Precision Planting Technology and Equipment in Advanced Crop Cultivation

Precision Planting Technology and Equipment in Advanced Crop Cultivation

Guest Editors

Xiaojun Gao
Qinghui Lai
Tao Cui

MDPI

Basel • Beijing • Wuhan • Barcelona • Belgrade • Novi Sad • Cluj • Manchester

Guest Editors
Xiaojun Gao
College of Engineering
Nanjing Agricultural University
Nanjing
China

Qinghui Lai
School of Energy and Environment Science
Yunnan Normal University
Kunming
China

Tao Cui
College of Engineering
China Agricultural University
Beijing
China

Editorial Office
MDPI AG
Grosspeteranlage 5
4052 Basel, Switzerland

This is a reprint of the Special Issue, published open access by the journal *Agriculture* (ISSN 2077-0472), freely accessible at: https://www.mdpi.com/journal/agriculture/special_issues/J881G82LXU.

For citation purposes, cite each article independently as indicated on the article page online and as indicated below:

Lastname, A.A.; Lastname, B.B. Article Title. *Journal Name* **Year**, *Volume Number*, Page Range.

ISBN 978-3-7258-3893-6 (Hbk)
ISBN 978-3-7258-3894-3 (PDF)
https://doi.org/10.3390/books978-3-7258-3894-3

Contents

agriculture

Article

Design and Testing of Electric Drive System for Maize Precision Seeder

Lin Ling [1,†], Yuejin Xiao [2,3,†], Xinguang Huang [2,3], Guangwei Wu [3,4], Liwei Li [3,4], Bingxin Yan [3,*] and Duanyang Geng [2,*]

1 College of Engineering and Technology, Southwest University, Chongqing 400716, China; mike-lin0923@outlook.com
2 School of Agricultural Engineering and Food Science, Shandong University of Technology, Zibo 255049, China; xiaoyj@nercita.org.cn (Y.X.); delicate-521@outlook.com (X.H.)
3 Intelligent Equipment Research Center, Beijing Academy of Agriculture and Forestry Sciences, Beijing 100097, China; wugw@nercita.org.cn (G.W.); lilw@nercita.org.cn (L.L.)
4 State Key Laboratory of Intelligent Agricultural Power Equipment, Beijing 100097, China
* Correspondence: yanbx@nercita.org.cn (B.Y.); dygxt@sdut.edu.cn (D.G.)
† These authors contributed equally to this work.

Abstract: To improve the expandability, seeding accuracy, and operating speed range of the electric drive system (EDS) of precision seeders, this study constructed an EDS based on a controller area network (CAN) bus and designed a motor controller based on a field-orientated control (FOC) algorithm. Full-factorial bench and field tests based on seed spacing (0.1, 0.2, and 0.3 m) and operating speed (3, 6, 9, 12, and 15 km/h) were carried out to evaluate the performance of the EDS. The results of bench tests showed that seeding quality varied inversely with operating speed and positively with seed spacing. The average quality of feed index (QFI) at 0.1, 0.2, and 0.3 m seed spacing in bench tests was 88.38%, 96.67%, and 97.36%, with the average coefficient of variation (CV) being 20.13%, 16.27%, and 13.20%. Analysis of variance confirmed that both operating speed and seed spacing had a significant effect on QFI and CV ($p < 0.001$). The analysis of motor rotational speed accuracy showed that the relative error of motor rotational speed above 410 rpm did not exceed 2.24%, and the relative error had less influence on the seeding quality. The average QFI was 85.93%, 95.91%, and 96.24%, with the average CV being 21.12%, 15.50%, and 16.49% at 0.1, 0.2, and 0.3 m seed spacing in field tests. The methods and results of this study can provide a reference for the design and optimization of the EDS in a maize precision seeder and provide an effective solution for the improvement of maize yields.

Keywords: CAN bus; field-orientated control; electric drive system; seeding quality; maize

Citation: Ling, L.; Xiao, Y.; Huang, X.; Wu, G.; Li, L.; Yan, B.; Geng, D. Design and Testing of Electric Drive System for Maize Precision Seeder. *Agriculture* **2024**, *14*, 1778. https://doi.org/10.3390/agriculture14101778

Academic Editor: Maohua Xiao

Received: 10 September 2024
Revised: 1 October 2024
Accepted: 2 October 2024
Published: 9 October 2024

1. Introduction

Land desertification and population growth have left most countries facing a severe food shortage in the future [1,2]. In this context, improving crop yields has become one of the important research hotspots at present [3–5]. Of the many ways to increase yields, improving seed spacing uniformity is especially critical [4,6,7]. The main reason for this is that seed spacing uniform effectively reduces competition for water, nutrients, and light during seed growth and development, which in turn results in higher yields and increased returns [8–12].

A precision seeder is the main carrier to achieve uniformity in seed spacing, and its seed spacing uniformity can effectively ensure a yield [13]. Traditional precision seeders use a mechanical drive system (MDS) to achieve the rotation of the seed meter plate. The MDS consists of a ground wheel, chain, and sprocket. Ground wheel slippage and chain jumping in MDS lead to fluctuations in the rotational speed of the seed meter plate, which in turn can lead to poor seeding quality in precision seeders [13–15]. In order to improve the impact of MDSs, EDSs with electric motors as drive elements have become a current research hotspot [2,16].

EDSs are common in developed countries [16]. The EDS developed by Horsch of Germany can achieve an operating speed of 15 km/h while ensuring seeding quality [17]. The vDrive motor developed by American Precision planting company can meet the operation demand of 16 km/h, which greatly improves the operation efficiency [18]. In Italy, the company Maschio launched the CHRONO high-speed seeder with an EDS that can achieve seeding in 15 km/h [19]. The above EDSs meet the high-speed requirements at the same time, with the use of a controller area network (CAN) bus based on the ISOBUS 11783 protocol, which can achieve the expansion of different rows of seeders, further improving the scope of their application. Meanwhile, the EDS with advanced motor control algorithm ensures seed spacing uniformity.

The widespread use of precision seeders with an MDS in developing countries is a serious constraint on increasing yields and is not conducive to improving farmers' incomes and ameliorating possible future food shortages. Although the use of EDSs in developing countries in applications is relatively lacking, their superiority has attracted the attention of many researchers. Yang et al. [20] designed an EDS for a four-row precision seeder and carried out experiments, and the results of these experiments showed that under the three operating speeds of 9, 11, and 12 km/h, the quality of feed index (QFI) of the EDS was increased by an average of 4.7% compared with that of the MDS, and the miss index (MI) was reduced by an average of 3.54%, which proved that the EDS was better adapted to seeding. Cay et al. [21] found that the EDS had a lower fuel consumption, with a 22% year-over-year reduction in fuel consumption. In the above studies, EDSs were designed and found to be advantageous, but the expandability of the system was not considered, and the designed EDSs were only adapted to seeders with the corresponding number of rows. Therefore, He et al. [22], Ding et al. [23], and Yang et al. [24], on the other hand, designed an expandable EDS based on a CAN bus to adapt to different rows of seeders. However, they neglected the improvement of seeding quality by the control algorithm. Also, the operating speeds in the above EDS were within 12 km/h, and high-speed operating situations were not considered. Du et al. [25] designed an electric drive unit, and bench experiments were conducted to compare the effects of square wave control and field-orientated control (FOC) algorithms on motor control and seeding quality, and the results proved that the FOC algorithm could achieve a higher seeding quality by virtue of its good rotational speed accuracy. However, they did not apply the algorithm to the actual EDS, and there is still a large upside in terms of the seeding quality.

The high cost of products of developed countries has prevented EDSs from spreading rapidly in developing countries. Moreover, the EDS in developing countries is deficient in terms of expandability, seeding accuracy, and operating speed. Therefore, the adaptability and high efficiency of the EDS cannot be further improved, which could be the main reason limiting the rapid promotion and application of EDSs in developing countries.

In order to solve the above problems, this study aims to achieve the following objectives: (1) construct an expandable EDS based on a CAN bus to adapt to maize seeders with different row units; (2) design a motor controller based on the FOC algorithm for precise control of the motor rotational speed; (3) conduct an testing study to validate the system performance.

2. Materials and Methods

2.1. System Components

The EDS structure is shown in Figure 1. The EDS consists of a master control unit, a seeder lift status detection unit, and several seeding control units. The master control unit includes an Android terminal (Nongxin Technology Co., Ltd., Beijing, China), a GNSS antenna, and a 4G antenna, which is mainly responsible for receiving satellite signals, network signals and human–machine interaction. The seeder lift status detection unit includes a seeder lift status detection controller and a travel switch, which is mainly responsible for detecting the lift and fall status of the seeder. The seeding control unit includes a permanent magnet synchronous motor (PMSM) (Nongxin Technology Co., Ltd., Beijing, China), a motor controller, and an air-suction seed meter (Shandong Dahua Machinery Co., Ltd., Jining, China), and the PMSM is coaxially connected to the seed meter. The motor controller is installed at the end of the motor. Communication between the motor controllers and the terminal via the CAN bus allows the EDS to be adapted to precision seeders with different row units. The 120 Ω termination resistors at both ends of the CAN bus are used for impedance matching to improve the CAN bus interference immunity.

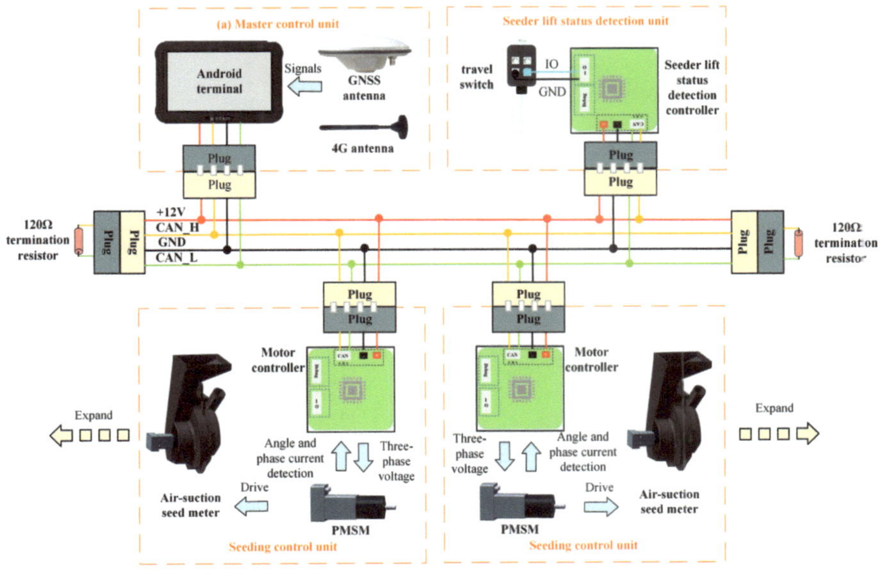

Figure 1. Structure of the EDS. Solid lines indicate connecting lines; dashed lines indicate system component units; notes are labeled near arrows and lines.

When the motor rotates, the seed meter plate connected to the motor rotates synchronously. The reduction ratio between the two is 32, and the diameter of the seed meter plate is 220 mm. The seeds are adsorbed on the seed meter plate (4.5 mm in diameter) under the action of negative pressure (4 to 6 kPa) and then pass through the double-sided seed-cleaning knives to scrape off the excess seeds. The seeds are finally transported to a seed feeding point, which operates without negative pressure, and placed into the seed tube under the influence of gravity and initial speed.

2.2. Working Principle

The working principle of the EDS is shown in Figure 2. The seeder lift status detection controller obtains the status of the tractor's three-point suspension bar according to the travel switch and then identifies the seeder lift status and sends it to the CAN bus. The Android terminal, as the main controller, acquires the operational speed of the seeder through the built-in board, external GNSS antenna, and 4G antenna in network RTK mode and calculates the target rotational speed of the motor according to the seeding parameters based on Equation (1). At the same time, the terminal also receives the seeder lift status, and, in combination with the motor start/stop status input by the user, it determines the enable state of each motor and sends it, together with the motor target rotational speed, to the CAN bus.

$$R_m = \frac{50vi}{3n_p d} \quad (1)$$

where R_m is the motor target rotational speed, rpm; v is the operating speed, km/h; i is the reduction ratio between the motor and the seed meter, 32; n_p is the number of holes in the seed meter plate, 26; and d is the seed spacing, m.

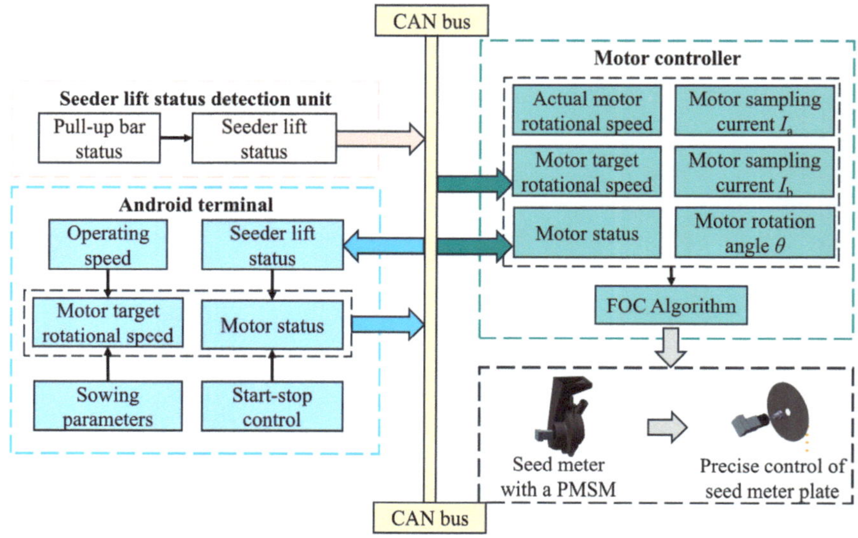

Figure 2. Working principle of the EDS.

2.3. Design of Communication Protocol

The communication protocol of the EDS is developed with reference to the ISO 11783 protocol. The ISO 11783 protocol is an international standard for communication between agricultural machinery and equipment [26]. The protocol data unit (PDU) is the basic unit of transmission in ISO 11783 [24]. The PDU consists of seven parts including the priority (P), reserved bit (R), data page (DP), protocol data unit format (PF), protocol specific data unit (PS), source address (SA), and data domain [27]. With reference to ISO 11783 regulations and the EDS architecture, the PDU information of the EDS is shown in Table 1.

Table 1. PDU information.

Equipment	P	R	DP	PF	PS	SA	Parameter Group PCN	PDU Identification
Controller	6	0	0	0xFF	0x22	0x80	00FF22	18FF2280
Android terminal	2	0	0	0xFF	0x23	0x26	00FF23	18FF2326

The data domain hosts the actual application data. According to the PDU identification, the EDS uses single-frame mode to transmit data, and the protocol is shown in Table 2. Among them, the row number is used to distinguish the type of controller; '0' means the seeder lift status detection sensor, '1' means the motor controller in the first row, '2' means the motor controller in the second row, and other sequential analogous types of controller.

Table 2. Data domain protocols.

Device	Byte1	Byte2	Byte3	Byte4	Byte5	Byte6	Byte7	Byte8
Controller	/	Row number	Motor/seeder lift status	Motor actual rotational speed		Motor target rotational speed		/
Android terminal	/	Row number	Motor status	Motor target rotational speed		/	/	/

2.4. Field-Orientated Control Algorithm

The schematic diagram of the FOC algorithm is shown in Figure 3. The currents I_a and I_b of PMSM are obtained by a current sampling process, and the current I_c can be obtained based on Kirchhoff's Law, as shown in Equation (2).

$$I_c = -(I_a + I_b) \tag{2}$$

The two currents are transformed by Clark's transformation to obtain the orthogonal axis currents I_α and I_β for the α-β coordinate system, and I_α, I_β are transformed by Park transformation to obtain the orthogonal axis currents I_q and I_d for the q-d coordinate system. I_q is the torque current component, and I_d is the excitation current component. Therefore, the target value of I_d, I_{d_ref}, is usually set to zero for torque maximization. The difference between the actual rotational speed, speed_act, obtained by the sensor and the target rotational speed, speed_ref, is fed into the PI controller to find the quadrature axis current I_q reference value I_{q_ref}. The actual (I_{q_act}, I_{d_act}) and reference (I_{q_ref}, I_{d_ref}) values of I_q and I_d are fed into the PI controller to find the quadrature axis voltages U_q and U_d, respectively. U_q and U_d are subjected to inverse Park transformation to obtain U_α and U_β for the α-β coordinate system, which are subjected to space voltage vector pulse width modulation (SVPWM) to obtain target voltage vectors (U_a, U_b, U_c) to achieve the FOC of the PMSM.

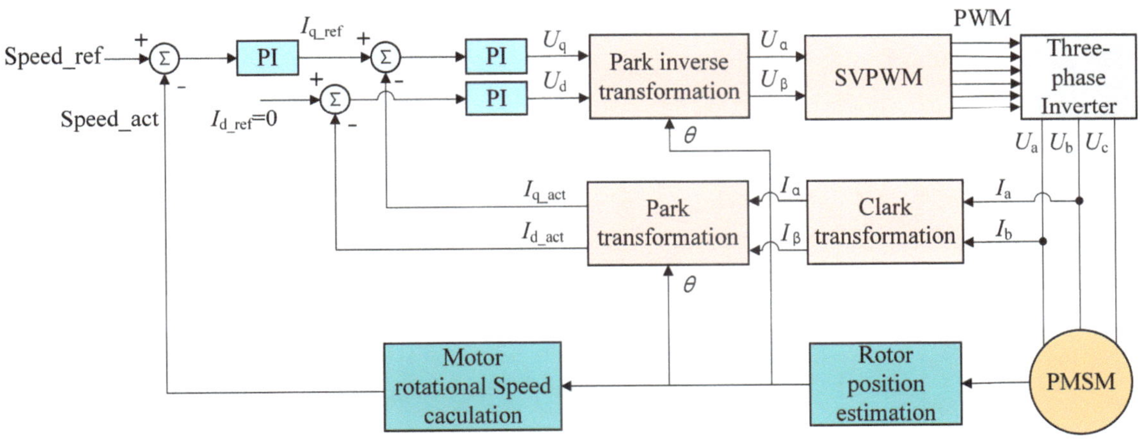

Figure 3. Schematic diagram of the FOC algorithm.

2.4.1. Coordinate Transformation

Coordinate transformation is used to solve the problem of the difficult calculation of AC quantities such as the voltage, current, and magnetic chain in the stator. The currents (I_a, I_b, I_c) in a three-phase stationary coordinate system (abc) are first projectively transformed into the two-phase orthogonal currents I_α and I_β in the two-phase stationary coordinate system (α-β) [25], as shown in Figure 4.

The relationship between the currents is shown in Equation (3).

$$\begin{bmatrix} I_\alpha \\ I_\beta \end{bmatrix} = \frac{2}{3} \begin{bmatrix} 1 & -\frac{1}{2} & -\frac{1}{2} \\ 0 & \frac{\sqrt{3}}{2} & -\frac{\sqrt{3}}{2} \end{bmatrix} \begin{bmatrix} I_a \\ I_b \\ I_c \end{bmatrix} \tag{3}$$

For the convenience of describing the motion of the rotor, I_α and I_β are projected onto the two-phase rotating coordinate system (q-d), as shown in Figure 5. Where θ is the rotation angle of the rotor, its value can be obtained by the sensor.

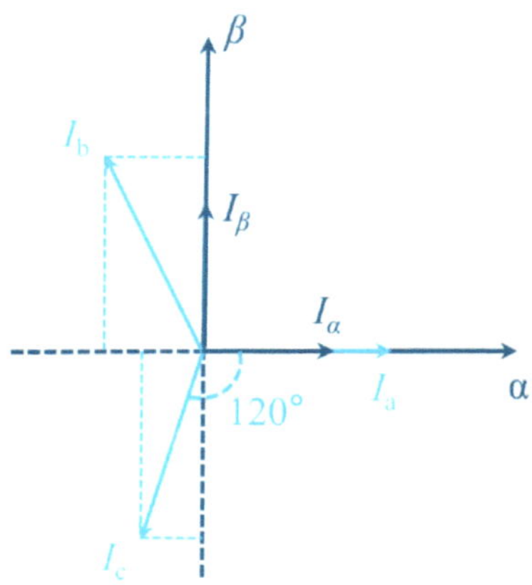

Figure 4. Two-phase stationary coordinate system.

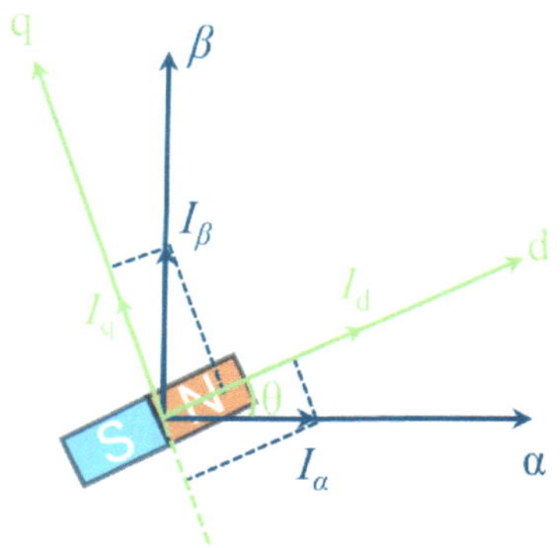

Figure 5. Two-phase rotating coordinate system.

The U_q and U_d in the qd coordinate system can be calculated by Equation (4), which is also known as the Park transformation.

$$\begin{bmatrix} U_d \\ U_q \end{bmatrix} = \begin{bmatrix} \cos\theta & \sin\theta \\ -\sin\theta & \cos\theta \end{bmatrix} \begin{bmatrix} U_a \\ U_\beta \end{bmatrix} \tag{4}$$

2.4.2. Space Voltage Vector Pulse Width Modulation

SVPWM outputs different space voltage vectors to drive the motor rotation by controlling the conduction of the MOSFET (Q1~Q6) in the three-phase inverter circuit (Figure 6) [28]. The three-phase inverter circuit consists of three sets of bridge arms, each of which contains two MOSFET arranged in upper and lower positions. There are eight combinations of upper and lower bridge arm switching, corresponding to eight space voltage vectors, including six non-zero vectors, U_1, U_2, U_3, U_4, U_5, and U_6, and two zero vectors, U_0 and U_7.

The magnitudes and positions of the eight space voltage vectors in the stationary coordinate system are shown in Figure 7, where six of the vectors divide the plane into six sectors (I, II, III, IV, V, and VI).

In each sector, two adjacent voltage vectors, as well as the zero vector, are selected to synthesize an arbitrary voltage vector according to the principle of volt–second balancing [29], as shown in Equation (5).

$$U_{ref} * T = U_x * T_x + U_y * T_y + U_0 * T_0 \tag{5}$$

where U_{ref} is the desired voltage vector, V; T is the PWM period, s; U_x and U_y are used to synthesize the two spatial voltage vectors of U_{ref}, corresponding to two of the six spatial voltage vectors mentioned above; and U_0 refers to the two zero vectors U_0 and U_7; and T_x, T_y, and T_0 are the action times corresponding to U_x, U_y, and U_0 in one PWM period, respectively.

The SVPWM allows the synthesis of space vectors of arbitrary direction and magnitude to achieve the FOC of PMSM [28].

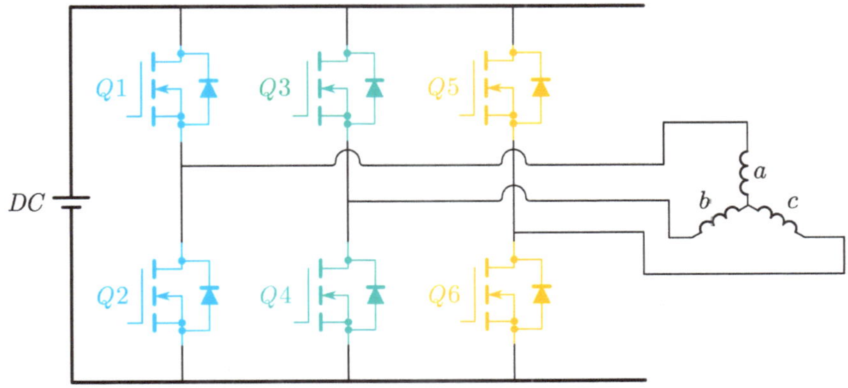

Figure 6. Three-phase inverter circuit. MOSFETs of the same color indicate the same bridge arm.

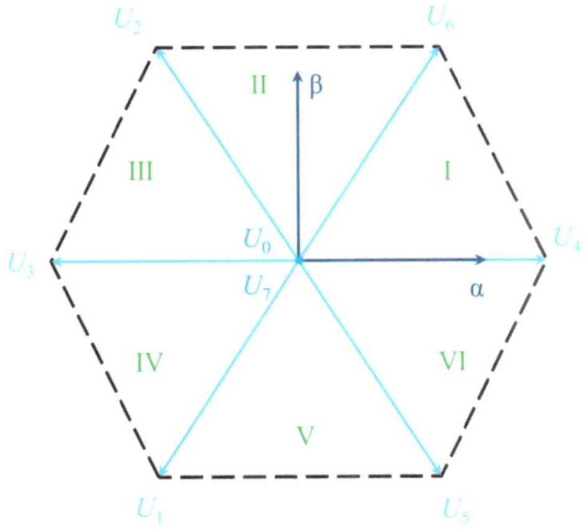

Figure 7. Space voltage vector. The triangular areas surrounded by neighboring arrows are sectors (I, II, III, IV, V, and VI).

2.5. Design of Motor Controller

The motor controller, tailored for the FOC in PMSM, comprises key components, as depicted in Figure 8. It incorporates two power supply modules (SCT2400, Silicon Content Technology Co., Ltd., Beijing, China) regulated to output 10 V (for gate drivers) and 3.3 V (for the MCU, magnetic encoder, and CAN transceiver). The heart of the system is an MCU (GD32C103CBT6, Zhaoyi Innovation Technology Group Co., Ltd., Beijing, China), leveraging USART, SWDIO for debugging, CAN (SN65HVD230, Texas Instruments Inc., Dallas, TX, USA) for communicating with Android terminal, TIMER0 for PWM generation, ADC for phase current sensing from operational amplifiers (OP-amp) (LVM324IPWR, Texas Instruments Inc., Dallas, TX, USA), and SPI for rotor angle acquisition from the magnetic encoder (MT6825, Magtek Inc., Shanghai, China). The PWMs were generated by the MCU drive gate drivers (FD6288Q, Texas Instruments Inc., Dallas, TX, USA), adjusting the MOSFETs (BSC0702LS, Infineon Technologies, Neubiberg, Germany) of the three-phase inverter circuit to achieve PMSM control.

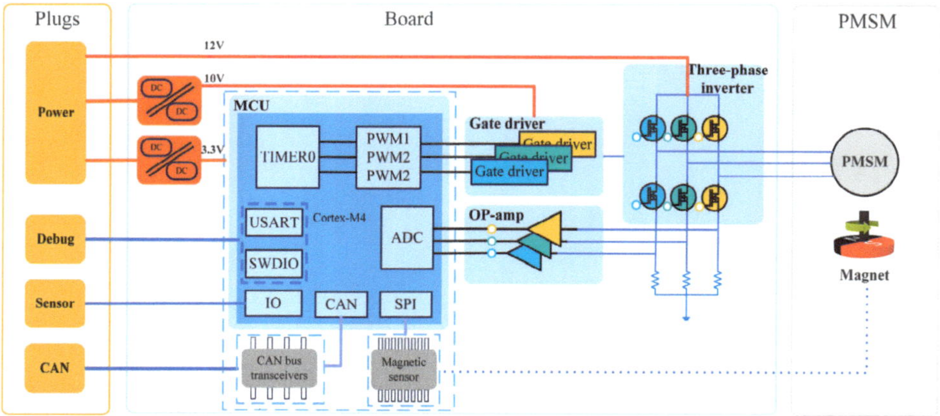

Figure 8. Controller structure and function.

The FOC process involves the MCU acquiring target speeds, rotor angle, and phase currents via CAN, SPI, and ADC. Based on these and the FOC algorithm (Figure 3), TIMER0 generates PWM signals, which, through SVPWM and gate drivers, produce a three-phase voltage to precisely control the PMSM.

It is worth mentioning that, in order to reduce the development period, the seeder lift status detection controller follows the motor controller, and the seeder lift status detection controller detects the triggering signal of the travel switch through the IO port and then obtains the lift status of the seeder.

3. Tests

In order to fully validate the seeding performance of the EDS, bench tests and field tests were carried out. The bench tests were used to evaluate the seed space uniformity of the EDS under a static environment and, at the same time, analyze the influence of the precision of motor rotational speed on seeding quality based on the relative error of the motor rotational speed. The field tests were used to validate the actual seeding performance of the EDS and to analyze the difference between bench tests and field tests regarding the seeding quality. NK815 maize seed was used in both bench tests and field tests. The seed had a half-horse tooth shape, and the 1000-seed weight was 312.05 g.

3.1. Bench Tests

Seed spacing and operating speed are important factors that affect seeding quality. Agronomy varies from region to region, and the seed spacing need is not consistent. In order to ensure the adaptability of the EDS, the range of seed spacing was set at 0.1–0.3 m. The operating speed greatly affects the operating efficiency, but due to the influence of the load of the seeder, the ground conditions, and the performance of the seed meter, the current seeder operating speed is generally within 15 km/h. Based on the range of seed spacing and operating speed, bench tests factor levels were set, as shown in Table 3.

Table 3. Bench tests factor levels.

Test Factor	Level
Seed spacing/m	0.1, 0.2, 0.3
Operating speed/(km/h)	3, 6, 9, 12, 15

Bench tests were conducted in September 2023. The bench tests equipment is shown in Figure 9 and included a seed meter detector [30], an air-suction seed meter (Shandong Dahua Machinery Co., Ltd., Jining, China), a PMSM (Nongxin Technology Co., Ltd., Beijing, China), a motor controller, a CAN tool (CANalyst-II, Chuangxin Technology, Zhuhai, China), and a terminal (Nongxin Technology Co., Ltd., Beijing, China). The air-suction seed meter is installed on the seed meter detector, which provides negative pressure (6kPa) via the seed meter detector, and the PMSM is connected coaxially to the seed meter plate. The outer seed cleaning knife of the seed meter is adjusted to the 9th position (14 positions in total), and the inner seed cleaning knife is adjusted to the 6th position (9 positions in total). The CAN tool is connected to the CAN bus between the motor controller and the terminal and is used to read CAN bus data.

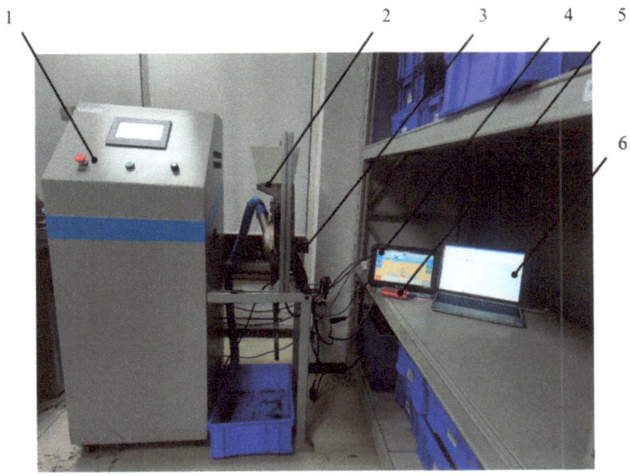

Figure 9. Bench tests equipment. 1. Seed meter detector, 2. air-suction seed meter, 3. PMSM, 4. Android terminal, 5. CAN tool, 6. computer.

Before the tests, the seed meter detector and the terminal were configured with the same operating speed and seed spacing according to Table 3, respectively. The seeding quality was recorded by the seed meter detector, and this was repeated three times at each factor level. The target number of seeds detected by the seed meter detector was set to 1000. At the same time, the motor rotational speed returned by the motor controller (return frequency of 20 Hz) through the CAN bus was read in real time by the CAN tool during the process, which is used to analyze the precision of motor rotational speed control.

3.2. Field Tests

Field tests were conducted to further explore the actual seeding performance of the EDS. The same factor levels as those used in the bench tests were selected for the field tests (Table 3). The field tests were conducted in October 2023. Before the tests, the soil (brown loam) was tilled by a rotary tiller working at a depth of 0.15 m, to ensure that the soil was soft. The tests equipment is shown in Figure 10, including a 2104 tractor (Case IH, Racine, WI, USA), a six-row precision seeder (2BMYFQ-6D, Shandong Dahua Machinery Co., Ltd., Jining, China), and an EDS. The EDS was installed on the seeder. The double-sided seed cleaning knife configuration was consistent with that used in the bench tests. The parameters of the seeder are shown in Table 4.

Figure 10. Field tests equipment. ① Antenna mounting method; ② Terminal mounting method; ③ Seeder lift detection unit mounting method; ④ PMSM and controller mounting method.

Table 4. Parameters of the seeder.

Name	Parameter
Overall dimensions/mm	2000 × 5900 × 1800
Weights/kg	1500
Working width/m	3.9
Number of rows	6
Row spacing/m	0.65
Seeding depth/mm	50
Fertilizer opener type	Double disk
Suppression wheel	V-shaped
Hitch way with tractor	Three-point suspension
Profiling mechanism	Machinery

Before the tests, six PMSMs were divided into three groups, and the three groups of motors were configured to 0.1, 0.2, and 0.3 m seed spacing, respectively. The seeding depth of each seeding unit was set at 0.05 m. The tractor rear hydraulic output was adjusted so that the rotational speed of the hydraulic motor of the seeder fan was increased as needed to ensure that the vacuum pressure of the seed meter was maintained at not less than 6.0 kPa. After the tests, 100 seed spacings per row were collected by digging out the seeds (Figure 11), translating to 200 seed spacings per group.

Figure 11. Seed digging and measuring.

3.3. Evaluation Index

The quality of feed index (QFI), miss index (MI), multiple index (MUL), and coefficient of variation (CV) specified in ISO 7256-1:1984, "Sowing equipment—Test methods Part 1: Single seed drills" [31], were used as evaluation indexes in both the bench tests and the field tests. The difference is that the evaluation indexes in the bench tests were given by direct calculations from the seed meter detector, and the results of the field tests were manually calculated based on seed spacings.

4. Analysis and Discussion of Results

4.1. Analysis and Discussion of Bench Tests

4.1.1. Seeding Quality of Bench Tests

The seeding quality at three seed spacings (0.1, 0.2, and 0.3 m) and five operating speeds (3, 6, 9, 12, and 15 km/h) is shown in Figure 12. With the increase in operating speed, the overall trend of QFI decreased, and the trend of MI, MUL and CV increased; with the increase in seed spacing, the trend of QFI decreased, and the trend of CV decreased.

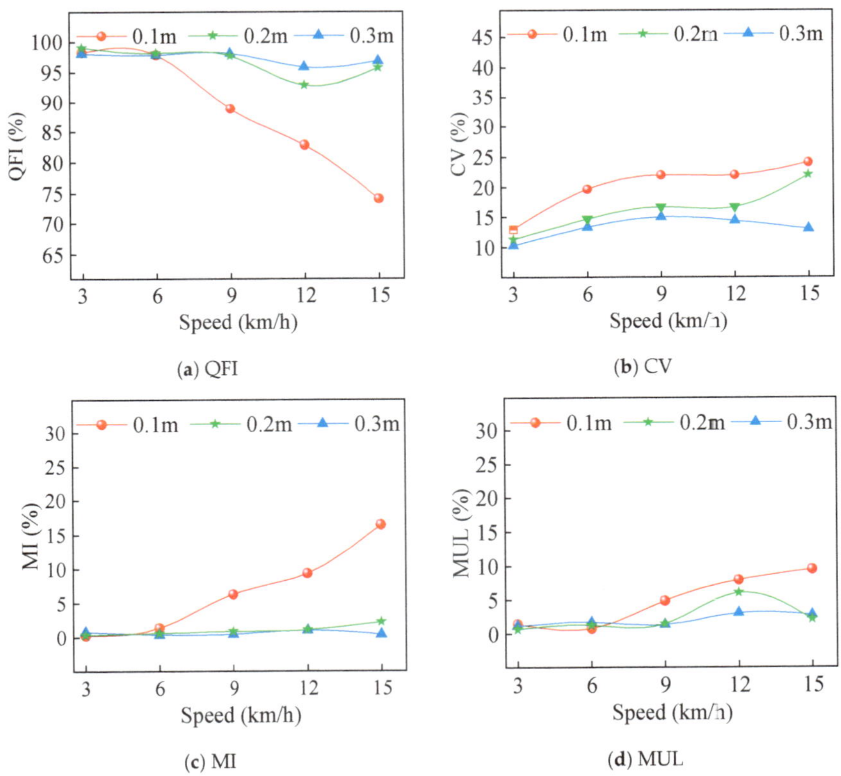

Figure 12. Seeding quality of bench tests.

Table 5 shows the results of bench tests. In the range of 3–15 km/h, the QFI varied from 74.10% to 98.30%, 92.87% to 99.00%, and 95.90% to 98.07% at 0.1, 0.2, and 0.3 m seed spacing, respectively. The average QFI was 88.38%, 96.67%, and 97.36%, respectively; the extreme difference (ED) in the QFI was 24.20%, 6.13%, and 2.23%, respectively. The CV varied from 13.00% to 24.00%, 11.33% to 22.00%, and 10.33% to 15.00% at 0.1, 0.2, and 0.3 m seed spacing, respectively; the average CV was 20.13%, 16.27%, and 13.20%, respectively; and the ED in the CV was 11.00%, 10.67%, and 4.67%, respectively.

Table 5. Results of the bench tests.

Seed Spacing/m	Operating Speed/(km/h)	QFI/%	Average QFI/%	ED of QFI/%	CV/%	Average CV/%	ED of CV/%
0.1	3	98.30	88.38	24.20	13.00	20.13	11.00
	6	97.83			19.67		
	9	88.90			22.00		
	12	82.77			22.00		
	15	74.10			24.00		
0.2	3	99.00	96.67	6.13	11.33	16.27	10.67
	6	98.10			14.67		
	9	97.67			16.67		
	12	92.87			16.67		
	15	95.70			22.00		
0.3	3	98.07	97.36	2.23	10.33	13.20	4.67
	6	97.87			13.33		
	9	98.13			15.00		
	12	95.90			14.33		
	15	96.83			13.00		

Based on the extreme difference data for QFI, it can be seen that the effect of increasing seed spacing on the decrease in QFI is not linear, but rather an accelerated deterioration process. The extreme difference in QFI at 0.1 m and 0.2 m seed spacing is 2.75 times and 10.84 times of that at 0.3 m seed spacing, respectively. The cause of this condition may be related to the rotational speed of seed meter plate and seed bouncing [2,16,32]. The decrease in seed spacing and the increase in operating speed lead to an increase in the rotational speed of the seed meter plate, which will result in less time for the seed adsorption and clearing process [16], thereby causing a decrease in QFI. Moreover, the increase in the rotational speed leads to an increase in the initial velocity of the seed when it is put into the seed tube, and the seed bouncing in the seed tube becomes violent, which leads to a decrease in QFI and an increase in CV.

4.1.2. Analysis of Variance

To further explore the effect of operating speed and seed spacing on seeding quality, an analysis of variance (ANOVA) was conducted, and the results of the ANOVA are shown in Table 6.

Table 6. ANOVA of bench tests results.

Index	Factor	SS	Df	MS	F	p
Seed spacing	QFI	748.950	2	374.475	655.695	0.000
	CV	362.133	2	181.067	94.744	0.000
Operating speed	QFI	670.056	4	167.514	293.313	0.000
	CV	343.200	4	85.800	44.895	0.000

Table 6 shows that both operating speed and seed spacing had a significant effect on QFI and CV ($p < 0.001$). From the F-values, it can be seen that both seed spacing and operating speed had a greater effect on QFI than on CV, and seed spacing had a greater effect on QFI and CV than operating speed had on QFI and CV.

4.1.3. Accuracy Analysis of Motor Rotational Speed

The decrease in seeding quality may be related to seed spacing and operating speed on the one hand, but there may also be a correlation with motor rotational speed accuracy [15,33]. In order to clarify the effect of motor rotational speed accuracy on seeding quality, the motor rotational speed accuracy in the tests was analyzed. The motor target rotational speed, relative error of motor rotational speed, QFI, and CV value correspondences during the tests are shown in Figure 13.

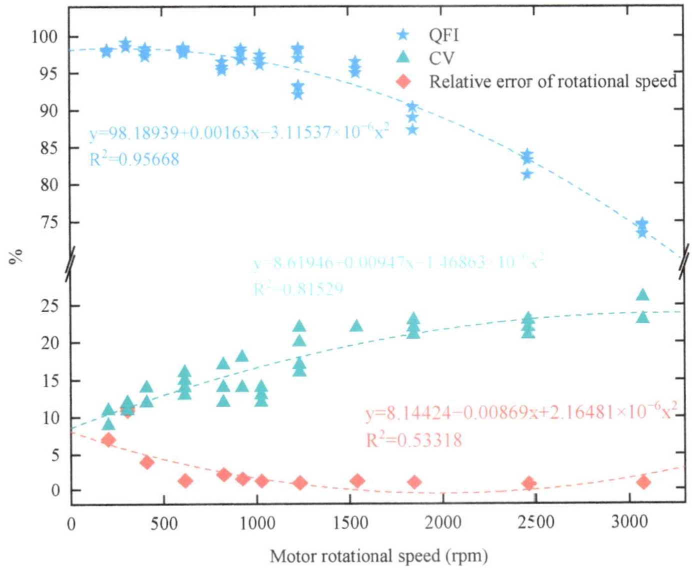

Figure 13. Correspondence between the relative error of motor rotational speed, QFI and CV.

Figure 13 shows that as the motor rotational speed increases, the QFI and the relative error of the motor rotational speed gradually decrease and the CV gradually increases. Among them, the relative error of the motor rotational speed does not exceed 2.24% when the motor rotational speed is more than 410 rpm; when the motor rotational speed is less than 410 rpm, the relative error of the motor rotational speed does not exceed 11.34%.

The reason for this condition is that at low motor rotational speeds, the resolution of the magnetic encoder is insufficient to accurately measure speed changes [34]. At the same time, the phase currents are small and the current sampling process is susceptible to noise. However, the QFI and CV corresponding to the low motor rotational speeds are at a high level. It can be assumed that the relative error of the motor rotational speed is low, having little effect on the seeding quality. The EDS's motor rotational speed accuracy can meet the demands for precision seeding of maize within the ranges of 3–15 km/h and 0.1–0.3 m seed spacing. Therefore, the reduction in QFI and the increase in CV are mainly due to the seed meter and seed bouncing. The seed adsorption and clearing time decreases at higher motor rotational speeds, leading to a reduction in QFI. At the same time, excessive rotational speed in the seed meter plate will make the seed bouncing in the seed tube become more violent, resulting in a decrease in the seeding quality and an increase in CV.

4.2. Analysis and Discussion of Field Tests

4.2.1. Seeding Quality of Field Tests

The seeding quality of field tests at three seed spacings (0.1, 0.2, and 0.3 m) and five operating speeds (3, 6, 9, 12, and 15 km/h) is shown in Figure 14. With the increase in operation speed, the overall trend of the QFI was decreasing, and the MI, MUL, and CV increased. With the increase in seed spacing, the QFI first increased rapidly and then stabilized, and the MI, MUL, and CV first decreased rapidly and then stabilized. The overall trend is similar to that of the indexes in the bench tests.

Table 7 shows the results of field tests. In the range of 3–15 km/h, the QFI varied from 66.37% to 98.02%, 91.79% to 98.52%, and 92.68% to 99.50% at 0.1, 0.2, and 0.3 m seed spacing, respectively. The average QFI was 85.93%, 95.91%, and 96.24%, respectively; the extreme difference in the QFI was 31.65%, 6.73%, and 6.82%, respectively. The CV varied from 17.94% to 24.02%, 10.43% to 20.65%, and 12.19% to 20.8% at 0.1, 0.2, and 0.3 m seed spacing, respectively; the average CV was 21.12%, 15.50%, and 16.49%, respectively; and the extreme difference in the CV was 6.08%, 10.22%, and 8.61%, respectively.

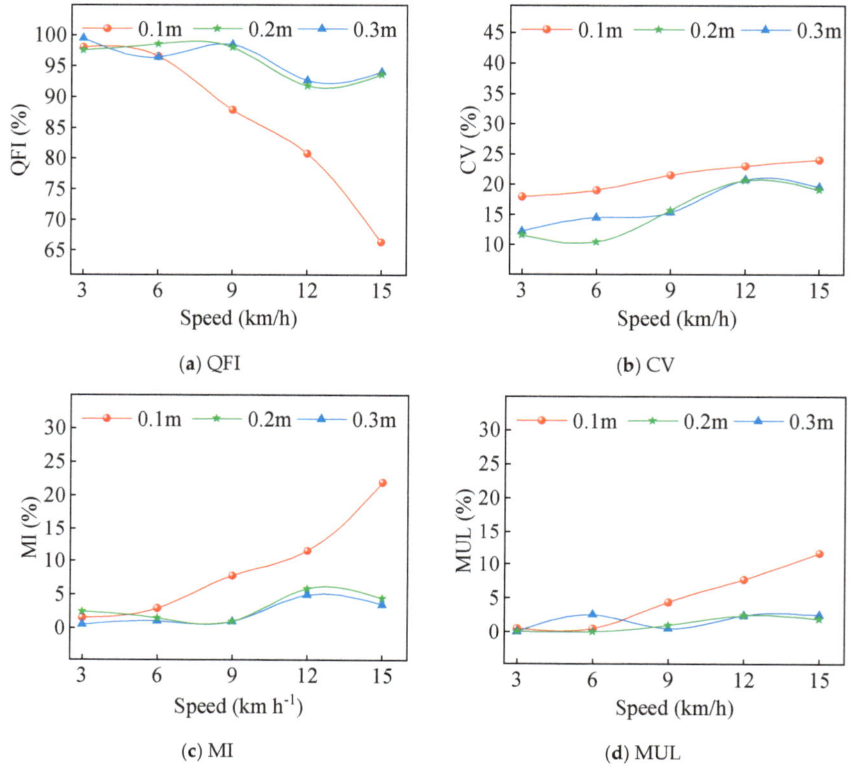

Figure 14. Seeding quality in field tests.

Table 7. Results of the field tests.

Seed Spacing/m	Operating Speed/(km/h)	QFI/%	Average QFI/%	ED of QFI	CV	Average CV	ED of CV
0.1	3	98.02	85.93	31.65	17.94	21.12	6.08
	6	96.59			19.05		
	9	87.92			21.54		
	12	80.77			23.05		
	15	66.37			24.02		
0.2	3	97.56	95.91	6.73	11.61	15.50	10.22
	6	98.52			10.43		
	9	98			15.69		
	12	91.79			20.65		
	15	93.66			19.13		
0.3	3	99.5	96.24	6.82	12.19	16.49	8.61
	6	96.45			14.49		
	9	98.51			15.38		
	12	92.68			20.8		
	15	94.06			19.58		

4.2.2. Comparison of Bench and Field Tests Results

Based on the extreme difference data of QFI in field tests, it is clear that smaller seed spacing has a significant effect on QFI. The extreme difference in QFI at 0.1 m seed spacing was 4.64 times that at 0.3 m seed spacing. In contrast, the extreme differences in QFI at 0.2 m and 0.3 m were similar, which may be related to the method used to obtain the seeding quality. Seeds were monitored in the bench tests by a monitoring sensor mounted on the seed tube, while, in the field tests, seeds were dug out manually. There was relatively little seed bouncing within the seed tube in the bench tests, whereas in the field tests seed bouncing was present throughout the seed tube and in the seed furrow. Seed bouncing affects seed spacing [35]; the effect of seed bouncing is more pronounced when seed spacing is small and less pronounced when seed spacing is large. As a result, the extreme differences in QFI are close at seed spacings of 0.2 and 0.3 m.

More seed bouncing in the field tests resulted in an overall decrease in seeding quality, with the average QFI at 0.1 m, 0.2 m, and 0.3 m seed spacing decreasing by 2.45%, 0.76%, and 1.12% compared to the average QFI in the bench tests and the average CV increasing by 0.98%, −0.77%, and 3.29% compared to the increase in the bench tests.

Overall, the EDS can meet the needs of a wide operating speed range (3–15 km/h) and a wide seed spacing range (0.1–0.3 m), and the motor controller designed based on the FOC algorithm can achieve high control accuracy. However, the seed meter performance and seed bouncing are important constraints that could limit further improvement in seeding quality, especially at high operating speeds. Therefore, it is necessary to optimize the seed meter based on the demand for high speeds and to improve seed bouncing in the seed tube and the furrow, which will help improve the yield in the future.

5. Conclusions

In this study, an EDS for a maize precision seeder based on a CAN bus was designed and a seeding controller based on FOC algorithm was developed to achieve the FOC of the PMSM. The seeding quality change rule for the EDS under different seed spacings and operating speeds was studied to determine the seeding performance of the system. Specific conclusions are as follows:

1. To improve the expandability, seeding accuracy, and operating speed range, an EDS for a maize precision seeder was designed based on the CAN bus and FOC algorithm. A CAN bus communication protocol designed based on ISO 11783 standard can be applied to different row seeders. The seeding controller based on the FOC algorithm can effectively ensure the seeding accuracy and speed range.
2. To explore the performance of the EDS and the change rule for seeding quality, bench tests were carried out. The results of the bench tests showed that seeding quality varied inversely with operating speed and positively with seed spacing. over a range of seed spacings (0.1, 0.2, and 0.3 m) and operating speeds (3, 6, 9, 12, and 15 km/h). The average QFI at 0.1, 0.2, and 0.3 m seed spacing in bench tests was 88.38%, 96.67%, and 97.36%, with the average CV being 20.13%, 16.27%, and 13.20%.
3. To explore the effect of tests factors on seeding quality, ANOVA and rotational speed accuracy were conducted based on bench tests. ANOVA showed that both operating speed and seed spacing have a significant effect on QFI and CV ($p < 0.001$). Both seed spacing and operating speed have a greater effect on QFI than on CV, and the effect of seed spacing on QFI and CV is greater than the effect of operating speed on QFI and CV. The analysis of motor rotational speed accuracy showed that the relative error of motor rotational speed above 410 rpm does not exceed 2.24% and the rotational speed control error has less influence on the seeding quality.
4. To further determine the performance of the system, field tests were conducted. The results of the field tests showed that the average QFI was 85.93%, 95.91%, and 96.24% at 0.1, 0.2, and 0.3 m seed spacing, and the average CV was 21.12%, 15.50%, and 16.49% in the range of operating speeds of 3, 6, 9, 12, and 15 km/h. Compared with the bench tests, the average QFI at 0.1 m, 0.2 m, and 0.3 m seed spacing decreasing by

2.45%, 0.76%, and 1.12%, and the average CV increased by 0.98%, −0.77%, and 3.29% respectively.

Synthesizing the results of the above tests, the EDS meets the demand for expandability, seeding accuracy, and operating speed range. The seed meter performance and seed bouncing can limit seeding quality at high operating speeds, and further optimization of the seed meter and seed guide mechanism will be needed in the future.

Author Contributions: Conceptualization, D.G.; Data curation, X.H.; Formal analysis, L.L. (Lin Ling), Funding acquisition, B.Y.; Investigation, L.L. (Lin Ling) and Y.X.; Methodology, L.L. (Lin Ling); Project administration, B.Y.; Resources, B.Y. and D.G.; Software, Y.X.; Supervision, G.W.; Validation, B.Y.; Visualization, L.L. (Lin Ling) and L.L. (Liwei Li); Writing—original draft, L.L. (Lin Ling); Writing—review and editing, L.L. (Lin Ling) and Y.X. All authors have read and agreed to the published version of the manuscript.

Funding: This research was funded by Next Generation Artificial Intelligence National Science and Technology Major Project (No. 2022ZD0115800), Modern AgricuituraIndustrial System of Shandong Province (No. SDAIT-02-12), and National Natural Science Foundation of China (No. 32301705).

Data Availability Statement: Data are contained within the article. The data presented in this study can be requested from the authors.

Conflicts of Interest: The authors declare no conflicts of interest.

References

1. Zhang, Q.; Dong, W.; Wen, C.; Li, T. Study on Factors Affecting Corn Yield Based on the Cobb-Douglas Production Function. *Agric. Water Manag.* **2020**, *228*, 105869. [CrossRef]
2. Ling, L.; Wu, G.; Wen, C.; Xiao, Y.; Fu, W.; Dong, J.; Ding, J.; Meng, Z.; Yan, B. Influence of Speed Measurement Method on Performance of an Electric-Drive Maize Precision Planter. *Biosyst. Eng.* **2024**, *238*, 175–187. [CrossRef]
3. Ding, Y.; He, X.; Yang, L.; Zhang, D.; Cui, T.; Li, Y.; Zhong, X.; Xie, C.; Du, Z.; Yu, T. Low-Cost Turn Compensation Control System for Conserving Seeds and Increasing Yields from Maize Precision Planters. *Comput. Electron. Agric.* **2022**, *199*, 107118. [CrossRef]
4. Yang, S.; Zhai, C.; Gao, Y.; Dou, H.; Zhao, X.; He, Y.; Wang, X. Planting Uniformity Performance of Motor-Driven Maize Precision Seeding Systems. *Int. J. Agric. Biol. Eng.* **2022**, *15*, 101–108. [CrossRef]
5. Wang, H.; Yang, L.; Zhang, D.; Cui, T.; He, X.; Xiao, T.; Li, H.; Du, Z.; Xie, C. Comparative Investigation and Evaluation of Electric-Drive Seed-Metering Systems across Diverse Speed Ranges for Enhanced High-Precision Seeding Applications. *Comput. Electron. Agric.* **2024**, *222*, 108976. [CrossRef]
6. He, X.; Zhang, D.; Yang, L.; Cui, T.; Ding, Y.; Zhong, X. Design and Experiment of a GPS-Based Turn Compensation System for Improving the Seeding Uniformity of Maize Planter. *Comput. Electron. Agric.* **2021**, *187*, 106250. [CrossRef]
7. Mishler, B.D. Quantifying Seed Uniformity and Yield Advantage of Precision Planter Technologies through Use of Field Tests and Machine Data. Master's Thesis, Kansas State University, Manhattan, KS, USA, 2021.
8. Nørremark, M.; Griepentrog, H.W.; Nielsen, J.; Søgaard, H.T. The Development and Assessment of the Accuracy of an Autonomous GPS-Based System for Intra-Row Mechanical Weed Control in Row Crops. *Biosyst. Eng.* **2008**, *101*, 396–410. [CrossRef]
9. Yang, L.; Yan, B.; Zhang, D.; Zhang, T.; Wang, Y.; Cui, T. Research Progress on Precision Planting Technology of Maize. *Trans. Chin. Soc. Agric. Mach.* **2016**, *47*, 38–48. [CrossRef]
10. Griepentrog, H.W. Seed Distribution Over The Area. In Proceedings of the AgEng. 1998: International Conference on Agricultural Engineering, Oslo, Norway, 24–27 August 1998; p. 98-A-059.
11. Panning, J.; Kocher, M.; Smith, J.; Kachman, S. Laboratory and Field Testing of Seed Spacing Uniformity for Sugarbeet Planters. *Appl. Eng. Agric.* **2000**, *16*, 7–13. [CrossRef]
12. Valentin, M.T.; Białowiec, A.; Karayel, D.; Jasinskas, A.; Ciolkosz, D.; Lavarias, J.A. Investigation of the Performance of a Cylindrical Hopper and Metering Device of a Carrot Seeder. *Sci. Rep.* **2023**, *13*, 813. [CrossRef] [PubMed]
13. Kamgar, S. Eslami Design, Development and Evaluation of a Mechatronic Transmission System for Upgrading Performance of a Row Crop Planter. *Int. J. Agron. Plant Prod.* **2013**, *4*, 480–487.
14. Wang, Y.; Zhang, W.; Qi, B.; Xia, Q. Comparison of Field Performance of Different Driving Systems and Forward Speed Measuring Methods for a Wet Direct Seeder of Rice. *Agronomy* **2022**, *12*, 1655. [CrossRef]
15. He, X.; Cui, T.; Zhang, D.; Wei, J.; Wang, M.; Yu, Y.; Liu, Q.; Yan, B.; Zhao, D.; Yang, L. Development of an Electric-Driven Control System for a Precision Planter Based on a Closed-Loop PID Algorithm. *Comput. Electron. Agric.* **2017**, *136*, 184–192. [CrossRef]
16. Wang, Y.; Zhang, W.; Qi, B.; Ding, Y.; Xia, Q. Research on Control System of Corn Planter Based on Radar Speed Measurement. *Agronomy* **2024**, *14*, 1043. [CrossRef]
17. MAESTRO SV/SX. Available online: https://www.horsch.com/en-ca/products/planting/maestro/maestro-sv/sx (accessed on 10 September 2024).

18. vDrive | Drive Systems. Available online: https://www.precisionplanting.com/products/planters/vdrive (accessed on 10 September 2024).
19. CHRONO—International—Corporate Website. Available online: https://www.maschiogaspardo.com/en/web/international/chrono (accessed on 10 September 2024).
20. Li, Y.; Xiantao, H.; Tao, C.; Dongxing, Z.; Song, S.; Zhang, R.; Mantao, W. Development of Mechatronic Driving System for Seed Meters Equipped on Conventional Precision Corn Planter. *Int. J. Agric. Biol. Eng.* **2015**, *8*, 1–9. [CrossRef]
21. Cay, A.; Kocabiyik, H.; May, S. Development of an Electro-Mechanic Control System for Seed-Metering Unit of Single Seed Corn Planters Part II: Field Performance. *Comput. Electron. Agric.* **2018**, *145*, 11–17. [CrossRef]
22. He, X.; Ding, Y.; Zhang, D.; Yang, L.; Cui, T.; Zhong, X. Development of a Variable-Rate Seeding Control System for Corn Planters Part I: Design and Laboratory Experiment. *Comput. Electron. Agric.* **2019**, *162*, 318–327. [CrossRef]
23. Ding, Y.; Yang, L.; Zhang, D.; Cui, T.; Li, Y.; Zhong, X.; Xie, C.; Ding, Z. Novel Low-Cost Control System for Large High-Speed Corn Precision Planters. *Int. J. Agric. Biol. Eng.* **2021**, *14*, 151–158. [CrossRef]
24. Yang, S.; Wang, X.; Gao, Y.; Zhao, X.; Dou, H.; Zhao, C. Design and Experiment of Motor Driving Bus Control System for Corn Vacuum Seed Meter. *Nongye Jixie Xuebao/Trans. Chin. Soc. Agric. Mach.* **2019**, *50*, 57–67. [CrossRef]
25. Du, Z.; Yang, L.; Zhang, D.; Cui, T.; He, X.; Xiao, T.; Xing, S.; Xie, C.; Li, H. Development and Testing of a Motor Drive and Control Unit Based on the Field-Oriented Control Algorithm for the Seed-Metering Device. *Comput. Electron. Agric.* **2023**, *211*, 108024. [CrossRef]
26. Paraforos, D.S.; Sharipov, G.M.; Griepentrog, H.W. ISO 11783—Compatible Industrial Sensor and Control Systems and Related Research: A Review. *Comput. Electron. Agric.* **2019**, *163*, 104863. [CrossRef]
27. Gao, Y.; Wang, X.; Yang, S.; Zhai, C.; Zhao, X.; Zhao, C. Development of CAN-Based Sowing Depth Monitoring and Evaluation System. *Nongye Jixie Xuebao/Trans. Chin. Soc. Agric. Mach.* **2019**, *50*, 15–28. [CrossRef]
28. Ramesh, P.; Umavathi, M.; Bharatiraja, C.; Ramanathan, G.; Athikkal, S. Development of a PMSM Motor Field-Oriented Control Algorithm for Electrical Vehicles. *Mater. Today Proc.* **2022**, *65*, 176–187. [CrossRef]
29. Wang, X.; Liu, N.; Na, R. Simulation of PMSM Field-Oriented Control Based on SVPWM. In Proceedings of the 2009 IEEE Vehicle Power and Propulsion Conference, Dearborn, MI, USA, 7–10 September 2009; pp. 1465–1469.
30. Hao, Y.; Cui, T.; Bora, G.; Zhang, D.; Wei, J.; He, X.; Wang, M.; Yang, L. Development of an Instrument to Measure Planter Seed Meter Performance. *Appl. Eng. Agric.* **2017**, *33*, 31–40. [CrossRef]
31. *ISO 7256-1:1984*; Sowing Equipment—Test Methods Part 1: Single Seed Drills (Precision Drills). ISO: Geneva, Switzerland, 1984.
32. Yan, B.; Wu, G.; Xiao, Y.; Mei, H.; Meng, Z. Development and Evaluation of a Seed Position Mapping System. *Comput. Electron. Agric.* **2021**, *190*, 106446. [CrossRef]
33. Wang, S.; Zhao, B.; Yi, S.; Zhou, Z.; Zhao, X. GAPSO-Optimized Fuzzy PID Controller for Electric-Driven Seeding. *Sensors* **2022**, *22*, 6678. [CrossRef]
34. Wang, Y.; Xie, W.; Chen, H.; Day-Uei, L.D. High-Resolution Time-to-Digital Converters (TDCs) with a Bidirectional Encoder. *Measurement* **2023**, *206*, 112258. [CrossRef]
35. Parish, R.L.; Bracy, R.P. An Attempt to Improve Uniformity of a Gaspardo Precision Seeder. *Horttech* **2003**, *13*, 100–103. [CrossRef]

agriculture

MDPI

Article

Design and Experiment of Compound Transplanter for Sweet Potato Seedling Belt

Wei Yan [1,2], Wenyi Zhang [1], Minjuan Hu [1], Yao Ji [1], Kun Li [1], Zhaoyang Ren [1,2] and Chongyou Wu [1,*]

[1] Nanjing Institute of Agricultural Mechanization, Ministry of Agriculture and Rural Affairs, Nanjing 210014, China; yanwei@caas.cn (W.Y.); zhangwenyi@caas.cn (W.Z.); huminjuan@caas.cn (M.H.); jiyao@caas.cn (Y.J.); likun03@caas.cn (K.L.); 82101222108@caas.cn (Z.R.)

[2] Graduate School, Chinese Academy of Agricultural Sciences, Beijing 100083, China

* Correspondence: wuchongyou@caas.cn

Abstract: To address the issues of high labor intensity, excessive manpower requirements, low planting spacing qualification rates, low planting depth qualification rates, and low operational efficiency associated with sweet potato transplanting, a sweet potato seedling belt transplanter has been designed. This machine can perform multiple processes: precision tillage and ridge shaping, orderly seedling feeding from rolls, the efficient separation of seedlings from the belt, flexible gripping and shaping, precise soil covering and the mechanism of exposing seedling tips. A three-factor, three-level orthogonal test was carried out using the forward speed of the machine, the pitch of the screw belt and the rotational speed of the screw as the influencing factors of the performance test, and the qualified rate of planting spacing and the qualified rate of planting depth as the evaluation indexes. The test results indicated that the significance order of the factors affecting the qualification rate for planting spacing with the optimal combination of factors was as follows: a forward speed of $0.3\ m\cdot s^{-1}$, a ribbon spacing of 60 mm, and a screw speed of 160 rpm. Field trials confirmed that under optimal conditions, the average qualification rate for planting spacing was 90.37%, meeting relevant technical standards and agronomic requirements.

Keywords: sweet potato; seedling belt transplanting; horizontal planting; seedling belt separation; mulching and matting tips

Citation: Yan, W.; Zhang, W.; Hu, M.; Ji, Y.; Li, K.; Ren, Z.; Wu, C. Design and Experiment of Compound Transplanter for Sweet Potato Seedling Belt. *Agriculture* **2024**, *14*, 1738. https://doi.org/10.3390/agriculture14101738

Academic Editor: John Fielke

Received: 5 September 2024
Revised: 29 September 2024
Accepted: 1 October 2024
Published: 2 October 2024

1. Introduction

China is the largest producer of sweet potatoes globally, with an annual planting area of approximately 2.2 million hectares, accounting for about 30% of the world's total, and an annual output of 48 million tons, approximately 53% of global production [1]. In recent years, with rising living standards and growing health awareness, the demand for fresh sweet potatoes has increased significantly, placing higher quality requirements on sweet potato production and consumption [2]. With the rapid development of social and economic as well as rural labor transfer, labor costs increase year by year, and the mechanized transplanting technology and equipment needs in sweet potato production areas urgently need more and more machinery to replace the dependence on many workers [3]. Currently, sweet potato transplanting accounts for about 23% of the labor used in the entire production process. Consequently, a substantial amount of transplanting is still carried out by labor, which is costly and uneven in quality, resulting in poor product marketability and low planting efficiency, thereby hindering the healthy and high-quality development of China's sweet potato industry [4]. Due to the requirement of bare-root transplanting, the morphology of sweet potato seedlings is complex, with tangled vines and inconsistent shapes, which complicates the mechanization of horizontal transplantation. Currently, transplantation primarily relies on manual labor or semi-automatic methods involving manual single-seedling placement. Semi-automatic transplanting requires two to three workers per row, with an average planting spacing pass rate of 85%, a planting depth pass rate of 80%, and an average transplanting machine operating efficiency of 35 plants per minute. The transplanting link has become an urgent issue in sweet potato production.

In order to solve the above problems, the author developed a sweet potato belt transplanter based on artificial seedling feeding, which solved the problem of a low qualified rate of planting distance and planting depth of transplanter. In terms of employment, the single-line operation has changed from the original two to one. In the production process, it is found that although the working efficiency of the transplanter is improved by 10% compared with the existing transplanter, it is difficult to popularize it. On this basis, in order to improve the working efficiency of the transplanter, an innovative technical solution has been proposed, incorporating "seedling belt feeding from rolls + automatic seedling belt separation + adaptive speed matching for the belt + flexible grip adjustment + stable shaping of seedlings". This technological approach involves breakthroughs in key techniques such as tension stability control, adaptive speed matching, seedling feeding from rolls, the automatic separation of seedlings, flexible gripping, and the shaping of seedlings. The design of the sweet potato seedling belt transplanter effectively reduces production costs, lessens labor intensity, enhances product quality, and increases economic efficiency. This development is significant for stabilizing planting scales, improving comprehensive benefits, enhancing market competitiveness, and promoting the green and efficient development of the sweet potato industry in China. Furthermore, it also provides an effective solution for the efficient transplantation of seedling cuttings.

2. Sweet Potato Seedling Transplanting Agronomy

2.1. Sweet Potato Seedling Mechanization Cultivation Model

In terms of cultivation methods, both ridge and flat systems are practiced in sweet potato production, predominantly with ridges. Various ridge configurations are used depending on production scenarios [5]. To address the compatibility issues related to suitable mechanized production operations, this proposal encompasses various stages, including ridging, transplanting, tillage, vine shredding, and harvesting. The approach emphasizes cost-effectiveness while integrating diverse operational scenarios and existing tractor availability across different regions, fostering the integration of agricultural machinery and agronomy. The recommended mechanized cultivation modes include single-ridge single-row, single-ridge double-row, and large-ridge double-row cultivation. From the perspective of mechanization, a single-ridge single-row cultivation model is recommended, as illustrated in Figure 1.

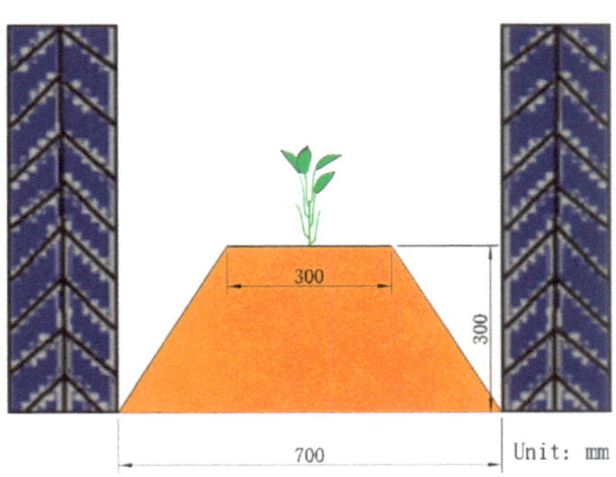

Figure 1. Single-ridge single-row cultivation model.

This model allows for the full mechanization of tillage, planting, management, and harvesting using a single tractor. It offers advantages such as high economic efficiency, simplicity in setup, broad adaptability, and low investment requirements, making it suitable for medium- and small-plot operations in most regions. In flat lowland areas, it can be paired with medium-sized four-wheeled tractors, while in hilly or small plots, walking tractors or mini tillers can be utilized. However, for larger fields, the machinery can be configured for multi-row operations to enhance efficiency.

2.2. Sweet Potato Seedling Transplanting Methods

Due to the tuber formation characteristics of sweet potatoes, sweet potato seedlings cannot be transplanted in soil-covered plug trays like vegetables. Instead, they must undergo a seedling cultivation process followed by cutting and then bare-root transplantation. Sweet potato mainly has five traditional planting methods: horizontal planting, diagonal planting, boat-bottom shaped planting, straight planting, and vine-pressing planting [6], as shown in Figure 2. The horizontal planting method requires the seedling to be planted in the soil part of the horizontal posture in the basic conditions of the potato, where each section is mostly occupied by rooting potato, with a less empty section having more potatoes, potato size uniformity, and other advantages. The oblique planting method requires the seedling to be planted in the soil part of the vertical direction at a certain angle, where the number of single-plant potatoes compared with the horizontal planting method is less than the number of larger potatoes in the upper nodes, and the lower nodes have smaller potatoes or even no potatoes. The ship-bottom-shaped planting method requires the seedling to be planted in the soil part of the middle of the pressure, where the head and tail ends of the two thus resemble the bottom of a ship. The seedling is implanted into the soil section more, causing more nodes to be close to the soil surface, which is conducive to the potatoes and leads to higher yields; however, at the central part of the seedling in the soil, the deeper nodes tend to be small and have fewer potatoes, or they can even become empty nodes. The direct planting method requires that the seedling be inserted directly into the soil. Due to the deeper direct insertion into the soil, only a few nodes are distributed in the topsoil layer of the potato species; the general number of single-plant potatoes is small; the potato has more than a few kinds of nodes in the upper part of the plant; expansion is fast; and the rate of large potatoes is high. The pressure vine planting method requires the top of the potato seedlings to be buried in the soil at both ends. The leaves are all exposed to the ground and results in more potatoes that are large; the stems and leaves are not easy to grow. It has high yield advantages, but poor drought resistance, and requires more planting labor with only a small area of high-yield cultivation.

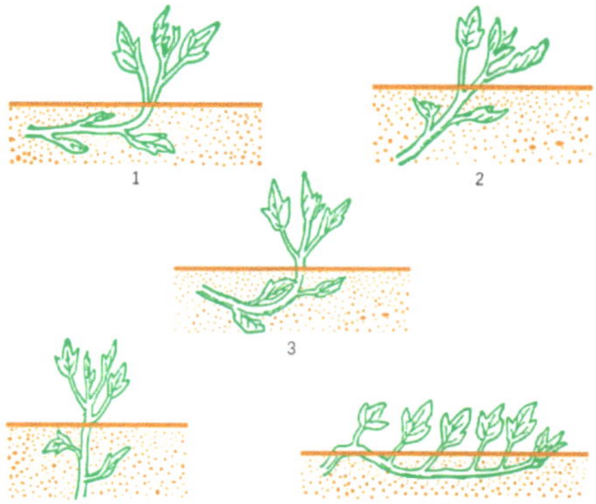

Figure 2. Five traditional methods of planting sweet potato: 1. horizontal planting; 2. inclined planting; 3. boat-shaped planting; 4. vertical planting; and 5. layering planting.

A comparative analysis of these five different planting methods, in conjunction with high-yield cultivation techniques, indicates that the current predominant practices involve horizontal, inclined, and vertical planting methods. Horizontal and inclined planting methods are primarily suitable for fresh sweet potato seedlings, while vertical planting is mainly used for waxy sweet potato seedlings. Horizontal planting has advantages over inclined planting, including higher tuber yield, uniformity in tuber size, better marketability, and an overall higher productivity, making it the leading method for fresh sweet potato seedling transplantation. For horizontal planting, the seedling length should be approximately 30 cm, with a horizontal planting depth of about 15 cm and a planting depth of 5~8 cm.

3. Design of the Sweet Potato Seedling Belt with Compound Transplanter

3.1. General Structure of the Machine

The overall structure of the sweet potato seedling belt with compound transplanter is schematically shown in Figure 3, which mainly consists of key components such as sliding rail, suspension device, hydraulic oil pipe, hydraulic cylinder, seedling roll belt, seat, angle adjustment, seedling tray, motor, support wheel, suppression wheel, mulching screw, furrow opener, seedling conveyor belt, frame, tension roller, monopoly shaping plate, soil crushing roller, rotary tillage cutter shaft and so on.

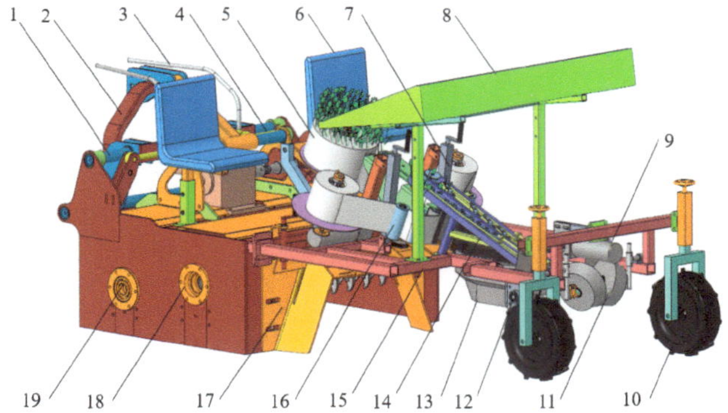

Figure 3. Structural diagram of the sweet potato seedling belt with compound transplanter: 1. sliding rail; 2. suspension device; 3. hydraulic oil pipe; 4. hydraulic cylinder; 5. seedling roll belt; 6. seat; 7. angle adjustment; 8. seedling tray; 9. motor; 10. support wheel; 11. suppression wheel; 12. mulching screw; 13. furrow opener; 14. seedling conveyor belt; 15. frame; 16. tension roller; 17. monopoly shaping plate; 18. soil crushing roller; and 19. rotary tillage cutter shaft.

3.2. Working Principle

During operation, the machine is connected to the rear of a tractor via a three-point hitch and coupled to the tractor's power output shaft using a universal joint. The rotary tillage part cuts and pulverizes the soil and throws it to the crusher rollers, which further crush the soil before the ridge shaping plate shapes it, completing the ridge-making operation. Manually, the rolled seedling belt is loaded onto the carrying device, and the dual-layer seedling belt is initially separated and bonded to the separation roller. A motor drives the separation roller to separate and collect the seedling belt. Once separated, the seedlings belt enters a conveyor belt driven by a motor, which transports the seedlings diagonally downward into the prepared seedling ditch. Once the roots touch the bottom of the ditch, the seedlings continuously bend and take a horizontal position. After the seedlings detach from the conveyor belt, the soil covering device performs soil covering and exposes the seedling tips to complete the seedling transplantation. Simultaneously, the bidirectional offset mechanism can be adjusted, allowing the equipment to operate along the edges of the field, between ridges, and in "S"-shaped operations.

3.3. Main Technical Parameters

According to the appropriate mechanization mode of operation and sweet potato horizontal planting agronomic requirements in sweet potato production, the design of the transplanting machine needs to meet the operation requirements of a ridge height of 300 mm and a distance between the center of the two ridges (ridge spacing) of 900 mm. According to the requirements of the planting of fresh potatoes, a planting spacing of 200~300 mm, a planting depth of 50~80 mm, and a planting spacing qualification rate of

≥90% are needed. In order to adapt to small-field operations, the machine needs to be compact and have low power consumption. The main technical parameters of sweet potato seedling with the compound transplanter are shown in Table 1.

Table 1. Main parameters of sweet potato seedling with compound transplanter.

Item	Parameter
Dimension of whole machine/(mm × mm × mm)	2450 × 1100 × 1150
Power requirements/kW	≥36
Ridge height/mm	300
Ridge spacing/mm	900
Planting spacing/mm	200~300
Planting depth/mm	50~80
Number of working rows	1
Qualified rate of planting spacing/%	≥90

4. Design of Key Component Parameters

4.1. Design of Key Parameters for Seedling Belt Carrying Device

4.1.1. Design of Seedling Belt Loading Program

Currently, sweet potato seedling transplanting machines primarily operate as semi-automatic systems with manual seedling feeding. The single-row operational efficiency of the equipment is limited by the manual feeding efficiency, resulting in relatively low overall productivity. Without improvements in feeding efficiency, significant enhancements in the machine's single-row operation efficiency are unlikely. To address these challenges, this section focuses on improving seedling feeding efficiency by innovatively proposing a seedling belt feeding technology solution. This approach constrains irregular and unordered sweet potato seedlings to the seedling belt in a standardized, orderly manner, allowing for specification and organization of the seedlings. The feeding method transitions from individual manual feeding to feeding entire rolls of the seedling belt. These seedling belt carrying devices can accommodate 50, 100, 200, or more seedlings as needed, leading to substantial improvements in feeding efficiency. Based on the previously measured geometric parameters and mechanical characteristics of sweet potato seedling plants, the seedling belt design was developed, as illustrated in Figure 4.

Figure 4. Design of the sweet potato seedling belt carrying device.

The designed seedling belt carrying device consists of two layers with a width of 150 mm, featuring designated planting positions at specified intervals. To facilitate manual planting, the upper openings of these positions have a width of 40 mm, while the lower openings measure 20 mm. The seedlings are placed from the upper opening to the lower opening, with roots positioned at the bottom of the lower opening, ensuring a spacing of 200 mm between seedlings. The preliminary selection of the seedling belt's materials includes paper-based and polymer-based dual-layer belts. The bonding methods for the seedlings belts can be continuous, intermittent, or dot-style. Experimental research has shown that under the same temperature, pressure, and sealing time conditions, the dot-style structure is easier to separate. Therefore, the bonding points of the seedling belt are designed as semicircular structures, with bonding methods chosen based on the materials used—either adhesive or point-sealing techniques. In conclusion, the differing sizes of the upper and lower openings and the semicircular design of the bonding points facilitate manual seedling placement. Additionally, the dual-layer seedling belt requires only minimal separation force during detachment. The specific structural design of the seedling belt carrying device is shown in Figure 5.

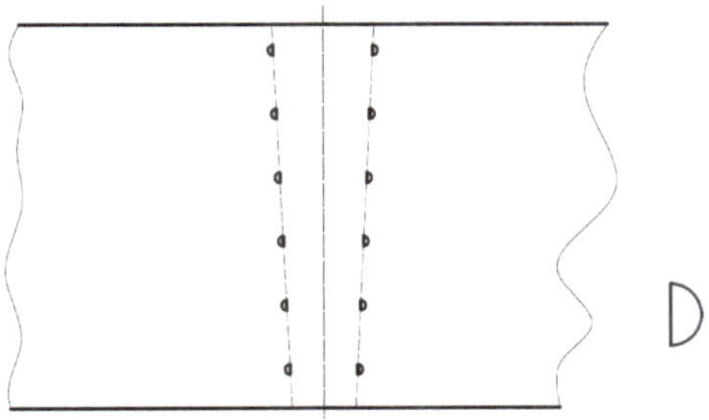

Figure 5. Structural configuration of the sweet potato seedling belt carrying device.

4.1.2. Experimental Research on Seedling Belt Carrying Device

To verify the suitability of different materials for the seedling belt carrying device and to determine the optimal material, experiments were conducted on paper-based carrier belts and polymer-composite carrier belts for sweet potato seedlings.

The paper-based carrier belt is constructed by bonding two layers of paper with adhesive, allowing for the manual insertion of the sweet potato seedlings into the designated slots of the belt. Experimental results indicate that sweet potato seedlings tend to lose moisture and wilt after a period when placed in the paper-based carrier belt. Additionally, the paper belt can easily break when exposed to water, making it unsuitable for watering or moisture-retaining operations, ultimately failing to maintain the freshness of the seedlings. Furthermore, the separation of the two layers of the paper carrier belt post-bonding is challenging, the bonding process is complex, and the cost of paper belts is relatively high, as shown in Figure 6.

Figure 6. Paper seedling carrier.

In contrast, the polymer-composite carrier belt is bonded using a thermal pressing technique to secure the two layers at specified points and distances. This method also allows for the manual insertion of the seedlings into the carrier belt's designated slots before rolling into rolls of polymer-composite carrier belts. To ensure the vitality of the seedlings, these polymer-composite rolls can be placed in a water tray or nutrient solution; however, care must be taken not to submerge them for too long, as prolonged exposure can lead to root bending, negatively affecting the subsequent growth of the tubers. Experimental findings demonstrate that the polymer-composite belts exhibit high strength, excellent flexibility, and low cost. When planting at a density of 3000~4000 seedlings per acre, the cost of the carrier belt is estimated to be between 40 and 50 RMB per acre, which aligns well with user production requirements. Consequently, polymer-composite carrier belts are preferred, as illustrated in Figure 7.

Figure 7. Polymer seedling carrier.

4.2. *Design of Key Parameters for Offset Tillage and Ridge Formation*

4.2.1. Design of the Bidirectional Offset Device

To address issues such as the mismatched wheel spacing of tractors, incomplete coverage in hilly or small plots, and difficulties in switching rows during ridge cultivation operations [7], a hydraulic bidirectional offset mechanism has been designed. This mechanism allows the equipment to operate close to field edges and rows, as well as execute "S"-shaped movements. Furthermore, it enables the equipment to align with the ridge edges during the end-of-row operations, minimizing the distance between neighboring ridges, improving equipment adaptability, and enhancing plot utilization.

The hydraulic bidirectional offset mechanism derives its power from the tractor. During operation, a hydraulic oil line connects to the tractor's hydraulic interface. Depending on the specific offset operation required, a manual directional valve controls the flow of hydraulic oil in and out of the left and right hydraulic cylinders, enabling lateral movement of the hydraulic rod, which is connected to the frame. This configuration allows the bidirectional offset mechanism to pivot around the three-point hitch to achieve lateral offset, as depicted in the schematic in Figure 8.

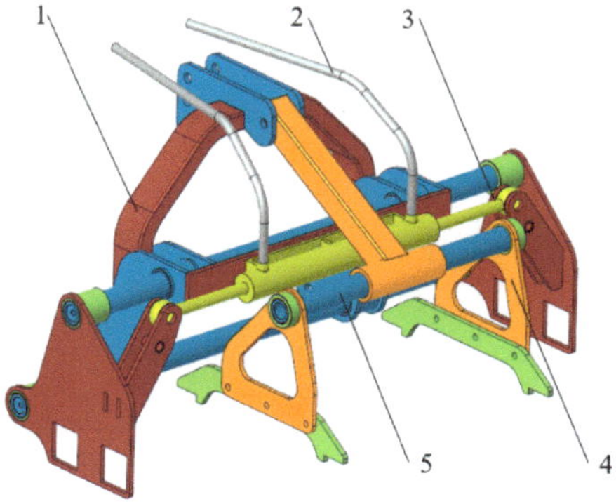

Figure 8. Structural diagram of the bidirectional offset mechanism: 1. linking device; 2. hydraulic oil hose; 3. hydraulic cylinder; 4. slide guide; and 5. connection plate.

Based on the parameters of tractors with power ratings between 50 and 70 horsepower, the left and right offset length of the mechanism is set at 20 cm, facilitating edge and aligned operations [8]. The hydraulic cylinder's body length is approximately 60 cm, while the stroke of the single-sided hydraulic rod is 20 cm. The load capacity of the hydraulic cylinder mainly arises from the sliding friction between the unit during translational movement and the frame. Based on trials of the initial prototype, the equipment weight is around 320 kg, translating to a gravitational force of approximately 3200 N. The installation position of the bidirectional offset mechanism is centrally located within the width of the three-point hitch with no imbalanced forces set for operation. The load force can be calculated using the following formula:

$$F = \mu G \tag{1}$$

In the formula, F is load force, N; μ is the coefficient of sliding friction; and G is the weight of the individual unit, N.

According to the Mechanical Design Handbook, the inner diameter of a hydraulic cylinder can be calculated using the following formula:

$$D = \sqrt{\frac{4F}{\pi P \times 10^6} + d^2} \tag{2}$$

In the formula, D—inner diameter, m; d—rod diameter, m; and P—actual load pressure, Mpa.

During actual operation, when the coefficient of sliding friction μ is 0.1, the rated output pressure of a hydraulic system on a 50~70 horsepower tractor is 16 MPa. Based on the Mechanical Design Handbook, if the gear ratio φ is taken as 1.46, d can be determined using the following formula:

$$d = D\sqrt{\frac{\varphi - 1}{\varphi}} \tag{3}$$

From this, Formula (3) can be expressed as follows:

$$D = \sqrt{\frac{146F}{25\pi P \times 10^6}} \tag{4}$$

By solving Equation (4), the result is calculated to be 6.09 mm. This indicates that a hydraulic cylinder inner diameter of 6.09 mm is sufficient to meet the operational requirements for bidirectional offset. Based on commonly used hydraulic cylinder specifications, a double-rod hydraulic cylinder with equal speed and stroke is selected. The main working parameters of the hydraulic cylinder are shown in Table 2.

Table 2. Hydraulic cylinder working parameters.

Name	Inner Diameter/mm	Rod Diameter/mm	Piston Stroke/mm
Double-rod hydraulic cylinder	40	20	200

4.2.2. Design of the Precision Tillage Device

Soil fragmentation is a critical factor affecting the quality of ridge shaping and transplanting [9,10]. To enhance soil fragmentation and improve the quality of ridge shaping and transplanting, this study utilizes a dual-structured "rotary tillage + soil crushing" precision tillage device. The dual-rotation mechanism is also effective in minimizing the effect of root stubble in the soil on the quality of starting and transplanting, as demonstrated in Figure 9.

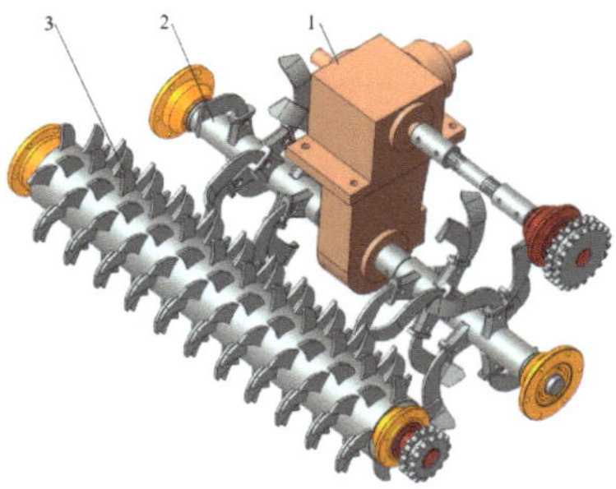

Figure 9. Structural schematic of the precision tillage device: 1. gearbox; 2. rotary tillage device; 3. soil crushing device.

The power for the fine tillage device is provided by the tractor, with a universal joint connecting the tractor's rear power output shaft to the gearbox of the rotary tiller. The gearbox, through gear transmission, supplies power to the rotary tillage device, while the side axle of the gearbox provides power to the soil crushing mechanism. During operation, the rotary tillage device throws the soil and root residues into the crushing device, where the soil clumps and residues are further cut into finer pieces. To optimize the soil crushing rate, the crushing blades of the device are arranged in an interleaved spiral configuration. Based on the preliminary transplanting test results, the lateral spacing for the crushing blades is designed to be 30 mm, satisfying the demands of the transplanting operations.

4.2.3. Design of the Ridge Shaping Device

The ridge shaping device primarily consists of a ridge shaping plate and an adjustment mechanism, as shown in Figure 10.

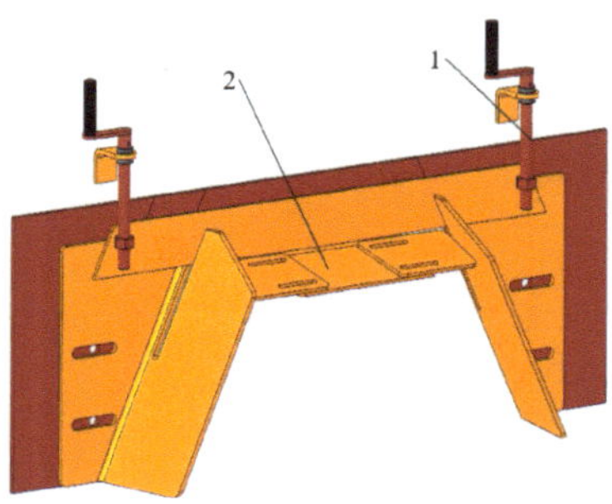

Figure 10. Structural diagram of the ridge shaping device: 1. ridge shaping plate and 2. adjustment mechanism.

To ensure soil permeability and ridge compaction, the formation plate is designed with a "trumpet" shape, with dimensions configured according to mechanized operational modes. The entrance of the ridge shaping plate is designed for a ridge height of 350 mm, a bottom width of 800 mm, and a top width of 350 mm, while the exit plate is designed for a ridge height of 300 mm, a bottom width of 700 mm, and a top width of 300 mm. Furthermore, to enhance the adaptability of the equipment, the ridge shaping plate can be adjusted laterally, and its height can be modulated via an adjustment lever.

4.3. Design of Key Parameters for Seedling Separation

4.3.1. Design of the Seedling Separation System

The seedling separation process refers to the stable separation and recovery of the double-layer seedling belt, allowing the sweet potato seedlings to be extracted from the roll and transferred in a single, orderly manner to the seedling transport belt. The key components of the seedling separation apparatus mainly include the active separation mechanism, the seedling holding mechanism, and the seedling belt tensioning mechanism, as shown in Figure 11.

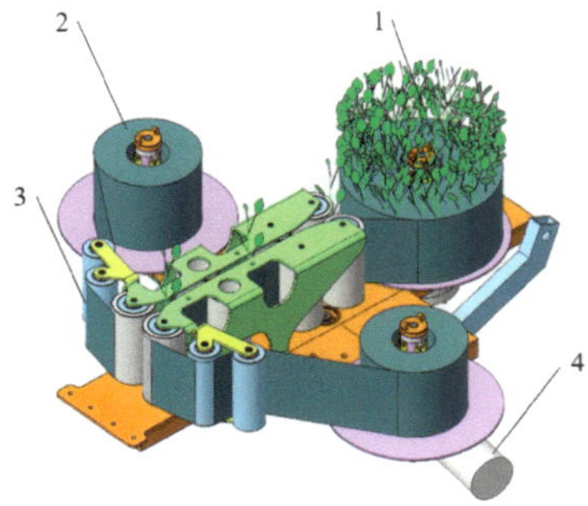

Figure 11. Structural diagram of the seedling separation device: 1. sweet potato seedling belt; 2. seedling separation device; 3. tensioning device; and 4. drive motor.

During operation, a motor drives the separation mechanism to perform a rotational motion. Through this rotational movement, the double-layer seedling belt is separated and collected into a roll. The seedling holding mechanism rotates around the axis of rotation to facilitate the seedling separation operation. To ensure an orderly and precise transfer of the separated seed potatoes to the transport device, it is essential to maintain continuous tension in the seedling belt throughout the process. This requires that both the active separation mechanism and the seedling holding mechanism rotate without self-rotation during their movements under the tensioning condition. Additionally, to enable the rapid loading and unloading of the rolled seedling belt and facilitate the quick detachment of the collected rolls by the separation mechanism, research on rapid replacement devices needs to be conducted.

4.3.2. Design and Selection of the Constant Tension Device

Seedling separation refers to the process of stably separating and recovering the double-layered seedling belt, allowing the sweet potato seedlings to be extracted from the carrier and conveyed individually and orderly onto the seedling delivery belt.

The seedling carrying mechanism produces a rotational motion under the influence of the separation mechanism. In the absence of a damping device, the rotational axis may easily self-rotate due to inertia, preventing the seedling belt from remaining in a tensioned state. To eliminate this issue and ensure that the seedling belt maintains constant tension during the separation process, a magnetic particle brake is designed, as illustrated in Figure 12.

Figure 12. Magnetic particle brake.

The magnetic particle brake utilizes a magnetic particle clutch, which transmits torque using magnetic particle clutches under the effect of electromagnetic force. When current flows through the coil, a magnetic field is generated, magnetizing the internal powder and causing it to attract itself, forming a strong magnetic coupling surface. As the current increases, the powder is activated in the magnetic field, which increases the friction of the clutch and enhances torque transmission between the input and output shafts. Conversely, when the current decreases or is cut off, the magnetization of the powder decreases, reducing the coupling force and thereby decreasing or disconnecting torque transmission. Therefore, by adjusting the current intensity, one can accurately control the friction and torque output of the clutch, enabling adjustments to the tension or rotational speed.

The constant tension is related to the characteristics of the seedling belt material, calculated using a specific formula:

$$F = k\sigma_s bh \tag{5}$$

In the formula, k is the tension coefficient; σ_s is the yield strength, MPa; b is the width of the seedling carrier, mm; and h is the thickness of the seedling carrier, mm. Here, $k < 1$; when $k = 1$, the tension in the carrier reaches the yield limit. After field adjustments, k is set to 0.4; $\sigma_s = 21$ MPa = 21 N·mm^{-2}; $b = 150$ mm, $h = 0.07$ mm, leading to a calculated constant tension $F = 88$ N.

According to the definition of torque, the maximum torque M_{max} for the film unwind roll can be calculated using the following formula:

$$M_{max} = \frac{F \times D_{max}}{2} \tag{6}$$

In the formula, M is the torque of the film unwind roll, N·m; F is the tension in the seedling carrier, N; and D is the diameter of the film unwind roll, m. With the maximum seedling carrier roll diameter D_{max} being 0.5 m, the calculated maximum torque M_{max} for the film unwind roll is 22 N·m. Therefore, when selecting a dual-shaft magnetic particle clutch, its rated torque must not be less than 22 N·m. The FL-25-S dual-shaft magnetic particle clutch is selected, with its performance parameters listed in Table 3 below.

Table 3. Magnetic particle clutch performance parameters.

Name	Parameter
Product model	FL-25-S
Rated torque/(N·m)	25
Rated voltage (V)	24
Rated speed (rpm)	1400
Rated current (A)	1.5

This paper selects the FL-25-S dual-shaft magnetic particle clutches, which allows adjustable torque ranging from 0 to 25 N·m and a rotational speed of 0~1400 rpm, meeting the requirements for maintaining constant tension in the seedling belt during roll operations.

Experimental studies indicate that the electromagnetic damping device maintains stable tension during the rotation of the seedling belt, keeping it in a tensioned state. The damping device ensures constant tension, which can also be adjusted based on need, possessing advantages of convenience, stability, and reliability.

4.3.3. Design of the Quick-Change Device

Through research, it has been found that the core shaft of the seedling roll production companies is commonly a cylindrical shape with a certain wall thickness, in accordance with relevant standards. The square coordinate fit for the core shaft is relatively rare. When using a cylindrical core shaft, a hinged quick replacement device has been designed to achieve the rapid switching of the rolled seedling belt and separation mechanism after coiling, while also preventing the rolling of the seedling belt and the coiled collection belt, as shown in Figure 13.

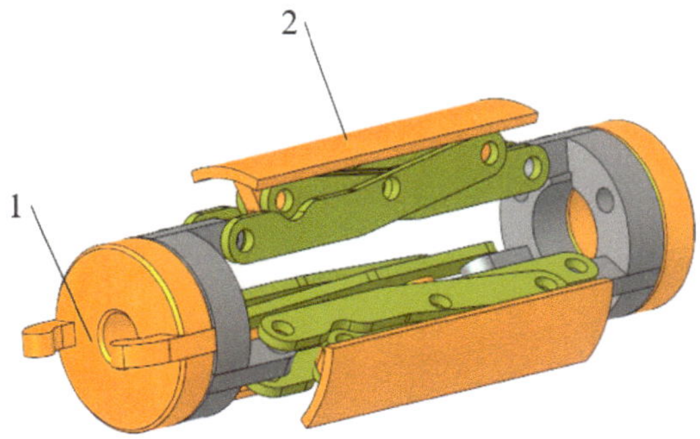

Figure 13. Structural diagram of the quick-change device: 1. locking mechanism and 2. support device.

The inner diameter of the seedling core shaft is 60 mm, and to standardize dimensions, the inner diameter of the collection belt shaft is also set to 60 mm. The design of the quick replacement device allows for a contraction and expansion range of 59~61 mm. During operation, the cylindrical core shaft is fitted into the replacement device, and by rotating the adjustment mechanism, the hinged structure is raised to facilitate quick replacement, which also prevents self-rotation. Conversely, the hinged structure can be contracted for rapid disassembly.

4.4. Design of Key Parameters for Horizontal Positioning and Shaping

4.4.1. Principle of Horizontal Positioning and Shaping

The process of achieving a horizontal posture of the seedlings refers to the horizontal positioning of the portion of the seed potatoes that is planted in the soil, as illustrated in Figure 14. During this horizontal positioning process, it is crucial to avoid damage to the seedling caused by the gripping components, as well as to prevent bending damage to the seedling roots when they touch the ground. To minimize gripping damage to the seedlings, the gripping component employs a flexible design made with double-layer sponge belts. Based on the previous experiments on seedling bending, it was determined that when the bending angle of the seedlings exceeds 90 degrees, bending damage is likely to occur. To avoid this, after achieving a horizontal posture, it is optimal for the angle between the seedling tip and the horizontal direction to be around 80 degrees. Furthermore, considering the physical characteristics of the seedlings, the width of the gripping component should not be excessively wide, as a wider grip may lead to the phenomenon where the stem separates from the transport belt while the leaf petiole and leaves remain attached. Through experimental research on different widths of the seedling belt, a width of 30 mm has been determined as suitable.

After the seedlings are separated from the seedling belt, they are conveyed downward at an angle by the grabbing component until the seedling roots touch the ground and bend to meet the horizontal planting requirement. To ensure a smooth transition of the seedlings onto the transport belt, the transport belt must simultaneously grip the seedlings before separation. Additionally, during the angled downward transportation of the seedlings, to guarantee the stability of the gripping posture and prevent the seedlings from falling during transport, an elastic pressing device should be added to the inner side of the transport belt to maintain stable gripping on the seedlings.

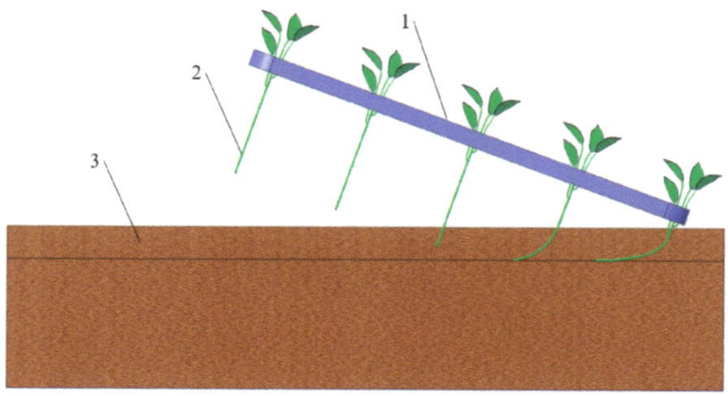

Figure 14. Schematic diagram of horizontal positioning and shaping: 1. seedling delivery belt; 2. sweet potato seedlings; and 3. ridge body.

4.4.2. Motion Analysis for Horizontal Positioning and Shaping

The movement relationships during the downward transportation of the seedlings on the transport belt are illustrated in Figure 15.

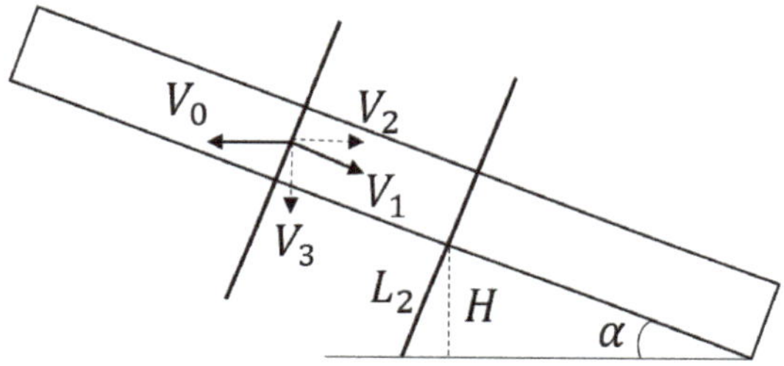

Figure 15. Motion analysis diagram for horizontal positioning and shaping.

In the figure, V_0 represents the forward speed of the machinery, V_1 is the speed of the transport belt, V_2 is the horizontal component speed of the transport belt, and V_3 is the vertical component speed of the transport belt. The length from the seedling root to the edge of the transport belt is L_2, and the distance from the edge of the transport belt to the bottom of the ditch is H. Once the machinery is operating stably, for the seedlings to achieve a horizontal posture when the roots touch the ground, the seedlings must exhibit a tendency to move in the opposite direction of the machinery's forward motion. The following conditions must be met:

$$V_2 = V_1 \cos\alpha > V_0 \tag{7}$$

If the specified planting distance for the seedlings is L_0, and the spacing between seedlings on the belt is L_1, the time required to plant the next seedling after the previous one is Δt. The transplanting machine must maintain stable planting spacing throughout the process, then

$$\Delta t = \frac{L_1 \sin\alpha}{V_3} \tag{8}$$

V_3 is the vertical speed of the transport belt, then

$$V_3 = V_1 \sin\alpha \tag{9}$$

Substituting Equation (9) into (8) yields

$$\Delta t = \frac{L_1}{V_1} \tag{10}$$

Based on the characteristics of horizontal posture formation, the planting distance can be expressed as

$$L_0 = L\cos\alpha - (V_2 - V_0)\Delta t \tag{11}$$

After simplification,

$$V_1 = \frac{L_1}{L_0} \times V_0 \tag{12}$$

When the seedlings separate from the transport belt, the conditions to achieve a horizontal posture must also be satisfied:

$$V_1 \sin\alpha\Delta t \geq H \tag{13}$$

That is,

$$V_1 \sin\alpha\Delta t \geq L_2 \cos\alpha \tag{14}$$

After simplification,

$$L_1 \sin\alpha \geq L_2 \cos\alpha \tag{15}$$

$$\cos\alpha \geq \sin\alpha \tag{16}$$

In summary, the angle between the transport belt and the horizontal is $\alpha \leq 45°$; this paper designates the angle between the transport belt and the horizontal to be 25°.

4.5. Design of Key Parameters for Planting Upright Seedlings

4.5.1. Design of the Ditching and Depth-Fixing Part

The main function of the grooving and depth setting parts is to open the seedling groove to prepare for the potato seedling to land on the bed, mainly including the grooving piece, connecting plate, etc. According to the potato seedling and seedling conveyor belt, the groove opener is designed as a large upper and small lower structure, as shown in Figure 16. The ditching and depth-fixing part is firmly attached to the frame through connecting plates. According to the agronomic requirements for sweet potato seedling planting, the ditch depth is designed to be 80 mm, with an adjustable ditch depth range of 0~80 mm. According to the physical characteristics of the potato seedlings, in order to minimize the collision of the seedlings with the side wall of the groove opener when they are transported diagonally downward, the width of the groove is designed to be 50 mm. To allow the seedling transport component to be submerged into the ditch opener, the upper opening of the ditch device is designed to be 190 mm wide. In order to make the trenching and depth setting parts have better soil breaking performance, after trenching to form a "clean zone" inside the trenching and depth setting parts, the design of the two trenching pieces into the soil at an angle of 35° can effectively reduce the impact of the soil, weeds, etc., on the transplanting [11–13], and the position of its position before and after is adjustable, with an adjustable range of 0~50 mm.

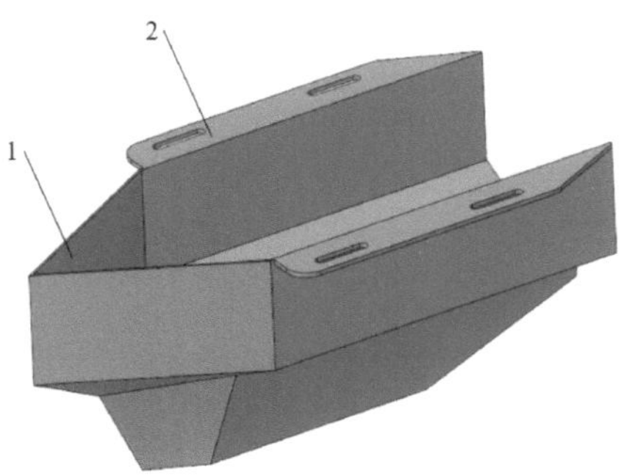

Figure 16. Structural diagram of the ditching and depth-fixing mechanism: 1. opening blade and 2. connection plate.

4.5.2. Design of Key Parameters for Soil Covering and Exposing Seedling Tips

Once the seedlings have achieved a horizontal posture and are about to be released from the seedling transport belt, soil covering of the seedling roots is required while ensuring that the seedling tips remain exposed above the soil surface, referred to as the "soil covering and exposing seedling tips". To achieve this operation, an active screw soil covering and tip exposure mechanism has been designed.

The active screw soil covering and exposing seedling tip mechanism mainly consists of a driving motor, connecting plate, screw mechanism, transmission shaft, chain case, and position adjustment plate, as shown in Figure 17. The motor drives the screw mechanism to perform the soil covering and exposes the seedling tips via chain transmission. The position of the screw mechanism can be adjusted both vertically and horizontally using the adjustment plate.

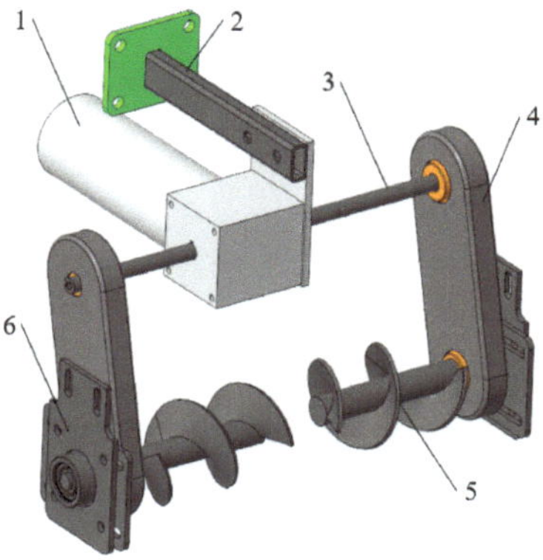

Figure 17. Active spiral soil covering and exposing seedling tip components: 1. drive motor; 2. connection plate; 3. drive shaft; 4. chain cover; 5. spiral mechanism; and 6. position adjustment plate.

Parameter Design of the Screw Mechanism

The soil covering process requires soil particles to move outward along the axial direction, meaning that the velocity vector of the particles must move from the inside toward the outside (with a friction angle relative to the normal of the screw surface) [14–17]. The conditions for this are as follows:

$$\tan\alpha \leq 1/\tan\varphi \tag{17}$$

In the formula, α is the screw angle (°) and φ is the friction angle between the soil and the steel plate (°). Since the screw angle varies at different points along the blade, with the maximum screw angle at the minimum radius, it is sufficient for the screw angle at the inner diameter to meet the conditions.

The theoretical volume of soil moved by the screw mechanism in the ditch can be considered as the volume of the furrow created, which can be modeled as a rectangular prism. The volume of soil required to cover the ditch can be expressed as

$$V = bvh \tag{18}$$

where b is the working width (m); v is the forward speed (m·s^{-1}); and h is the tillage depth (m). Based on the agronomic requirements for the sweet potato seedling transplanter and to improve transplanting efficiency, the operational speed of the tractor is set to be between 0.1 and 0.4 m·s^{-1}. With a ditch width of 50 mm and a ditch depth of 80 mm, the volume of soil required to cover the ditch is calculated to be approximately 4×10^{-4}~1.6×10^{-3} m^3·s^{-1}.

Referencing the design methods for open screw conveyors, to ensure stable soil movement and prevent clogging, the screw covering mechanism must meet the following requirements:

The conveying capacity of the screw mechanism must exceed the soil input rate, or sludging may occur.

The soil must not exhibit vertical jumping or rolling perpendicular to the direction of conveyance.

The soil must be conveyed axially, meaning that both axial forces and axial velocity must be greater than 0.

Based on the design principles of screw conveyors, the formulas for the design parameters of the screw mechanism are as follows:

Spiral outer diameter:

$$D \geq K \sqrt[2.5]{\frac{V}{\varphi \varepsilon \rho}} \tag{19}$$

Shaft diameter:

$$d = (0.2 \sim 0.35)D \tag{20}$$

Spiral speed:

$$n \leq n_{max} = \frac{A}{\sqrt{D}} \tag{21}$$

Screw pitch:

$$S_{max} = (0.5 \sim 2.2)D \tag{22}$$

In the formula, V is the amount of mud transported, $m^3 \cdot s^{-1}$; K is the comprehensive characteristic coefficient of the mud; φ is the filling coefficient; ε is the slope coefficient; ρ is the bulk density of the conveyed material, $t \cdot m^{-3}$; and A is the comprehensive characteristic coefficient of the material.

Since the left and right soil covering screws are arranged symmetrically, the mud transport capacity of each screw mechanism should be half of the total soil required to fill the seedling ditch, and thus is set to range between 2×10^{-4} and 0.8×10^{-3} $m^3 \cdot s^{-1}$. The topsoil for sweet potato seedling transplanting has a certain fluidity, referring to the design standards for screw conveyors, and the estimated comprehensive characteristic coefficient for the soil K is taken as 0.0415. The filling factor φ is set to 0.4; since the spirals are positioned horizontally, the inclination coefficient ε is set to 1.0; the comprehensive characteristic coefficient A for the soil is taken as 75; and the bulk density ρ of the soil is set at 1.8 $t \cdot m^{-3}$.

From Equation (19), the outer diameter of the screw is calculated to be $D \geq 72$ mm. Considering the overall mass of the machine and its compatibility with the tractor, the outer diameter is determined to be $D = 100$ mm. From Equation (20), the screw shaft diameter is determined to be between 20 and 35 mm. To minimize the entanglement of grass on the shaft and to enhance soil transport capacity, the screw shaft diameter is chosen as $d = 30$ mm. The pitch from Equation (22) is determined to be in the range of 50 to 220 mm. To ensure smooth soil transport within the screw and to prevent mud accumulation, the soil transport amount of each pitch section must exceed the combined mud input from the current and previous pitch sections. Therefore, this study adopts a variable-pitch, constant-diameter screw, where the pitch increases along the helices towards the furrow, with the largest pitch closest to the furrow. Based on the structure and configuration dimensions of the machine, the maximum operational rotational speed from Equation (21) is calculated to be 237 rpm.

5. Performance Testing and Analysis

To achieve efficient and sustainable sweet potato seedling planting, the performance of the sweet potato seedling belt transplanter is evaluated.

5.1. Testing Conditions and Equipment

In May 2023, a sweet potato seedling transplanting test was conducted in Qingzhou City, Shandong Province, where the test site was an idle winter field, consisting of sandy soil with a moisture content of approximately 16% to 18% (0~100 mm). The sweet potato seedlings used in the test were 300 to 350 mm in length and 4 to 6 in diameter. The test equipment primarily consisted of a Dongfanghong 504 tractor, sweet potato seedling belt transplanter, moisture meter, calculators, ruler, tape measure, tachometer, etc.

5.2. Experimental Design and Methods

To improve planting quality, achieve horizontal planting of the sweet potato seedlings, and enhance yield, quality, and uniformity of sweet potato seedling tubers, several key parameters affecting the whole machine's performance and operational effectiveness were selected as test factors: the forward speed of the machine v, the screw speed n, and the spacing of the ribbons τ (the distance between the centerline of the screw shaft and the end of the conveyor belt, hereinafter referred to as the pitch of the screw belt). The planting spacing Z was taken as the evaluation index to characterize the operation quality of the machine. According to the above experimental plan, single-factor tests were conducted to identify the factors influencing the qualified rate of planting spacing and their respective ranges. A three-factor, three-level orthogonal test ($L^9(3^4)$) [18,19] was designed, with factor levels shown in Table 4.

Table 4. Factors and levels of the orthogonal test.

Test Level	The Forward Speed of the Machine v/m·s^{-1}	Spacing of the Ribbons τ/mm	The Screw Speed n/rpm
1	0.20	40	120
2	0.30	60	140
3	0.40	80	160

During the tests, different combinations of forward speed, spacing of the ribbons, and screw speeds were adjusted to calculate the qualified rate of planting spacing Z, thus analyzing and evaluating the operational performance of the machine. Each experimental group was repeated three times to take an average value [20]. The assessment criteria for the experiments were based on GB/T 5262 standards [21]. Three measurement areas were randomly selected within the test site, where each area must contain two adjacent operational widths. In each area, one row was chosen, and 120 consecutive planting spacings were measured [22]. A standard planting spacing of D = 200 mm was established, with measured spacings considered qualified if they fell within D(1 ± 10%). The percentage of qualified spacings relative to the total measured count was calculated to determine the qualified rate of spacings [23,24], based on Equations (23) and (24):

$$Z_i = \frac{N_{zi}}{N} \times 100 \tag{23}$$

$$Z = \frac{\sum_{i=1}^{n} z_i}{n} \times 100 \tag{24}$$

In the above formulas, the following notations are used:
N_{zi}—The number of qualified plant spacings in the testing area;
N—Total number of samples determined in the detection area, N = 120;
Z_i—The qualified rate of planting spacing in the detection area (%);
Z—The qualified rate of planting spacing (%);
n—The number of detection areas, n= 3.

5.3. Test Results and Analysis

The experimental results obtained from the orthogonal performance tests are presented in Table 5, where A, B and C represent the levels of the parameters v, τ and n, respectively.

Table 5. Results and orthogonal test.

Test Number		Test Factor *A*	*B*	*C*	Test Index Planting Spacing Qualification Rate Z/%
1		1	1	1	85.56
2		1	2	2	91.78
3		1	3	3	87.57
4		2	1	2	91.17
5		2	2	3	95.33
6		2	3	1	87.86
7		3	1	3	91.21
8		3	2	1	91.95
9		3	3	2	88.42
Z	k_1	88.30	89.31	88.46	
	k_2	91.45	93.02	90.46	
	k_3	90.53	87.95	91.37	
	R	3.15	5.07	2.91	

An analysis of the range values for each factor in Table 5 indicates that [25], regarding the evaluation indicators, the order of significance for each factor's influence is as follows: B, A, C, with the optimal combination of factor levels identified as $A_2B_2C_3$.

Using SPSS 21.0 data processing software, a variance analysis was performed on the experimental results, as shown in Table 6 [26]. As can be seen from Table 6, the value of the sum of squares of the error terms is much smaller than the sum of squares of the influencing factors, indicating that the interaction between the test factors does not have a significant effect on the test assessment indicators [27,28]. When analyzing the qualified rate of planting spacing Z, it is evident from the F-value that the experimental factors A, B and C have a significant impact on the evaluation indicator Z. From $F_B > F_A > F_C$, it is evident that the experimental factor B has the greatest impact on the evaluation indicators, followed by factor A, while factor C has the least influence, which is consistent with the results of the range analysis.

Table 6. Analysis of variance.

Evaluation Indicators	Source of Variation	Sum of Squares	Degrees of Freedom	Sum of Mean Squares	*F*	*p*
Z	calibration model	70.35	6	11.73	49.06	0.020 *
	A	15.72	2	7.86	32.90	0.029 *
	B	41.30	2	20.65	86.42	0.011 *
	C	13.32	2	6.67	27.87	0.035 *
	error	0.48	2	0.24		

Note: $R^2 = 0.995$; * is significant ($p < 0.05$).

Combining the results of the range analysis and the variance analysis, it can be concluded that the best parameter combination for transplanting performance was $A_2B_2C_3$, which corresponds to a forward speed of 0.3 m·s^{-1}, a ribbon spacing of 60 mm, and a screw speed of 160 rpm.

5.4. Field Experiment

In order to verify the working performance of the transplanter in the above results, a field trial was conducted in May 2023 at the sweet potato planting base in Qingzhou City, Shandong Province, China. Prior to the experiments, the machine's operational parameters were adjusted to the optimal combination: a forward speed of 0.3 m·s^{-1}, a ribbon spacing of 60 mm, and a screw speed of 160 rpm. A total of six sets of repeated experiments were conducted, with each set containing 120 measured planting spacings. The results of these experiments are presented in Table 7 and in Figure 18.

Table 7. Results of the field test.

Test Serial Number	Planting Spacing Qualification Rate Z/%
1	90.21
2	89.63
3	91.32
4	89.74
5	90.96
6	90.38
mean	90.37

Figure 18. Field experiment results: (**a**) field experiment equipment; (**b**) experiment result; (**c**) growth situation; and (**d**) growth situation.

The test results show that the average planting distance of the sweet potato bare seedling horizontal transplanter is 90.37%, which is in line with the relevant agricultural machinery industry technology and standards and local agronomic production requirements. The horizontal planting of potato seedlings can be realized, which is beneficial to improve the uniformity and yield of sweet potato.

6. Discussion

In this paper, the effects of machine advance speed, spacing between spirals and the screw speed on the planting spacing qualification rate were studied. Further optimization and improvement of the machine is needed in later experiments for indexes such as the planting depth qualification rate and planting spacing variation coefficient.

Additionally, it was observed during the experiments that, with the planting spacing fixed, there is a need to match the seedling delivery speed with the forward speed. However, as the roll diameter of the seedling conveyor increases, the relationship between the seedling separation line speed and the forward speed changes continuously, which affects the planting spacing to some extent.

The sweet potato seedling belt transplanter has preliminarily achieved automation in sweet potato seedling transplanting; however, manual intervention is still required to thread the seedlings onto the conveyor belt. Future research could focus on developing key technologies for automatic seedling bundling.

7. Conclusions

The designed sweet potato seedling belt replica transplanter initially realizes automatic and horizontal planting of sweet potato. The machine saves the labor of one to two people, and at an operating speed of 0.3 $m{\cdot}s^{-1}$, the machine plants 90 plants per minute, which improves the efficiency by more than 40% compared with the semi-automatic transplanting machine.

The main parameters affecting the working performance and operational effectiveness of the whole machine were obtained. Tests on the main parameters affecting the working performance and operational effectiveness of the whole machine were carried out. The test results show that the primary and secondary orders of the Z significance of the qualified rate of planting spacing for the impact evaluation index are B, A and C, the optimal level combination of influencing factors is $A_2B_2C_3$, the forward speed of the machine is 0.3 $m{\cdot}s^{-1}$, the spacing of the ribbons is 60 mm, and the screw speed is 160 rpm.

Transplanters meet sweet potato production requirements. The field test results show that under the optimal combination of factor levels, the Z-average rate of planting spacing was 90.37%, which met the relevant technical standards and agronomic requirements.

Author Contributions: Conceptualization, W.Y. and M.H.; methodology, W.Y.; software, K.L.; validation, W.Y., M.H. and C.W.; formal analysis, W.Y.; investigation, K.L.; resources, Y.J.; data curation, W.Y.; writing—original draft preparation, W.Y.; writing—review and editing, W.Y. and Z.R.; visualization, C.W.; supervision, Z.R.; project administration, W.Z. All authors have read and agreed to the published version of the manuscript.

Funding: This research was funded by the Key R&D Plan of Jiangsu Province (Grant No. BE2021311) and China Agriculture Research System of MOF and MARA (Grant No. CARS-10-B19).

Institutional Review Board Statement: Not applicable.

Data Availability Statement: The datasets used and/or analyzed during the current study are available from the corresponding author on reasonable request.

Conflicts of Interest: The authors declare no conflicts of interest.

References

1. Li, Q.; Zhao, H.; Jin, Y.L.; Zhu, J.C.; Ma, D.F. Analysis and perspectives of sweetpotato industry contributing to national food security in China. *Jiangsu JOAS* **2022**, *38*, 1484–1491.
2. Ma, D.F.; Li, Q.; Cao, Q.H.; Niu, F.X.; Xie, Y.P.; Tang, J.; Li, H.M. Development and prospect of sweetpotato industry and its technologies in China. *Jiangsu JOAS* **2012**, *28*, 969–973.
3. Hu, L.L.; Ji, F.L.; Wang, B.; Ling, X.Y.; Hu, Z.C.; Yu, X.T. Latest developments of sweet potato mechanical transplanting in China. *J. Chin. Agric.* **2015**, *36*, 289–291, 317.
4. Yan, W.; Zhang, W.Y.; Hu, M.J.; Ji, Y.; Li, K.; Qi, B. Present situation of research and expectation on plant mechanization of sweet potato in China and abroad. *J. Chin. Agric.* **2018**, *39*, 12–16.
5. Sarkar, P.; Upadhyay, G.; Raheman, H. Active-passive and passive-passive configurations of combined tillage implements for improved tillage and tractive performance. *Span. J. Agric. Res.* **2021**, *19*, e02R01. [CrossRef]
6. Ma, D.F.; Liu, Q.C.; Zhang, L.M. China Sweet Potato. *Jiangsu PSTP* **2021**, *3*, 337–340.
7. Upadhyay, G.; Raheman, H. Performance of combined offset disc harrow (front active and rear passive set configuration) in soil bin. *J. Terramech.* **2018**, *78*, 27–37. [CrossRef]
8. Zhang, L.M.; Ma, D.F. *Main Cultivation Modes of Sweet Potato in China*, 1st ed.; China Agricultural Science and Technology Press: Beijing, China, 2012; pp. 102–108.
9. Zhao, H.; Liu, X.X.; Pan, Z.G.; Li, L.; Sun, Y. Agronomic characteristics and mechanized planting technology of sweet potato. *J. Chin. Agric* **2021**, *42*, 21–26.

10. Shao, Y.Y.; Xuan, G.T.; Hou, J.L.; Hu, Z.C.; Wang, Y.X.; Liu, Y. Design and Simulation of sweet potato mulched transplanting mechanism with "boat"-shape. *ASABE* **2018**, *39*, 2–6.
11. Zhao, S.H.; Gu, Z.Y.; Yuan, Y.W.; Lu, J.Q. Bionic Design and Experiment of Potato Curved Surface Sowing Furrow Opener. *Trans. CSAM* **2021**, *52*, 32–42.
12. Zhao, S.H.; Yang, L.L.; Zhang, X.; Hou, L.T.; Yuan, Y.W.; Yang, Y.Q. Design and Experiment of Zigzag Opener for Double-row No-tillage Seeding on Soybean Ridge. *Trans. CSAM* **2022**, *53*, 74–84.
13. Zeng, S.; Tang, H.T.; Luo, X.W.; Ma, G.H.; Wang, Z.M.; Zang, Y.; Zhang, M.H. Design and experiment of precision rice hill-drop drilling machine for dry land with synchronous fertilizing. *Trans. CSAE* **2012**, *28*, 12–19.
14. Qi, J.T.; Meng, H.W.; Kan, Z.; Li, C.S.; Li, Y.P. Analysis and test of feeding performance of dual-spiral cow feeding device based on EDEM. *Trans. CSAE* **2017**, *33*, 65–71.
15. Yang, W.W.; Luo, X.W.; Wang, Z.M.; Zhang, M.H.; Zeng, S.; Zang, Y. Design and experiment of track filling assembly mounted on wheeled-tractor for paddy fields. *Trans. CSAE* **2016**, *32*, 26–31.
16. Meng, H.W.; Gao, Z.J.; Kan, Z.; Lin, H. Design and experiment on dairy cow precision-feeding device based on equal-diameter and variable-pitch. *Trans. CSAE* **2011**, *27*, 103–107.
17. Dai, F.; Zhang, S.L.; Song, X.F.; Zhao, W.Y.; Ma, H.J.; Zhang, F.W. Design and Test of Combined Operation Machine for Double Width Filming and Covering Soil on Double Ridges. *Trans. CSAM* **2020**, *51*, 108–117.
18. Chen, K. *Experimental Design and Analysis*, 2nd ed.; Tsinghua University Press: Beijing, China, 2015; pp. 172–188.
19. Shao, Y.Y.; Liu, Y.; Xuan, G.T.; Hu, Z.C.; Han, X.; Wang, Y.X.; Chen, B.; Wang, W.Y. Design and Test of Multifunctional Vegetable Transplanting Machine. *IFAC* **2019**, *52*, 92–97. [CrossRef]
20. Shi, Y.Y.; Luo, W.W.; Hu, Z.C.; Wu, F.; Gu, F.W.; Chen, Y.Q. Design and Test of Equipment for Straw Crushing with Strip-laying and Seed-belt Classification with Cleaning under Full Straw Mulching. *Trans. CSAM* **2019**, *50*, 58–67.
21. *GB/T 5262*; Measuring Methods for Agricultural Machinery Testing Conditions-General Rules. SAC: Beijing, China, 2008.
22. Dai, F.; Zhao, W.Y.; Song, X.F.; Xin, S.L.; Liu, F.J.; Xin, B.B. Operating Parameter Optimization and Experiment of Device with Elevating and Covering Soil on Plastic-film. *Trans. CSAM* **2017**, *48*, 88–96.
23. Yan, W.; Hu, Z.C.; Wu, N.; Xu, H.B.; You, Z.Y.; Zhou, X.X. Parameter optimization and experiment for plastic film transport mechanism of shovel screen type plastic film residue collector. *Trans. CSAE* **2017**, *33*, 17–24.
24. Yan, W.; Hu, M.J.; Li, K.; Wang, J.; Zhang, W.Y. Design and experiment of horizontal transplanter for sweet potato seedlings. *Agriculture* **2022**, *12*, 675. [CrossRef]
25. Hu, L.L.; Wang, B.; Wang, G.P.; Yu, Z.Y.; You, Z.Y.; Hu, Z.C.; Wang, B.K.; Gao, X.M. Design and experiment of type 2ZGF-2 duplex sweet potato transplanter. *Trans. CSAE* **2016**, *32*, 8–16.
26. Xiang, W.; Wu, M.L.; Guan, C.Y.; Xu, Y.J. Design and experiment of planting hole forming device of crawler transplanter for rape (*Brassica napus*) seedlings. *Trans. CSAE* **2015**, *31*, 12–18.
27. Xu, H.B.; Hu, Z.C.; Zhang, P.; Gu, F.W.; Song, W.L.; Wang, C.C. Optimization and Experiment of Straw Back-Throwing Device of No-Tillage Drill Using Multi-Objective QPSO Algorithm. *Agriculture* **2021**, *11*, 986. [CrossRef]
28. Yang, H.G.; Yan, J.C.; Wei, H.; Wu, H.C.; Wang, S.Y.; Ji, L.L.; Xu, X.W.; Xie, H.X. Gradient Cleaning Method of Potato Based on Multi-Step Operation of Dry-Cleaning and Wet Cleaning. *Agriculture* **2021**, *11*, 1139. [CrossRef]

agriculture

Article

Design and Testing of Soybean Double-Row Seed-Metering Device with Double-Beveled Seed Guide Groove

Huajiang Zhu [1], Sihao Zhang [1], Wenjun Wang [1,*], Hongqian Lv [1], Yulong Chen [1], Long Zhou [1], Mingwei Li [1] and Jinhui Zhao [2]

[1] School of Agricultural Engineering and Food Science, Shandong University of Technology, Zibo 255000, China; 22503030045@stumail.sdut.edu.cn (H.Z.); 22403010040@stumail.sdut.edu.cn (S.Z.); 23403010309@stumail.sdut.edu.cn (H.L.); cyl06471@sdut.edu.cn (Y.C.); zhoulong21@sdut.edu.cn (L.Z.); lmw271314@sdut.edu.cn (M.L.)

[2] State Key Laboratory of Agricultural Equipment Technology, Beijing 100083, China; zhaojinhui@caams.org.cn

* Correspondence: wjwang2016@163.com

Abstract: During the operation of a shaped hole seed-metering device, poor seed-filling quality and inconsistent seed-casting points lead to poor seed spacing uniformity, especially in a one-chamber double-row seed-metering device. To solve this problem, a soybean double-row seed-metering device with double-beveled seed guide groove was designed to ensure a high single-seed rate and seed-casting point consistency. Through the theoretical analysis of the working process of the seed-metering device, dynamic and kinematic models of the seeds were established, and the main structural parameters of the seed discharge ring, triage convex ridge, shaped hole, and seed guide groove were determined. The main factors affecting the seeding performance were obtained as the following: the inclination angle of the triage convex ridge, the radius of the shaped hole, and the depth of the seed guide groove. A single-factor test was carried out by discrete element simulation to obtain the inclination angle of the triage convex ridge α_3 = 29°, the radius of the shaped hole r_1 = 4.16–4.5 mm, and the depth of the seed guide groove l_1 = 0.49–1.89 mm. A two-factor, five-level, second-order, orthogonal rotation combination test was conducted to further optimize the structural parameters of the seed-metering device. The two test factors were the radius of the shaped hole and the depth of the seed guide groove, and the evaluation indices were the qualified rate, replay rate, and missed seeding rate. The results showed that the optimal combinations of the structural parameters were the radius of the shaped hole r_1 = 4.33 mm and the depth of the seed guide groove l_1 = 1.20 mm. Subsequent bench testing demonstrated that the seed discharge's qualified rate was above 94% at operating speeds of 6–10 km/h, and the seeding performance was stable. The final results of the soil trench test showed that the seed-metering device exhibited a qualified rate of 93.31%, replay rate of 2.04%, and missed seeding rate of 4.65% at an operating speed of 8 km/h. This research outcome may serve as a valuable reference and source of inspiration for the innovative design of precision seed-metering devices.

Keywords: soybean; double-row planting; seed-metering device; guide filling; discrete element method

Citation: Zhu, H.; Zhang, S.; Wang, W.; Lv, H.; Chen, Y.; Zhou, L.; Li, M.; Zhao, J. Design and Testing of Soybean Double-Row Seed-Metering Device with Double-Beveled Seed Guide Groove. *Agriculture* **2024**, *14*, 1595. https://doi.org/10.3390/agriculture14091595

Academic Editors: Xiaojun Gao, Qinghui Lai and Tao Cui

Received: 20 July 2024
Revised: 7 September 2024
Accepted: 9 September 2024
Published: 13 September 2024

1. Introduction

Precision seeding represents a pivotal developmental trajectory and prevailing trend within the realm of contemporary seeding technology, embodying a commitment to enhancing agricultural efficiency. A seed-metering device is the core component of a precision seeding system, directly influencing the seeding quality and working efficiency [1,2]. Precision seed-metering devices are categorized into mechanical and pneumatic types based on the principle of operation [3]. A mechanical seed-metering device features a simple structure, an ease of maintenance, and low processing costs, making it widely utilized in China.

Researchers have extensively researched on how different filling ways and shaped hole structures affect the filling effect of seed-metering devices [4–9]. Zhang et al. [10] aimed to improve the probability of seeds entering the holes of disc seeders, addressing the issue of empty planting by designing a height difference at the edges of the holes. Meng et al. [11] developed a fixed disturbance ring to enhance the activity of seed populations, which subsequently facilitated the design of a fixed disturbance-assisted soybean seeder. To identify the pivotal parameters of a seed-metering device, Li et al. [12] analyzed the movement process and state of the seeds. Li et al. [13] adopted a centrifugal cone disc-pushing seed-filling method to improve the stability of seeding quality, adjusting the seed orientation through friction and vibration. Zhang et al. [14] developed a pneumatic, mechanical, combination single-disc double-row seeder that minimizes seed damage while simultaneously reducing production costs. Li et al. [15] investigated the structure of a seed disc and seed-cleaning device, leading to the design of a double-row pneumatic disc seeder that significantly enhanced the seed-filling rate. Guo et al. [16] improved seeding quality by utilizing a seed guide groove and an auxiliary air-assisted seed-filling shaped hole structure. Chen et al. [17] employed discrete element simulation software to dissect the factors and their combinations influencing seeding performance, ultimately guiding the design of a convex scoop seed-metering device. Dun et al. [18] studied a shaped hole structure. Huang et al. [19] improved the filling force and seeding stability by altering the structure of a seed-filling shaped hole. Their study provided a reference for the design of shaped hole structures. Foreign researchers designed a secondary seed-casting device based on the principle of orderly seeding, with seeds discharged from the same position [20–23]. Maschio Company (Campodarsego, Italy) uses airflow to ensure uniform plant spacing when seed-casting [24]. Researchers in China have also been researching the uniformity of seed-casting. A seed-guiding mechanism designed by Dong et al. [25] significantly enhanced the seed-casting performance of a seed-metering device under high-speed operating conditions. Chen et al. [26] designed a belt-type seed delivery device aimed at improving the uniformity of the seed-casting process. Li et al. [27,28] designed a linear seed-pushing mechanism for an air-aspirated seed-metering device. This mechanism served to prevent the seed from changing position due to colliding with the seed tube's interior. Their proposed end toggle mechanism to disengage the seed and the straight-line seed-casting method were effective in resolving lateral seed displacement from the seed-casting effect. The above research directions mainly focused on two aspects of realizing orderly seed-filling and precise seed-casting. These studies effectively improved the performance of seed-metering devices. The problems of the above seed-metering devices mainly include the following: (1) the seed is in an unconstrained state before it enters the shaped hole, and the stability of seed-filling is poor and (2) by solely relying on its own gravity to disperse the seed, the device leads to irregular seed-casting points, thereby impacting the uniformity of spacing among the sown seeds.

Therefore, a soybean double-row seed-metering device with double-beveled seed guide groove was designed. The precision seed-metering device enables the systematic and orderly introduction of seeds into individual holes, thus ensuring the singular presence of one seed per hole. Furthermore, the seeding wheel guarantees the achievement of a consistent seed-casting point.

2. Materials and Methods

2.1. Overall Structure and Working Principle of the Seed-Metering Device

2.1.1. Overall Structure

Figure 1 illustrates the design of a soybean double-row seed-metering device, featuring a double-beveled seed guide groove. It consists of a front cover, front seed chamber, seed discharge ring, rear seed chamber, drive disc, seed guard, rear shell, drive shaft, and other components.

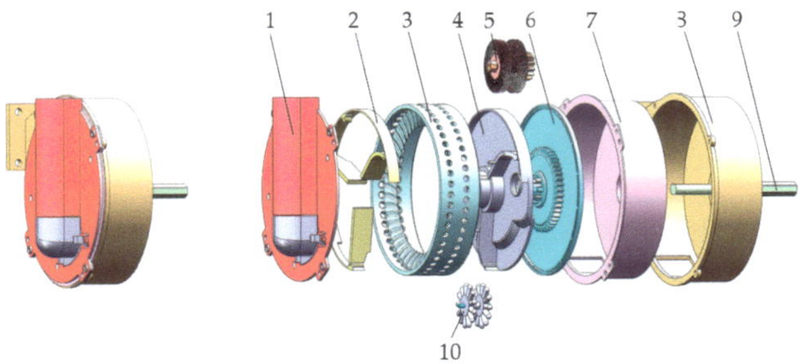

Figure 1. Diagram of the structural principles of a soybean double-row seed-metering device with double-beveled seed guide groove, where 1 is the front cover; 2 is the front seed chamber; 3 is the seed discharge ring; 4 is the rear seed chamber; 5 is the brush wheel; 6 is the drive disc; 7 is the seed guard; 8 is the rear shell; 9 is the drive shaft; and 10 is the seeding wheel.

The drive shaft initiates the rotation of the drive disc, which in turn propels the seed discharge ring into circular motion for the purpose of sowing. Simultaneously, the brush wheel, powered by the drive disc, rotates to execute seed-clearing operations. The teeth of the seeding wheel are strategically positioned within the contoured holes of the seed discharge ring, which, upon being rotated by the latter, performs the crucial task of seed-casting. Except for the above three sets of rotating parts, all other parts are non-rotating fixed parts. The triage convex ridge on the seed discharge ring has the capacity to divide the seed population into two distinct portions, which can then be filled independently, which improves the seed-filling efficiency. The seed guide grooves on the triage convex ridge effectively guide the seeds into the shaped holes using the restraining effect of the grooves on the seeds to enhance the seed-filling rate. The teeth of the seeding wheel push the seed to reach the seed-casting point out of the shaped hole, keeping the seed-casting point consistent.

2.1.2. Working Principle

The working process of the seed-metering device is mainly divided into four areas: seed-filling area I, seed-clearing area II, seed transport area III, and seed-casting area IV. The working process is shown in Figure 2.

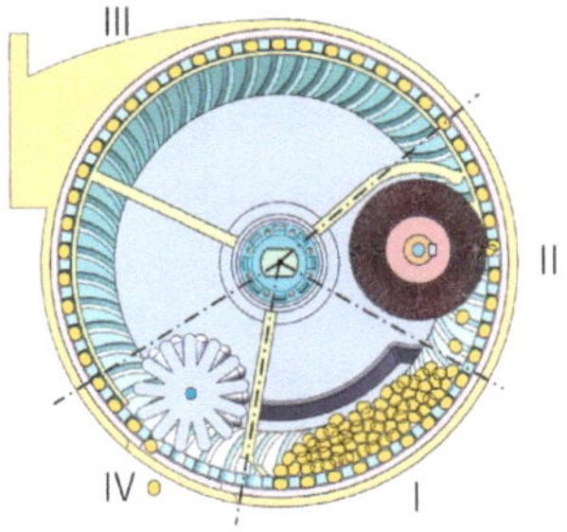

Figure 2. Schematic diagram of the division of the working process area of the seed-metering device, where I is seed-filling area; II is seed-clearing area; III is seed transport area; IV is seed-casting area.

When the seed-metering device operates, the seed falls into seed-filling area I. The connection between seed-filling area I and seed-casting area IV is divided by a partition on the front and rear seed discharge chambers. The partition serves to prevent the seed from entering seed-casting area IV directly from seed-filling area I. The seeds within the population traverse along the seed guide groove toward the shaped holes. The constraints and propulsion provided by the seed guide groove facilitate a smoother entry of the seeds into the shaped holes, thereby enabling effective guidance. This mechanism results in a significant increase in the seed-filling rate compared to seed-metering devices that rely solely on the seeds' gravity for filling. The seeds in the shaped holes follow the seed discharge ring as it rotates counterclockwise. When the seed reaches seed-clearing area II, the excess seed falls back to seed-filling area I under the action of the brush wheel and gravity. Afterward, the seeds waiting to be discharged follow the movement of the seed-discharging ring and pass through seed transport area III to seed-casting area IV. The teeth of the seeding wheel extend into the shaped holes on the seed discharge ring. When the seeds reach the seed-casting point, the teeth of the seed discharge ring immediately push the seeds out of the shaped holes, completing the seed-casting operation. Since the position of the seeding wheel remains fixed, the location of the seed-casting point also remains constant, ensuring consistency in the seed-casting point.

2.2. Key Component Design and Related Parameter Analysis

2.2.1. Overall Structure of the Seed Discharge Ring

The seed discharge ring, as depicted in Figure 3, constitutes the core functional element of the seed-metering device, playing a pivotal role in its operation. The seed discharge ring structure comprises a triage convex ridge, seed guide groove, shaped hole, and limiting convex ridge. Triage convex ridges, seed guide grooves, and shaped holes form the seed-filling structure. Double rows of seed guide grooves are staggered on both sides of the triage convex ridge of the seed discharge ring. The shaped holes are designed with rounded corners where they come into contact with the inner circle of the seed discharge ring. This not only effectively reduces seed damage but also facilitates the removal of excess seed from the shaped holes by the brush wheel. The seed discharge ring was designed with a limiting convex ridge on one side, fitting into a slot on the edge of the drive disc to form a whole.

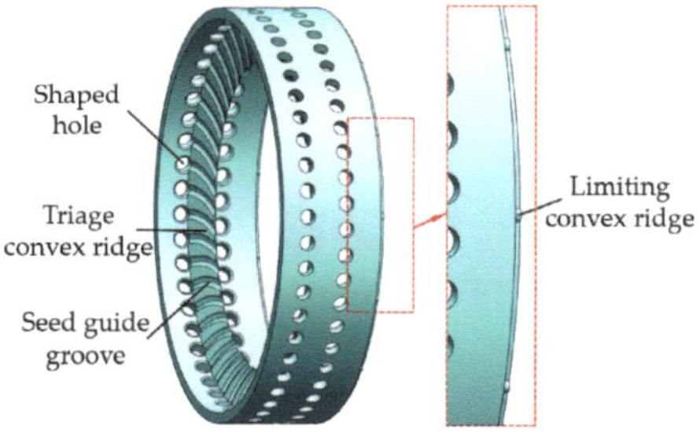

Figure 3. Schematic diagram of the structure of the seed discharge ring.

2.2.2. Analysis of the Force Acting on Soybean Seeds

During the seed-filling process, the origin is defined by the center of mass of the seed, with the x-axis aligned to the tangent direction of the seed discharge ring's rotation, the y-axis to its normal direction, and the z-axis perpendicular to the plane of the seed discharge ring. The three-dimensional coordinate system $oxyz$ is shown in Figure 4.

In the xoy plane, forces acting on the seed include friction from the slanting surface of the seed guide, thrust perpendicular to the surface, friction of the seed cluster on the seed in the tangential direction of the triage convex ridge, and gravity. The equation of equilibrium is as follows.

$$F_{c1}\cos\alpha_1 + f_{m1}\sin\alpha_1 = f_{m2}\cos\alpha_2 \tag{1}$$

where F_{c1} is the thrust perpendicular to the seed guide groove, N; f_{m1} is the friction force along the incline, N; f_{m2} is the population friction along the tangent to the triage convex ridge, N; α_1 is the angle between the slant of the seed guide groove and the x-axis, (°); and α_2 is the angle between the tangent to the triage convex ridge and the x-axis, (°).

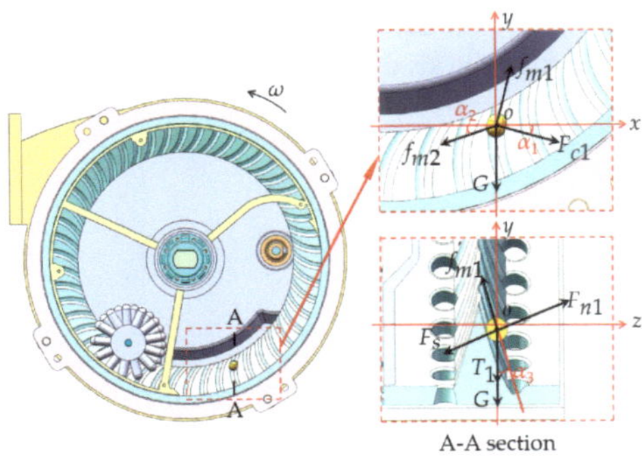

Figure 4. Schematic diagram of the forces on the seeds during seed-filling.

In the *yoz* plane, forces acting on the seed include the crowding force on the seed cluster perpendicular to the incline, supporting force perpendicular to the incline, combined force along the *y*-axis, and gravity. Forces on the seed are in equilibrium in the z-axis direction. The equation of equilibrium is as follows.

$$F_{n2}\cos\alpha_1 = F_s\cos\alpha_1 + f_{m1}\sin\alpha_1 \tag{2}$$

The f_{m1} in Equation (2) is given in Equation (3) below.

$$f_{m1} = \mu F_{n1} \tag{3}$$

where F_{n1} is the support force perpendicular to the incline, N; F_s is the crowding force from the seed cluster perpendicular to the inclined plane, N; and μ is the sliding friction factor between the seed and the seed discharge ring.

To facilitate the seamless downward progression of the seed within the designated guide groove, the resultant force acting upon the seed must be oriented consistently downward, conforming to the inclined plane of the triage convex ridge. The force equation is as follows.

$$G\cos\alpha_3 + T_1\cos\alpha_3 > f_{m1} \tag{4}$$

where G is the gravitational force acting on the seed, N; T_1 is the combined force along the *y*-axis, N; and α_3 is the inclination angle of the triage convex ridge, (°). The T_1 in Equation (4) is given in Equation (5) below.

$$T_1 = F_{c1}\sin\alpha_1 + f_{m2}\sin\alpha_2 - f_{m1}\cos\alpha_1 \tag{5}$$

A comprehensive analysis was conducted on both the internal architecture of the seed-metering device and the configuration of the seed discharge ring. It was found that when the inclination angle of the triage convex ridge α_3 was taken as 0°, the crowding pressure on the seeds on both sides of it by the seed population increased with time. Consequently, the seeds may be prone to breakage or erratic movements, ultimately compromising the quality of the seed-filling process. However, according to Equation (4), the larger the inclination angle of the triage convex ridge α_3, the smaller the $G\cos\alpha_3$ and $T_1\cos\alpha_3$. Therefore, the lower the guiding force is on the seeds, and thus the less favorable the seed-filling. Further tests are necessary to definitively determine the optimal range of the inclination angle of the triage convex ridge α_3.

As soybean seeds rotate concurrently with the seed discharge ring, transitioning from the seed transport area to the seed-casting area, they relinquish the supportive force provided by the seed guard, enabling their steady ejection under the impetus of the seeding wheel. At this time, the seed is subjected to forces, including the thrust of the shaped hole F_{c2}, the thrust of the seeding wheel F_{c3}, centrifugal force J, and gravity G, which make up the combined force T_2. The combined force T_2 provides the seed with a plane velocity v of motion, which can be decomposed into a horizontal velocity v_x and a vertical velocity v_y. A schematic diagram of the seed-casting process is shown in Figure 5.

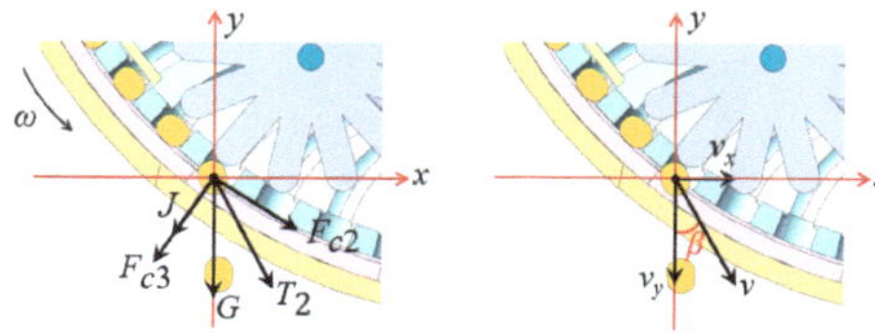

Figure 5. Schematic diagram of seed force and speed in the process of seed-casting.

A Cartesian coordinate system xoy was established in the plane of the rotating seed discharge ring, with the center of gravity of the soybean seeds designated as the origin. The trajectory of the soybean seed is as follows.

$$\begin{cases} x = v_x t_1 \\ y = v_y t_1 + \frac{g t_1^2}{2} \end{cases} \tag{6}$$

where x is the displacement of the soybean seed in the horizontal direction, mm; y is the displacement of the soybean seed in the vertical direction, mm; v_x is the velocity of the soybean seed in the horizontal direction, m/s; v_y is the velocity of the soybean seed in the vertical direction, m/s; and t_1 is the seed-casting time, s. The v_x, and v_y in Equation (6) are given in Equations (7) and (8) below.

$$v_x = v \sin \beta \tag{7}$$

$$v_y = v \cos \beta \tag{8}$$

The v in Equations (7) and (8) is given in Equation (9) below.

$$v = \frac{n_p \pi r_1}{30} \tag{9}$$

where β is the angle of the seed-casting, (°); n_p is the rotational speed of the seed discharge ring, r/min; and r_1 is the radius of the shaped hole, mm.

According to Equations (6)–(9), the equations for the trajectory of the soybean seed in the xoy plane were obtained as follows.

$$\begin{cases} x = \frac{n_p \pi r_1 \sin \beta}{30} t_1 \\ y = \frac{n_p \pi r_1 \cos \beta}{30} t_1 + \frac{g t_1}{2} \end{cases} \tag{10}$$

Under constant structural parameters of both the seed discharge ring and seeding wheel, the rotational velocity n_p of the ring governs the seed trajectory. Rigorous mechanical characterization is imperative for the rational design of the seed discharge ring, facilitating the precise analysis and calculation of its crucial parameters.

2.2.3. Design of the Seed Discharge Ring

The diameter of the seed discharge ring is instrumental in determining not only its overall size but also the magnitude of linear velocity and centrifugal force during its operation. Hence, its design necessitates meticulous consideration of specific operational conditions. It was found that the diameter of the seed discharge ring was directly proportional to the number of seed guide grooves and shaped holes [29]. Therefore, after determining the forward speed of the seeder and the seed spacing, the diameter of the seed discharge ring can be increased accordingly to reduce the rotation speed of the seed discharge ring.

The diameter range of the seed discharge ring is usually 140–260 mm [30]. For this study, the diameter of the seed discharge ring was taken as 240 mm and the width of the seed discharge ring was taken as 52 mm. The distance between the centers of the shaped holes on both sides of the seed discharge ring was designed to be 20 mm. 3D-printed technology was utilized for the fabrication of seed discharge rings using resin materials.

(1) Design of the triage convex ridge

The structure of the triage convex ridge is shown in Figure 6. The triage convex ridge ensures that the seeds on either side are filled independently, without interference, during the seed-filling operations of the seed-metering device. An optimal angle of inclination for the triage convex ridge is beneficial for seed-filling operations. However, if the angle of inclination is excessively large, the height of the ridge becomes constrained, which impairs the ability of the seed cluster to separate into both sides simultaneously, thereby reducing its guiding capability. This reduces the ability of the seeds to fill the shaped holes.

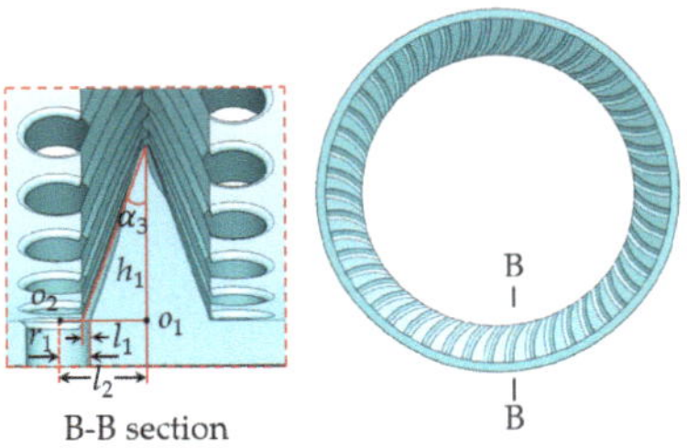

Figure 6. Schematic diagram of the structure of the triage convex ridge.

The formula for the inclination angle of the triage convex ridge was obtained from Figure 6.

$$\alpha_3 = \arctan\left(\frac{l_2 - (r_1 - l_1)}{h_1}\right) \tag{11}$$

where l_1 is the depth of the seed guide groove, mm; l_2 is the distance from o_2 to o_1, mm; h_1 is the height of the triage convex ridge, mm; o_1 is the midpoint of the distance between the centers of the shaped holes on both sides; and o_2 is the center of the shaped holes on one of the sides

The distance between points o_2 and o_1 is half the distance between the centers of the shaped holes on both sides, and therefore l_2 was taken as 10 mm. The triage convex ridge is limited by the positions of the brush wheel and the seeding wheel. Hence, the maximum height of the triage convex ridge was designed to be 25 mm, ensuring uninterrupted rotation of the brush and seeding wheels. The minimum height, exceeding the largest seed dimension, was set at 15 mm to facilitate seed separation on both sides and ensure guidance into the shaped holes via the seed guide groove. Therefore, the height of the triage convex ridge h_1 was taken in the range of 15–25 mm. The presence of diversion ridges effectively segmented the seed population into two distinct groups, each facilitated independently for seed-filling. As the rotary seed discharge ring rotates, the seed guide grooves adjacent to the diversion ridges meticulously direct the seeds toward their respective shaped holes. The inherent diversity among soybean varieties results in variations in their physical attributes, particularly in their dimensions. Consequently, larger seeds experience enhanced guiding forces during their trajectory, increasing their likelihood of successfully accessing the shaped holes. In contrast, smaller seeds encounter less pronounced guiding forces, which leads to a comparatively lower seed-filling efficiency. The determination of the range of height of the triage convex ridge provides the basis for the determination of the range of inclination angle of the triage convex ridge in a single-factor test.

(2) Design of the shaped holes

- Structure and dimensions of the shaped hole

The structure of the shaped hole has a great influence on a seed-metering device's seeding performance. A right-angled configuration at the shaped hole entrance may inflict damage on seeds during both the filling and cleaning processes. Therefore, the shaped holes were designed with chamfered corners with a radius of 1 mm, as shown in Figure 7.

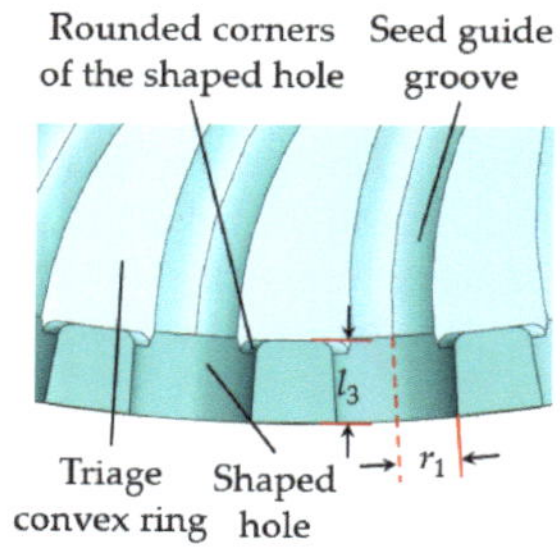

Figure 7. Schematic diagram of the structure of the shaped hole.

The radius of a shaped hole being too large will lead to multiple smaller seeds entering, while the radius of a shaped hole being too small will result in larger seeds being unable to enter. The depth of the shaped hole was designed to hold one seed and only one seed to ensure a smooth seed-filling and seed-clearing process [31]. It was essential to adhere to the following constraints for the individual dimensions.

$$\begin{cases} 0.75d_{\max} > r_1 > 0.5d_{\max} \\ \overline{X}_W > l_3 > d_{\min} \end{cases} \tag{12}$$

where $d_{\max}$ is the maximum size of the seed, mm; $\overline{X}_W$ is the average of the width of the seed, mm; $d_{\min}$ is the minimum size of the seed, mm; and l_3 is the depth of the shaped hole, mm.

The widely cultivated Zhonghuang 37 soybean seed was employed in the test procedure. A sample of 100 seeds were randomly extracted from the seed cluster, and their dimensions (length L, width W, and thickness T) were meticulously measured. The measurements were recorded and subsequently processed. The outcomes of the processing are summarized in Table 1.

Table 1. Statistical results of the soybean seeds' geometric size.

Parameter	Length L/mm	Width W/mm	Thickness T/mm
Maximum value	10.18	8.18	7.38
Minimum value	7.06	6.4	5.22
Average value	8.19	7.49	6.55

Table 1 shows that the average values of the dimensions of the soybean seeds were 8.19 mm in length, 7.49 mm in width, and 6.55 mm in thickness. Upon substituting the dimensions into Equation (12), it was determined that the range of values for the radius of the shaped hole was 4.1–6.14 mm. Thus, when the seed size was excessively large ($L > 6.14$ mm), there was an elevated risk of missed seeding occurring. Conversely, if the seed size was overly small ($L < 4.1$ mm), there was an increased likelihood of replay phenomena taking place. The 1 mm gap between the seed discharge ring and the seed guard serves to prevent friction, with the seed that fills the shaped holes subsequently dropping down to the seed guard. Therefore, the depth of the shaped hole was reduced by 1 mm. The value of l_3 was taken to be within the range of 5.56–6.49 mm, and the depth of the shaped hole was rounded up to l_3 = 6 mm. The range of the shaped hole's radius was subsequently determined through a rigorous single-factor test.

- Distribution of the shaped hole

The diameter of the seed discharge ring directly determines the number of shaped holes. When the operating speed and plant spacing are constant, the decrease in the rotation speed of the seed discharge ring can increase the number of shaped holes. Equation (13) shows the relationship between the rotation speed of the seed discharge ring, the number of shaped holes, and the forward speed of the seeder.

$$v = \frac{6zmn}{10000} \tag{13}$$

where v is the forward speed of the seeder, km/h; z is the seed spacing, calculated by taking 10 cm; n is the rotation speed of the seed discharge shaft, r/min; and m is the number of shaped holes, piece.

Based on Equation (13), it was evident that the drive shaft's rotational speed inversely correlated with the number of shaped holes, given constant seeder forward speed and plant spacing. Maintaining a constant diameter of the seed discharge ring, a reduced drive shaft speed translates to an increased number of shaped holes, enhancing seed-filling efficiency. However, an overly high number of holes may impede seed-filling. Therefore, determining the optimal maximum number of holes and the minimum spacing between them is crucial for ensuring smooth and efficient seed-filling. This simplifies the seed-filling process of a seed-metering device, as shown in Figure 8.

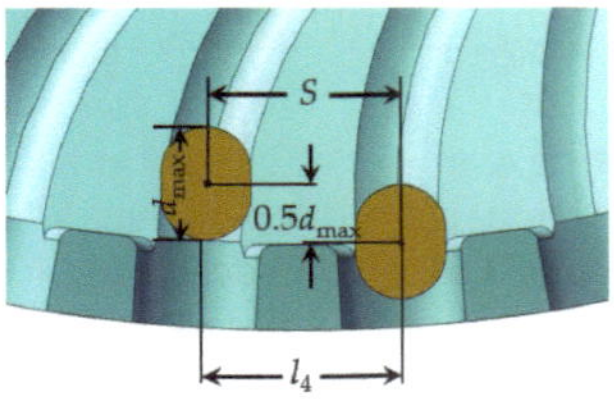

Figure 8. Schematic diagram of distribution distance of the shaped hole.

We assumed that the seed underwent horizontal projectile motion before entering the shaped hole from the seed guide groove. Then, the conditions for successful seed-filling would entail that the portion of the seed entering the shaped hole exceeded half of the seed itself, which corresponded to a drop height of more than $0.5d_{max}$. The fall time required for successful seed-filling was $t_2 = \sqrt{\frac{d_{max}}{g}}$, and the displacement in the horizontal direction $S = v_0 t_2$. According to the reference literature [17], Equations (14)–(16) are necessary conditions for smooth seed-filling.

$$l_4 \geq S + 0.5d_{max} \tag{14}$$

$$v_0 = \frac{\omega D}{2} \tag{15}$$

$$l_4 \approx \frac{\pi D}{m} \tag{16}$$

where S is the distance the seed traveled in the horizontal direction, mm; l_4 is the distance between the shaped holes, mm; v_0 is the linear velocity of the seed discharge ring, rad/s; ω is the angular velocity of the seed discharge ring, rad/s; and D is the diameter of the seed discharge ring, mm.

The following relationship was obtained from Equations (14)–(16).

$$m \le \frac{60\pi}{\pi n \sqrt{\frac{d_{\max}}{g} + \frac{30 d_{\max}}{D}}} \tag{17}$$

Equation (17) indicates that the number of shaped holes is directly proportional to the diameter of the seed discharge ring and inversely proportional to both the rotational speed of the drive shaft and the seed size. When the distance between the shaped holes is the smallest, the number of shaped holes is the largest and the following relationship was obtained.

$$l_4 \approx \frac{\sqrt{3}+1}{2} d_{\max} \tag{18}$$

$$m l_4 \approx m \frac{\sqrt{3}+1}{2} d_{\max} \approx \pi D \tag{19}$$

Collation yielded the following relationships.

$$m \approx \frac{2\pi D}{\left(\sqrt{3}+1\right)} d_{\max} \tag{20}$$

$$v_0 t \le l_4 \tag{21}$$

Substituting Equation (15) into Equation (21) yielded Equation (22).

$$n \le \frac{\sqrt{g d_{\max}}\left(\sqrt{3}+1\right) \times 30}{\pi D} \tag{22}$$

Given a seed discharge ring diameter of 240 mm and a maximum seed size of 8.19 mm, the calculated maximum rotational speed for the seed discharge ring, as determined by Equation (22), was 58.43 r/min. Calculation according to Equation (20) gave $m \approx 67.39$ when the neighboring shaped holes were next to each other. To have enough space between the shaped holes during seed-filling, the number of shaped holes on one side of the triage convex ridge was taken as 50. At a seeder forward speed of 10 km/h, the seed discharge ring speed was 33.33 r/min, less than 58.43 r/min, and met the design requirements.

(3) Design of the seed guide groove

The reasonable structure of the seed guide groove not only guides the seed into the shaped hole but also enhances the disturbance of the seed cluster. As the seed discharge ring rotates, the seed guide groove and seed cluster exert pressure on the seed in the seed guide groove. This force ensures that the seed outside the seed guide groove cannot influence the seed in the seed guide groove to enter the shaped hole. This ensures that the seed-filling is effective.

A Cartesian coordinate system XOY was established, with the origin positioned at the center O of the seed discharge ring, as depicted in Figure 9. To guarantee seamless seed entry into the shaped hole, the absolute trajectory of the seed must align precisely with the tangent of the circle defined by the top of the seed guide groove. At time t_4, point A_1 moves to point A_2, and the seed moves from point A_1 to point A_3.

After time t_4, point A_1 rotates to point A_3, at which point the velocity at this position is as follows.

$$v_1 = v_2 / \cos \gamma_2 \tag{23}$$

$$\begin{cases} v_2 = r_3 \omega \\ \gamma_2 = k \omega t_3 \\ r_3 = r_2 / \cos \gamma_2 \end{cases} \tag{24}$$

where v_1 is the velocity of the seed at point A_3, m/s; v_2 is the velocity component of the seed in the direction tangent to point A_3, m/s; γ_2 is the angle between line segment OA_1

and line segment OA_3, (°); r_2 is the radius of the base circle of the seed guide groove, mm; r_3 is the radius of the circle at point A_3, mm; k is the angular rate coefficient for soybeans, which was taken as 0.1 to 0.9; and t_3 is the time of movement of the soybean seed, s.

According to Equations (23) and (24), the following results were obtained.

$$v_1 = r_2\omega / \cos^2(k\omega t_3) \tag{25}$$

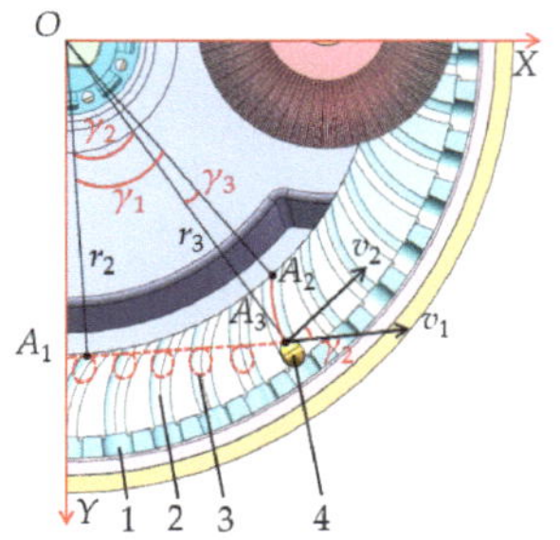

Figure 9. Schematic diagram of the curve equation of the guide groove on the seed-metering ring, where 1 is the shaped hole; 2 is the seed guide groove; 3 is seeds in the process of moving; 4 is the position the soybean seed needs to reach at time t_4; the dotted line A_1A_3 segment is the absolute motion trajectory of the soybean seed; and the Arc A_2A_3 segment is the relative motion trajectory of the soybean seed.

After time t_4, the absolute motion trajectory of the soybean seed following the movement of the seed discharge ring is represented by the following equation.

$$l_{A_1A_3} = \int_0^{t_4} \frac{r_2\omega}{\cos^2(k\omega t_3)} dt_3 = r_2 \tan(k\gamma_1)/k \tag{26}$$

where t_4 is the actual movement time of the soybean seed, s, and γ_1 is the angle between line segment OA_1 and line segment OA_2, (°).

The soybean seeds have a defined trajectory as the seed discharge ring rotates. To maximize the ability of the seed guide groove to guide the seed-filling, the curve equation of the seed guide groove was made to coincide with the relative motion trajectory of the seed. The expression of the curve equation is as follows.

$$\begin{cases} x_{A_3} = r_2 \cos\gamma_3 \\ y_{A_3} = r_2 \sin\gamma_3 \end{cases} \tag{27}$$

$$\gamma_3 = \gamma_1 - \gamma_2 \tag{28}$$

$$r_2 = r_3 \cos\gamma_2 \tag{29}$$

where γ_3 is the angle between line segment OA_2 and line segment OA_3, (°).

The following equation was obtained by combining Equations (27)–(29).

$$\begin{cases} x_{A_3} = r_2(\cos\gamma_1 + \tan\gamma_2 \sin\gamma_1) \\ y_{A_3} = r_2(\sin\gamma_1 - \tan\gamma_2 \cos\gamma_1) \end{cases} \tag{30}$$

$$\tan\gamma_2 = l_{A_1A_3}/r_2 \tag{31}$$

The result of the simplification of Equation (30) is as follows.

$$\begin{cases} x_{A_3} = r_2\left(\cos\gamma_2 + \frac{\tan(k\gamma_1)\sin\gamma_1}{k}\right) \\ y_{A_3} = r_2\left(\sin\gamma_2 + \frac{\tan(k\gamma_1)\cos\gamma_1}{k}\right) \end{cases} \quad (\gamma_1 \in (0,\ \zeta)) \tag{32}$$

where x_{A_3} is the coordinate of point A_3 on the x-axis in the absolute coordinate system, mm; y_{A_3} is the coordinate of point A_3 on the y-axis in the absolute coordinate system, mm; and ζ is the upper bound of the range, (°).

The expression for r_3 is as follows.

$$r_3 = r_2\sqrt{1 + (\tan(k\gamma_1)/k)^2} \tag{33}$$

Equation (32) is the relative motion trajectory of the soybean seeds. The radius of the circumference in which the shaped hole is located must exceed r_3, so the value of γ_1 was in the range $37.17° < \gamma_1 < 42.02°$. The optimal depth range for the seed guide groove was established to prioritize its role in guiding seed-filling over its tangential function in pushing the seed. This prevents the excessive depth of the seed guide groove from compromising the efficiency of seed-filling. The depth of the seed guide groove l_1 must be less than $0.5T_{min}$, and according to Table 1, T_{min} = 5.22 mm, so the range of values for l_1: $l_1 < 2.61$ mm.

2.3. Test Conditions and Methods

2.3.1. Modeling and Parameter Setting

The operation of the seed-metering device was simulated using Altair EDEM 2021 discrete element simulation software. Soybean seeds were modeled using 5-ball collocation based on the known dimensional parameters of the soybean seeds [32]. The Hertz–Mindlin non-slip contact model was used for the seed-to-seed and seed-to-component contact models [33–36]. The parameters of the simulation model are presented in Table 2.

Table 2. Simulation of model parameters.

Item	Parameters	Numerical Values
Soybean seed	Density/(kg·m^{-3})	0.4
	Shear modulus/MPa	11
	Poisson's ratio	1053
	Moisture content/%	10.2
Resin	Density/(kg·m^{-3})	1454.80
	Shear modulus/MPa	120
	Poisson's ratio	0.35
Soybean–soybean	Static friction coefficient	0.5
	Rolling friction coefficient	0.01
	Coefficient of restitution	0.6
Soybean–resin	Static friction coefficient	0.32
	Rolling friction coefficient	0.04
	Coefficient of restitution	0.35

In the parameter settings of EDEM, the 3D particle sizes were set to be generated according to a normal distribution based on the distribution of the three-axis sizes of the seeds. The coefficient of variation for the seed population was 0.157, with a total seed count of 1500. The total time of the simulation was 10 s, where the step size of the simulation was 0.0000314 s, and the data were saved at intervals of 0.01 s. The simulation's boundary conditions were the following: (1) During the simulation, the operating speed was 8 km/h. The seed discharge ring, the drive disc, and the drive shaft rotated together counterclockwise in the same direction and their rotation speed was set to 160 deg/s. The brush wheel rotated clockwise at −525.71 deg/s. The seeding wheel rotated counterclockwise at a speed of 615.38 deg/s. And the remaining seed-metering device components were fixed. (2) The simulation material, including the front cover, as well as the front and rear seed chambers, were all constructed of resin. The seed discharge ring and seeding wheel were also composed of resin material. The bristles of the brush wheel were manufactured from nylon. The seed guard was constructed from high-strength steel. Lastly, the rear shell of the device was made of cast iron, selected for its strength and resistance to wear and tear. (3) The computational domains were the following: x-axis (maximum: 210 mm, minimum: 90 mm); y-axis (maximum: 265 mm, minimum: −20 mm); and Z-axis (maximum: 120 mm, minimum: −160 mm). The scope of the computational domain was combined with the seed-metering device structure to meet the needs of the simulation. The simulation process is shown in Figure 10.

Figure 10. Seed-metering device model and discrete element simulation process.

2.3.2. Single-Factor Test

Following the structural design and theoretical analysis of the triage convex ridge, shaped hole, and seed guide groove, a targeted single-factor test was undertaken to refine the structural parameters. This endeavor sought to quantify the influence of diverse structural features on the performance index of the seed-metering device. In the single-factor test, the test factors were the inclination angle of the triage convex ridge α_3, the radius of the shaped hole r_1, and the depth of the seed guide groove l_1. The operational speed during the single-factor test was set at 8 km/h. When investigating the effect of the inclination angle of the triage convex ridge α_3 on seeding performance, each of the other factors took a fixed value: the radius of the shaped hole r_1 = 4.4 mm and the depth of the seed guide groove l_1 = 1.5 mm. When investigating the effect of the radius of the shaped hole r_1 on seeding performance, each of the other factors took a fixed value: the inclination angle of the triage convex ridge α_3 = 20.5° and the depth of the seed guide groove l_1 = 1.5 mm. When investigating the effect of the depth of the seed guide groove l_1 on seeding performance, each of the other factors took a fixed value: the inclination angle of the triage convex ridge α_3 = 20.5° and the radius of the shaped hole r_1 = 4.4 mm.

According to GB/T 6973-2005 [37] Testing methods of single seed drills (precision drills), the qualified rate, replay rate, and missed seeding rate were taken as test indices. The seeds discharged from the seed-casting point during the simulation were counted, and the qualified rate, replay rate, and missed seeding rate were calculated. When only one seed was found to have been discharged from the shaped hole, it was marked as passing. When two or more seeds were found to have been discharged from the shaped hole, they were marked as reseeding. When no seed was found to have been discharged from the shaped hole, it was marked as missed seeding.

The test methods for the qualified rate, replay rate, and missed seeding rate were the following: In the first step, the simulation was played back to record the total number of shaped holes that were traversed through the seed-casting point, starting from the first seed-filled shaped hole until the end of the simulation. This total is referred to as the total number of shaped holes turned through the seed-casting point. The number of shaped holes that dispensed only one seed while passing through the seed-casting point was designated as the qualified count. The number of shaped holes that dispensed two or more seeds was classified as the replay count, while the number of shaped holes that dispensed no seeds was termed the missed seeding count. In the second step, the recorded qualified count, replay count, and missed seeding count were divided by the total count to calculate the qualified rate, replay rate, and missed seeding rate, respectively. In the third step, the qualified rates, replay rates, and missed seeding rates from the three repeated simulation tests were averaged to yield the final test results.

2.3.3. Second-Order Orthogonal Rotation Combination Test

Based on the results of the single-factor test, the radius of the shaped hole and the depth of the seed guide groove were identified as pivotal test factors for further investigation. The range of values for the radius of the shaped hole was 4.16–4.5 mm and the range of values for the depth of the seed guide groove was 0.49–1.89 mm. A two-factor, five-level, second-order orthogonal rotation combination test was conducted with the qualified rate, replay rate, and missed seeding rate as test indicators. The measurement of test indicators was conducted in accordance with the methodology employed in the single-factor test, utilizing the two-factor level coding presented in Table 3. These parameters provided the basis for the successful implementation of the orthogonal test and were processed using the data analysis software Design-Expert 13.0.

Table 3. Test factor coding.

Code	Radius of the Hole X_1/mm	Depth of Seed Guide Groove X_2/mm
−1.414	4.09	0.2
−1	4.16	0.49
0	4.33	1.19
1	4.5	1.89
1.414	4.57	2.18

2.3.4. Bench Test

Bench tests are conducted to verify the reasonableness of the optimized results of parameters. The radius of the shaped hole was taken as 4.33 mm, and the depth of the seed guide groove was taken as 1.20 mm. The main working parts and the rear shell inside the 3D-printed seed-metering device formed the seed-metering device, which was mounted on a bench, as shown in Figure 11. According to GB/T 6973-2005 Testing methods of single seed drills (precision drills), the qualified rate, replay rate, and missed seeding rate were taken as test indicators. Bench test verification was carried out on the JSP-12 computer vision-measuring device performance test bench at 5 operating speeds of 6, 7, 8, 9, and 10 km/h. Each operating speed was tested three times and the results were recorded. Theoretical seed spacing was 10 cm. If the distance between two adjacent seeds in the row was greater than 5 cm and less than or equal to 15 cm, it was considered passing. Seed spacing less than or equal to 5 cm was considered to be reseeding. Seed spacing greater than 15 cm was considered to be a missed seeding.

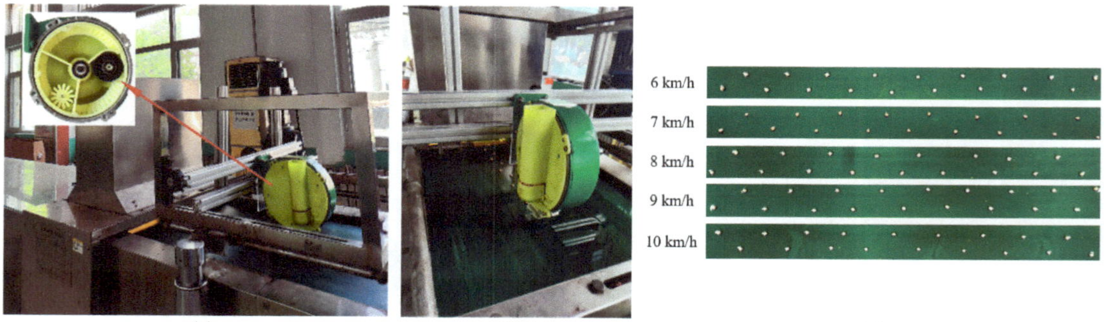

(a) (b) (c)

Figure 11. Bench validation test. (**a**) Bench test device; (**b**) Test process; (**c**) Test effect.

2.3.5. Soil Trench Test

Using two-factor optimum parameter combination, a soil trench validation test was carried out at an operating speed of 8 km/h to investigate the seeding performance of the seed-metering device on the soil. The soil trench was 30 m long, 2.8 m wide, and 25 cm deep. The tests were carried out from 16 April to 26 April 2024 on the soil trench test bench in the laboratory of Shandong University of Technology. The physical properties of the soil were measured before the test. The soil moisture content was 12.6% at 0–10 cm and 14.3% at 10–20 cm. The soil compactness was 1.98 MPa at 0–10 cm and 2.5 MPa at 10–20 cm. The physical properties of the soil in the soil trench tests fell within the range of those encountered in field operations [38,39]. The bench frame with the seed-metering device was fixedly installed on the rear suspension beam of the soil trench trolley, as shown in Figure 12. The test indicators included the qualified rate, replay rate, and missed seeding rate, which were measured in the same way as in the bench test.

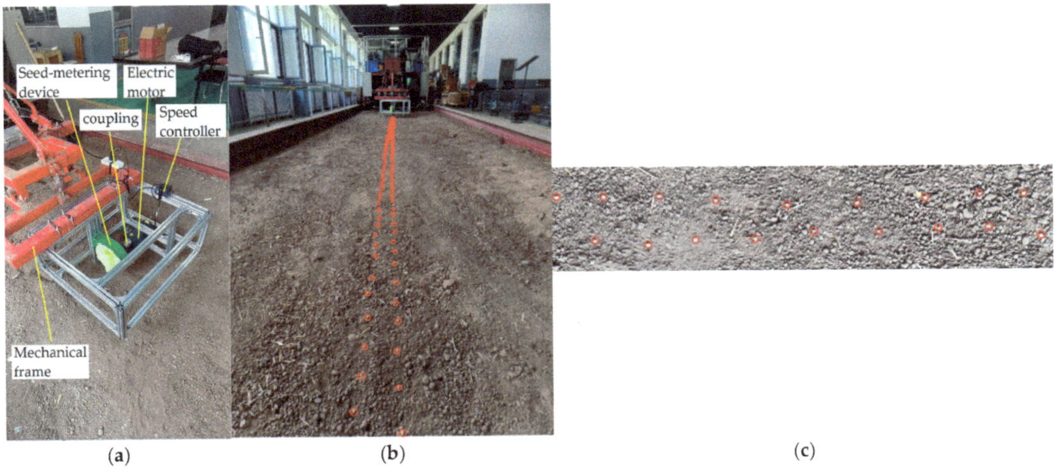

Figure 12. Soil trench validation test. (**a**) Soil trench test device; (**b**) Test process; (**c**) Test effect. where the red circles in (**b**,**c**) indicate the position of the seeds after landing.

3. Results and Discussion

3.1. *Single-Factor Test Results and Analyses*

The results of the single-factor test are shown in Table 4.

Table 4. Single-factor test results.

Factor	Level	Qualified Rate/%	Replay Rate/%	Missed Seeding Rate/%
Inclination angle of the triage convex ridge $\alpha_3/°$	12	91.32	8.48	0.2
	16.25	93.96	5.07	0.97
	20.5	95.13	3.99	0.88
	24.75	96.3	1.85	1.85
	29	96.49	1.07	2.44
Radius of the shaped hole r_1/mm	4.1	93	1.22	5.78
	4.2	96.56	1.22	2.22
	4.3	95.67	2.55	1.78
	4.4	95.89	3.78	0.33
	4.5	95.11	4.67	0.22
	4.6	93.9	5.34	0.76
	4.7	93.34	6.11	0.55

Table 4. *Cont.*

Factor	Level	Qualified Rate/%	Replay Rate/%	Missed Seeding Rate/%
Depth of the seed guide groove l_1/mm	0.5	95.03	1.07	3.9
	1	96.98	1.36	1.66
	1.5	96.3	1.85	1.85
	2	94.64	4.78	0.59
	2.5	92.98	6.82	0.2

3.1.1. Effect of the Inclination Angle of the Triage Convex Ridge on the Test Index

It was known that l_2 = 10 mm, h_1 = 15–25 mm, r_1 = 4.4 mm, and l_1 = 1.5 mm. Substituting this into Equation (11), the range of values of the inclination angle of the triage convex ridge α_3 was calculated to be 11.97–29.57°. The inclination angles of the triage convex ridge were taken as 12°, 16.25°, 20.5°, 24.75°, and 29°. In order to more clearly show the change in the trend of the effect of the inclination angle of the triage convex ridge on the test index, the results of the test are shown in Figure 13.

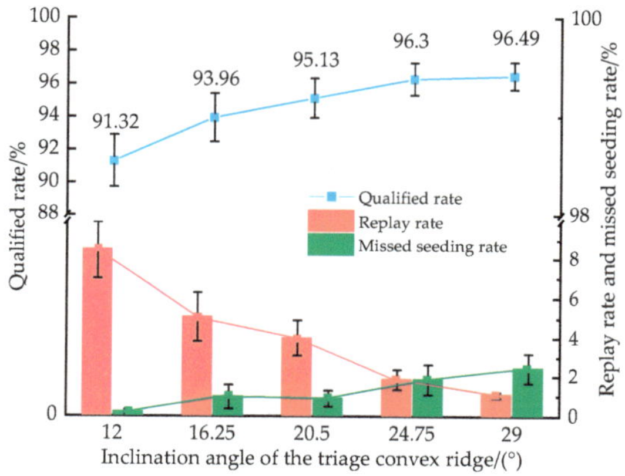

Figure 13. Effect curve of the inclination angle of the triage convex ridge on the test index.

Figure 13 shows that the qualified rate increased with the increased inclination angle of the triage convex ridge. When the inclination angle of the triage convex ridge increased to 24.75°, the changed trend of the qualified rate slowed down significantly.

The replay rate decreased with the increased inclination angle of the triage convex ridge and was lowest when the inclination angle of the triage convex ridge increased to 29°. The increase in the inclination angle of the triage convex ridge led to a reduction in the squeezing force on the seeds on both sides, which reduced the likelihood of multiple seeds entering the shaped hole.

The missed seeding rate gradually increased with the inclination angle of the triage convex ridge, peaking at 29°. At angles higher than this, the downward guiding force of the seed guide groove was diminished, thereby affecting the quality of seed filling. The optimal seeding performance was observed when the inclination angle was set at 29°, which was then chosen for conducting single-factor tests regarding the radius of the shaped hole and the depth of the seed guide groove.

Based on the observed trend in the qualified rate, as illustrated in Figure 13, it could be inferred that an increase in the inclination angle of the triage convex ridge resulted in improved seeding performance. Consequently, the maximum value of 29° was selected as the inclination angle of the triage convex ridge for subsequent testing.

3.1.2. Effect of the Radius of the Shaped Hole on the Test Index

Simulation pre-tests were performed for the radius range of the shaped hole due to the large radius range of the shaped hole. The pre-test results are shown in Figure 14a.

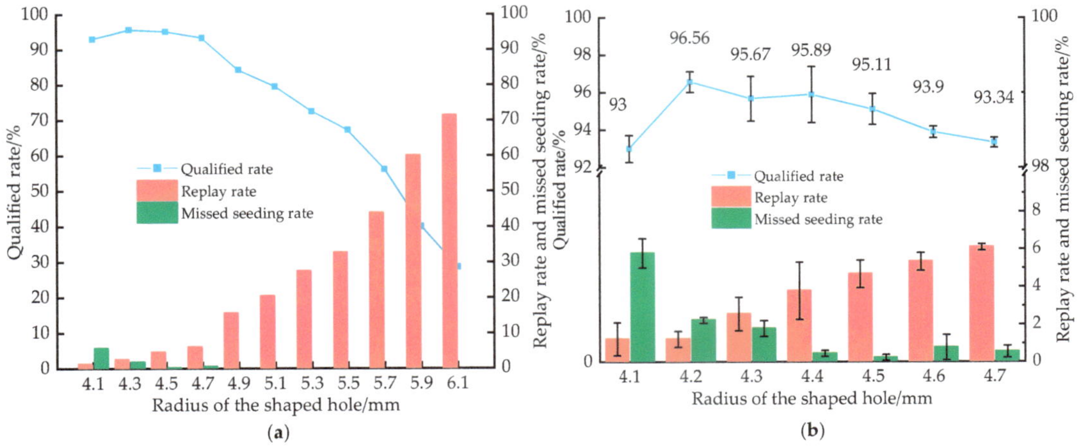

Figure 14. Effect curve of the radius of the shaped hole on the test index. (**a**) Pre-test results; (**b**) Test results.

The analysis of the test results, presented in Figure 14a, indicated that the radius of the shaped hole r_1 had to range from 4.1 to 4.71 mm to achieve a qualified rate exceeding 93%. Conversely, at a radius of 4.9 mm, the qualified rate was observed to be 84.33%. When the radius of the shaped hole was larger than 4.9 mm, the qualified rate continued to decrease, and the requirement for the precision operation of the seeder could not be met. The range of values for the radius of the shaped hole r_1 was determined to be 4.1–4.7 mm, and 4.1, 4.2, 4.3, 4.4, 4.5, 4.6, and 4.7 mm were selected for the single-factor test. The test outcomes are depicted in Figure 14b.

By analyzing the test results in Figure 14b, the following regression equation was obtained for the effect of the radius of the shaped hole on the seeding performance.

$$Y_1 = 99.167X_1^3 - 1337.8X_1^2 + 6004.4X_1 - 8871 \quad (34)$$

$$Y_2 = -37.5X_1^3 + 494.85X_1^2 - 2165.1X_1 + 3144.1 \quad (35)$$

$$Y_3 = -61.667X_1^3 + 842.44X_1^2 - 3834.8X_1 + 5816.8 \quad (36)$$

where Y_1, Y_2, and Y_3 are the qualified rate, replay rate, and missed seeding rate, respectively, %, and X_1 is the radius of the shaped hole, mm.

The qualified rate exhibited a pronounced variation with the augmentation of the radius of the shaped hole, attaining a peak value of 96.56% at a radius of precisely 4.2 mm. Subsequently, a decline was observed as the radius surpassed 4.2 mm, reaching 4.3 mm, whereupon a marginal yet discernible increase occurred between 4.3 mm and 4.4 mm. This slight elevation was attributed to the mitigation of seed collisions at the entry point, thereby facilitating enhanced seed ingress. Notably, this transient increase did not undermine the overarching trend of the radius of the shaped hole's influence on the qualified rate, which then resumed its continuous downward trajectory.

As the radius of the shaped hole increased, the replay rate augmented, with a deceleration in this trend observed when the radius surpassed 4.5 mm. As the radius of the shaped holes increased to a certain extent, the probability of multiple seeds entering the shaped holes continued to rise. However, the collisions among the seeds at the entrance of the shaped holes became more intense, resulting in a deceleration of this upward trend.

With the increase in the radius of the shaped holes, the missed seeding rate initially decreases. When the radius reaches 4.5 mm, the missed seeding rate begins to rise slightly The reason for this is that the shaped hole size was excessive, exceeding 4.5 mm. This caused some seeds to collide at the mouth of the shaped hole and prevented them from entering the shaped hole.

According to regression Equations (34)–(36), when the radius of the shaped holes ranged from 4.1 to 4.5 mm, the qualified rate exceeded 95%, and the replay rate was below 5%. Furthermore, when the radius of the shaped holes ranged from 4.11 to 4.7 mm, the missed seeding rate was less than 5%. The highest qualified rate of 96.56% was achieved with a radius of the shaped hole of 4.2 mm.

3.1.3. Effect of the Depth of the Seed Guide Groove on the Test Index

In the single-factor test, the depth of the seed guide groove l_1 was taken as 0.5, 1, 1.5, 2, and 2.5 mm, where the seed guide grooves did not affect each other. To more clearly show the change in trend of the effect of the depth of the seed guide groove on the test index, the results of the test are depicted in Figure 15.

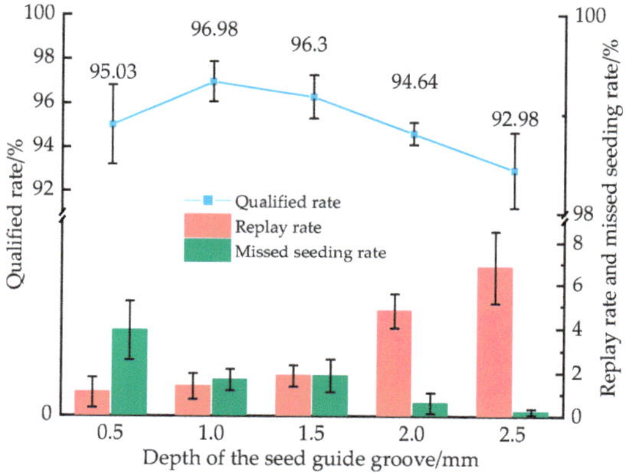

Figure 15. Effect curve of the depth of the seed guide groove on the test index.

By analyzing the test results in Figure 15, the following regression equation was obtained for the effect of the depth of the seed guide groove on the seeding performance.

$$Y_1 = 1.7533X_2^3 - 10.233X_2^2 + 16.085X_2 + 89.336 \quad (37)$$

$$Y_2 = -0.7267X_2^3 + 4.9671X_2^2 - 6.3948X_2 + 3.196 \quad (38)$$

$$Y_3 = -1.04X_2^3 + 5.3229X_2^2 - 9.7586X_2 + 7.49 \quad (39)$$

where Y_1, Y_2, and Y_3 are the qualified rate, replay rate, and missed seeding rate, respectively, %, and X_2 is the depth of the seed guide groove, mm.

The qualified rate increased with increasing depth of the seed guide groove, reaching a maximum value of 96.98% when the depth of the seed guide was increased to 1 mm, and then gradually decreased. As the depth of the seed guide groove increases, the capacity of the groove to direct seed movement toward the shaped hole is enhanced, thereby facilitating a more streamlined seed entry into the shaped hole. However, the depth of the seed guide groove was too large, causing multiple seeds to enter the same shaped hole, thereby reducing the qualified rate. The highest qualified rate of 97% was achieved with a depth of seed guide groove of 1.09 mm.

The replay rate increased slowly as the depth of the seed guide groove significantly increased when it increased to 1.5 mm. The lowest replay rate of 0.89% was achieved with a depth of the seed guide groove of 0.78 mm.

The missed seeding rate initially decreased with the depth of the seed guide groove; however, it augmented at 1–1.5 mm before resuming its downward trend beyond 1.5 mm. This increase in the missed seeding rate may be due to the increased ability of the seed guide groove to constrain the seed, resulting in some of the seeds colliding at the entrance of the shaped hole, thus failing to enter the shaped hole.

According to regression Equations (37)–(39), when the depth of the seed guide groove ranged from 0.49 to 1.89 mm, the qualified rate exceeded 95%. Additionally, when the depth of the seed guide groove was between 0.23 and 1.4 mm, the replay rate was below 2%. Finally, when the depth of the seed guide groove exceeded 1.01 mm, the missed seeding rate was below 2%.

3.2. Second-Order Orthogonal Rotation Combination Test Results and Analysis

Design-Expert 13.0 software was used to design a second-order orthogonal rotary combination test to investigate the effect of the radius of the shaped hole and the depth of the seed guide groove on the test indices. The test results are presented in Table 5.

Table 5. Second-order orthogonal rotating combination test design and results.

No.	X_1	X_2	Qualified Rate Y_1/%	Replay Rate Y_2/%	Missed Seeding Rate Y_3/%
1	−1	−1	91.23	0.29	8.48
2	1	−1	95.71	1.85	2.44
3	−1	1	96.39	2.92	0.69
4	1	1	95.03	4.39	0.58
5	−1.414	0	94.74	1.17	4.09
6	1.414	0	95.91	3.8	0.29
7	0	−1.414	92.69	0.58	5.27
8	0	1.414	95.32	4.39	0.29
9	0	0	96.2	1.46	2.34
10	0	0	96.49	2.15	1.36
11	0	0	96.3	1.27	2.43
12	0	0	96.49	1.75	1.76
13	0	0	96.79	1.66	1.55

Note: X_1 and X_2 are the factor-coded values for the radius of the shaped hole and the depth of the seed guide groove, respectively. The interaction between the radius of the shaped hole and the depth of the seed guide groove was significant. The effects of two factors on the test indices had a significant impact.

Regression analysis was conducted on the test results presented in Table 5, yielding the subsequent regression equations.

$$Y_1 = -0.5832X_1^2 - 1.24X_2^2 - 1.46X_1X_2 + 0.5968X_1 + 1.02X_2 + 96.45 \quad (40)$$

$$Y_2 = 0.3829X_1^2 + 0.3829X_2^2 - 0.0225X_1X_2 + 0.8437X_1 + 1.32X_2 + 1.66 \quad (41)$$

$$Y_3 = 0.2916X_1^2 + 0.5866X_2^2 + 1.48X_1X_2 - 1.44X_1 - 2.09X_2 + 1.89 \quad (42)$$

Analysis of variance was performed on the experimental data using Design-expert 13.0 software. The results are presented in Table 6.

The p-values of the models for qualified rate, replay rate, and missed seeding rate were all less than 0.001, and the differences were extremely significant. The respective F values for the lack of fit terms were 2.30, 0.2888, and 2.29, with corresponding p-values of 0.2187, 0.8323, and 0.2203, all exceeding the significance threshold of 0.05. The insignificant variations in the lack-of-fit terms indicate that the second-order regression equation, as fitted by the model, closely aligned with reality, offering a robust prediction of the test results. In the regression model for the qualified rate, replay rate, and missed seeding rate, the two factors and their interaction terms had extremely significant effects on the qualified rate, replay rate, and missed seeding rate.

Table 6. Analysis of variance.

Source	Qualified Rate/%		Replay Rate/%		Missed Seeding Rate/%	
	F Value	*p*-Value	*F* Value	*p*-Value	*F* Value	*p*-Value
Model	80.08	<0.0001	56.02	<0.0001	35.75	<0.0001
X_1	35.90	0.0005	74.41	<0.0001	47.14	0.0002
X_2	105.86	<0.0001	182.10	<0.0001	98.90	<0.0001
X_1 X_2	107.40	<0.0001	0.0265	0.8754	24.96	0.0016
${X_1}^2$	29.81	0.0009	13.33	0.0082	1.68	0.2360
${X_2}^2$	135.45	<0.0001	13.33	0.0082	6.80	0.0351
Lack of Fit	2.30	0.2187	0.2888	0.8323	2.29	0.2203

Note: $0.01 < p < 0.05$ is generally significant, $0.001 < p < 0.01$ is significant, and $p < 0.001$ is extremely significant.

Response surface plots were generated in Design-expert 13.0 software to analyze the relationship between the effects of the two factors on the test indices, as shown in Figure 16.

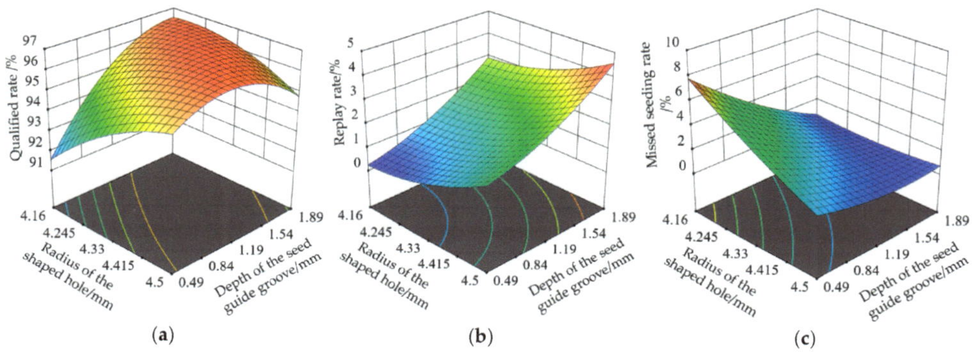

Figure 16. Response surface diagram: (**a**) the radius of the shaped hole and the depth of the seed guide groove on the qualified rate; (**b**) the radius of the shaped hole and the depth of the seed guide groove on the replay rate; and (**c**) the radius of the shaped hole and the depth of the seed guide groove on the missed seeding rate.

From Figure 16, it was evident that there was a significant interaction between the two factors. When the radius of the shaped hole was relatively small, both the qualified rate and replay rate increased with the depth of the seed guide groove, while the missed seeding rate decreased. This phenomenon can be attributed to the fact that an increased depth of the seed guide groove enhances the contact area with the seeds, thereby allowing for more effective guidance of the seeds into the shaped holes. Conversely, when the radius of the shaped hole was larger, seeds could more easily enter the shaped holes, resulting in an increase in both the qualified rate and replay rate, along with a decrease in the missed seeding rate. However, if the depth of the seed guide groove became excessively large, the guiding force exerted on the seeds may have become too strong, leading to a significant increase in the replay rate and, consequently, a decrease in the qualified rate.

When the depth of the seed guide groove was shallow, the qualified rate increased as the radius of the shaped hole expanded. This occurred because larger seeds encountered difficulties when attempting to enter smaller shaped holes, and a larger radius facilitated their entry, thereby enhancing seeding effectiveness. However, with a deeper seed guide groove, the qualified rate decreased as the radius increased. Although a larger radius allowed for easier entry, the deeper seed guide groove enhanced guidance, resulting in multiple seeds entering the same shaped hole more frequently. This phenomenon led to a higher replay rate and a subsequent decline in the qualified rate.

To explore the optimal parameter combination for each test factor and ensure that the conditions for optimization fell within a reasonable range of design parameters, an optimization solution was employed that aimed to maximize the qualified rate while minimizing the replay rate and missed seeding rate. Equation (43) presents the objective function along with the constraint range for the working parameters.

$$\begin{cases} \max Y_1(X_1, X_2) \\ \min Y_2(X_1, X_2) \\ s.t.\begin{cases} -1.414 \leq X_1 \leq 1.414 \\ -1.414 \leq X_2 \leq 1.414 \end{cases} \end{cases} \quad (43)$$

The optimization process was executed utilizing the "Optimization" module within Design-expert 13.0 software, yielding optimized parameters: the radius of the shaped hole was 4.33 mm and the depth of the seed guide groove was 1.20 mm. These conditions facilitated a qualified rate of 96.48%, a replay rate of 1.69%, and a missed seeding rate of 1.83%.

3.3. Bench Test Results and Analysis

A comparative analysis of the bench and simulation test results was conducted to validate the seeding quality achieved by the seed-metering device during actual operational conditions. The results of the test were averaged over three times, and the results are shown in Table 7 and Figure 17.

Table 7. Bench test results.

Factor	Level	Qualified Rate/%		Replay Rate/%		Missed Seeding Rate/%	
		Test Value	Simulation Value	Test Value	Simulation Value	Test Value	Simulation Value
Operating speed /(km·h^{-1})	6	96.18	96.65	1.24	1.81	2.58	1.54
	7	95.34	96.02	1.02	1.93	3.64	2.05
	8	95.63	95.61	1.08	2.44	3.29	1.95
	9	95.44	95.45	0.81	2.11	3.75	2.44
	10	94.91	95.33	0.54	1.7	4.55	2.97

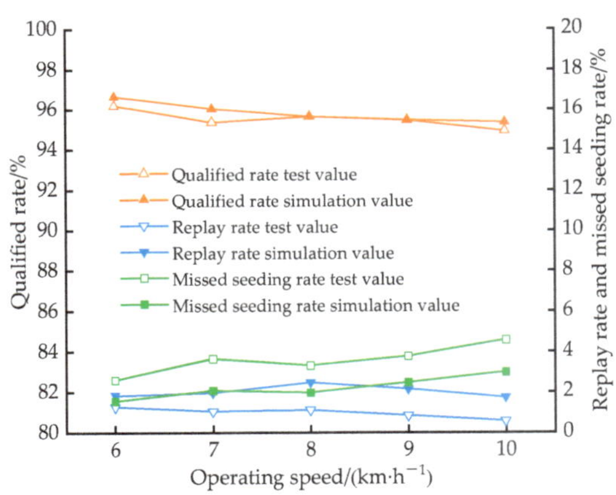

Figure 17. Bench test results.

According to Table 7 and Figure 17, the test results from the bench and simulation were found to be closely aligned. The bench tests, conducted at five operating speeds—6, 7, 8, 9, and 10 km/h—yielded relatively favorable outcomes, with all tests achieving a qualified rate exceeding 94%, demonstrating stable seeding performance. At a forward speed of 8 km/h, the bench test recorded a qualified rate of 95.12%, a replay rate of 1.08%, and a missed seeding rate of 3.80%. This result closely corresponds to the data obtained from the regression model calculations, indicating the reliability of the optimized parametric combination.

According to Table 7 and Figure 17, it was observed that with an increase in the operating speed, both the qualified rate and the replay rate decreased, and the missed seeding rate increased. An increase in the forward speed of the seeder may have induced vibrations in the seed-metering device, potentially impeding the efficient entry of seeds into the shaped hole. Concomitantly, as the working speed intensified, the seed-metering device's velocity escalated, curtailing the seed filling duration. This decrease in filling time resulted in a diminished seed-filling rate and a consequential rise in the missed seeding rate.

During the bench tests, the seed-metering device was mounted on a JSP-12 computer vision-measuring device performance test bench, with only the vibrations generated by the device's own operation acting as a source of disturbance. In comparison to actual field operations, the system exhibited superior performance, facilitated more effective filling of the shaped holes, and achieved greater uniformity in seed spacing.

3.4. Soil Trench Test Results and Analysis

The soil trench test results are shown in Table 8.

Table 8. Soil trench test results.

No.	Qualified Rate/%	Replay Rate/%	Missed Seeding Rate/%
1	93.50	2.28	4.22
2	93.11	1.71	5.18
3	93.33	2.12	4.55
Average value	93.31	2.04	4.65

According to Table 8, the seed-metering device operating at a speed of 8 km/h achieved a qualified rate of 93.31%, a replay rate of 2.04%, and a missed seeding rate of 4.65%. These findings suggest that the seed-metering device exhibits stable seeding performance, satisfying the criteria for precision sowing operations. The seeding quality of the seed-metering device in the soil trench test was reduced compared to the bench test. It may be that the vibration caused by the forward movement of the soil trench trolley during the soil trench test resulted in a more intense vibration of the seed-metering device, which reduced the quality of seed-filling and seed-casting. In addition, the bouncing roll of the seed as it hit the ground also reduced seeding quality. In the soil trench test, the soil trench trolley operated relatively smoothly and exhibited low vibration. In contrast, during field operations, the seeder may have experienced significant vibrations due to complex ground conditions, which could have adversely affected seeding outcomes. Therefore, the results of the soil trench test may have deviated from the empirical results obtained from the combined field tests.

4. Conclusions

(1) A soybean double-row seed-metering device with a double-beveled seed guide groove was designed. Through the theoretical analysis of the working process of the seed-metering device, dynamic and kinematic models of the seeds were established. The main factors affecting the seeding performance were obtained as the following: the inclination angle of the triage convex ridge, the radius of the shaped hole, and the depth of the seed guide groove. The ranges of values for the inclination angle of the triage convex ridge, the radius of the shaped hole, and the depth of the seed guide groove were determined to be 11.97–29.57°, 4.1–4.7 mm, and 0–2.61 mm, respectively.

(2) A discrete element simulation was employed to conduct a single-factor test, using the qualified rate, replay rate, and missed seeding rate as evaluation indicators. The test results were obtained as the following: the inclination angle of the triage convex ridge $\alpha_3 = 29°$, the radius of the shaped hole $r_1 = 4.16$–4.5 mm, and the depth of the seed

guide l_1 = 0.49–1.89 mm. The qualified rates of the tests were all greater than 95%, meeting the operational requirements of precision seeding.

(3) A two-factor, five-level, second-order orthogonal rotation combination test was conducted with the radius of the shaped hole and the depth of the seed guide groove as test factors. A regression model between the evaluation indices of the radius of the shaped hole and the depth of the seed guide groove was developed. The test results showed that the radius of the shaped hole and the depth of the seed guide groove had a significant effect on each evaluation index. According to the regression model, the optimal parameter combination was 4.33 mm for the radius of the shaped hole and 1.20 mm for the depth of the seed guide groove. The qualified rate was 96.48%, the replay rate was 1.69%, and the missed seeding rate was 1.83%.

(4) Bench tests were carried out, and the results showed that the test results from the bench and simulation were consistent. At the operating speed of 6–10 km/h, the qualified rate of seeding was above 94%. A soil trench test was conducted, and the results showed that the seed spacing qualified rate of the seed-metering device was 93.31% at an operating speed of 8 km/h. The replay rate was 2.04% and the missed seeding rate was 4.65%. This met the requirements of precision sowing operation. These research results can provide a reference for the innovative design of precision seed-metering devices.

Author Contributions: Data curation, H.Z. and S.Z.; funding acquisition, W.W.; investigation, Y.C. and M.L.; methodology, L.Z. and J.Z.; project administration, W.W.; supervision, L.Z.; validation, H.Z. and H.L.; writing—original draft, H.Z.; writing—review and editing, W.W. All authors have read and agreed to the published version of the manuscript.

Funding: This work was supported by the National Natural Science Foundation of China (52005307), High-quality Development project for the Ministry of Industry and Information (2023ZY02009), and Opening Fund of the State Key Laboratory of Agricultural Equipment Technology (NKL-2023-010).

Institutional Review Board Statement: Not applicable.

Data Availability Statement: The data used to support the findings of this study are available from the corresponding author upon request.

Conflicts of Interest: The authors declare no conflicts of interest.

References

1. Gao, X.J.; Zhou, Z.Y.; Xu, Y.; Yu, Y.B.; Su, Y.; Cui, T. Numerical simulation of particle motion characteristics in quantitative seed feeding system. *Powder Technol.* **2020**, *367*, 643–658. [CrossRef]
2. Yang, L.; Yan, B.X.; Zhang, D.X.; Zhang, T.L.; Wang, Y.X.; Cui, T. Research Progress on Precision Planting Technology of Maize. *Trans. Chin. Soc. Agric. Mach.* **2016**, *47*, 38–48.
3. Yuan, Y.W.; Bai, H.J.; Fang, X.F.; Wang, D.C.; Zhou, L.M.; Niu, K. Research Progress on Maize Seeding and Its Measurement and Control Technology. *Trans. Chin. Soc. Agric. Mach.* **2018**, *49*, 1–18.
4. Li, Y.H.; Yang, L.; Han, Y.; Zhang, D.X.; Cui, T.; Yu, Y.M. Design and Experiment of Spoon-clamping Type Metering Device for Faba Beans. *Trans. Chin. Soc. Agric. Mach.* **2018**, *49*, 108–116.
5. Li, Y.H.; Wei, Y.N.; Yang, L.; Zhang, D.X.; Cui, T.; Zhang, K.L. Design and Experiment of Mung Bean Precision Seed-metering device with Disturbance for Promoting Seed-filling. *Trans. Chin. Soc. Agric. Mach.* **2020**, *51*, 43–53.
6. Liu, H.X.; Liu, J.X.; Tang, S.F.; Xu, X.M. Design on opposed inclined-plate high-speed precision seed-metering device and its working mechanism analysis. *Trans. Chin. Soc. Agric. Eng.* **2016**, *32*, 24–31.
7. Correia, T.P.; Sousa, S.F.; Silva, P.R.; Dias, P.P.; Gomes, A.R. Sowing performance by a metering mechanism of continuous flow indifferent slope conditions. *J. Agric. Eng.* **2016**, *36*, 839–845.
8. Wang, Y.C.; Sun, H.; Li, B.Q.; Han, X.; Chen, H.T. Design and Experiment of Centralized Belt Type Soybean Seed-metering Device. *Trans. Chin. Soc. Agric. Mach.* **2019**, *50*, 74–83.
9. Jia, H.L.; Chen, Y.L.; Zhao, J.L.; Guo, M.Z.; Huang, D.Y.; Zhuang, J. Design and key parameter optimization of an agitated soybean seed metering device with horizontal seed-filling. *Int. J. Agric. Biol. Eng.* **2018**, *11*, 76–87. [CrossRef]
10. Zhang, F.X.; Qing, C.L.; Wang, X. Study on Leak in Seeding Prevention and Seed Protection of Disc Precision Soybean Metering Device. *J. Agric. Mech. Res.* **2022**, *44*, 65–70.
11. Meng, H.; Wang, X.; Li, X.H.; Li, H.; Yu, Y.C. Design and experiment of soybean seed metering device with constant disturbance aided filling soybean. *J. China Agric. Mech. Chem.* **2023**, *44*, 18–25.

12. Li, W.; Xu, H.; Li, X.H.; Li, H.; Yu, Y.C. Design and Experiment of Soybean Side Nest Seed Meter Based on EDEM. *J. Agric. Mech. Res.* **2023**, *45*, 164–169.
13. Li, M.T.; Li, T.Y.; Guan, X.D.; Zhao, G.K.; Zhou, F.J. Design on Push Structure of Centrifugal Cone Seed-metering Device and Its Filling Mechanism Analysis. *Trans. Chin. Soc. Agric. Mach.* **2018**, *49*, 77–85.
14. Zhang, X.L.; Song, J.; Li, X.Y.; Xie, F.X.; Zhang, L. Design and simulation analysis of pneumatic mechanical combined single disc and double row seed meter. *J. China Agric. Mech. Chem.* **2020**, *41*, 30–34.
15. Li, B.H.; Li, H.; Qi, X.D.; Xu, S.Y.; Bai, M.C. Design and Test of Double—row Pneumatic Precision Metering Device for Brassica Chinensis. *J. Agric. Mech. Res.* **2022**, *44*, 71–77.
16. Guo, P.; Zheng, X.S.; Wang, D.W.; Hou, J.L.; Zhao, Z. Design and Experiment of Precision Seed-metering device with Pneumatic Assisted Seed-filling for Peanut. *Trans. Chin. Soc. Agric. Mach.* **2024**, *55*, 64–74.
17. Chen, Y.L.; Jia, H.L.; Wang, J.X.; Wang, Q.; Zhao, J.L.; Hu, B. Design and Experiment of Scoop Metering Device for Soybean High-speed and Precision Seeder. *Trans. Chin. Soc. Agric. Mach.* **2017**, *48*, 95–104.
18. Dun, G.Q.; Yu, C.L.; Yang, Y.Z.; Ye, J.; Du, J.X.; Zhang, J.T. Parameter simulation optimization and experiment of seed plate type hole for soybean breeding. *Trans. Chin. Soc. Agric. Eng.* **2019**, *35*, 62–73.
19. Huang, Y.X.; Li, P.; Dong, J.X.; Chen, X.H.; Zhang, S.L.; Liu, Y. Design and Experiment of Side-mounted Guided High Speed Precision Seed-metering device for Soybean. *Trans. Chin. Soc. Agric. Mach.* **2022**, *53*, 44–53+75.
20. Yazgi, A. Effect of seed tubes on corn planter performance. *Appl. Eng. Agric.* **2016**, *32*, 783–790.
21. Kocher, M.F.; Coleman, J.M.; Smith, J.A.; Kachman, S.D. Corn seed spacing uniformity as affected by seed tube condition. *Appl. Eng. Agric.* **2011**, *27*, 177–183. [CrossRef]
22. Yang, L.; Yan, B.X.; Cui, T.; Yu, Y.M.; He, X.T.; Liu, Q.W.; Liang, Z.J.; Yin, X.W.; Zhang, D.X. Global overview of research progress and development of precision maize planters. *Int. J. Agric. Biol. Eng.* **2016**, *9*, 9–26.
23. Liao, Y.T.; Li, C.L.; Liao, Q.X.; Wang, L. Research Progress of Seed Guiding Technology and Device of Planter. *Trans. Chin. Soc. Agric. Mach.* **2020**, *51*, 1–14.
24. Available online: https://www.maschio.com/en/web/international (accessed on 7 February 2020).
25. Dong, J.X.; Gao, X.J.; Zhang, S.L.; Huang, Y.X.; Zhang, C.Q.; Shi, J.T. Design and Test of Guiding Seed Throwing Mechanism for Maize Posture Control and Driving Metering Device. *Trans. Chin. Soc. Agric. Mach.* **2023**, *54*, 25–34.
26. Chen, X.G.; Zhong, L.M. Design and test on belt-type seed delivery of air-suction metering device. *Trans. Chin. Soc. Agric. Eng.* **2012**, *28*, 8–15.
27. Li, Y.H.; Yang, L.; Zhang, D.X.; Cui, T.; Zhang, K.L.; Xie, C.J.; Yang, R.M. Analysis and test of linear seeding process of maize high speed precision metering device with air suction. *Trans. Chin. Soc. Agric. Eng.* **2020**, *36*, 26–35.
28. Li, Y.H.; Yang, L.; Zhang, D.X.; Cui, T.; He, X.T.; Du, Z.H.; Wang, D.C. Performance analysis and structure optimization of the maize precision metering device with air suction at high-speed condition. *Trans. Chin. Soc. Agric. Eng.* **2022**, *38*, 1–11.
29. Wang, D.W.; Ji, R.Q.; He, X.N.; Guo, P.; Shi, Y.X.; Zhang, C.X. Drive-guided Combination Slot-assisted Seed-attached Air-absorbing Peanut High-speed Precision Seed Meter. *Trans. Chin. Soc. Agric. Mach.* **2023**, *54*, 59–70+149.
30. Zhao, L.J.; Yan, S.S.; Wang, Y.J.; Zhang, Y.X.; Han, Y.M.; Cai, X.H.; Xu, C.L. Experiment and Design on Air Suction Single and Double Row Precision Universal Metering Device. *J. Agric. Mech. Res.* **2019**, *41*, 136–141.
31. Gai, M.M.; Yang, J.; Li, X.H.; Qu, Z.; Li, H.; Yu, Y.C. Design and test of a guided horizontal disc soybean precision seed-metering device. *Jiangsu Agric. Sci.* **2023**, *51*, 200–206.
32. Yang, D.X. *Particle Modelling of Soybean Seeds and the Simulation Analysis and Experimental Study of the Seed-Throwing and Pressing*; Jilin University: Changchun, China, 2021.
33. Qu, Z.; Liu, L.; An, X.; Lu, Q.; Ding, L.; Yu, Y.C. Optimization Design and Experiment of Soybean Hole Wheel Seed Metering Device Besed on EDEM. *J. Agric. Mech. Res.* **2023**, *45*, 103–110.
34. Niu, Y.Y.; Xu, M.C.; Qu, Z.; An, X.; Li, H.; Yu, Y.C. Design and test of EDEM-based self-disturbing internal spoon type soybean precision seed-metering device. *Jiangsu Agric. Sci.* **2022**, *50*, 202–207.
35. Dun, G.Q.; Yang, Y.Z.; Guo, Y.L.; Liu, X.L.; Yu, C.L.; Du, J.X.; Zhang, J.T. Analysis of EDEM simulation of different soybean seed-filling characteristics. *J. Henan Agric. Univ.* **2019**, *53*, 93–98.
36. Xu, J.; Cai, Z.S.; Gan, Y.Q.; Zhang, L.L. Optimization study on seed-clearing process for declined disc-scoop-type soybean seed metering device based on EDEM. *J. Northeast Agric. Univ.* **2018**, *49*, 79–88.
37. *GB/T 6973-2005*; Testing Methods of Single Seed Drills (Precision Drills). AQSIQ-SAC: Beijing, China, 2005.
38. Zhang, S.J.; Qiu, Z.J.; Wang, F.H.; Zhao, H.M.; Zhang, H.H. Design and Test on the Field Soil Moisture and Compaction Acquisition Instrument. *Trans. Chin. Soc. Agric. Mach.* **2010**, *41*, 75–79.
39. Xu, S.; Kan, Y.C. Correlation of Soil Compactness and Water Content Under Different Fertility Levels. *Chin. Agric. Sci. Bu.* **2022**, *38*, 94–100.

agriculture

MDPI

Article

Simulation and Optimization of a Rotary Cotton Precision Dibbler Using DEM and MBD Coupling

Long Wang [1,2], Xuyang Ran [1,2], Lu Shi [1,2], Jianfei Xing [1,2], Xufeng Wang [1,2], Shulin Hou [3] and Hong Li [1,2,*]

1 College of Mechanical and Electronic Engineering, Tarim University, Alar 843300, China; 120140002@taru.edu.cn (L.W.); ranxuyang98@126.com (X.R.); fengleaf77@126.com (L.S.); 120200012@taru.edu.cn (J.X.); wxf@taru.edu.cn (X.W.)

2 Xinjiang Production and Construction Corps (XPCC) Key Laboratory of Utilization and Equipment of Special Agricultural and Forestry Products in Southern Xinjiang, Alar 843300, China

3 College of Engineering, China Agricultural University, Beijing 100083, China; h01520@cau.edu.cn

* Correspondence: 120140003@taru.edu.cn

Abstract: Investigating the seeding mechanism of precision seeders is of great significance for improving the quality of cotton sowing operations. This paper designs a rotary type-hole cotton precision mulching dibbler. The main factors influencing the entry of cotton seeds into the seed wheel holes during the seeding process are then theoretically analyzed. Following this, an accurate discrete element model of coated cotton seeds is established and combined with a discrete element method (DEM) and multi-body dynamics (MBD)-coupled simulation model of the seed drill for seed picking and planting. Simulation experiments on the seeding performance of the precision dibbler were performed to study the influence of the seed wheel structure and motion parameters on the picking and planting performance under different speeds. The optimal parameter combination for the seed wheel is obtained through optimization experiments, and a precision dibbler is manufactured for bench testing. The bench test results are consistent with the simulation test results. At the precision dibbler rotation speed of 16 r/min, the qualified index reaches a maximum value of 93.28%, the skip sowing index increases with the precision dibbler rotation speed, and the re-sowing index decreases as the speed increases. These optimization results significantly improved seeding precision and efficiency and are of great significance for the reliability and effectiveness of cotton sowing operations.

Keywords: discrete element method; multi-body dynamics; coupling simulation; dibbler

Citation: Wang, L.; Ran, X.; Shi, L.; Xing, J.; Wang, X.; Hou, S.; Li, H. Simulation and Optimization of a Rotary Cotton Precision Dibbler Using DEM and MBD Coupling. *Agriculture* **2024**, *14*, 1411. https://doi.org/10.3390/agriculture14081411

Academic Editor: Maohua Xiao

Received: 27 July 2024
Revised: 16 August 2024
Accepted: 18 August 2024
Published: 20 August 2024

1. Introduction

Coupling the discrete element method (DEM) with multi-body dynamics (MBD) to conduct dynamic simulations of complex mechanical structures containing granular materials can effectively improve the dynamic performance of mechanisms and optimize mechanical structures and motion parameters [1–3]. This coupling approach can determine the motion behavior of materials and equipment and also predicts the operational performance of equipment [4]. The DEM calculates the loads of bulk granular materials within the geometric bodies of the equipment. The determined values are then input into the MBD to obtain the relative motion of the bodies. Following this, the motion is transferred back to the DEM, which calculates the behavior of the particles and the loads on the geometric bodies based on the motion. This mutual transfer enables coupled calculations [5].

The application of MBD-DEM co-simulation techniques for analyzing the motion state of granular materials in complex structures was first applied in the field of engineering. Lu [6] and Coetzee [4] designed MBD-DEM-coupled simulation models for civil engineering and geotechnical engineering, respectively. Applications have also been presented in aerospace engineering [7] and mining engineering [8]. Ji [9] employed DEM-MBD-coupled simulations to analyze the impact of the landing method and cushioning mechanism on the landing process of the lunar lander. Mohajeri [10] simulated the grabbing process of iron ore under bonding and stress using MBD-DEM-coupled simulations. Shi [11,12] analyzed the influence of compaction operations on the mechanical quality and damage of ballast based on the DEM-MBD-coupled method. Lommen [5] established a framework for the development, validation, and application of combined MBD and DEM simulations. Xiao [13,14] employed one-way-coupled MBD-DEM simulations to determine the influence of damping particles on the dynamic response of gear transmission. Chung [15] and Wu [16] developed bidirectional-coupled MBD-DEM dynamic models for gear and bearing systems, with damping particles filled in the gear cavities, to analyze the system's dynamic response and conduct experimental validation. The South African company VR Steel optimized the bucket of a dragline through co-simulation using ADAMS-EDEM, increasing the bucket fill rate while reducing the driving forces of the equipment [17].

Recent studies have applied the coupling simulation analysis of DEM and MBD to agricultural engineering [18]. Xu [19] established a simulation analysis model for precision seeders based on DEM-MBD coupling to analyze the soil covering and compaction process during planting operations. There was a strong agreement between the simulation and experimental results, validating the feasibility and applicability of the coupled simulation method. Yan [20] realized simulation analysis of the soybean seeding monomer working process by coupling EDEM and RecurDyn. The simulation results agreed with the test results, proving the accuracy of the coupling method. Liu [21] employed coupled simulation analysis using EDEM and RecurDyn to analyze the load on the rotary tiller shaft of a micro-tiller. Lai [22] conducted a DEM-MBD-coupled simulation experiment to analyze the impact of different seeding chain tension, seeder structure, and operating parameters on the working performance of a chain spoon-type ginseng precision seeder. Kim [23] established a soil-tool coupling simulation model based on DEM-MBD coupling and conducted simulation experiments combined with field measurements. The results indicated the ability of the simulations to effectively and accurately replace on-site testing. Dong [24] used DEM-MBD coupling to simulate the vibration-deep loosening operation, analyzing the soil disturbance and velocity distribution during the operation. However, despite these advances, the application of DEM-MBD coupling specifically to cotton precision dibblers remains relatively limited. In light of this, we believe that exploring the use of DEM-MBD coupling for the design and optimization of cotton precision dibblers could provide new insights into improving seeding performance. The previous literature indicates that the DEM-MBD simulation method can effectively analyze the motion characteristics of seeds in a seeder and determine the main factors affecting the seeding performance of the seeder. Applying DEM-MBD coupling to the design and optimization of cotton precision dibblers differs from previous agricultural applications. This method allows for more accurate prediction of seeding performance by simulating the unique interactions between cotton seeds and the dibbler components, thereby improving the precision and efficiency of cotton sowing operations.

This study aims to improve the structural design of seeding machinery, optimizing the existing configurations to enhance the efficiency of seed picking and sowing processes. Based on the DEM-MBD-coupled simulation model for a cotton precision dibbler, this research analyzed the main factors influencing the seeder's performance and explored the effects of different rotational speeds and structural and motion parameters of the seed wheel on seeding performance. A consistent trend observed between the simulations and bench tests validates the feasibility of the coupled simulation approach and the effectiveness of the simulation parameters. This work provides a reference for the structural analysis and optimization of the seeding unit.

2. Structure Analysis and Modeling of the Dibbler

Figure 1 presents the rotary type-hole cotton precision dibbler structure designed for the mulching planting mode of cotton in the northwest region of China. The main components of the structure include a seed tube, waist belt, duckbill, seed wheel, disrupter, and a moving plate. The seed tube is designed to guide the cotton seeds falling from the seed box along the correct path into the seed chamber. The outer side of the belt is used to secure the duckbill of the rotary type-hole dibbler, while the inner side is employed to install the type-hole wheel. The cotton seeds fall into the duckbill through the type-hole wheel by gravity. The duckbill, which directly contacts the soil, is a critical component that ensures the cotton seeds are ultimately deposited into the soil. The type-hole wheel is an essential component of the rotary type-hole dibbler, responsible for precisely metering the seeds that enter the seed chamber. The disrupter stirs the seeds within the seed chamber, making them easier for the type-hole wheel to capture. The moving plate connects and secures all components, facilitating rotational seeding.

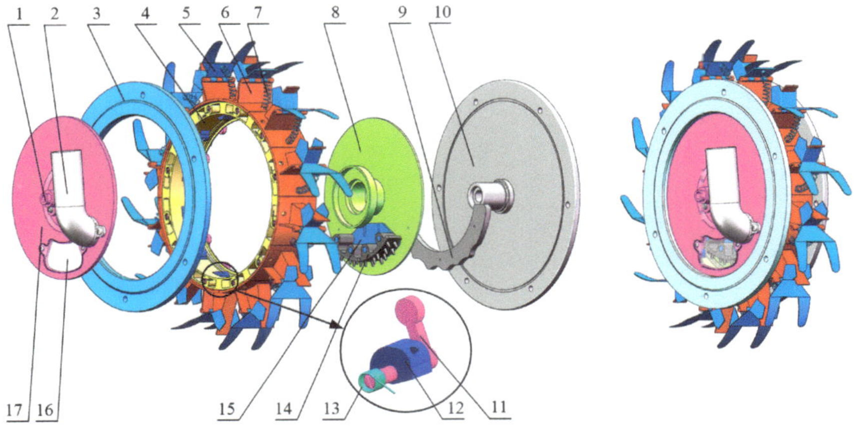

Figure 1. Structure of the rotary type-hole dibbler. (1) Shaft, (2) seed tube, (3) dynamic ring, (4) waist belt, (5) duckbill dynamic plate, (6) spring, (7) duckbill fixed plate, (8) fixed plate, (9) gear plate, (10) moving plate, (11) shifter, (12) type-hole wheel, (13) torsion spring, (14) disrupter, (15) disrupter support frame, (16) sight hole cover plate, and (17) end cover.

2.1. Working Principle of the Rotary Type-Hole Dibbler

The working process of a dibbler can be divided into three key stages: seed picking and cleaning; seed storage and dropping, and hole forming and seeding (Figure 2). During operation, the dibbler rolls on the mulch and soil at speed v under the drive of the main frame of the seeding machine. The cotton seeds in the seed box then enter the seed chamber through a seed-dropping pipe. The dibbler rotates as the main frame advances. When the type-hole wheel enters the seed picking and cleaning area, the gear plate moves the shifter, which connects with the type-hole wheel, causing the holes on the type-hole wheel to rotate in and out in the seed chamber as the belt rotates, achieving seed picking and cleaning. As the dibbler continues to rotate, the type-hole wheel enters the seed storage and dropping area, and the shifter disengages from the gear plate. The type-hole wheel then returns to its initial position under the action of the torsion spring, and the hole enters the seed storage chamber. As the dibbler continues to rotate, the hole opens downwards, and the cotton seeds slide out under the effect of gravity, passing through the seed storage chamber and entering the duckbill. At this point, the duckbill is in a closed state under the action of the spring. As the dibbler continues to rotate, it enters the seeding area and the duckbill reaches the bottom of the dibbler and inserts into the mulch and soil. The duckbill blade moves upwards under the pressure of the soil, causing the duckbill to open, and the cotton seeds are sown into the soil, thus completing the seeding operation.

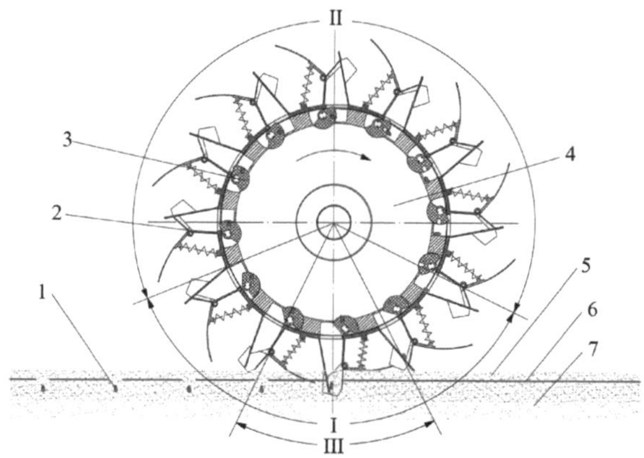

Figure 2. Working principle of the rotary type-hole dibbler. (I) Seed picking and cleaning, (II) seed storage and dropping, (III) hole forming and seeding. (1) cotton seeds, (2) duckbill, (3) type-hole wheel, (4) dibbler, (5) soil on plastic film, (6) plastic film, and (7) soil under plastic film.

2.2. Analysis of the Seed-Picking Movement of the Type-Hole Wheel

The successful seed picking of the dibbler-type hole wheel depends on the movement of the cotton seeds along the surface of the hole wheel. In particular, the cotton seeds can slide, roll, or exhibit a combination of both movements. The dibbler rotates and transfers at a certain speed to the lower layer of the cotton seeds that are in contact with the inner surface of the belt. This speed is less than or equal to the belt speed. The lower cotton seed layer then transfers a certain speed to the upper layer of cotton seeds, thereby transferring its speed to the cotton seeds inside the seed box. The relative speed of the cotton seeds is key in investigating the successful picking of seeds by the type-hole wheel. Without relative movement, successful seed picking cannot occur. Moreover, if the relative speed is too low, the frequency of successful seed picking is low, resulting in low sowing efficiency, while if the relative speed is too high, the type-hole wheel may not pick seeds in time, leading to missed sowing. Therefore, a precision dibbler must have a good seed-picking performance to clearly define the maximum limit speed at which cotton seeds can pass through the type holes of the type-hole wheel.

The motion trajectory equation of the hole wheel during the seed-picking process is described as follows:

$$\begin{cases} S = A\sin(2\pi f t) \\ f = \frac{360nz}{\tau} \end{cases} \tag{1}$$

where S is the harmonic motion displacement of the type-hole wheel (mm); A is the vibration amplitude of the type-hole wheel (mm); f is the vibration frequency of the type-hole wheel (Hz); n is the rotational speed of the seeder (r/min); z is the number of teeth on the gear plate; t is the time (min); and τ is the corresponding angular degree of the vibrating motion region of the seed-picking wheel (°).

By taking the first derivative of the motion trajectory equation of the hole wheel with respect to time t, the motion equation for harmonic motion velocity v_0 of the hole wheel can be obtained as follows:

$$v_0 = \frac{\mathrm{d}S}{\mathrm{d}t} = 2\pi A\cos(2\pi f) \tag{2}$$

In addition, the hole wheel moves in a circular motion along with the belt. The motion equation for its circular motion $v_r{'}$ is given by Equation (3):

$$v_r' = \omega R = 2\pi n R \tag{3}$$

The velocity of the hole wheel v' is the vector sum of its harmonic motion velocity v_0 and circular motion velocity v_r, expressed as Equation (4).

$$v' = \sqrt{v_0^2 + v_r'^2} = \sqrt{4\pi^2 n^2 R^2 + 4\pi^2 A^2 f^2 (\cos(2\pi f t))^2} \tag{4}$$

When the cotton seed is in an inverted position and needs to enter the type hole, the maximum distance of movement is required. We assume that the cotton seed approaches the type hole with its germ end, the coordinate origin is at the top right point of the type hole, the X-axis is the tangential direction of the seeding wheel rotation, and the Z-axis is the normal direction of the seeding wheel. When the center of gravity point O' moves directly above the edge of the type hole, a gap is generated between the lower side of the cotton seed's center of gravity and the edge of the type hole. Under the influence of gravity, the cotton seed begins to flip counterclockwise until it completely enters the type hole. As the cotton seed enters the type hole, its motion is similar to free falling and rotational motion, and the trajectory of the motion is approximately parabolic, as shown in Figure 3.

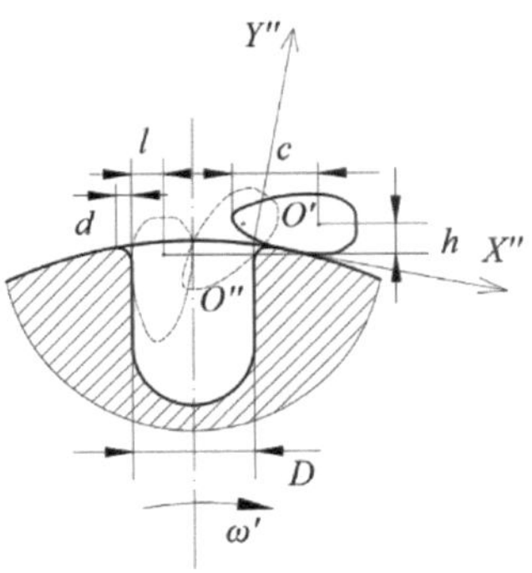

Figure 3. Motion state of the cotton seed falling into the type hole.

When the cotton seed enters the type hole at maximum speed v_{max}, the motion trajectory of the cotton seed center is given by Equation (5):

$$\begin{cases} D + d - l = v_{\max} t_0 \\ h = \frac{1}{2} g t_0^2 \end{cases} \tag{5}$$

where l is the horizontal distance between the cotton seed center of mass and the type hole wall (mm); t_0 is the time taken by the cotton seed center of mass (from the beginning) to fall into the type hole (s); h is the vertical distance that the cotton seed center of mass passes through during the falling process into the type hole (mm); d is the chamfer width (mm); and g is the acceleration due to gravity (m/s^2).

From Equation (5), the required time t_0 for the cotton seed to fall into the type hole can be obtained as follows:

$$t_0 = \sqrt{\frac{2h}{g}} \tag{6}$$

Substituting Equation (6) into Equation (5) gives the terminal velocity v_{max} at which the cotton seed fills the type hole as Equation (7):

$$v_{\max} = (D + d - l)\sqrt{\frac{g}{2h}} \tag{7}$$

The relative movement velocity v of the cotton seed with respect to the type-hole wheel is the vector sum of the type-hole wheel movement velocity v' and the relative velocity of the cotton seed v_r with respect to the dibbler, expressed as Equation (8):

$$v = \sqrt{v'^2 + v_r^2} = \sqrt{4\pi^2 n^2 R^2 + 4\pi^2 A^2 f^2 (\cos(2\pi f t))^2 + v_r^2} \tag{8}$$

To ensure the seeding accuracy of the furrow opener, the movement velocity v of the cotton seed with respect to the type-hole wheel must be less than the terminal velocity v_{max} of the cotton seed filling the type hole:

$$\sqrt{4\pi^2 n^2 R^2 + 4\pi^2 A^2 f^2 (\cos(2\pi f t))^2 + v_r^2} < (D + d - l)\sqrt{\frac{g}{2h}} \tag{9}$$

Equation (9) reveals that when parameters such as the type hole diameter, hole chamfer size, cotton seed dimensions, cotton seed friction coefficient, cotton seed layer height inside the dibbler, and belt radius are constant, the successful filling of cotton seeds into the type-hole wheel is related to its vibration amplitude, the vibration frequency, and the rotation speed of the dibbler.

In the device design process, we used the equations derived in this paper to calculate the key parameters. These equations helped determine the optimal rotation speed of the seeder, the number of gear plate teeth, and the oscillation amplitude of the type-hole wheel to ensure efficient performance in actual operation.

2.3. Establishment of a Discrete Element Model for Coated Cotton Seeds

The coated cotton seeds are similar to ovals. This study utilized reverse engineering technology to obtain an accurate contour model of the cotton seeds. A Capture MINI scanner (3D Systems) was employed to perform a three-dimensional (3D) scan of the outer contour of the cotton seed, and Geomagic Wrap 3D 2017 (3D Systems) was used to process the 3D scanning data. Figure 4 presents the process adopted for the cotton seed contour model. The physical coated cotton seed is depicted in Figure 4a. The 3D scanning data were employed to derive the point clouds of the coated cotton seed. The point cloud data were preprocessed to remove noise points from the original point cloud and compress it. It was then encapsulated to form a polygon model. Automatic curvature repair was applied to the missing areas in the polygon model to obtain the contour model of the coated cotton seed, as shown in Figure 4b.

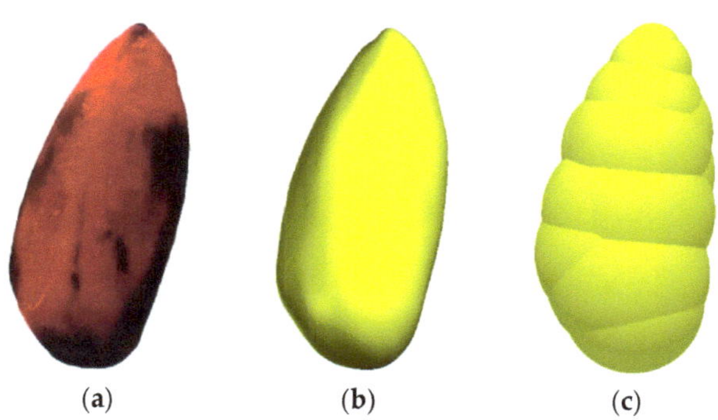

Figure 4. Processing of the coated cotton seed contour model: (**a**) coated cotton seed, (**b**) contour model, (**c**) multi-sphere aggregation particle model.

In the discrete element simulation software, spherical or other shaped particles are required to establish the discrete element model of the target materials. However, actual materials are mostly irregular in shape. In the EDEM discrete element software, the bonded particle [25] and multi-sphere [26] methods can be used to establish material models. The bonded particle model requires complex computations and a long simulation time due to the large number of particles used. The multi-sphere aggregation model has lower computational and simulation time requirements. Based on this, we adopt the multi-sphere aggregation model for the simulation experiments. The material contour model is imported into the EDEM software, and according to the model contour, several spherical particles with different diameters are overlapped and stacked to form the model. Figure 4c presents the cotton seed model established by this method, comprising 13 spherical particles with different diameters.

2.4. Establishment of the DEM-MBD Coupling Simulation Model

We implemented the MBD-DEM method using RecurDyn V9R2 (FunctionBay) and EDEM 2020 (DEM-Solutions). RecurDyn can simulate the complex motion of the planter, while EDEM simulates the motion between particles and components. Combining the RecurDyn and EDEM simulations can achieve real-time data transfer, enabling the bidirectional coupling of the cotton seed particles and furrow wheel planter. During the joint simulation, RecurDyn transfers the motion of the belt and seeding wheel components in the planter to EDEM. EDEM then calculates the forces and moments exerted by the cotton seed particles on the moving components of the planter and returns relevant information to RecurDyn. In the next time step, RecurDyn calculates the new displacements, velocities, accelerations, etc. of the components based on the new force, torque information, and planter motion, which are then transferred back to EDEM and implemented on the cotton seed particles. The data exchange process iterates back and forth to achieve bidirectional coupling calculations between RecurDyn and EDEM.

To ensure the reasonable and efficient simulation and calculation of the furrow wheel planter, the planter model is simplified by considering only the components in direct contact with the cotton seeds in the seed chamber, including end caps, fixed plates, belts, seeding wheels, and pushrods. The furrow wheel planter is modeled in 3D using SolidWorks 2018 (Dassault Systemes). The material properties, connections, forces, and contacts are set up in the multibody dynamics software RecurDyn. In the model, the end caps and fixed plates are stationary, the belt rotates clockwise around the axis, the seeding wheel rotates around the axis, and the motion is controlled by torsion springs and limit stops.

The direct target of the furrow wheel planter is the coated cotton seeds. The virtual simulation process uses the Hertz–Mindlin (no slip) model to calculate the interaction relationships between the coated cotton seeds. According to actual operation conditions, cotton seeds are first introduced into the seed chamber from the seed drop tube. Therefore, the mechanism does not move during the seed-filling process. A particle factory is set on the seed drop tube using pre-calibrated discrete element particles of coated cotton seeds with contact parameters. A total of 3000 cotton seed particles are generated at a rate of 1000 seeds/s and in the direction of gravity, with a volume proportional to a normal distribution.

The intrinsic and contact parameters of materials are crucial in the discrete element simulation process. The intrinsic parameters include material density, Poisson's ratio, elastic modulus, etc., while the contact parameters include the collision coefficient, sliding friction coefficient, and rolling friction coefficient between materials. Table 1 reports the values of the intrinsic and contact parameters of cotton seeds and resin materials, determined based on previous experimental results and references [27,28].

The total simulation time is set as the same value for EDEM and RecurDyn in the coupling interface, and the time step in RecurDyn is fixed as an integer multiple of that in EDEM. Here, the time steps in RecurDyn and EDEM are set as 0.01 s and 2.5×10^{-6} s, respectively. Once the parameters are set, the coupled simulation commences. After coupling, cotton seed particles also appear in RecurDyn, as shown in Figure 5. The movements of the components of the seeder and the cotton seed particles are simultaneously displayed in the two software, with arrows indicating the rotation direction of the conveyor belt.

Table 1. Simulation model parameters.

Parameter	Value
Density of cotton seed (g/cm^3)	0.981
Poisson's ratio of cotton seed	0.27
Shear modulus of cotton seed (MPa)	1.4×10^7
Density of resin (g/cm^3)	1.18
Poisson's ratio of resin	0.38
Shear modulus of resin (MPa)	177
Seed-to-seed static friction coefficient	0.130
Seed-to-seed dynamic friction coefficient	0.225
Seed-to-seed collision restoration coefficient	0.185
Seed-to-resin static friction coefficient	0.49
Seed-to-resin dynamic friction coefficient	0.21
Seed-to-resin collision restoration coefficient	0.25

Figure 5. EDEM-RecurDyn-coupled simulation: (**a**) RecurDyn simulation; (**b**) EDEM simulation. The direction of the arrow indicates the direction of dibbler movement.

In EDEM, a cotton seed particle that has been successfully picked and discharged by the type-hole wheel is selected to generate the motion trajectory (Figure 6). The cotton seed enters the seed chamber from the seed drop tube, follows the rotation of the conveyor belt, and revolves with the seed group. It then moves from the bottom to the top of the conveyor belt, falls back to the bottom, and enters the hole of the type-hole wheel. When the hole of the type-hole wheel rotates back to the seed storage bin, the cotton seed reaches the seed storage bin and is discharged under the action of gravity.

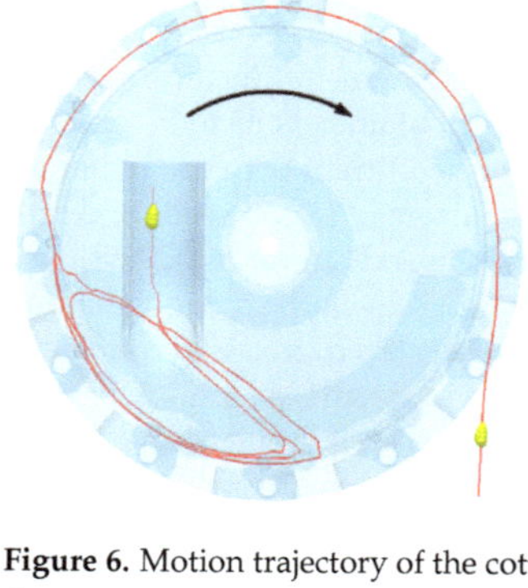

Figure 6. Motion trajectory of the cotton seed. Yellow represent cotton seed, and red lines represent the movement trajectory of cotton seed.

3. Simulation and Bench Tests

3.1. Simulation Test

The theoretical analysis in Section 2.2 reveals that the type-hole wheel structure has a certain disturbance effect on the seed population, and increasing the disturbance of the population can improve the probability of successful seed filling in the type-hole. To study the effect of the type-hole wheel structure on the seed population inside the seed cavity, this study designed three type-hole wheel structures, namely, a circular-type-hole wheel, a convex-type-hole wheel, and a corrugated-type-hole wheel. The basic structural parameters of these three type-hole wheels are the same, with a hole diameter of 7.0 mm and a depth of 10.0 mm. The convex and corrugated seed wheels have a surface wave amplitude of 2.0 mm, and a rounded corner is made at the upper edge of the shape hole with a radius of 0.5 mm (Figure 7). The dibbler speed, number of gear plate teeth, and vibration amplitude of the type-hole wheel are set as 20 r/min, 6, and 10 mm, respectively, and the working speed of the dibbler is controlled by setting the rotation speed of the belt.

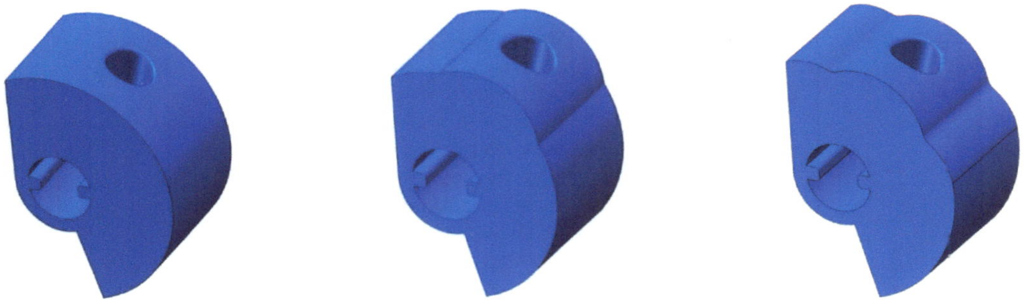

(a) (b) (c)

Figure 7. Different type-hole wheel structures designed in this study: (**a**) circular-type-hole wheel, (**b**) convex-type-hole wheel, (**c**) corrugated-type-hole wheel.

The vibration frequency and oscillation amplitude of the type-hole wheel influence the cotton seed filling effect on the shape hole. In addition, the oscillation frequency corresponds to the number of teeth on the gear plate during the movement of the type-hole wheel in the seeding and clearing area, and the oscillation amplitude corresponds to the tooth surface fluctuation amplitude of the gear plate. To investigate the impact of the vibration frequency and oscillation amplitude of the type-hole wheel on the seeding performance of the dibbler, we design a simulation test plan based on an optimized design of the combined experimental method. The test plan is used to determine the combination of type-hole wheel motion parameters that can achieve the best seeding performance. The type-hole wheel structure is corrugated, and based on preliminary experiments, the type-hole wheel oscillation amplitude is selected as 8–12 mm and the number of gear plate teeth is 4–8 (Figure 8).

(a) (b) (c)

Figure 8. Gear plates of dibbler structures designed in this study: (**a**) Four teeth, (**b**) six teeth, (**c**) eight teeth.

To reduce the impact of the rotation speed on the seed population of the dibbler, the seeder rotation speed of 10–30 r/min was selected and the levels of seeding experiment factors were encoded (Table 2).

Table 2. Encoding of the seeding experiment factor levels.

Level	Rotation Speed of Dibbler X_1 (r/min)	Number of Teeth on the Gear Plate X_2	Oscillation Amplitude of the Type-Hole Wheel X_3 (mm)
−1	10	4	7
0	20	6	10
1	30	8	13

The simulation planting operation status has three key forms, namely single seeding (1 seed), double seeding (≥2 seeds), and missed seeding (no seed). To facilitate the observations of cotton seed movements, during the EDEM simulation process, only cotton seed particles within the seed disc thickness in the seed metering device are selected during seed retrieval. Moreover, the transparency of all mechanical components is set to 0.1 (Figure 9). Figure 9a depicts the qualified sowing state, where the type-hole wheel typically takes out a single cotton seed from the group of seed particles and transfers it to the seed storage bin through the type hole. Figure 9b shows the re-sowing state, where the type-hole wheel takes out multiple cotton seeds from the group of seed particles and transfers them to the seed storage bin through the seed-type hole. The cotton seed status in the type holes under multiple double-seeding states reveals that smaller cotton seeds are more prone to double-seeding. This is because when the size of cotton seeds around the type hole is small, it is easier for two or more cotton seeds to enter the hole under their own gravity and the compression of other cotton seeds. This makes it difficult for the seed cleaning plate to remove them from the type hole. Furthermore, in the double seeding state, the two cotton seeds in the type hole are mostly overlapping. Figure 9c presents the skip sowing state, where the type-hole wheel fails to smoothly remove the cotton seed from the group of seed particles. The cotton seed status around the type hole under multiple missed seeding states indicates that larger cotton seeds are more likely to experience missed seeding. This is because when the size of the cotton seeds around the type hole is large, the cotton seeds have difficulty entering the type hole under their own gravity and the compression of other cotton seeds, leading to unsuccessful seed retrieval. In addition, when a cotton seed is about to enter the type hole, it escapes from the hole due to the compression of the surrounding cotton seeds, resulting in failed seed retrieval.

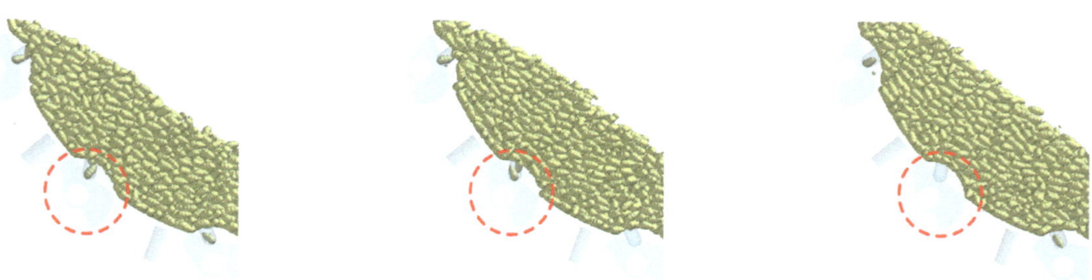

(**a**) (**b**) (**c**)

Figure 9. EDEM virtual simulation of seed sowing operation status: (**a**) qualified sowing state, (**b**) re-sowing state, (**c**) skip sowing state. The regions between red dashed lines represent seed sowing operation status.

According to the national standards GB/T 6976-2005 'Test Methods for Single-Granule (Precision) Seeders' and JB/T10293-2013 'Technical Conditions for Single-Granule (Precision) Seeders', this study selects the qualified index Y_1 (Y_1 represents the degree of matching between the number of seeds successfully sown by the seeder in each seeding cycle and the theoretical number of seeds), re-sowing index Y_2 (Y_2 refers to the proportion of instances where more than one seed is sown at a single seeding point during a seeding cycle), and skip sowing index Y_3 (Y_3 indicates the proportion of instances where no seeds are sown at a seeding point during the seeding process) as simulation test indicators to evaluate the quality of the seed metering device operation. These indices are calculated as follows:

$$Y_1 = \frac{N - n_1 - n_2}{N} \times 100\% \tag{10}$$

$$Y_2 = \frac{n_1}{N} \times 100\% \tag{11}$$

$$Y_3 = \frac{n_2}{N} \times 100\% \tag{12}$$

where N is the total number of cotton seeds counted in the experiment; n_1 is the number of cotton seeds that have a distance with adjacent seeds less than 0.5 times the theoretical seed distance; and n_2 is the number of cotton seeds that have a distance with adjacent seeds greater than 1.5 times the theoretical seed distance.

3.2. Bench Test

The JPS-12 seed metering device developed by the Heilongjiang Agricultural Machinery Research Institute was employed for the seeding performance test (Figure 10). During the test, the designed hole-type cotton seed metering device is installed on the test stand and the seedbed conveyor belt moves in the opposite direction to the seed metering device to simulate the actual field operation process. Cotton seeds are collected by the seed metering device and seeded on the seedbed conveyor belt coated with sticky oil to prevent the cotton seeds from bouncing on the seedbed conveyor belt. The image acquisition system of the test stand detects the cotton seeds in real-time to determine various seeding performance indicators and subsequently outputs the test results.

Figure 10. Seed planting performance evaluation test; (1) the rotary type-hole dibbler, (2) the motor, (3) the transmission shaft, (4) the conveyor belt, and (5) the fixed bracket. The red boxes and arrow represent enlarged view of the dibbler.

To validate the accuracy of the optimized simulation test results, the optimal parameter combination obtained from the simulation seeding test was used to create a gear plate, with a tooth number of 5.64 and a type-hole wheel swing amplitude of 9.71 mm. The components of the seeder were 3D printed using light-sensitive resin. Transparent resin materials were used for the seed-dropping tube and end cover to observe the seed movement inside the seeder. Xinlu Zhong 67 deseeded cotton seeds were selected for the experimental sample. The seeding device speeds were set to 12 r/min, 16 r/min, 20 r/min, 24 r/min, and

28 r/min for the seeding tests. In each test group, the number of seeding measurements was set to 250 (except for the start-up and stop phases), and each test group was repeated three times.

4. Results and Discussion

4.1. Impact of the Type-Hole Wheel Structure on the Seeding Performance

Figure 11 presents the vector diagram of the compression force of cotton seeds at 5.6 s when the rotational speed of the planter is 20 r/min. At this same time, the corrugated-type-hole wheel exhibits the maximum cotton seed compression force, reaching 0.317 N, followed by the convex-type-hole wheel (0.299 N) and the circular-type-hole wheel (0.212 N). The cotton seeds in the seed chamber of the circular-type-hole wheel generally move clockwise around the center. The cotton seeds at the bottom of the cotton seed group move to the upper right under the drive of the belt, and after reaching the top, they slide down to the left along the upper surface under the action of gravity and friction. The cotton seeds in the middle of the cotton seed group move clockwise under the drive of the upper and lower layer cotton seeds. The movement of the upper and lower layer cotton seeds in the seed chamber of the convex and corrugated-type-hole wheels is similar to that of the circular wheel. However, the movement of the cotton seeds in the middle of the cotton seed group exhibits an irregular trend, with a chaotic movement direction in some areas. Compared to the circular-type-hole wheel, the movements caused by the convex and corrugated-type-hole wheels are more irregular, indicating that they exert the greatest disturbance to the cotton seed particles, which is conducive to the filling of type holes by the cotton seeds.

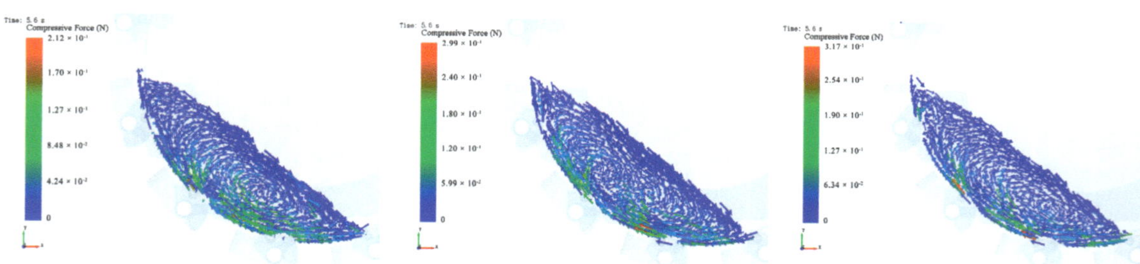

(a) (b) (c)

Figure 11. Vector plot of the cotton seed compression force at 5.6 s: (**a**) circular-type-hole wheel, (**b**) convex-type-hole wheel, (**c**) corrugated-type-hole wheel.

To further analyze the influence of the seeding wheel structure on the seeding process, the average velocity curves of cotton seed particles within the time steps of 3 to 15 s are shown in Figure 12. The cotton seed particle velocity with the circular-type-hole wheel is markedly lower than that of the convex and corrugated types. The fluctuation magnitude of the cotton seed particle velocity, from the largest to smallest, is as follows: convex-type-hole wheel (0.0284 m/s); circular-type-hole wheel (0.0276 m/s); and corrugated-type-hole wheel (0.0242 m/s). The relative velocity of cotton seed particles with the circular-type-hole wheel is low and the velocity fluctuates greatly. Although the convex-type-hole wheel can increase the relative velocity of the cotton seed particles, its velocity also exhibits large fluctuations. In contrast, the corrugated-type-hole wheel enhances the relative velocity of cotton seed particles while maintaining velocity fluctuations at a low level. This indicates the ability of the corrugated-type-hole wheel to improve the dibbler filling performance.

Therefore the structure of the type-hole wheel determines the mobility and compression force inside the dibbler. A well-designed type-hole wheel structure can effectively reduce the compression force between seeds, and increase the mobility of seeds, thereby improving filling efficiency. If the type-hole wheel structure is not optimized, the seeds may be subjected to excessive compression force, leading to seed damage or inadequate filling, which will further affect the quality and efficiency of seeding.

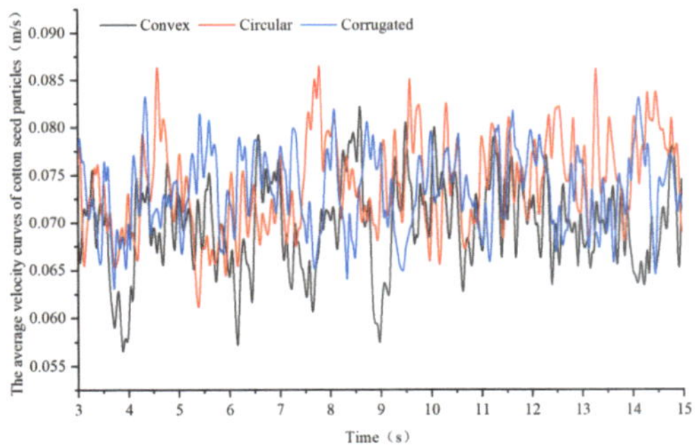

Figure 12. Average velocity of cotton seed particles under the action of different type-hole wheels with different structures.

4.2. Influence of the Dibbler Rotation Speed on Seeding Performance

To analyze the relationship between the speed of the type-hole wheel and the cotton seeds during the seed-picking process, two cotton seeds successfully filled into the type holes were randomly selected. When the dibbler rotation speed was 10 r/min, the selected seed numbers were 648 and 1472; when the planter rotation speed was 20 r/min, the selected seed numbers were 1088 and 1724; when the planter rotation speed was 30 r/min, the selected seed numbers were 564 and 116. The cotton seeds were marked in EDEM to obtain the speed of the cotton seeds and the corresponding type-hole wheels during the seed-picking process. The data were plotted using Origin 2018 (OriginLab Corporation, Indianapolis, United States) to present the speed change trends of the cotton seeds and the corresponding type-hole wheels at different rotation speeds, as shown in Figure 13.

Figure 13a,b reveal that when the dibbler rotation speed is 10 r/min, the duration of the type-hole wheel in the seed picking and cleaning area is approximately 2.65 s, with the speed fluctuating between 0.12 and 0.23 m/s. The time taken for cotton seeds no. 648 and no. 1472 to fill into the holes is 6.84 s and 10.52 s, with speeds of 0.0947 m/s and 0.1190 m/s, respectively. At this time, the corresponding speeds of the type-hole wheel are 0.1559 m/s and 0.1374 m/s, respectively. Under a rotation speed of 10 r/min, the speed fluctuation range of successfully picked cotton seeds is between 0.05 and 0.27 m/s, with the speed of the cotton seeds before filling into the holes generally higher than that of the type-hole wheel.

Figure 13c,d show that when the dibbler rotation speed is 20 r/min, the duration of the type-hole wheel in the seed picking and cleaning area is approximately 1.33 s, with the speed fluctuating between 0.25 and 0.45 m/s. The time taken for cotton seeds no. 1088 and no. 1724 to fill into the holes is 9.61 s and 12.64 s, with speeds of 0.1933 m/s and 0.1809 m/s, respectively. At this time, the corresponding speeds of the type-hole wheel are 0.2804 m/s and 0.2799 m/s, respectively. Under a rotation speed of 20 r/min, the speed fluctuation range of successfully picked cotton seeds is between 0.13 and 0.48 m/s, and the of the cotton seeds before filling into the holes is generally still higher than that of the type-hole wheel.

Figure 13e,f reveal that when the dibbler rotation speed is 30 r/min, the duration of the type-hole wheel in the seed picking and cleaning area is approximately 0.88 s, with the speed ranging from 0.37 to 0.66 m/s. The time taken for cotton seeds no. 564 and no. 116 to fill into the holes is 9.26 s and 12.28 s, with speeds of 0.3016 m/s and 0.4054 m/s, respectively. At this time, the corresponding speeds of the type-hole wheel are 0.4198 m/s and 0.4954 m/s, respectively. Under a rotation speed of 30 r/min, the speed fluctuation range of successfully picked cotton seeds is between 0.11 and 0.73 m/s, and the speed of the cotton seeds before filling into the holes is mostly lower than that of the type-hole wheel.

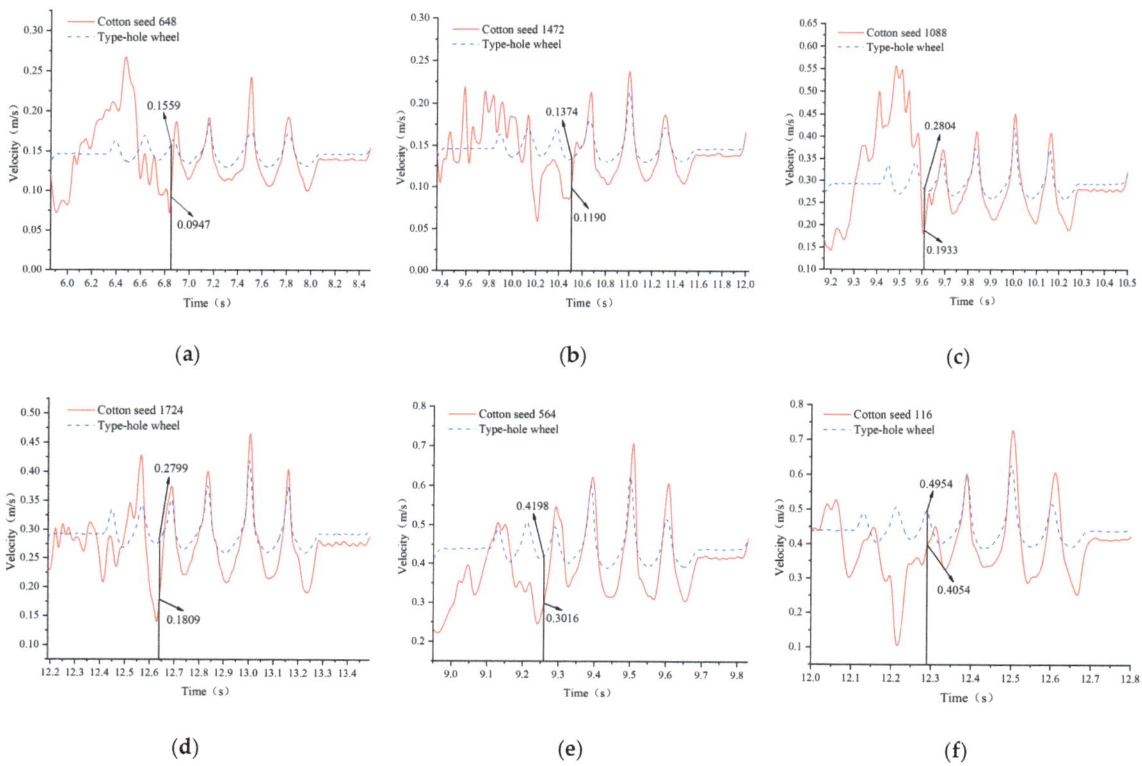

Figure 13. Speed of cotton seeds and type-hole wheels at different rotation speeds: (**a**) cotton seed 648 with the dibbler rotation speed at 10 r/min, (**b**) cotton seed 1472 with the dibbler rotation speed at 10 r/min, (**c**) cotton seed 1088 with the dibbler rotation speed at 20 r/min, (**d**) cotton seed 1724 with the dibbler rotation speed at 20 r/min, (**e**) cotton seed 564 with the dibbler rotation speed at 30 r/min, and (**f**) cotton seed 116 with the dibbler rotation speed at 30 r/min.

In summary, as the dibbler rotation speed increases, the duration of the type-hole wheel in the seed picking and cleaning area decreases, while the speed fluctuation range of the cotton seeds increases. Under the same dibbler rotation speed, the variation in the type-hole wheel speed is constant, but the cotton seed speed exhibits irregular random fluctuations. At different planter rotation speeds, a higher rotation speed results in a larger instantaneous speed of the cotton seeds. Before filling into the holes of the type-hole wheel, the speed changes are relatively chaotic and fluctuate greatly. However, after filling the holes, the speed of the cotton seeds follows a similar trend to that of the type-hole wheel speed, but with a larger speed fluctuation range than the type-hole wheel. After filling the holes, in addition to moving with the type-hole wheel, the cotton seeds also move within the holes. An improved planting effect is achieved when the relative speed between the cotton seeds and the type-hole wheel is smaller. Therefore, by adjusting the dibbler rotation speed, the relative speed of the cotton seeds within the dibbler can be changed, thereby improving the effectiveness of filling the cotton seeds into the holes.

4.3. Effects of the Motion Parameters of the Type-Hole Wheel on Seeding Performance

Based on the actual production situation, the qualified index Y_1, re-sowing index Y_2, and skip-sowing index Y_3 were selected as the experimental indicators for planting experiments. Table 3 reports the design of the Box–Behnken response surface experiment and the corresponding results.

Table 3. Design and results of the seeding experiment.

No.	Rotation Speed of Dibbler X_1 (r/min)	Number of Teeth on the Gear Plate X_2	Oscillation Amplitude of Type-Hole Wheel X_3 (mm)	Qualified Index Y_1 (%)	Re-Sowing Index Y_2 (%)	Skip Sowing Index Y_3 (%)
1	−1	1	0	89.69	7.25	3.06
2	0	−1	−1	89.16	5.39	5.45
3	0	−1	1	89.22	4.16	6.62
4	−1	0	1	90.43	6.52	3.05
5	0	1	−1	88.48	6.34	5.18
6	0	0	0	91.65	3.88	4.47
7	0	0	0	91.57	4.32	4.11
8	1	−1	0	86.58	4.49	8.93
9	0	1	1	86.96	6.89	6.15
10	0	0	0	91.52	4.21	4.27
11	1	0	−1	87.77	4.54	7.69
12	1	0	1	85.31	4.85	8.84
13	−1	0	−1	90.96	6.26	2.78
14	0	0	0	92.19	3.75	4.06
15	−1	−1	0	90.45	6.38	3.17
16	1	1	0	86.48	5.23	8.29
17	0	0	0	91.96	3.65	4.39

We employed Design-Expert 12.0 (Stat-Ease Inc., Minia Bonis, Minneapolis, MN, USA) to conduct polynomial regression analysis on the seeding experiment results. The second-order polynomial regression models for the qualified index, re-sowing index, and skip-sowing index are expressed in Equations (13)–(15):

$$Y_1 = 91.78 - 1.92X_1 - 0.48X_2 - 0.56X_3 + 0.17X_1X_2 - 0.48X_1X_3 - 0.40X_2X_3 - 1.66X_1^2 - 1.82X_2^2 - 1.50X_3^2 \quad (13)$$

$$Y_2 = 3.96 - 0.91X_1 + 0.66X_2 - 0.01X_3 - 0.03X_1X_2 + 0.01X_1X_3 + 0.45X_2X_3 + 0.86X_1^2 + 1.01X_2^2 + 0.72X_3^2 \quad (14)$$

$$Y_3 = 4.26 + 2.71X_1 - 0.19X_2 + 0.45X_3 - 0.13X_1X_2 + 0.22X_1X_3 - 0.05X_2X_3 + 0.67X_1^2 + 0.93X_2^2 + 0.66X_3^2 \quad (15)$$

Tables 4–6 report the results of the regression model variance analysis. The p-values of the regression models for the three indices are less than 0.0001, and the p-values of the lack-of-fit terms exceed 0.05. This indicates that the regression models are highly significant, the lack-of-fit of the models is not significant, and the regression models have a high fitting degree.

Table 4. Analysis of variance for the qualified index.

Source of Variance	Sum of Squares	Degrees of Freedom	F	p	Significance
Model	74.67	9	42.97	<0.0001	**
X_1	29.61	1	153.33	<0.0001	**
X_2	1.81	1	9.35	0.0184	*
X_3	2.48	1	12.82	0.0090	**
X_1X_2	0.1089	1	0.5640	0.4771	
X_1X_3	0.9312	1	4.82	0.0641	
X_2X_3	0.6241	1	3.23	0.1152	
X_1X_1	11.57	1	59.93	0.0001	**
X_2X_2	13.95	1	72.25	<0.0001	**
X_3X_3	9.51	1	49.24	0.0002	**
Residual	1.35	7			
Lack of fit	1.02	3	4.14	0.1017	
Error	0.3291	4			
Sum	76.02	16			

Note: ** denotes an extremely significant impact of the parameter ($p < 0.01$); and * represents a significant impact of the parameter ($p < 0.05$). The same as Tables 5 and 6.

Table 5. Analysis of variance for the re-sowing index.

Source of Variance	Sum of Squares	Degrees of Freedom	F	p	Significance
Model	21.69	9	15.29	0.0008	**
X_1	6.66	1	42.26	0.0003	**
X_2	3.50	1	22.19	0.0022	**
X_3	0.0015	1	0.0096	0.9247	
X_1X_2	0.0042	1	0.0268	0.8746	
X_1X_3	0.0006	1	0.0040	0.9516	
X_2X_3	0.7921	1	5.03	0.0599	
X_1X_1	3.12	1	19.82	0.0030	**
X_2X_2	4.33	1	27.46	0.0012	**
X_3X_3	2.18	1	13.81	0.0075	**
Residual	1.10	7			
Lack of fit	0.7647	3	3.01	0.1574	
Error	0.3387	4			
Sum	22.79	16			

Table 6. Analysis of variance for the skip sowing index.

Source of Variance	Sum of Squares	Degrees of Freedom	F	p	Significance
Model	69.15	9	270.38	<0.0001	**
X_1	58.81	1	2069.37	<0.0001	**
X_2	0.2775	1	9.77	0.0167	*
X_3	1.58	1	55.75	0.0001	**
X_1X_2	0.0702	1	2.47	0.1599	
X_1X_3	0.1936	1	6.81	0.0349	*
X_2X_3	0.0100	1	0.3519	0.5717	
X_1X_1	1.90	1	66.76	<0.0001	**
X_2X_2	3.65	1	128.49	<0.0001	**
X_3X_3	1.83	1	64.30	<0.0001	**
Residual	0.1989	7			
Lack of fit	0.0753	3	0.8126	0.5498	
Error	0.1236	4			
Sum	69.35	16			

Table 4 reveals that the interaction terms (X_1X_2, X_1X_3, and X_2X_3) between the rotation speed of the dibbler, the number of teeth on the gear plate, and the oscillation amplitude of type-hole wheel do not exert a significant influence ($p > 0.05$) on the qualified index model, while all other factors have a significant impact on the qualified index model. Table 5 indicates that the oscillation amplitude of the type-hole wheel does not significantly affect the re-sowing index model ($p > 0.05$), suggesting that the oscillation amplitude of the type-hole wheel has a minor impact on the re-sowing index. In addition, the interaction terms (X_1X_2, X_1X_3, and X_2X_3) between the rotation speed of the dibbler, the number of teeth on the gear plate, and the oscillation amplitude of the type-hole wheel also do not significantly influence ($p > 0.05$) the re-sowing index model, while all other factors have a significant impact on the re-sowing index model. Table 6 shows that the interaction term (X_1X_2) between the rotation speed of the dibbler and the number of teeth on the gear plate and the interaction term (X_2X_3) between the number of teeth on the gear plate, and the oscillation amplitude of type-hole wheel do not significantly impact ($p > 0.05$) the skip sowing index model, while all other factors have a significant impact on the skip sowing index model.

Removing the non-significant terms from the regression models of the qualified index, re-sowing index, and skip-sowing index results in the following simplified equations, respectively:

$$Y_1 = 91.78 - 1.92X_1 - 0.48X_2 - 0.56X_3 - 1.66X_1^2 - 1.82X_2^2 - 1.50X_3^2 \quad (16)$$

$$Y_2 = 3.96 - 0.91X_1 + 0.66X_2 - 0.01X_3 + 0.86X_1^2 + 1.01X_2^2 + 0.72X_3^2 \quad (17)$$

$$Y_3 = 4.26 + 2.71X_1 - 0.19X_2 + 0.45X_3 + 0.22X_1X_3 + 0.67X_1^2 + 0.93X_2^2 + 0.66X_3^2 \quad (18)$$

After simplification, the goodness of fit R^2 for the regression models of the qualified index, re-sowing index, and skip-sowing index are 0.9603, 0.9960, and 0.9166, respectively. This indicates that the predicted values of the regression model equations have a good fit with the actual values. Thus, the regression models for the seed selection qualified index, re-broadcast index, and missed broadcast index have high reliability.

To obtain the optimal parameter combination for the rotation speed of the dibbler, the number of teeth on the gear plate, and the oscillation amplitude of the type-hole wheel, this study takes the maximum value of the qualified index, and the minimum values of the re-sowing index and skip sowing index as the optimization objectives. The optimization objectives and constraints for the rotation speed of the dibbler, the number of teeth on the gear plate, and the oscillation amplitude of the type-hole wheel can be formulated as Equation (19):

$$\begin{cases} \max Y_1(X_1, X_2, X_3) \\ \min[Y_2(X_1, X_2, X_3), Y_3(X_1, X_2, X_3)] \\ 10 \le X_1 \le 30 \\ 4 \le X_2 \le 8 \\ 7 \le X_3 \le 13 \end{cases} \quad (19)$$

The optimization module of Design-Expert 12.0 is used to optimize the constrained objective function, obtaining the best parameter combination can be obtained. At the dibbler rotation speed of 16.34 r/min, the number of teeth on the gear plate is 5.64, the oscillation amplitude of the type-hole wheel is 9.71 mm, and the seeding performance of the seed planter reaches is optimized. At this point, the qualified index is 92.23%, the re-sowing index is 3.63%, and the skip-sowing index is 4.18%.

4.4. Optimization and Validation of Seeder Performance in Bench Tests

The qualified index, re-sowing index, and skip-sowing index were selected as the evaluation criteria for seeding performance. Table 7 presents the bench test results.

Table 7. The bench test results.

Rotation Speed of the Dibbler (r/min)	Qualified Index (%)	Re-Sowing Index (%)	Skip Sowing Index (%)
12	91.78	6.64	1.58
16	93.28	4.35	2.37
20	92.95	3.98	3.07
24	92.17	3.04	4.79
28	90.71	2.82	6.47

The experimental results presented in Table 7 show that as the dibbler rotation speed increases, the qualified index initially increases and then decreases, while the re-sowing index exhibits a decreasing trend and the skip-sowing index exhibits an increasing trend. As the rotation speed increases, the movement of seeds inside the seeder accelerates, improving filling efficiency and increasing the qualified index. However, when the speed increases further, the effect of centrifugal force becomes more pronounced, causing some seeds to fail to enter the seeding holes properly, which leads to a decrease in filling efficiency and a subsequent decline in the qualified index. When the rotation speed of the dibbler reaches 16 r/min, the qualified index reaches its maximum value of 93.28%, with the skip-sowing index at 2.37% and the re-sowing index at 4.35%. The qualified index from the bench test is slightly higher than that from the simulation test. This discrepancy may be attributed to the vibrations caused by the rotation of the motor and the chain gear transmission in the bench test, which intensifies the seed movement within the seeder and facilitates seed filling in the seed extraction wheel. The relative error between the bench test and simulation test results is not significant, indicating that the simulation test results are relatively accurate.

5. Conclusions

A simulation model of a type-hole wheel precision cotton dibbler was established based on the DEM-MBD-coupled algorithm to analyze the impact of the type-hole wheel structure and the motion parameters of the dibbler on its seed extraction and seeding performance. The main conclusions are as follows:

(1) Three different structures of the type-hole wheel were designed: circular; convex; and corrugated. The effects of the type-hole wheel structural designs and operating speed on the dibbler seeding performance were analyzed. The results showed that the average movement speed of the cotton seed particles in the seed chamber increased with the dibbler rotation speed. Among the three structures, the corrugated-type-hole wheel caused the greatest disturbance to the cotton seed population within the dibbler. In particular, the corrugated-type-hole wheel type increased the relative speed of the cotton seed particles and the seed population speed fluctuated less compared to the other type-hole wheel types, indicating it to be more conducive to filling the holes with the seeds.

(2) A combination design experiment was conducted to study the optimal motion parameter combination for the type-hole wheel and establish a mathematical model relating the seeding performance indicators to the experimental parameters. The rotation speed of the dibbler, the number of teeth on the gear plate, and the oscillation amplitude of the type-hole wheel were selected as experimental factors, and the qualified index, re-sowing index, and skip-sowing index were used as evaluation criteria. The regression mathematical model was optimized, yielding the following optimal parameter combination: dibbler rotation speed of 16.34 r/min; the number of gear plate teeth of 5.64; and type-hole wheel oscillation amplitude of 9.71 mm.

(3) A bench test of the seeder's seeding performance was conducted. The test results indicated that when the rotation speed of the dibbler was 16 r/min, the qualified index reached a maximum value of 93.28%, with a re-sowing index of 4.35% and a skip-sowing index of 2.37%. The results from the bench test were consistent with those from the simulation test, showing that the re-sowing index decreased with the increase in rotation speed, while the skip-sowing index increased with the dibbler rotation speed.

This study presents significant advancements in the seeding performance of the dibbler, achieved by optimizing both the structure of the type-hole wheel and the motion parameters of the device. The findings demonstrate that variations in type-hole wheel design considerably influence seed movement and speed fluctuations. Although these parameters exhibited strong performance under controlled experimental conditions, they may be influenced by real-world factors, including terrain variations, differences in soil conditions, and the maintenance of equipment during operational use. Therefore, to ensure sustained optimal performance across diverse operating environments, regular maintenance of the dibbler is recommended, along with the adjustment of parameters tailored to specific conditions. Future research should consider additional variables to further refine and optimize the dibbler's design, including the impact of seed type and shape, soil conditions, and fluctuations in the operational environment, thereby ensuring robust performance across a range of field conditions.

Author Contributions: Conceptualization, L.W. and H.L.; methodology, L.W. and X.R.; software, L.W. and X.R.; validation, L.W., X.W. and J.X.; formal analysis, L.S.; investigation, L.S. and X.R.; data curation, S.H.; writing—original draft preparation, L.W.; writing—review and editing, L.W. and H.L.; project administration, L.W. and H.L.; funding acquisition, L.W. All authors have read and agreed to the published version of the manuscript.

Funding: This research was supported by the Bintuan Science and Technology Program (2022CB001-06, 2023AB005-01) and the President Fund from Tarim University (TDZKYS202302).

Institutional Review Board Statement: Not applicable.

Data Availability Statement: The data presented in this study are available upon request from the authors.

Acknowledgments: The authors would like to thank their schools and colleges, as well as the funding providers of the project. All support and assistance are sincerely appreciated.

Conflicts of Interest: The authors declare no conflicts of interest.

References

1. Zhang, X.; Wu, B.; Niu, L.K.; Xiong, X.Y.; Dong, Z.X. Dynamic characteristics of two-way coupling between flip-flow screen and particles based on DEM. *Chin. Coal Soc.* **2019**, *44*, 1930–1940.
2. Gan, J.Q.; Zhou, Z.Y.; Yu, A.B.; Ellis, D.; Attwood, R.; Chen, W. Co-simulation of multibody dynamics and discrete element method for hydraulic excavators. *Powder Technol.* **2023**, *414*, 118001. [CrossRef]
3. Fang, J.; Zhao, C.F.; Lu, X.Y.; Xiong, W.H.; Shi, C. Dynamic behavior of railway Vehicle-Ballasted track system with unsupported sleepers based on the hybrid DEM-MBD method. *Constr. Build. Mater.* **2023**, *394*, 132091. [CrossRef]
4. Coetzee, C.J.; Els, D.N.J.; Dymond, G.F. Discrete element parameter calibration and the modelling of dragline bucket filling. *J. Terramech.* **2010**, *47*, 33–44. [CrossRef]
5. Lommen, S.; Lodewijks, G.; Schott, D.L. Co-simulation framework of discrete element method and multibody dynamics models. *Eng. Comput.* **2018**, *35*, 1481–1499. [CrossRef]
6. Lu, Z.; Lu, X.L.; Lu, W.S.; Masri, S.F. Studies of the performance of particle dampers under dynamic loads. *J. Sound Vib.* **2010**, *329*, 5415–5433. [CrossRef]
7. Ahmad, N.; Ranganath, R.; Ghosal, A. Modeling and experimental study of a honeycomb beam filled with damping particles. *J. Sound Vib.* **2017**, *391*, 20–34. [CrossRef]
8. Barrios, G.K.P.; Tavares, L.M. A preliminary model of high pressure roll grinding using the discrete element method and multi-body dynamics coupling. *Int. J. Miner. Process.* **2016**, *156*, 32–42. [CrossRef]
9. Ji, S.Y.; Liang, S.M. DEM-FEM-MBD coupling analysis of landing process of lunar lander considering landing mode and buffering mechanism. *Adv. Space Res.* **2021**, *68*, 1627–1643. [CrossRef]
10. Mohajeri, M.J.; de Kluijver, W.; Helmons, R.L.J.; van Rhee, C.; Schott, D.L. A validated co-simulation of grab and moist iron ore cargo: Replicating the cohesive and stress behaviour of bulk solids. *Adv. Powder Technol.* **2021**, *32*, 1157–1169. [CrossRef]
11. Shi, S.W.; Gao, L.; Xiao, H.; Xu, Y.; Yin, H. Research on ballast breakage under tamping operation based on DEM-MBD coupling approach. *Constr. Build. Mater.* **2021**, *272*, 1810–1822. [CrossRef]
12. Shi, S.W.; Gao, L.; Cai, X.P.; Yin, H.; Wang, X.L. Effect of tamping operation on mechanical qualities of ballast bed based on DEM-MBD coupling method. *Comput. Geotech.* **2020**, *124*, 103574. [CrossRef]
13. Xiao, W.Q.; Huang, Y.X.; Jiang, H.; Lin, H.; Li, J.L. Energy dissipation mechanism and experiment of particle dampers for gear transmission under centrifugal loads. *Particuology* **2016**, *27*, 40–50. [CrossRef]
14. Xiao, W.Q.; Li, J.N.; Pan, T.L.; Zhang, X.; Huang, Y.X. Investigation into the influence of particles' friction coefficient on vibration suppression in gear transmission. *Mech. Mach. Theory* **2017**, *108*, 217–230. [CrossRef]
15. Chung, Y.C.; Wu, Y.R. Dynamic modeling of a gear transmission system containing damping particles using coupled multi-body dynamics and discrete element method. *Nonlinear Dyn.* **2019**, *98*, 129–149. [CrossRef]
16. Wu, Y.R.; Chung, Y.C.; Wang, I.C. Two-way coupled MBD-DEM modeling and experimental validation for the dynamic response of mechanisms containing damping particles. *Mech. Mach. Theory* **2021**, *159*, 104257. [CrossRef]
17. Curry, D.R.; Deng, Y. Optimizing Heavy Equipment for Handling Bulk Materials with Adams-EDEM Co-simulation. In Proceedings of the 7th International Conference on Discrete Element Methods (DEM), Dalian, China, 15 December 2017.
18. Hu, J.P.; Pan, J.; Chen, F.; Yue, R.C.; Yao, M.J.; Li, J. Simulation Optimization and Experiment of Finger-clamping Seedling Picking Claw Based on EDEM RecurDyn. *Chin. Soc. Agric. Mach.* **2022**, *53*, 75–85+301.
19. Xu, T.Y.; Zhang, R.X.; Wang, Y.; Jiang, X.M.; Feng, W.Z.; Wang, J.L. Simulation and Analysis of the Working Process of Soil Covering and Compacting of Precision Seeding Units Based on the Coupling Model of DEM with MBD. *Processes* **2022**, *10*, 1103. [CrossRef]
20. Yan, D.X.; Xu, T.Y.; Yu, J.Q.; Wang, Y.; Guan, W.; Tian, Y.; Zhang, N. Test and Simulation Analysis of the Working Process of Soybean Seeding Monomer. *Agriculture* **2022**, *12*, 1464. [CrossRef]
21. Liu, Y.; Liu, Y.P.; Zhang, T. Load Analysis of Rotary Cutter Shaft for Power Tiller Based on DEM and MBD Theory. *J. Agric. Sci. Technol.* **2020**, *22*, 79–86.
22. Lai, Q.H.; Jia, G.X.; Su, W.; Zhao, L.J.; Qiu, X.B.; Lu, Q. Design and Test of Chain Spoom Type Precision Seed Metering Device for Ginseng Based onDEM-MBD Coupling. *Chin. Soc. Agric. Mach.* **2022**, *53*, 91–104.
23. Kim, Y.S.; Lee, S.D.; Baek, S.M.; Baek, S.Y.; Jeon, H.H.; Lee, J.H.; Siddique, M.A.; Kim, Y.J.; Kim, W.S.; Sim, T.; et al. Development of DEM-MBD coupling model for draft force prediction of agricultural tractor with plowing depth. *Comput. Electron. Agric.* **2022**, *202*, 107405. [CrossRef]

24. Dong, X.Q.; Su, C.; Zheng, H.N.; Han, R.Q.; Li, Y.L.; Wan, L.P.C.; Song, J.N.; Wang, J.C. Analysis of soil disturbance process by vibrating subsoiling based on DEM-MBD coupling algorithm. *Transact. Chin Soc. Agric. Eng.* **2022**, *38*, 34–43.
25. Yu, Q.X.; Liu, Y.; Chen, X.B.; Sun, K.; Lai, Q.H. Calibration and Experiment of Simulation Parameters for Panax notoginseng Seeds Based on DEM. *Chin. Soc. Agric. Mach.* **2020**, *51*, 123–132.
26. Abbaspour-Fard, M.H. Theoretical validation of a multi-sphere discrete element model suitable for biomaterials handling simulation. *Biosyst. Eng.* **2004**, *88*, 153–161. [CrossRef]
27. Wang, L.; He, X.W.; Hu, C.; Guo, W.S.; Wang, X.F.; Xing, J.F. Measurement of physical parameters of coated cotton seed and parameter calibration of discrete element. *Chin. Agric. Univ.* **2022**, *27*, 71–82.
28. Wang, L.; Hu, C.; He, X.W.; Guo, W.S.; Hou, S.L. A general modelling approach for coated cotton-seeds based on the discrete element method. *Inmateh-Agric. Eng.* **2021**, *63*, 219–228.

agriculture

MDPI

Article

Strip Tillage Improves Productivity of Direct-Seeded Oilseed Rape (*Brassica napus*) in Rice–Oilseed Rape Rotation Systems

Chaosu Li [1,2], Ming Li [2], Tao Xiong [2], Hongkun Yang [1], Xiaoqin Peng [3], Yong Wang [3], Haiyan Qin [4], Haojie Li [2], Yonglu Tang [2] and Gaoqiong Fan [1,*]

1 Crop Ecophysiology and Cultivation Key Laboratory of Sichuan Province, Sichuan Agricultural University, Chengdu 611130, China; 2020101005@stu.sicau.edu.cn (C.L.); 14442@sicau.edu.cn (H.Y.)
2 Crop Research Institute of Sichuan Academy of Agricultural Sciences, Environment-Friendly Crop Germplasm Innovation and Genetic Improvement Key Laboratory of Sichuan Province, Chengdu 610066, China; limimg1005@gmail.com (M.L.); xiongtao27@163.com (T.X.); lhjie@163.com (H.L.); ttyycc88@163.com (Y.T.)
3 Sichuan Academy of Agricultural Machinery Sciences, Chengdu 610066, China; scsnjypxq@163.com (X.P.); luojun0803@126.com (Y.W.)
4 Sichuan Agricultural Technology Extension Station, Chengdu 610041, China; scnj@vip.163.com
* Correspondence: fangaoqiong@sicau.edu.cn; Tel.: +86-28-86290870

Abstract: Oilseed rape (*Brassica napus*) is a crucial global oil crop. It is generally cultivated in rotation with rice in southern China's Yangtze River Basin, where the wet soil and residue retention after rice harvest significantly hinder its seedling establishment. Hence, this study developed a strip-tillage seeder for oilseed rape seeding following rice harvest. Additionally, seedling establishment, soil infiltration and evaporation post-seeding, soil moisture change, oilseed yield, and weed occurrence under strip tillage (ST) were compared with conventional shallow rotary-tillage (SR) and deep rotary-tillage (DR) seeding practices. Compared to SR and DR, the results demonstrated that ST had a higher seeding efficiency and 53.8% and 80.2% lower energy consumption, respectively. ST also enhanced seedling growth and oilseed yield formation more effectively than the competitor tillage treatments, with an oilseed yield increase exceeding 6%. Additionally, ST improved water infiltration and reduced soil water evaporation, resulting in higher topsoil (0–20 cm) moisture during the critical growth stages. Furthermore, ST reduced soil disturbance, significantly decreasing the density of the dominant weed, *Polypogon fugax*. Overall, ST seeding technology has the potential to improve the productivity of oilseed rape in rice–oilseed rape rotation systems, and its yield superiority is mainly due to seedling establishment improvement and soil moisture adjustment.

Keywords: strip tillage; oilseed rape; seedling establishment; yield; soil moisture; weed density

Citation: Li, C.; Li, M.; Xiong, T.; Yang, H.; Peng, X.; Wang, Y.; Qin, H.; Li, H.; Tang, Y.; Fan, G. Strip Tillage Improves Productivity of Direct-Seeded Oilseed Rape (*Brassica napus*) in Rice–Oilseed Rape Rotation Systems. *Agriculture* **2024**, *14*, 1356. https://doi.org/10.3390/agriculture14081356

Academic Editors: Tao Cui, Vito Armando Laudicina, Xiaojun Gao and Qinghui Lai

Received: 20 June 2024
Revised: 31 July 2024
Accepted: 8 August 2024
Published: 14 August 2024

1. Introduction

Oilseed rape (*Brassica napus*) is a vital global oil crop and the largest in China, cultivated on 7.5 million hectares, representing nearly one-fifth of global cultivation and production [1,2]. Despite this, China's significant consumption creates a supply–demand imbalance [1,3]. To address this, the Chinese government has implemented policies to boost domestic production. Nevertheless, mechanization, particularly in sowing and harvesting, remains low, discouraging farmers from planting oilseed rape [4]. Currently, in the Yangtze River Basin, the main cultivation area of oilseed rape in southern China, primary production relies on labor-intensive and inefficient artificial seedling transplanting [2]. Although direct seeding for oilseed rape, compared to traditional manual transplanting, saves labor and cost, the high soil moisture after rice harvest, continuous autumn rainfall, and significant residue retention often result in poor establishment of direct-seeded oilseed rape in this region [5].

Several studies have aimed to improve the establishment of mechanically direct-seeded oilseed rape in high-soil-moisture conditions [6–8]. The main procedure involves soil disturbance and straw mixing, followed by mechanical seeding and fertilization. These steps can be performed separately or simultaneously using a combine seeder. Large clods may be created if the topsoil is too wet when tilled, resulting in high soil roughness affecting subsequent mechanical seeding operations [9]. High moisture levels or excessive straw return can also block mechanical seeding structures, adversely impacting seeding practice and seedling establishment [6,8]. Additionally, in wet clay soils, severe soil disturbance can exacerbate waterlogging for subsequent dryland crops in rice-based systems [5,10].

When exploring alternatives, strip tillage, where rotary tillage occurs only in a narrow strip to facilitate the seeding of the next crop, reduces soil disturbance and straw blockage and adapts well to fields with high soil moisture and residue mulching, particularly for wheat seeding [11]. The tilled soil in the strips contributes to water infiltration and crop root development [12,13]. Removing the residues from tilled strips has a positive effect on mitigating soil temperature changes [14,15]. Moreover, retaining crop residues between tilled strips reduces soil evaporation, allowing increasing moisture storage in the soil profile used by crop roots [14]. In most cases, strip tillage minimizes soil compaction and optimizes the seedbed environments for seed germination and crop seedling growth [16,17]. However, strip tillage adoption is regionally specific, and its positive effect on crop yield depends on multiple factors, such as tillage strip width, soil properties, and climatic conditions [13]. Nevertheless, there is evidence that strip tillage may benefit multiple-cropping production with *graminaceous* crops [13]. Most of the existing strip-tillage practices for oilseed rape production have been conducted in dryland systems, and the yield response varies according to different experimental conditions [18–20]. In contrast, strip tillage under complex conditions with high soil moisture and full residue retention has rarely been reported. Moreover, plowing or rotary tillage has been used as an effective way to control weeds, and strip tillage may increase weed occurrence and thus affect crop yields [21,22]. Therefore, this paper developed a strip-tillage seeder for oilseed rape seeding and conducted a field comparison with conventional shallow and deep rotary-tillage practices. This study aimed to (1) evaluate the seedling establishment of oilseed rape with a strip-tillage seeder; (2) determine the impact of strip tillage on oilseed rape growth and yield; and (3) estimate the effects of strip tillage on soil moisture change and weed occurrence.

2. Materials and Methods

2.1. The Machine Structure and Working Principle of the Strip-Tillage Oilseed Rape Seeder

A design diagram of the innovative strip-tillage oilseed rape seeder is shown in Figure 1. It comprises the following key components: a frame, strip rotary blades, a seed and fertilizer box, a seeding regulation system, a fertilizer distribution system, and a ground wheel. The seeder accommodates 6 or 8 seeding rows, corresponding to 6 or 8 blade groups on the rotary axis. Each group contains 4 blades for preparing the seedbed.

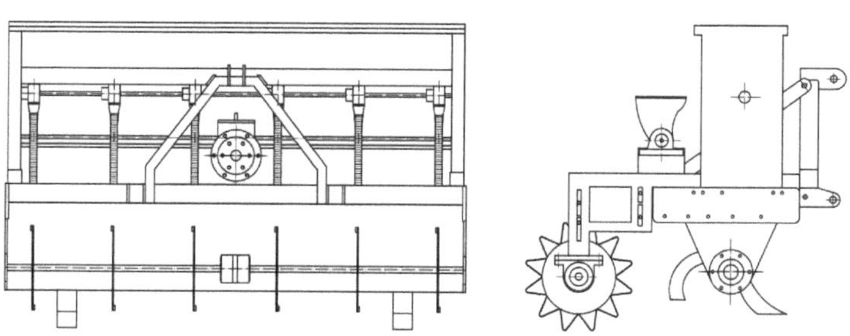

Figure 1. Main view (**left**) and side view (**right**) of the strip-tillage oilseed rape seeder.

The seeder uses a three-wire suspension system connected to the tractor, with the sowing and fertilizer devices driven by the ground wheel behind the seeder. The tractor's movement drives the rotation of the ground wheels, which in turn drive the seeding and fertilizing shafts via a chain. As the blades rotate, the fertilizer dropped in front of them is mixed effectively with soil and straw. Since the seeding tube's pipe mouth is directly opposite the corresponding blade group, the rape seeds fall onto the rotary-tillage belt after the strip rotary-tillage operation. No-tillage soil is maintained between seeding rows to minimize straw blockage. When the tractor stops or the seeder is raised to turn, the ground wheels cease operation, halting seeding and fertilizing automatically. To reduce mud and straw adhesion, the ground wheels behind the seeder are not traditional rollers but two narrow, serrated wheels. This design prevents slippage and minimizes mud accumulation. A picture of the field operation is shown in Figure 2.

Figure 2. Field operation of the strip-tillage oilseed rape seeder.

2.2. Comparison of Different Oilseed Rape Seeding Regimes

2.2.1. Field Trial Design

After developing the above machine, a comparative experiment was conducted under field conditions using conventional rotary-tillage seeding techniques. The experiment took place in Guanghan City, Sichuan Province, located in southwest China during the 2021/2022 and 2022/2023 growing seasons. In 2021/2022, the topsoil (0–20 cm) was sandy loam, while in 2022/2023 it was closer to silty loam. The topsoil moisture before tillage was 52.5% and 45.5% in 2021 and 2022, respectively. The previous crop, rice, was harvested in late September. In 2021, a semi-feed combine harvester was used, leaving straw chopped and dispersed on the soil surface. In 2022, a full-feed combine harvester was employed, and the straw was additionally crushed by a straw crusher before land preparation. The current local rice yield is approximately 10,000 kg ha^{-1} (14% grain moisture content), and the residue amount returned to the field is close to 9000 kg ha^{-1} each year. The seeding dates were 9 October 2021 and 7 October 2022, respectively. In the 2021/2022 season, the experiment included two seeding methods: the innovative seeding practice with the above strip-tillage seeder (ST) and conventional shallow rotary tillage (SR). Each treatment had three replications and was randomly arranged. In the 2022/2023 season, a deep rotary-tillage (DR) seeding practice was added, and three seeding practices were implemented. Each plot covered an area of 120 m^2.

For the ST seeding practice, an 8-row seeder was used in 2021 and a lighter 6-row seeder was used in 2022 under no-tillage and residue-retention conditions. For the SR seeding treatment, an integrated seeding machine performed rotary tillage for land preparation, sowing, and fertilization in one pass under no-tillage conditions, with a till depth of 7–8 cm. The DR seeding practice employed a traditional deep rotary tiller for tillage at a depth of 18–20 cm before seeding, and seeds were sown using the same seeder as in the SR seeding practice.

In the 2021/2022 season, the variety Chuanyou 81 was sown at a rate of 2.9 kg ha^{-1}. In the 2022/2023 season, the variety Chuanyou 83 was used, with seeding rates of 2.6, 2.7, and 2.7 kg ha^{-1} for the ST, SR, and DR treatments, respectively. Throughout the growth period, each treatment received 150 kg nitrogen (N) ha^{-1}, with 60% applied as compound fertilizer

(N-P-K, 15-15-15) during sowing and the remainder as urea at bolting. No irrigation was applied, and diseases and pests were well controlled according to technical requirements. During the 2021/2022 season, given the requirement to investigate the occurrence of weeds, chemical weeding was not conducted during the entire growth stage, while in the 2022/2023 season, chemical weeding was performed. Table 1 reports the temperature and rainfall during the period in which the experiment was conducted.

Table 1. Temperature and rainfall during the experiment.

Year	Factor	October	November	December	January	February	March	April
2021/2022	Mean temperature (°C)	17.3	11.3	8.3	7.7	7.4	17.0	18.3
	Rainfall (mm)	114.8	3.2	1.7	3.1	24.2	18.2	70.7
2022/2023	Mean temperature (°C)	18.3	15.2	7.2	6.6	10.0	14.0	18.9
	Rainfall (mm)	27.6	1.8	0.5	0.1	6.3	21.4	32.3

2.2.2. Sowing Efficiency and Energy Consumption

In the 2022/2023 growing season, the time and fuel consumption for each operation during land preparation and seeding were recorded, including tillage, seeding, fertilization, etc.

2.2.3. Topsoil Disturbance

After seeding in 2022/2023, two representative quadrats were selected in each plot. Each quadrat covered one row and was 35 cm in length. The soil clods in the quadrats were categorized into four groups based on diameter (>20 cm, 10–20 cm, 2–10 cm, and <2 cm). The number of clods and the averaged maximum diameters of the first three groups were measured.

2.2.4. Soil Infiltration and Topsoil Evaporation after Seeding

In 2022/2023, the double-ring method was used to measure soil infiltration after seeding [23]. The rings had a height of 25 cm and diameters of 50 cm (outer) and 25 cm (inner). For the ST treatment, the center of the inner and outer rings was on the midline of the rotary-tilled strip, and for the SR and DR treatments, the topsoil was fully tilled, and the center of the rings was close to the location of the seeds. They were inserted 20 cm into the soil and filled with water using two Mariotte bottles. The water level dropped rapidly in the first 30 min, then the rate slowed down, and data were recorded every 30 min after the slowdown. The water level was consistently maintained at 3 cm in both rings throughout the experiment. A consistent infiltration rate was assumed after three successive measurements showed identical values. The cutting-ring method was used to measure soil evaporation [24]. After seeding, topsoil was excavated using a cutting ring, with a height of 5 cm and a volume of 100 m^2. The ring was sealed with plastic film, and the total weight (soil and cutting ring) was recorded and returned to the sampling position. The cutting rings were removed and weighed after three days to calculate soil evaporation per unit area.

2.2.5. Seedling Emergence

Following seedling emergence, the number of emerged seedlings was assessed. In 2021/2022, six sample sites per plot were randomly investigated. Each site included two rows 2 m in length. In 2022/2023, three quadrats containing three rows 3 m in length were randomly selected to investigate the number of emerged seedlings. The coefficient of variation for the number of seedlings for each plot was also calculated.

2.2.6. Agronomic Characters Measurement at Seedling Stage

At 30 days post-seeding, fifteen representative seedlings were selected from each plot to measure the stem base diameter and the number of green leaves per plant. Then, all seedlings sampled in each plot were mixed and oven-dried at 105 °C for 20 min and then at 75 °C to a constant weight to determine the aboveground dry mass per seedling and the dry mass per m^2. Dry mass per m^2 was the product of dry mass per seedling and the number of seedlings per m^2.

2.2.7. Yield and Yield Components Investigation

Before harvesting at maturity, five and nine representative plants were taken from each plot in the 2021/2022 and 2022/2023 seasons, respectively, and these plants were used to investigate the number of siliques per plant, the number of seeds per silique, and 1000-seed dry weights [25]. At maturity, surviving plants in the center of the plot, undamaged by sampling, were harvested to evaluate the yield. In 2021/2022, the harvested area was 8 m^2 per plot. In 2022/2023, the harvested area encompassed three locations, each with two rows measuring 3 m in length. The number of surviving plants in each quadrat was investigated before harvest. After sun-drying and manual threshing, the seeds were weighed and moisture contents were measured. Seed yield was calculated based on 8% seed moisture.

2.2.8. Weed Occurrence Survey

Weed occurrence was assessed at 90 days post-seeding in the 2021/2022 season. Two quadrats, each 1 m^2, were examined in each plot. The species and stem numbers of weeds in each quadrat were recorded, and weed density per m^2 was calculated.

2.2.9. Soil Moisture Change

In the 2022/2023 season, every 30 days after seeding, soil from the 0–40 cm depth was sampled at three plot locations and divided into layers of 0–10, 10–20, 20–30, and 30–40 cm. The soil from each layer was mixed evenly, and its moisture content was determined by the oven-drying method, where samples were dried at 105 °C to a constant weight.

2.3. Data Analysis

Data were processed using Microsoft Excel 2013. Analysis was conducted using IBM SPSS 22.0 software, and the data were presented as averages. The significant differences among treatments were analyzed using ANOVA, the independent two-sample *t*-test was used to determine the difference between the two treatments in the 2021/2022 season, and Duncan's multiple range test was used to identify the differences among the three treatments in the 2022/2023 season.

3. Results

3.1. Variance Analysis of Seed Yield and Yield Components

The variance analysis revealed that oilseed yield was mainly affected by the seeding practice rather than the year (Table 2). The difference in emerged seedlings per m^2 between the years was noticeable, affecting surviving plant number per m^2. Moreover, the seeding practice mainly affected the number of siliques per plant, while the number of seeds per silique and 1000-grain weights were significantly affected by the year.

Table 2. Variance analysis of oilseed yield, yield components, and benefits under the two experimental years and the different seeding practices.

Factor	Oilseed Yield	Seedling Number per m^2	Surviving Plants per m^2	Number of Siliques per Plant	Number of Seeds per Silique	1000-Seed Dry Weight
Year (Y)	3.1	191.4 **	90.4 **	0.3	42.6 **	10.0 *
Seeding practice (S)	14.9 **	5.5 *	18.0 **	9.7 **	2.0	0.4
Y × S	4.3	7.4 *	26.8 **	0.2	0.002	2.1

The data presented in the table are *F* values. "*" means significantly different ($p < 0.05$); "**" means extremely significantly different ($p < 0.01$).

3.2. Yield and Yield Components

Consistent trends in oilseed yield among seeding treatments were observed across both years, with the ST practice consistently outperforming the other practices (Table 3). In 2021/2022, the ST treatment yielded 6.9% more than SR, though the difference was not significant. In 2022/2023, ST yielded 48.9% and 107.2% more than SR and DR, respectively, and a significant difference was observed among the treatments. The seed yield increase of ST was mainly due to the substantial increase in the number of siliques per plant.

Table 3. Effect of different oilseed rape seeding practices on oilseed yield and yield components.

Year	Seeding Practice	Oilseed Yield (kg ha^{-1})	Surviving Plants per m^2	Number of Siliques per Plant	Number of Seeds per Silique	1000-Seed Dry Weight (g)
2021/2022	ST	2838 ± 168 a	28.9 ± 1.1 b	284.5 ± 27.7 a	13.6 ± 1.6 a	4.22 ± 0.10 a
	SR	2655 ± 496 a	41.1 ± 3.6 a	196.9 ± 69.1 a	12.3 ± 1.0 a	4.13 ± 0.07 a
2022/2023	ST	2899 ± 385 a	24.0 ± 1.1 b	286.6 ± 8.9 a	17.8 ± 0.8 a	3.82 ± 0.09 a
	SR	1947 ± 218 b	24.5 ± 1.0 b	225.0 ± 36.2 ab	16.5 ± 1.1 a	3.99 ± 0.27 a
	DR	1399 ± 211 b	27.2 ± 1.5 a	138.0 ± 67.1 b	17.3 ± 0.7 a	3.84 ± 0.12 a

ST, strip-tillage seeding; SR, shallow rotary-tillage seeding; DR, deep rotary-tillage seeding. Data are presented as means with standard deviations (means ± SDs), and data in the same column from the same year followed by different lowercase letters are significantly different ($p < 0.05$).

3.3. Seeding Efficiency, Energy Consumption, and Soil Disturbance

The efficiency and energy consumption of different seeding methods varied due to differing pre-seeding operations and soil disturbance intensity (Table 4). ST increased seeding efficiency by 28.6% and 62.5% and reduced energy consumption by 53.8% and 80.2% compared to the SR and DR treatments, respectively. Different tillage practices created varied soil clod distributions (Table 4). Clods with extra-large (>20 cm) and large (10–20 cm) diameters dominated in the DR and SR treatments, while the ST treatment mainly produced clods with medium diameters.

Table 4. Effects of different oilseed rape seeding practices on seeding efficiency, energy consumption, and soil clod distributions during the 2022/2023 season.

Seeding Practice	Seeding Efficiency (h ha^{-1})	Fuel Consumption (L ha^{-1})	NCED	ACDED (cm)	NCLD	ACDLD (cm)	NCMD	ACDMD (cm)
ST	3.0	18.7	0 b	-	0 c	-	71.5 ± 8.9 a	5.6 ± 0.5 b
SR	4.2	40.5	1.9 ± 3.3 b	26.2 ± 0.5 a	22.9 ± 5.7 a	15.1 ± 1.8 a	19.0 ± 8.7 b	5.5 ± 1.1 b
DR	8.0	94.5	9.5 ± 3.3 a	27.1 ± 4.5 a	15.2 ± 3.3 b	16.2 ± 2.0 a	3.8 ± 6.6 b	7.5 ± 0.7 a

ST, strip-tillage seeding; SR, shallow rotary-tillage seeding; DR, deep rotary-tillage seeding; NCED, number of clods with extra-large diameters (>20 cm) per m^2; ACDED, average clod diameter for extra-large diameters; NCLD, number of clods with large diameters (10–20 cm) per m^2; ACDLD, average clod diameter for large diameters; NCMD, number of clods with medium diameters (2–10 cm) per m^2; ACDMDA, average clod diameter for medium diameters. Data for soil clods are presented as means with standard deviations (means ± SDs), and data in the same column followed by different lowercase letters are significantly different ($p < 0.05$).

3.4. Seedling Establishments

With nearly the same seeding rates, the difference in seedling number emerged per m^2 between treatments did not reach a significant level in the two experimental years (Table 5). However, the ST treatment exhibited better field distribution uniformity in the 2021/2022 growing season.

Table 5. Effect of different oilseed rape seeding practices on seedling establishments.

Year	Seeding Practice	Seedling Number per m^2	CVSD (%)
2021/2022	ST	47.0 ± 3.0 a	11.1 ± 1.7 b
	SR	56.6 ± 4.0 a	27.0 ± 8.3 a
2022/2023	ST	27.0 ± 2.8 a	11.0 ± 2.0 a
	SR	26.8 ± 2.8 a	11.5 ± 5.8 a
	DR	31.5 ± 2.7 a	10.2 ± 4.9 a

ST, strip-tillage seeding; SR, shallow rotary-tillage seeding; DR, deep rotary-tillage seeding; CVSD, coefficient of variation of seedling distribution. Data are presented as means with standard deviations (means ± SDs), and data in the same column in the same year followed by different lowercase letters are significantly different ($p < 0.05$).

3.5. Agronomic Characters at Seedling Stage

Most agronomic character values for oilseed rape seedlings decreased with the increase in tillage intensity at 30 days post-seeding (Table 6). In 2021/2022, significant differences existed in the main individual parameters between the two seeding practices, while dry mass per m^2 values remained similar. However, in 2022/2023, all tested agronomic parameters were significantly higher under ST compared to SR and DR, and dry mass per m^2 increased by 65.3% and 95.1%, respectively.

Table 6. Effects of different oilseed rape seeding practices on agronomic characters at seedling stage (30 days post-seeding).

Year	Seeding Practice	Diameter of Stem Base (cm)	Number of Green Leaves per Plant	Dry Mass per Plant (g)	Dry Mass per m^2 (g)
2021/2022	ST	4.55 ± 0.12 a	5.93 ± 0.29 a	1.27 ± 0.19 a	60.3 ± 13.1 a
	SR	3.73 ± 0.26 b	5.36 ± 0.23 b	0.88 ± 0.10 b	50.4 ± 8.4 a
2022/2023	ST	4.75 ± 0.36 a	4.92 ± 0.20 a	1.50 ± 0.22 a	40.2 ± 2.4 a
	SR	3.63 ± 0.27 b	4.19 ± 0.20 b	0.92 ± 0.16 b	24.3 ± 3.2 b
	DR	3.45 ± 0.20 b	4.00 ± 0.02 b	0.65 ± 0.08 b	20.6 ± 3.9 b

ST, strip-tillage seeding; SR, shallow rotary-tillage seeding; DR, deep rotary-tillage seeding. Data are presented as means with standard deviations (means ± SDs), and data in the same column in the same year followed by different lowercase letters are significantly different ($p < 0.05$).

3.6. Weed Occurrence

Seeding methods significantly influenced weed occurrence in the field (Table 7). In this study, the dominant weed was *Polypogon fugax*. At 90 days post-seeding with ST, the *Polypogon fuga* densities decreased by 85.2% and 83.7% compared to the SR and DR treatments, respectively.

Table 7. Effects of different oilseed rape seeding practices on field weed density at 90 days post-seeding in the 2021/2022 season.

Seeding Practice	Total Weed Density (Seedlings m^{-2})	*Polypogon Fugax* Density (Seedlings m^{-2})
ST	529 ± 18 b	525 ± 18 b
SR	3560 ± 966 a	3553 ± 967 a
DR	3235 ± 1342 a	3224 ± 1334 a

ST, strip-tillage seeding; SR, shallow rotary-tillage seeding; DR, deep rotary-tillage seeding. Data are presented as means with standard deviations (means ± SDs), and data in the same column followed by different lowercase letters are significantly different ($p < 0.05$).

3.7. Soil Infiltration and Evaporation Post-Seeding

Soil infiltration and evaporation measurements post-seeding showed that the stable infiltration rate in the ST treatment was significantly higher than in the other two treatments, while evaporation was lower (Figure 3). Over three days post-seeding, topsoil evaporation in the ST treatment decreased by 35.1% and 32.7% compared to the SR and DR treatments, respectively.

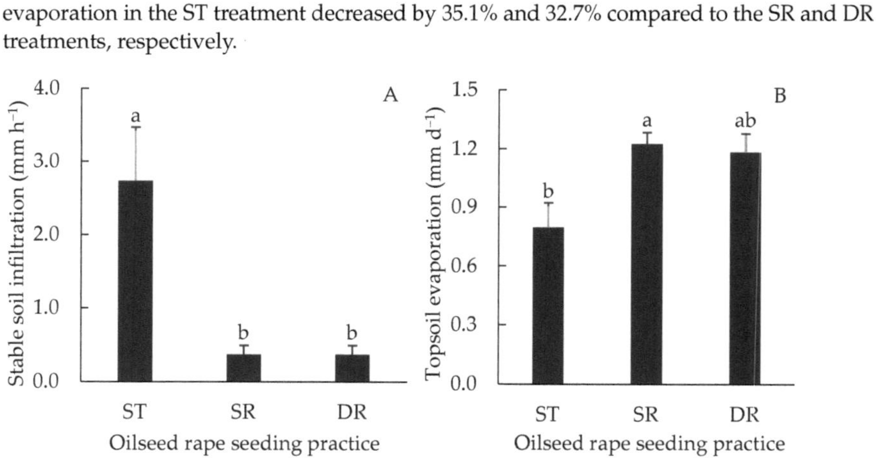

Figure 3. Effects of different oilseed rape seeding practices on stable soil infiltration (**A**) and topsoil evaporation (**B**) post-seeding in the 2022/2023 season. ST, strip-tillage seeding; SR, shallow rotary-tillage seeding; DR, deep rotary-tillage seeding. Bar graphs show the means, and the error bars represent standard deviations. On the bar chart, different lowercase letters indicate significant differences between different oilseed rape seeding practices for each parameter ($p < 0.05$).

3.8. Soil Moisture Change during Oilseed Rape Growth

The ST treatment maintained higher topsoil moisture (0–20 cm) throughout the oilseed rape growth period (Figure 4). The average soil moisture during the whole growth stages at depths of 0–10 cm and 10–20 cm in the ST treatment increased by 11.6% and 24.6% and by 8.0% and 16.6%, respectively, compared to the SR and DR treatments. However, soil moisture differences in deep soil (20–40 cm) between the treatments were reduced.

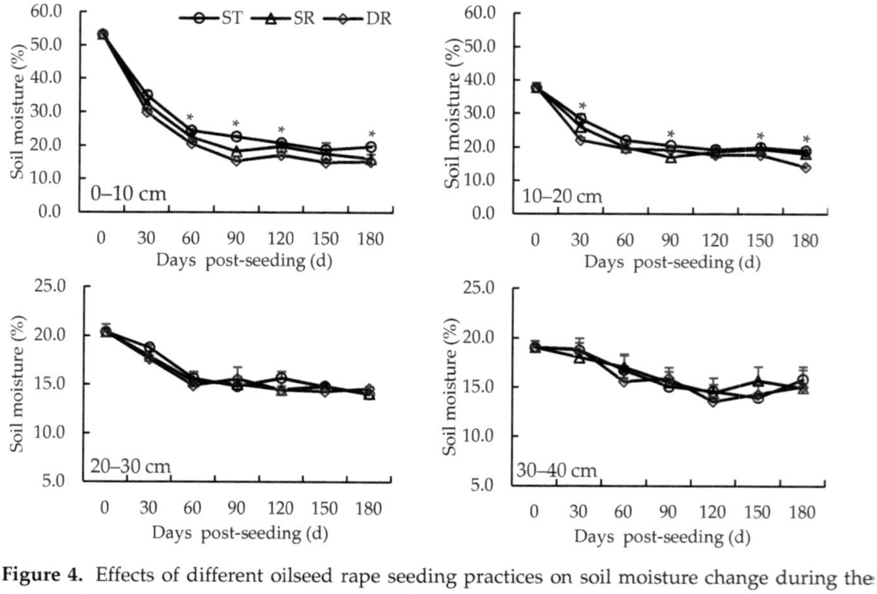

Figure 4. Effects of different oilseed rape seeding practices on soil moisture change during the 2022/2023 season. The moisture data in the figure show the means and the error bars represent standard deviations. ST, strip-tillage seeding; SR, shallow rotary-tillage seeding; DR, deep rotary-tillage seeding. "*" indicates a significant difference in soil moisture among the three treatments at a specific sampling time.

4. Discussion

The low quality of mechanized direct seeding is an important factor limiting oilseed rape yield improvement in rice–oilseed rape rotation systems in southern China [5,6]. This study reveals that, compared with conventional rotary tillage, strip-tillage practices can improve the seedling establishment and yield of oilseed rape. Many studies have confirmed that tillage systems have a pronounced impact on crop yield [26,27]. Strip tillage combines the advantages of no tillage and full tillage and thus positively affects crop yields in most regions, especially increasing dryland system production in cooler temperate areas [13]. This study confirmed that strip tillage positively affects oilseed rape production in subtropical–humid regions.

The positive effects of strip tillage on crop growth and yield are related to improvements in temperature and moisture conditions in seedbeds [12–15]. Some studies have proved that strip tillage in cooler areas can increase seedbed temperature, promoting crop emergence and growth [14,15]. However, the average temperature exceeds 15 °C when oilseed rape is seeded in southern China, and temperature may be not a limiting factor for oilseed rape establishment in this region. In contrast, in this study, the enhancement of seedling establishment and oilseed yield with strip tillage may have been caused by improved seedbed moisture conditions. High soil moisture after rice harvest and continuous rainfall make fields prone to waterlogging [5]. In this study, the soil water infiltration rate with strip tillage was higher than that in the shallow- and deep-tillage systems, which reduced the adverse effects of excessive soil moisture on seed germination and seedling growth. Other studies also supported this result, indicating that reduced tillage preserves soil capillary structure and increases water infiltration [28,29]. In addition, increasing soil surface roughness after intensive tillage leads to rapid topsoil moisture loss, thus affecting water absorption by seeds and seedlings [9]. This study also found that strip tillage with less soil disturbance led to fewer large soil clods and less soil water evaporation, resulting in a relatively ideal soil moisture state in the seedbeds, which positively affected oilseed rape growth. Compared with other conventional tillage practices, the oilseed yield increase with strip tillage in 2021/2022 was lower than that in 2022/2023. The rainfall in the former season was much higher than in the latter, especially in winter and spring, which further confirmed the positive effect of soil moisture retention with strip tillage on oilseed rape yield improvement. Due to the improvement in soil moisture conditions, most agronomic parameters with strip tillage at the oilseed rape seedling stage were superior to those of the competitor tillage regimes, and the number of siliques per plant at maturity was significantly improved. In conclusion, the positive effects of strip tillage on oilseed rape production in southern China are mainly due to soil moisture adjustment, which promotes surface water infiltration to alleviate waterlogging and reduces water evaporation to improve crop drought resistance.

Weeds significantly impact crop growth and yield, yet there is no consensus regarding their occurrence under conservation tillage conditions [30]. This study identified *Polypogon fugax* as the dominant weed species in oilseed rape fields, and its density and total weed density were lower in the strip-tillage treatment than in the other traditional tillage treatments. The occurrence of weeds in the field is influenced by factors including the number and spatial distribution of weed seeds in the soil and the tillage method used for previous crops [31]. In China's Yangtze River Basin, rotary tillage is typically employed before rice transplanting to accommodate transplanting, resulting in most weed seeds being buried deep in the soil. Once the topsoil is disturbed, the weeds rapidly germinate when air and water conditions around the seeds are favorable. In this study, *Polypogon fugax* was primarily found in rotary-tillage belts (Figure 5), contrasting with the relatively uniform distribution of weeds in fully rotary-tilled fields. Therefore, in rice–dryland crop rotation systems, ST seeding practices may potentially reduce weed occurrence during dryland crop growth seasons. In addition, strip tillage reduces fuel consumption in seeding operations by more than 50% (Table 4); thus, coupled with other innovative nutrient managements [32], it will be an option for future oilseed rape sustainable production in rice–oilseed rotation systems.

Figure 5. Weeds mainly occurred in tilled rows under the strip-tillage seeding practice in the oilseed rape fields.

5. Conclusions

Mechanical direct seeding for oilseed rape in rice–oilseed rape rotation systems in southern China faces great challenges due to high soil moisture and residue retention. In this study, the innovative oilseed rape strip-tillage (ST) seeding practice minimized soil disturbance, facilitated efficient seedling establishment, and enhanced seedling growth and seed yield, which was confirmed to benefit oilseed rape production. The seed yield improvement is attributable to the preservation of soil moisture and the suppression of weed germination. ST technology for oilseed rape seeding needs to be evaluated in multiple environments to further improve its adaptability to complex environments following rice harvest.

Author Contributions: Conceptualization, C.L. and G.F.; methodology, C.L., M.L., H.Y., H.Q. and G.F.; investigation, C.L., M.L., T.X. and H.Q.; writing—original draft preparation, C.L.; writing—review and editing, C.L., H.Y., Y.T. and G.F.; visualization, C.L., X.P. and Y.W.; funding acquisition, C.L., Y.T. and H.L. All authors have read and agreed to the published version of the manuscript.

Funding: This research was funded by the Agricultural Technology Collaborative Extension Project of Sichuan Province, China; the Science and Technology Innovation Team of Oilseed Rape of Sichuan Province, China; and the Natural Science Foundation of Sichuan Province (2024NSFSC1223).

Institutional Review Board Statement: Not applicable.

Data Availability Statement: Data are contained within the article.

Acknowledgments: The authors would like to express their gratitude to EditSprings for the expert linguistic services provided.

Conflicts of Interest: The authors declare no conflicts of interest.

References

1. Bonjean, A.P.; Dequidt, C.; Sang, T.; Limagrain, G. Rapeseed in China. *OCL* **2016**, *236*, D605. [CrossRef]
2. Hu, Q.; Hua, W.; Yin, Y.; Zhang, X.; Liu, L.; Shi, J.; Zhao, Y.; Lu, Q.; Chen, C.; Wang, H. Rapeseed research and production in China. *Crop J.* **2017**, *5*, 127–135. [CrossRef]
3. Li, F.; Guo, K.; Liao, X. Risk assessment of china rapeseed supply chain and policy suggestions. *Int. J. Environ. Res. Public Health* **2023**, *20*, 465. [CrossRef] [PubMed]
4. He, A.; Yuan, B.; Jin, Z.; Man, J.; Peng, S.; Zhang, L.; Liu, H.; Nie, L. Comparative study on annual yield, water consumption, irrigation water use efficiency and economic benefits of different rice-oilseed rape rotation systems in Central China. *Agric. Water Manag.* **2021**, *247*, 106741. [CrossRef]
5. Li, L.; Li, J.; Wei, C.; Yang, C.; Zhong, S. Effect of mechanized ridge tillage with rice-rape rotation on paddy soil structure. *Agriculture* **2022**, *12*, 2147. [CrossRef]

6. Liao, Y.; Gao, L.; Liao, Q.; Zhang, Q.; Liu, L.; Fu, Y. Design and test of side deep fertilizing device of combined precision rapeseed seeder. *Trans. Chin. Soc. Agric. Mach.* **2020**, *51*, 65–75. [CrossRef]
7. Wu, M.; Guan, C.; Tang, C.; Chen, S.; Luo, H.; Wang, G.; Xie, F.; Li, X.; Yang, W. Experimental research on 2BYF-6 type no-tillage rape combine seeder in paddy stubble field. *Trans. Chin. Soc. Agric. Eng.* **2007**, *23*, 172–175. [CrossRef]
8. Wei, G.; Zhang, Q.; Liu, L.; Xiao, W.; Sun, W.; Liao, Q. Design and experiment of plowing and rotary tillage buckle device for rapeseed direct seeder. *Trans. Chin. Soc. Agric. Mach.* **2020**, *51*, 38–46. [CrossRef]
9. Keller, T.; Arvidsson, J.; Dexter, A.R. Soil structures produced by tillage as affected by soil water content and the physical quality of soil. *Soil Tillage Res.* **2007**, *92*, 45–52. [CrossRef]
10. Ding, J.; Li, F.; Le, T.; Xu, D.; Zhu, M.; Li, C.; Zhu, X.; Guo, W. Tillage and seeding strategies for wheat optimizing production in harvested rice fields with high soil moisture. *Sci. Rep.* **2021**, *11*, 119. [CrossRef]
11. Li, C.; Tang, Y.; McHugh, A.D.; Wu, X.; Liu, M.; Li, M.; Xiong, T.; Ling, D.; Tang, Q.; Liao, M.; et al. Development and performance evaluation of a wet-resistant strip-till seeder for sowing wheat following rice. *Biosyst. Eng.* **2022**, *220*, 146–158. [CrossRef]
12. Jaskulska, I.; Jaskulski, D. Strip-till one-pass technology in central and eastern Europe: A MZURI pro-til hybrid machine case study. *Agronomy* **2020**, *10*, 925. [CrossRef]
13. Dou, S.; Wang, Z.; Tong, J.; Shang, Z.; Deng, A.; Song, Z.; Zhang, W. Strip tillage promotes crop yield in comparison with no tillage based on a meta-analysis. *Soil Tillage Res.* **2024**, *240*, 106085. [CrossRef]
14. Licht, M.A.; Al-Kaisi, M. Strip-tillage effect on seedbed soil temperature and other soil physical properties. *Soil Tillage Res.* **2005**, *80*, 233–249. [CrossRef]
15. Tabatabaeekoloor, R. Soil characteristics at the in-row and inter-row zones after strip-tillage. *Afr. J. Agric. Res.* **2011**, *6*, 6598–6603. [CrossRef]
16. Chen, Q.; Zhang, X.; Sun, L.; Ren, J.; Yuan, Y.; Zang, S. Influence of tillage on the Mollisols physicochemical properties, seed emergence and yield of maize in northeast China. *Agriculture* **2021**, *11*, 939. [CrossRef]
17. Mikha, M.M.; Hergert, G.W.; Qiao, X.; Maharjan, B. Soil chemical properties after 12 years of tillage and crop rotation. *Agron. J.* **2020**, *112*, 4395–4406. [CrossRef]
18. David, B.; Lucie, B.; Perla, K.; Pavel, C.; Katerina, P.; Vlastimil, M.; Jan, V. Growth and yield of winter oilseed rape under strip-tillage compared to conventional tillage. *Plant Soil Environ.* **2021**, *67*, 85–91. [CrossRef]
19. Saldukaitė, L.; Šarauskis, E.; Zabrodskyi, A.; Adamavičienė, A.; Buragienė, S.; Kriaučiūnienė, Z.; Savickas, D. Assessment of energy saving and GHG reduction of winter oilseed rape production using sustainable strip tillage and direct sowing in three tillage technologies. *Sustain. Energy Technol. Assess.* **2022**, *51*, 101911. [CrossRef]
20. Jankowski, K.J.; Sokólski, M.; Szatkowski, A.; Załuski, D. The effects of tillage systems on the management of agronomic factors in winter oilseed rape cultivation: A case study in North-Eastern Poland. *Agronomy* **2024**, *14*, 437. [CrossRef]
21. Hobbs, P.R.; Sayre, K.; Gupta, R. The role of conservation agriculture in sustainable agriculture. *Philos. Trans. R. Soc. B Biol. Sci.* **2008**, *363*, 543–555. [CrossRef] [PubMed]
22. Hendrix, B.J.; Young, B.G.; Chong, S. Weed management in strip tillage corn. *Agron. J.* **2004**, *96*, 229–235. [CrossRef]
23. Ren, Z.; Zhang, G.; Wang, B.; Shi, Y. Effects of double-ring diameter on soil infiltration rate. *J. Soil Water Conserv.* **2012**, *26*, 94–99.
24. Liu, M.; Wang, S.; Fan, J.; Fu, W.; Du, M. Rapid in-situ determination of soil evaporation with cutting ring method. *Chin. J. Soil Sci.* **2021**, *52*, 55–61.
25. Cai, G.; Yang, Q.; Chen, H.; Yang, Q.; Zhang, C.; Fan, C.; Zhou, Y. Genetic dissection of plant architecture and yield-related traits in *Brassica napus*. *Sci. Rep.* **2016**, *6*, 21625. [CrossRef] [PubMed]
26. Su, Y.; Gabrielle, B.; Makowski, D. A global dataset for crop production under conventional tillage and no tillage systems. *Sci. Data* **2021**, *8*, 33. [CrossRef] [PubMed]
27. Gawęda, D.; Haliniarz, M. Grain yield and quality of winter wheat depending on previous crop and tillage system. *Agriculture* **2021**, *11*, 133. [CrossRef]
28. Li, H.; He, J.; Gao, H.; Chen, Y.; Zhang, Z. The effect of conservation tillage on crop yield in China. *Front. Agric. Sci. Eng.* **2015**, *2*, 179–185. [CrossRef]
29. Wardak, D.L.R.; Padia, F.N.; de Heer, M.I.; Sturrock, C.J.; Mooney, S.J. Zero tillage has important consequences for soil pore architecture and hydraulic transport: A review. *Geoderma* **2022**, *422*, 115927. [CrossRef]
30. Kumar, V.; Mahaja, G.; Dahiya, S.; Chauhan, B.S. Challenges and opportunities for weed management in no-till farming systems. In *No-Till Farming Systems for Sustainable Agriculture*; Dang, Y., Dalal, R., Menzies, N., Eds.; Springer: Berlin/Heidelberg, Germany; Cham, Switzerland, 2020. [CrossRef]
31. Radicetti, E.; Mancinelli, R. Sustainable weed control in the agro-ecosystems. *Sustainability* **2021**, *13*, 8639. [CrossRef]
32. Pramanick, B.; Mahapatra, B.S.; Datta, D.; Dey, P.; Singh, S.P.; Kumar, A.; Paramanik, B.; Awasthi, N. An innovative approach to improve oil production and quality of mustard (*Brassica juncea* L.) with multi-nutrient-rich polyhalite. *Heliyon* **2023**, *9*, e13997. [CrossRef] [PubMed]

agriculture

MDPI

Article

Improving Sowing Uniformity of a Maize High-Speed Precision Seeder by Incorporating Energy Dissipator

Rui Liu, Guangwei Wu, Jianjun Dong, Bingxin Yan and Zhijun Meng *

Intelligent Equipment Research Center, Beijing Academy of Agriculture and Forestry Sciences, Beijing 100097, China; liur@nercita.org.cn (R.L.); wugw@nercita.org.cn (G.W.); dongjj@nercita.org.cn (J.D.); yanbx@nercita.org.cn (B.Y.)

* Correspondence: lmengzj@nercita.org.cn; Tel.: +86-178-0113-7458

Abstract: To enhance the sowing uniformity of the vacuum seeder in the high-speed working state, a flexible energy-dissipation receiving device was designed. We analyzed the angle and velocity of seed ejection from the seed-metering device. Additionally, we explored the rheological properties of four different sodium alginate (SA) solutions. Combined with high-speed camera technology, the movement characteristics of four kinds of energy dissipators were revealed, and it was determined that the fabrication material of the energy dissipator is colloid with an SA percentage of 10%. The influence of the thickness of the energy dissipator body, impact velocity, and impact angle on the pre- and post-impact velocity difference and end-of-motion transverse displacement value was investigated. The quadratic regression equation between experimental factors and experimental indexes was established, and it was determined that the thickness of the energy dissipator was 7 mm. Field experiment results showed that the working speed was 12~16 km·h^{-1}, the leakage rate was less than 6.83%, the multiple rates were less than 0.97%, the qualified rate was stable at more than 92.4%, and the qualified grain distance variation rate was stable at less than 16.57%. The designed energy-dissipation device is beneficial to improve the overall working performance of high-speed precision seeders. In the future, if the reliability and long-term performance of the energy-dissipation device are further improved, it will be able to meet the requirements for precision seeding under high-speed conditions.

Keywords: maize; high-speed precision; seed transport system; energy dissipation; seeding quality

Citation: Liu, R.; Wu, G.; Dong, J.; Yan, B.; Meng, Z. Improving Sowing Uniformity of a Maize High-Speed Precision Seeder by Incorporating Energy Dissipator. *Agriculture* **2024**, *14*, 1237. https://doi.org/10.3390/agriculture14081237

Academic Editor: Simone Bergonzoli

Received: 9 July 2024
Revised: 22 July 2024
Accepted: 25 July 2024
Published: 26 July 2024

1. Introduction

With the acceleration of China's urbanization process, agricultural production has scaled up significantly. Business entities are now pursuing high-speed and high-efficiency production methods. As a result, high-speed precision seeders have become one of the crucial components in the mechanized production of maize [1–3]. The precision of high-speed maize seeders depends not only on the accuracy of individual key components but also on their ability to work together effectively. Seeds are transported from the seed box to the seed furrow through a series of processes, including filling, carrying, clearing, dropping, and guiding. These processes are conducted sequentially by the vacuum seed-metering device and the seed-guiding device [4–8]. The vacuum seed-metering device picks up the seeds precisely and carries them in a circular motion, while the seed-guiding device restrains the seeds from moving orderly into the seed furrow. The difficulty of cooperation between the two is that the seeds will start from the seed discharge port of the seed-metering device at a certain angle of ejection due to inertia. The position and direction of the seeds entering the seed-guiding device are different with different working speeds. After entering the seed-guiding device, the seeds will collide and bounce. Thus, the seeds cannot continue to enter the seed furrow in a single-seed orderly state, which reduces the sowing accuracy.

In Europe and the USA, high-speed precision planters have the seed-guiding device closely connected to the seed-metering device at the seed discharge port. Additionally, a binding force is applied to guide the seeds smoothly and orderly into the seed-guiding device. The Proesm K precision seeder, developed by [9], transports seeds from the seed box to the seed furrow by vacuum seed-metering device and seed-guiding tube. A vertical, lateral scraper is installed at the seed discharge port of the seed-metering device to force the seeds into the seed-guiding tube in a vertical direction. Chrono 700 high-speed precision seeder, developed by [10], consists of a vacuum seed-metering device and a pneumatic seed-guiding tube. The seed-metering device is inclined at 15 degrees, and the upper part of the seed-guiding tube is positioned close to the seed discharge port. Seeds separated from the seed-metering device enter the seed-guiding tube in a straight line under the action of gravity. The 3505 Planter precision seeder, developed by [11], consists of a vacuum seed-metering device and a belt-type seed-guiding device. Two oppositely rotating seed wheels are installed at the seed discharge port of the seed-metering device, which transfers the seeds to the conveyor belt of the belt-type seed-guiding device. The Tempo R precision planter developed by [12], is equipped with a pneumatic seed-metering device and an air-fed seed-guiding tube, with the help of a pressure-relief wheel and a positive airflow to guide the seed into the seed-guiding tube.

Currently, research in China mainly focuses on either seed-metering devices or seed-guiding devices. Due to the lack of a cooperative theoretical basis and advanced technology, the domestic seed-metering device and seed-guiding tube cannot be closely connected under high-speed working conditions. The difference in seed shape, high vibration frequency of the machine, and variations in seed injection angle into the seed-guiding device cause seeds to easily collide with the inner wall of the seed-guiding device. This prevents the formation of an orderly seed flow, which reduces the seeding uniformity and affects the seeding quality [13,14]. A few researchers in China have added collaborative parts to the seed discharge port of the seed-metering device or the upper part of the seed-guiding tube, which guides the seeds to enter the seed-guiding tube along the same position and direction. The linear seed-pushing device combined with the asymptotic gas-blocking device is only suitable for seed-metering disks with guide grooves, which have a narrow range of applications [15]. The uniformity and stability of seed guidance are poor at high rotational speeds of V-shaped grooved wheels, so it is not suitable for the high-speed working of the seeder [16]. Under the condition of not changing the structure of the seed-metering device or seed-guiding device and not adding a secondary auxiliary seeding device, the technology of stable and orderly seeding is studied to solve the problems of poor uniform grain distance, high leakage, and missed multiple rates caused by collision when seeds enter the seed-guiding device. This is of great significance to realize high-speed and precision seeding.

To eliminate the dynamic energy head that collides with the wall surface of the seed-guiding device and ensure that all seeds can enter the seed-guiding device in an orderly state, therefore improving the uniformity of the sowing of the vacuum seeder in the high-speed working state. A method of realizing the energy dissipation using shear-thinning fluid as the undertaking part was put forward. The dynamic relationship of the seed undertaking process was defined, and the key factors and numerical range affecting the energy-dissipation effect were determined. In addition, the high-speed camera technology was used to analyze the energy-dissipation process, to determine the best working parameters of the energy-dissipation parts. Finally, the feasibility and effectiveness of the energy-dissipation are verified by field experiments.

2. Materials and Methods

2.1. Seed Transport System of High-Speed Vacuum Seeder

As shown in Figure 1, the seed transport system of a high-speed vacuum seeder is mainly composed of a case, upper seed clearing knife, lower clearing knife, seed disk, forced clearing knife, energy-dissipation device, and positive-pressure airflow-assisted blowing seed-guiding device. When the sowing system is working, the single seed can be carried out under the synergistic effect of negative pressure, seed disk, and clearing knife. In the seed-throwing section, a single seed hits the energy-dissipation device under the action of inertia force and gravity. Because the kinetic energy of the seed is absorbed by the energy dissipator, the seed will fall into the barotropic airflow-assisted blow seed-guiding device along the wall or after a slight bounce. When the seed reaches the seed-guiding section, it is accelerated under the airflow blowing and then leaves the seed-guiding device to complete the whole seeding process.

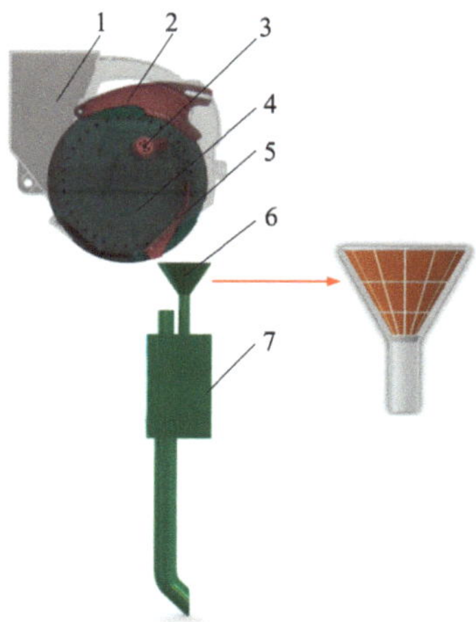

Figure 1. Seed transport system of high-speed vacuum seeder. (1) Case; (2) Upper seed clearing knife; (3) Lower clearing knife; (4) Seed disk; (5) Forced clearing knife; (6) Energy-dissipation device; (7) Positive-pressure airflow-assisted blowing seed-guiding device.

2.2. *Effect of Seed-Throwing Process on Qualified Rate of Grain Distance*

2.2.1. Motion Analysis in the Initial Stage of Seed-Throwing

As shown in Figure 2, a rectangular coordinate system is established with the horizontal direction as the x-axis and the vertical direction as the y-axis. Because the speed relative to the seed disk is zero when the seeds move circularly with the seed disk at a uniform speed, the instantaneous speed of the seeds leaving the seed-metering disk at the seed discharge port obeys the following Formula (1):

$$\begin{cases} v_{tx1} = 2\pi n r_1 \sin\theta_1 \\ v_{ty1} = 2\pi n r_1 \cos\theta_1 \\ \theta_1 = (k_t - 1)\frac{2\pi}{Z} \\ n = \frac{1\times 10^3 v}{3.6LZ} \end{cases} \tag{1}$$

where v is the forward speed of the planter, km·h^{-1}; v_{t1} is the speed of the seed as it leaves the seed disk, m·s^{-1}; L is theoretical grain distance, mm; Z is the number of seed disk holes of the seed disk; n is the rotation speed of seed disk, r·s^{-1}; r is the radius of the circle where the center of the seed disk hole is located, m; θ_1 is the angle of rotation of seeds in the seed discharge area, degree; k_t is the number of seed disk holes that seeds rotate in the seed discharge area.

Simplify Formula (1) to obtain the following Formula (2):

$$v_{tx1} = \frac{1 \times 10^3 \pi v r_1}{1.8LZ} \sin\left(\frac{2\pi k_t - 2\pi}{z}\right) \tag{2}$$

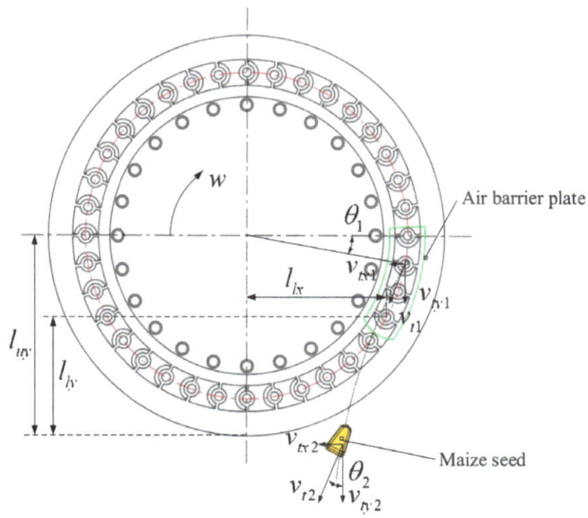

Figure 2. Schematic diagram of motion analysis of seeds after leaving seed disk.

According to the analysis of Formula (2), the horizontal speed of seeds in the initial stage of seed-throwing is related to the forward speed of the seeder, the circle radius of seed disk holes, the number of seed disk holes rotated by seeds in the discharge area, the theoretical grain distance and the number of seed disk holes of seed disk. When the structure of the seed-metering device is determined, the horizontal velocity of the seed at the initial stage of seed-throwing is only positively correlated with the forward speed of the seeder and negatively correlated with the theoretical grain distance. Seeds with a certain initial velocity enter the seed-guiding device with a parabolic trajectory when thrown in an unconstrained state. The higher the speed of the seeder, the higher the horizontal velocity of the seeds, and the greater the transverse displacement, which in turn increases the probability of collision with the wall of the seed-guiding device.

Ideally, the seeds will break away from the seed disk when they reach the upper area of the air barrier plate, but it is impossible for the seeds to break away from the seed disk immediately when the actual working speed is at 8~16 km·h^{-1}. In this study, the upper and lower limit areas of the air barrier plate were taken to calculate the speed of seeds leaving the seed disk and entering the seed-guiding device. Based on Formula (1) and the formula for accelerated motion, when the seeds are detached from the seed disk at the upper and lower limit positions, the speed is 0.183 m·s^{-1} $\leq v_{t1} \leq$ 0.321 m·s^{-1}. Subsequently, the seed does parabolic motion with initial velocity v_{t1}. The seed starts to move to the upper part of the seed-guiding device from the upper limit position (flush with the lower end surface of the seed disk), ignoring air resistance, the combined velocity of the seed at the end of the motion is 1.84 m·s^{-1} $\leq v_{t2} \leq$1.98 m·s^{-1}; according to the displacement formula, we can obtain l_{ux} = 0 m, l_{uy} = 0.14 m. By the formula of the sine function, we can obtain that the angle θ_t of the seed entering the seed-guiding device is 0 degrees.

Similarly, it can be obtained that when the seed moves from the lower limit position, the combined velocity of the seed at the end of the motion is 1.01 m·s^{-1} $\leq v_{t2} \leq$ 1.055 m·s^{-1}, the displacement of seed movement is 0.00796 m $\leq l_{lx} \leq$ 0.0125 m, l_{ly} = 0.07 m, the angle θ_2 of seed entering the seed-guiding device is 6.5~10.1 degrees. The angle value of the seed entering the seed-guiding device under actual working conditions cannot be the same as the theoretical value, so it is necessary to expand the width of the experimental value to investigate the motion law of the seed energy-dissipation process. Based on the above analysis, the angle of seeds entering the seed-guiding device θ_2 ϵ [0°, 30°] and the speed of seeds entering the seed-guiding device 1 m·s^{-1} $\leq v_{t2} \leq$ 3 m·s^{-1} were selected for experimental study.

2.2.2. Analysis of Seed Collision Motion

Maize seeds can be classified as horse-tooth shape, spheroid shape, and ellipsoid shape according to their geometrical shapes, as shown in Figure 3. Due to the irregular shape of the seed, different parts of the seed impact the wall of the seed-guiding device, the direction of rotation in the seed guide tube is different, and the seed collision equation of motion obeys the following Formula [17–20]:

$$\begin{cases} m\frac{d^2x}{dt^2} = F_n + F_\tau + G \\ I\frac{dw_p}{dt} = F_n b\cos\psi - F_\tau b\sin\psi \\ F_n = \frac{4}{3}Y^* R^{1/2}\zeta_n^{3/2} \\ F_\tau = -S_\tau \xi_\tau \\ G = mg \end{cases} \tag{3}$$

where F_n is the normal contact force, N; F_τ is the tangential contact force, N; G is the weight of the seed, N; I is the moment of inertia of the seed, kg·m^2; w_p is the angular velocity of the seed rotation, rad·s^{-1}; b is the length of the long axis of the seed, m; ψ is the angle between the collision position of the seed and the long axis, degree; x is the displacement of the seed center, m; Y^* is the equivalent Young's modulus of the seed, Pa; R is the equivalent radius of the seed, m; ζ_n is the normal coincidence, m; ζ_τ is the tangential coincidence, m.

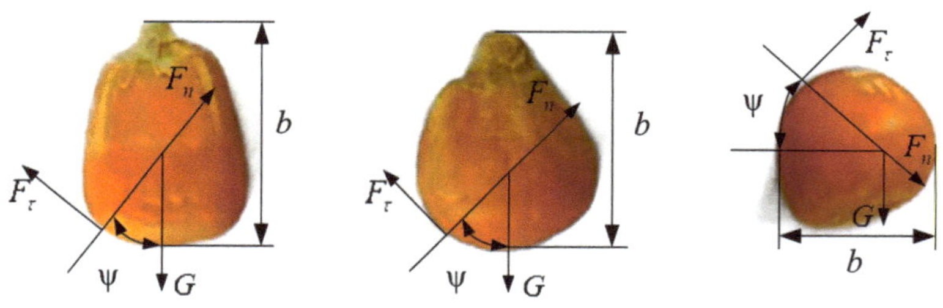

Figure 3. Maize seeds of different shapes. (**a**) Horse-tooth shape; (**b**) Spheroid shape; (**c**) Ellipsoid shape.

When the seed collides with the wall of the seed-guiding device, energy is absorbed by the deformation of the seed [21]; therefore, the kinetic energy equation after the collision is shown in Formula (4):

$$E = \frac{1}{2}mv_p^2 - \int_0^q F_n v_n dq \tag{4}$$

where v_p is the velocity before the seed impacts the wall surface, m·s^{-1}; q is the normal momentum when the seed recovers deformation, N·s v_n is the normal relative velocity between particles at the collision point, m·s^{-1}.

Combined with Formulae (3) and (4), it can be seen that the irregular shape of seeds, inconsistent collision points, different rotational movements, and different kinetic energy losses after seeds bounce up result in different trajectories for seed bounce movements. To evaluate the order of seeds entering the seed-guiding device, it is necessary to analyze the changes in seed velocity and movement after collision.

2.3. Design of Energy-Dissipation Device and Principle of Seed Energy Dissipation

2.3.1. Design of Energy-Dissipation Device

Viscous fluid dampers are highly sensitive to velocity changes and can effectively absorb vibration and impact energy. They have been successfully used in bridges, engineering machinery, and rolling stock to absorb vibration shock energy and reduce equipment damage caused by shock [22–25]. The results of the study show that when the relative motion speed of the viscous fluid due to external excitation is small, the energy dissipated by the shear-thinning fluid is much larger than that dissipated by the shear-thickening fluid. Conversely, when the relative motion speed of the viscous fluid due to external excitation is large, the energy dissipated by the shear-thickening fluid is greater than that dissipated by the shear-thinning fluid [26]. Because the theoretical velocity of seeds entering the seed-guiding device is small, the external excitation to viscous fluid is also small. Therefore, the energy-dissipation characteristics of different types of viscous fluid are comprehensively compared and analyzed. Based on this analysis, we choose shear-thinning fluid as the material of energy-dissipation parts in this study.

The energy-dissipation device consists of a funnel and an energy dissipator. To facilitate the impacted seeds to quickly fall to the seed-guiding device, the inclination angle of the funnel is determined to be 70 degrees. The outlet of the funnel is directly connected to the air intake chamber of the seed-guiding device, so their diameters are identical. Since the energy-dissipation material is a viscoelastic fluid, grid strips are arranged on the inner wall of the funnel to prevent deformation under bumpy conditions. The energy-dissipation materials are laid in blocks within the grid.

2.3.2. Theoretical Model of Seed Energy Dissipation

The process of the seed impacting the energy-dissipation material after departure from the seed disk is a decelerating motion. As a viscous fluid, the shear-thinning fluid provides resistance to the seeds. From the initial moment of impact to the point of lowest velocity, the resistance experienced by the seeds is primarily due to liquid drag force and viscous shear stress [27,28]. Taking maize seeds as the analysis object, Formula (5) can be obtained according to the momentum theorem.

$$\begin{cases} mv_{t2} - mv_{t3} = \int_0^{t_e} F_e + F_{vis} \\ F_e = \dfrac{1}{2}\rho_l C_e A_e v_{t2}^2 \\ F_{vis} = f_l A_l \\ A_l = 2\pi a_e x_l \\ f_l = \dfrac{\eta_a v_{t2}}{x} \end{cases} \tag{5}$$

where v_{t2} is the velocity of the seed impacting the flexible material, m·s^{-1}; v_{t3} is the minimum velocity of the seed after impacting the flexible material, m·s^{-1}; t_e is the time of the seed impacting the flexible material, s; F_e and F_{vis} are the fluid drag force and viscous force exerted by the fluid on the seed, respectively, N; ρ_l is the density of the fluid, kg·m^{-3}; C_e is the coefficient of resistance; A_e is the characteristic area, i.e., the seed entering the flexible material part of the horizontal projection, m^2; f_l is the shear stress exerted by the seed on the flexible body, N·m^{-2}; A_l is the characteristic area of f_l acting on the surface of the seed, m^2; a_e is the equivalent radius of the seed, m; x is the length of the seed entering into the flexible material, m.

Simplify Formula (5) to obtain the following Formula (6):

$$v_{t3} = v_{t2} - \frac{\rho_l C_e A_e t_e v_{t2}^2 + 4\pi a_e t_e \eta_a v_{t2}}{2m} \tag{6}$$

Formula (6) represents the energy-dissipation equation after the seed impacts the flexible material. We can clearly find that the velocity after the seed impacts the flexible material is negatively correlated with the velocity of the seed impacting the flexible material and the viscosity of the fluid, respectively, i.e., the larger the velocity of the impact, the larger the value of the viscosity of the fluid and the larger the kinetic energy dissipated by the seed. The kinetic energy dissipated by the seed is related to the horizontal projection area of the seed entering flexible materials. Because maize seeds are irregular objects, the attitude of the seed impacting flexible materials will also affect the energy dissipation of the seed.

2.4. Preparation of Experimental Materials

Sodium alginate (SA) dissolves in water to form a viscous colloid with good softness and uniformity. The viscosity of its aqueous solution increases with the increase in SA concentration [29,30]. Therefore, SA is selected as dispersed phase particles and deionized water as a dispersed medium to prepare shear-thinning fluid. A certain amount of SA (Shanghai Macklin Biochemical Co., Ltd., Shanghai, China) is dissolved in deionized water at 25 °C and then stirred in a magnetic stirring water bath (Spring instrument. Co., Ltd., Changzhou, China) until the SA was uniformly distributed. Under the same experimental conditions, the solutions were prepared with SA mass shares of 5%, 10%, 15% and 20%, respectively.

2.5. Test Bench

To study the feasibility of using SA to make an energy-dissipation effect, experiments were carried out on a self-made impact device test bench, as shown in Figure 4. The test equipment mainly consists of an L-PRI 1000 high-speed camera (produced by AOS Technologies AG, with a frame rate of 100 frames per second during the test), bracket, energy dissipator, and computer. During the test, the high-speed camera's frame rate is set to 100 frames per second. Each experiment records the movement of the seeds with the high-speed camera, saves the video in. raw4 format, and imports it into the motion analysis software TEMA Classic 2021. The first step is to select the marker tracking point and set the algorithm and parameters for the point. The second step is to select the seed motion segment to identify and interpret the tracking point. The third step is to calibrate the seed motion segment in two dimensions. The fourth step is to display the tracking data and obtain the 2D-pixel coordinates of the maize seed in the image sequence. Finally, the speed, displacement and acceleration of the seeds were calculated by TEMA Classic software.

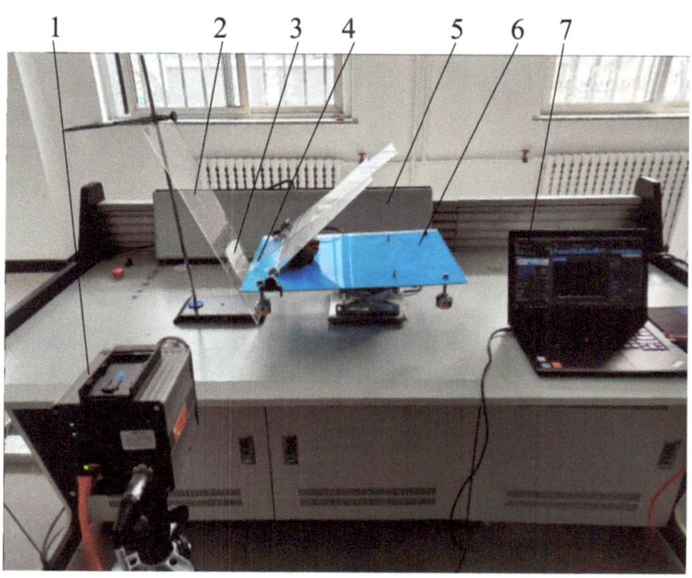

Figure 4. Impact test bench. (1) L-PRI 1000 high-speed camera; (2) Acrylic plate bracket; (3) Energy dissipator; (4) Protractor; (5) Fill light; (6) Adjusting bracket; (7) Computer.

Under natural conditions, SA aqueous solution loses water, which directly affects its viscosity and elasticity if water loss is excessive. Therefore, for the test, the SA solution is made into an energy dissipator with dimensions of 150 × 150 × 2 mm and wrapped with 0.02 mm thick polyethylene material. (which meets the requirements of GB/T1040.3-2006 [31]). In practical applications, the energy-dissipation material is laid on a funnel with a fixed inclination angle, forming an energy-dissipation device. The inclination angle of the fixed energy-dissipation material bracket on the impact test bench is set to match that of the funnel, which is 70 degrees. In the test, the maize seeds are released manually by adjusting the height. The seeds slid down a smooth acrylic plate to achieve the desired impact speed. The movement angle of the seeds is changed by rotating the adjusting bracket. Zhengdan 958 maize seeds are selected, with a moisture content of 13% and a thousand-grain mass of 375 g.

2.6. Field Experiments

To further verify the effect of the energy dissipator on sowing quality enhancement under high-speed conditions, the 2BQX-6J maize/soybean high-speed vacuum precision seeder, with and without an energy-dissipation device, was used as the test carrier. The field test was carried out at the Beijing Agricultural Machinery Experimental Station. The supporting power for the test was provided by a John Deere 1654 tractor (power 165 hp). The fan of the seed-metering device was driven by a hydraulic motor, with the fan pressure set at 4.5 kPa with an error of 0.5 kPa. The theoretical grain distance was set to 25 cm, and the soil type of the test site was sandy loam. Ungraded Zhengdan 958 maize seed was used in the test, and the field test site is shown in Figure 5. The length of the test area was set to 100 m, with the middle 20 m designated as the data collection area. The working speeds were set to 12 km·h^{-1}, 14 km·h^{-1}, and 16 km·h^{-1}, respectively.

Figure 5. Field experiments.

2.7. Experimental Scheme and Evaluation Index

2.7.1. Evaluation Index

The pre-and post-impact velocity difference and the end-of-motion transverse displacement value are used as the bench test indexes. The average value of 20 seeds in each group is taken to evaluate the test results. The calculation method of the experiment index is shown in Formula (7):

$$\begin{cases} y_1 = \dfrac{\sum\limits_{j=1}^{n} (v_{cj} - v_{mj})}{n} \\ y_2 = \dfrac{\sum\limits_{j=1}^{n} x_z}{n} \end{cases} \tag{7}$$

where y_1 is the pre- and post-impact velocity difference, m·s^{-1}; y_2 is the value of transverse displacement at the end of the movement, mm; v_{cj} is the maximum velocity before the seed impacts the flexible body, m·s^{-1}; v_{mj} is the minimum velocity after the seed impacts the flexible body, m·s^{-1}; x_z is the horizontal position of the end of the seed movement (about to touch the tabletop), mm.

According to the requirements of JB/T10293-2013 [32] Technical Conditions for Single-Seed (Precision) Seeder, the field experiment was carried out with qualified rate, leakage rate, multiple rate, and qualified grain distance variation rate as experimental indexes. The calculation method of the experiment index is shown in Formula (8):

$$\begin{cases} y_3 = \frac{n_1}{N} \times 100\% \\ y_4 = \frac{n_2}{N} \times 100\% \\ y_5 = \frac{n_0}{N} \times 100\% \\ y_6 = \sqrt{\frac{\sum n_i X_i}{n'_2} - \overline{X^2}} \times 100\% \end{cases} \quad (8)$$

Specifically,

$$\begin{cases} X_i = \frac{x_m}{x_t} \\ n'_1 = \sum n_i(X_i \in \{0 \sim 0.5\}) \\ n'_2 = \sum n_i(X_i \in \{> 0.5 \sim \leq 1.5\}) \\ n'_3 = \sum n_i(X_i \in \{> 1.5 \sim \leq 2.5\}) \\ n'_4 = \sum n_i(X_i \in \{> 2.5 \sim \leq 3.5\}) \\ n'_5 = \sum n_i(X_i \in \{> 3.5 \sim +\infty\}) \\ n_2 = n'_1 \\ n_1 = n'_1 + n'_2 + n'_3 + n'_4 + n'_5 - 2n_2 \\ n_0 = n'_3 + 2n'_4 + 3n'_5 \\ N = n'_2 + 2n'_3 + 3n'_4 + 4n'_5 \end{cases} \quad (9)$$

where y_3 is the qualified rate, %; y_4 is the multiple rate, %; y_5 is the leakage rate, %; y_6 is the qualified grain distance variation rate,%; X_i is the scale value, x_m is measured seed spacing, mm; x_t is theoretical seed spacing, mm; n_i is the frequency.

2.7.2. One Factor and Multifactor Experimental Design

To study the rheological behavior of SA solution in a steady shear flow field, we used a DHR-2 rheometer (TA Instrument Co., Ltd., New Castle, DE, USA) to obtain the macroscopic response characteristics of SA solution under stress or strain conditions. To investigate the changes in the fluid under dynamic shear, time was taken as the independent variable, while the energy storage modulus and energy-dissipation modulus were taken as the dependent variables. Oscillatory tests were performed using the rheometer, setting the low shear rate to 0.1/s and the high shear rate to 100/s, with each time period lasting 20 s. The mass fractions of SA in the dispersed phase were 5%, 10%, 15% and 20%.

Based on the theoretical analysis of seed energy dissipation, the mass fraction of dispersed phase in the fluid, impact velocity, and impact angle are selected as experiment factors. The pre-and post-impact velocity difference and the transverse displacement at the end of the movement are selected as experiment indexes. A single-factor experimental study on the process of seed energy dissipation is carried out. The level codes of experimental factors are shown in Table 1.

Table 1. Factor level of single-factor experiment.

Level	Factors		
	Mass Fraction of Dispersed Phase in Fluid/%	**Impact Velocity/m·s^{-1}**	**Impact Angle/°**
1	10	1	0
2	15	2	15
3		3	30

2.7.3. Orthogonal Experimental Design

To improve the energy-dissipation effect of seed, it is necessary to optimize the structure of the energy-dissipation body. Based on the single-factor experiment, the thickness of the energy dissipator, impact velocity and impact angle are selected as experimental factors to carry out a quadratic regression orthogonal experiment with three factors and three levels. The level codes of experimental factors are shown in Table 2.

Table 2. Factor level of quadratic regression orthogonal experiment.

Level	Factors		
	Thickness of Energy Dissipator *E*/mm	**Impact Velocity** *F*/m·s^{-1}	**Impact Angle** *W*/°
−1	4	1	0
0	6	2	15
1	8	3	30

3. Results and Discussion

3.1. Analysis of Rheological Property of Shear-Thinning Fluid

Figure 6 shows the variation curves of fluid viscosity values with the shear rate for different SA mass fractions. It can be seen from the figure that when the mass fraction of SA is fixed value in the test level, the viscosity of fluid decreases with the increase of shear rate, which shows a continuous shear-thinning phenomenon. When the mass fraction of SA is less than 10%, and the shear rate is greater than 10/s, the fluid's viscosity is very low, indicating a low critical shear rate. At a shear rate of 0.1/s, the viscosity values are 1003.3 Pa·s and 2257.5 Pa·s for SA mass fractions of 15% and 20%, respectively. The viscosity decreases significantly when the shear rate is increased to 100/s, indicating that a proper increase in SA is beneficial to increase the fluid's critical shear rate.

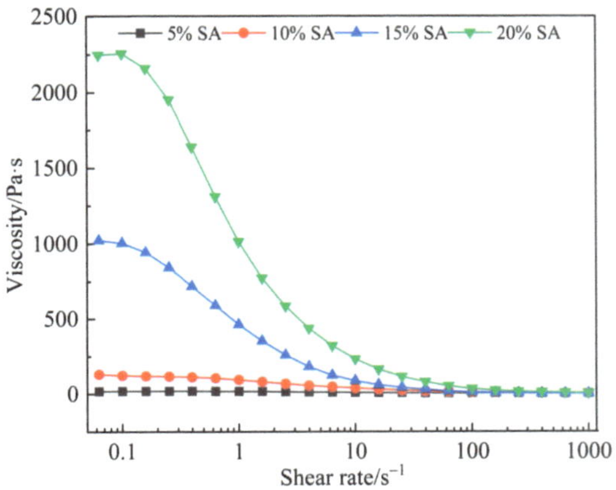

Figure 6. Rheological curves of fluid under different mass fractions.

The SA solution is a viscoelastic fluid, making the study of its viscoelasticity particularly important. According to the theory of viscoelasticity, elasticity represents the solid behavior of the system, while viscosity represents the liquid behavior. The elastic and viscous strengths of the system are expressed by the storage modulus (G′) and the dissipation modulus (G″), respectively [33]. When maize seeds impact on energy-dissipation parts made of SA solution at a certain velocity, the fluid movement process can be roughly divided into three stages under the action of shear force: the no-collision period under low shear rate, the collision period under high shear rate and recovery period under low shear rate.

Figure 7a–d record the values of energy storage modulus (G′) and energy-dissipation modulus (G″) for fluids with SA mass fractions of 5%, 10%, 15%, and 20%, respectively, across the three stages. As can be seen from the figure, the values of energy storage modulus and energy-dissipation modulus increase with the increase of the mass fraction of SA in the four sets of tests. Specifically, the values of G′ and G″ for 5% SA fluid are smaller, and the energy storage modulus of the fluid is smaller than the energy-dissipation modulus in the three time periods. Please note that G″ is larger than G′ during both the collision and recovery stages in the same set of tests, which indicates that the 5% SA fluid exhibits liquid-like behavior and has poor shape retention. For 10% SA fluid, the energy storage modulus is smaller than the energy-dissipation modulus throughout all three stages, but the values of G′ and G″ increase significantly, indicating enhanced viscous and elastic properties. Therefore, it demonstrates that the 10% SA fluid can absorb energy and recover from deformation. For 15% SA fluid, the energy storage modulus is larger than the energy consumption modulus in the time period of 0~20 s and 40~60 s, which indicates that the fluid is a kind of solid and has a certain shape-retaining ability without shear action. The energy consumption modulus is larger than the energy storage modulus in the time period of 20~40 s, which indicates that the fluid mainly exhibits viscous characteristics when it is subjected to shear force. Thus, it proves that 15% SA fluid mainly consumes energy. The value of G′ for 15% SA fluid is approximately equal to three times 10% SA fluid, and the value of G″ for 15% SA fluid is approximately equal to twice 10% SA fluid for the collision-free period at a low shear rate. In the recovery period at a low shear rate, the value of G′ for 15% SA fluid is approximately equal to two times 10% SA fluid, and the difference between the two G″ values is very small, which indicates that the ability of 15% SA fluid to recover deformation in the absence of shear is slightly better than that of 10% SA fluid. The values of G′ and G″ are approximately equal for 15%SA fluid and 10% SA fluid for the collision period at a high shear rate. A comprehensive analysis of the rheological properties of the 10% SA and 15% SA fluids shows that the energy-dissipation modulus of both is approximately equal from the shear stage to the recovery stage, but there is a difference in specific rheological properties of both during the recovery stage. The G′ and G″ variation rules of 20% SA fluid and 15% SA fluid are the same in all three time periods, but the energy storage modulus of 20% SA fluid in the absence of shear is significantly increased, which indicates that the elastic properties of the fluid are better.

When seeds come into contact with the energy-dissipation body, the fluid flows under the influence of shear force, resulting in a damping force that dissipates external energy. In this paper, the shear-thinning fluid is used as the energy-dissipation material, which is laid in the middle of the seed-guiding device and the seed-metering device. In the ideal state, the seeds lose all their kinetic energy upon impact with the fluid, the fluid does not deform after impact, and the fluid exhibits excellent shape retention, which ensures that the seeds enter the seed-guiding device along the vertical direction. Fluid needs to have an appropriate balance of viscosity and elasticity. If the elasticity of the fluid is too low, the ability to recover from deformation is poor, which can easily cause sticking. Conversely, if the fluid's elasticity is too high, it can cause the seeds to rebound, failing to meet the requirements for energy dissipation. In addition, the analysis shows that the 5% SA fluid is mainly in a liquid-like state, with a large amount of subsidence after seed impact, which makes it impossible to maintain the shape of the energy dissipator. Conversely, the 20% SA fluid is more elastic, with a sharp rebound after seed impact. Therefore, 5% SA and 20% SA fluids are not suitable as materials for energy-dissipation devices. Thus, 10% SA and 15% SA fluids will be selected for experimental studies on seed energy dissipation.

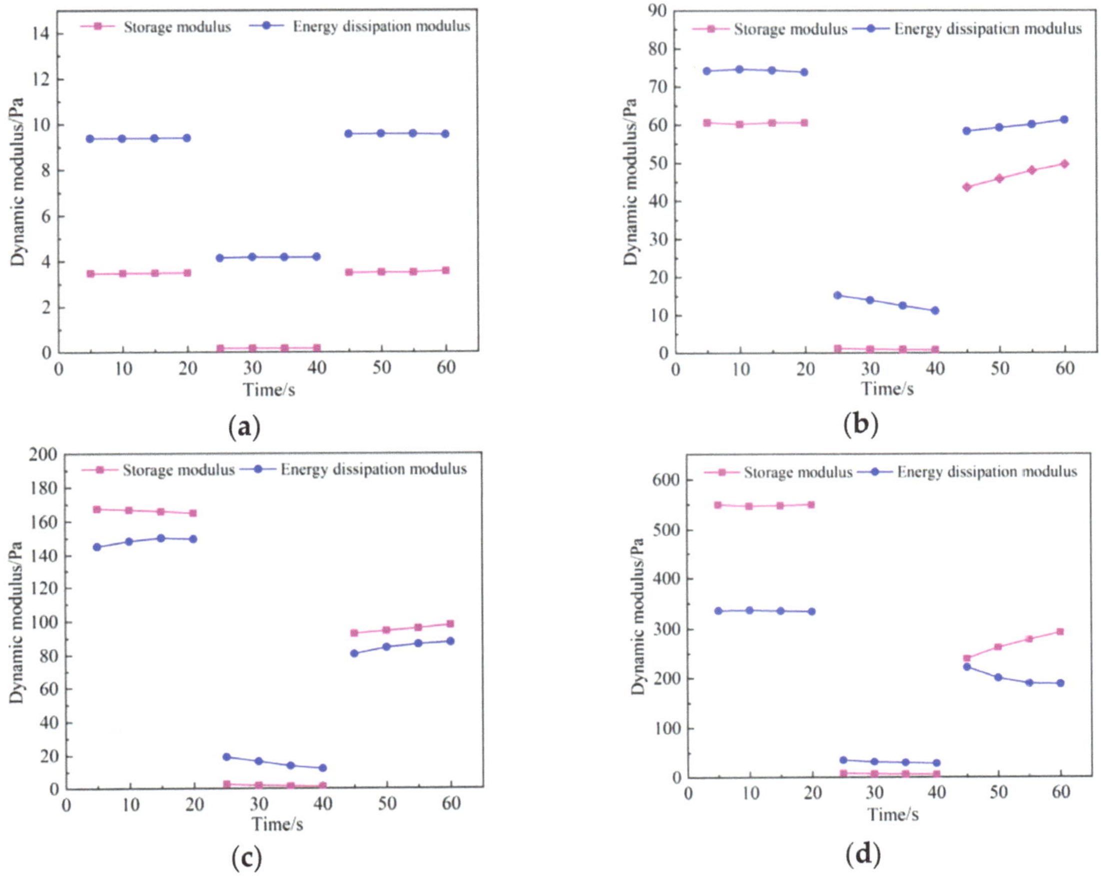

Figure 7. Viscoelasticity of fluid under different mass fractions. (**a**) 5% SA; (**b**) 10% SA; (**c**) 15% SA; (**d**) 20% SA.

3.2. Analysis of the Seed Energy-Dissipation Process

3.2.1. Feasibility of the Seed Energy Dissipation

When the mass fraction of SA in the dispersed phase is 10%, the impact velocity is 2 m·s^{-1} and the impact angle is 30 degrees, maize seeds might impact flexible material in four postures: the largest side in the plane of seed contacts flexible material (broad-side impact), the tip of seed contacts flexible material (tip impact), the side of seed contacts flexible material (side impact) and the tail of seeds contacts flexible materials (bottom impact), as shown in Figure 8a–d in turn. Additionally, the movement posture of seeds changes obviously after impacting the energy-dissipation material. The movement of seeds after contact with flexible material is different, with different postures of seeds impacting flexible materials. This phenomenon is consistent with the theoretical analysis of the seed energy-dissipation equation.

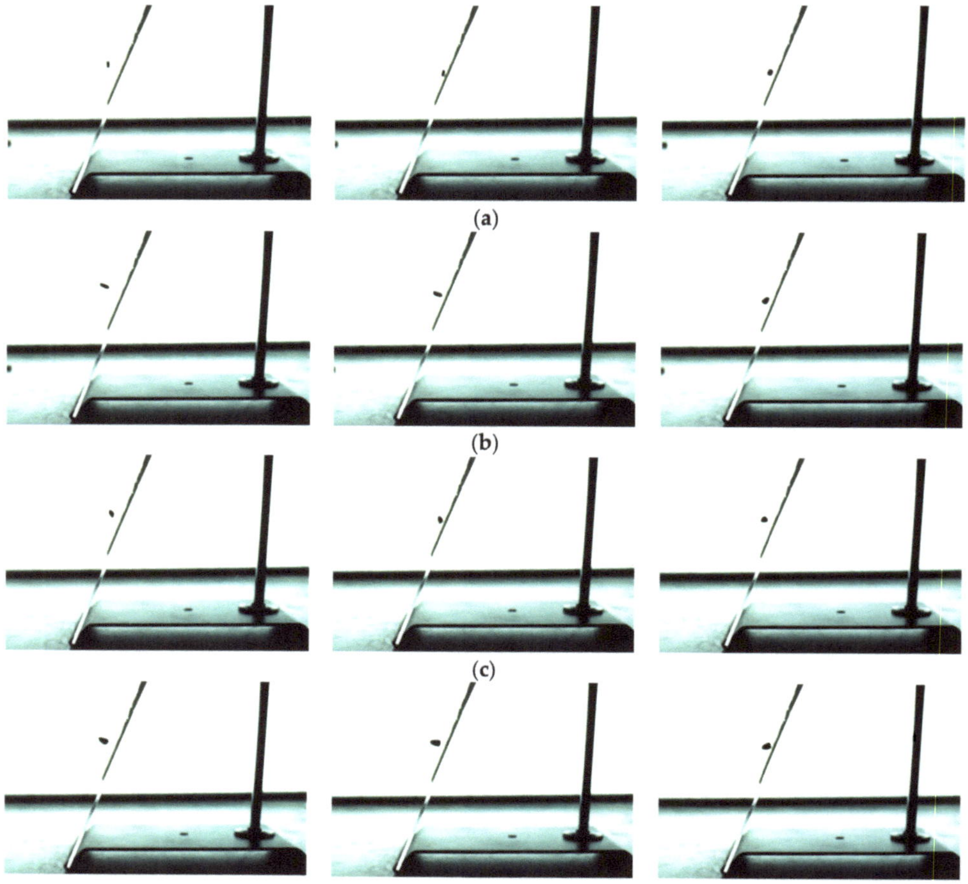

Figure 8. Different postures of seeds impacting the energy dissipator. (**a**) Broad-side impact; (**b**) Tip impact; (**c**) Side impact (**d**) Bottom impact.

The velocity and transverse displacement of the seed under four impact postures as a function of time are shown in Figures 9–12.

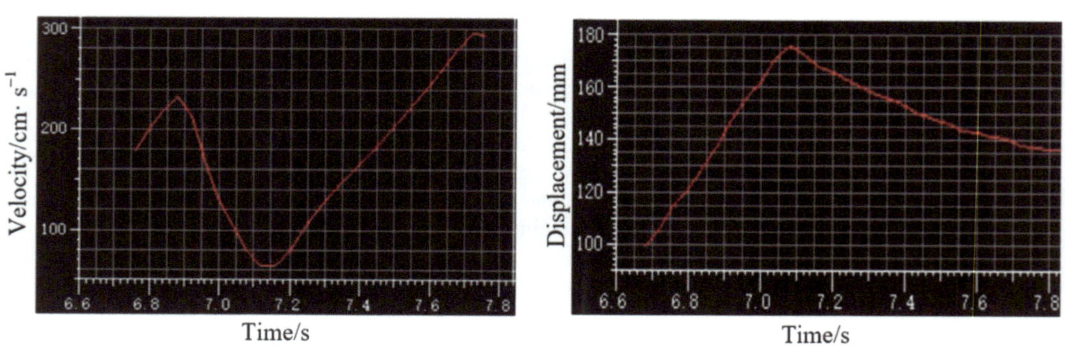

(**a**) (**b**)

Figure 9. Motion analysis of broad-side impact. (**a**) Time-velocity curve; (**b**) Time-transverse position curve.

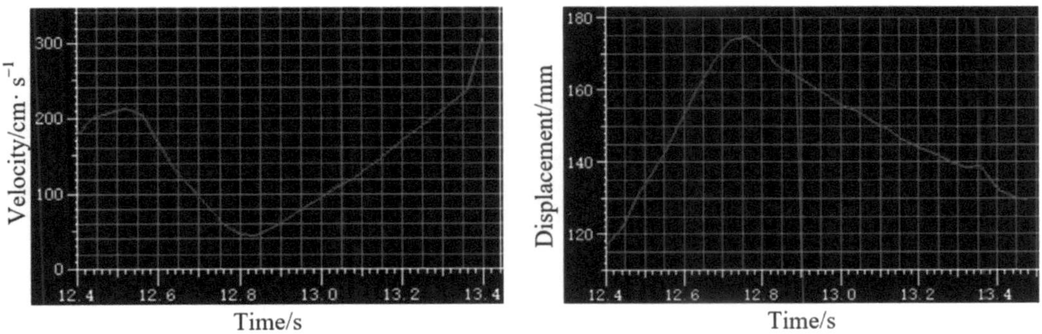

(**a**) (**b**)

Figure 10. Motion analysis of tip impact. (**a**) Time-velocity curve; (**b**) Time-transverse position.

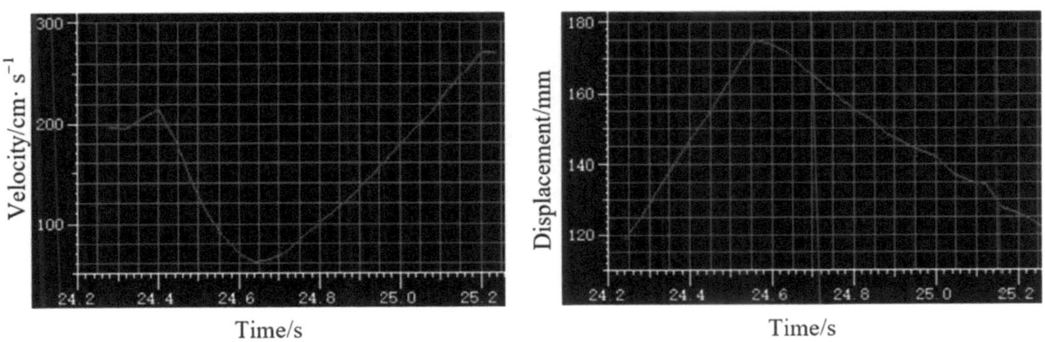

(**a**) (**b**)

Figure 11. Motion analysis of side impact. (**a**) Time-velocity curve; (**b**) Time-transverse position curve.

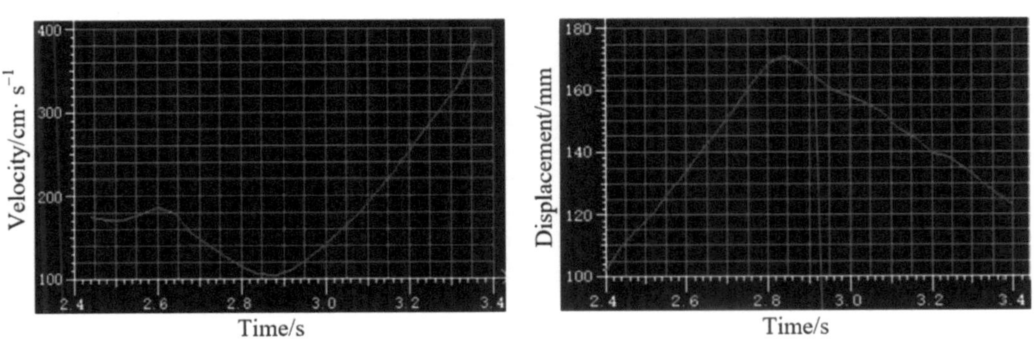

(**a**) (**b**)

Figure 12. Motion analysis of bottom impact. (**a**) Time-velocity curve; (**b**) Time-transverse position curve.

By analyzing the time-velocity curves in Figures 9–12, it can be seen that before and after the seeds hit the flexible materials, the seed velocity experienced three stages: acceleration, deceleration, and re-acceleration over time. Before the seeds contact the flexible material, due to the small height of the flexible material, the seed velocity increases slightly under gravity. After the seed comes into contact with the flexible material, the shear force is exerted on the flexible material, causing the material to flow and rapidly dissipate the seed's kinetic energy of the seed, resulting in the seed's velocity rapidly decreasing to a minimum. When the seed separates from the flexible material, it accelerates under the action of gravity, causing its velocity to increase. The seed's velocity in the second stage is greatly reduced, which proves that the flexible materials designed in this paper effectively reduce the seed's velocity. It also can be found that the reduction in kinetic energy varies with different impact postures. Specifically, the seed's velocity decreased from 2.32 m·s^{-1} to 0.63 m·s^{-1} for broad-side impact, from 2.18 m·s^{-1} to 0.41 m·s^{-1} for tip impact, from 2.11 m·s^{-1} to 0.60 m·s^{-1} for side impact, and from 1.92 m·s^{-1} to 1.01 m·s^{-1} for bottom impact. The greater the pre- and post-impact velocity difference, the more effective the energy absorption of the energy dissipator. The seed velocity curve rapidly changes from a minimum in the second stage to continuous growth in the third stage, indicating that the seeds are not stuck in the flexible material.

By analyzing the time-transverse position curves in Figures 9–12, it can be seen that before and after the seed impacts the flexible material, the seed's transverse displacement goes through two stages: an increase followed by a decrease over time. Due to the horizontal impact velocity, the seed's transverse displacement gradually increases before contacting the material. After the seed contacts the flexible material, the seed has a reverse horizontal velocity, and its transverse displacement gradually decreases, indicating the elasticity of the flexible material, which ensures that the seed quickly separates from the flexible material after impact. In the experiment, the energy-dissipation material bracket is fixed. Considering the arrangement position of the test device, it can be seen that the greater the transverse displacement of the seed's moving end, the smaller the distance between the seed and the bracket's end face, indicating a smaller bounce amplitude after impact. The transverse displacement of the seed's moving end is the largest for broad-side impact, resulting in the smallest bounce amplitude. The transverse displacement values of the moving end of seeds are different under the four impact postures, but the difference is very small, which shows that seeds can enter the seed-guiding device along the same position. The above analysis verified the feasibility of using an SA solution to make an energy-dissipation device.

3.2.2. Influence of Experimental Factors on Energy-Dissipation Process

The influence of different mass fractions of fluid dispersed phase on the pre-and post-impact velocity difference is shown in Figure 13. When the impact angle is a fixed value in the test range, the energy dissipator can effectively dissipate the seed's kinetic energy at each velocity level. The 10% SA fluid has a better kinetic energy-dissipation ability than the 15% SA fluid because the 10% SA fluid exhibits stronger flow performance under the same shear force, producing a larger damping force and thus dissipating more energy. At the same impact velocity, the pre- and post-impact velocity difference is smallest at an impact angle of 0 degrees and largest at an impact angle of 30 degrees. This may be because maize seeds are irregular objects; a larger impact angle increases the contact area with the energy dissipator during oblique motion, resulting in a larger trailing force from the fluid and greater kinetic energy dissipation.

The influence of different mass fractions of fluid dispersed phase on the transverse displacement values at the end of the seed movement is shown in Figure 14. At the same impact velocity and angle, the transverse displacement values of seeds impacting the 10% SA fluid are larger than those impacting the 15% SA fluid, further supporting the good energy absorption effect of the 10% SA fluid. Based on the above experimental results, 10% SA fluid is determined to be the fabrication material for the energy dissipator. However, within the test range, although the velocity difference of seeds after impacting the energy dissipator is significantly reduced, the post-impact seed velocity remains high. For example, when the impact velocity is 3 m·s^{-1} and the impact angle is 0 degrees, the pre-and post-impact velocity is 1.14 m·s^{-1}, indicating the need to further optimize the energy dissipator's structure.

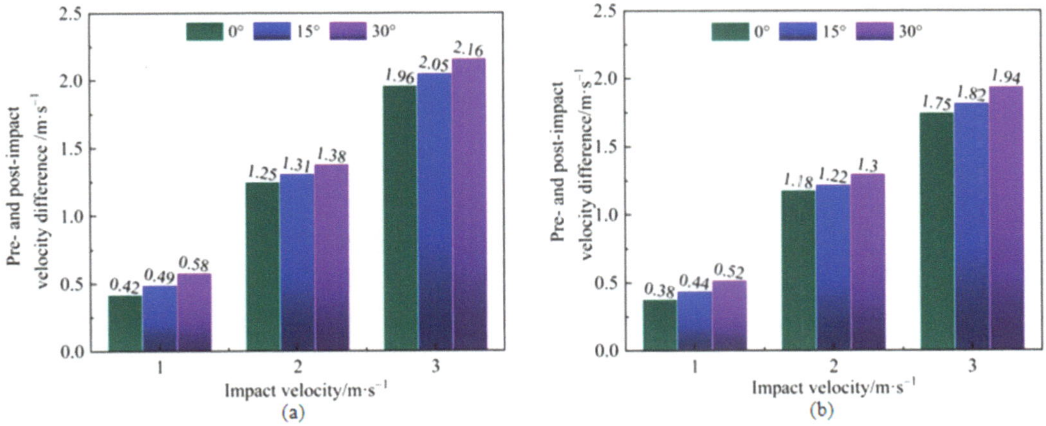

Figure 13. Influence of different mass fractions of fluid dispersed phase on the pre- and post-impact velocity difference. (**a**) 10% SA; (**b**) 15% SA.

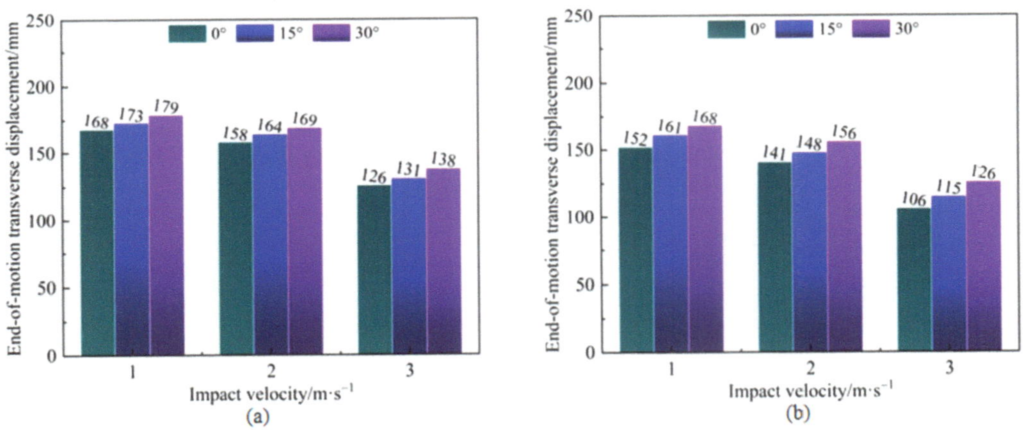

Figure 14. Influence of different mass fractions of fluid dispersed phase on the transverse displacement. (**a**) 10% SA; (**b**) 15% SA.

3.3. Optimization of Energy Dissipator

To further reduce the bounce degree of seeds after impacting the energy dissipator, based on the previous test and analysis results, a regression orthogonal test design was carried out according to Box-Behnken central combination test design theory. The test scheme and results are shown in Table 3.

3.3.1. Analysis of Variance and Establishment of Regression Model

Using Design-Expert 10 software (Stat-Ease Ltd., Godward St NE, Minneapolis, MN, USA) to carry out regression analysis on the experiment data in Table 3, the regression equations between the pre- and post-impact velocity difference y_1, end-of-motion transverse displacement value y_2, energy dissipator thickness E, impact velocity F, and impact angle W as shown in Formula (10).

$$\begin{cases} y_1 = 1.8 + 0.23E + 0.078F + 0.81W + 0.045EF - 0.15E^2 - 0.15F^2 - 0.057W^2 \\ y_2 = 170.06 + 2.5E - 18.75F + 6.25W - 8.18F^2 + 4.83W^2 \end{cases} \quad (10)$$

Table 3. Design and results of the orthogonal experiment.

Number	Thickness of Energy Dissipator E/(mm)	Impact Velocity F/(m·s^{-1})	Impact Angle W/(degree)	Pre- and Post-Impact Velocity Difference y_1/(m·s^{-1})	Value of Transverse Displacement at the End of the Movement y_2/(mm)
1	−1	0	−1	1.44	167
2	−1	0	1	1.62	178
3	1	1	0	2.45	145
4	1	0	1	1.73	185
5	−1	1	0	2.18	141
6	0	1	−1	2.33	142
7	0	0	0	1.78	169
8	0	0	0	1.86	172
9	0	−1	−1	0.68	180
10	−1	−1	0	0.65	179
11	0	0	0	1.81	166
12	0	1	1	2.47	156
13	1	−1	0	0.74	184
14	0	0	0	1.82	175
15	0	0	0	1.74	171
16	0	−1	1	0.91	191
17	1	0	−1	1.59	171

The results of the variance analysis of the regression equation are shown in Table 4. The models of pre- and post-impact velocity difference and end-of-motion transverse displacement value are extremely significant. However, the lack of fit of regression equations is insignificant. The determination coefficients R^2 of y_1 and y_2 are 0.9469 and 0.8574, respectively, which indicates that the predicted value of the regression equation has an excellent correlation with the actual value. The primary and secondary factors affecting the pre- and post-impact velocity difference and end-of-motion transverse displacement value are impact velocity F, impact angle W, and energy dissipator thickness E, in turn.

Table 4. Variance analysis of regression equations.

Source of Variation	y_1/(m·s^{-1})				y_1/(mm)			
	SS	*DF*	*F*	*P*	*SS*	*DF*	*F*	*P*
Model	5.54	9.00	441.51	<0.0001 **	3453.24	9.00	57.70	<0.0001 **
E	0.05	1.00	34.48	0.0006 **	50.00	1.00	7.34	0.03 *
F	5.20	1.00	3731.64	<0.0001 **	2812.50	1.00	412.74	<0.0001 **
W	0.06	1.00	42.71	0.0003 **	312.50	1.00	45.86	0.0003 **
EF	8.1×10^{-3}	1.00	5.81	0.00467 **	0.25	1.00	0.03	0.8535
EW	4.0×10^{-4}	1.00	0.29	0.608	2.25	1.00	0.33	0.5835
FW	2.0×10^{-3}	1.00	1.45	0.267	2.25	1.00	0.33	0.5835
E^2	0.09	1.00	67.75	<0.0001 **	0.13	1.00	0.33	0.8945
F^2	0.09	1.00	65.51	<0.0001 **	281.39	1.00	0.02	0.0004 **
W^2	0.01	1.00	9.9	0.0162 *	98.02	1.00	41.29	0.0068 **
Lack of fit	1.67×10^{-3}	3.00	0.28	0.84	2.50	3.00	14.39	0.971
Pure error	8.1×10^{-3}	4.00	441.51	<0.0001				
Total	5.55	16.00						

Note: * indicates significance at 0.05 level; ** denotes significance at 0.01 level.

3.3.2. Analysis of the Influence Effect of Energy Dissipator Thickness

To analyze the relationship between the energy dissipator thickness and the energy-dissipation effect, contour plots between the elements and the indicators were obtained using the Design-Expert 10 software, as shown in Figure 15. At the same velocity level, the pre-and post-impact velocity difference increases with the energy dissipator thickness. However, when the energy dissipator increases from 6 mm to 8 mm, the growth rate of the pre-and post-impact velocity difference decreases. This may be because the seed's impact velocity is small. The applied shear force cannot make the energy dissipator flow and then cannot dissipate more kinetic energy. Therefore, within the range of experimental values, there is an optimal thickness of the energy dissipator, which can dissipate the kinetic energy of seeds to the maximum extent. At a fixed impact velocity, the interaction between the thickness of the energy dissipator and the impact angle has a minimal effect on the pre-and post-impact velocity difference. When the impact angle or impact velocity is fixed, the end-of-motion transverse displacement is positively correlated with the energy dissipator thickness.

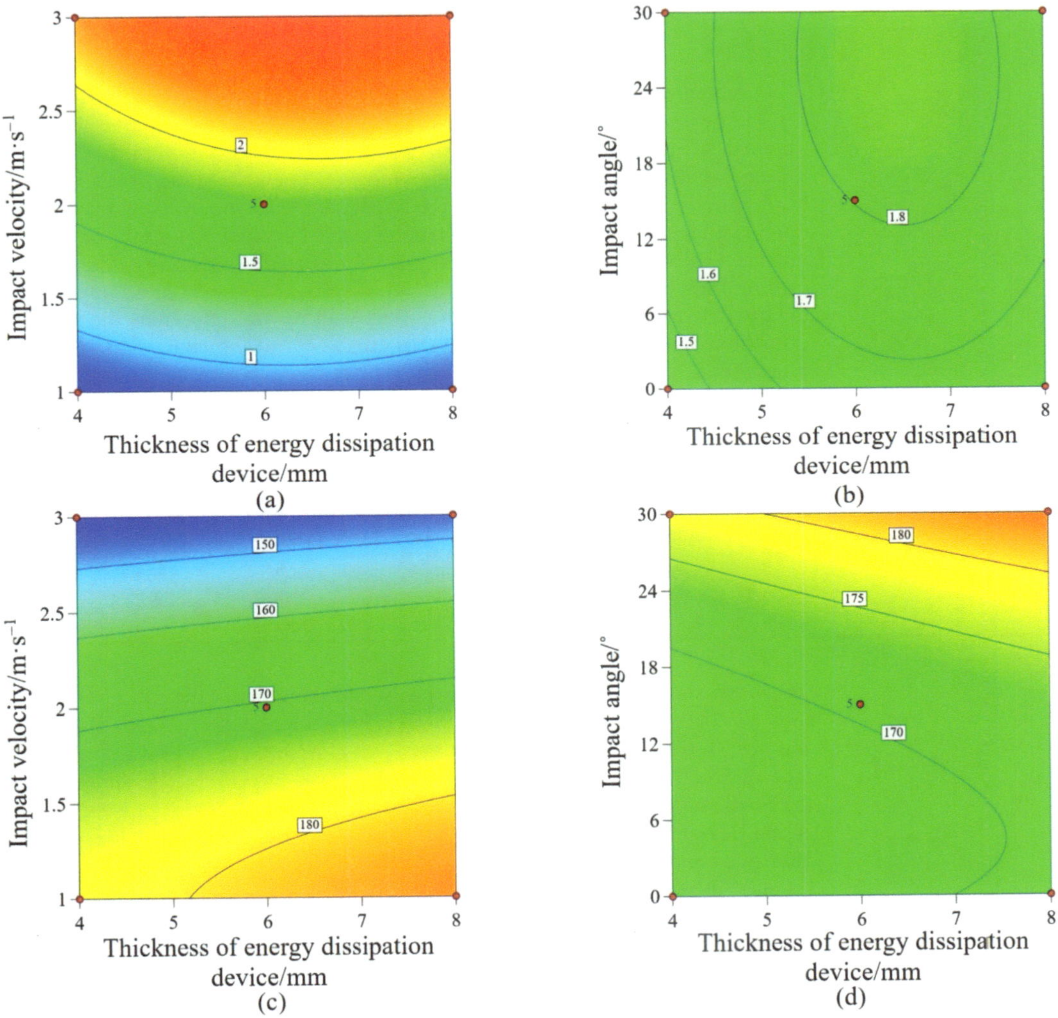

Figure 15. Influence effect of two-factor interaction on evaluation index. (**a**) Influence effect of impact velocity and thickness on pre- and post-impact velocity difference; (**b**) Influence effect of impact angle and thickness on pre- and post-impact velocity difference; (**c**) Influence effect of impact velocity and thickness on end-of-motion transverse displacement; (**d**) Influence effect of impact angle and thickness on end-of-motion transverse displacement.

3.3.3. Parameter Optimization and Experimental Validation

To determine the optimal thickness of the energy-dissipation flexible bearing body, the maximum pre-and post-impact velocity difference and end-of-motion transverse displacement are used as the evaluation index. The regression model is solved by multi-objective optimization, and the optimization objective function and constraint conditions are shown in Formula (11).

$$\begin{cases} maxy_1(E,F,W)\begin{cases} 4 \le E \le 8 \\ 1 \le F \le 3 \\ 0 \le W \le 30 \end{cases} \\ maxy_2(E,F,W)\begin{cases} 4 \le E \le 8 \\ 1 \le F \le 3 \\ 0 \le W \le 30 \end{cases} \end{cases} \quad (11)$$

Combined with the test results, it is observed that at each level of impact velocity, the pre-and post-impact velocity difference and end-of-motion transverse displacement are the smallest at an impact angle of zero. Therefore, the optimization calculation module of Design-Expert is used to determine the optimum thickness of the energy-dissipation flexible bearing body at three speeds when the impact angle is zero, as shown in Table 5. The optimal thickness of the energy-dissipation flexible bearing body for the optimized range of impact velocities is 6.3~6.9 mm. Because the thickness of the energy-dissipation flexible bearing body needs to meet the minimum bounce of the seed at all impact velocities, the predicted value at the maximum impact velocity was chosen as the optimal value. Considering the production difficulty, the thickness of the energy-dissipation flexible bearing body is determined to be 7 mm.

Table 5. Parameter optimization and experimental validation results.

Impact Velocity/(m·s^{-1})	Impact Angle/(°)	Thickness/(mm)		y_1/(m·s^{-1})		y_2/(mm)	
		Predicted Value	Measured Value	Predicted Value	Measured Value	Predicted Value	Measured Value
1	0	6.3	7.0	0.68	0.71	180.8	173.0
2	0	6.6	7.0	1.46	1.48	169.7	164.0
3	0	6.9	7.0	2.36	2.41	142.1	137.0

Verification tests were carried out to verify the accuracy of the optimization results, as shown in Table 5. The results showed that when the impact velocity was 1~3 m·s^{-1}, the pre- and post-impact velocity difference was 0.71~2.41 m·s^{-1}, with a seed velocity absorption ratio of 71~80.3%, indicating a better energy-dissipation effect. The difference between the predicted value and the actual value of the pre-and post-impact velocity difference is small. For the end-of-motion transverse displacement, the actual value is less than the predicted value, with an error of less than 8 mm. The main reason is that maize seeds are irregular in shape, and the postures of the seeds at the end of the movement are different. During image processing, the center of the seeds is taken as the measuring point, so the displacement values are different.

3.4. Analysis of Field Experiment Results

Comparative test results are shown in Figure 16. According to the field test results at different working speeds, the data analysis showed that the numerical differences for the same test index were very small at the same working speed level, which indicated that the seed transport system of the high-speed vacuum seeder could work continuously and stably. At all speed levels, the leakage rate was less than 6.83%, the multiple rate was less than 0.97%, the qualified rate was stable at more than 92.4%, and the qualified grain distance variation rate was stable at less than 16.57%. When working parameters were the same, the grain spacing uniformity was significantly better than the planter without the energy-dissipation device. For example, at a working speed of 14.13 km·h^{-1}, the leakage rate decreased by 2.4%, the multiple rates decreased by 0.4%, the qualified index increased by 2.8%, and the qualified grain distance variation rate decreased by 3.63%. The results showed the same tendency with Li's research [34]. This proves that the energy-dissipation device designed in this paper is beneficial for improving the overall working performance of high-speed precision seeders.

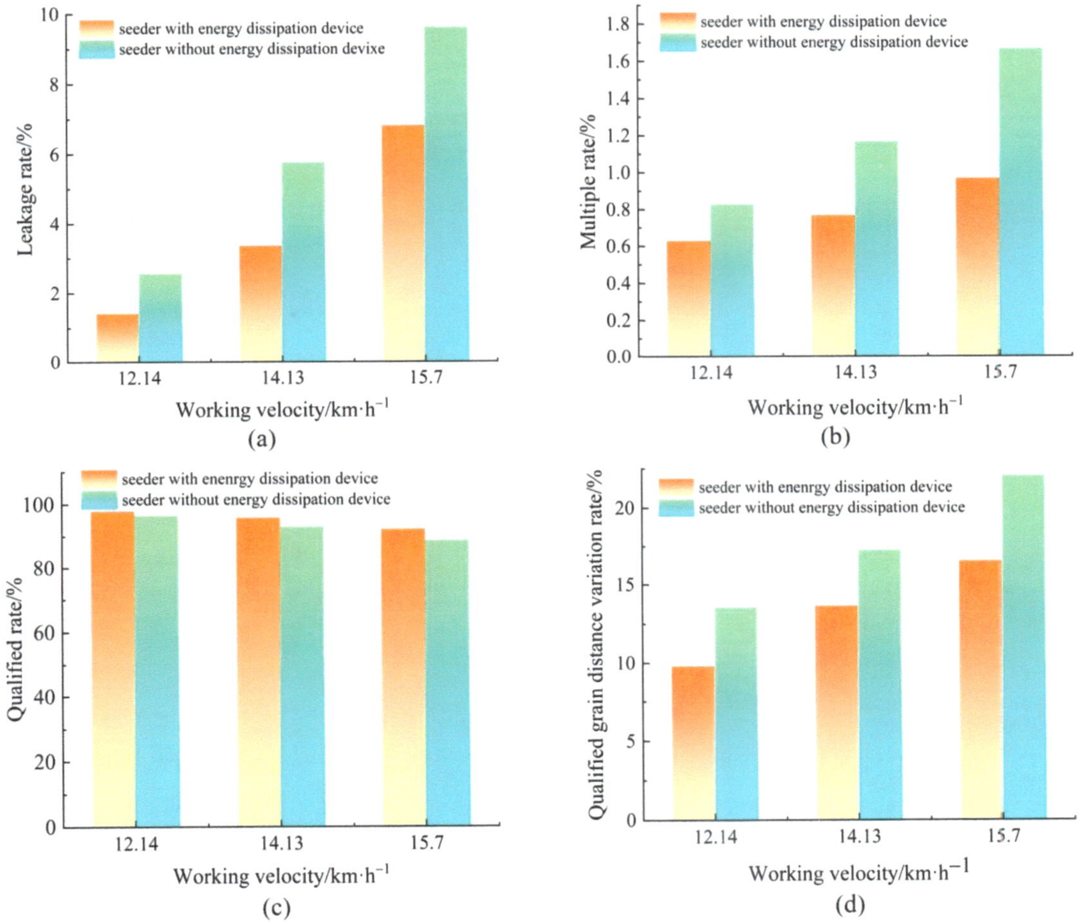

Figure 16. Comparative test results. (**a**) Leakage rate at different velocities; (**b**) Multiple rates at different velocity; (**c**) Qualified rate at different velocities; (**d**) Qualified grain distance variation rate at different velocities.

4. Conclusions

(1) Seed-throwing in an unconstrained state and seed impact with the wall of the seed-guiding device results in poor sowing uniformity of the seeder under high-speed working.

(2) The mass fraction of the fluid dispersed phase significantly influenced the energy dissipation and flexible undertaking. The results of rheological performance experiments show that the rheological properties of 10% SA and 15% SA fluids are similar. Both fluids exhibit good energy absorption and strong deformation recovery ability.

(3) The motion posture, velocity, and displacement of the seed energy-dissipation process are evaluated, and it is determined that the material of the energy dissipator is the colloid with an SA percentage of 10%.
(4) The thickness of the energy dissipator was determined to be 7 mm through regression orthogonal experiments and parameter optimization. The results of the validation experiments show that when the impact velocity was 1~3 $m{\cdot}s^{-1}$, the pre-and post-impact velocity difference was 0.71~2.41 $m{\cdot}s^{-1}$, the absorption ratio of seed velocity was 71~80.3%, indicating a better energy-dissipation effect.
(5) The field experiment results prove that the high-speed vacuum seed transport system could work continuously and stably. When the working speed was 12~16 $km{\cdot}h^{-1}$, the leakage rate was less than 6.83%, the multiple rate was less than 0.97%, the qualified rate was stable at more than 92.40%, and the qualified grain distance variation rate was stable at less than 16.57%, which proves that the energy-dissipation device is beneficial for improving the overall working performance of high-speed precision seeder. In the future, the dynamics of seed movement, as they fall into the seed furrow, will be studied in detail, with the aim of enhancing the application of the energy-dissipation device in practical agricultural production.

Author Contributions: Conceptualization, G.W.; methodology, R.L.; software, R.L.; validation, Z.M.; formal analysis, R.L.; investigation, B.Y. and R.L.; resources, B.Y. and Z.M.; data curation, R.L.; writing—original draft preparation, R.L.; writing—review and editing, Z.M and G.W.; visualization, J.D. and R.L.; supervision, G.W.; project administration, J.D.; funding acquisition, Z.M. All authors have read and agreed to the published version of the manuscript.

Funding: This research was funded by the National Key Research and Development Program of China (2021YFD2000402) National Natural Science Foundation (32301705).

Institutional Review Board Statement: Not applicable.

Data Availability Statement: Data are contained within the article. The data presented in this study can be requested from the authors.

Conflicts of Interest: The authors declare no conflicts of interest.

References

1. Kus, E. Evaluation of some operational parameters of a vacuum single-seed planter in maize sowing. *J. Agric. Sci. Tarim Bilim. Derg.* **2021**, *27*, 327–334.
2. Yatskul, A.L.; Lemiere, J.P. Establishing the conveying parameters required for the air-seeders. *Biosyst. Eng.* **2018**, *166*, 1–12. [CrossRef]
3. Yang, L.; He, X.T.; Tao, C.; Zhang, D.X.; Song, S.; Rui, Z.; Mantao, W. Development of mechatronic driving system for seed meters equipped on conventional precision corn planter. *Int. J. Agric. Biol. Eng.* **2015**, *8*, 1–9.
4. Ding, Y.Q.; He, X.T.; Yang, L.; Zhang, D.X.; Cui, T.; Li, Y.H.; Zhong, X.J.; Xie, C.J.; Du, Z.H.; Yu, T.C. Low-cost turn compensation control system for conserving seeds and increasing yields from maize precision planters. *Comput. Electron. Agric.* **2022**, *199*, 107118. [CrossRef]
5. Han, D.; Zhang, D.; Jing, H.; Yang, L.; Cui, T.; Ding, Y.; Wang, Z.; Wang, Y.; Zhang, T. DEM-CFD coupling simulation and optimization of an inside-filling air-blowing maize precision seed-metering device. *Comput. Electron. Agric.* **2018**, *150*, 426–438. [CrossRef]
6. Pareek, C.M.; Tewari, V.K.; Machavaram, R. Multi-objective optimization of seeding performance of a pneumatic precision seed metering device using integrated ANN-MOPSO approach. *Eng. Appl. Artif. Intell.* **2023**, *117*, 105559. [CrossRef]
7. Liu, Q.W.; He, X.T.; Yang, L.; Zhang, D.X.; Cui, T.; Qu, Z.; Yan, B.X.; Wang, M.T.; Zhang, T.L. Effect of travel speed on seed spacing uniformity of corn seed meter. *Int. J. Agric. Biol. Eng.* **2017**, *10*, 98–106.
8. Shi, S.; Zhou, J.; Liu, H.; Fang, H.; Jian, S.; Zhang, R. Design and experiment of pneumatic precision seed-metering device with guided assistant seed-filling. *Trans. Chin. Soc. Agric. Mach.* **2019**, *50*, 61–70. (In Chinese with English Abstract)
9. Sola. Proesm-K. Available online: https://solagrupo.com/es/pr/sembradoras-monograno/prosem-k-variant-manual-fija-Prosem%20K%20Variant%20manual%20fija-11 (accessed on 21 June 2024).
10. Maschio. Chrono700. Available online: https://www.maschio.com/en/web/international/chrono-700 (accessed on 21 June 2024).
11. Kinze. 3505Planter. Available online: https://www.kinze.com/planters/3505-planter/ (accessed on 21 June 2024).

12. Vaderstad. TempoR. Available online: https://www.vaderstad.com/ca-en/planting/tempo-planter/powershoot/ (accessed on 21 June 2024).
13. Liu, R.; Liu, L.J.; Li, Y.J.; Liu, Z.J.; Zhao, J.H.; Liu, Y.Q.; Zhang, X.D. Numerical simulation of seed-movement characteristics in new maize delivery device. *Agriculture* **2022**, *12*, 1944. [CrossRef]
14. Liu, R.; Liu, L.J.; Li, Y.J.; Liu, Z.J.; Zhao, J.H.; Liu, Y.Q.; Zhang, X.D. Design and experiment of corn high speed air suction seed metering device with disturbance assisted seed-filling. *Trans. Chin. Soc. Agric. Mach.* **2022**, *53*, 50–59. (In Chinese with English Abstract)
15. Li, Y.; Yang, L.; Zhang, D.; Cui, T.; Zhang, K.; Xie, C.; Yang, R. Analysis and test of linear seeding process of maize high speed precision metering device with air suction. *Trans. Chin. Soc. Agric. Eng.* **2020**, *36*, 26–35. (In Chinese with English Abstract)
16. Zhao, S.; Chen, J.; Wang, J.; Chen, J.; Yang, C.; Yang, Y. Design and Experiment on V-groove dialing round type guiding-seed device. *Trans. Chin. Soc. Agric. Mach.* **2018**, *49*, 146–158. (In Chinese with English Abstract)
17. Gao, Z.; Lu, C.Y.; Li, H.W.; He, J.; Wang, Q.J.; Huang, S.H.; Li, Y.X.; Zhan, H.M. Measurement method of collision restitution coefficient between corn seed and soil based on the collision dynamics theory of mass point and fixed surface. *Agriculture* **2023**, *12*, 1611. [CrossRef]
18. Li, H.; Li, Y.; Tang, Z.; Xu, L.; Zhao, Z. Numerical simulation and analysis of vibration screening based on EDEM. *Trans. Chin. Soc. Agric. Eng.* **2011**, *27*, 117–121. (In Chinese with English Abstract)
19. Li, X.P.; Zhang, W.T.; Xu, S.D.; Ma, F.L.; Du, Z.; Ma, Y.D.; Liu, J. Calibration of collision recovery coefficient of corn seeds based on high-speed photography and sound waveform analysis. *Agriculture* **2023**, *13*, 1677. [CrossRef]
20. Wojtkowski, M.; Pecen, J.; Horabik, J.; Molenda, M. Rapeseed impact against a flat surface: Physical testing and DEM simulation with two contact models. *Powder Technol.* **2010**, *198*, 61–68. [CrossRef]
21. Li, J.; Su, N.; Hu, J.; Yang, Z.; Sheng, L.; Zhang, X. Numerical analysis and experiment of abrasive flow machining microhole structure based on CFD-DEM coupling. *Trans. Chin. Soc. Agric. Eng.* **2018**, *34*, 80–88. (In Chinese with English Abstract)
22. Chen, Y.; Geng, R.; Ma, L. Design and selection of fluid viscous devices for shock control of bridges. *China Civ. Eng. J.* **2007**, *40*, 55–61. (In Chinese with English Abstract)
23. Elsinawi, A.H.; Jhemi, A.; Alhamaydeh, M. Adaptive seismic isolation of structures using MR-fluid dampers. In Proceedings of the 2013 5th International Conference on Modeling, Simulation and Applied Optimization, ICMSAO 2013, Hammamet, Tunisia, 28–30 April 2013.
24. Jia, J.; Shen, X.; Du, J.; Wang, Y.; Hua, H. Design and experimental research on fluid viscous dampers. *Chin. J. Mech. Eng.* **2008**, *44*, 194–198. (In Chinese with English Abstract) [CrossRef]
25. Park, B.; Lee, Y.; Park, M.; Ju, Y.K. Vibration control of a structure by a tuned liquid column damper with embossments. *Eng. Struct.* **2018**, *168*, 290–299. [CrossRef]
26. Zhang, C.-H.; Wang, Y.; Du, J.-Y.; Wen, Z.-D. Effect of power-law fluid damping in a shock isolation system. *J. Ship Mech.* **2015**, *19*, 975–981. (In Chinese with English Abstract)
27. Ma, J.; Liu, D.; Chen, X. Experimental study of oblique impact between dry spheres and liquid layers. *Phys. Rev. E Stat. Nonlinear Soft Matter Phys.* **2013**, *88*, 033018. [CrossRef] [PubMed]
28. Ma, J.; Liu, D.; Chen, Z.; Chen, X. Agglomeration characteristics during fluidized bed combustion of salty wastewater. *Powder Technol.* **2014**, *253*, 537–547. [CrossRef]
29. Lee, K.Y.; Mooney, D.J. Alginate: Properties and biomedical applications. *Prog. Polym. Sci.* **2012**, *37*, 106–126. [CrossRef]
30. Zhu, J.; Marchant, R.E. Design properties of hydrogel tissue-engineering scaffolds. *Expert Rev. Med. Devices* **2011**, *8*, 607–626. [CrossRef]
31. Li, W.C.; Qiao, K.; Zheng, Y.D.; Yan, Y.; Xie, Y.J.; Liu, Y.; Ren, H.M. Preparation, mechanical properties, fatigue and tribological behavior of duble crosslinked high strength hydrogel. *J. Mech. Behav. Biomed. Mater.* **2022**, *126*, 105009. [CrossRef] [PubMed]
32. Ling, L.; Wu, G.W.; Wen, C.K.; Xiao, Y.J.; Fu, W.Q.; Dong, J.J.; Ding, J.H.; Meng, Z.J.; Yan, B.X. Influence of speed measurement method on performance of an electric-drive maize precision planter. *Biosyst. Engineering.* **2024**, *238*, 175–187. [CrossRef]
33. Wittmer, J.P.; Xu, H.; Benzerara, O.; Baschnagel, J. Fluctuation-dissipation relation between shear stress relaxation modulus and shear stress autocorrelation function revisited. *Mol. Phys.* **2015**, *113*, 2881–2893. [CrossRef]
34. Li, Y.; Yang, L.; Zhang, D.; Cui, T.; He, X.; Du, Z.; Wang, D. Performance analysis and structure optimization of the maize precision metering device with air suction at high-speed condition. *Trans. Chin. Soc. Agric. Eng.* **2022**, *38*, 1–11.

agriculture

Article

Design and Experiment of the Profiling Header of River Dike Mower

Mingsheng Li [1,*], Yulin Yan [1], Lin Tian [2], Xingzheng Chen [1] and Fanyi Liu [1]

[1] College of Engineering and Technology, Southwest University, Chongqing 400715, China; yanyulin2023@email.swu.edu.cn (Y.Y.); chenxzh@swu.edu.cn (X.C.); liufanyi2018@swu.edu.cn (F.L.)
[2] College of Engineering, China Agricultural University, Beijing 100091, China; tianlin2022@126.com
* Correspondence: lms040709321@swu.edu.cn

Abstract: Drawing upon advancements in profiling technology, this paper presents an innovative lateral profiling mechanism for the header to improve mowing efficiency and the ability to adapt to terrain for river dike mowers. It delves into the imitation principle and forced situations. Furthermore, a novel lawn protection boot design has been introduced, capable of adjusting mowing heights with swift transitions. The structural integrity of this boot has been optimized through rigorous finite element analysis. Meanwhile, the rolling shaft and cutter have been carefully selected and designed, with a mechanical model of the cutter established to examine its motion and force characteristics. In addition, hydraulic circuits tailored to fulfill the required functions of the header have been devised, and key hydraulic components have been appropriately selected. Key components are subjected to finite element analysis by using ANSYS to verify and optimize their structural strength. Prototype testing and field trials are subsequently conducted, revealing that the mower can achieve a mowing speed of 0.85 m/s on flat ground and a 25-degree slope, thereby fulfilling the design requirements for mowing speed. The imitation mechanism adapts to different embankment terrains. Notably, the lawn protection boots offer adjustable mowing heights of 10.4 cm, 12 cm, and 14 cm, respectively, with a height adjustment range of approximately 2 cm for each position, meeting the requirement for adjusting mowing heights. In addition, the transition time between different positions of the lawn protection boots is less than 5 min, achieving rapid switching and operational efficiency. Furthermore, a mowing uniformity test is conducted by using a header equipped with profiling functionality. The results reveal that the mowing effect of the profiling header meets design requirements, demonstrating its effectiveness and reliability in agricultural applications.

Keywords: mower; profiling header; lawn protection boot; simulation analysis; experiment

Citation: Li, M.; Yan, Y.; Tian, L.; Chen, X.; Liu, F. Design and Experiment of the Profiling Header of River Dike Mower. *Agriculture* **2024**, *14*, 1188. https://doi.org/10.3390/agriculture14071188

Academic Editor: Xiaojun Gao

Received: 20 June 2024
Revised: 15 July 2024
Accepted: 16 July 2024
Published: 19 July 2024

1. Introduction

China, renowned for its extensive river networks, attributes substantial investments dedicated to river flood control annually, particularly in the field of embankment construction and reinforcement [1,2]. The total length of these constructed embankments in China stands at an impressive 413,676 km, among which a significant portion of these embankments, specifically those rated at Grade 5 or higher, spans an outstanding 275,495 km. Thus, the rigorous management and construction of these river embankments play a pivotal role in effectively mitigating the risks of flooding disasters and embankment breaches, therefore fostering national economic development and safeguarding lives and properties of the populace [3].

Notably, research has demonstrated the crucial role of vegetation in stabilizing river embankments [4]. To mitigate the deleterious effects of rainfall erosion and subsequent damage to embankments, grass-covering techniques have traditionally been the frontline approach for embankment preservation [5–8]. Nevertheless, if vegetation becomes overly tall and dense, it can hinder the timely identification of seepage, rodent burrows, ant nests, and breaches, thereby posing serious threats to flood prevention efforts. Consequently, the implementation of regular maintenance and lawn care practices has been paramount [9]. However, the existence of challenges such as excessive embankment slopes, loose soil, and uneven terrain has posed significant difficulties in maintaining the integrity of these vegetable barriers [10].

Therefore, the embankment mower, a specialized type of mower characterized by large header dimensions, long cutting width, and high mowing efficiency, can serve as a useful tool for embankment maintenance. Although the range of available mowers is diverse, few have met the stringent requirements for embankment mowing with imitation capabilities [11–14]. Furthermore, existing mowers have often struggled with issues such as sluggish height adjustments, cumbersome operational processes, inconvenient replacement of lawn protection boots, and damage to embankment lawns during mowing operations [15–18]. As for the application of imitation technology in mowers, researchers are mainly attracted by two distinct types: mechanical structure imitation and hydraulic imitation. On the one hand, mechanical structure imitation primarily relies on sliding panels, mechanical lifting devices, and special mechanical structures to achieve imitation. For one thing, the imitation of sliding panels falls under the category of passive imitation, employing a straightforward structure that adjusts the distance between the cutter and the ground by swapping out sliding panels. As an example, the R0212 mower, designed by the German company Neumeister, incorporates sleeve wrenches and threaded rods to precisely adjust the height of the cutting unit, thereby fulfilling the imitation goal of altering the distance from the sliding panel [19]. For another, the imitation through lifting devices involves elevating the components of the mower to achieve the desired imitation. Many of these products rely on the balancing effect of springs to achieve floating imitation. Wu et al., for instance, have developed a small, self-propelled mower trimmer that utilizes the deformation of profiling springs to achieve floating imitation on uneven terrain [20]. Hydraulic profiling, on the other hand, represents an active profiling method, leveraging hydraulic mechanisms to lift the cutting device and achieve profiling.

The present paper is dedicated to addressing the challenges posed by embankment mowers, which suffer from inadequacy in profiling capability, incompetence in adjusting mowing heights, incapability in lawn protection boot switching, and dilemmas in embankment lawns. The study, with a focus on enhancing operational efficiency, proposes an innovative imitation solution alongside an optimized design for lawn protection boots to ensure the cleanliness of embankment lawns, elevate the imitation capabilities of river dike mowers, and safeguard embankment safety. Thorough optimization has been conducted on various components, including the header bracket, cutters, and transmission components, leading to the creation of a profiling-capable header and the implementation of adjustable mowing heights. Therefore, these advancements culminate in the successful prototyping of the entire machine. To further validate the effectiveness of the novel design, extensive field experiments have been conducted on the embankments of the Yellow River. These tests concentrate on evaluating the imitation capabilities, precision of mowing height adjustment, and protection of the lawn by the optimized header. The ultimate objective of this research is to provide both technical support and a robust theoretical framework that can be leveraged to further advance the field of mower imitation technology.

2. Materials and Methods

2.1. Machine Structure and Operational Fundamentals

The river dike mower is composed primarily of four distinct components: the walking chassis, the header support, the profiling mechanism, and the mowing device, as depicted in Figure 1. Firstly, the tracked chassis serves as the foundation, supporting the weight of the entire machine and facilitating the movement of the header for grass cutting. Secondly, the header support, in conjunction with the lifting hydraulic cylinder, allows for precise adjustment of the header's elevation, ensuring optimal cutting performance. Thirdly, the profiling capabilities of the mower are categorized into lateral and longitudinal profiling.

Noticeably, lateral profiling is primarily achieved through a meticulously designed mechanical structure, while longitudinal profiling is facilitated by the controlled movement of the lifting hydraulic cylinder. Fourthly, the mowing device, situated on the header, is the heart of the mowing process, with internal blades rotating at high speeds to shear the grass through centrifugal force. In this study, we also focus on the development and application of "Lawn Protection Boots," a specialized footwear designed to minimize damage to grass surfaces during maintenance and other activities on turf areas. Finally, the lawn protection boots serve as a crucial interface between the mower's header and the lawn surface, which not only protects the integrity of the lawn but also allows for fine-tuning of the mowing height. The engine chosen for this mower is a water-cooled, four-stroke, three-cylinder model known as the 3TNV88, manufactured by Yanmar Engine (Qingdao, China) Co. Table 1 summarizes the key structural parameters that govern the performance and functionality of this agricultural machinery.

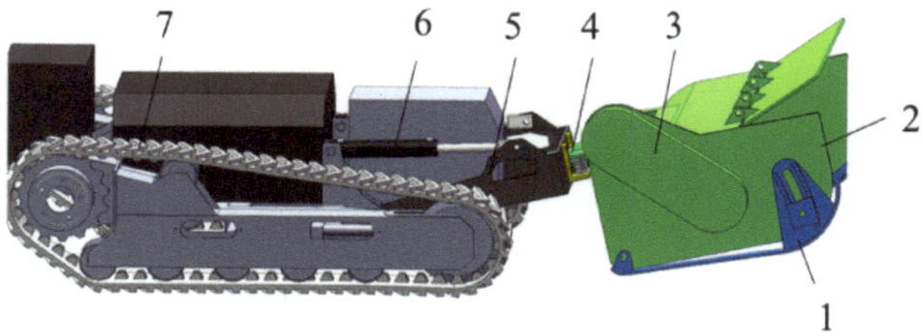

Figure 1. The overall structure of the river dike mower: 1. Lawn protection boots; 2. Mowing device; 3. Hydraulic motor; 4. Profiling mechanism; 5. Header support; 6. Lifting hydraulic cylinder; 7. Crawler chassis.

Table 1. Key structural parameters of river dike mowers.

Parameter Name	Parameter Value
Engine type	Yanmar 3TNV88
Engine power (kW)	20
Dimensions length/width/height (mm)	2900/1570/1000
Mass (kg)	880
Number of cutter circumferences (blade)	4
Number of cutter axes (blade)	9
Mowing range (mm)	1300
Hydraulic motor type	BMR50

2.2. *Design of Key Components*

2.2.1. Design of Lifting Mechanism

The lifting mechanism of the header incorporates three components: the base, the hydraulic cylinder, and the header support. The function of this mechanism lies in its ability to efficiently raise and lower the header. Notably, during the lifting and lowering process, the mower does not perform mowing operations. During mowing, the hydraulic cylinder incorporates a strategic amount of float, thus enabling longitudinal profiling and enhancing mowing precision. This structure is clearly depicted in Figure 2a.

Let us consider the geometric relationships within this mechanism by assuming that the angle between the header support AC and the horizontal direction is a, and the angle between the hydraulic cylinder BC and the vertical direction is a_2. Furthermore, when the piston rod of the hydraulic cylinder is fully retracted, the length of BC is L_{BC0}. Additionally, let L_{BC0} be the length at which the piston rod starts to extend from the inner end of the hydraulic cylinder. Based on these geometric relationships, a comprehensive geometric model is constructed, as illustrated in Figure 2b.

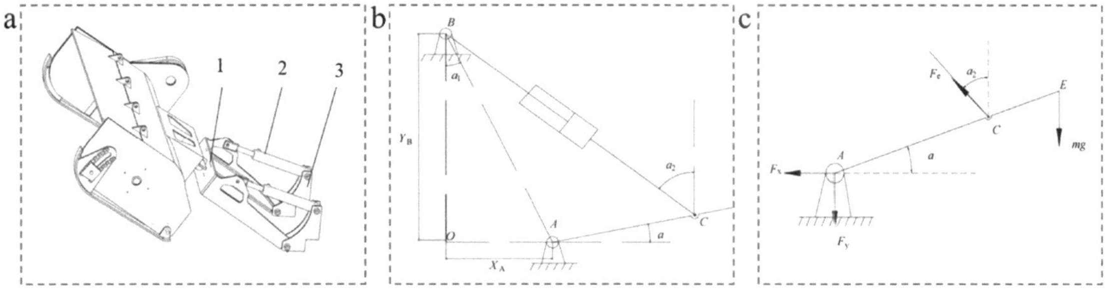

Figure 2. Design of the header lifting mechanism: (**a**) Structure configuration of the header lifting mechanism: 1. Header support, 2. Lifting hydraulic cylinder, 3. Base; (**b**) Sketch map of the header lifting mechanism; (**c**) Force analysis diagram of the header support.

In triangle ABC, there exists the following:

$$\begin{cases}(L_{BC0} + x) \cdot \sin \alpha_2 = L_{AB} \cdot \sin \alpha_1 + L_{AC} \cdot \cos \alpha \\ L_{AB} \cdot \cos \alpha_1 = (L_{BC0} + x) \cdot \cos \alpha_2 + L_{AC} \cdot \sin \alpha\end{cases} \quad (1)$$

The following can be derived from Formula (2):

$$\alpha = \arcsin \frac{(L_{BC0} + x)^2 - L_{AB}{}^2 - L_{AC}{}^2}{2L_{AB} \cdot L_{AC}} + \alpha_1 = f(x) \quad (2)$$

in the following formula: $L_{AB} = \sqrt{x_A^2 + y_B^2}$, $\alpha_1 = \arctan\frac{x_B}{y_B}$.

L_{AB} is the distance between the upper and lower articulation points on the support, L_{AC} is the distance between the lower articulation point on the base and the articulation point on the header support, α_2 is the angle between the line connecting the upper articulation point on the base as well as the lower articulation point on the hydraulic cylinder and the vertical direction, and $f(x)$ is the function of the hydraulic cylinder's extension displacement x.

The force situation of the header support in the vertical direction is elaborated in Figure 2c. The force and moment equilibrium equations of the header support are as follows:

$$\begin{cases}F_e \cdot \cos \alpha_2 - mg - F_y = 0 \\ F_e \cdot \sin \alpha_2 - F_x = 0 \\ F_e \cdot L_{AC} \cdot \cos(\alpha_2 - \alpha) - L_{AD} \cdot mg \cdot \cos \alpha = 0\end{cases} \quad (3)$$

In the formula, F_x is the horizontal force at the articulation point between the header support and the base, F_y is the vertical force at the articulation point, m is the mass of the header, F_e is the force exerted by the hydraulic cylinder, L_{AC} is the distance between the articulation point of the header support and the base and the articulation point on the hydraulic cylinder, and L_{AD} is the distance between the articulation point of the header support and the base and the connection point between the support and the header.

2.2.2. Design of Mowing Device

The Y-type swing blade stands out due to its diverse and notable characteristics, which includes its exceptional impact resistance, double-sided blade, wear resistance, formidable crushing ability, and uniform force distribution. This blade type has found widespread application in a range of mowers. Given its distinguished performance profile, the Y-type cutter, depicted in Figure 3a, has been selected as the primary blade for this study.

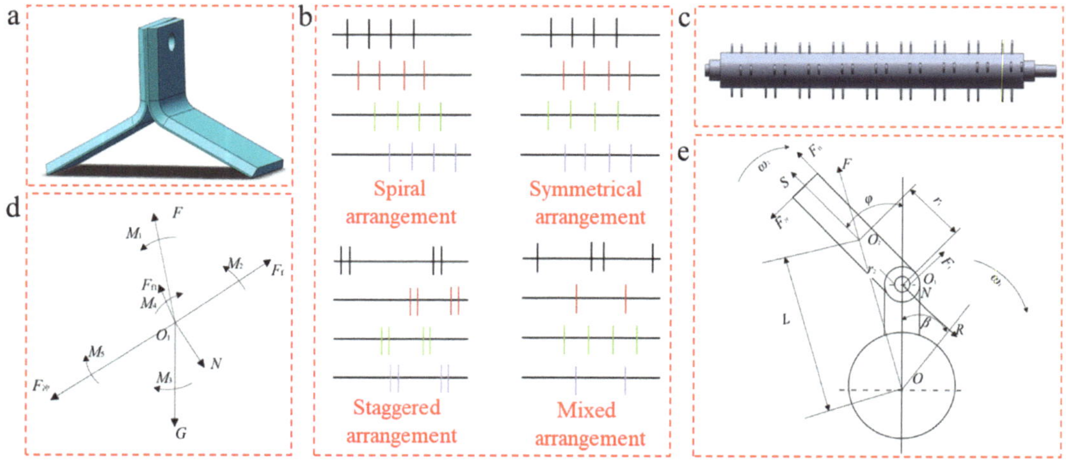

Figure 3. Design of the mowing device: (**a**) Schematic diagram of cutter structure; (**b**) Arrangement of cutters; (**c**) Structure of the rollers; (**d**) Torque diagram of the cutter; (**e**) Force diagram of the cutter.

When configuring the blade arrangement on the roller, it is imperative to ensure that no cuts are missed during the mowing process and that there is no potential for blockages or interference among the blades. The typical blade arrangements include spiral, symmetrical, staggered, and symmetrical staggered configurations, as exemplified in Figure 3b. Among these four arrangements, the symmetrical layout offers uniformity in blade spacing, resulting in even stress distribution across the roller during rotation. Additionally, this arrangement simplifies the assembly process. Consequently, the roller of this mower has been designed with a symmetrical blade arrangement.

To determine the optimal number of cutters, it is of great significance to take careful consideration of the roller's length and arrangement. An insufficient number of cutters can lead to missed cuts, compromising mowing efficiency. Conversely, an excess of blades can increase the torque and startup power of the roller, potentially resulting in blockages and entanglement issues. Hence, the number of cutters needs to be designed rationally [21]. Typically, the number of cutters is calculated using the following formula:

$$n = L_T \delta \tag{4}$$

In the formula, n is the total number of cutters (piece), δ is the cutter density (piece/mm), straight blades are generally taken as 0.05 to 0.07 piece/mm, T-shaped blades are generally taken as 0.01 piece/mm, and L_T is the distribution length of cutters on the roller, mm.

By substituting L_T = 1400 mm into the equation, the number of cutters can be calculated as n = 36 pieces. The result is exhibited in Figure 3c.

The roller speed is a crucial parameter in the performance of the mower. Once the roller speed attains an optimal level, the cutters can execute an efficient cutting process [22]. Furthermore, the roller speed is intricately linked to the determination of the appropriate speed for the power mechanism, thereby emphasizing the necessity for a judicious choice of the roller speed [23]. Generally, the roller speed can be accurately determined through the application of Equation (5).

$$n_1 = \frac{30(v_a + v_m)}{\pi R_s} \tag{5}$$

In the formula, v_a is the cutting speed of the blade (m/s), v_m is the operating speed of the mower (m/s), and R_s is the rotational radius of the cutter root edge (m).

Research has embodied that to achieve effective cutting, the minimum linear velocity at the cutter root edge should not be less than 10 m/s, while the cutting speed of the cutter edge generally ranges from 20 to 50 m/s [23]. Take v_a = 25 m/s as an example. In this case, the forward speed of the mower is taken as v_m = 0.8 m/s, and the rotational radius at the root is R_s = 120 mm. Substituting these values into Equation (5), we obtain n_1 = 2054 r/min. Thus, the speed of the shaft is taken as 2100 r/min.

During the operational phrase, a force analysis of the cutter's motion state has been conducted, as illustrated in Figure 3d,e. The forces acting on the cutter mainly consist of the gravitational force G of the cutter, the centripetal force S generated by the rotation of the cutter around the pin axis, the centripetal force F generated by the rotation of the cutter around the central axis, the normal pressure force N exerted on the connection point between the pin axis and the cutter, as well as the impact force F_t exerted by vegetation and soil on the front of the cutter during collision cutting and the frictional force F_{f1} between vegetation and the cutter. The formulas for quantifying the impact force and frictional force, particularly when contaminants such as weeds and soil collide with and rub against the cutter, are as follows:

$$F_t t = M\Delta v \tag{6}$$

$$F_1 = \mu F_t \tag{7}$$

In the formula, M is the weight of impurities at the time of colliding (kg), Δv is the linear velocity difference at the time of colliding (m/s), t is the colliding time (s), and μ is the coefficient of friction between the cutter and the impurities.

At the rotational center point O_1 of the cutter, the following exists:

$$m_1 r_1 R \omega_1^2 \sin\phi + F_f r_2 - m_1 g \sin(\omega_1 t + \phi) - F_t t - \frac{1}{2} F_{f_1} \beta = 0 \tag{8}$$

The critical angle at which the cutter is relatively stationary to the center O_1 of the pinhole is $\varphi \approx 1.4°$. By substituting the various parameters of the cutter into Equation (8) we can accurately determine the precise pressure applied by the pin on the cutter, along with other pertinent values, at any specific moment in time.

2.2.3. Design of the Profiling Mechanism

The lateral profiling mechanism and its primary operational process are outlined in Figure 4a–c. When both ends of the header encounter minor bumps or pits in the terrain, the imbalance between the header and the ground surface triggers a rotational movement around the pivot axis. This rotational stability is maintained by a pair of springs, one in compression and the other in tension, jointly maintaining the overall equilibrium of the header. Specifically, if the left end of the header encounters a minor protrusion as the pivot axis rotates, the left spring assumes a tensile state, while the right spring compresses. Similarly, when the right end of the header encounters a bump or depression, the profiling mechanism responds with the corresponding lateral adjustment. This adaptive profiling is achieved through the concurrent stretching and compression of the springs at both ends, along with the rotational movement of the pivot axis, effectively delivering lateral profiling functionality at both extremities of the header.

The longitudinal profiling mechanism and its primary operational process are visualized in Figure 4d–f. As the mower traverses terrain with significant elevation, the hydraulic cylinder initiates a retraction action. Conversely, when the mower encounters a deep depression in the soil, the hydraulic cylinder extends outward, thereby enabling profiling functionality in the longitudinal direction. This adaptive behavior ensures that the mower maintains optimal contact with the ground, regardless of the terrain's irregularities.

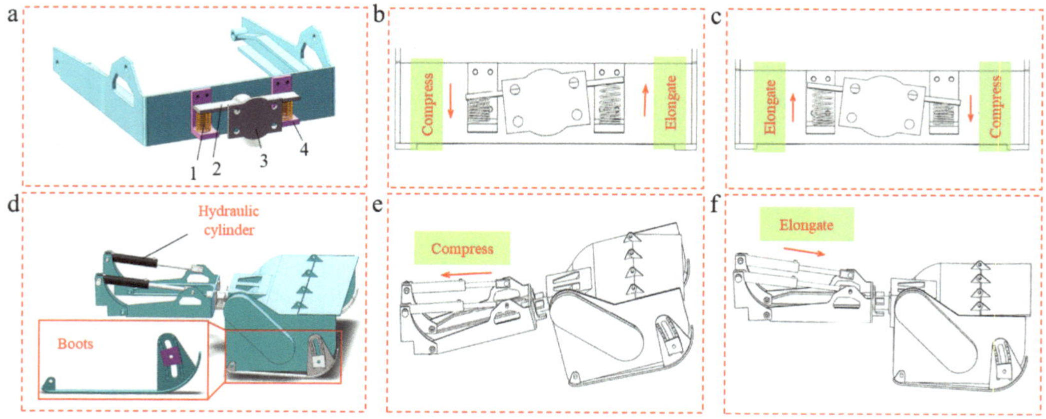

Figure 4. Design of the profiling mechanism: (**a**) Schematic diagram of the transverse profiling mechanism: 1. Spring, 2. Upper guard plate, 3. Header connecting plate, 4. Lower support; (**b**,**c**) Operational status of the transverse profiling mechanism; (**d**) Schematic diagram of longitudinal profiling mechanism; (**e**,**f**) Operational status of the longitudinal profiling mechanism.

2.2.4. Design of Hydraulic System

The design principle of hydraulic system for controlling the header lifting is illustrated in Figure 5. Initially, the electromagnetic valves 5 and 6 are set in their appropriate positions. As the header encounters a raised road surface, the hydraulic cylinder piston rod retracts. This retraction triggers a sequence of events: the oil flows through the appropriate side of electromagnetic valve 5, then passes through check valve 8, and ultimately returns to the oil tank through the hydraulic cylinder. Conversely, upon encountering a depressed road surface, the piston rod extends. At this point, oil pumped from the hydraulic cylinder passes through throttle valve 7, then flows through the right position of the electromagnetic valve 6, and finally returns to the oil tank. These two circuits are pivotal in achieving the longitudinal profiling functionality required for traversing embankment terrain. When it becomes necessary to lift the header, electromagnetic valve 5 is actuated to the left position. This allows the oil to flow from the hydraulic pump 2, through electromagnetic valve 5, then through check valve 8, and into the hydraulic cylinder. Once the oil passes through the cylinder, it returns to the oil tank. Conversely, for lowering the header, triggered by the activation of the stroke switch, electromagnetic valve 6 is shifted to its left position, facilitating their respective limit positions. This ensures that the lifting or lowering of the header is precisely controlled and halted at the desired heights.

The principal hydraulic components of the lawn mower comprise hydraulic motors dedicated to propulsion and cutter rotation, a hydraulic pump, cylinders, a reservoir, and valves, of which the meticulous calculation and selection of these components are paramount to ensure optimal performance [24].

When it comes to hydraulic motors, the mower features both a traveling hydraulic motor and a blade rotation hydraulic motor. While the selection and calculation for the motor that rotates the cutter have already been addressed in a prior section, the focus now shifts to the selection and calculation of the propulsion hydraulic motor [25,26].

The torque required for the crawler drive is precisely known to be 1963.5 N·m, with a reduction ratio of 6 for the crawler section. As a result, the output torque of the hydraulic motor only needs to exceed M_k = 327.25 N·m to fulfill the power requirements. The displacement V of the traveling hydraulic motor can typically be determined using the following standard formula:

$$V = \frac{2\pi M_k}{p\eta} \tag{9}$$

In the formula, p is the working pressure of the mower's hydraulic system, and we assume it to be 20 MPa, and η is the volumetric efficiency of the high-speed hydraulic motor and we adopt a value of 0.9.

The maximum speed of the hydraulic motor is generally calculated by the following formula:

$$v_{max} = 2\pi r_d n_{max} \cdot 60 \tag{10}$$

In the formula, v_{max} is the maximum speed of the mower, and we assume it to be 12 km/h, n_{max} is the rated maximum speed of the hydraulic motor (r/min), and r_d is the selected drive radius of the crawler, and we set it to be 160 mm.

Put each value into the calculation to obtain $V \approx 120$ mL/r, and $n_{max} \approx 200$ r/min. Both the selected hydraulic motor displacements should meet 120 mL/r; and the maximum rated speed is 200 r/min. Based on various parameters of the mower's movement, the hydraulic motor model BMR50 is selected as the traveling hydraulic motor.

Furthermore, other types of hydraulic components, including pumps, tanks, and valve components, are selected according to the required functionalities and parameters. This selection is made according to comprehensive operational analysis, considering both the practicality and cost-effectiveness of the various options. The detailed breakdown of the selection process for the various hydraulic components can be found in Table 2.

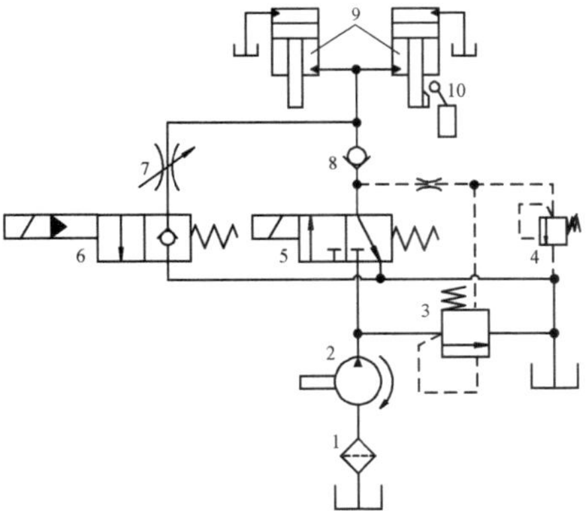

Figure 5. Design of hydraulic system for header elevation control: 1. Oil filter; 2. Positive displacement hydraulic pump; 3. Differential pressure relief valve; 4. Safety valve; 5. Two-position three-way solenoid directional valve; 6. Two-position two-way solenoid directional valve; 7. Throttle valve; 8. Check valve; 9. Header lifting hydraulic cylinder; 10. Stroke switch.

Table 2. Main hydraulic components selection of mower.

Hydraulic Components	Parameters
The cutter rotation hydraulic motor	M4MF37-32 series axial piston hydraulic motor
The walking hydraulic motor	BMR50 series hydraulic motor
Hydraulic pump	Closed-loop triple gear pump: M4PV28-28S228AR3BJR-557 + 2 HMPZA12RTXGSGAPMM1
Check valve	DIF-L20H1
Oil tank	250L

2.3. *Experiment Scheme*

2.3.1. Simulation Experiment

As the lawn protection boot serves as the principal component responsible for transmitting forces between the header and the ground, this present study employs the software

ANSYS2020 Workbench to perform a rigorous static analysis and optimization of the key structures of the boot. The mesh size is set to 2.437 mm. Furthermore, the material properties utilized in the study are outlined in Table 3.

Table 3. Material property settings.

Type of Material	Density P/kg·m^{-3}	Modulus of Elasticity E/MPa	Poisson's Ratio	Strength of Yield σ_s/MPa
Q235	7850	200,000	0.3	235

To access the dynamic behavior of the mower, which is the pivotal working unit, motion simulation analysis is conducted using the software ADAMS2020 [27]. Specifically, the roller shaft is configured with a rotational constraint, simulating a rotational speed of 2600 rpm, equivalent to the nominal operating speed of the mower. The simulation duration is set to 0.01 s, with a total of 5000 steps, ensuring a detailed and accurate representation of the mower's operational dynamics.

2.3.2. Field Experiment

The experiment site is located on the river dike of the Yellow River, specifically in Mengjin County, Luoyang City, Henan Province, China, positioned at 34 degrees north latitude and 112 degrees east longitude. The experiment was performed beneath a clear sky. The river dike, an artificially constructed embankment, exhibits a slope ranging from 10 to 30 degrees and boasts a width of 3 to 5 m. The grass species within this locale are mainly composed of ryegrass, carpet grass, bermudagrass, and step grass, maintaining an average height of 23.2 cm and an average diameter of 40 mm.

- Test contents: The specific experiments include verifying the mower's mowing speed [28], validating mowing productivity [29], testing the uniformity of mowing with and without profiling conditions, and analyzing the effect of the lawn protection boot at three different height adjustment positions during mowing.
- Test equipment: The key component is the prototype mower, other vital equipment includes a smartphone, a tape measure for accurate measurements (DELI with a range of 10 m and precision markings at the millimeter level), an angle gauge for assessing slopes (SOUTHERN JZC-B2 and measuring accuracy 1 Degree), a stopwatch for timing (CASIO HS-70W accurate to 1/100 s), and a marker bar for designation.
- Test methods: For the conduct of the experiment, a five-point sampling technique, also recognized as the diagonal method, is employed. This technique involves the selection of a uniformly spaced grassy area along the bank and the subsequent division of this area into five small rectangular sections arranged diagonally. The centers of four sample points are positioned equidistant from the center point on each diagonal. Each designated sample point encompasses an area of 0.25 square meters. The experiments are conducted on flat terrain and on slopes of 25 degrees, and the duration required for mowing each sample point is meticulously recorded [30,31]. The following is about validation test design:

(1) The experimental design for the mowing speed and productivity:

Mowing speed and productivity are fundamental performance indicators of the mower during normal operational conditions. Only when these two parameters meet the design requirements can the mower be considered as compliant with the required performance criteria. Consequently, it is necessary to conduct experimental verification and analysis on both parameters.

(2) The experimental design for a mowing height test in three gears of a lawn protection boots:

To ascertain the optimal mowing height for each gear, three distinct sections of embankment grassland are selected, ensuring similar terrain conditions and uniform grass height density across all test sites. The lawn protection boots are installed at gears one, two, and three, respectively, for mowing tests. After every two meters of mowing, the remaining stubble height is measured five times to ensure accurate data collection. The collected data are then organized into a comparative chart, allowing for a visual analysis of the remaining stubble heights across the three lawn protection boot positions.

(3) The experimental design for comparing the uniformity of mowing with and without contour-following conditions on the header:

Two embankment grasslands, each 10 m in length and exhibiting slight unevenness, are selected as test sites. Both grasslands possess identical grass density, height, and other relevant parameters, facilitating a fair comparison between the two conditions. Initially, the mower without the profiling mechanism is tested on Grassland One. Following the mowing process, the stubble height is measured at 2-m intervals, totaling 5 sampling points. The same procedure is repeated with a mower equipped with the profiling mechanism on Grassland Two, with the stubble height at 2-m intervals to obtain another set of data.

3. Results and Discussion

3.1. Analysis of Simulation Results

The header exhibits a gravity load of 160 kg, bearing a lawn protection boot force of 1600 N. Given the structural characteristics of the lawn protection boots, one end has borne 1000 N while the other has borne 600 N, as manifest in Figure 6a. As illustrated in Figure 6b,c, the maximum stress on the sliding boots has reached 137.57 MPa, accompanied by a maximum deformation of 0.35 mm, indicating notable regions of stress concentration. To alleviate stress concentration and streamline the design, a thorough optimization of the lawn protection boot's structure is required. A range of optimization methods, including shape optimization, structural topology optimization, and dimension optimization, are available. For this specific optimization, dimension optimization and shape optimization are chosen. The optimization process involves making strategic modifications to the design. Small blocks are excised from the locking points of the lawn protection boots and replaced with three pairs of locking holes. This approach reduces the weight and overall dimensions of the lawn protection boots, thereby simplifying the installation process. The simulation analysis of the optimized lawn protection boot structure reveals intriguing results. As presented in Figure 6d–i, the highest stress points consistently occur at the juncture between the bottom support plate and the base plate of the boots across gears one, two, and three. In gears one and two, the maximum deformation occurs at the apex of the upper support plate, while in gear three, the point of maximum deformation shifts to the highest point of the boot base. As the gear levels progress, both the maximum stress and total deformation exhibit a decreasing trend. Across all three gear levels, the peak stress recorded is 44.52 MPa. This value falls well within the yield strength of the material Q235 used for the lawn protection boots, which is rated at 235 MPa. Thus, the optimized design meets the required standards, ensuring durability and reliability in agricultural applications.

After enhancing the design of the lawn protection boots, a substantial reduction in stress and deformation is observed. According to the comparison of stress and deformation cloud maps before and after the improvements [32], the maximum stress exerted on the boots decreases significantly, from 135.57 MPa to 44.52 MPa, representing a notable drop of 91.05 MPa. Similarly, the maximum deformation also decreases markedly, dropping from 0.35 mm to 0.11 mm, a momentous reduction of 0.24 mm. These findings clearly indicate that the stress concentration within the lawn protection boots has been significantly mitigated, highlighting the effectiveness of the enhancements made.

The kinematic simulation model of the mower device is depicted in Figure 7a. As evident in Figure 7b–d, the mower undergoes a start-up phase ranging from 0 to 0.005 s, during which there are significant fluctuations in the blade speed. Subsequently, the blade speed gradually stabilizes and eventually converges to approximately 2600 rpm, closely matching the roller speed. Similarly, the highest and lowest points of the blade exhibit comparable fluctuations. Although the trajectory of the blade tip initially appears irregular, it ultimately stabilizes into a circular motion centered around the roll axis. This observation suggests that the cutters on the roller are arranged and distributed in a uniform and reasonable manner, fulfilling the requirements for high-speed mower operation.

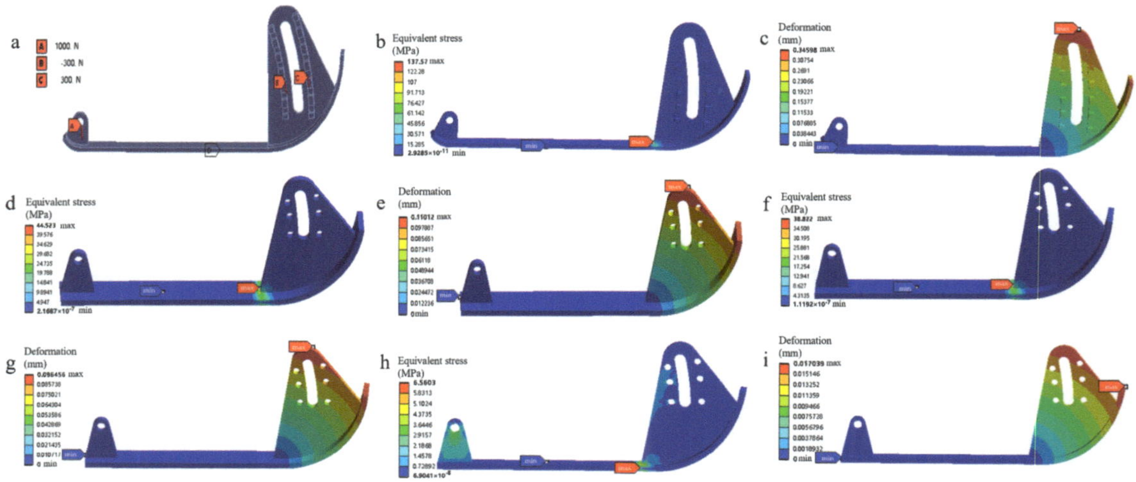

Figure 6. Static analysis of the lawn protection boots: (**a**) Load and constraint settings of the lawn protection boot; (**b**,**c**) Equivalent stress and deformation contour plots of the lawn protection boot; (**d–i**) Equivalent stress and deformation contour plots of the optimized lawn protection boot in gears one through three.

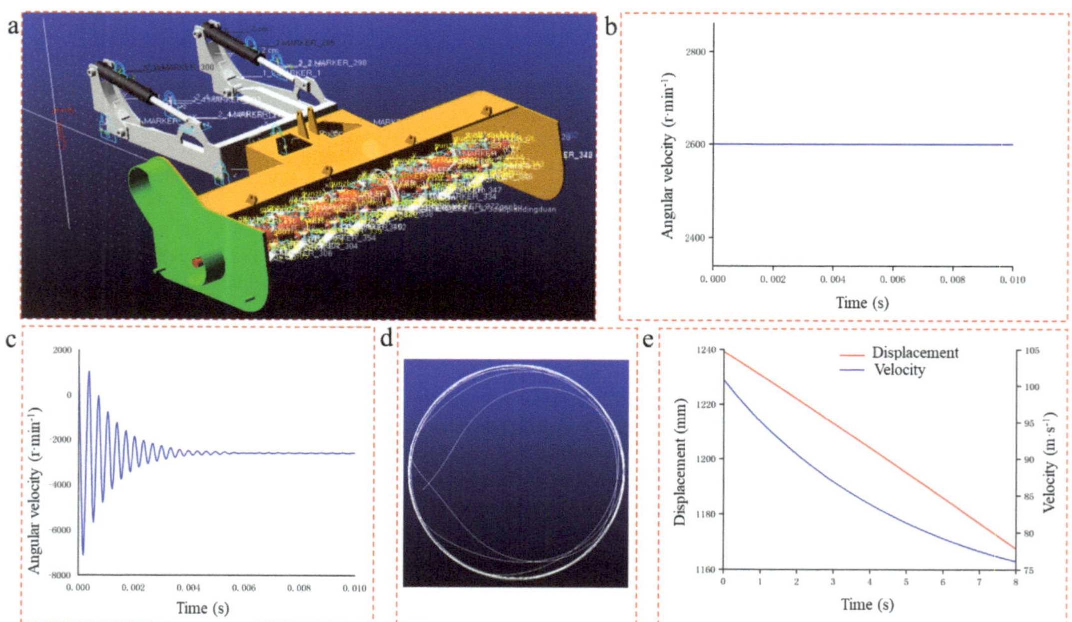

Figure 7. Kinematic simulation analysis of the header: (**a**) Results of simulation model construction; (**b**) Variation of roller shaft angular velocity; (**c**) Variation of the cutter center angular velocity; (**d**) Motion trajectory curve of blade tip; (**e**) Velocity and displacement curves at the centroid of the header.

To further validate the rationality of the mower lifting mechanism design, a simulation analysis of the mower lifting mechanism can be conducted [33]. This simulation thoroughly examines the motion characteristics of the header during the lifting process, thereby verifying the functionality of the lifting mechanism. The simulation modeling process involves importing the model by adding constraints, kinematic subs, and drivers. Drawing upon the findings of previous conclusion, which indicate that the hydraulic cylinder stroke is 160 mm, and the lifting time should not exceed 10 s, we assume a lifting speed of 20 mm/s for the hydraulic cylinder every 20 mm of upward movement. Consequently, the simulation is executed for a duration of 8 s, divided into 200 discrete steps. By observing the lifting situation of the header, as presented in Figure 7e, we can deduce that during the lifting process, the maximum speed reached is 100.8 mm/s, while the minimum speed is 76.1 mm/s. This comprehensive simulation analysis confirms that the design and parameter selection of the header lifting mechanism align with the intended working conditions and satisfy the designated design requirements.

3.2. *Analysis of Field Experiment Results*

3.2.1. Mower Mowing Speed and Productivity Experiment Verification

Experimental principle: The operational speed of the mower is calculated according to the following formula, and the average of five measurements is taken at the end:

$$v = \frac{\sum L}{\sum t} \tag{11}$$

In the formula, v is the machine working speed (m/s), $\sum L$ is the total length of the area, which is determined by the experiment (m), and $\sum t$ is time spent (s).

Specific steps: Five 10-m sites are selected on flat ground with a 25° slope. Under normal mower operating conditions, the time taken for mowing is recorded at each sample point. Finally, the speed values of the five sample points on flat ground and the 25° slope are calculated using Formula (11), as presented in Table 4.

Table 4. Mower mowing speed on flat ground and a 25° slope.

Samples	1	2	3	4	5
Level mowing speed/m·s^{-1}	0.63	0.77	0.83	1	0.91
25° slope mowing speed/m·s^{-1}	0.67	0.83	1	1	0.83

The mowing productivity of a mower serves as a key parameter for assessing the effectiveness of its header design, which in turn serves as an indirect indicator of the success of the mower's profiling. The following outlines the methodology for verifying the mowing productivity of a mower:

Experimental principle of mowing productivity of mower:

$$p = \frac{A}{t} \tag{12}$$

In the formula, p is mowing productivity (m^2/s), A is Area of mowing (m^2), and t is mowing time (s).

Specific steps: Five plots of land are selected, including flat ground and featuring a 25° slope, as depicted in Figure 8d. Each plot measures 10 m in length and width, facilitating the conduct of precise mowing experiments. During these experiments, the width of the mower's cutting swath is measured, and the time taken for each mowing session is recorded. These data are then utilized to compute the mowing productivity by using the Formula (12).

The measured width of the mower's cutting swath is determined to be 1.3 m. By substituting these values into Formula (12), the mowing productivity on flat ground and the 25° slope is calculated, respectively, as presented in Table 5.

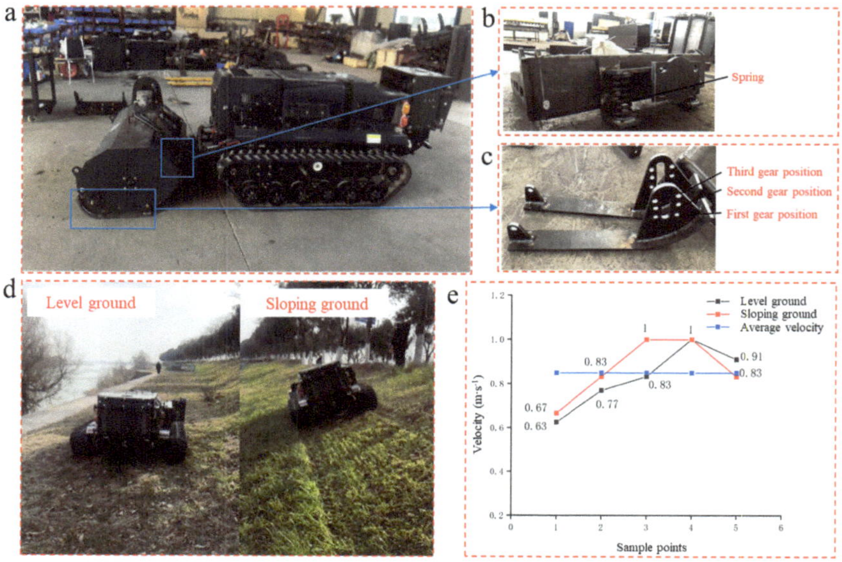

Figure 8. Machine prototype and field experiment: (**a**) Overall machine structural configuration; (**b**) Lateral profiling mechanism; (**c**) Lawn protection boot; (**d**) Flat and sloping ground tests; (**e**) Mowing speed analysis.

Table 5. Mowing productivity of the mower on flat ground and a 25° slope.

Samples	1	2	3	4	5
Flat mowing productivity/$m^2{\cdot}h^{-1}$	2925	3600	3900	4680	4255
25° slope mowing productivity/$m^2{\cdot}h^{-1}$	3120	4680	4680	4680	3900

The results of the processed prototype mower are presented in Figure 8a–c. Its operating efficiency is evaluated on both flat ground and a 25° slope. The experiment area and the corresponding results are depicted in Figure 8d,e. Notably, the average working speed of the mower on both flat ground and slopes is recorded as 0.85 m/s. This speed falls squarely within the designed mowing speed range of 0.27 m/s to 2.7 m/s, thus fulfilling the designated requirements.

A comparative graph of mowing speeds derives from speed curve graphs plotted for mowing on flat ground with a 25° slope. This is clearly manifested in Figure 8e. From the graph, it can be observed that under normal operating conditions, the mower exhibits nearly identical mowing speed on both flat and sloped terrain. Notably, the speed stabilizes at 0.85 m/s. This speed falls within the predefined mowing speed range of 1 km/h to 10 km/h, thus fulfilling the established design criteria.

The values presented in Table 5 reveal an average mowing productivity of 3872 m^2/h on flat ground and an even higher rate of 4212 m^2/h on a 25° slope. Notably, both of these averages exceed the minimum stipulated requirement of 3000 m^2/h for the mower's designated mowing productivity. Therefore, it can be confidently concluded that the header effectively meets the established design requirements for mowing operations.

The mower's productivity when mowing on a 25° slope is higher compared to flat ground. This is primarily because of its ability to adjust to changes in the terrain's elevation, allowing it to maintain a consistent cutting effect. Additionally, the force of gravity on the slope provides the mower with extra power, resulting in more efficient mowing on sloped surfaces.

3.2.2. Experiment Verification of a Mowing Height Test in Three Gears of Lawn Protection Boots

To assess the functional efficiency of the lawn protection boots, extensive tests have been conducted at three different gear positions, each tailored to varying cutter heights relative to the ground: gear position 1 set at a cutter height of 6 cm, gear position 2 at 8 cm, and gear position 3 at 10 cm.

The results of these tests are depicted in Figure 9a,b. It is observed that the differences in mowing height among the three gears increase at almost the same rate. Specifically, the average stubble height for Gear 1 is recorded as 10.4 cm, Gear 2 exhibited 12 cm, and Gear 3 yielded 14 cm. Notably, the height disparities among the three gears are approximately 2 cm, and the total time required for gear adjustments does not exceed 5 min, thus meeting the design requirement for gear modulation in the boots.

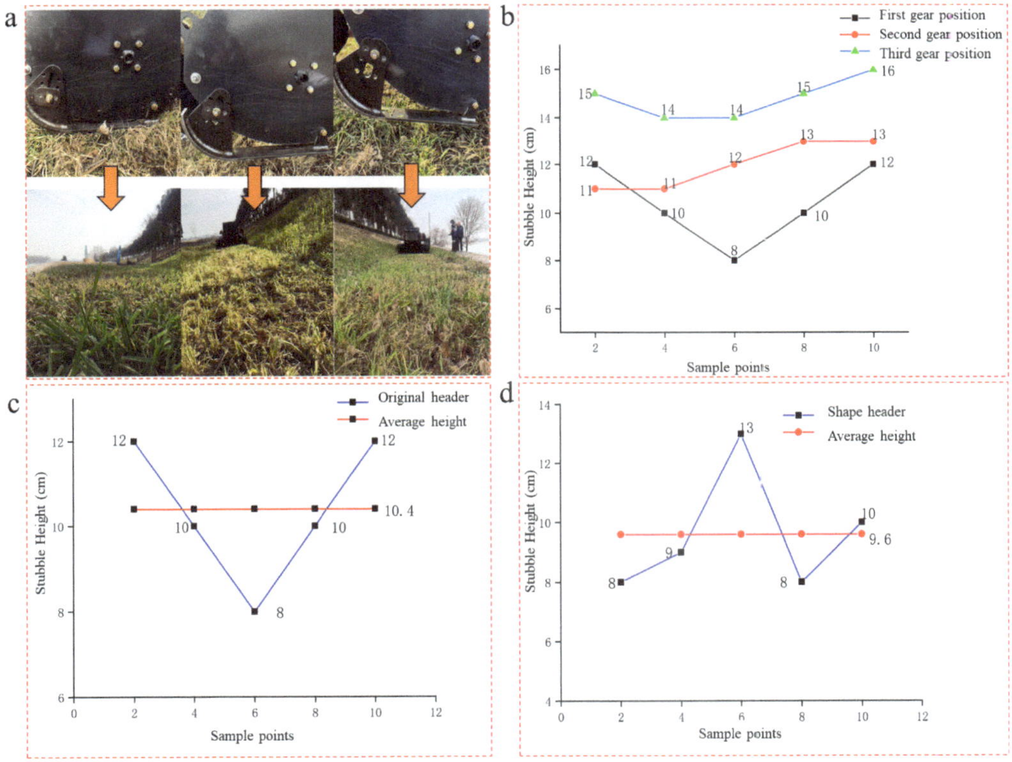

Figure 9. Comprehensive performance experiment of header: (**a**) Effect of three-speed change experiment of lawn protection boot; (**b**) Stubble height after three-speed change of lawn protection boot; (**c**) Stubble height of conventional header operation; (**d**) Stubble height for profiling header operation.

3.2.3. Comparison Experiment of Mowing Uniformity with and without Profiling of the Header

Furthermore, to evaluate the operational effectiveness of the profiling header, a comparative analysis is conducted between a traditional header and the profiling header, as manifested in Figure 9c,d. The stubble heights for platforms devoid of profiling and those equipped with profiling are calculated using Equation (13), yielding values of μ_N = 10.4 cm and μ_Y = 9.6 cm, respectively. In this case, 'N' and 'Y' serve as abbreviations, representing platforms without a profiling header and with a profiling header, respectively.

$$\mu = \frac{\sum_{i=1}^{n} x_i}{n} \tag{13}$$

In the formula, μ is the mean stubble height; $\sum_{i=1}^{n} x_i$ is the sum of stubble height; and n is the number of sampled data points.

Then, utilizing the variance calculation Formula (14), we can compute the variance to obtain a value of σ_N^2 = 2.24, and σ_Y^2 = 3.34.

$$\sigma^2 = \frac{\sum_{i=1}^{n} (x_i - \mu)^2}{n} \tag{14}$$

Here, σ^2 is the variance; $\sum_{i=1}^{n}(x_i - \mu)^2$ is the sum of squares of the differences between each data point and the mean value; and n is the number of sampled data points. Then, using Formula (15), we can obtain the standard deviation of σ_N = 1.497 and σ_Y = 1.828. Finally, we employ the coefficient of variation using Formula (16), and then acquire CV_N = 14.39% and CV_Y = 19.04%. Here, σ is standard deviation; and CV is coefficient of variation.

$$\sigma = \sqrt{\frac{\sum_{i=1}^{n} (x_i - \mu)^2}{n}} = \sqrt{\sigma^2} \tag{15}$$

$$CV = \frac{\sigma}{\mu} \times 100\% \tag{16}$$

After the comparison between the performance data of the profiling headers and non-profiling (traditional) headers, it is evident that the profiling header exhibits a slightly higher standard deviation of 1.828 compared to 1.497 for the non-profiling header. In addition, the coefficient of variation of the profiling header, standing at 19.04%, is also slightly higher than that of the 14.39% of the non-profiling header. These findings indicate that during the mowing process, the performance of the profiling header exhibits relatively greater fluctuations and variability.

A higher standard deviation signifies that the profiling header demonstrates greater variance in the measured data. This may reflect pronounced differences in its performance across diverse conditions or at various times. Moreover, the elevated coefficient of variation further corroborates the relatively unstable performance of the profiling header, which can be attributed to a range of factors, including variations in terrain, crop type, or operating parameters.

The performance data of the profiling header exhibits certain fluctuations; however, it remains within the design specifications for crucial indicators such as mowing height. The fact that the deviation amount remained within the permissible range underscores the overall reliability and practicality of the system.

4. Conclusions

(1) This simulation verifies the functionality of the lifting mechanism by closely analyzing the motion characteristics of the header during the lifting process. Based on this information, we can deduce that the lifting process achieved a peak velocity of 100.8 mm/s and a minimum velocity of 76.1 mm/s. This comprehensive simulation analysis confirms that the design and parameter selections of the header lifting mechanism satisfy the specified design requirements and align with the intended working conditions.

(2) Significantly, the mower maintains a consistent operating speed of 0.85 m/s on both flat and sloping terrain. This velocity falls precisely within the designated range of 0.27 m/s to 2.7 m/s, which is the intended speed for mowing. The findings indicated a significantly greater rate of 4212 m^2/h on a 25° incline, while the average mowing efficiency on flat terrain was 3872 m^2/h. Notably, both of these averages exceed the

minimum stipulated requirement of 3000 m^2/h for the mower's designated mowing productivity, thus fulfilling the designated requirements.

(3) An experiment was conducted to determine the maximum level of gear adjustment for the lawn protection boots. Notably, the height differences between the three gears are only about 2 cm, which meets the design requirement for gear modulation in the boots. When assessing the uniformity of mowing with and without a profiling header, the validation indicates that the former exhibits a slightly greater standard deviation (1.828) compared to the latter (1.497). Essentially, the profiling header adapts to the transformation of river dikes within an acceptable range of variation and meets the design criteria for important factors such as the evenness of mowing.

Author Contributions: Conceptualization, M.L.; methodology, M.L.; software, M.L. and Y.Y.; validation, M.L. and Y.Y.; formal analysis, M.L. and Y.Y.; investigation, M.L., Y.Y. and L.T.; resources, M.L. and L.T.; data curation, M.L.; writing—original draft preparation, M.L., Y.Y. and L.T.; writing—review and editing, X.C. and F.L.; visualization, M.L. and L.T.; supervision, X.C., F.L. and M.L.; project administration, M.L. All authors have read and agreed to the published version of the manuscript.

Funding: This research has been funded by the Chongqing Special Key Project for Technological Innovation and Application Development, under the grant number cstc2021jscx-gksb0003.

Institutional Review Board Statement: Not applicable.

Data Availability Statement: The datasets used and/or analyzed during the current study are available from the corresponding author upon reasonable request.

Conflicts of Interest: The authors declare that they have no conflicts of interest pertaining to the content, execution, or publication of this research.

References

1. *GB 50286-2013*; Code for Design of Embankment Slope Engineering. Ministry of Water Resources of the People's Republic of China: Beijing, China, 2013.
2. *2019 National Water Conservancy Development Statistics Bulletin*; Ministry of Water Resources of the People's Republic of China: Beijing, China, 2019.
3. Bátori, Z.; Kiss, P.J.; Tölgyesi, C.; Deák, B.; Valkó, O.; Török, P.; Erdős, L.; Tóthmérész, B.; Kelemen, A. River embankments mitigate the loss of grassland biodiversity in agricultural landscapes. *River Res. Appl.* **2020**, *36*, 1160–1170. [CrossRef]
4. Löbmann, M.T.; Geitner, C.; Wellstein, C.; Zerbe, S. The influence of herbaceous vegetation on slope stability—A review. *Earth-Sci. Rev.* **2020**, *209*, 103328. [CrossRef]
5. Cole, L.J.; Stockan, J.; Helliwell, R. Managing riparian buffer strips to optimise ecosystem services: A review. *Agric. Ecosyst. Environ.* **2020**, *296*, 106891. [CrossRef]
6. Vannoppen, W.; Poesen, J.; Peeters, P.; De Baets, S.; Vandevoorde, B. Root properties of vegetation communities and their impact on the erosion resistance of river dikes. *Earth Surf. Process. Landf.* **2016**, *41*, 2038–2046. [CrossRef]
7. Scheres, B.; Schüttrumpf, H. Investigating the erosion resistance of different vegetated surfaces for ecological enhancement of sea dikes. *J. Mar. Sci. Eng.* **2020**, *8*, 519. [CrossRef]
8. van den Hoven, K.; Grashof-Bokdam, C.J.; Slim, P.A.; Wentholt, L.; Peeters, P.; Depreiter, D.; Koelewijn, A.R.; Stoorvogel, M.M.; van den Berg, M.; Kroeze, C.; et al. Greening the dike revetment with historic sod transplantation technique in a living lab. *J. Flood Risk Manag.* **2024**, *17*, e12968. [CrossRef]
9. Zanetti, C.; Macia, J.; Liency, N.; Vennetier, M.; Mériaux, P.; Provensal, M. Roles of the riparian vegetation: The antagonism between flooding risk and the protection of environments. In Proceedings of the 3rd European Conference on Flood Risk Management FLOODrisk, Lyon, France, 17–21 October 2016; Volume 7, p. 13015.
10. van der Meer, J.; Steendam, G.J.; Mosca, C.A.; Guzzo, L.B.; Takata, K.; Cheong, N.S.; Eng, C.K.; Lj, L.A.; Ling, G.P.; Siang, C.W.; et al. Wave overtopping tests to determine tropical grass species and topsoils for polder dikes in a tropical country. *Coast. Eng. Proc.* **2020**, *36*, 31. [CrossRef]
11. Orlyanskaya, I.A.; Jhalnin, E.V.; Orlyansky, A.V.; Petenev, A.N.; Kulaev, E.V. Comparative efficiency of segment-finger and rotary mowers for grass cutting. In Proceedings of the AIP Conference Proceedings, Stavropol, Russia, 27–30 September 2021; Volume 2661, p. 070002.
12. Nishimura, Y.; Yamaguchi, T. Grass Cutting Robot for Inclined Surfaces in Hilly and Mountainous Areas. *Sensors* **2023**, *23*, 528. [CrossRef] [PubMed]
13. Liao, J.C.; Chen, S.H.; Zhuang, Z.Y.; Wu, B.W.; Chen, Y.J. Designing and Manufacturing of Automatic Robotic Lawn Mower. *Processes* **2021**, *9*, 358. [CrossRef]

14. Xiao, M.H.; Zhao, Y.; Wang, H.; Xu, X.; Bartos, P.; Zhu, Y. Design and Test of Hydraulic Driving System for Undercarriage of Paddy Field Weeder. *Agriculture* **2024**, *14*, 595. [CrossRef]
15. Kang, C.Q.; Ng, P.K.; Liew, K.W. The Conceptual Synthesis and Development of a Multifunctional Lawnmower. *Inventions* **2021**, *6*, 38. [CrossRef]
16. Khodke, K.R.; Kukreja, H.; Kotekar, S. Literature Review of Grass Cutter Machine. *Int. J. Emerg. Technol. Eng. Res.* **2018**, *6*, 97–101.
17. Diwakaran, S.; Kumar, M.D.V.; Mohanreddy, P.S.; Rishika, C.; Sreenivasulu, P.; Sivasubramanian, M. Design of an Autonomous Mower with Height Adjustable Cutting Motor. In Proceedings of the 2023 4th International Conference on Signal Processing and Communication (ICSPC), Coimbatore, India, 23–24 March 2023; pp. 348–352.
18. Kang, C.Q.; Ng, P.K.; Liew, K.W. A TRIZ-Integrated Conceptual Design Process of a Smart Lawnmower for Uneven Grassland. *Agronomy* **2022**, *12*, 2728. [CrossRef]
19. Lei, X.H.; Qi, Y.N.; Zeng, J.; Yuan, Q.C.; Chang, Y.H.; Lyu, X. Development of unilateral obstacle-avoiding mower for Y-trellis pear orchard. *Int. J. Agric. Biol. Eng.* **2022**, *15*, 71–78. [CrossRef]
20. Wu, B.; Wang, D.; Wang, G.; Fu, Z.; Guo, Z.; Gong, Z.; Li, T. Design and Simulation of the Copying Spring of Selfpropelled Mowers. In Proceedings of the 2015 ASABE Annual International Meeting, New Orleans, LA, USA, 26–29 July 2015; p. 152188918.
21. Zhang, Y.; Tian, L.; Cao, C.; Zhu, C.; Qin, K.; Ge, J. Optimization and validation of blade parameters for inter-row weeding wheel in paddy fields. *Front. Plant Sci.* **2022**, *13*, 1003471. [CrossRef] [PubMed]
22. Sheng, Y.Y.; Tian, C.Y.; Xiao, M.H.; Yao, W.Q.; Wang, Q.Q.; Shi, X.Y. Design and Analysis of a Portable Greenhouse Weeding Machine. *J. Comput. Methods Sci. Eng.* **2019**, *19*, 229–241. [CrossRef]
23. Tian, F.; Xia, K.; Wang, J.; Song, Z.; Yan, Y.; Li, F.; Wang, F. Design and experiment of self-propelled straw forage crop harvester. *Adv. Mech. Eng.* **2021**, *13*, 16878140211024455. [CrossRef]
24. Nkakini, S.; Branly, Y. Design, Fabrication and Evaluation of a Spiral Blade lawn mower. *Eur. Int. J. Sci. Technol.* **2014**, *3*, 165–172.
25. Zhao, L.; Wang, J.; Zhang, Z.W. Research on Parameter Design of Multi-axis Hydrostatic Transmission Vehicle. *MATEC Web Conf.* **2017**, *139*, 00215. [CrossRef]
26. Cao, L.; Miao, S. Design of Chinese Cabbage Harvester. In Proceedings of the 2020 IEEE International Conference on Mechatronics and Automation (ICMA), Beijing, China, 13–16 October 2020; pp. 243–248.
27. Guan, P.; Yang, J. Design and test of a new type of coupling weeder. *Int. J. Mech.* **2023**, *17*, 16–24. [CrossRef]
28. Zhang, L.; Yao, C.; Ying, W.; Luo, S.; Ying, F. Theoretical analysis and design of roller mower straight blade. *J. Mech. Sci. Technol.* **2024**, *38*, 3597–3606. [CrossRef]
29. Pirchio, M.; Fontanelli, M.; Frasconi, C.; Martelloni, L.; Raffaelli, M.; Peruzzi, A.; Gaetani, M.; Magni, S.; Caturegli, L.; Volterrani, M.; et al. Autonomous Mower vs. Rotary Mower: Effects on Turf Quality and Weed Control in Tall Fescue Lawn. *Agronomy* **2018**, *8*, 15. [CrossRef]
30. Iwano, Y.; Hasegawa, T.; Tanaka, A.; Iizuka, K. Development of the trimmer-type mowing system against a slope. In Proceedings of the 2016 International Conference on Advanced Mechatronic Systems (ICAMechS), Melbourne, VIC, Australia, 16 January 2017; pp. 23–28.
31. Fontanelli, M.; Chiaro, N.D.; Gagliardi, L.; Frasconi, C.; Raffaelli, M.; Peruzzi, A.; Luglio, S.M. Measuring the operative performance of autonomous mowers on slopes. In Proceedings of the 2023 IEEE International Workshop on Metrology for Agriculture and Forestry (MetroAgriFor), Pisa, Italy, 6–8 November 2023; pp. 404–408.
32. Yan, J.; Gao, L.; Wang, L.; Gong, X. Design, finite element analysis and utility of automatic weeder for port. *Coast. Res.* **2019**, *93*, 901–904. [CrossRef]
33. Peigang, J.; Yinnan, L.; Yang, L. Design and Simulation Analysis of Weeding Machine Chassis Based on CATIA and ADAMS. In Proceedings of the 2023 International Conference on Electronics and Devices, Computational Science (ICEDCS), Marseille, France, 22–24 September 2023; pp. 158–162.

agriculture

MDPI

Article

Analysis of Vibration Characteristics of Tractor–Rotary Cultivator Combination Based on Time Domain and Frequency Domain

Yuanyuan Gao [1,2], Yifei Yang [1], Shuai Fu [3], Kangyao Feng [1], Xing Han [1], Yongyue Hu [1], Qingzhen Zhu [1,2] and Xinhua Wei [1,2,*]

1 School of Agricultural Engineering, Jiangsu University, Zhenjiang 212013, China; gaoyy0910@ujs.edu.cn (Y.G.); yyfjsdxyx@stmail.ujs.edu.cn (Y.Y.); kangyaofeng@stmail.ujs.edu.cn (K.F.); hanx1010@stmail.ujs.edu.cn (X.H.); hy_0512@stmail.ujs.edu.cn (Y.H.); qingzhen_zhu@ujs.edu.cn (Q.Z.)
2 Key Laboratory of Modern Agricultural Equipment and Technology, Ministry of Education, Zhenjiang 212013, China
3 Weichai Lovol Intelligent Agricultural Technology Co., Ltd., Weifang 261200, China; fushuai@lovol.com
* Correspondence: 1000003563@ujs.edu.cn

Abstract: A good planting bed is a prerequisite for improving planting quality, while complex ground excitation often leads to machine bouncing and operation vibration, which then affects the operation effect. In order to improve the quality of rotary tillage operations, it is necessary to study the effects of various vibration excitations on the unit during tractor rotary tillage operations and analyze the vibration interaction relationship among the tractor, the three-point suspension mechanism, and the rotary tiller. For this purpose, multiple three-way acceleration sensors were installed at different positions on the rotary tiller unit of a Lexing LS1004 tractor(Lexing Agricultural Equipment Co. Ltd., Qingdao, China) to collect vibration data at different operating speeds and conduct vibration characteristic analysis between different components. The test results showed that when the unit moved forward at 2.1 km/h, 3.6 km/h, and 4.5 km/h, respectively, the vibration acceleration of the tractor, the three-point suspension mechanism, and the rotary tiller increased with the increase in speed, and there was indeed interaction between them. The vertical acceleration change during the test in the three-point suspension mechanism was the most significant (5.914 m/s^2) and was related to the increase in the speed of the vehicle and the vibration transfer of the rotary tiller. Meanwhile, the vertical vibration acceleration of the tractor's symmetrical structure was not similar, suggesting the existence of structural assembly problems. From the perspective of frequency domain analysis, the resonant frequency at the cab of the tractor was reduced in a vertical vibration environment, with relatively low frequencies (0~80 Hz) and small magnitudes, which might be beneficial to the driver's health. The rotary tillage group resonated around 350 Hz, and this characteristic can be used to appropriately increase the vibration of the rotary tiller to reduce resistance. The tractor cab resonated around 280 Hz, which must be avoided during field operations to ensure driver health and reduce machine wear. The research results can provide a reference for reducing vibration and resistance during tractor rotary tillage operations, as well as optimizing and improving the structure of rotary tillers and tractors.

Keywords: seed bed preparation; tractor rotary tillage; ground excitation; vibration interaction relationship; time–frequency characteristics; power spectral density

Citation: Gao, Y.; Yang, Y.; Fu, S.; Feng, K.; Han, X.; Hu, Y.; Zhu, Q.; Wei, X. Analysis of Vibration Characteristics of Tractor–Rotary Cultivator Combination Based on Time Domain and Frequency Domain. *Agriculture* **2024**, *14*, 1139. https://doi.org/10.3390/agriculture14071139

Academic Editors: Xiaojun Gao, Qinghui Lai and Tao Cui

Received: 13 June 2024
Revised: 10 July 2024
Accepted: 11 July 2024
Published: 13 July 2024

1. Introduction

In agricultural production, land preparation is a key point. Through land preparation operations, the soil in the field can be made more suitable for the growth of crops. Traditional farming machinery, such as disc harrows and share plows, require greater tractor traction. In contrast, tractor rotary tillage units have become the main method for land preparation due to their wide range of applicable operations and lower power load requirements. Therefore, the research on the operation performance of a rotary tiller will help to improve the final operation quality and crop yield [1–3].

Due to the complex field operating environment, agricultural machinery often faces severe vibrations caused by multisource excitations during operation, which reduces the reliability of the machinery and leads to a decline in work quality [4,5]. However, the appropriate vibration can help reduce the resistance of rotary tillage operations and tractor power consumption. For this reason, it is necessary to understand the vibration characteristics of the tractor during rotary tillage operations. At present, there have been a large number of studies on the working quality of tractors and rotary tillers at home and abroad. In terms of tractors, Han et al. designed a coordinated control system between engine load characteristics and plowing depth to stabilize the engine load by adjusting the plowing depth, allowing the tractor to work within the range of high power and low fuel consumption [6]. An approach that combines multi-objective particle swarm optimization (MOPSO) and wavelet decomposition algorithms has been used to collect input vibration signals efficiently, making it extremely convenient for tractor bumping tests [7]. Watanabe et al. established a three-degree-of-freedom nonlinear tractor dynamics model and used a delayed feedback (DF) control algorithm to eliminate complex vibrations in tractor motion [8]. Singh et al. implemented the online transmission of raw acceleration data in all directions of the tractor cab seat and driver's head based on the Internet of Things module and obtained its power spectral density (PSD) [9]. In addition to tractors, the mounted machine directly affects the quality of operation. Therefore, some scholars have studied the vibration characteristics of land preparation machinery [10,11]. For example, Lai et al. used EDEM 2017, ADAMS 2013, and other simulation software to derive the load model of the rotary tiller gearbox and then carried out stress–strain analysis to find out the design defects of the gearbox [12]. Guo et al. studied the vibration acceleration signals of the hood and handrails of the micro-tiller under four conditions and found that the anisotropic vibration of the hood and handrails was less than that of the static conditions when the engine was running at high speed, because the blades cutting the soil increased the damping of the whole machine [13]. Then, the soil's absorption of certain frequency energies caused the tiller to experience violent vibrations in the handle due to unbalanced inertial force, providing a reference for reducing resistance and energy consumption during rotary tillage operations. Du et al. considered the influence of tractor wheel subsidence and structural deformation of rotary tiller unit on tillage depth and designed a system that could realize the automatic monitoring of the tillage depth of a suspended rotary tiller, which would allow for a comprehensive evaluation of the quality of tillage depth [14]. Chen took the micro-tiller as their research object, tested the vibration of each structure of the micro-tiller under eight working conditions, and analyzed the time domain and frequency domain, thereby reducing the work resistance of the micro-tiller [15]. She then verified the established dynamic model of the whole machine through modal analysis and other means and proposed optimization measures. In the field of the effectiveness of land leveling technology, laser-controlled leveling and GNSS-controlled leveling technologies have been used to enhance the practical benefits of land leveling, contributing valuable research ideas for the future direction of reducing mechanical energy loss during leveling technology [16].

It can be seen from the studies above that research on the vibration characteristics of tillage and land preparation equipment has mostly focused on the tractor, and there is a lack of research on the analysis of the vibration characteristics of the tractor and rotary tiller combination during the operation [17–19]. Since rotary tillers and tractors are related to each other when working in the field, they are stimulated to vibrate and affect each other's vibration [20]. To this end, this paper takes a rotary tiller, a tractor, and the three-point suspension mechanism used to fixedly connect the two as the research objects, collecting and analyzing the vibration conditions of the tractor, three-point suspension mechanism, and rotary tiller, exploring the possible interaction between the units at a stable speed and studying the impact of vibration on the normal operation of the unit, in order to provide a basis for improving the reliability of tractors and rotary tiller, as well as improving and optimizing the structure of the implements to reduce vibration and resistance from soil.

2. Materials and Methods

2.1. Tractor and Rotary Tiller Structure

In this test, a Lexing LS1004 tractor (Lexing Agricultural Equipment Co. Ltd., Qingdao, China) was used to pull the rotary tiller. Its main performance parameters are shown in Table 1. The Dongfanghong 1GQN-250G rotary tiller (YTO GROUP CORPORATION, Luoyang, China) was selected for plowing operations, as shown in Figure 1, and it was connected to the tractor with a three-point suspension mechanism. Its main performance parameters are shown in Table 2 [21].

Table 1. Main performance parameters of tractor.

Rotating Speed/rpm	PTO/rpm	Overall Size/mm
2100	540/750/1000	4540 × 2000 × 2650

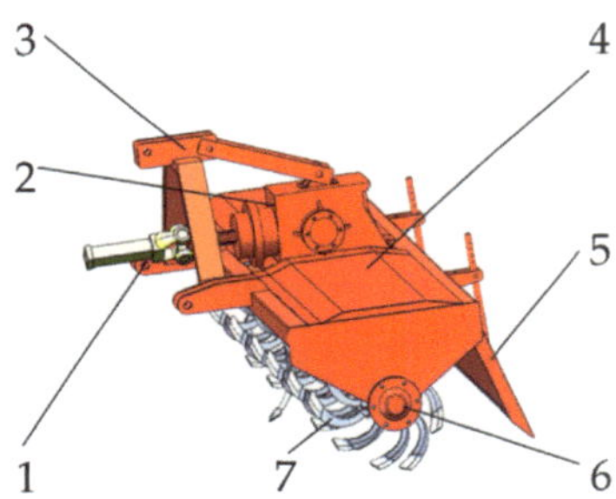

Figure 1. Main structure of rotary tiller: (1) Transmission system, (2) Gear box, (3) Linkage frame, (4) Cover, (5) Soil leveling board, (6) Rotary tiller shaft, (7) Rotary tiller blade.

Table 2. Main performance parameters of rotary tiller.

Matching Horsepower/Hp	Cultivated Width/cm	Cultivation Depth/cm	PTO/rpm
90~105	250	2~16	720/540

The rotary tiller is mainly composed of a transmission system, a gear box, a linkage frame, a cover, a flat soil support plate, a rotary tiller shaft, and a rotary tiller blade. The tractor power was inputted into the gear box through the universal joint of the transmission system, driving the cutter shaft to rotate and the rotary tillage blade to perform soil plowing and soil crushing operations. The soil leveling board was used to prevent soil from splashing and further breaking up the soil during the operations.

2.2. Analysis of the Main Vibration Sources of the Unit

The Dongfanghong 1GQN-250G rotary tiller adopts an intermediate bevel gear transmission. The power of the tractor's power output shaft was transmitted to the intermediate gearbox through the universal joint transmission shaft. The specific transmission path is shown in Figure 2, and the direction of power transmission has been marked with red arrows in the figure. It can be seen from the figure that when a tractor-mounted rotary tiller was operating, in addition to the tractor engine and the possible uneven field surface, its main vibration sources included the rotary tiller gear box, the rotary tiller blade shaft, and the impact load generated by the rotary tiller blade cutting the soil [22].

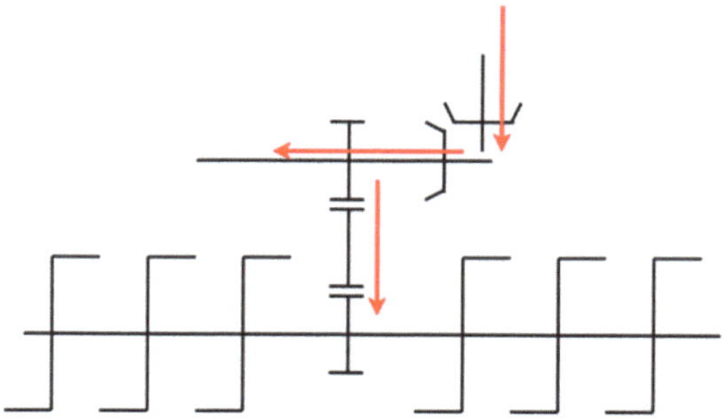

Figure 2. Rotary tiller power transmission route.

When the unit was performing rotary tillage operations, we set the tractor's power take off (PTO) speed to 540 r/min. At this time, the theoretical engine speed was about 1680 r/min. In the test, the first-stage bevel gear transmission ratio of the rotary tiller gearbox was 15:22, and the theoretical rotation speed of the rotary tiller's cutter shaft was 252 r/min [23]. The theoretical vibration frequency of the engine can be obtained according to Equation (1), and the theoretical vibration frequency of PTO and rotary blade shaft can be obtained according to Equation (2) [24]. The theoretical vibration frequencies of each vibration source are shown in Table 3.

$$f_1 = \frac{2}{60c} n_i i \tag{1}$$

where f_1 is the theoretical vibration frequency of the engine, Hz; c is the number of engine strokes; n_i is the theoretical engine speed, r/min; i represents the number of cylinders.

$$f = \frac{n}{60} \tag{2}$$

where f is the theoretical vibration frequency of PTO and rotary blade shaft, Hz; n is the theoretical speed of the kinematic mechanism, r/min.

Table 3. Theoretical vibration frequency of the main vibration source of the unit.

Vibration Source	Theoretical Vibration Frequency/Hz
PTO	8~10
Tractor engine	54~58
Rotary tiller shaft	4~5

2.3. Laboratory Equipment

The vibration acceleration acquisition and analysis system of the tractor rotary tillage unit mainly consists of a dynamic signal analyzer, three-way acceleration sensors, post-processing software, a voltage inverter, a 12 V DC power supply and a laptop, as shown in Figure 3. Among them, the Spider-80Xi analyzer (Crystal Instruments Company, Santa Clara, CA, USA) has 32 input channels and is powered by a DC power supply with a voltage of 10 V to 22 V; the three-way acceleration sensor model is BWJ13533, and the sensitivity of each sensor is shown in Table 4. The inverter model is BOKAI 3000 WDC/AC (Zhongshan Xinyihao E-commerce Co. Ltd, Zhongshan, China) inverter, which was used to convert 12 V DC power into 220 V AC power to power the dynamic signal analyzer.

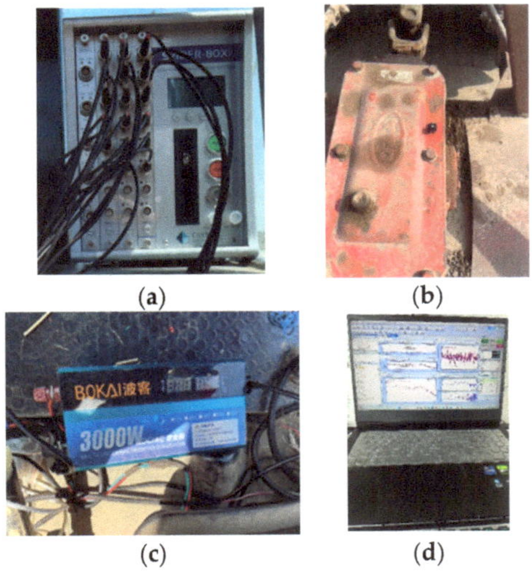

Figure 3. Collection and analysis system: (**a**) Spider-80Xi analyzer, (**b**) Three-way acceleration sensor (at the rotary tiller), (**c**) Inverter, (**d**) Upper computer interface.

Table 4. Sensor sensitivity in all directions.

Sensor Number	Sensitivity in All Directions/mV·g^{-1}		
	X	Y	Z
1	50.24	50.39	50.53
2	49.64	50.27	50.16
3	49.10	49.88	49.60
4	49.35	50.41	51.05
5	49.21	50.33	49.62
6	49.86	50.05	49.28
7	48.82	50.14	50.26
8	49.40	48.66	49.01

In order to facilitate subsequent test data processing, during the test process, the unit's traveling direction was set as the X direction, the vertical forward direction as the Y direction, and the vertical direction to the ground as the Z direction. In this way, the vibration test data at different positions of the unit were uniformly divided into three directions, X, Y, and Z.

2.4. Experimental Program

During the tractor rotary tillage operations, the vibration signals of the rotary tiller, tractor, and linkage were collected by the three-way acceleration sensors and transmitted to the road spectrum test and analysis system through the Spider-80Xi dynamic signal analyzer. Then, the characteristic values (RMS) of the acceleration signal were extracted, and the time domain analysis and PSD analysis were carried out to obtain the sequence of effects of the main excitation sources on the vibration magnitude of the unit components [25].

The experiment was conducted in a rice stubble field. Since the ground of the paddy field was relatively flat after beating and drying, the influence of uneven ground on the vibration of the machine was considered as a minor factor in this experiment. In order to reduce the impact of the tractor engine on the collected signals, sensors were arranged at different positions of the unit as a blank control in the experiment. As shown in Figure 4, three-way acceleration sensors were installed on the rotary tiller, tractor, and three-point suspension mechanism to obtain vibration data. The installation position should be as close as possible to the excitation source to obtain accurate measurement data [26].

For this purpose, the position of each measuring point was set as follows: measuring point 1 (center of the rotary cultivator), measuring point 2 (suspended lower left pull rod), measuring point 3 (suspended right lower pull rod), measuring point 4 (tractor left rear wheel), measuring point 5 (tractor right rear wheel), measurement point 6 (tractor cab), measurement point 7 (tractor left front wheel), and measurement point 8 (tractor right front wheel). Finally, a total of 8 measurement points and 3 working conditions were used to conduct vibration characteristics test experiments. Among them, measuring points 4, 5, 7, and 8 were all arranged on the front and rear axles of the tractor chassis. The three working conditions were, respectively, the operating status of the rotary tiller unit when traveling at speeds of 2.1 km/h, 3.6 km/h, and 4.5 km/h. In addition, in order to reduce the influence of external factors on the test results [27], the vibration characteristic parameters of the rotary tiller and tractor cab with zero speed were collected. Meanwhile, the operating parameters of the rotary tiller were kept unchanged during the experiment to test the effect of working speed and soil excitation on the vibration characteristics of the machine.

Figure 4. Installation location of each sensor.

2.5. Signal Acquisition and Analysis

According to the above-mentioned measurement requirements, the sensors were first installed in a specific position and their three-axis orientations were recorded. They were then connected to the dynamic signal analyzer via a wire harness to check whether the individual channel data were normal. Meanwhile, in order to ensure the accuracy of the experimental waveform, the sampling frequency should be at least twice the frequency of the analyzed signal [28]. Before the experiment, the Spider-80Xi dynamic signal analyzer sampling rate was set to 2560 Hz, continuous sampling mode was performed, and 24 input channels were turned on. After the settings were completed, the vibration signal collection test was performed. While the three working speeds were stable, the vibration acceleration data was collected. Under stable working conditions, the sampling time of each group was 40 s. Based on the acceleration data obtained under stable sampling time, the vibration characteristics of the tractor rotary tillage operation were analyzed.

2.5.1. Time Domain Analysis

The root mean square (RMS) is often used to measure the fluctuation of random signal near the mean value, which can directly reflect the strength of the signal [29]. Therefore, the vibration signal data were processed, and the RMS was calculated to provide a basis for subsequent vibration analysis. The calculation formula of the RMS is shown in Equation (3).

$$RMS = \sqrt{\frac{1}{N}\sum_{k=1}^{N} x_k^2} = \sqrt{\frac{x_1^2 + x_2^2 + x_3^2 + \cdots + x_k^2}{N}} \tag{3}$$

where x_k is the vibration acceleration value, m/s^2; N is the number of signals collected.

With a view to reduce the impact of the location distribution of measuring points on the measurement results and obtain more accurate measurement data, the sensors fixed on the unit components were arranged symmetrically along the center line of the unit. Among the 8 measuring points, measuring points 2 and 3 were symmetrical about the center line of the unit; measuring points 4 and 5 on the tractor's rear axle were symmetrical about the tractor's center line; and measuring points 7 and 8 on the tractor's front axle were symmetrical about the tractor's center line. The measurement data of measuring point 1 and measuring point 6 were not additionally processed.

2.5.2. Frequency Domain Analysis

In order to further analyze the vibration signal, the time domain signal is generally converted into the frequency domain signal through Fast Fourier Transformation (FFT) [30]. This method is often used to analyze the signal spectrum in digital signal processing. The PSD of the rotary tiller and the tractor can be obtained by analyzing their frequency domain data [31]. PSD is a measure of the mean square value of random variables. It eliminates the impact of frequency resolution on the amplitude of the spectral function and represents the distribution of signal power at each frequency point. For any random signal, the PSD expression can be determined through Equation (4) [32].

$$\lim_{T\to\infty}\frac{1}{T}\int_0^T |\sigma(t)|^2 dt = \lim_{T\to\infty}\frac{1}{T}\int_{-\infty}^{+\infty} |F_\sigma(\omega)|^2 df = \int_{-\infty}^{+\infty} S_\sigma(f) df \tag{4}$$

where $\sigma(t)$ is a random signal; $F_\sigma(\omega)$ is $\sigma(t)$ after Fourier transform; $S_\sigma(f)$ represents the distribution of signal average power spectrum in frequency domain.

For the purpose of studying the influence of vehicle speed on the vibration characteristics of the rotary tillers and tractors, the vibration acceleration data of measuring point 1 on the rotary tiller and measuring point 6 on the tractor under three working conditions were processed. We then obtained the power spectrum peak value and corresponding peak frequency of each working condition of the two points.

3. Results and Discussion

3.1. Time Domain Analysis of Rotary Tillage Unit

The vertical RMS values of the rotary tiller and tractor with zero speed are listed in Table 5. The RMS values of the vertical vibration acceleration at different positions of the unit under the three working conditions are shown in Table 6, and the RMS values of the vertical acceleration on both sides of the rotary tillage unit are shown in Table 7.

Table 5. Vertical vibration acceleration of the cab of rotary tiller and tractor under no-load condition.

The Location of the Measurement Point	RMS/$m\cdot s^{-2}$
Rotary tiller	20.865
Cab	1.201

Table 6. RMS of vertical vibration acceleration at different positions of the rotary tillage unit under three working conditions.

Speed/ km·h^{-1}	RMS of Vertical Vibration Acceleration at Different Positions of the Unit/m·s^{-2}				
	Rotary Tiller	Three-Point Suspension Mechanism	Tractor Rear Wheels	Cab	Tractor Front Wheels
2.1	21.532	12.507	2.104	1.218	6.500
3.6	24.950	20.594	4.820	1.333	4.884
4.5	26.399	24.026	5.306	1.346	7.726
Average value	24.294	19.042	4.077	1.299	6.370
Standard deviation	2.499	5.914	1.726	0.070	1.425

Table 7. RMS of vertical acceleration on both sides of the rotary tillage unit under three working conditions.

Speed/ km·h^{-1}	RMS of Vertical Acceleration on Both Sides of the Rotary Tillage Unit/m·s^{-2}					
	Linkage Left Lower Tie Rod	Linkage Right Lower Tie Rod	Tractor Left Rear Wheel	Tractor Right Rear Wheel	Tractor Left Front Wheel	Tractor Right Front Wheel
2.1	20.382	4.632	4.129	0.078	10.973	2.018
3.6	24.463	16.725	9.566	0.074	7.192	2.575
4.5	27.509	20.543	10.529	0.082	12.848	2.603
Average value	24.118	13.967	8.075	0.078	10.338	2.399
Standard deviation	3.576	8.306	3.451	0.004	2.881	0.330

The following can be seen from the data in Tables 5–7:

(1) The vertical amplitude of the rotary tiller and tractor cab in the working state was higher than in the no-load state, and the amplitude increased with the speed, indicating that the unit had other excitation sources besides the rotary tiller. However, this change would be obvious at a higher speed. When the vehicle speed was 2.1 km/h, the amplitude increased by only 0.667 m/s^2 and 0.017 m/s^2, respectively, compared with zero speed, while the amplitude increase was 5.534 m/s^2 and 0.145 m/s^2, respectively, at the speed of 4.5 km/h. During the tractor rotary tillage operations, the average vertical acceleration generated by the rotary tiller at different vehicle speeds was 24.294 m/s^2, and the average vertical acceleration generated by the three-point suspension mechanism was 19.042 m/s^2, which was much larger than the vertical acceleration everywhere of the tractor. This showed that the rotary tiller was the most important excitation source in the unit. In addition, an analysis of the vibration conditions at different positions of the tractor showed that the vertical vibration of the front wheel (6.370 m/s^2) was greater than the tractor's rear wheel (4.077 m/s^2) and the cab (1.299 m/s^2). In addition to the different distances from the rotary tiller, the reason might also be related to the front and rear wheel load, stubble thickness, front and rear wheel vibration amplitude, and other factors.

(2) Investigating the changes in the vertical vibration acceleration of the different parts, it could be seen that the influence of working speed on each measurement part from large to small was as follows: three-point suspension mechanism > rotary tiller > tractor rear wheels > tractor front wheels > cab. Among them, the standard deviation of the three-point suspension mechanism (5.914 m/s^2) was much larger than that of other parts, indicating that speed changes had the greatest impact on the three-point suspension mechanism. The reason may be that the three-point suspension mechanism, the tractor, and the rotary tiller used pins and other movable connections, so it was greatly affected by inertia. The vertical vibration change in the rotary tiller (2.499 m/s^2) was second only to the three-point suspension mechanism, indicating that the vehicle speed was an important factor in the vibration changes in the rotary

tiller operation, which, in turn, affected the quality of the rotary tillage operation. While the cab's vertical vibration acceleration changed the least (0.070 m/s^2), presumably because the tractor's vibration damping device had good vibration damping performance near the cab.

(3) The vertical vibration amplitude and vibration acceleration standard deviation of the rotary tiller and the three-point suspension mechanism were much larger than those of the tractor components, indicating that the biggest excitation source for the two came from their interaction, which meant their interaction led to a severe vibration. The vertical acceleration amplitude of the rear wheels of the tractor (4.077 m/s^2) was generally smaller than that of the front wheels (6.370 m/s^2), which was related to the fact that the mass of the rear wheel was greater than that of the front wheel. However, the acceleration changes in the rear wheels (1.726 m/s^2) were greater than those in the front wheel (1.425 m/s^2). It was speculated that the rear wheels were affected by the violent vibration of the three-point suspension device, which accelerated the change in its own vibration amplitude.

(4) From working condition 1 to working condition 3, the overall vertical vibration acceleration of the rotary tiller and tractor increased, but the growth rate slowed down. The vibration acceleration of each part of the tractor increased as the vehicle sped up as a whole but the vibration acceleration of the tractor's front wheel decreased by 1.616 m/s^2 from 2.1 km/h to 3.6 km/h, and then increased again when it reached 4.5 km/h. It was speculated that when the vehicle speed was 3.6 km/h, the load distribution of the tractor changed, and the load on the left front wheel became smaller.

(5) The durability of agricultural machinery could be detected using vibration signals [33]. The vibration of the symmetrical structure of the wheeled tractor should be similar, but the vibration acceleration on the left side of each measured part in Table 5 was greater than that of the right side. This might have been caused by problems with the tractor parts and structural assembly, which, in turn, led to abnormal vibration transmission paths of the front and rear wheels and the three-point suspension mechanism symmetry point [34]. The vibration acceleration of the tractor's front left wheel only decreased by 3.781 m/s^2 when the speed was 3.6 km/h, resulting in a decrease in the vibration acceleration of the tractor's front wheel, as in Table 7. At this time, the tractor's vibration damping mechanism could not achieve a more balanced vibration suppression effect on the left and right sides, and further inspection of the quality of the tractor's assembly and parts was required.

(6) During the experiment, the three-point suspension mechanism produced the maximum amplitude, influencing the rotary tiller and tractor connected to its axle pin to produce greater vibration, which, in turn, affected the operation of the unit. For the tractor, the vibration was transmitted from the rear axle to the cab, causing a higher vibration of the seat, resulting in reduced comfort of the tractor seat, which was not conducive to the driver's long-term operation of the tractor. At the same time, appropriately increasing the amplitude of the rotary tiller could reduce the cutting resistance of the rotary tiller blade into the soil, which is conducive to improving the quality of rotary tillage operation [35], but excessive vibration may also aggravate blade fatigue damage. For this reason, the resonance frequency of the rotary tiller should be avoided as much as possible.

(7) Judging from the results of time domain analysis alone, higher speed means larger amplitude for each component, but this inference still needs further verification.

3.2. Frequency Domain Analysis of Rotary Tillage Unit

In order to obtain the change in the energy occupied by the center of mass vibration signal of the rotary cultivator and tractor in the unit frequency band along with the frequency, the collected time domain data was processed. Since measuring point 6 was installed in the cab, the vibration it measured could also reflect the vibration characteristics to the operator to a certain extent. For this reason, FFT transformation was performed on the vertical vibration data of measuring point 6 at three speeds, and the frequency domain data in the range of 0~80 Hz were obtained, as shown in Figure 5. The power spectrum peaks and corresponding peak frequencies at the center of mass of the rotary tiller and tractor under different working conditions are shown in Table 8, and the PSD is shown in Figure 6. For non-stationary random signals, the above processing could not reflect the characteristics of the signal frequency changing with time. In order to process non-stationary random signals, it was necessary to jointly represent the time domain and frequency domain of the signal in two dimensions, which means, time–frequency analysis. The signal analyzer in Matlab 2022a could be used to perform time–frequency analysis on the original time domain data collected. The local time–frequency analysis of the gearbox and cab under working condition 1 is shown in Figure 7.

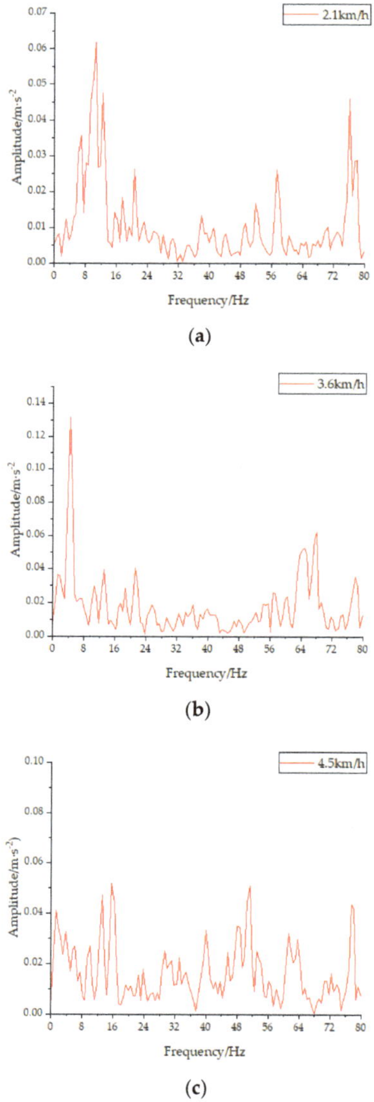

Figure 5. Vertical FFT diagrams of the cab in the low frequency range (0−80 Hz): (**a**) The speed at 2.1 km/h, (**b**) The speed at 3.6 km/h, (**c**) The speed at 4.5 km/h.

Table 8. Peak power spectrum of center of mass of rotary tiller and tractor.

Measuring Point Location	Rotary Tiller Gearbox			Tractor Cab		
Speed/km·h^{-1}	2.1	3.6	4.5	2.1	3.6	4.5
Peak axial	X	Y	Y	X	X	X
Peak/(m·s^{-2})^2	15.728	7.213	7.958	0.216	0.115	0.794
Frequency/Hz	351	466	548	270	275	274

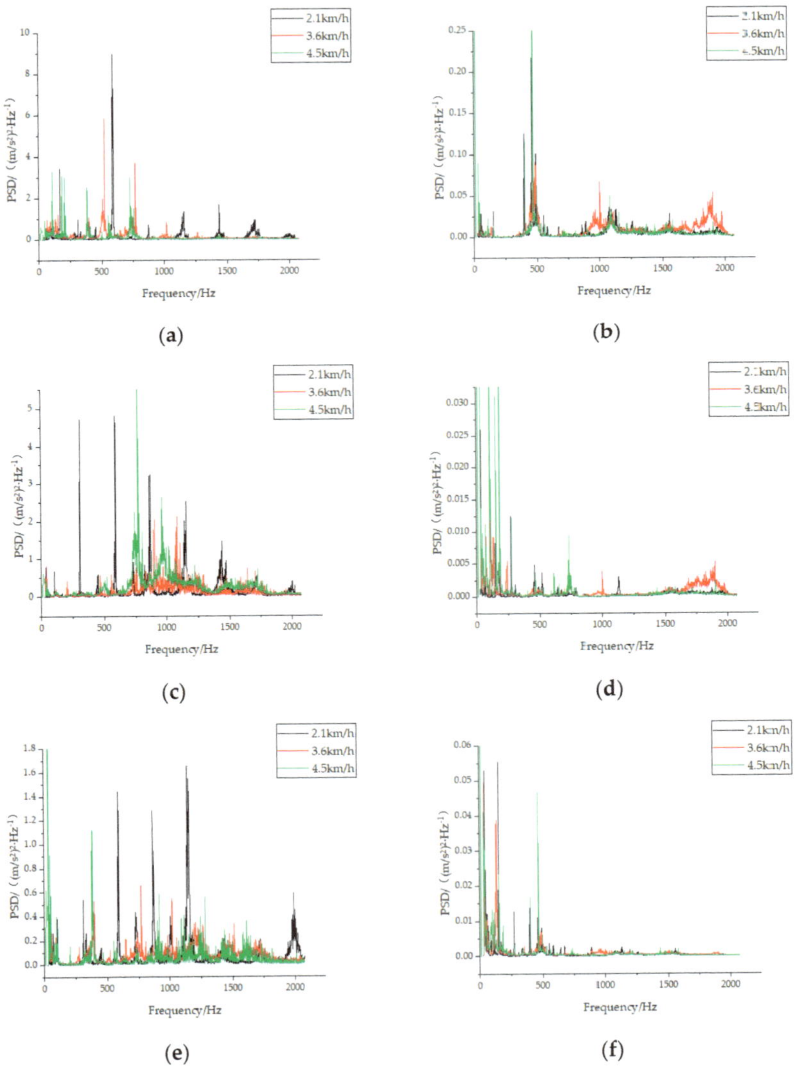

Figure 6. PSD drawings of gearbox and cab in various directions: (**a**) Gearbox, X direction, (**b**) Cab, X direction, (**c**) Gearbox, Y direction, (**d**) Cab, Y direction, (**e**) Gearbox, Z direction, (**f**) Cab, Z direction.

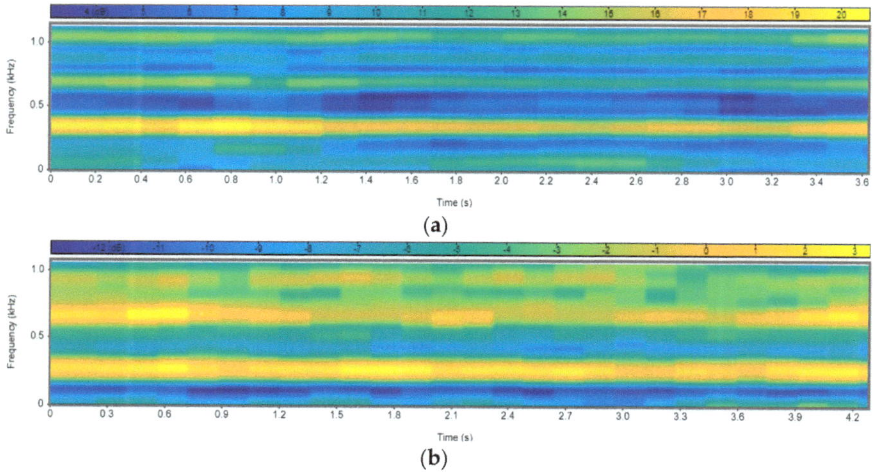

Figure 7. Local time−frequency analysis of working condition 1: (**a**) Gearbox and (**b**) Cab.

Based on the data charts above, the relationship between vibration acceleration and frequency changes near the gearbox (rotary tiller) and tractor cab (tractor) was analyzed, and the following conclusions were drawn:

(1) At the three working speeds, the amplitude of the tractor cab appeared as small peaks in the frequency range of 0~8 Hz, and with the increase in working speed, the peak frequency of the amplitude reduced from 6.8 Hz to 4.9 Hz. As the human body is more sensitive to vertical vibration frequencies in the range of 4–8 Hz, it would not be appropriate to work for a long time in that environment. The change in the peak frequency of the seat vibration in the low frequency range (1~16 Hz) was consistent with the content in the literature [36]. Therefore, maintaining high-speed operation of the unit in actual operation may avoid the sensitive frequency interval of the human body and maintain human health.

(2) For the tractor, the peaks of vibration energy first occurred intensively in the low-frequency range of 0~100 Hz. Under the three working speeds, the power spectrum density in the X, Y, and Z directions all produced the first peak near 33 Hz. Regardless of the speed and direction, in the frequency range of 30~50 Hz, the energy occupied by the tractor's vibration signal began to gather significantly, especially at the highest speed, indicating that the first-order natural frequency of the tractor may exist in the continuous range of 30~50 Hz. In the wide range of 100~500 Hz, the power spectrum density values of the tractor in each direction varied, and the overall vibration energy density was higher. Combined with the corresponding frequency of the power spectrum peak of the rotary tiller in Table 8, it could be seen that the tractor had a resonance band of a certain width in this range. Unlike the tractor, the vibration signal power distribution of the rotary tiller in the range of 0~100 Hz was not significant, and the vibration energy in all directions was less than in the range above 250 Hz. Considering that its design purpose was to loosen the topsoil, it is beneficial for the rotary tiller to have much greater vibration energy than the tractor cab when working.

(3) For the movement of the rotary tiller in the X direction, the peak power spectrum of working condition 1 (15.728 $(m/s^2)^2$) was much larger than that of working conditions 2 and 3. Figure 7a is the time–frequency diagram of the X-direction vibration of the rotary tiller gearbox under working condition 1. After processing with the band stop filter, it could be seen that there was a relatively obvious vibration energy distribution at various frequencies, among which the energy distribution at a frequency of about 350 Hz was the most significant, and the energy concentration did not change over

time. More concentrated energy distributions could also be seen when the frequencies were near 700 Hz, 1050 Hz, and 1400 Hz (that was, integer multiples of 350 Hz). In the short period, the other frequency components except 350 Hz were not significant, indicating that the energy of the rotary tiller under working condition 1 was the most concentrated at the frequency of 350 Hz. This was consistent with the frequency corresponding to the gearbox vibration peak in Table 8 (351 Hz). It was inferred that the center of mass of the rotary cultivator resonated near the frequency of 350 Hz.

(4) Figure 7b is a time–frequency diagram of the vibration of the tractor cab in the X direction under working condition 1. The figure showed that the frequency corresponded to the maximum vibration at about 280 Hz, and the energy there remained unchanged during the sampling time. In Table 8, the tractor centroid power spectrum peak at a frequency of 274 Hz under working condition 3 increased significantly (0.794 $(m/s^2)^2$), and the peak frequencies under the three working conditions were all in the 270–280 Hz range. Therefore, it can be speculated that there was a resonance frequency in this interval.

(5) The X-direction resonance frequency of the rotary tiller was around 350 Hz, which was higher than the X-direction resonance frequency of the tractor cab (280 Hz). For tractors, people would want to avoid resonance, but this is different for rotary tillers. For the entire unit, improving operating efficiency is to better perform the rotary tillage operations. Since increasing the vibration of the rotary tiller can reduce the adhesion of soil to the rotary tiller blades [37], the operating speed can be increased to increase the amplitude of the rotary tiller operation, thereby reducing the resistance of the soil to the rotary tiller operation and improving operating efficiency and quality.

(6) For the movement of the gearbox in the Y direction, the PSD diagram showed that its vibration energy was concentrated in the high-frequency range of 750~1000 Hz, which was much higher than the resonance frequencies in the X and Z directions. Therefore, when considering avoiding resonance of the rotary cultivator, only the vibration in the X and Z directions need to be considered. The tractor cab in the Y direction generated a large vibration acceleration near 78 Hz. The vibration energy was concentrated in the lower frequency range of 75~135 Hz under various working conditions, but the overall vibration acceleration of the Y-axis was low, causing little impact.

(7) The Z-direction rotary tiller gearbox had a wide frequency range of violent vibrations, and large vibrations occurred between 430 and 1200 Hz under three working conditions. It was speculated that this was related to the fact that the rotary tiller would jump up and down in the vertical direction due to the rotary tiller blade encountering resistance when entering the soil and the vibration acceleration of the blade axis changing more obviously in the Z direction. The maximum vibration acceleration frequencies generated by the tractor in the Z direction were 66 Hz, 78 Hz, and 272 Hz, respectively, which were more concentrated than those in the X and Y directions.

(8) The results of the frequency domain analysis showed that high speed did not always lead to high amplitude. Sometimes, the PSD values of the rotary tiller in each direction at low speed were higher than those at high speed. In the low-frequency range relevant to human health, the low-speed amplitude was also greater than the high-speed amplitude.

(9) When using the signal analyzer in Matlab 2022a for the time–frequency analysis shown in Figure 7, a band-stop filter was also used to preprocess the original signal data, and the frequency components of the resulting time–frequency diagram were still relatively complex. In order to obtain a time–frequency diagram with clearer vibration energy distribution and better time–frequency aggregation, the next step could be to use continuous wavelet transform and other analysis methods with good time–frequency resolution to extract the characteristics of the vibration signal.

4. Conclusions

This paper took the rotary tiller, tractor, and three-point suspension mechanism as research objects and explored the interaction between the units at different speeds through vibration characteristic analysis. Based on the results of the vibration test experiments of the rotary tillage unit under three working speed conditions, combined with time domain and frequency domain analysis, the following conclusions can be drawn:

(1) The vibration amplitude of the rotary tiller needs to be appropriately increased to reduce the resistance to soil penetration, but the interaction between the tractor and the rotary tiller causes the vibration to be transmitted to the tractor synchronously, which is harmful to the tractor. Therefore, while the tractor is mounted with a rotary tiller and operates at high speed and efficiency, the vibration reduction capacity of the tractor also needs to be optimized to achieve higher quality of work and more stable operation.

(2) Generally speaking, the vibration acceleration at each measurement point increased with the increase in working speed, but this was not necessarily the case in certain frequency domain intervals. In some frequency domain intervals, the amplitude did not always increase with increasing speed. It might be that the low-frequency range is less affected at low speed, while the amplitude change in the high-frequency range at higher speeds still needs to be tested.

(3) The reason for why the vibration amplitude of the unit increased with the tractor speed might be related to the more violent collision between the unit and the stubble surface caused by the increase in speed. However, in the low-frequency and low-amplitude range, which has an important impact on human health, the vertical resonance frequency of the cab seat decreased with the increase in amplitude (from 6.8 Hz to 4.9 Hz). In the future, the change law of the vertical amplitude of the seat under high-speed operation of the unit can be studied to help the driver operate the tractor more healthily.

This paper obtained the interaction relationship among the components of the unit by collecting and analyzing the vibration data of the tractor, three-point suspension mechanism, and rotary tiller under different working conditions, and it also speculated on the possible structural assembly problems of the tractor within the experiment. The vibration data analysis method used in this paper can be further used to explore the interaction relationship and vibration characteristics between the tractor and the suspension agricultural implement system. In the future, the vibration analysis method can be further used to study the influence of factors such as field surface flatness and of PTO speed on the combined operation effect of the tractor and rotary tiller.

Author Contributions: Conceptualization, Y.G. and X.W.; methodology, Y.G., Y.Y. and S.F.; software, Y.Y. and K.F.; validation, Y.Y.; formal analysis, X.H. and S.F.; investigation, Y.Y. and K.F.; resources, Y.G. and X.W.; data curation, Y.Y.; writing—original draft preparation, Y.Y. and S.F.; writing—review and editing, Y.G. and Y.Y.; visualization, Y.H. and Q.Z.; supervision, S.F.; project administration, Y.G. All authors have read and agreed to the published version of the manuscript.

Funding: This study was financially supported by Jiangsu Provincial Natural Science Foundation (Grant No. BK20210776), the National Natural Science Foundation of China (Grant No. 32201672), the Priority Academic Program Development of Jiangsu Higher Education Institutions (Grant No. PAPD-2023-87), and Jiangsu Province and Education Ministry Co-sponsored Synergistic Innovation Center of Modern Agricultural Equipment (Grant No. XTCX1002).

Institutional Review Board Statement: Our studies did not involve humans or animals.

Data Availability Statement: The data presented in this study are available on request from the corresponding author.

Conflicts of Interest: Author Shuai Fu was employed by the company Weichai Lovol Intelligent Agricultural Technology Co., Ltd. The remaining authors declare that the research was conducted in the absence of any commercial or financial relationships that could be construed as a potential conflict of interest.

References

1. Qin, J.W. Current status and suggestions for the use of tillage machinery standards. *Agric. Mach. Qual. Superv.* **2020**, *01*, 20–21.
2. Chen, W.X. Discussion on the technical characteristics of rotary tiller and land preparation technology. *Guangxi Agric. Mech.* **2022**, *05*, 22–25.
3. Wang, Z.; Zhou, J.; Wang, X. Research on Energy Management Model for Extended-range Electric Rotary-tilling Tractor. *Trans. Chin. Soc. Agric. Mach.* **2023**, *54*, 428–438. [CrossRef]
4. Chen, S.R.; Zhou, Y.P.; Tang, Z.; Lu, S.N. Modal vibration response of rice combine harvester frame under multi-source excitation. *Biosyst. Eng.* **2020**, *194*, 177–195. [CrossRef]
5. Zhai, C.; Long, J.; Taylor, R.; Weckler, P.; Wang, N. Field scale row unit vibration affecting planting quality. *Precis. Agric.* **2020**, *21*, 589–602. [CrossRef]
6. Han, J.Y.; Yan, X.X.; Tang, H. Method of controlling tillage depth for agricultural tractors considering engine load characteristics. *Biosyst. Eng.* **2023**, *227*, 95–106. [CrossRef]
7. Sun, L.; Liu, M.; Wang, Z.; Wang, C.; Luo, F. Research on Load Spectrum Reconstruction Method of Exhaust System Mounting Bracket of a Hybrid Tractor Based on MOPSO-Wavelet Decomposition Technique. *Agriculture* **2023**, *13*, 1919. [CrossRef]
8. Watanabe, M.; Sakai, K. Delayed feedback control for chaotic vibration in nonlinear impact dynamics of bouncing agricultural tractor. *Sci. Rep.* **2023**, *13*, 10695. [CrossRef]
9. Singh, A.; Nawayseh, N.; Samuel, S.; Dhabi, Y.K.; Singh, H. Real-time vibration monitoring and analysis of agricultural tractor drivers using an IoT-based system. *J. Field Robot.* **2023**, *40*, 1723–1738. [CrossRef]
10. Sun, Y.; Ke, S.; Wang, G.; Liu, Z. Control Strategy Simulation Analysis of a New Micro-cultivator MR Elastomer Vibration Isolation System. *Adv. Eng. Sci.* **2020**, *52*, 218–225. [CrossRef]
11. Liao, Y.T.; Qi, T.X.; Liao, Q.X.; Zeng, R.; Li, C.L.; Gao, L.P. Vibration characteristics of pneumatic combined precision rapeseed seeder and its effect on seeding performance. *J. Jilin Univ. (Eng. Technol. Ed.)* **2022**, *52*, 1184–1196. [CrossRef]
12. Lai, Q.H.; Yu, Q.X.; Dong, J.Y. Dynamic analysis of rotary tiller gearbox based on EDEM, ADAMS and ANSYS. *J. Intell. Fuzzy Syst.* **2019**, *36*, 1153–1160. [CrossRef]
13. Guo, L.; Jian, C.; Xie, H.J.; Wang, S.M. Vibration test and analysis of mini-tiller. *Int. J. Agric. Biol. Eng.* **2016**, *9*, 97–103. [CrossRef]
14. Du, X.W.; Yang, X.L.; Pang, J.; Ji, J.T.; Jin, X.; Chen, L. Design and Test of Tillage Depth Monitoring System for Suspended Rotary Tiller. *Trans. Chin. Soc. Agric. Mach.* **2019**, *50*, 43–51. [CrossRef]
15. Chen, Y. Random Vibration Characteristics Analysis and Experimental Research of Self-Propelled Micro Tillage Machine. Ph.D. Thesis, Chongqing University of Technology, Chongqing, China, 2017.
16. Chen, G.; Hu, L.; Luo, X.; Wang, P.; He, J.; Huang, P.; Zhao, R.; Feng, D.; Tu, T. A review of global precision land-leveling technologies and implements: Current status, challenges and future trends. *Comput. Electron. Agric.* **2024**, *220*, 108901. [CrossRef]
17. Shao, X.D.; Yang, Z.H.; Song, Z.H.; Liu, J.H.; Yuan, W. Analysis of influence of tractor rotary tillage load on power take-off driveline. *Trans. Chin. Soc. Agric. Mach.* **2022**, *53*, 332–339. [CrossRef]
18. Liu, W.; Zheng, E.L.; Zhou, Y.Q.; Yao, H.P.; Zhu, Y. Vibration characteristics analysis of wheeled tractor/implement system under plough operation condition. *J. Nanjing Agric. Univ.* **2021**, *44*, 1002–1012. [CrossRef]
19. Zhu, S.H.; Xu, G.; Yuan, J.Q.; Ma, J.F.; Yi, L.; Li, K. Influence of implement's mass on vibration characteristics of tractor-implement system. *Trans. Chin. Soc. Agric. Eng.* **2014**, *30*, 30–37. [CrossRef]
20. Hu, H.B.; Li, Z.Y. Discussion on the working status and efficiency of tractor rotary tillage unit. *Trans. Chin. Soc. Agric Mach.* **1989**, *03*, 96–99.
21. Liu, B.J.; Geng, Y.F. Research & development on dongfanghong 1gqn series rotary tiller. *Shandong Agric. Mach.* **2002**, *8*, 5–9. [CrossRef]
22. Lin, J.X.; Liao, Q.X.; Zhang, Q.S.; Zhang, Q.S.; Kang, Y.; Zhang, J.Q. Vibration characteristics analysis and structural improvement of the shovel type seedbed preparation machine suitable for rapeseed mechanical direct seeding. *Trans. Chin. Soc. Agric. Eng.* **2023**, *39*, 39–48. [CrossRef]
23. Wei, J.B.; Wang, Y.; Zhang, C.L.; Jiang, Y.B.; Guan, L. Application analysis of matching tractor PTO shaft and agricultural ma-chinery. *China South. Agric. Mach.* **2023**, *54*, 45–48. [CrossRef]
24. Wang, J.X.; Shuai, S.J. *Automotive Engine Fundamentals*, 2nd ed.; Tsinghua University Press: Beijing, China, 2011; 128p.
25. Pang, J.; Li, Y.M.; Ji, J.T.; Xu, L.Z. Vibration excitation identification and control of the cutter of a combine harvester using triaxial accelerometers and partial coherence sorting. *Biosyst. Eng.* **2019**, *185*, 25–34. [CrossRef]
26. Gao, Z.P.; Xu, L.Z.; Li, Y.M.; Wang, Y.D.; Sun, P.P. Vibration measure and analysis of crawler-type rice and wheat combine harvester in field harvesting condition. *Trans. Chin. Soc. Agric. Eng.* **2017**, *33*, 48–55. [CrossRef]
27. Qian, P.; Lu, T.; Shen, C.; Chen, S. Influence of vibration on the grain flow sensor during the harvest and the difference elimination method. *Int. J. Agric. Biol. Eng.* **2021**, *14*, 149–162. [CrossRef]
28. Yao, Y.C.; Song, Z.H.; Du, Y.F.; Zhao, Y.X.; Mao, E.R.; Liu, F. Analysis of vibration characteristics and its major influenced factors of header for corn combine harvesting machine. *Trans. Chin. Soc. Agric. Eng.* **2017**, *33*, 40–49. [CrossRef]

29. Yan, J. Study on Vibration Characteristics of Tractor Based on Excitation of Agricultural Terrain Roughness. Ph.D. Dissertation, Inner Mongolia Agricultural University, Hohhot, China, 2020.
30. Wu, X. Application of fast fourier transform in signal processing. *Inf. Rec. Mater.* **2021**, *22*, 184–186. [CrossRef]
31. Zhang, H.Q.; Wang, C.; Cui, Y. Comparative analysis study of conventional time-frequency analysis methods. *Chang. Inf. Commun.* **2020**, *08*, 55–56.
32. Geng, L.X.; Li, K.; Pang, J.; Jin, X.; Ji, J.T. Test and analysis of vibration characteristics of transplanting machine based on time frequency and power spectral density. *Trans. Chin. Soc. Agric. Eng.* **2021**, *37*, 23–30. [CrossRef]
33. Ma, Z.; Zhang, Z.L.; Zhang, Z.H.; Song, Z.Q.; Liu, Y.B.; Li, Y.M.; Xu, L.Z. Durable Testing and Analysis of a Cleaning Sieve Based on Vibration and Strain Signals. *Agriculture* **2023**, *13*, 2232. [CrossRef]
34. Zhang, M.J.; Jin, J.F.; Chen, Y.Y.; Chen, Y.K. Vibration symmetry characteristics of wheeled tractor structure. *J. Jilin Univ. (Eng. Technol. Ed.)* **2023**, *53*, 2136–2142. [CrossRef]
35. Wang, W.M.; Wang, T.Y.; Guo, B.; Chen, X.; Zhou, F.J. Design and test of key components of rotary cultivator based on vibration drag reduction principle. *Trans. Chin. Soc. Agric. Mach.* **2019**, *50*, 35–45. [CrossRef]
36. Zhou, Z.; Griffin, M.J. Response of the seated human body to whole-body vertical vibration: Biodynamic responses to sinusoidal and random vibration. *Ergonomics* **2014**, *57*, 693–713. [CrossRef] [PubMed]
37. Liu, G.; Xia, J.; Zheng, K.; Cheng, J.; Wang, K.; Zeng, R.; Wang, H.; Liu, Z. Effects of vibration parameters on the interfacial adhesion system between soil and metal surface. *Soil Till. Res.* **2022**, *218*, 105294. [CrossRef]

agriculture

MDPI

Article

Experimental and Numerical Analysis of Straw Motion under the Action of an Anti-Blocking Mechanism for a No-Till Maize Planter

Qingyi Zhang [1,2,3], Huimin Fang [1,2,3,*], Gaowei Xu [4], Mengmeng Niu [5] and Jinyu Li [1]

1 School of Agricultural Engineering, Jiangsu University, Zhenjiang 212013, China; zhangqingyi@ujs.edu.cn (Q.Z.); 2212316040@stmail.ujs.edu.cn (J.L.)
2 Key Laboratory for Theory and Technology of Intelligent Agricultural Machinery and Equipment, Jiangsu University, Zhenjiang 212013, China
3 Jiangsu Province and Education Ministry Co-Sponsored Synergistic Innovation Center of Modern Agricultural Equipment, Zhenjiang 212013, China
4 Department of Automotive Engineering, Shandong Jiaotong University, Jinan 250357, China; 202107@sdjtu.edu.cn
5 Field Operation Technology and Equipment Innovation Center, Shandong Academy of Agricultural Machinery Sciences, Jinan 250100, China; moonniu@126.com
* Correspondence: fanghuimin@ujs.edu.cn; Tel.: +86-155-0860-2886

Abstract: To address the low clearance rate issue of the anti-blocking mechanism for maize no-till planters in the Huang-Huai-Hai Plain of China, experiments and simulations were conducted to analyze the individual and collective movements of straw under the action of the round roller-claw anti-blocking mechanism. A tracer-based measurement method for straw displacement was applied firstly. Experimental results showed that the straw forward displacement could be characterized by the average horizontal displacements of longitudinal and lateral tracers, while the straw side displacement could be characterized by the lateral displacement of the longitudinal tracer. The straw forward displacement was 58.95% greater than the side displacement. Forward, side, and total displacements of straw increased as the mechanism's forward speed increased from 3 km/h to 7 km/h, with corresponding rates of increase at 233.98%, 43.20%, and 162.47%, respectively. Furthermore, a model of straw–soil–mechanism interaction was constructed in EDEM 2022 software. The relative error between experimental and simulated straw clearance rates was 11.20%, confirming the applicability of the simulation model for studying straw–soil–mechanism interaction. Based on the simulation model, three straw tracers of different lengths were selected to study the motion behavior of straw. It was inferred that despite differences in straw length, the movement behaviors of the three straw tracers under the influence of the anti-blocking mechanism were similar. Additionally, longer straws exhibited greater displacements in all directions. This paper serves as a reference for studying straw motion behavior influenced by anti-blocking mechanisms.

Keywords: no-till planting; anti-blocking mechanism; straw displacement; tracer technology; discrete element method; straw clearance rate; straw motion behavior

Citation: Zhang, Q.; Fang, H.; Xu, G.; Niu, M.; Li, J. Experimental and Numerical Analysis of Straw Motion under the Action of an Anti-Blocking Mechanism for a No-Till Maize Planter. *Agriculture* **2024**, *14*, 1001. https://doi.org/10.3390/agriculture14071001

Academic Editors: Xiaojun Gao, Qinghui Lai and Tao Cui

Received: 15 May 2024
Revised: 19 June 2024
Accepted: 23 June 2024
Published: 26 June 2024

1. Introduction

The Huang-Huai-Hai Plain in China is characterized by semi-arid and semi-humid climates, with the predominant agricultural practice being the rotation of winter wheat and summer maize crops. With the continuous improvement in crop yields in recent years, there has been a corresponding increase in the production of straw [1]. However, an increasing amount of straw is being either discarded or openly burned, leading to significant resource wastage and a range of associated issues such as atmospheric pollution and traffic disruptions [2].

Incorporating crop residue cover, as a fundamental tenet of conservation agriculture, offers multifaceted benefits. It not only enhances straw utilization efficiency and mitigates pollution [3] but also mitigates soil erosion [4], enhances the physico-chemical properties of soil [5], and boosts crop yield [6]. However, the presence of straw residue covers poses challenges to maize planting in the Huang-Huai-Hai Plain. Maize planters are prone to blockages, leading to poor planting possibility and sowing quality [7]. Anti-blocking maize planters play a crucial role in ensuring efficient seeding and high sowing quality [8]. An effective anti-blocking mechanism is essential for mitigating straw blockages and enhancing the sowing reliability of maize planters.

Various types of anti-blocking mechanisms are designed to perform tasks such as chopping residues ahead and along the planting path, pushing residues sideways, or burying residues in a strip ahead of furrow opening. Recent developments in anti-blocking mechanisms include an oblique anti-blocking device [9], a round roller-claw anti-blocking mechanism [10], a spiral discharge anti-blocking and row-sorting device [11], a bionic shifting and diffluence straw anti-blocking device [12], and a combined stubble burying and anti-blocking device [13]. These anti-blocking mechanisms find applications in rapeseed direct planting as well as corn no-tillage planting. Theoretical [14,15], experimental [16,17], and numerical [18,19] methods are employed to assess the performance of anti-blocking mechanisms in terms of soil disturbance and straw cleaning rate. For instance, Li et al. [14] analyzed the performance of symmetrical spiral row-sorting of the straw device with a theoretical method, Jiang et al. [16] tested the performance of a seedbed preparation machine before transplanting of rapeseed with a combined transplanter in the field, and Zhu et al. [18] optimized the structural parameters of the bionic shifting and diffluence straw anti-blocking mechanism via EDEM 2022 software. However, these studies focused on the design and optimization of the anti-blocking mechanism. There is a lack of attention to the interaction between straw and the anti-blocking mechanism.

The interaction between straw and agricultural implements is crucial for developing effective anti-blocking mechanisms, and the interaction is usually characterized by straw movement [20]. Researchers employ various methods to analyze straw movement, including theoretical analysis, advanced testing, numerical simulations, and experiments. Gu et al. [21] demonstrated the effectiveness of the "laminar flow splitter" due to its property of round-flow. Gao et al. [22] designed an anti-blocking mechanism, combining a driven divider with a passive residue separating device based on the theory of the boundary layer in fluid mechanics. Liao et al. [23] tracked the movement and throwing trajectory of straws, cut down by the saw-tooth anti-blocking mechanism of a no-tillage planter, with the help of high-speed photography. Niu et al. [15] studied individual and group straw movement numerically. Fang et al. [20] evaluated anti-blocking devices through experiments and simulations. Some studies utilized high-speed cameras [9] and tracer technology [10] to observe straw movement but lacked detailed explanations. Understanding straw motion requires clarity on how anti-blocking devices control and guide it, an aspect currently under-researched in straw–soil–mechanism interaction studies.

In order to address the dilemma of our unclear understanding of straw motion under the influence of anti-blocking mechanisms, a controlled soil bin experiment and discrete element simulation were conducted to achieve the following objectives: (i) propose a measurement method for straw displacement and investigate the effects of working parameters of the anti-blocking mechanism on straw displacement via an experiment; (ii) validate the simulation model of straw–soil–mechanism interaction, and analyze the motion behavior of straw based on simulation.

2. Materials and Methods

2.1. Description of Anti-Blocking Mechanism

The anti-blocking furrow opener, depicted in Figure 1a, is constructed from steel and affixed to a maize planter, specifically tailored for no-tillage sowing in the Huang-Huai-Hai Plain. A pivotal component of this furrow opener is the round roller-claw anti-blocking mechanism, illustrated in Figure 1b. This mechanism has the capability to rotate horizontally under power while advancing alongside the anti-blocking furrow opener. The geometrical characteristics of the anti-blocking mechanism, as outlined in Table 1, were employed for its shape and design.

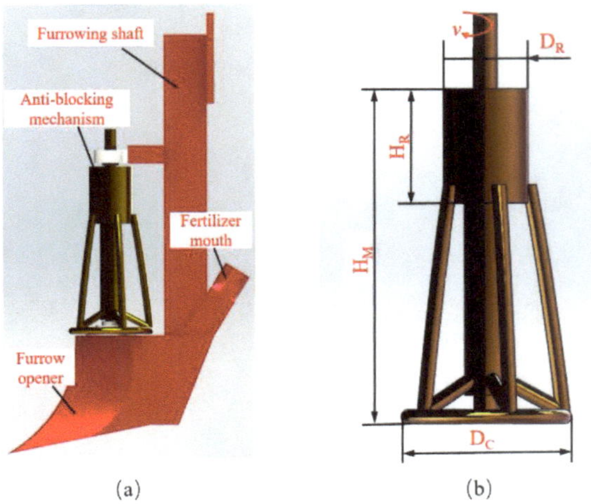

Figure 1. Experimental mechanism: (**a**) furrow opener; (**b**) anti-blocking mechanism.

Table 1. Characteristics of a round roller-claw anti-blocking mechanism.

Parameters	Values
Height of whole mechanism H_M/(mm)	234
Height of round roller H_R/(mm)	80
Biggest diameter of claw D_C/(mm)	120
Diameter of round roller D_R/(mm)	60

2.2. Experimental Design

2.2.1. Description of Soil Bin Experiment

The indoor experiments, depicted in Figure 2, were carried out in the soil bin at the Shandong Academy of Agricultural Machinery Sciences, located in Shandong Province, People's Republic of China (China). The dimensions of soil bin are 60 m × 2.5 m × 1 m (length × width × depth) and it is filled with sufficient soil, enabling the testing of the maize planter. The maize planter was propelled by a PTO shaft of the soil bin tester, while two sets of gear reducers facilitated the rotational motion of the anti-blocking mechanism. The soil density was managed at 1.70 g/cm^3 with the average moisture content of the 0–5 cm soil layer as 10.07%, and 13.51% for the next 5 cm soil depth. The soil hardness was 0.96 MPa.

The straw was uniformly distributed across the soil surface, creating a coverage area of straw measuring 1 m in width and 20 m in length, ensuring that the anti-blocking mechanism moved forward in the middle of the straw row. The wheat straw used in the experiments was harvested from fields in Shandong Province, China, with an average mulching quantity of 610 kg/mu.

Figure 2. Experimental site in soil bin: (**a**) experimental equipment; (**b**) straw distribution.

2.2.2. Description of Straw Tracer

Five groups of straw tracers were positioned in the test sample area, which were painted purple, blue, red, green, and black, respectively. Each group of straw tracers was spaced 1 m apart to avoid motion interference. Two straw tracers with same color were utilized, with one aligned parallel to the forward motion of the anti-blocking mechanism, termed the longitudinal straw tracer. The other straw tracer, placed perpendicular to it, was called the lateral straw tracer. Every set of longitudinal and lateral straw tracers shared the same color, where the lateral straw tracers were distinguished by special markings at both ends. Three sets of straw tracers (purple, red, and black) were positioned along the forward trajectory of the anti-blocking mechanism, while the remaining two sets (blue and green) were placed laterally, offset by 30 mm from the forward path. The arrangement of the straw tracers is illustrated in Figure 3.

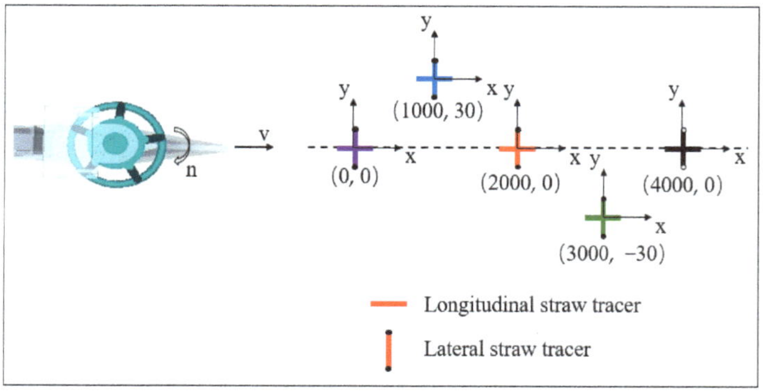

Figure 3. Schematic view of straw tracer placements. Note: v represents the forward velocity, km/h; n represents the rotational speed of the anti-blocking mechanism, r/min.

2.2.3. Measurement

(A) Straw displacement

Straw displacement refers to the movement of straw in both horizontal and lateral directions. It is determined by calculating the absolute difference between the initial and final positions of straw tracers.

(B) Straw clearance rate

The straw clearance rate quantifies the change in the quantity of straw within a 60 mm range on either side of the anti-blocking mechanism forward path. It is calculated as the ratio of the difference in straw weight before and after the anti-blocking mechanism's operation to the initial straw weight within this specified range.

2.3. Simulation Design

2.3.1. Description of Simulation Model

The simulation model of straw–soil-mechanism interaction was developed using EDEM 2022 (Version 8.0.0, Altair Engineering, Inc., Troy, MI, United States). Hard spheres of 5 mm radius were used to represent soil particles, and the contact model between soil particles was based on the Hertz–Mindlin bonding model. Hard spheres with a radius of 3 mm were used to represent straw particles, with a space of 5 mm separating adjacent straw particles.

Considering the actual length of the straw in the field and the size of the straw particles, three lengths of 36, 76, and 116 mm, respectively, were used to model the straw in this study. The 7, 15, and 23 spheres, respectively, were connected to form the 36, 76, and 116 mm straw, respectively. There were 60,000 soil particles and 2400 straw particles, with three lengths of 36 mm, 76 mm, and 116 mm, respectively, generated randomly. According to the previous work [10], the material properties and interaction parameters used in this simulation were as detailed in Table 2. Figure 4 illustrates this simulation model.

Table 2. Simulation parameters.

Parameters	Values
Density of soil/(gm^{-3})	1.85
Density of straw/(gm^{-3})	0.24
Density of steel/(gm^{-3})	7.87
Poisson ratio of soil	0.38
Poisson ratio of straw	0.4
Poisson ratio of steel	0.3
Shear modules of soil/(Pa)	1×10^6
Shear modules of straw/(Pa)	1×10^6
Shear modules of steel/(Pa)	7.9×10^{10}
Recovery coefficient of soil–soil	0.6
Recovery coefficient of soil–steel	0.6
Recovery coefficient of straw–steel	0.3
Static fiction coefficient of soil–soil	0.6
Static fiction coefficient of soil–steel	0.6
Static fiction coefficient of straw–steel	0.3
Rolling fiction coefficient of soil–soil	0.4
Rolling fiction coefficient of soil–steel	0.05
Rolling fiction coefficient of straw–steel	0.01

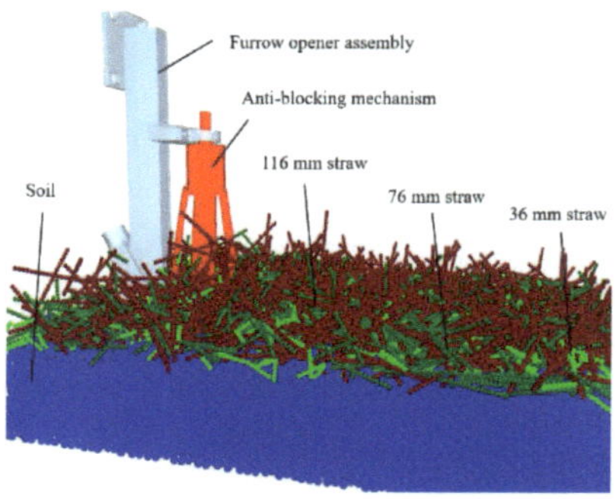

Figure 4. Simulation model of straw–soil-mechanism interaction.

2.3.2. Description of Straw Tracer

In the simulation, 2400 straw particles were randomly generated on the soil surface. Straws of different lengths were chosen as the straw tracer to assess and analyze the displacement of a single piece of straw by monitoring its movement. As straw was randomly distributed on the soil surface, it was essential to compare the differences in the motion behavior of straw at the same location. There are two principles for selecting the three straw tracers. Firstly, tracers should align closely with the forward path of the anti-blocking mechanism, maintaining consistency in the X coordinate values. Secondly, maintaining consistent relative heights of the tracers ensures uniformity in the Y coordinate values, that is, the Y coordinate values are as consistent as possible. Based on these principles, the selected straw tracers and their relative positions in the anti-blocking mechanism are as shown in Figure 5.

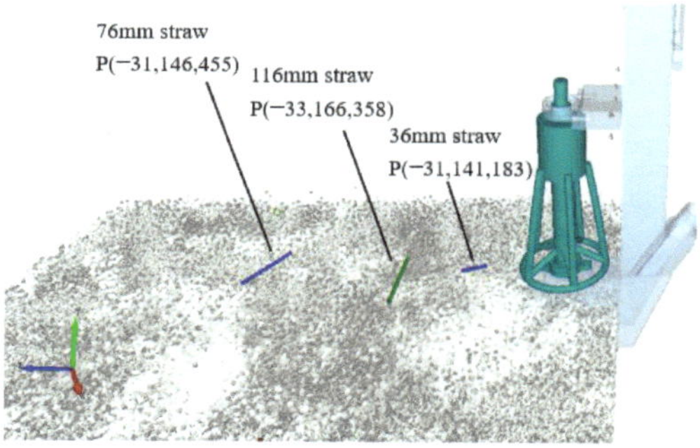

Figure 5. The straw tracers in the simulation and their positions.

2.3.3. Measurement

(A) Straw movement

The positions of three straw tracers were continuously monitored at every instance, utilizing the established model for full coverage of straw distribution. Straw movement refers to the positional changes in straw in both horizontal and lateral directions. The simulation characterizes straw movement by observing the fluctuations in tracer positions over time.

(B) Straw clearance rate

We chose a 60 mm span on either side of the anti-blocking mechanism as the designated area for analyzing straw clearance in the simulation. The straw clearance rate is defined as the ratio of the difference in the straw quantity before and after simulation to the initial straw quantity.

2.4. Experimental and Numerical Treatment Design

Three sets of experiments and simulation were designed and conducted in this study. In the first group of experiments, the orientation and position of the straw tracers were evaluated to determine their impact on straw displacement based on the soil bin test. Additionally, the method for measuring straw displacement was established. Subsequently, the effect of the operational parameters of the anti-blocking mechanism (forward and rotational speed) on straw displacement (forward, side, and total displacement) was investigated using the established measurement technique. This experimental set included three sets of forward speeds (3, 5, and 7 km/h) and four sets of rotational speeds (260, 400, 530, and 740 r/min).

The second experiment focused on using the straw clearance rate to characterize straw side movement. The verification of the simulation model was assessed by comparing the difference in the straw clearance rate between simulation and experimental results. This experiment involved five sets of forward speeds (3, 5, 7, 8, and 9 km/h) while maintaining a constant rotational speed of 400 r/min.

In the simulation, three different straw lengths (36 mm, 76 mm, and 115 mm) positioned approximately at the same location were chosen for motion tracking based on the simulation model. This numerical treatment aimed to analyze the motion behavior of straw under constant conditions of rotational speed (400 r/min) and forward speed (7 km/h).

2.5. Statistical Analysis

Detection of significant differences among the results was performed through analysis of variance using SPSS ver.23 [24]. A significance level of 0.05 in a Duncan test was adopted in this study [25].

3. Results and Discussions

3.1. Measurement Method of Straw Displacement

3.1.1. Effect of Tracer Orientations on Straw Displacement

Longitudinal and lateral orientations were observed for all straw tracers during the experiment. Forward and side straw displacements, measured using longitudinal and lateral tracers, are depicted in Figure 6. The round roller-claw anti-blocking mechanism was tested at rotational speeds of 260, 400, 530, and 740 r/min, respectively, along with three forward speeds of 3, 5, and 7 km/h.

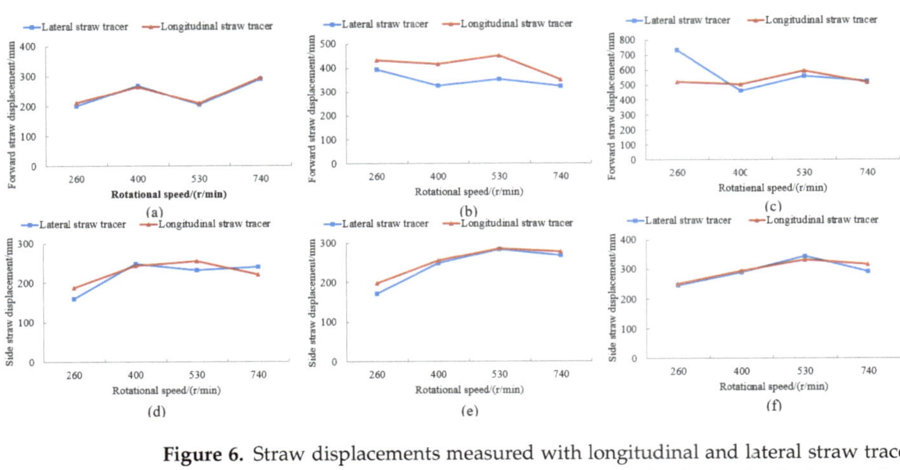

Figure 6. Straw displacements measured with longitudinal and lateral straw tracers: (**a**,**d**) straw forward and side displacements under forward speeds of 3 km/h; (**b**,**e**) straw forward and side displacements under forward speeds of 5 km/h; (**c**,**f**) straw forward and side displacements under forward speeds of 7 km/h.

Figure 6a–c depicts the forward displacement of straw measured by tracers placed both longitudinally and laterally under varying operational conditions. The discrepancy between the measurements was minimal at a forward speed of 3 km/h (Figure 6a), exhibiting an average absolute difference of 6.2 mm and an average relative difference of 2.8% However, at a forward speed of 5 km/h (Figure 6b), the disparity peaked, with an average absolute difference of 63.1 mm and an average relative difference of 18.3%. Notably, at a forward speed of 7 km/h, a notable difference emerged only when the anti-blocking mechanism operated at a rotational speed of 260 rpm, resulting in an absolute difference of 212 mm, possibly attributable to considerable error induced by straw dragging [26]. Additionally, the average absolute error under varying operating parameters was 29.22 mm, with a relative error of 5.83%. The side displacement of straw, as depicted in Figure 6d–f, was measured using tracers positioned both longitudinally and laterally across various operational settings. The comparison revealed a slight disparity between the measurements obtained from the two placement strategies, showing an average absolute difference of 13.93 mm and a relative difference of 5.83%.

The analysis suggested that tracers employing the two placement strategies manifest noteworthy errors in assessing forward displacement, which ought to be the mean of both. Nonetheless, the discrepancy in side displacement measurement was insignificant. Hence, longitudinal straw tracers are recommended for side displacement measurement, aligning with the findings of Fang et al. [26], that is, when studying the straw displacement under the action of a rotating component, the straw forward displacement can be obtained by considering the average horizontal displacements of longitudinal and lateral tracers, and the straw side displacement can be obtained by considering the lateral displacement of the longitudinal tracer.

3.1.2. Effect of Tracer Positions on Straw Displacement

Using straw tracers positioned variably to measure both forward and side displacements of straw, the effects of the tracer position on straw displacements were ascertained, as depicted in Figure 7.

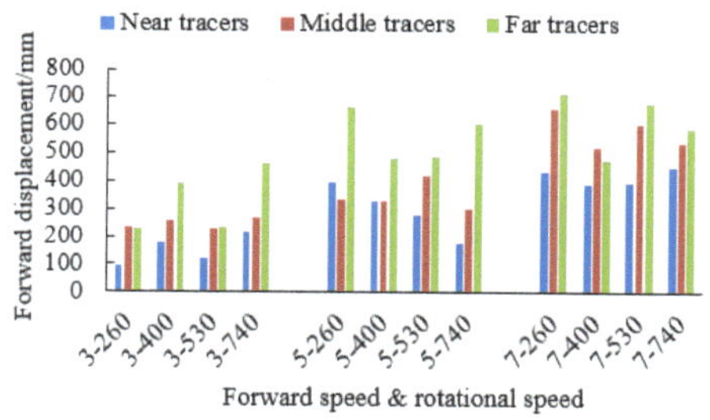

(**a**)

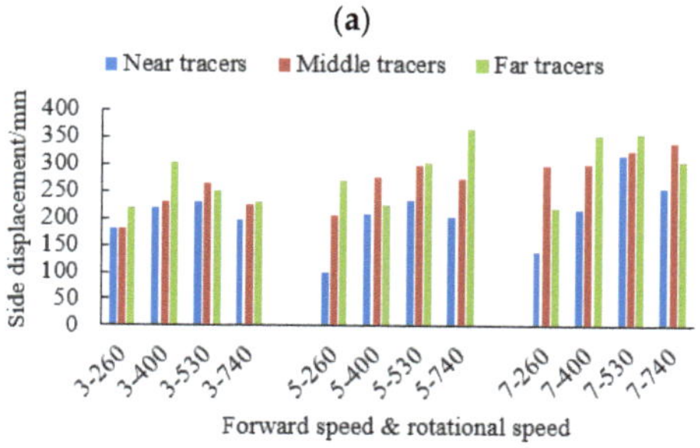

(**b**)

Figure 7. Straw displacements measured with straw tracers at different positions: (**a**) the forward straw displacement measured with straw tracers at different positions; (**b**) the side straw displacement measured with straw tracers at different positions.

The anti-blocking mechanism induces horizontal movement in straw as it advances. Additionally, the rotating mechanism throws straw tangentially, resulting in a combined motion in both horizontal and lateral directions. A comparison of straw displacements in the horizontal and lateral directions reveals that lower forward and rotational speeds of the anti-blocking mechanism result in smaller side displacements of the nearby straw tracer. Furthermore, the side displacement of straw was generally less than the forward displacement, aligning with previous research findings [27]. This phenomenon primarily occurs due to the straw's low side component velocity upon being thrown tangentially by the mechanism, coupled with constraints from surrounding straw.

Straw tracers placed at various positions experience distinct forces from the anti-blocking mechanism and surrounding straw, leading to variations in straw displacement. Both forward and side displacements followed a consistent trend, where the straw tracer at the distant position showed the highest displacement, followed by the middle position, and finally, the nearby position had the smallest displacement. This is attributed to the motion of straw under the influence of the anti-blocking mechanism resembling fluid flow; hence, straw at distant positions experiences larger displacements as it is situated at the periphery of the fluid. Using the displacement measured by the middle-position straw tracer as the control, the average forward displacement of the distant-position straw tracer exceeded that of the middle position by 28.37%, while the middle-position tracer's displacement was 25.75% higher than the nearby tracer. In the lateral direction, the side displacement measured by the straw tracer at the distant position was 5.95% higher than that measured by the straw tracer at the middle position, and the side displacement of the middle-position tracer was 22.07% higher than that of the nearby tracer. Concerning various forward speeds of the anti-blocking mechanism, the forward displacements measured by the distant-position straw tracer were 34.03%, 61.94%, and 5.98% higher than those of the middle position, while the middle-position tracer's displacements were 37.66%, 14.18%, and 27.59% higher than those of the nearby tracer at forward speeds of 3, 5, and 7 km/h. In terms of side displacement, the straw tracer in the middle position showed displacements 8.26%, 28.89%, and 26.26% higher than those of the nearby tracer, respectively, at forward speeds of 3, 5, and 7 km/h. Side displacements measured by the distant-position straw tracer were 11.34% and 10.79% higher than those measured by the middle-position straw tracer at forward speeds of 3 and 5 km/h, respectively. At a forward speed of 7 km/h, the distant position straw tracer's displacements were marginally (1.94%) less than those of the middle-position straw tracer.

Furthermore, upon comparing the average displacements of the straw tracers at distant and nearby positions with the displacement obtained by the straw tracer in the middle position, it was found that the relative errors in forward and side displacements under various operating parameters of the anti-blocking mechanism were 21.25% and 12.35%, respectively. At a forward speed of 3 km/h for the anti-blocking mechanism, the difference in forward displacement between them exhibited no discernible pattern with variations in the rotational speed of the anti-blocking mechanism. This observation was consistent for side displacement as well, with relative errors in forward and side displacements at 23.06% and 9.75%, respectively. When the forward speed of the anti-blocking mechanism was 5 km/h, the mean forward displacement of the straw tracers at distant and nearby positions was 29.95% higher than that of the straw tracer in the middle position, while the mean side displacement was 11.14% lower than that of the straw tracer in the middle position. At a forward speed of 7 km/h for the anti-blocking mechanism, the average forward displacement of the straw tracers at distant and nearby positions was 10.73% less than that of the straw tracer in the middle position. Moreover, the side displacement of the two exhibited no apparent correlation with alterations in the rotational speed of the anti-blocking mechanism, with a relative error of 16.16%.

Hence, to comprehensively account for the impact of straw tracers at distinct positions on straw displacement and to better reflect the actual dynamics of straw movement, this research adopted the mean displacements of straw tracers at varied positions as representative values for straw displacement. The straw tracers were generally not placed in different positions in previous studies [26,27], and this study provides a straw displacement measurement method to reduce measurement errors caused by the position of straw tracers. Specifically, the mean forward displacement of straw tracers at distant, middle, and nearby positions will serve as the horizontal displacement metric for straw, while the mean side displacement of straw tracers in the same positions will function as the corresponding side displacement metric.

3.2. Straw Displacement Influenced by the Working Parameters of the Mechanism

Following the harvest of the wheat crop, the straw is distributed across the field, after which maize seeding is conducted directly. If the straw cannot be pushed ahead and sideways, the maize planter is prone to blockages. Assessing the impact of operational parameters on straw displacement is crucial for evaluating the effectiveness of the anti-blocking mechanism [10]. Figure 8 displays the forward, side, and total straw displacement under various rotational speeds (260, 400, 530, and 740 rpm) and forward speeds (3, 5, and 7 km/h) of the anti-blocking mechanism. Apart from instances where the straw forward displacement was marginally less than the side displacement at a forward speed of 3 km/h and a rotational speed of 530 rpm, the forward displacement consistently exceeded the side displacement across all operational parameters of the anti-blocking mechanism. The average absolute value of displacement was 155 mm, with a relative difference of 58.95%.

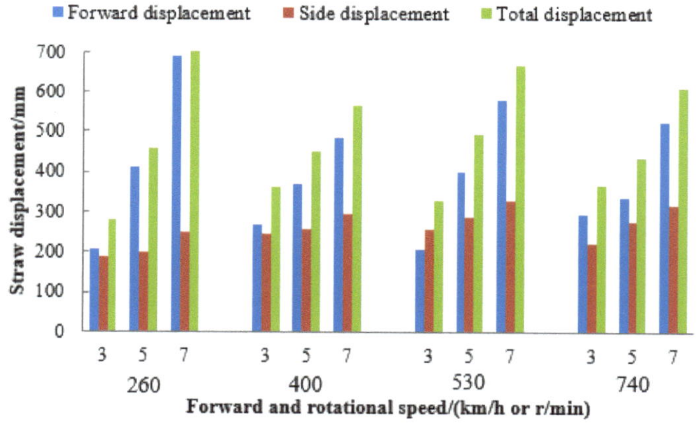

Figure 8. The forward straw displacement measured in different forward and rotational speed conditions.

Forward displacement of straw increased with the forward speed of the mechanism, which maintained consistency across all four rotational speeds of the anti-blocking mechanism. a lower rotational speed of the anti-blocking mechanism correlated with a higher increase in the straw forward displacement rate. At a rotational speed of 260 rpm, the forward displacement increased by 100.49% when the forward speed changed from 3 km/h to 5 km/h, and by 233.98% when the forward speed increased from 3 km/h to 7 km/h. Nevertheless, no consistent pattern emerged in the straw forward displacement variations with rotational speed at a constant forward speed of the mechanism.

Regarding side displacement, straw displacement increased with forward speed, consistent across all four rotational speeds of the anti-blocking mechanism. A higher rotational speed of the anti-blocking mechanism correlated with a greater increase in straw side displacement rate. At a rotational speed of 740 rpm, side displacement increased by 25.38% as the forward speed changed from 3 km/h to 5 km/h, and by 43.20% as the forward speed increased from 3 km/h to 7 km/h. Nevertheless, no consistent pattern emerged in straw side displacement variations with rotational speed at a constant forward speed of the mechanism.

Concerning total displacement, straw displacement rose with an increasing forward speed of the mechanism at constant rotational speeds of the anti-blocking mechanism, consistent across all four rotational speeds. Since forward displacement of straw consistently exceeded side displacement, the trend in total straw displacement mirrored that of forward displacement. Thus, a lower rotational speed of the anti-blocking mechanism resulted in a higher rate of increase in total straw displacement. At a rotational speed of 260 rpm, total displacement increased by 64.23% as the forward speed changed from 3 km/h to 5 km/h, and forward displacement increased by 162.47% as the forward speed increased from 3 km/h to 7 km/h. Nevertheless, there was no discernible pattern in the variations in total straw displacement with rotational speed at a constant forward speed of the mechanism.

3.3. Verification of the Numerical Model and Straw Motion Analysis

3.3.1. Verification of the Simulation Model Based on the Straw Clearance Rate

The straw clearance rate might be more suitable to describe straw side movement because it represents the movement behavior of a group of pieces of straw in a defined specific area, while the straw side displacement was measured only by specific straw tracers. The experimental and simulation results of straw clearance rates with different forward speeds are shown in Figure 9, where the rotational speed of the anti-blocking mechanism is 740 rpm.

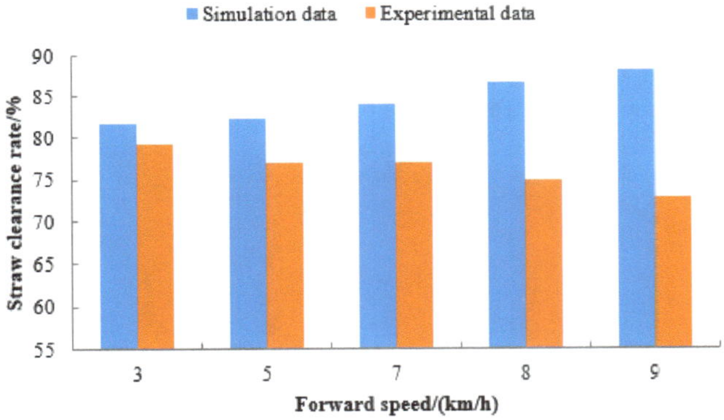

Figure 9. Comparison of straw clearance rates between simulation and experimental results.

The straw clearance rate decreased with the increase in the forward speed of the anti-blocking mechanism in the experiment. The forward speed was too fast for the straw to be ejected from the furrow in time when it moved outward with the anti-blocking mechanism. This also indicated that the straw would remain in the furrow when the ratio of circumferential velocity to forward velocity was large. However, the decrease in the proportion of straw clearance rate with the increase in the forward speed of the mechanism was not significant; the straw clearance rate only increased by 8.17% when the forward speed increased from 3 km/h to 9 km/h.

In the simulation, the straw clearance rate increased with the increase in the forward speed of the anti-blocking mechanism. Additionally, the straw clearance rate only increased by a small margin (7.84%) when the forward speed increased from 3 km/h to 9 km/h. There was a different trend in straw clearance rate variation between the simulation and experimental results, which might be because the straw in the simulation was rigid and did not intertwine and clump together while moving with the anti-blocking mechanism. But during the experiment, where the straw could become entangled and compressed when the forward speed was high, this phenomenon made the straw difficult to eject. The difference in straw clearance rates the between simulation and experimental results might indicate that rigid straw was ejected more easily.

Although the simulation straw clearance rates under various operating parameters were higher than the experimental data, the discrepancies were relatively minor. The average relative error between the experimental and simulation straw clearance rates was 11.20%, with the minimum error (2.99%) occurring at a forward speed of 3 km/h for the anti-blocking mechanism, and the maximum error (20.95%) occurring at a forward speed of 9 km/h. The relative error increased with the increase in the forward speed of the mechanism, indicating that the assumption of rigid straw in the simulation had a greater impact when the forward speed of the mechanism was high. However, the small differences in their values also indicated that the simulation model was suitable for studying the interaction between straw and the anti-blocking mechanism, especially at a lower forward speed (less than 7 km/h, where the error is 9.11%). Therefore, subsequent simulations of straw motion were based on a forward speed of 7 km/h for the anti-blocking mechanism.

3.3.2. Analysis of Individual Straw Movement Behaviors

Three straw tracers, each of different lengths, were selected to track their motion during the simulation. At the beginning of the simulation, all three straws were stationary and were located on the ground in front of the anti-blocking mechanism. Subsequently, under the combined effect of the anti-blocking mechanism and the surrounding straw, they ascended along the rotation direction of the mechanism at a certain speed. Afterwards, the straws were ejected under the increasing centrifugal force and decreasing support force of the straw bed. The changes in the position of the straw in the side, forward, and vertical directions are shown in Figure 10. From the figure, it is evident that although the lengths of the straws are different, their motion behaviors under the action of the anti-blocking mechanism are similar.

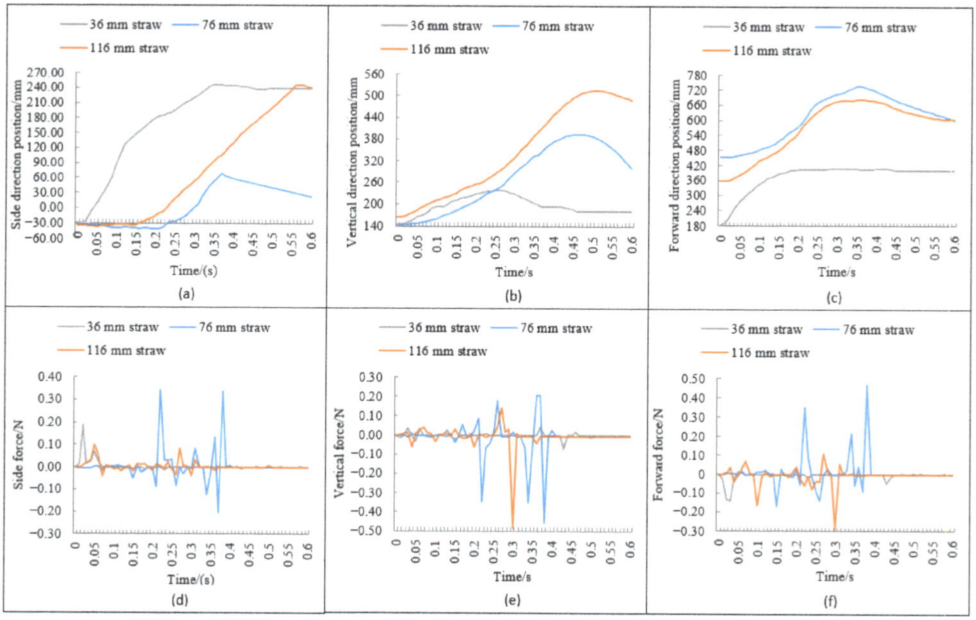

Figure 10. The movement and force of selected straws: (**a–c**) the movement trajectory of straw in the side, vertical, and forward directions over time; (**d–f**) the force exerted on straw in the side, vertical, and forward directions over time.

In the side direction, the straws moved along the rotation direction of the anti-blocking mechanism and were subsequently ejected. This phenomenon was similar to the motion behavior of straw under the action of rotating mechanisms [27]. The side displacements of the longer and shorter straws were similar, while that of the medium-length straw was the smallest. The shortest straw (36 mm) rapidly moved approximately 280 mm laterally under the influence of the anti-blocking mechanism and then remained nearly stationary thereafter. The longest straw (116 mm) did not exhibit as drastic a movement as the shortest straw (36 mm) but had approximately equal side displacements. The smallest side displacement of the medium-length straw might be due to the inhibiting effect of the surrounding dense straw.

In the vertical direction, the straws exhibited climbing behavior under the influence of the anti-blocking mechanism. This proved that the mechanism with a small-diameter upper and large-diameter lower structure would lift the straw during rotation [15]. Moreover, the climbing behavior was more pronounced and the climbing duration longer for the longer straw. The shortest straw only climbed approximately 95 mm before descending, while the medium-length and the longest straw climbed 248 mm and 348 mm, respectively. The vertical motion of the straw tracers confirmed the functionality of the anti-blocking mechanism in guiding the straw to climb.

In the horizontal direction, the straw moved forward under the influence of the anti-blocking mechanism. All straws of different lengths moved forward under the push of the mechanism initially. Subsequently, the medium-length straw and the longest straw exhibited backward behavior after losing the guiding effect of the mechanism. The continuous backward movement behavior indicated that the straw formed a smooth flow under the action of the mechanism. Additionally, the longer the straw, the greater the distance it travelled; for example, the 116 mm straw moved 325 mm in the forward direction.

4. Conclusions

A tracer-based measurement method was applied to investigate the impact of operational parameters of the anti-blocking mechanism on straw displacement. Then, a straw–soil-mechanism interaction model was developed using EDEM 2022 software. The model's feasibility was confirmed from the perspective of straw clearance rates. Furthermore, the model was used to study the individual motion behavior of straw over time under the influence of the anti-blocking mechanism. The research results showed that the tracer method could be used to study the straw displacement under the action of the anti-blocking mechanism. The experimental results also indicated that the forward displacement, side displacement, and total displacement of the straw increased with the increase in the forward speed of the mechanism at a constant speed of rotation. The straw–soil–mechanism interaction model could be used to study the motion behavior of straw. Straws in different positions had similar movement behaviors under the action of the anti-blocking mechanism. The motion behavior of straw under the anti-blocking mechanism offers insights into the interactions among straw, soil, and anti-blocking mechanisms.

Author Contributions: Conceptualization, Q.Z. and H.F.; methodology, Q.Z. and M.N.; software, Q.Z. and H.F.; validation, Q.Z., H.F. and G.X.; formal analysis, M.N. and J.L.; investigation, Q.Z., H.F., G.X. and M.N.; resources, H.F.; data curation, H.F.; writing—original draft preparation, Q.Z. and J.L.; writing—review and editing, H.F., G.X. and M.N.; visualization, Q.Z.; supervision, H.F.; project administration, H.F.; funding acquisition, H.F. All authors have read and agreed to the published version of the manuscript.

Funding: This research was funded by the China Postdoctoral Science Foundation, grant number 2023M741433, the Research Foundation for Talented Scholars of Jiangsu University, grant number 22JDG041, the Priority Academic Program Development of Jiangsu Higher Education Institutions, grant number PAPD-2023-87, and the Jiangsu University Agricultural Engineering Department Project, grant number NGXBTD20240307.

Institutional Review Board Statement: Not applicable.

Data Availability Statement: The analyzed datasets are available from the corresponding author on reasonable request.

Conflicts of Interest: The authors declare no conflicts of interest.

References

1. Li, W.; Zhang, C.; Ma, T.; Li, W. Estimation of summer maize biomass based on a crop growth model. *Emir. J. Food Agric.* **2021**, *33*, 742–750. [CrossRef]
2. Memon, M.S.; Chen, S.; Niu, Y.; Zhou, W.; Elsherbiny, O.; Liang, R.; Du, Z.; Guo, X. Evaluating the efficacy of sentinel-2B and landsat-8 for estimating and mapping wheat straw cover in rice-wheat fields. *Agronomy* **2023**, *13*, 2691. [CrossRef]
3. Guo, Y.; Cui, M.; Xu, Z. Spatial characteristics of transfer plots and conservation tillage technology adoption: Evidence from a survey of four provinces in China. *Agriculture* **2023**, *13*, 1601. [CrossRef]
4. Melland, A.R.; Antille, D.L.; Dang, Y.P. Effects of strategic tillage on short-term erosion, nutrient loss in runoff and greenhouse gas emissions. *Soil Res.* **2016**, *55*, 201–214. [CrossRef]
5. Tunio, M.H.; Gao, J.; Talpur, M.A.; Lakhiar, I.A.; Chandio, F.A.; Shaikh, S.A.; Solangi, K.A. Effects of different irrigation frequencies and incorporation of rice straw on yield and water productivity of wheat crop. *Int. J. Agric. Biol. Eng.* **2020**, *13*, 138–145. [CrossRef]
6. Khan, I.; Iqbal, B.; Khan, A.A.; Inamullah; Rehman, A.; Fayyaz, A.; Shakoor, A.; Farooq, T.H.; Wang, L. The interactive impact of straw mulch and biochar application positively enhanced the growth indexes of maize (*Zea mays* L.) crop. *Agronomy* **2022**, *12*, 2584. [CrossRef]
7. Ahmad, F.; Adeel, M.; Qiu, B.; Ma, J.; Shoaib, M.; Shakoor, A.; Chandio, F.A. Sowing uniformity of bed-type pneumatic maize planter at various seedbed preparation levels and machine travel speeds. *Int. J. Agric. Biol. Eng.* **2021**, *14*, 165–171. [CrossRef]
8. Zhang, X.; Li, H.; Du, R.; Ma, S.; He, J.; Wang, Q.; Chen, W.; Zheng, Z.; Zhang, Z. Effects of key design parameters of tine furrow opener on soil seedbed properties. *Int. J. Agric. Biol. Eng.* **2016**, *9*, 67–80.
9. Yao, W.; Diao, P.; Miao, H.; Li, S. Design and experiment of anti-blocking components for shallow stubble clearing based on soil bin test. *Agriculture* **2022**, *12*, 1728. [CrossRef]
10. Fang, H.; Niu, M.; Zhu, Z.; Zhang, Q. Experimental and numerical investigations of the impacts of separating board and anti-blocking mechanism on maize seeding. *J. Agric. Eng.* **2022**, *LIII*, 1273.
11. Li, Y.; Lu, C.; Li, H.; He, J.; Wang, Q.; Huang, S.; Gao, Z.; Yuan, P.; Wei, X.; Zhan, H. Design and experiment of spiral discharge anti-blocking and row-sorting device of wheat no-till planter. *Agriculture* **2022**, *12*, 468. [CrossRef]
12. Zhu, H.; Zhang, X.; Hong, Y.; Bai, L.; Zhao, H.; Ma, S. Design and experiment of bionic shifting and diffluence straw anti-blocking device. *Agric. Res. Arid Areas* **2023**, *41*, 318–328.
13. Du, W.; Zhou, G.; Zhang, Q.; Bian, Q.; Liao, Q.; Liao, Y. Design and experiment of the anti-blocking device combined stubble burying for rapeseed direct seeding. *Trans. Chin. Soc. Agric. Eng.* **2024**, *40*, 60–70.
14. Li, Y.; Lu, C.; Li, H.; Wang, Z.; Gao, Z.; Wei, X.; He, D. Design and experiment of symmetrical spiral row-sorting of the straw device based on kinematics analysis. *Agriculture* **2022**, *12*, 896. [CrossRef]
15. Niu, M.; Fang, H.; Chandio, F.A.; Shi, S.; Xue, Y.; Liu, H. Design and experiment of separating guiding anti-blocking mechanism for no-tillage maize planter. *Trans. Chin. Soc. Agric. Mach.* **2019**, *50*, 52–58.
16. Jiang, L.; Tang, Q.; Wu, J.; Yu, W.; Zhang, M.; Jiang, D.; Wei, D. Design and test of seedbed preparation machine before transplanting of rapeseed combined transplanter. *Agriculture* **2022**, *12*, 1427. [CrossRef]
17. Yao, W.; Zhao, D.; Miao, H.; Cui, P.; Wei, M.; Diao, P. Design and experiment of oblique anti-blocking device for no-tillage planter with shallow plowing stubble clearing. *Trans. Chin. Soc. Agric. Mach.* **2022**, *53*, 42–52.
18. Zhu, H.; Wu, X.; Qian, C.; Bai, L.; Ma, S.; Zhao, H.; Zhang, X.; Li, H. Design and experimental study of a bi-directional rotating stubble-cutting no-tillage planter. *Agriculture* **2022**, *12*, 1637. [CrossRef]
19. Wang, L.; Bian, O.; Liao, Q.; Wang, B.; Liao, Y.; Zhang, O. Burying stubble and anti-blocking deep fertilization composite device for rapeseed direct planting in high stubble and heavy soil rice stubble field. *Trans. Chin. Soc. Agric. Mach.* **2023**, *54*, 83–94.
20. Fang, H.; Shi, S.; Qiao, L.; Niu, M.; Xu, G.; Jian, S. Numerical and experimental study of working performance of round roller claw type anti-blocking mechanism. *J. Chin. Agric. Mech.* **2018**, *39*, 1–9.
21. Gu, Y.; Zhang, Y.; Song, J. A study on "laminar flow splitter" as blocking proofing device for mulching no-tillage planters. *Trans. Chin. Soc. Agric. Mach.* **1994**, *25*, 46–51.
22. Gao, N.; Zhang, D.; Yang, L.; Cui, T. Design of anti-blocking mechanism combined driven divider with passive residue separating device. *Trans. Chin. Soc. Agric. Mach.* **2014**, *45*, 85–91.
23. Liao, Q. Analysis on the saw-tooth anti-blocking mechanism for no-tillage planter by the high-speed photograph technology. *Trans. Chin. Soc. Agric. Mach.* **2005**, *26*, 46–49.
24. George, D.; Mallery, P. *IBM SPSS Statistics 23 Step by Step: A Simple Guide and Reference*, 14th ed.; Routledge, Taylor & Francis Group: London, UK, 2016; pp. 169–176.
25. Fang, H.; Niu, M.; Wang, X.; Zhang, Q. Effects of reduced chemical application by mechanical-chemical synergistic weeding on maize growth and yield in East China. *Front. Plant Sci.* **2022**, *13*, 1024249. [CrossRef]

26. Fang, H.; Zhang, Q.; Chandio, F.A.; Guo, J.; Sattar, A.; Arslan, C.; Ji, C. Effect of straw length and rotavator kinematic parameter on soil and straw movement by a rotary blade. *Eng. Agric. Environ. Food* **2016**, *9*, 235–241. [CrossRef]
27. Fang, H.; Ji, C.; Ahmed, A.T.; Zhang, Q.; Guo, J. Simulation analysis of straw movement in straw-soil-rotary blade system. *Trans. Chin. Soc. Agric. Mach.* **2016**, *47*, 60–67.

agriculture

MDPI

Article

Design and Experimentation of Rice Seedling Throwing Apparatus Mounted on Unmanned Aerial Vehicle

Peichao Yuan [1], Youfu Yang [1], Youhao Wei [1], Wenyi Zhang [2,*] and Yao Ji [1,*]

[1] Nanjing Institute of Agricultural Mechanization, Ministry of Agriculture and Rural Affairs, Nanjing 210014, China; 82101212127@caas.cn (P.Y.); 82101215527@caas.cn (Y.Y.); 82101225605@caas.cn (Y.W.)
[2] Graduate School of Chinese Academy of Agricultural Sciences, Beijing 100083, China
* Correspondence: zhangwenyi@caas.cn (W.Z.); jiyao@caas.cn (Y.J.)

Abstract: In order to further exploit the production advantages of rice throwing, this paper proposes a systematically designed throwing device suitable for integration with unmanned aerial vehicles (UAVs). The device primarily comprises a seedling carrying and connection system, a seedling pushing mechanism, and a seedling guiding device. The operational principles and workflow of the device are initially elucidated. Subsequently, an analysis of factors influencing rice throwing effectiveness is conducted, with throwing height, working speed, and the bottom diameter of the seedling guide tube identified as key factors. Seedling spacing uniformity and seedling uprightness are designated as performance indicators. A three-factor, three-level response surface experiment is conducted, yielding regression models for the experimental indicators. Through an analysis of the response surface, the optimal parameter combination is determined to be a throwing height of 142.79 cm, a working speed of 55.38 r/min, and a bottom diameter of the seedling guide tube of 43.51 mm. At these parameters, the model predicts a seedling spacing uniformity of 88.43% and a seedling uprightness of 88.12%. Field experiments validate the accuracy of the optimized model results. Experimental data indicate that under the optimal operational parameters calculated by the regression model, the seedling spacing uniformity is 86.7%, and the seedling uprightness is 84.2%. The experimental results meet the design requirements, providing valuable insights for UAV rice-throwing operations.

Keywords: agricultural machinery; unmanned aerial vehicle; transplanting device; experiment

Citation: Yuan, P.; Yang, Y.; Wei, Y.; Zhang, W.; Ji, Y. Design and Experimentation of Rice Seedling Throwing Apparatus Mounted on Unmanned Aerial Vehicle. *Agriculture* **2024**, *14*, 847. https://doi.org/10.3390/agriculture14060847

Academic Editors: Dainius Steponavičius and Caiyun Lu

Received: 16 April 2024
Revised: 21 May 2024
Accepted: 27 May 2024
Published: 28 May 2024

1. Introduction

Rice constitutes a primary staple crop in China. Rice seedlings transplanted through the process of seedling throwing exhibit growth advantages such as rapid growth and prolific tillering. This practice contributes to the enhancement of rice yield, demonstrating considerable potential for further development [1–4]. To further exploit the advantages of rice throwing, mechanized rice transplanters have rapidly evolved from disorderly to orderly operations [5]. Domestically, orderly rice transplanters mainly include the following three methods: pneumatic, roller-type, and clamping, which transfer seedlings into the field through seedling guides. Pneumatic transplanters occasionally encounter issues with nozzle failure, resulting in unstable operation. Roller-type transplanters may cause damage to seedlings. Clamping transplanters are prone to drift issues and require higher field preparation standards [6–8]. Additionally, the terrain in hilly and mountainous areas is characterized by small agricultural plots and steep slopes. Ground machinery operation is challenging, and field-to-field transitions pose difficulties. The mechanization level of rice cultivation in domestic hilly areas is less than 10% [9–11].

To address these issues, this study aims to leverage the advantages of UAVs for seedling throwing operations. UAV operations are characterized by rapid execution speed and excellent maneuverability [12]. Using UAVs allows for liberation from the structural constraints inherent in ground-based machinery. This facilitates the adjustment of seedling throwing height, thereby enhancing planting effectiveness. In hilly and mountainous terrain, UAVs experience lesser influence from topography. This is another advantage that distinguishes it from ground-based machinery.

Research on agricultural UAVs includes aspects such as agricultural monitoring and disaster assessment using remote sensing technology [13–16]. In terms of UAV operations, applications such as fertilization, pesticide spraying, and seeding have been explored [17–20]. There has been limited research on UAV rice throwing. Therefore, investigating the application of UAVs for solid material handling is of significant reference value to this study [21,22]. In the fertilization stage of UAV operations, flight altitude and spreading speed significantly affect fertilization effectiveness [23–25]. In the seeding stage, agricultural UAVs mainly employ the following two forms of seeding: disorderly broadcasting and orderly row broadcasting [26]. Orderly row broadcasting typically requires guiding channels to control the seed distribution. Song Cancan et al. [27] have studied the wind distribution around multi-rotor UAVs and designed a pneumatic rice seeding device. This device is regulated by a groove wheel to control the amount of seed discharge. The high-speed airflow generated by the fan accelerates the expulsion of rice seeds. Subsequently, the rice seeds are dispersed through separate and independent diversion channels. This method enhances the controllability and uniformity of rice seed sowing. Liu Wei et al. [28] studied the stability of the displacement of a sowing device under a fixed sowing speed condition. He developed a rice seeding device capable of simultaneously sowing five rows. The seeding rate of this device can be adjusted according to the operational requirements.

In conclusion, operational speed, flight altitude, and guiding devices are important factors influencing the orderly operation of UAVs. In recent years, domestic agricultural UAVs have seen rapid development, with continuous expansion of their operational capabilities and significant enhancement of their performance. With the increasing market ownership of UAVs, a deep material technology foundation has been laid for orderly seedling-throwing by UAVs. This paper presents the design of a UAV-mounted sequential rice seedling throwing apparatus. The operational parameters of this device were optimized through bench testing. The validation of the device was conducted through field trials. The aim is to provide design references for expanding the production applications of UAVs and complement existing seedling-throwing machinery types.

2. Materials and Methods

2.1. Overall Structure and Workflow

The UAV-mounted rice-throwing device comprises the seedling carrying and connection system, the seedling pushing mechanism, and the seedling guiding mechanism, as illustrated in Figure 1. The MG-1P agricultural UAV is chosen as the power platform. This UAV is manufactured by DJI, located in Shenzhen, China. It has a rated payload of 10 kg and a maximum flight time of approximately 9 min when fully loaded. The seedling carrying and connection system is placed on a specially designed landing gear, with the seedling pushing mechanism mounted at the bottom of the seedling carrying and connection system. The seedling guiding mechanism is installed beneath the seedling carrying and connection system and the seedling pushing mechanism. The devices that the UAV needs to suspend are all manufactured using 3D printing technology (manufactured by Wenext, headquartered in Shenzhen, China). Their total mass is approximately 5.2 kg, and when fully loaded with seedlings, the overall weight does not exceed 9.8 kg.

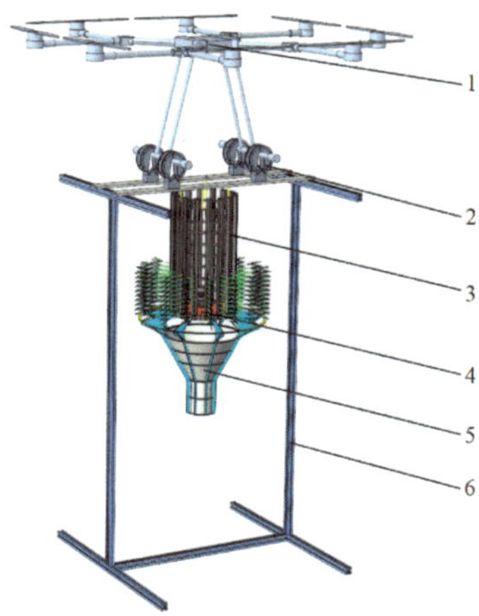

Figure 1. Overall structure of UAV transplanting system: 1. unmanned aerial vehicle; 2. connection mechanism; 3. seedling carrying box; 4. seedling pushing mechanism; 5. seedling guiding device; 6. docking frame.

Based on the operational status during UAV operation, the process can be divided into four stages. First, during the preparation stage, well-nurtured rice bowl seedlings are preloaded into the seedling carrying mechanism before UAV operation. The entire device is then placed on the docking frame. Second, during the seedling retrieval stage, the UAV is remotely controlled to fly above the rice seedling carrying and connection system, hover at an appropriate altitude, and then connect with the UAV using specific mechanical claws. Third, during the operation stage, the UAV suspends the rice seedling throwing apparatus and flies to the designated field. The UAV flies at a constant speed along the planned route. Simultaneously, the remote-controlled seedling-pushing mechanism starts operating. When the first seedling in the seedling slot is ejected, the remaining seedlings fall successively with the assistance of counterweight blocks. Rice seedlings are continuously pushed into the seedling guiding device at a constant speed. Subsequently, the seedlings fall into the field under the combined influence of their own gravity and the airflow generated by the UAV's rotors. The planting process is illustrated in Figure 2. Finally, after the seedling throwing is completed, the UAV flies back above the docking frame. Simultaneously, using a remote control, it opens the mechanical claw to detach from the seedling throwing apparatus. Afterward, the UAV flies toward another seedling-loaded throwing apparatus, repeating the above process in a cycle.

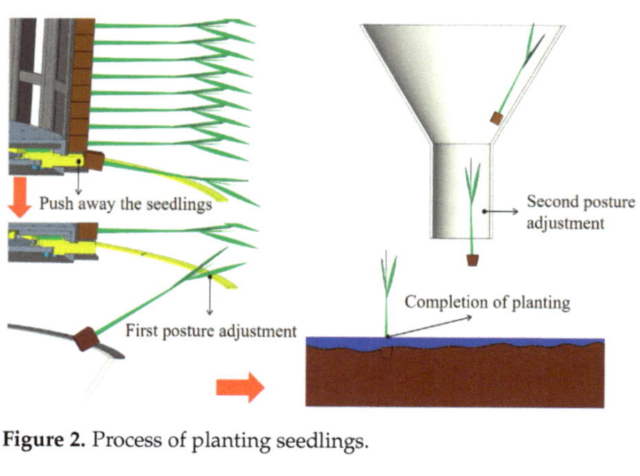

Figure 2. Process of planting seedlings.

2.2. Working Principle

2.2.1. Rice Seedling Carrying and Connection System

The seedling carrying and connection system is employed for both loading seedlings and connecting to the UAV. As shown in Figure 3, it consists of two main parts as follows: the seedling carrying box and the connection mechanism. In this study, the traditional tray-in-a-box approach was abandoned, and a ring-stacked seedling carrying box was introduced to facilitate the installation of the UAV. The height of the seedling carrying box is 900 mm. It consists of 16 seedling slots and can carry up to 960 rice seedlings. To avoid the entanglement of leaves, the seedling slots are evenly distributed on a circle with a diameter of 300 mm. The advantage of the circular distribution is that the consistent position of the seedlings can be easily adjusted. Even if the seedlings are pushed out at different positions, there is no cumulative positional error. At the bottom of each seedling tank, there is a 12 mm diameter hole for the push rod and a seedling outlet, and a top cover is installed on the top of the seedling carrying box. It serves to improve the overall structural strength and provides a mounting position for the connecting mechanism. The connection mechanism comprises four specially designed claws arranged in a rectangular distribution and two cantilever beams. The specialized clamp utilizes a parrot-beak-like comb-shaped claw, altering the force distribution of the clamp. It transforms the lateral clamping force acting on the mechanical claw into longitudinal pressure on the "parrot beak" structure. This pressure further differentiates into shear force on the bolts connecting the claws, ensuring the stability of the connection. The connection mechanism is connected to the top cover of the seedling carrying box via four copper screws.

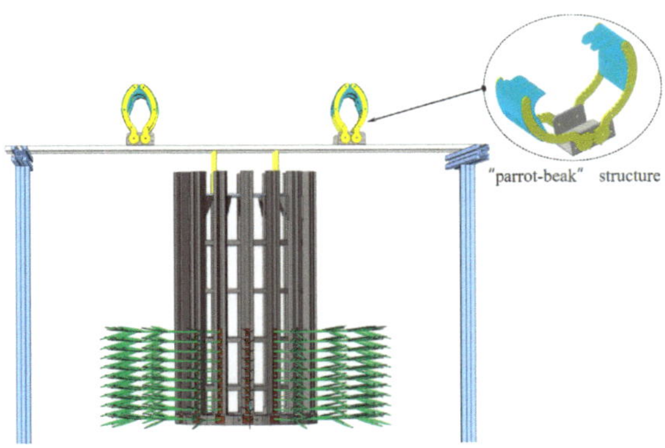

Figure 3. Rice seedling carrying and connection system.

2.2.2. Seedling Pushing Mechanism

The seedling pushing mechanism is utilized to sequentially push rice seedlings out of the seedling carrying box in a predetermined order. Its internal structure is illustrated in Figure 4. The motor base is mounted on the back of the chassis. Within the recess of the chassis, there are 16 identical-sized push rods and connecting rods installed. The push rods are secured onto sliders (MGN7C, manufactured by HIWIN, headquartered in Taiwan, China), ensuring linear movement. At the end face of each push rod, there is a layer of elastic material adhered to prevent damage to the rice bowl substrate during seedling pushing. The distance between the follower crankshaft and the primary crankshaft remains constant. While the length of the connecting rods maintains a fixed proportion to the center distance of the primary crankshaft. The disk is the core component of the seedling-pushing mechanism. It is connected to the follower crankshaft at the bottom, connected to the connecting rods at the top, and linked to the primary crankshaft in the middle section. The motor (42GA775, manufactured by Vantel, headquartered in Jiangsu, China) is connected to the center of the chassis via a primary crankshaft. When the motor drives the main crankshaft to rotate one full circle, it sequentially pushes out 16 rice seedlings.

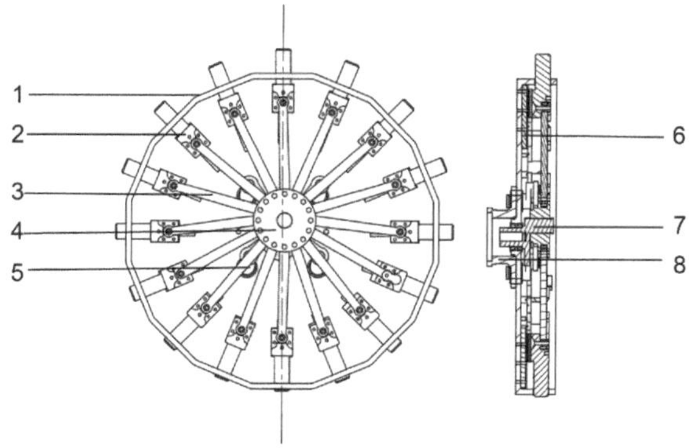

Figure 4. Composition of the Pushing Seedling Institution section: 1. seedling pushing chassis; 2. push rod; 3. connecting rod; 4. disk; 5. follower crankshaft; 6. slide; 7. primary crankshaft; 8. motor base.

The seedling pushing mechanism consists of two types of four-bar linkages combined. The primary crankshaft, follower crankshaft, and disk form a double-crank mechanism. The disk, connecting rod, and push rod form a crank-slider mechanism. In the double-crank mechanism, the motion of each point on the disk follows a circular trajectory consistent with the end of the primary crankshaft. Thus, this point and the center of the hidden trajectory can serve as the crank in the crank-slider mechanism. With this design approach, it is possible to achieve a single motor input and multiple pushrod outputs.

Taking the end actuator of the seedling pushing mechanism as the research subject, the schematic diagram of its mechanism motion is illustrated in Figure 5, where a Cartesian coordinate system is established. The rotated center of the primary crankshaft after translation is the origin *O*. The direction of the push rod axis is the x-axis and the vertical direction is the y-axis. Here, *ab* represents the primary crankshaft, with *a* length of l_1, and *bc* represents the connecting rod with a length of l_2. When the primary crankshaft rotates at an angle α, the angle between the push rod axis and the connecting rod is β, at which point the distance of the slider from the rotation center is *L*.

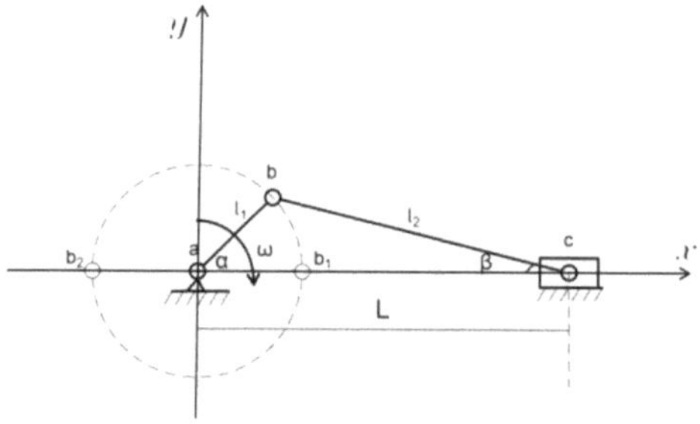

Figure 5. Sketch of the actuating end mechanism of the rice-pushing mechanism.

During the crank rotation, the primary crankshaft and connecting rod are twice in the same straight line, the crank positions are ab_1 and ab_2, which correspond exactly to the two limit positions of the slider, and their distances from the center of rotation are L_{max} and L_{min}, respectively, which can be computed to derive the push rod strokes as follows:

$$s = L_{max} - L_{min} = 2 \times l_1 \tag{1}$$

The height of the rice seedling bowl used in this article is 18 mm. To ensure that the rice seedlings can be fully pushed out, according to Equation (1), the length of the main shaft should be at least 9 mm. Without affecting the working performance, and to ensure a more compact coordination of each part, the dimensions of each component are finally determined. The center distance between the primary crankshaft and the follower crankshaft is 12 mm. The length of the connecting rod is 72 mm. The diameter of the push rod is 12 mm. The diameter of the disk is 50 mm. The diameter of the outer circle of the seedling pushing chassis is 254 mm.

The closed vector equation of the mechanism is as follows:

$$\begin{cases} \vec{L}_1 + \vec{L}_2 = \vec{L} \\ l_1 e^{-i\alpha} + l_2 e^{i\beta} = L \end{cases} \tag{2}$$

Expanding Equation (2) using Euler's formula as follows:

$$\begin{cases} L = l_1 \cos\alpha + l_2 \cos\beta \\ l_1 \sin\alpha - l_2 \sin\beta = 0 \end{cases} \tag{3}$$

Differentiating Equation (2) yields the velocity equation for the slider L as follows:

$$-l_1\omega_1 i e^{-i\alpha} + l_2\omega_2 i e^{i\beta} = V_l \tag{4}$$

Expanding Equation (4) using Euler's formula as follows:

$$\begin{cases} -l_1\omega_1 i \cos\alpha + l_1\omega_1 \sin\alpha + l_2\omega_2 i \cos\beta - l_2\omega_2 \sin\beta = V_l \\ V_l = l_1\omega_1 \sin\alpha - l_2\omega_2 \sin\beta \end{cases} \tag{5}$$

where L represents the distance from the slider to the center of rotation; l_1 denotes the distance from the crankshaft center; l_2 stands for the length of the connecting rod; α represents the angle turned by the crankshaft; β indicates the angle of the connecting rod; ω_1 represents the angular velocity of the crankshaft; and ω_2 represents the angular velocity of the connecting rod.

From Equation (3), it can be observed that when the end displacement of the push rod is maximal, α = 0° and β = 180°, or when the end displacement of the push rod is minimal, α = 180° and β = 0°. Substituting these values into Equation (3) reveals that the velocity of the push rod at these points is zero. This reduces the risk of damaging the seedling matrix. Additionally, the positions of the push rod at any two symmetric locations within the transplanting mechanism are identical, but their trends of variation are opposite. Within the 180° division, the values of α and β for any given phase of the push rod are unequal. It indicates that the motion states of the push rod correspond differently to each phase. This validates the feasibility of gradual transplanting.

The operation process of a single push rod is illustrated in Figure 6a, which depicts the following three sequential positions: the origin, midpoint, and apex positions. These positions correspond to the beginning, middle, and end stages of seedling transplantation, respectively, indicating the relative positions of the seedlings and the transplanting mechanism. As depicted in Figure 6b, the origin position aligns with the bottom end face of the seedling trough, while the apex position aligns with the seedling ejection point. Building upon this foundation, sixteen push rods initiate reciprocating motions in the same sequential order, as shown in Figure 6c. When push rod 1 reaches the apex position, the seedling is completely ejected from the trough, marking the completion of the seedling pushing process. Meanwhile, push rod 2 reaches the midpoint position with its velocity directed outward along the axis, indicating that the seedling is about to be ejected. Conversely, push rod 16, the last push rod in the rotation direction, also reaches the midpoint position with its velocity directed inward along the axis. This indicates that the seedlings have been pushed out before push rod 1.

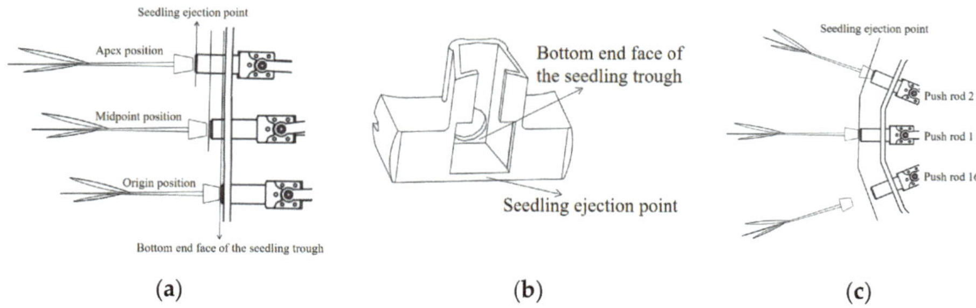

Figure 6. Timing diagram of pushing rice planting movement: (**a**) different timing positions for the same putter; (**b**) position interpretation; (**c**) same timing position for different putters.

2.2.3. Seedling Guiding Device

The seedling guiding apparatus is utilized to adjust the posture of rice seedlings during transplantation, ensuring that the root portion of the seedlings faces downward and descends in an upright manner, as depicted in Figure 7. The blocking ring structure restricts the movement of the seedling tip, facilitating the seedling bowl to drop first. The conical section of the seedling guide pipe concentrates the landing position of rice seedling bowls. The cylindrical narrowing section of the guide pipe enhances the uprightness of the seedlings. In order to reduce the overall weight of the device, it comprises an external framework and internal seedling guide pipe. The external framework is securely fastened to the seedling carrier frame using bolts. The internal seedling guide pipe is positioned within the external framework, enabling adjustment of the rice seedling drop channel. Since the external framework bears the primary load, the internal seedling guide pipe can be made as thin as possible. By maintaining seedling guidance functionality, the overall weight of the guide tube can be significantly decreased.

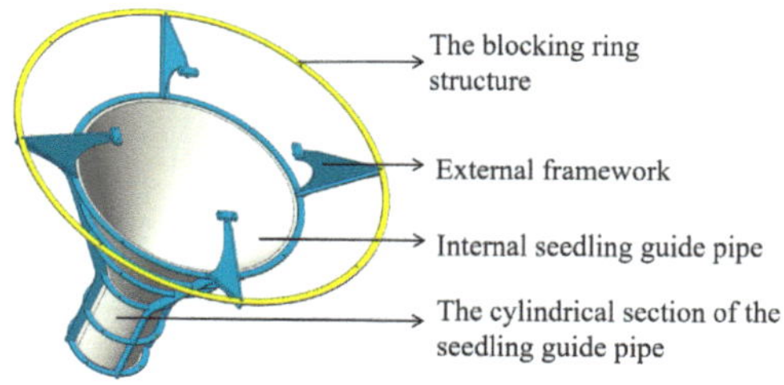

Figure 7. Seedling Guiding Device.

3. Working Performance Test

3.1. Platform Test

The platform test in this study utilized a square soil trough provided by the Nanjing Agricultural Mechanization Research Institute of the Ministry of Agriculture and Rural Affairs. The dimensions of the trough are 500 cm × 100 cm × 30 cm. According to the requirements for rice seedling throwing, the soil in the trough needs to undergo soaking, pulping, and settling treatment. Rice seedlings selected for throwing should be at the three-leaf stage, with the height of the seedlings ranging between 160 and 180 mm. The experimental platform is depicted in Figure 8. The UAV was securely fastened to the platform and rotated at idle speed during the experiments. The seedling throwing apparatus could be directly placed on the square frame. A motor was used to drive a synchronous belt for moving the platform.

Figure 8. Test stand.

3.1.1. Test Conditions

To investigate the effects of seedling throwing height, primary crankshaft working speed, and bottom diameter of the seedling guide tube on the rice seedling throwing performance. These factors were chosen as the three variables in the experiment. The spacing pass rate and the uprightness pass rate were selected as the evaluation criteria. The expressions for each criterion are as follows:

$$U = \frac{V_0}{V} \times 100\% \tag{6}$$

$$I = \frac{M_0}{V} \times 100\% \tag{7}$$

where U represents the spacing pass rate; I denotes the qualified rate of rice seedling uprightness; V_0 stands for the number of qualified seedlings with appropriate plant spacing within the measured area; M_0 represents the number of upright seedlings counted within the measured area; and V represents the total number of seedlings measured within the measured area.

In this experiment, the standard plant spacing for transplanting is 150 mm. To maintain stable plant spacing, the rotational speed of the main shaft is kept in linkage with the flight speed of the UAV. Each rotation of the primary crankshaft will push out 16 rice seedlings. The flight speed of the UAV at this point is shown in Equation (8). Design-Expert 10 was utilized to design a three-factor, three-level quadratic orthogonal experimental design. After verification, it was determined that the throwing height should be selected between 120 and 160 cm; the primary crankshaft working speed should range from 45 to 75 r/min; and the bottom diameter of the seedling guide tube should be between 40 and 60 mm. Table 1 presents the influencing factors and their levels.

$$v = \frac{16qn}{60} \tag{8}$$

where v represents the flight speed of the UAV, m/s; q represents the standard plant spacing, m; and n represents the rotational speed of the primary crankshaft, rpm.

Table 1. Factor and level design.

Level	Height X_1/cm	Working Speed X_2 (r/min)	Diameter X_3/mm
−1	120	45	40
0	140	60	50
1	160	75	60

3.1.2. Results and Discussion

The experimental results are presented in Table 2. Using Design-Expert 10 software, a multivariate regression fitting analysis was conducted to establish quadratic polynomial regression models for the spacing pass rate, Y_1, and the uprightness pass rate, Y_2, as shown in Equations (9) and (10). Subsequently, variance analysis was performed on the regression equations, with results detailed in Table 3.

$$Y_1 = 87.92 - 0.42X_1 - 1.24X_2 - 2.76X_3 + 0.4X_1X_2 - 0.15X_1X_3 - 0.57X_2X_3 - 0.9{X_1}^2 - 1.42{X_2}^2 - 3.17{X_3}^2 \quad (9)$$

$$Y_2 = 87.84 - 0.93X_1 - 0.28X_2 - 1.58X_3 - 0.025X_1X_2 - 0.13X_1X_3 + 0.22X_2X_3 - 2.16{X_1}^2 - 1.46{X_2}^2 - 1.36{X_3}^2 \quad (10)$$

Table 2. Test results.

Test Number	X_1	X_2	X_3	Spacing Pass Rate Y_1/%	Uprightness Pass Rate Y_2/%
1	1	1	1	80.6	82.2
2	1	−1	0	86.6	83.2
3	−1	1	0	83.8	85.3
4	0	1	0	88.4	88.2
5	0	1	−1	86.1	86.2
6	−1	1	1	82.2	83.7
7	−1	−1	0	87.8	85.6
8	0	1	0	87.6	87.2
9	0	1	1	78.8	83.1
10	1	1	0	84.2	82.8
11	0	1	0	88.2	87.8
12	0	−1	−1	86.7	87.4
13	0	1	0	88.2	87.4
14	1	1	−1	85.8	85.2
15	−1	1	−1	86.8	86.2
16	0	−1	1	81.7	83.4
17	0	1	0	87.2	88.6

Table 3. Analysis of variance for the qualified rates of plant spacing and seedling uprightness.

Variance Source	Degree of Freedom	Sum of Squares		Mean Square		F Value		*p* Value	
		Y_1	Y_2	Y_1	Y_2	Y_1	Y_2	Y_1	Y_2
Model	9	135.47	67.87	15.05	7.54	32.47	21.7	<0.0001	0.0003
X_1	1	1.44	6.85	1.44	6.85	3.12	19.7	0.1208	0.003
X_2	1	12.25	0.61	12.25	0.61	26.42	1.74	0.0013	0.2285
X_3	1	61.05	19.85	61.05	19.85	131.68	57.12	<0.0001	0.0001
X_1X_2	1	0.64	0.0025	0.64	0.0025	1.38	0.0072	0.2785	0.9348
X_1X_3	1	0.09	0.063	0.09	0.063	0.19	0.18	0.6728	0.6842
X_2X_3	1	1.32	0.2	1.32	0.2	2.85	0.58	0.1351	0.4701
${X_1}^2$	1	3.39	19.6	3.39	19.6	7.32	56.41	0.0304	0.0001
${X_2}^2$	1	8.52	8.94	8.52	8.94	18.38	25.74	0.0036	0.0014
${X_3}^2$	1	42.38	7.76	42.38	7.76	91.4	22.33	<0.0001	0.0021
Residual	7	3.25	2.43	0.46	0.35				
Lack of fit	3	2.24	1.12	0.75	0.37	2.96	1.14	0.1609	0.4346
Pure terror	4	1.01	1.31	0.25	0.33	0.41			
Cor total	16	138.72	70.3						

X_1, X_2, and X_3 represent the levels of throwing height, working speed, and bottom diameter of the seedling guide tube, respectively. Y_1 denotes the spacing pass rate, while Y_2 represents the uprightness pass rate.

As shown in Table 3, the significance levels in both regression models are highly significant; all lack-of-fit terms are greater than 0.05, indicating a high degree of fit for the regression models. Therefore, the working parameters of the transplanting device can be optimized using this model.

From the analysis of Table 3, it is evident that the working speed of the throwing mechanism and the diameter of the seedling guide tube have the most significant impact on the spacing pass rate in the model, with their interaction also being highly significant. Similarly, for the uprightness pass rate, the throwing height and the diameter of the seedling guide tube have the most significant impact on the model, with their interaction also being highly significant. To further analyze the interactive effects of throwing height. the working speed of the throwing mechanism, and the bottom diameter of the seedling guide tube on the spacing pass rate and the uprightness pass rate, response surface plots were generated using Design-Expert 10 software, as depicted in Figure 9.

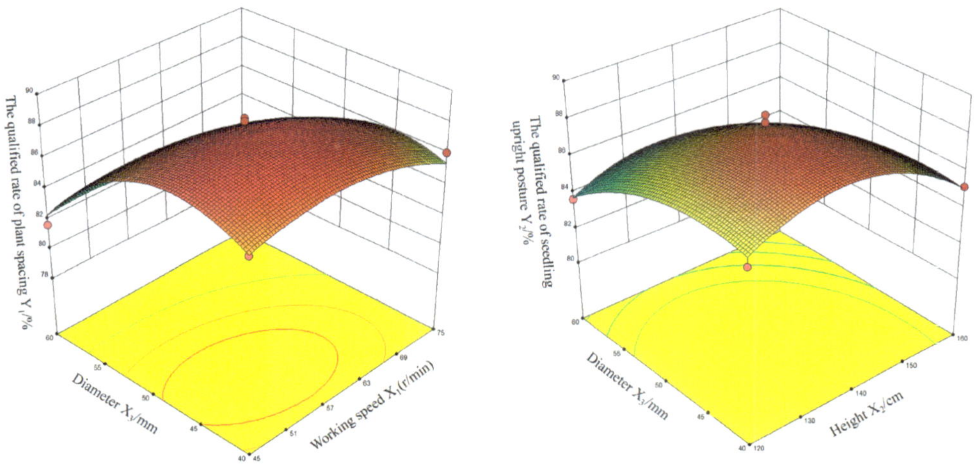

(a) (b)

Figure 9. Response surface analysis: (**a**) response surface of the spacing pass rate under different working speeds and bottom diameters of the seedling guide tube; (**b**) response surface of the uprightness pass rate under different throwing heights and bottom diameters of the seedling guide tube.

To achieve the best transplanting effect and obtain the optimal working parameters for each influencing factor, Design-Expert 10 software was utilized to optimize and solve for the throwing height, working speed, and diameter of the bottom opening of the seedling guide tube. The constraint conditions are as shown in Equation (11). When the throwing height is 142.79 cm, the working speed is 55.38 r/min, and the diameter of the seedling guide tube is 43.51 mm, the spacing pass rate is 88.43%, and the uprightness pass rate is 88.12%.

$$\begin{cases} \begin{cases} \max Y_1(X_1, X_2, X_3) \\ \max Y_2(X_1, X_2, X_3) \end{cases} \\ s.t. \begin{cases} 120 < X_1 < 160 \\ 45 < X_2 < 75 \\ 40 < X_3 < 60 \end{cases} \end{cases} \quad (11)$$

3.2. Field Test

To validate the actual planting effect of the UAV orderly throwing device, field experiments were conducted in rice fields at the Baima Experimental Base in Nanjing City, Jiangsu Province. Based on the optimal operating parameters obtained from the platform test, the UAV flight speed was set to 2.2 m/s, the seedling throwing height was set to 143 cm, the primary crankshaft working speed was 55 r/min, and the bottom diameter of the seedling guide tube was 44 mm. The experimental process is depicted in Figure 10.

(a) (b) (c)

Figure 10. Test procedure: (**a**) disengagement from the docking frame; (**b**) flight toward the operating area; (**c**) transplanting effectiveness.

Through experimentation, the spacing pass rate was determined to be 86.7%, while the uprightness pass rate was 84.2%. The experimental findings align with the anticipated design objectives, albeit marginally lower than the optimized model's predictions. This variance is attributed to disparities between the laboratory settings and the field environment. In practical rice paddy operational settings, the effectiveness of throwing seedlings is subject to the influence of environmental wind conditions. Furthermore, distinctions in soil composition between the laboratory and the experimental field contribute to non-uniform soil hardness and varying water depths across different areas of the experimental site. These factors collectively contribute to a slight decrease in the compliance rate observed.

4. Discussion

The basic principle of the seedling throwing machine is to allow rice seedlings to fall into the field relying on their own gravity. During this process, the posture of the seedlings and the speed at which they enter the mud have a significant impact on the planting effect. Currently, most rice seedling throwing machines use the chassis of transplanters, which greatly limits the throwing height. In actual operation, insufficient speed when entering the mud leads to the problem of seedling drifting. Therefore, pneumatic seedling throwing machines have been developed to increase the speed of seedlings entering the mud. Due to the different principles of seedling throwing, the methods for adjusting the posture of different seedling-throwing machines also vary. However, they all aim to ensure a better upright posture of the seedlings. In addition, ground-based seedling throwing machines are inconvenient in complex terrain environments such as hilly areas with steep slopes.

This study employs UAV as the platform for seedling throwing. UAVs can navigate away from the ground, thus exhibiting better adaptability to different plots. UAVs can also easily control height and are almost unrestricted within an effective throwing height. Based on the above advantages, there is a need for more in-depth research on the effect of seedling throwing using UAVs.

From the response surface of the spacing pass rate, it can be observed that an increase in working speed and an enlargement diameter of the seedling guide tube will lead to a decrease in the spacing pass rate. The spacing pass rate is higher when the working speed is between 50 and 70 r/min and the diameter of the seedling guide tube is between 45 and 55 mm. This is because a higher working speed results in a greater throughput of seedlings per second within the seedling guide tube, increasing the risk of seedling jamming and leading to significant variations in plant spacing. A decrease in the diameter of the seedling guide tube causes seedlings to be closer to the central axis of the bottom opening, thereby reducing the degree of variation in plant spacing. However, when the diameter of the seedling guide tube is too small, it may lead to poor seedling dropping, resulting in a decrease in the spacing pass rate.

From the response surface of the uprightness pass rate, it can be seen that an increase in the diameter of the seedling guide tube and an increase in throwing height will result in a decrease in the uprightness pass rate. The best seedling uprightness is achieved when the diameter of the seedling guide tube is between 40 and 55 mm and the throwing height is between 130 and 150 cm. This is because a smaller diameter of the seedling guide tube allows for better adjustment of seedling posture. Lower throwing heights increase the risk of seedling tilting, while higher throwing heights result in greater variability in seedling posture in the air, affecting seedling uprightness.

Within the above parameters, although the UAV operates quickly in a single line, the overall efficiency is limited. For example, when the primary crankshaft speed is 60 r/min, the UAV's flight speed is 2.4 m/s. Without considering the replacement of seedling trays, the theoretical operational efficiency is approximately 0.25 hm^2/h. This efficiency gap compared to ground machinery is significant. To improve efficiency, future research will focus on the following:

1. Investigating multi-row seedling throwing operations by UAVs. Within the limits of payload and flight stability, adding more rows significantly improves the efficiency of seedling throwing;
2. Researching rapid seedling loading methods for UAVs. Although this study proposed a design for a quick-change seedling box, it still requires the manual loading of seedlings into the box. Currently, there is a lack of automated row-wise seedling loading systems to enhance loading efficiency. In the future, efforts can be made to further supplement logistical equipment for UAV operations, gradually improving the completeness of UAV throwing systems;
3. Studying uneven rotor airflow distribution in multi-row operations. After UAVs implement multi-row seedling throwing, the airflow distribution from the rotors beneath the UAV varies. This results in inconsistent wind conditions affecting different parts of the throwing devices. To ensure uniform transplanting, further research is needed on the distribution of rotor airflow in multi-row UAV operations.

5. Conclusions

This paper, based on analyzing the current situation of seedling throwing, presents a novel seedling-throwing scheme design and experimental analysis. The following conclusions have been drawn:

1. In response to the challenges of difficult terrain in hilly areas for ground-based agricultural machinery and suboptimal rice throwing effects, a UAV-based orderly seedling throwing apparatus was designed. The main structural components and operational procedures of the device are introduced. The working principles of key components, such as the seedling carrying and connection system, seedling pushing mechanism, and seedling guiding device, are analyzed;
2. To further enhance the operational performance of the seedling throwing apparatus, an analysis of factors affecting rice throwing effectiveness was conducted. The throwing height, working speed, and diameter of the seedling guide tube outlet were considered as influencing factors. Performance indicators, including the spacing pass rate and the uprightness pass rate, were used. A three-factor, three-level response surface experimental design was employed, and a regression model for the experimental indicators was obtained. An analysis of the response surface results identified the optimal parameter combination as follows: a throwing height of 142.79 cm, a working speed of 55.38 r/min, and a bottom diameter of the seedling guide tube of 43.51 mm. At these optimal parameters, the model predicted a seedling spacing compliance rate of 88.43% and a seedling erectness compliance rate of 88.12%;
3. To validate the accuracy of the model optimization results, field experiments were conducted. Experimental data showed that under the optimal operating parameters calculated using the regression model, the spacing pass rate was 86.7%, and the uprightness pass rate was 84.2%. The experimental results met the design requirements, providing valuable insights for UAV-based rice seedling throwing.

Author Contributions: Conceptualization, P.Y. and Y.J.; methodology, P.Y., Y.Y. and Y.J.; software, P.Y. and Y.W.; validation, W.Z. and P.Y.; formal analysis, P.Y., Y.J. and W.Z.; investigation, P.Y. and

W.Z.; resources, P.Y., Y.J., Y.Y. and W.Z.; data curation, P.Y., Y.Y. and Y.W.; writing—original draft preparation, P.Y.; writing—review and editing, P.Y. and Y.Y.; visualization, P.Y. and Y.W.; supervision, W.Z. and Y.J.; project administration, W.Z. and Y.J.; funding acquisition, W.Z. and Y.J. All authors have read and agreed to the published version of the manuscript.

Funding: This research was funded by the Science and Technology Innovation Project of the Chinese Academy of Agricultural Sciences (Grant No. CAAS-SAE-202301).

Institutional Review Board Statement: Not applicable.

Informed Consent Statement: Not applicable.

Data Availability Statement: The data presented in this study are available in the article.

Acknowledgments: The authors thank the editor and anonymous reviewers for providing helpful suggestions for improving the quality of this manuscript.

Conflicts of Interest: The authors declare no conflicts of interest.

References

1. Zhang, H.; Hu, Y.; Yang, J.; Dai, Q.; Huo, Z.; Xu, K.; Wei, H.; Gao, H.; Guo, B.; Xing, Z.; et al. Development and Prospect of Rice Cultivation in China. *Sci. Agric. Sin.* **2021**, *54*, 1301–1321.
2. Tian, X.; Qin, F.; Li, M. Empirical analysis of the difference of rice production costs in China. *Trans. CSAE* **2005**, *21*, 247–250.
3. Wu, G.; Yu, G.; Xiang, X.; Wang, L. Design and test of rice potted-seedling transplanting mechanism with three transplanting arms. *Trans. CSAE* **2017**, *33*, 15–22.
4. Xia, Q.; Zhang, W.; Ji, Y.; Li, K. Research status and trend of mechanical throwing seedling technology and equipment in China. *J. Chin. Agric. Mech.* **2019**, *40*, 201–208.
5. Li, Z.; Ma, X.; Li, X.; Chen, L.; Li, H.; Yuan, Z. Research Progress of Rice Transplanting Mechanization. *Trans. CSAM* **2018**, *49*, 1–20.
6. Tang, Y.; Wang, Y.; Zhang, D.; Luo, X. Research on Four-Wheel Drive Pneumatic Throwing Transplanter of Paddy Seedling. *J. Chin. Agric. Mech.* **2009**, *3*, 72–75.
7. Wang, L.; Wang, P.; Li, Y.; Song, J. Experiment Study on 2ZPY-H530 Rice Potted-seedling Transplant. *J. China Agric. Univ.* **2002**, *7*, 21–24.
8. Qian, M.; Yu, G.; Jiang, C.; Zhao, Y. Work Principle and Parameter Optimization of Rice-seedling Transplanter with Non-circular Gears. *Trans. CSAM* **2014**, *45*, 64–69.
9. Li, Y.; Xu, J.; Wang, M.; Liu, D.; Sun, H.; Wang, X. Development of autonomous driving transfer trolley on field roads and its visual navigation system for hilly areas. *Trans. CSAE* **2019**, *35*, 52–61.
10. Gong, X.; Fu, Q.; Sun, A.; Guan, Y.; Wang, B.; Li, M. Rice planting potential in plain-hill-wetland area estimated by nature-society water cycle model. *Trans. CSAE* **2019**, *35*, 138–147.
11. Jiang, Y.; Sun, Z.; Wang, R.; Xia, C.; Ye, Q.; Guo, Y. Design and performance test of the omnidirectional leveling system for crawler work machine in hilly areas. *Trans. CSAE* **2023**, *39*, 64–73.
12. Liu, W.; Zhou, Z.; Xu, X.; Gu, X.; Zou, S.; He, W.; Luo, X.; Huang, J.; Lin, J.; Jiang, R. Evaluation method of rowing performance and its optimization for UAV-based shot seeding device on rice sowing. *Comput. Electron. Agric.* **2023**, *207*, 107718. [CrossRef]
13. Ni, J.; Yao, L.; Zhang, J.; Cao, W.; Zhu, Y.; Tai, X. Development of an unmanned aerial vehicle-borne crop-growth monitoring system. *Sensors* **2017**, *17*, 502. [CrossRef]
14. Li, J.; Lan, Y.; Zhou, Z.; Zeng, S.; Huang, C.; Yao, W.; Zhang, Y.; Zhu, Q. Design and test of operation parameters for rice air broadcasting by unmanned aerial vehicle. *Int. J. Agric. Biol. Eng.* **2016**, *9*, 24–32.
15. Yeh, J.; Lin, K.; Yuan, L.; Hsu, J. Automatic counting and location labeling of rice seedlings from unmanned aerial vehicle images. *Electronics* **2024**, *13*, 273. [CrossRef]
16. Amarasingam, N.; Gonzalez, F.; Salgadoe, A.S.A.; Sandino, J.; Powell, K. Detection of White Leaf Disease in Sugarcane Crops Using UAV-Derived RGB Imagery with Existing Deep Learning Models. *Remote Sens.* **2022**, *14*, 6137. [CrossRef]
17. Li, J.; Lan, Y.; Shi, Y. Research progress on airflow characteristics and field pesticide application system of rotary-wing UAV. *Trans. CSAE* **2018**, *34*, 104–118.
18. Gao, X.; You, Z.; Wu, H.; Wang, S.; Cao, M. Design and Experiment of Green Manure Seed Broadcast Device Based on Unmanned Aerial Vehicle Platform. *Trans. CSAM* **2022**, *53*, 76–85.
19. Zhang, Q.; Zhang, K.; Liao, Q.; Liao, Y.; Wang, L.; Shu, C. Design and experiment of rapeseed aerial seeding device used for UAV. *Trans. CSAE* **2020**, *36*, 138–147.
20. Zhang, H.; Lan, Y.; Wen, S.; Chen, C.; Xu, T.; Chen, S. Modelling approach of spray retention on rice in plant protection using unmanned aerial vehicle. *Trans. CSAE* **2022**, *38*, 40–50.
21. Villette, S.; Piron, E.; Miclet, D. Hybrid centrifugal spreading model to study the fertiliser spatial distribution and its assessment using the transverse coefficient of variation. *Comput. Electron. Agric.* **2017**, *137*, 115–129. [CrossRef]
22. Chojnacki, J.; Berner, B. The influence of air stream generated by drone rotors on transverse distribution pattern of sown seeds. *J. Res. Appl. Agric. Eng.* **2018**, *63*, 9–12.

23. Tang, Y.; Fu, Y.; Guo, Q.; Huang, H.; Tan, Z.; Luo, S. Numerical simulation of the spatial and temporal distributions of the downwash airflow and spray field of a co-axial eight-rotor plant protection UAV in hover. *Comput. Electron. Agric.* **2023**, *206*, 107634. [CrossRef]
24. Song, C.; Liu, L.; Wang, G.; Han, J.; Zhang, T.; Lan, Y. Particle deposition distribution of multi-rotor UAV-based fertilizer spreader under different height and speed parameters. *Drones* **2023**, *7*, 425. [CrossRef]
25. Song, C.; Zhou, Z.; Zang, Y.; Zhao, L.; Yang, W.; Luo, X.; Jiang, R.; Ming, R.; Zang, Y.; Zi, L.; et al. Variable-rate control system for UAV-based granular fertilizer spreader. *Comput. Electron. Agric.* **2020**, *180*, 105832. [CrossRef]
26. Xiao, H.; Li, Y.; Yuan, L.; Zhang, Z. Application and Prospect of China Agricultural Unmanned Aerial Vehicle in Rice Production. *J. Guangxi Agric. Sci.* **2021**, *48*, 139–147.
27. Song, C.; Zhou, Z.; Jiang, R.; Luo, X.; He, X.; Ming, R. Design and parameter optimization of pneumatic rice sowing device for unmanned aerial vehicle. *Trans. CSAE* **2018**, *34*, 80–88.
28. Liu, W.; Zou, S.; Xu, X.; Gu, Q.; He, W.; Huang, J.; Huang, J.; Lyu, Z.; Lin, J.; Zhou, Z.; et al. Development of UAV-based shot seeding device for rice planting. *Int. J. Agric. Biol. Eng.* **2022**, *15*, 1–7. [CrossRef]

agriculture

Article

Research on a Multi-Lens Multispectral Camera for Identifying Haploid Maize Seeds

Xiantao He [1,2], Jinting Zhu [1], Pinxuan Li [1], Dongxing Zhang [1,2], Li Yang [1,2,*], Tao Cui [1,2], Kailiang Zhang [1,2] and Xiaolong Lin [1]

1 College of Engineering, China Agricultural University, Beijing 100083, China
2 Key Laboratory of Soil-Machine-Plant System Technology of Ministry of Agriculture, Beijing 100083, China
* Correspondence: yl_hb68@126.com; Tel.: +86-010-6273-7765

Abstract: Haploid breeding can shorten the breeding period of new maize varieties and is an important means to increase maize yield. In the breeding program, a large number of haploid seeds need to be screened, and this step is mainly achieved manually, which hinders the industrialization of haploid maize breeding. This article aims to develop a multispectral camera to identify the haploid seeds automatically. The camera was manufactured by replacing narrow-band filters of the ordinary CCD camera, and the RGB, 405 nm, 980 nm and 1050 nm images of haploid or diploid seeds were simultaneously captured (the characteristic wavelengths were determined according to color and high-oil markers of maize). The performance was tested using four maize varieties with the two genetic markers. The results show that the developed multispectral camera significantly improved the recognition accuracy of haploid maize seeds to 92.33%, 97.33%, 97% and 93.33% for the TYD1903, TYD1904, TYD1907 and TYD1908 varieties, respectively. The cameras in the near-infrared region (wavelengths of 980 nm and 1050 nm) achieved better performance for the varieties of high-oil marker, with an increase of 0.84% and 1.5%, respectively. These results demonstrate the strong potential of the multispectral imaging technology in the haploid seed identification of maize.

Keywords: maize; haploid seed identification; multispectral imaging; AlexNet-based model

Citation: He, X.; Zhu, J.; Li, P.; Zhang, D.; Yang, L.; Cui, T.; Zhang, K.; Lin, X. Research on a Multi-Lens Multispectral Camera for Identifying Haploid Maize Seeds. *Agriculture* **2024**, *14*, 800. https://doi.org/10.3390/agriculture14060800

Academic Editor: Bruno Bernardi

Received: 4 April 2024
Revised: 16 May 2024
Accepted: 18 May 2024
Published: 22 May 2024

1. Introduction

Maize (*Zea mays* L., also commonly known as corn) is one of the important food crops in the world, and its output has a significant influence on global food security [1–3]. The world population is gradually growing, while arable land is limited, resulting in an increasing demand for maize yield [4]. The cultivation of new varieties of maize is the key means to boost per-unit yield. There are many breeding methods to obtain new maize varieties [5]. Compared with other breeding methods, the haploid breeding technique has the following advantages: (1) shortening the breeding period; (2) overcoming the incompatibility of distant hybridization; (3) reducing land and labor use; (4) improving breeding efficiency, etc. [6,7]. However, the proportion of haploid seeds is low in haploid breeding, and only about 10% of the required haploid seeds can be gained in a single breeding cycle [8]. It is difficult to identify haploid seeds in mixed maize seeds (haploid and diploid), which restricts the engineering of haploid breeding of maize. Therefore, it is of great significance to develop automatic detection technology of haploid seeds.

Morphological identification (MI), cyto-anotomy identification (CI) and gene marking (GM) are the three main technologies to identify haploid maize seeds [9]. MI refers to judging whether maize is haploid or not by observing the morphological structure of maize at the seedling stage. Although this method is intuitive and effective, the detection cycle is long and the cost is high. CI is complicated and extremely destructive, and haploid seeds are identified by dissecting all seeds. GM is the most widely used and effective haploid identification method at present [10]. It marks the haploid and diploid seeds with different genotypes and phenotypes, and a large number of haploid corn seeds can be selected manually. The advantages of this method are that haploid seeds can be identified at the seed stage, and it is easier to achieve automatic haploid screening. Color marker and high-oil marker are two commonly used gene markers. Color marker uses the R1-nj gene to induce pigment differences in maize seeds, so that diploid embryos have purple marks, while haploid embryos do not [11]. The color difference in seed embryos enables human beings to intuitively identify haploid and diploid corn seeds. However, the expression of R1-nj varies greatly among different varieties of maize, resulting in a high recognition error rate by the human eye when the color difference of the embryo is faint. Haploid maize seeds can also be identified by high-oil marker, and the oil content of induced diploid embryos is significantly higher than that of haploid embryos [12]. However, some varieties exhibit overlapping oil content in haploid and diploid seeds, making it difficult to distinguish. The simultaneous use of color marker and high-oil marker is an effective means to improve the recognition accuracy of haploid seeds [13]. In practice, the haploid corn seeds marked by these two methods are mainly identified by artificial means, which takes a lot of manpower and time, and also cannot meet the needs of the breeding industry. Therefore, there is an urgent need to develop a rapid and non-destructive haploid seed identification and screening method.

Some detection techniques based on gene markers have been used for the automatic identification of haploid maize seeds, including RGB image, nuclear magnetic resonance (NMR), near-infrared spectroscopy (NIR), hyper-spectral imaging (HSI) and multispectral imaging (MSI). RGB image uses the RGB camera to capture the color marker of the seed embryo, and this method has not been satisfactory for haploid seeds with a small color difference [14–19]. NMR separates haploids based on oil marker, but NMR is cumbersome in operation, slow in sorting and costly in manufacturing [8,20–23]. NIR is widely used for the quantitative detection of liquids and powders, such as the contents of protein, oil, sugar and pigment, but the accuracy is easily affected by the size and shape of seeds when identifying haploid maize seeds [24–29]. Compared with NIR, HSI technology can obtain the external characteristics of seeds, such as color, size and texture, as well as the internal chemical composition distribution of the seeds with spectral and image information, which counteracts the effects of the size and shape of the seeds on the spectral information [30–34]. However, the data acquisition of HIS is slow and its cost is high. MSI is similar to HSI, and the key difference is that MSI only obtains spectral images under characteristic wavelengths [35–38]. The acquisition is fast, and the cost is greatly reduced. However, MSI needs to be customized according to the application scenario. At present, there is no report on MSI technology applied to the detection of haploid seeds of maize.

The main objectives of this study are as follows: (1) To design a four-channel multi-lens MSI camera, which can collect the information of color marking and high-oil marking of maize seeds simultaneously; (2) To research the preprocessing algorithm of the MSI image of haploid seeds, to establish the classification model of haploid seeds based on convolutional deep neural network, and to realize the identification of haploid seeds with color marking and high-oil marking; (3) To establish a detection system of haploid seeds of maize and to provide technical support for the following automatic sorting of haploid seeds.

2. Materials and Methods

2.1. The Design of Multispectral Camera for Haploid Maize Identification

The developed MSI camera is shown in Figure 1a, including an RGB camera and cameras with wavelengths of 405 nm, 980 nm and 1050 nm. The RGB camera can capture the shape and color of seeds. The 405 nm camera is in the range of 380 nm~470 nm of the violet light, which can strengthen the image of purple marks on the embryo. The 980 nm and 1050 nm cameras are in the band range of near-infrared, and their wavelengths are the strong absorption peaks of the C-H functional groups characterizing the oil contents, so as to achieve the identification of haploid seeds with high-oil marker. In this way, 4 MSI images of haploid seeds can be captured simultaneously. The manufacturing scheme of the MSI camera is shown in Figure 1b. The optical filter of the ordinary CCD camera (manufactured by Xingkai Security Technology Company (Shenzhen, China), camera diameter 15 mm, resolution 1280 × 720, signal-to-noise ratio ≥48 dB, frames 30 f/s, 5 V powered by the image processor) was removed and replaced with a narrow-band filter in front of the camera. It was successively replaced by the filter with a wavelength at 405 nm, 980 nm and 1050 nm (manufactured by Fuzhe Laser Technology Company (Shenzhen, China), with a diameter of 14–15 mm and a thickness of 1–1.2 mm). When the camera operates, the spectral information of the seed first passes through the lens, then reaches the filter. The filter blocks all spectra except for the characteristic wavelengths, and finally, the CCD sensor perceives the seed image information at the characteristic wavelengths. Consequently, the four images at RGB, 405 nm, 980 nm and 1050 nm are obtained and transmitted to the image processor via USB.

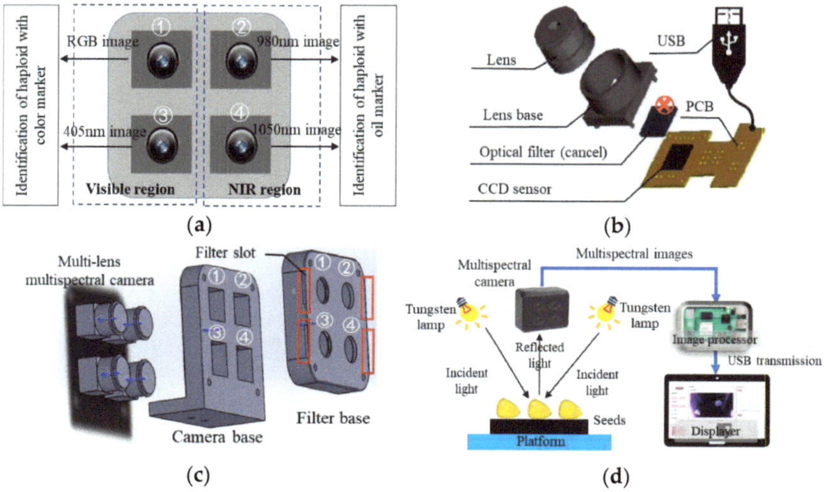

Figure 1. The design of muti-lens MSI camera for haploid seed identification. (**a**) Wavelength selection for MSI camera; (**b**) Manufacturing method of spectral camera based on CCD camera; (**c**) Fixing device of MSI camera; (**d**) Identification system of haploid seeds.

The designed fixing device of camera is shown in Figure 1c, including a camera base and a filter base. The filters were inserted and fixed through the slots on the side of the filter base. The diameter of the circular hole for fixing filter is 12 nm, which is good to install and replace the filter. The detection system of haploid seeds is shown in Figure 1d, which includes seed placement platform, light source, MSI camera, image processor (Raspberry PI 4b, Raspberry PI, Cambridge, UK) and displayer. The light source uses two symmetrically placed tungsten lamps (12 V, 20 W). When working, the tungsten lamp shines the light on maize seed; then, the light is reflected to the MSI camera. The MSI camera transmits the multispectral images to the image processor; then, the processor discriminates haploid and diploid seeds and sends the identification results on the display. The image processor uses the Rasberry Pi 4b, which has four USB ports to connect four cameras and facilitates the sequential capture of images via USB connections.

2.2. Samples and Sample Preparation

Maize samples were divided into two categories: color-marker section and high-oil-marker section. All samples were provided by a breeding company (China National Tree Seed Group Co., Ltd., Beijing, China). In the first section, two varieties (TYD1903 and TYD1904) were obtained using the traditional color inducer CAU5 as the paternal line. The embryo and endosperm of the diploid seeds were both labeled purple, while only the endosperm of the haploid seeds was marked purple. In the second section, the high-oil inducer material CHOI3 was used as the paternal line to produce two other varieties (TYD1907 and TYD1908). These varieties generated haploid and diploid seeds with both oil content and color markers. Specifically, the diploid seeds not only had a purple embryo marker but also a higher oil content compared to the haploid seeds. The total sample size was 400 seeds, with 100 seeds in each category.

2.3. Chemical Composition Determination

The pH-differential method was used for the determination of the anthocyanin content (AC, mg·L^{-1}). Three seeds from each haploid and diploid group of every variety were randomly chosen from the four maize varieties for AC determination. Cyanidin-3-O-glucoside (CAS No.: 7084-24-4) was employed as the standard sample, with a molar absorptivity (ε) of 26,900 M^{-1}cm^{-1} and a molar mass (MW) of 432 Da. Then, the AC value was determined using the formula $AC = a \cdot b/(\varepsilon \cdot l) \times MW \times 10^3$, where a represents the difference in absorptivity between two pH solutions, b denotes the dilution rate and l signifies the path length of the colorimetric vessel (cm). During testing, the values of b and l were established as 24 and 1 cm, respectively. The oil content rate (OCR, %) of haploid and diploid seeds was determined by the NMR-based oil testing instrument NM120-015V-I (Shanghai Newman Electronic Technology Co., Ltd., Shanghai, China).

2.4. Multispectral Image Collection

The detection system of haploid seeds (Figure 2) was established, and the MSI camera was pointed at the seed embryo, and its height was adjusted to ensure that the four cameras could collect seed images at the same time. The distance between the MSI camera and the acquisition platform was determined to be 150 mm. The focus of each camera was adjusted to obtain clear images. The black platform was use to acquire MSI images with clean background. The fixed rod of tungsten lamp on left and right side was at 60 degrees with the vertical rod of the holder, and the distance between the halogen tungsten lamp and the vertical rod was 280 mm. The MSI images of maize seeds were collected one by one, and 4 images of each were obtained at the same time. The images were stored by category for the subsequent establishment of identification model of haploid seeds.

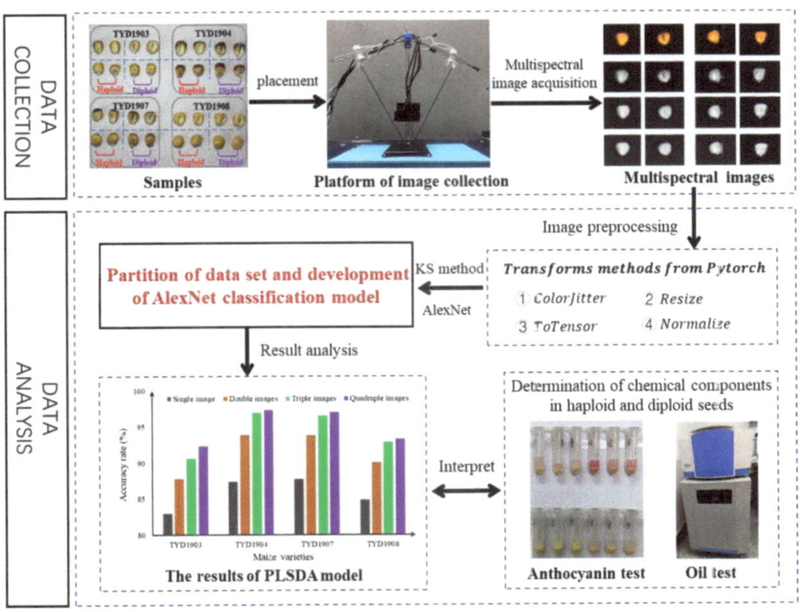

Figure 2. Schematic flow of the image collection and the haploid identification.

2.5. Image Preprocessing

Pytorch was applied to preprocess the obtained images of the maize seeds (100 images each variety). The process includes the following: (1) Transforms.Resize was used to scale the image sizes in proportion to AlexNet's input size of 224 × 224. (2) Transforms.ColorJitter was used to adjust the brightness, contrast, saturation and tone of the images with value of 0.1, thus obtaining an expanded 400 image sets for each variety. Considering that the position of seeds might affect haploid recognition, the rotation (90 degrees in one step) and translation (50 up, down, left and right) of image were conducted to obtain an expanded 800 image sets with orientation changed. Thus, additional 1200 images were obtained. (3) Transforms.ToTensor and Transforms.Normalize were used to normalize images. This process can reduce the occurrence of gradient explosion and gradient disappearance and accelerate the convergence of the model. KS method was used to divide the data set. For 1300 sets of each category, 700 sets were used as the model training and the other 600 sets were used as the model testing.

2.6. Classification Model of Haploid Seeds

AlexNet was applied to establish the classification models of haploid and diploid seeds. It is a classical convolutional neural network, whose network structure consists of 5 convolutional layers and three fully connected layers (i.e., 2 fully connected hidden layers and 1 fully connected output layer) [39]. Each convolution layer contains convolution kernel, bias term, activation function, and local response normalization (LRN) module. The first, second, and fifth convolutional layers are followed by a maximum pooling layer, and the last three layers are fully connected layers. The final output layer, softmax, converts the network output into probability values that are used to predict the category of the image. The nonlinear activation function, modified linear unit (ReLU), is used in the AlexNet. Compared with traditional activation functions (sigmoid and tanh), ReLU can effectively solve the problem of gradient disappearance while maintaining the fast computation speed, thus making training more efficient. LRN was used to suppress the response of neighboring neurons and avoid overfitting, and this can improve the generalization ability of the network.

$$b^i_{x,y} = \frac{a^i_{x,y}}{\left(\mathrm{k} + \alpha \sum_{j=\max(0, i-\frac{n}{2})}^{\min(N-1,\, i+\frac{n}{2})} \left(a^j_{x,y}\right)^2\right)^{\beta}} \tag{1}$$

where $b^i_{x,y}$ is the normalized value of the ith feature map at position (x, y), $a^i_{x,y}$ is the original value of the ith feature map at position (x, y), n is the neighborhood size, N is the number of feature maps, and k, α and β are hyperparameters with values of 2, 10^{-4} and 0.75, respectively.

3. Results and Discussion

3.1. Analysis of Seed Multispectral Images

The results depicted in Figure 3 illustrate the outcomes of analyzing both haploid and diploid seeds across color-marked and oil-marked varieties. In general, diploid seeds exhibited higher AC values compared to haploid seeds across all four maize varieties. This distinction can be attributed to the utilization of the R1-nj pigment gene during maize seed induction, resulting in the presence of a purple marker in diploid embryos, whereas haploid embryos lack this marker. Notably, TYD1904 demonstrated the highest AC, with the greatest disparity between haploid and diploid seeds. Conversely, TYD1908 exhibited the lowest AC, while TYD1903 displayed the smallest difference between haploid and diploid AC values. Distinct OCRs were noted between haploid and diploid seeds for TYD1907 and TYD1908, whereas TYD1903 and TYD1904 showed significant overlaps. This suggests that distinguishing between haploid and diploid seeds based on oil content alone is challenging for color-marked varieties. Furthermore, the average OCRs of diploid samples for TYD1907 and TYD1908 were noticeably higher than those of haploid samples.

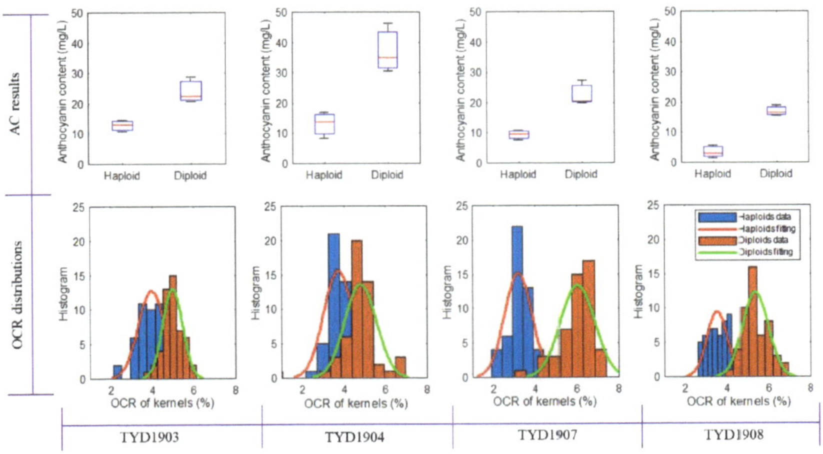

Figure 3. The result of anthocyanin and oil measurement of four maize varieties. AC: anthocyanin content; OCR: oil content rate.

The images of haploid and diploid seeds of four varieties in different wavelengths are shown in Figure 4. For the RGB images of all varieties, it can be observed that the embryo purple of TYD1904 is the deepest, while that of TYD1908 is the lightest, consistent with the measured anthocyanin content results. The anthocyanin content of TYD1903 and TYD1907 is roughly the same, but in the RGB images, the embryo purple of TYD1907 is more pronounced than that of TYD1903. This is because more anthocyanins of TYD1903 are distributed inside the seeds. For the 405 nm images, it can be seen that the embryo purple was enhanced, especially in the varieties of TYD1904 and TYD1907. Compared to the RGB and 405 nm images, the 980 nm and 1050 nm images contain the smaller purple difference between haploid and diploid seeds, but the differences from TYD1907 and TYD1908 are greater than that from TYD1903 and TYD1904. Overall, there are differences between the images of haploid and diploid seeds, which is the basis for identifying the haploid seeds from diploid seeds.

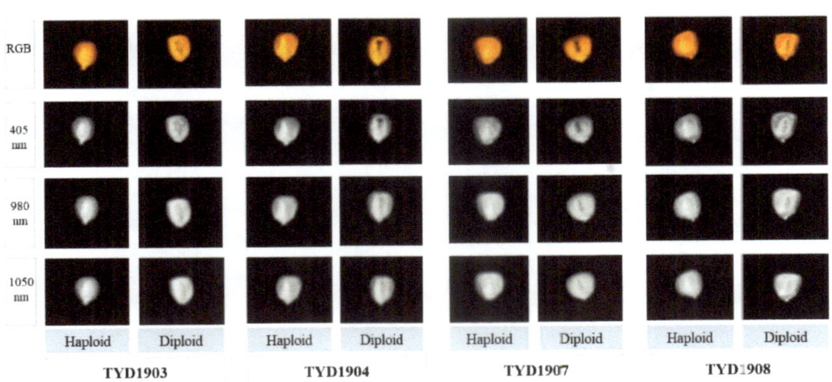

Figure 4. The images of haploid and diploid seeds at different wavelengths.

3.2. The Performance of AlexNet-Based Model

The recognition accuracy of haploid seeds of each variety under a single wavelength is shown in Table 1. The RGB camera had the highest recognition accuracy for haploid seeds with an average of 93.58%, followed by the 405 nm camera with an average of 92.42%. The performance of the 980 nm camera and the 1050 nm camera was comparatively poorer, ranging from 77.25% to 79.63%. Moreover, the performance of the 1050 nm camera was worse compared to the 980 nm camera. The reason for this is that as the wavelength of the filter increases, the intensity of the transmitted wavelength decreases under the same light source, leading to a weaker image signal at the wavelength of 1050 nm. The RGB and 405 nm cameras had significantly higher recognition accuracy for the varieties with an obvious purple marker (TYD1904 and TYD1907), reaching over 95%. The 980 nm and 1050 nm cameras had higher recognition accuracy for the varieties with a high-oil marker (TYD1907 and TYD1908) compared to the varieties with only a color marker (TYD1903 and TYD1904), reaching accuracies of 87.75% and 84.79%, respectively. This is because the 980 nm and 1050 nm cameras are in the near-infrared region and can identify both the anthocyanin difference and the oil difference between haploid and diploid seeds.

Table 1. Results of the AlexNet-based model for identifying haploid seeds with single image.

Varieties	RGB (%)	405 nm (%)	980 nm (%)	1050 nm (%)	Average (%)
TYD1903	90.33	88.00	78.00	75.67	83.00
TYD1904	96.67	95.83	79.83	77.00	87.33
TYD1907	95.83	95.00	81.17	79.00	87.75
TYD1908	91.50	90.83	79.50	77.33	84.79
Average (%)	93.58	92.42	79.63	77.25	85.72

The recognition accuracy of the four varieties under different combinations of cameras are shown in Figure 5 and Table 2. It can be observed that the accuracy of the combination of two cameras ranged from 78.67% to 97%, while the accuracy of the combination of three cameras ranged from 89% to 97.33%. The full combination of four cameras yielded the best performance with an accuracy range of 92.33% to 97.33%. It was evident that the recognition accuracy of the combination of multiple cameras was higher than that of a single camera, and with an increase in the number of cameras, the recognition accuracy increased (Figure 6).

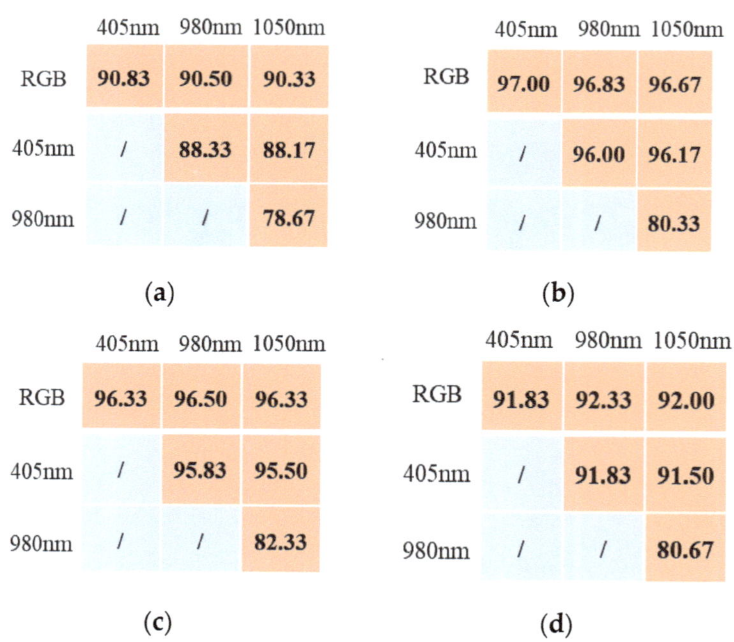

Figure 5. Results of the AlexNet-based model for identifying haploid seeds with two cameras. (**a**) TYD1903; (**b**) TYD1904; (**c**) TYD1907; (**d**) TYD1908.

Table 2. Results of the AlexNet-based model for identifying haploid seeds with three or four cameras.

Varieties	RGB-405 nm–980 nm (%)	RGB-405 nm–1050 nm (%)	RGB-980 nm–1050 nm (%)	405 nm–980 nm–1050 nm (%)	RGB-405 nm–980 nm–1050 nm (%)
TYD1903	91.50	91.17	90.83	89.00	92.33
TYD1904	97.33	97.17	97.00	96.33	97.33
TYD1907	96.50	96.50	96.67	96.33	97.00
TYD1908	92.83	93.17	93.00	92.50	93.33

The presented analysis indicates that combinations with the RGB camera performed better than those without the RGB camera. For instance, the combination of the RGB camera and 980 nm camera for the TYD1903 variety achieved 90.5%, which was higher than the 88.33% achieved by the combination of the 405 nm camera and the 980 nm camera. The reason for this result is that the RGB camera captured images with three channels (i.e., R, G and B channels), resulting in higher image clarity. Additionally, the color differences among the four varieties are more pronounced than the differences in oil content.

For varieties with a clear color marker, the accuracy of the RGB camera was already high. Adding other cameras could further improve recognition accuracy, but the improvement margin was limited. For example, the recognition accuracy for TYD1904 using the RGB camera was 96.67%, and it increased to 97.33% when other cameras were added, resulting in a marginal increase of 0.66%. However, for varieties with a light color marker, it is essential to combine other cameras with the RGB camera. For TYD1903, the recognition accuracy with the RGB camera alone was 90.33%, which increased to 92.33% when three other cameras were added, resulting in an increase of 2%.

Moreover, the 980 nm and 1050 nm cameras performed better for varieties with a high oil content. For instance, the accuracy of the combination of the RGB, 980 nm and 1050 nm cameras for TYD1907 and TYD1908 increased by 0.84% and 1.5%, respectively, while for TYD1903 and TYD1904, the accuracy increased only by 0.5% and 0.33%, respectively. This indicates that adding cameras in the near-infrared region can improve the recognition accuracy of haploid seeds, especially for the varieties with a high-oil marker.

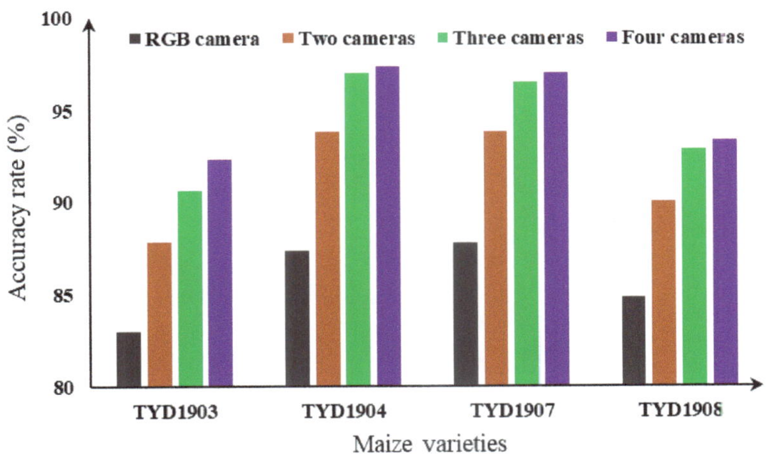

Figure 6. The results of identifying haploid seeds under different varieties and multispectral images.

3.3. Comprehensive Analysis

Through the above analysis, it is evident that the developed MSI camera has significant performance advantages in identifying haploid maize seeds compared to a single camera. Furthermore, it exhibits clear cost advantages in comparison to existing MSI cameras on the market. As shown in Table 3, the CCD camera costs USD 14.6, while the filters for 405 nm, 980 nm, and 1050 nm cost USD 16.7, USD 13.9 and USD 38.9, respectively. The total cost of the MSI camera is only USD 139.1, making it relatively affordable compared to market-available MSI cameras (which typically cost more than USD 1000).

Table 3. The price list of components used in the MSI camera.

Component Name	Number	Unit price (USD)	Total (USD)
CCD camera	4	14.6	139.1
Filter 405 nm	1	16.7	
Filter 980 nm	1	13.9	
Filter 1050 nm	1	38.9	
Fixing device	1	11.2	

Compared to the HSI camera and NIR and NMR spectrometers, MSI has significant advantages in real-time seed detection systems (Table 4). However, existing MSI cameras on the market usually have fixed wavelengths and are often mounted on drones for the acquisition of normalized difference vegetation index (e.g., NDVI), making them difficult to directly apply to the identification of haploid maize seeds. The wavelengths of the images captured by the developed multispectral camera were selected based on the characteristics of haploid maize seeds, and the filters can be easily replaced through the slots on the side of the filter base for expanding detection purposes in the future. In addition, the image augmentation technique was utilized in this study to expand the data set and construct an AlexNet-based model. To mitigate the potential for overfitting, we plan to acquire data sets from a broader variety of maize breeds in future studies. This proactive approach will serve to enhance the robustness of the model, particularly in its ability to accurately identify haploid seeds across a range of maize varieties.

Table 4. The comparison with different devices for detecting haploid seeds.

Detection Device	Price	Detection Speed	Accuracy	Remarks
MSI camera	Low	High	Medium	Images are associated with the characteristic wavelengths.
HSI camera	High	Slow	High	Slow data acquisition and processing speed.
NIR spectrometer	Medium	High	Medium	Generally used for detection of homogeneous liquids or powders
NMR spectrometer	High	Medium	Medium	The cost is very high, commonly used in medical fields

4. Conclusions

This study developed a multi-lens MSI camera for the identification of haploid maize seeds, and the performance was tested using four maize varieties with two different genetic markers. The conclusions of the research can be drawn as follows:

(1) Compared to the conventional RGB camera, the developed MSI camera significantly improved the recognition accuracy of haploid maize seeds. The accuracy of the MSI camera for identifying haploid seeds of four maize varieties (TYD1903, TYD1904, TYD1907 and TYD1908) was 92.33%, 97.33%, 97% and 93.33%, respectively, while the accuracy of the RGB camera was 90.33%, 96.67%, 95.83% and 91.5%, respectively. These results demonstrate the strong potential of MSI technology in haploid maize seed identification.

(2) The cameras in the NIR region (wavelengths of 980 nm and 1050 nm) showed better identification performance for the varieties with a high-oil marker. The accuracy of

the combination of the RGB, 980 nm and 1050 nm cameras for TYD1907 and TYD1908 increased by 0.84% and 1.5%, respectively, while for TYD1903 and TYD1904, the accuracy increased only by 0.5% and 0.33%, respectively. Therefore, adding cameras in the NIR region can improve the recognition accuracy of haploid seeds, especially for the varieties with a high-oil marker.

(3) The MSI camera was manufactured based on the ordinary CCD camera, and the overall cost was much lower than for existing MSI cameras on the market. Furthermore, the developed MSI camera is specifically tailored for the identification of haploid maize seeds, making it easier for this MSI camera to be widely applied in haploid maize breeding.

Author Contributions: Conceptualization, X.H., D.Z. and L.Y.; methodology, X.H., J.Z. and P.L.; software, X.H., J.Z. and P.L.; validation, X.H., J.Z. and T.C.; formal analysis, J.Z., P.L. and K.Z.; investigation, X.H. and J.Z.; resources, X.H. and J.Z.; data curation, X.H., J.Z. and P.L.; writing—original draft preparation, X.H., J.Z., X.L. and L.Y.; writing—review and editing, X.H., J.Z., P.L., X.L. and L.Y.; visualization, D.Z. and T.C.; supervision, L.Y. and K.Z.; project administration, X.H.; funding acquisition, X.H. and T.C. All authors have read and agreed to the published version of the manuscript.

Funding: This research was supported by The National Key Research and Development Program of China (2021YFD2000402, 2021YFD2000404) and the earmarked fund for CARS-02.

Institutional Review Board Statement: Not applicable.

Data Availability Statement: The data used in this study were self-tested and self-collected. As the control method designed in this paper is still being further improved, data cannot be shared at present.

Acknowledgments: We acknowledge the China National Tree Seed Group Corporation Limited for providing the samples for our study.

Conflicts of Interest: The authors declare no conflict of interest.

References

1. Erenstein, O.; Jaleta, M.; Sonder, K.; Mottaleb, K.; Prasanna, B.M. Global Maize Production, Consumption and Trade: Trends and R&D Implications. *Food Secur.* **2022**, *14*, 1295–1319. [CrossRef]
2. Ranum, P.; Peña-Rosas, J.P.; Garcia-Casal, M.N. Global Maize Production, Utilization, and Consumption. *Ann. N. Y. Acad. Sci.* **2014**, *1312*, 105–112. [CrossRef] [PubMed]
3. Rouf Shah, T.; Prasad, K.; Kumar, P. Maize—A Potential Source of Human Nutrition and Health: A Review. *Cogent Food Agric.* **2016**, *2*, 1166995. [CrossRef]
4. Tian, X.; Engel, B.A.; Qian, H.; Hua, E.; Sun, S.; Wang, Y. Will Reaching the Maximum Achievable Yield Potential Meet Future Global Food Demand? *J. Clean. Prod.* **2021**, *294*, 126285. [CrossRef]
5. Li, R.; Zhang, G.; Liu, G.; Wang, K.; Xie, R.; Hou, P.; Ming, B.; Wang, Z.; Li, S. Improving the Yield Potential in Maize by Constructing the Ideal Plant Type and Optimizing the Maize Canopy Structure. *Food Energy Secur.* **2021**, *10*, e312. [CrossRef]
6. Takele, M. Review on Haploid and Double Haploid Maize (*Zea mays*) Breeding Technology. *Int. J. Agric. Sci. Food Technol.* **2022**, *8*, 052–058. [CrossRef]
7. Meng, D.; Liu, C.; Chen, S.; Jin, W. Haploid Induction and Its Application in Maize Breeding. *Mol. Breed.* **2021**, *41*, 20. [CrossRef] [PubMed]
8. Melchinger, A.E.; Schipprack, W.; Würschum, T.; Chen, S.; Technow, F. Rapid and Accurate Identification of in Vivo-Induced Haploid Seeds Based on Oil Content in Maize. *Sci. Rep.* **2013**, *3*, 2129. [CrossRef]
9. De Oliveira Couto, E.G.; Davide, L.M.C.; de Oliveira Bustamante, F.; von Pinho, R.G.; Silva, T.N. Identificação de Milho Haploide Por Citometria de Fluxo, Marcadores Morfológicos e Moleculares. *Cienc. Agrotecnol.* **2013**, *37*, 25–31. [CrossRef]
10. Wang, Y.; Lv, Y.; Liu, H.; Wei, Y.; Zhang, J.; An, D.; Wu, J. Identification of Maize Haploid Kernels Based on Hyperspectral Imaging Technology. *Comput. Electron. Agric.* **2018**, *153*, 188–195. [CrossRef]
11. Chaikam, V.; Nair, S.K.; Babu, R.; Martinez, L.; Tejomurtula, J.; Boddupalli, P.M. Analysis of Effectiveness of R1-Nj Anthocyanin Marker for in Vivo Haploid Identification in Maize and Molecular Markers for Predicting the Inhibition of R1-Nj Expression. *Theor. Appl. Genet.* **2015**, *128*, 159–171. [CrossRef]
12. Liu, C.; Li, J.; Chen, M.; Li, W.; Zhong, Y.; Dong, X.; Xu, X.; Chen, C.; Tian, X.; Chen, S. Development of High-Oil Maize Haploid Inducer with a Novel Phenotyping Strategy. *Crop J.* **2022**, *10*, 524–531. [CrossRef]

13. Chaikam, V.; Molenaar, W.; Melchinger, A.E.; Boddupalli, P.M. Doubled Haploid Technology for Line Development in Maize: Technical Advances and Prospects. *Theor. Appl. Genet.* **2019**, *132*, 3227–3243. [CrossRef] [PubMed]
14. Zhang, J.; Wu, Z.; Song, P.; Li, W.; Chen, S.; Liu, J. Embryo Feature Extraction and Dynamic Recognition Method for Maize Haploid Seeds. *Nongye Gongcheng Xuebao/Trans. Chin. Soc. Agric. Eng.* **2013**, *29*, 199–203.
15. Altuntaş, Y.; Kocamaz, A.F.; Cömert, Z.; Cengiz, R.; Esmeray, M. Identification of Haploid Maize Seeds Using Gray Level Co-Occurrence Matrix and Machine Learning Techniques. In Proceedings of the 2018 International Conference on Artificial Intelligence and Data Processing (IDAP), Malatya, Turkey, 28–30 September 2018; pp. 8–12. [CrossRef]
16. Veeramani, B.; Raymond, J.W.; Chanda, P. DeepSort: Deep Convolutional Networks for Sorting Haploid Maize Seeds. *BMC Bioinform.* **2018**, *19*, 289. [CrossRef]
17. Song, P.; Zhang, H.; Wang, C.; Luo, B.; Zhang, J.X. Design and Experiment of a Sorting System for Haploid Maize Kernel. *Int. J. Pattern Recognit. Artif. Intell.* **2018**, *32*, 1855002. [CrossRef]
18. Altuntaş, Y.; Cömert, Z.; Kocamaz, A.F. Identification of Haploid and Diploid Maize Seeds Using Convolutional Neural Networks and a Transfer Learning Approach. *Comput. Electron. Agric.* **2019**, *163*, 104874. [CrossRef]
19. Sabadin, F.; Galli, G.; Borsato, R.; Gevartosky, R.; Campos, G.R.; Fritsche-Neto, R. Improving the Identification of Haploid Maize Seeds Using Convolutional Neural Networks. *Crop Sci.* **2021**, *61*, 2387–2397. [CrossRef]
20. Wang, H.; Liu, J.; Xu, X.; Huang, Q.; Chen, S.; Yang, P.; Chen, S.; Song, Y. Fully-Automated High-Throughput NMR System for Screening of Haploid Kernels of Maize (Corn) by Measurement of Oil Content. *PLoS ONE* **2016**, *11*, e0159444. [CrossRef]
21. Li, H.; Qu, Y.; Yang, J.; Cui, L.; Mao, X.; Liu, Z. Analysis on Single Kernel Weight and Oil Content of Different Grain Types in Maize Based on NMR. *Trans. Chin. Soc. Agric. Eng.* **2018**, *34*, 183–188. [CrossRef]
22. Ge, W.; Li, J.; Wang, Y.; Yu, X.; An, D.; Chen, S. Maize Haploid Recognition Study Based on Nuclear Magnetic Resonance Spectrum and Manifold Learning. *Comput. Electron. Agric.* **2020**, *170*, 105219. [CrossRef]
23. Qu, Y.; Liu, Z.; Zhang, Y.; Yang, J.; Li, H. Improving the Sorting Efficiency of Maize Haploid Kernels Using an NMR-Based Method with Oil Content Double Thresholds. *Plant Methods* **2021**, *17*, 2. [CrossRef] [PubMed]
24. Sendin, K.; Manley, M.; Williams, P.J. Classification of White Maize Defects with Multispectral Imaging. *Food Chem.* **2018**, *243*, 311–318. [CrossRef] [PubMed]
25. Li, W.; Li, J.; Li, W.; Liu, L.; Li, H.; Chen, C.; Chen, S. Near Infrared Spectroscopy Analysis Based Machine Learning to Identify Haploids in Maize. *Spectrosc. Spectr. Anal.* **2018**, *38*, 2763–2769. [CrossRef]
26. Cui, Y.; Ge, W.; Li, J.; Zhang, J.; An, D.; Wei, Y. Screening of Maize Haploid Kernels Based on near Infrared Spectroscopy Quantitative Analysis. *Comput. Electron. Agric.* **2019**, *158*, 358–368. [CrossRef]
27. Shen, F.; Huang, Y.; Jiang, X.; Fang, Y.; Li, P.; Liu, Q.; Hu, Q.; Liu, X. On-Line Prediction of Hazardous Fungal Contamination in Stored Maize by Integrating Vis/NIR Spectroscopy and Computer Vision. *Spectrochim. Acta-Part A Mol. Biomol. Spectrosc.* **2020**, *229*, 118012. [CrossRef]
28. Sendin, K.; Manley, M.; Marini, F.; Williams, P.J. Hierarchical Classification Pathway for White Maize, Defect and Foreign Material Classification Using Spectral Imaging. *Microchem. J.* **2021**, *162*, 105824. [CrossRef]
29. Chavez, R.A.; Cheng, X.; Herrman, T.J.; Stasiewicz, M.J. Single Kernel Aflatoxin and Fumonisin Contamination Distribution and Spectral Classification in Commercial Corn. *Food Control* **2022**, *131*, 108393. [CrossRef]
30. Golhani, K.; Balasundram, S.K.; Vadamalai, G.; Pradhan, B. A Review of Neural Networks in Plant Disease Detection Using Hyperspectral Data. *Inf. Process. Agric.* **2018**, *5*, 354–371. [CrossRef]
31. Weng, H.; Lv, J.; Cen, H.; He, M.; Zeng, Y.; Hua, S.; Li, H.; Meng, Y.; Fang, H.; He, Y. Hyperspectral Reflectance Imaging Combined with Carbohydrate Metabolism Analysis for Diagnosis of Citrus Huanglongbing in Different Seasons and Cultivars. *Sens. Actuators B Chem.* **2018**, *275*, 50–60. [CrossRef]
32. Wakholi, C.; Kandpal, L.M.; Lee, H.; Bae, H.; Park, E.; Kim, M.S.; Mo, C.; Lee, W.H.; Cho, B.K. Rapid Assessment of Corn Seed Viability Using Short Wave Infrared Line-Scan Hyperspectral Imaging and Chemometrics. *Sens. Actuators B Chem.* **2018**, *255*, 498–507. [CrossRef]
33. Kimuli, D.; Wang, W.; Wang, W.; Jiang, H.; Zhao, X.; Chu, X. Application of SWIR Hyperspectral Imaging and Chemometrics for Identification of Aflatoxin B1 Contaminated Maize Kernels. *Infrared Phys. Technol.* **2018**, *89*, 351–362. [CrossRef]
34. Kimuli, D.; Wang, W.; Lawrence, K.C.; Yoon, S.C.; Ni, X.; Heitschmidt, G.W. Utilisation of Visible/near-Infrared Hyperspectral Images to Classify Aflatoxin B1 Contaminated Maize Kernels. *Biosyst. Eng.* **2018**, *166*, 150–160. [CrossRef]
35. Yang, X.; Gao, S.; Sun, Q.; Gu, X.; Chen, T.; Zhou, J.; Pan, Y. Classification of Maize Lodging Extents Using Deep Learning Algorithms by UAV-Based RGB and Multispectral Images. *Agriculture* **2022**, *12*, 970. [CrossRef]
36. He, X.; Liu, L.; Liu, C.; Li, W.; Sun, J.; Li, H.; He, Y.; Yang, L.; Zhang, D.; Cui, T.; et al. ScienceDirect Discriminant Analysis of Maize Haploid Seeds Using Near-Infrared Hyperspectral Imaging Integrated with Multivariate Methods. *Biosyst. Eng.* **2022**, *222*, 142–155. [CrossRef]
37. Ma, F.; Yuan, M.; Kozak, I. Multispectral Imaging: Review of Current Applications. *Surv. Ophthalmol.* **2023**, *68*, 889–904. [CrossRef] [PubMed]

38. Zhang, W.; Zhu, L.; Zhuang, Q.; Chen, D.; Sun, T. Mapping Cropland Soil Nutrients Contents Based on Multi-Spectral Remote Sensing and Machine Learning. *Agriculture* **2023**, *13*, 1592. [CrossRef]
39. Lu, J.; Tan, L.; Jiang, H. Review on Convolutional Neural Network (CNN) Applied to Plant Leaf Disease Classification. *Agriculture* **2021**, *11*, 707. [CrossRef]

agriculture

MDPI

Article

Simulation and Optimization of a Pendulum-Lever-Type Hole-Seeding Device

Hengshan Zhou [1,2], Fei Dai [1,2,*], Ruijie Shi [2], Cai Zhao [1], Huan Deng [2], Haifu Pan [2] and Qinxue Zhao [2]

[1] State Key Laboratory of Aridland Crop Science, Gansu Agricultural University, Lanzhou 730070, China; zhouhs@st.gsau.edu.cn (H.Z.); zhaoc@gsau.edu.cn (C.Z.)

[2] College of Mechanical and Electrical Engineering, Gansu Agricultural University, Lanzhou 730070, China; shirj@gsau.edu.cn (R.S.); 17797691649@163.com (H.D.); 18198029608@163.com (H.P.); zhaoqx@st.gsau.edu.cn (Q.Z.)

* Correspondence: daifei@gsau.edu.cn

Abstract: The process of hole seeding on the mulch during full-film double-row furrow corn planting faces issues such as poor seed discharge and seed blockage. To address these challenges, a pendulum-lever-type hole-forming mechanism is designed, along with an adjustment device. By analyzing the working principles of the pendulum-lever-type hole seeder and the adjustment device, the structural parameters of the device are determined. Through theoretical analysis and simulation experiments, three-dimensional models of seeds and hole seeders are constructed. Based on MBD-DEM cosimulation, the trajectory of seed movement and the seeding process of the hole seeder are analyzed to elucidate the effects of the hole-former opening and the number of pendulum bearings on seeding quality. To improve the operational performance of the hole seeder, experiments are conducted using the hole seeder's rotating disc speed, lever angle of the hole-former, and the number of pendulum bearings as experimental factors, with the qualification index, miss-seeding index, and reseeding index as experimental indicators. A three-factor, three-level Box–Behnken central composite experiment is performed to obtain mathematical models of the relationships between the experimental factors and indicators. Using Design-Expert 12 software, the regression models are optimized for multiple objectives to obtain the optimal parameter combination: a seeder disc speed of 49 r/min (corresponding to a forward speed of 5.76 km/h), a lever angle of 131°, and four pendulum bearings. Under this optimal parameter combination, the qualification index is 91.70%, the miss-seeding index is 4.57%, and the reseeding index is 3.73%. Experimental validation of the seeding performance of the hole seeder under the optimal parameter combination is conducted. Bench tests show that the qualification index, miss-seeding index, and reseeding index are 90.53%, 5.60%, and 3.87%, respectively. Field tests demonstrate a qualification index of 89.13%, a miss-seeding index of 5.46%, and a reseeding index of 6.41%. The actual results are consistent with the optimized values, providing valuable insights for the design and performance optimization of hole seeders.

Keywords: hole-former; hole seeder; parameter optimization; DEM-MBD

Citation: Zhou, H.; Dai, F.; Shi, R.; Zhao, C.; Deng, H.; Pan, H.; Zhao, Q. Simulation and Optimization of a Pendulum-Lever-Type Hole-Seeding Device. *Agriculture* **2024**, *14*, 750. https://doi.org/10.3390/agriculture14050750

Academic Editors: Xiaojun Gao, Qinghui Lai and Tao Cui

Received: 18 April 2024
Revised: 6 May 2024
Accepted: 9 May 2024
Published: 11 May 2024

1. Introduction

In response to the prevalent "nine droughts in ten years" phenomenon in Northwest China's arid regions, the widespread adoption of the whole-film mulching and double-ridge furrow-sowing drought-resistant tillage technology is currently underway. This approach combines a unique ridge structure with film mulching furrow sowing, offering benefits such as water retention, soil moisture conservation, and temperature elevation [1,2]. This planting method has been widely adopted in the cultivation of crops such as corn and potatoes [3]. Production practice has demonstrated that the whole-film mulching and double-ridge furrow seeding technique increases crop yields by nearly 30% compared to conventional mulch cultivation [4]. Precision seeding on film is vital for achieving mechanized operations throughout the entire process of full-film double-ridge field crop production [3]. Film hole-seeding involves creating planting holes in the film at specific row and hole spacings, effectively minimizing damage to the film caused by seeding machinery, safeguarding the water and thermal systems of the double ridges covered by the film [5], enhancing rainwater collection and moisture retention, boosting seeding efficiency, and reducing labor and resource requirements [6]. Hence, enhancing the operational performance of hole seeders holds significant importance in advancing the mechanization of maize production in full-film double-ridge fields and promoting the evolution of precision seeding technology and equipment [7].

With the advancement of precision seeding technology, high-speed field operations have become the pivotal benchmark for evaluating seeder performance [8,9]. Gao et al. [10,11] explored the influence of particle count and cleaning element angle on seeding performance during high-speed seeding processes. Dong et al. [12] and Ma et al. [13] investigated maize seeding postures and proposed methods for adjusting seed guidance trajectories and filling postures to enhance the high-speed operation performance of seed-metering devices. However, these studies overlooked the impact of the hole-forming device on seeding effectiveness. Accelerating hole-seeder operation speeds leads to shortened seeding times, potentially resulting in hole seeders failing to smoothly dispense seeds or seeds not being dispensed promptly, causing seed blockages and damage. Research on the effect of hole-forming mechanisms on seeding outcomes during hole-seeder operations remains limited. While optimized duckbill hole-forming devices, as proposed by Wang [14], reduce soil plug formation, the incomplete opening of the soft soil pressure plate may impede seed dispersion. Gu et al. [15] suggested a robust opening method using levers and limit blocks to improve seeding success rates, but excessive duckbill opening could increase the damage caused to plastic film and the soil.

Addressing these issues, a pendulum-lever-type hole-seeding device was designed to enhance the operational performance of the seeder and promote the development of precision seeding technology. This study focused on widely used mechanical hole seeders, analyzing the seeding principles of hole-seeder hole-forming components, investigating hole-seeder bench test adjustment requirements, and optimizing structural parameters. Through cosimulation analysis using discrete element software EDEM 2020 [16] and multibody dynamics software RecurDyn V9R4 [17], joint simulations were conducted to identify optimal parameter combinations for hole seeders and key components of the hole-forming mechanism. Bench tests and field tests were conducted to validate the performance improvements.

2. Materials and Methods

2.1. Machine Structure and Working Principle

2.1.1. Machine Structure

The pendulum plate and lever hole-seeding device was mainly composed of a fixed plate, movable shaft, electric actuator, rotating disc, connecting bracket, fixed disc cover, pendulum plate, and hole-former, and its structure is shown in Figure 1.

The fixed plate and connecting bracket were coaxially mounted on the movable shaft with the hole seeder's rotating disc, fixed disc, and fixed disc cover; the hole-former was circumferentially distributed on the movable plate; the seed-discharging wheel dial was coaxially mounted with the seed-discharging wheel; the pendulum mechanism hinged to the lower portion of the fixed disc was adjacent to the limit lever; and the return tension spring was connected to the pendulum plate at one end and was hooked up to the fixed disc at the other end.

2.1.2. Working Principle

When the pendulum paddle hole-seeding device operates, according to the preset sowing depth and the size of the hole-seeding device, the length of the electric push rod is adjusted to change the installation height of the hole-seeding device (the vertical distance between the center of the hole-seeding device axis and the seedbed belt). The adjustment range of the electric push rod was 305~505 mm, and the installation height of the hole-seeding device ranged from 180~471 mm, to meet the agronomic technical requirements of different sowing depths. The adjusting trajectory is shown as a red line in Figure 2a.

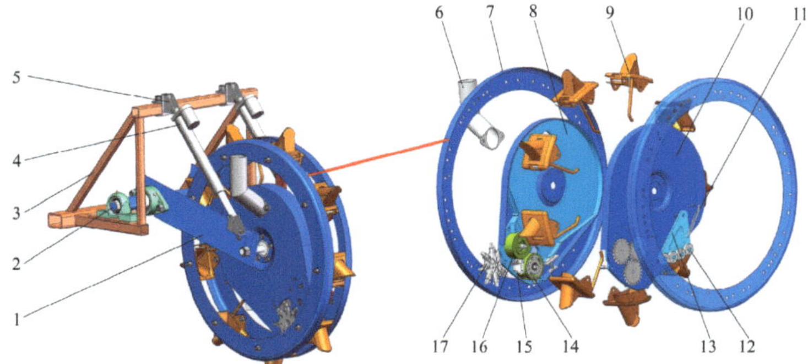

Figure 1. Structural diagram of pendulum-lever-type hole-seeding device: (1) fixed plate; (2) movable shaft; (3) bracket; (4) electric actuator; (5) connecting bracket; (6) seed delivery tube; (7) rotating disc; (8) fixed disc cover; (9) hole-former; (10) fixed disc; (11) return tension spring; (12) pendulum plate bearing; (13) pendulum plate; (14) seed-discharging wheel; (15) guard plate (16) seed brush wheel; (17) seed-discharging wheel dial.

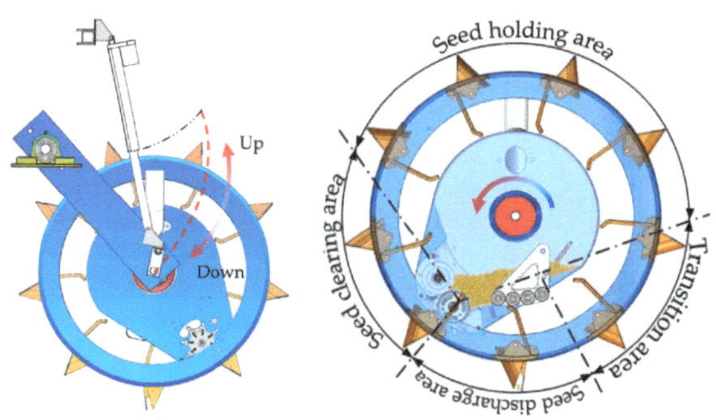

Figure 2. Working diagram of wobble plate lever hole-seeding device. (**a**) Adjustment devices; (**b**) hole seeder.

Within the hole-seeding chamber, the corn seed population flows dynamically under the influence of friction and gravity in the seed holding area of the hole seeder's fixed disk chamber. Seeds in the seed holding area rely on gravity and interseed force to enter the holes of the seed discharge wheel. The rotating sleeve contacts the dial of the seed discharge wheel, driving the rotation of both the seed discharge wheel and the seed brush wheel. The brush wheel rotates counterclockwise to clear any seeds that have not fully entered the holes. As the seed discharge wheel rotates to a specific position, seeds fall into the cavity of the hole-forming device under the combined influence of gravity and the seed-clearing plate. At this point, the lever slides around the outer circle of the pendulum plate bearing, while the bearing group acts on the upper end of the lever to keep the movable hole-former open. The seeds then fall into the soil, completing one sowing cycle, as illustrated in Figure 2b.

2.2. Design of Key Components for Hole-Seeding Devices

The hole-seeding device plays a crucial role in the film-sowing process and is primarily composed of the hole-former unit and the swinging disk mechanism. The size and timing of the hole-former's opening are critical factors influencing both the seeding performance of the hole seeder and the moisture retention effect of the full-film hole seeder. Excessive opening size or prolonged opening time may result in film tearing or picking, leading to noticeable soil disturbance around the holes, disrupting the ridge structure of the full-film double-row furrows and reducing soil temperature and water transfer capacity [18]. Therefore, it is essential to minimize the degree of film damage and soil disturbance caused by the hole-former while ensuring smooth seeding, thereby preserving the hydrothermal effect of the full-film double-row production system [19,20].

2.2.1. Design of the Hole-Forming Mechanism

The hole-former shape is mainly conical- and wedge-shaped, and research shows that cone-shaped hole-formers are more effective than wedge-shaped hole-formers in terms of hole formation and soil movement [21]. The structure of the conical hole-forming group, which is mainly composed of a fixed hole-former, base fixed plate, rotating sleeve, lever, return spring, and movable hole-former, is shown in Figure 3.

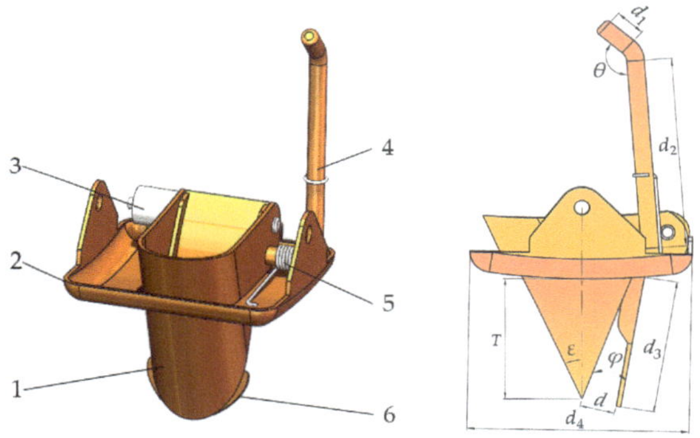

(**a**) 3D diagram (**b**) Schematic diagram

Figure 3. Hole-former structure: (1) fixed hole-former; (2) base fixed plate; (3) rotating sleeve; (4) lever; (5) return spring; (6) movable hole-former.

As shown in Figure 3b, to ensure that the seed drop is smooth and the hole-former works without clogging and film hanging, the hole-former opening degree d satisfies [22]

$$d = 1.2 \sim 1.5 d_{\text{max}} \quad (1)$$

where d is the hole-former opening, mm; d_{max} is the maximum geometric size of the seed, mm.

To ensure that the seed discharge wheel-type holes are filled with single grains of seeds, the larger Jinsui No. 4 corn seeds were selected [23]. Then, the range of values of hole-former openings d was determined as 15~18.75 mm.

The length of the hole-forming mechanism affects not only the sowing depth but also the effect of the tip scraping when the soil comes out. If the length of the hole-forming mechanism is too short, it is difficult to break the film into holes; if the length of the hole-forming mechanism is too long, it is easy to hang the film, destroying the shape of the film holes, affecting corn seedling emergence. The calculation formula for the length of the dynamic into the hole device is

$$d_3 = \frac{T}{\cos \varepsilon} \quad (2)$$

where d_3 is the length of the movable hole-former, mm; T is the sowing depth. According to the requirements of full-film double-ridge furrow agronomic technology [2], we took 50 mm; ε is the angle between the hole-forming edge line of the fixed hole-former and the line connecting the bottom and the center of the drum, °.

Referring to the literature [24], ε = 30°, which was brought into Equation (5) to obtain d_3 = 57.74 mm. After the hole-former opening was determined, the movable hole-former angle φ was determined from the geometrical relationship:

$$\varphi = 2\sin^{-1}\left(\frac{d}{2d_3}\right) \tag{3}$$

where φ is the angle of rotation of the dynamic hole-forming apparatus, °.

From this, the rotational angle φ of the movable hole-former was determined to be 14.93°~18.68°. The long arm of lever d_2 is the distance from the hole-forming part to the pendulum mechanism, which is 100 mm; to ensure the stability of the opening of the dynamic hole-forming mechanism, the value of the short arm of lever d1 should be larger than the radius of the pendulum bearings by 15 mm and smaller than the distance between the centers of the two adjacent bearings by 30 mm, which was taken as d_1 = 22 mm.

2.2.2. Design of Pendulum Mechanism

The pendulum mechanism comprises the pendulum plate, pendulum bearing, return tension spring, and limit lever, as depicted in Figure 4. During operation, the return tension spring keeps the pendulum disk mechanism close to the fixed disk limit lever. The lever slides along the outer circle of the pendulum disk bearing, ensuring the hole-forming mechanism remains open for seed deposition into the soil. Furthermore, the limiting device on the pendulum disk regulates the opening and closing of the hole-forming mechanism within an appropriate timeframe, facilitating smooth seeding.

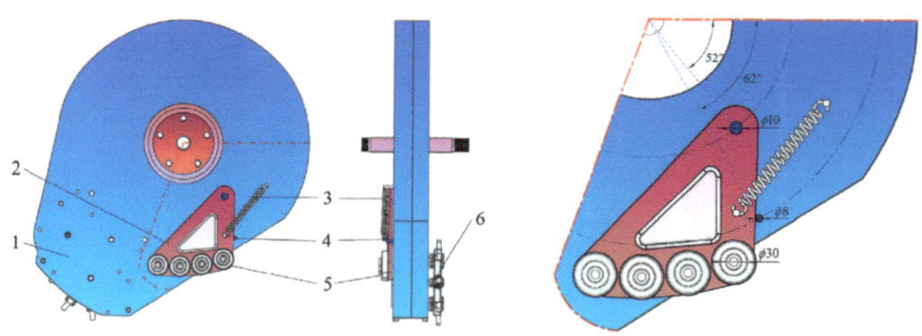

(**a**) Schematic structure (**b**) Schematic of installation dimensions

Figure 4. Pendulum mechanism: (1) fixed plate; (2) pendulum plate; (3) return tension spring; (4) limit lever; (5) pendulum bearing; (6) seed-discharging wheel dial.

In the pendulum device with reverse rotation protection, as the rotating disc moves backward, the lever glides along the outer perimeter of the bearing, prompting the pendulum disk mechanism to rotate around the pendulum pin in the same direction. Throughout this process, the movable hole-former maintains a tight closure against the fixed hole-former. Upon the lever's disengagement from the outer circle of the bearing, the pendulum disk mechanism reverts to its initial position due to the action of the return spring, as illustrated in Figure 5.

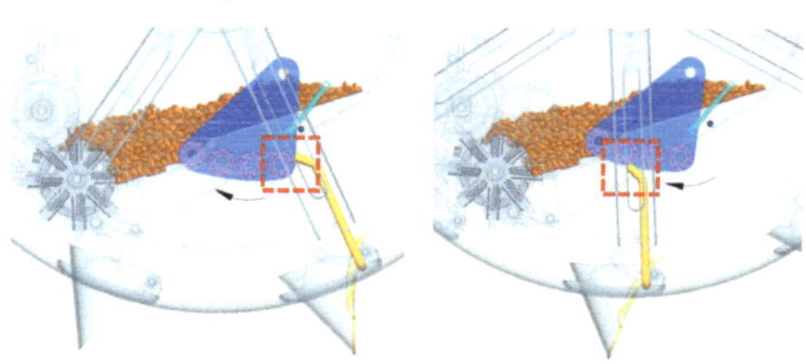

(T_1) (T_2)

Figure 5. Moments T_1 and T_2 during the reversal process of the hole seeder.

2.3. Design of the Adjustment Mechanism and Test Bench

2.3.1. Design of Adjustment Device

The adjustment device serves the purpose of controlling and regulating sowing depth and is adaptable for altering the bench height to accommodate different sizes of hole-seeder seeding tests. Comprising the bracket, bearing seat, electric actuator, and switch controller, as depicted in Figure 6, the adjustment device is an integral component of the pendulum-lever-type hole-seeding apparatus. During operation, in accordance with various test conditions and hole-seeder dimensions, the electric actuator's stroke is controlled to adjust the hole seeder's height position. Concurrently, the fixed plate rotates along the axis of the movable shaft, providing support and fixation for the hole seeder.

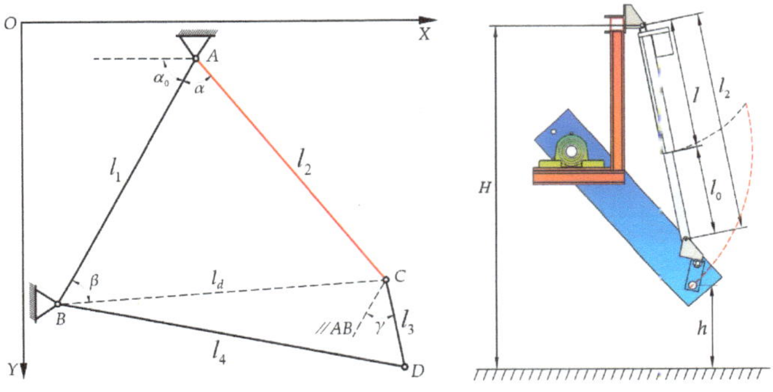

Figure 6. Adjustment trajectory analysis. (**a**) Sketch of the regulating mechanism. (**b**) Hole-seeding axis adjustment track.

Figure 6a shows a sketch of the four-bar mechanism of the mounting bracket, constructing the coordinate system *XOY*, the *X*-axis over point *A*, and the *Y*-axis over point *B*. According to the analytical method [25], the trajectory of the movement of point *D* on the connecting rod can be adjusted to the hole-seeding center-axis trajectory *D*(*x*,*y*):

$$\begin{cases} x_D = l_1\cos\alpha_0 + l_2\cos(\pi-\alpha_0-\alpha) + l_3\cos(\pi-\alpha_0-\gamma), \alpha \in [0,\pi] \\ y_D = l_2\sin(\pi-\alpha_0-\alpha) + l_3\sin(\pi-\alpha_0-\gamma), \alpha \in [0,\pi] \\ {x_D}^2 + (y_D - l_1\sin\alpha_0)^2 = {l_3}^2 \end{cases} \tag{4}$$

$$\begin{cases} \delta = \arccos\frac{{l_1}^2+d^2-{l_4}^2}{2l_1l_d} - \beta \\ \beta = \arcsin\frac{l_2\sin\alpha}{l_d} \\ l_d = \sqrt{{l_1}^2+{l_2}^2-2l_1l_2\cos\alpha} \end{cases} \tag{5}$$

where (x_D, y_D) are the coordinates of point *D*, mm; l is the initial length of the motorized actuator, mm; l_0 is the electric actuator adjustment stroke, mm; l_1 is the length between the bracket fixing points, mm; l_2 is the crank length, mm; l_3 is the length of connecting rod, mm; α_0 is the angle between the line connecting points *A* and *B* and the direction of *X* axis; α is the crank rotation angle, °; δ is the angle between connecting rod *AB* and *CD*, °; l_4 is the rocker length, mm; β is the angle between *AB* and *BC*, °.

The electric actuator stroke l_0 ranges from 0 to 200 mm, i.e., the length of l_2 varies from 305 to 505 mm. Considering the size of the hole-forming mechanism and the radius of the hole-seeding device [26], l_4 = 400 mm, l_1 = 320 mm, l_3 = 100 mm, and α_0 = 60°.

When the minimum adjustment stroke l_0 = 0 mm, the crank angle α = 7 °, calculated using formulas (1) and (2) in the coordinate system *XOY D*, with point coordinates for (328, 279); we adjusted the l_0 maximum to 200 mm, when the crank angle α = 43.2 °, for

D point coordinates for (213, 570). The hole-seeding installation height h and point D longitudinal coordinate relationship formula is

$$h = H - y_D > 0 \tag{6}$$

where h is the vertical distance between the axis of hole seeding to the seedbed belt, mm; H is the vertical distance between the bracket and motorized actuator connecting the axis to the seedbed belt, mm.

When H = 750 mm, the variation range of hole-seeding installation height h is 180~471 mm, which is in line with the range of the hole-seeding radius. The axial trajectory of the hole-seeding installation position is shown as a red line in Figure 6b.

2.3.2. Structure of the Test Bench

We used this regulating device in a test stand; the structure of the test bench mainly included a bench, conveyor belt, drive roller, and roller, as shown in Figure 7.

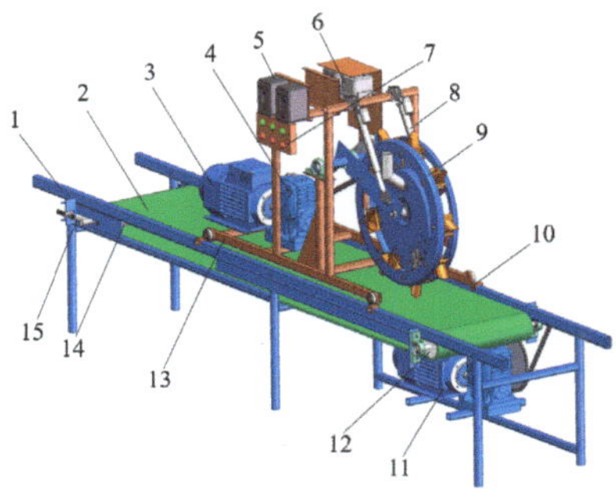

(**a**) 3D diagram

(**b**) Photograph of device

Figure 7. Hole-seeding device test bed: (1) bench; (2) conveyor belt; (3) hole seeder motor; (4) retarder; (5) inverter; (6) contactor; (7) central control panel; (8) electric push rod; (9) hole seeder; (10) bracket restrainer; (11) conveyor belt motor; (12) drive roller; (13) bracket; (14) roller; (15) conveyor belt tensioning device.

The transmission system of the seeding test bench of the hole-seeding device is shown in Figure 8, and the direction of the arrow is the direction of power transmission. The driving motor of the hole seeder is transmitted to the gearbox through a universal transmission shaft. After the gearbox changes direction, it is first transmitted to the movable shaft by belt transmission, and then transmitted to the hole seeder through the coaxial sprocket two-stage transmission. In the same way, the conveyor belt drive motor is transmitted to the conveyor belt drive roller by chain transmission after variable speed commutation of the gearbox.

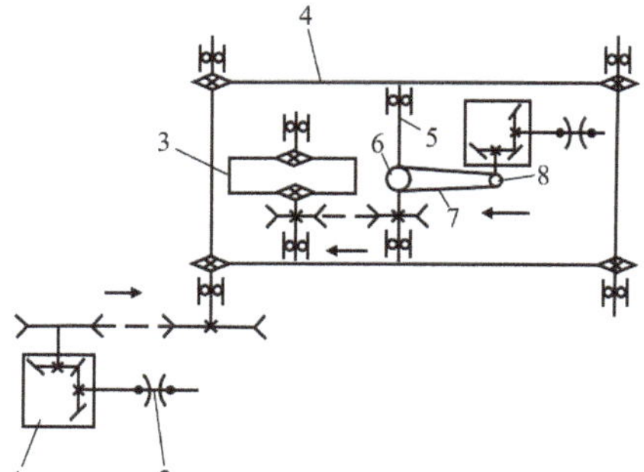

Figure 8. Schematic diagram of the test stand drive system: (1) retarder; (2) coupler; (3) rotating disc; (4) conveyor belt; (5) movable shaft; (6) large pulley; (7) V belt; (8) small pulley.

3. Seeding Process Simulation

3.1. Discrete Elemental Modeling of Corn Kernels

We randomly selected 500 seeds of Jinsui No. 4. Considering the influence of kernel shape on the seed-casting process, the kernels were categorized into flat and round according to the shape contour for counting, and the three-axis dimensions of the maize kernels were determined using digital vernier calipers, which were the length (*L*), width (*W*), and thickness (*T*) of the corn kernels (Figure 9).

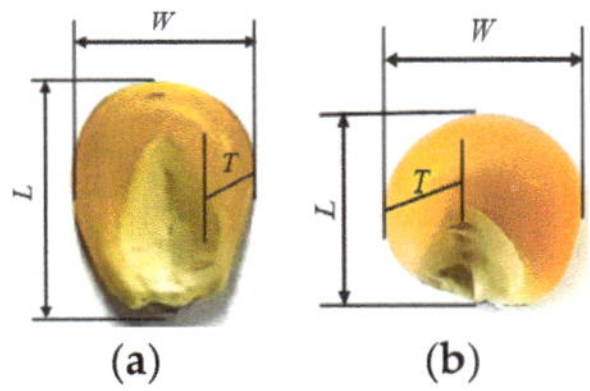

Figure 9. Corn kernel external dimensions. (**a**) Flattened; (**b**) orbicular.

The calculated triaxial geometric mean diameters of the seeds are shown in Table 1. Based on the three-axis geometric mean diameter and shape profile of the corn kernels, 3D modeling software was used to establish a 3D model of the kernel, and the model was imported into EDEM software in .stl format. To more accurately fit the motion of the corn kernels, spheres with unequal diameters were used to fill the kernel. We reduced the number of spherical particles required to fill a single grain to save simulation time, and we established 13 and 14 large-spherical corn grain models, as shown in Figure 10.

Table 1. Triaxial geometric mean diameter of seeds.

Shape	Average Length *L*/mm	Average Width *W*/mm	Average Thickness *T*/mm	Proportion
Flattened	11.65	7.62	4.85	77.4%
Orbicular	10.73	7.68	5.87	22.6%

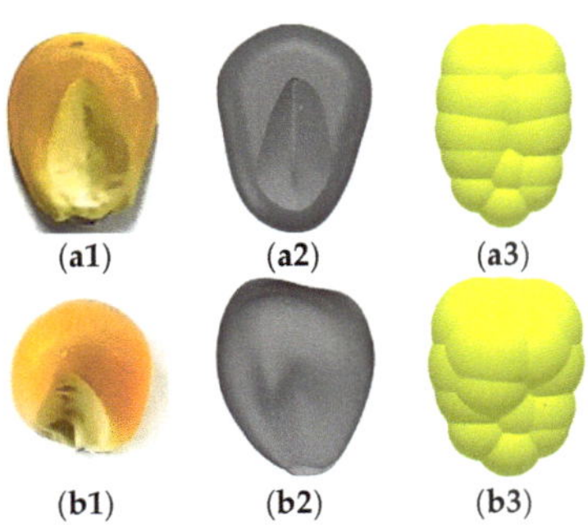

Figure 10. Discrete metamodel of corn kernels. (**a1**) Flat-shaped grain entity; (**a2**) 3D model of flat grain; (**a3**) discrete flat-grain model; (**b1**) circular grain entity; (**b2**) 3D model of circular grain; (**b3**) discrete circular-grain model.

3.2. Modeling of Hole-Seeding Applicator Rows

While ensuring successful sowing with the hole seeder, the 3D model created using SolidWorks 2022 software was simplified to reduce the computational complexity of the simulations. The simplified model was then imported into multibody dynamics simulation software RecurDyn in .x_t format to add constraints, forces, and contact relationships for simulating the trajectory of the hole seeder during seeding. The multibody dynamics model in RecurDyn was imported into EDEM for further analysis using a wall intermediate format file. Simulation tests of the hole-seeding process were conducted based on MBD-DEM. The necessary material and contact parameters for the simulation model in the seed discharge process were obtained by reviewing the literature, as shown in Tables 2 and 3 [27,28].

Table 2. Material parameters of the simulation model.

Material	Poisson's Ratio	Shear Modulus/MPa	Density/(kg·m^{-3})
Corn	0.357	2.17×10^2	1250
Steel	0.28	3.5×10^4	7850
Seed brush	0.40	1×10^2	1150
Rubber band	0.48	1×10^3	1380

Table 3. Contact parameters of the simulation model.

Form of Contact	Coefficient of Restitution	Coefficient of Static Friction	Coefficient of Rolling Friction
Corn–Corn	0.60	0.50	0.10
Steel–Corn	0.60	0.30	0.01
Seed Brush–Corn	0.45	0.50	0.01
Rubber Belt–Corn	0.711	0.784	0.035

After establishing a simplified model for hole-seeding simulation [29], the simulation parameters were set to generate a total of 1000 corn grains, with each type generated at a rate of 300 grains/s, as depicted in Figure 11a. The simulation time was set to 2 min, with a time step of 2.58×10^{-5} s. To observe in detail how the seed model underwent the processes of filling, carrying, and casting during hole-seeding discharge by the hole seeder, kernels with entry hole movement within the corn grain group were selected as the observation objects. The hole-seeding model was set to be displayed in the form of filled objects with an opacity of 0.1, and its trajectory is depicted in Figure 11b. Upon completion of the simulation, Figure 12 illustrates the seedbed with falling seeds.

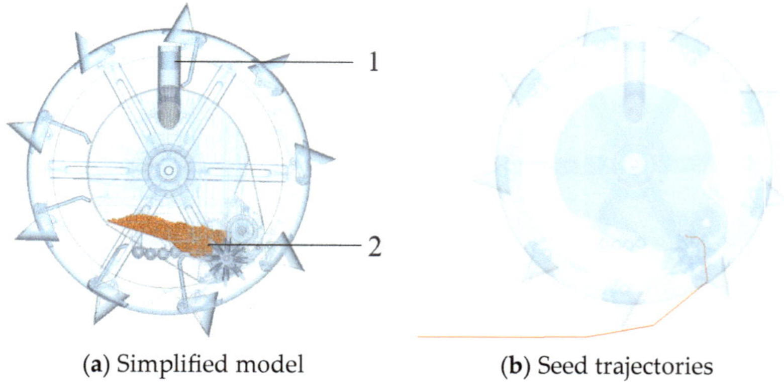

Figure 11. Simulation model. (1) Corn model generation cylinder; (2) corn particles.

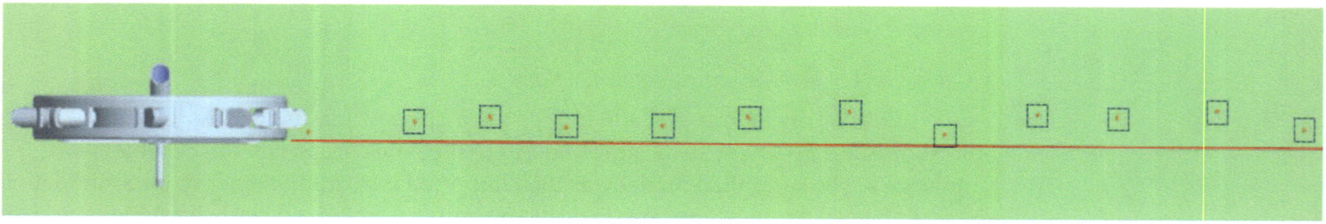

Figure 12. Schematic diagram of simulated seed drop.

3.3. Single-Factor Simulation Test

To determine the range of model operating parameters, single-factor simulation experiments were necessary. Based on the operating principles of the device and relevant operational parameters of the seeder, we separately determined the experimental value range of the seeder's dynamic disc speed, the dynamic forming plate lever angle, and the number of pendulum plate bearings.

3.3.1. Determination of Rotating Speed of Rotating Disc

The seed discharger is the core component of the hole seeder, and its seed supply speed directly determines the rotational speed of the hole seeding. To improve the operation speed of hole seeding, considering the actual seed supply speed of this mechanical seeder, the maximum speed of the hole seeder's rotating disc was set at 80 r/min, which means that the forward speed of the hole seeding was 9.40 km/h. To adapt to different operating conditions of the seeder, we verified the sowing performance at different speeds. Selecting a low speed of 40 r/min, a medium speed of 60 r/min, and a high speed of 80 r/min, the forward speeds of the seeder were 4.72 km/h, 7.06 km/h, and 9.40 km/h, respectively.

3.3.2. Determination of the Angle of the Lever

The angle of the movable hole-former lever primarily affects the size of the hole-former opening, which varies with the lever angle. During operation, Figure 13a,b illustrate that when the lever angle of the hole-forming mechanism is 110°, the opening of the movable hole-former is small, increasing the likelihood of 'stuck seeds' if seeds are not promptly dispensed, leading to seed damage and affecting subsequent seeding strokes. Additionally, when the hole-forming mechanism is used for mulch-hole seeding, a small lever angle narrows the opening of the movable hole-former, preventing seeds from falling into the soil promptly, resulting in decreased seeding quality and an increased empty hole rate. In Figure 13d, when the hole-former lever is at 150°, the turning angle of the movable hole-former λ is 22.72°, elevating the risk of film tearing and soil disturbance. Conversely, in Figure 13c, when the lever is at 140°, the turning angle λ = 18.71°, nearing the maximum range value of 18.69°. Considering these factors, the optimal range for the movable hole-former lever angle was determined to be 110° to 140°.

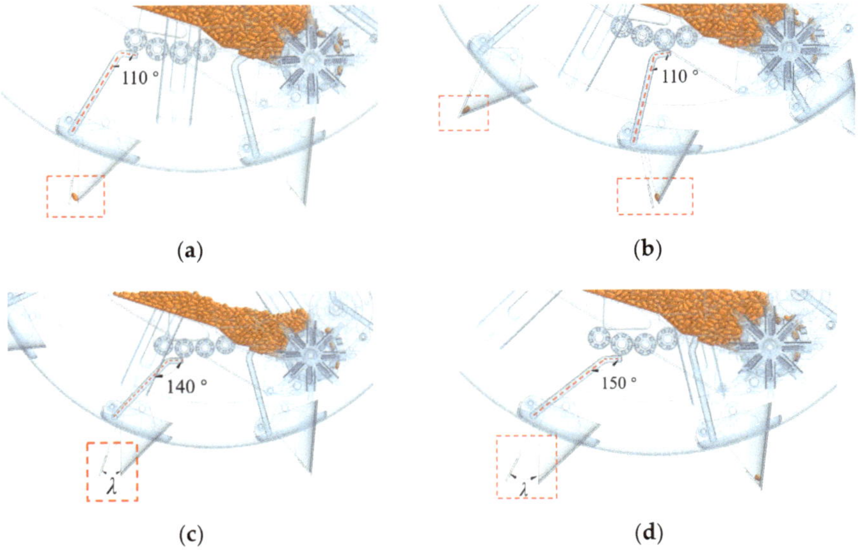

Figure 13. Movable hole-formers with different lever angles: (**a**) for 110° lever; (**b**) seed not dropped for 110°, (**c**) 140°, and (**d**) 150° levers.

3.3.3. Determination of the Number of Pendulum Bearings

The pendulum mechanism serves as the primary opening mechanism of the hole-former, ensuring that it operates within a reasonable timeframe. The number of pendulum bearings is a critical component for controlling the opening and closing times of the hole-forming device. If the number of pendulum bearings is insufficient, it may shorten the opening time of the moving hole-forming device, leading to issues such as seed jamming and the delayed discharge of corn kernels, which can significantly impact the effectiveness of sowing. Conversely, an excessive number of pendulum bearings extends the opening time of hole seeding but increases the risks of film tearing and the disruption of the structure of the whole-film mulching and double-ridge furrow, resulting in lower ground temperature around the seed holes and reduced moisture transfer capability. We calculated the seed drop time to satisfy the following conditions:

$$\begin{cases} \Delta t = \frac{\Delta\theta}{\omega} \\ \omega = \frac{v}{r} = 2\pi n \end{cases} \tag{7}$$

where $\Delta\theta$ is the angle between the drop point of the discharge wheel of the hole seeder and the line connecting the center and the vertical centerline of the hole seeder, (°); Δt is the time for the hole seeding to turn over, s; ω is the hole-forming apparatus's angular velocity, rad.

Considering the seed dispenser had a $\Delta\theta$ of 36° and speeds of 40 r/min, 60 r/min, and 80 r/min, the duration of hole seeding was 0.15 s, 0.1 s, and 0.075 s, respectively. When the speed of the hole seeding was increased from 40 to 80 r/min, the theoretical pulsation period of the seeding time reduced from 0.15 to 0.075 s.

Through simulation testing, we obtained the actual seed drop time of the hole seeder. We set the EDEM data-saving interval to 0.01 s and exported speed data for a single seed to create a two-dimensional line graph, as depicted in Figure 14. We analyzed the seed drop time and speed change graph for different dynamic disc speeds, and we determined the optimal number of pendulum bearings to install. We adjusted the opening and closing time of the hole-former and optimized the seed drop effect of the hole seeder.

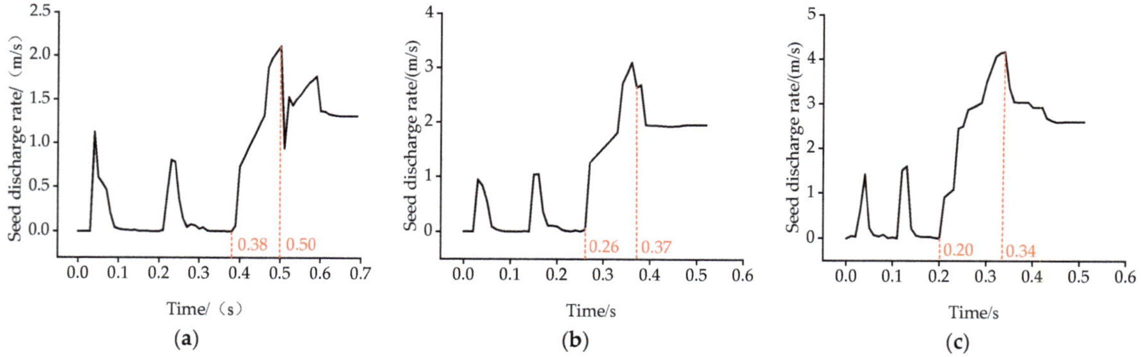

Figure 14. Seed drop time vs. speed at different rotational speeds. (**a**) Speed = 40 r/min; (**b**) speed = 60 r/min; (**c**) speed = 80 r/min.

In the seed movement process, the seed falls into the hole first, the two hole positions are rotated with the seeding wheel, and then the seed contacts the hole-forming device freely, colliding in the cavity of the hole-forming device, and finally reaching the bottom of the hole-forming device. Then, the hole-forming device opens and the seed falls to the seedbed belt. It can be seen from the Figure 14 that when the rotating speed of the hole seeder was 40 r/min, 60 r/min, and 80 r/min, the corn seeds fell freely from the seeding. The time required for this process was 0.12 s, 0.11 s, and 0.14 s, respectively. When the rotating speed of the moving disc was 80 r/min, the seed dropping time did not decrease but increased, because the rotating speed of the moving disc was too fast, and the seed was dragged by the hole-forming device, resulting in delayed falling. The maximum speed of the seeds was more than 4 m/s, and there was an obvious bounce phenomenon that occurred when the seed contacting the seedbed. Figure 15 is the change diagram of the seeding index Y of the hole seeder at different speeds when the number of pendulum bearings ranged from one to six.

$$Y = \frac{m}{M} \tag{8}$$

where Y is the seeding index, %; M is the theoretical number of seeding holes; m is the actual number of seeding holes.

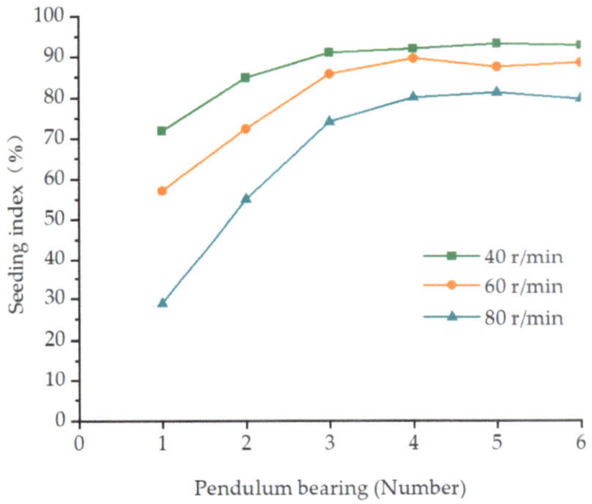

Figure 15. Plot of number of bearings versus row index at different speeds.

It can be seen from the Figure 15 that the number of pendulum bearings was one, the seeding index under different speeds was less than 80%. This is attributed to the brief opening duration of the hole-forming device with a single bearing. When seeds do not reach the bottom of the hole-forming device, the movable hole-former completes its cycle prematurely, resulting in fewer seeding holes. However, the seeding index was 71.94% at 40 r/min. The main reason was that the seeding period of the hole seeder was longer at a slower rotating speed, which prolonged the opening and closing time of the hole-forming device and the falling time, and the seeds could pass through before the closing of the hole-forming device. When the number of bearings was three or less, the seeding index of the hole seeder at different speeds increased significantly. When the number of bearings was more than three, the seeding index at the speed of 40 r/min tended to be stable. The seeding index at the speed of 60 r/min and 80 r/min increased by 4.82% and 5.94%, respectively, and then the seeding index fluctuated within a range of 3%. When the number of bearings was six, the seeding index did not change significantly. From our analysis, the optimal value range of the pendulum plate bearing was found to be between three and five.

3.4. Box–Behnken Experimental Design

To optimize the parameters of the hole-seeding device's oscillating disc mechanism, an experiment was conducted using a Box–Behnken central combination experimental design to determine the optimal parameter combinations of the factors affecting the seed discharge performance. In the test, we took the hole-seeding machine moving disc speed, the angle of the moving hole-forming machine lever, and the number of swing disc bearings as factors, and based on the results of the theoretical analysis and simulation analysis, we set the zero-level value, and we carried out a three-factor, three-level orthogonal test. The experimental factor coding is shown in Table 4. A total of 17 groups of tests were conducted, each group of tests was repeated three times, and the average value was taken as the final test results (X_1, X_2, and X_3 were the factor coding values).

Table 4. Test factor codes.

Levels	Test Factors		
	Rotational Speed of Rotating Disc X_1/(r·min^{-1})	Lever Angle X_2/(°)	Pendulum Bearings X_3/(Number)
−1	40	110	3
0	60	125	4
1	80	140	5

According to the parameter requirements of each test sequence number, we adjusted the rotational speed of the moving disc, angle of the lever, and number of bearings. We completed the modeling of the hole-seeding machine using SolidWorks, set the rotational speed of the moving disc in RecurDyn software, and carried out simulation tests based on MBD-DEM. We extracted the data of the number of grains in each group of tests and computed the qualification index, the miss-seeding index, and the replanting index. We recorded the test results. According to GB/T 6973-2005 "Test Methods for Single Grain (Precision) Seeders" [30], the test takes the seed discharge qualification index Y_1, miss-seeding index Y_2, and replanting index Y_3 as the test indices, which are continuously detected under stable working conditions of the hole-seeding machine, and the test results are shown in Table 5.

$$Y_1 = \frac{n_1}{N} \times 100\% \tag{9}$$

$$Y_2 = \frac{n_0}{N} \times 100\% \tag{10}$$

$$Y_3 = \frac{n_2}{N} \times 100\% \tag{11}$$

where N is the theoretical number of seeds; n_0 is the number of miss-seeded rows; n_1 is the number of single seed rows; n_2 is the number of repeated rows.

Table 5. Experimental project and results.

Test No.	Factors			Response Values		
	X_1	X_2	X_3	Y_1/%	Y_2/%	Y_3/%
1	−1	1	0	89.26	5.42	5.32
2	−1	0	1	90.4	4.54	5.06
3	−1	0	−1	89.98	5.29	4.73
4	−1	−1	0	72.49	25.23	2.28
5	0	1	1	88.02	7.73	4.25
6	0	1	−1	85.03	11.32	3.65
7	0	0	0	87.52	8.73	3.75
8	0	0	0	87.41	9.41	3.18
9	0	0	0	86.61	9.47	3.92
10	0	0	0	86.66	9.57	3.77
11	0	0	0	84.56	11.63	3.81
12	0	−1	1	67.5	29.59	2.91
13	0	−1	−1	64.62	32.73	2.65
14	1	1	0	79.68	17.97	2.35
15	1	0	−1	77.6	20.26	2.14
16	1	0	1	72.56	25.68	1.76
17	1	−1	0	54.33	44.04	1.63

4. Results and Analysis

4.1. Simulation Test Results

4.1.1. Regression Modeling and Testing

Regression fitting and variance analysis were carried out using Design-Expert 12 data processing software. The results are shown in Table 6, and the regression equations of the qualification index Y_1, miss-seeding index Y_2, and replanting index Y_3 with the coded values of the test factors were obtained as follows:

$$Y_1 = 86.55 - 7.25X_1 + 10.38X_2 + 1.42X_3 + 2.15X_1X_2 + 1.15X_1X_3 + 0.0275X_2X_3 - 3.13{X_1}^2 - 9.48{X_2}^2 - 0.7323{X_3}^2 \quad (12)$$

$$Y_2 = 9.76 + 8.43X_1 - 11.14X_2 - 1.61X_3 - 1.57X_1X_2 - 1.17X_1X_3 - 0.1125X_2X_3 + 3.50{X_1}^2 + 9.90X_2 + 0.6790{X_3}^2 \quad (13)$$

$$Y_3 = 3.69 - 1.19X_1 + 0.7625X_2 + 0.1963X_3 - 0.5800X_1X_2 + 0.0125X_1X_3 + 0.0850X_2X_3 - 0.3667{X_1}^2 - 0.4242{X_2}^2 + 0.1032{X_3}^2 \quad (14)$$

Table 6. Variance analysis of the regression coefficients.

Test Index	Sources of Variance	Square of Sum	Degree of Freedom	Mean Square	*F*-Value	*p*-Value
Qualification index	Model	1765.11	9	196.12	207.28	<0.0001 **
	X_1	99.94	1	99.94	105.62	<0.0001 **
	X_2	6.54	1	6.54	6.91	0.0340 *
	X_3	7.58	1	7.58	8.01	0.0254 *
	X_1X_2	18.40	1	18.40	19.45	0.0031 **
	X_1X_3	5.34	1	5.34	5.64	0.0493 *
	X_2X_3	0.0030	1	0.0030	0.0032	0.9565
	${X_1}^2$	41.38	1	41.38	43.73	0.0003 **
	${X_2}^2$	378.18	1	378.18	399.70	<0.0001 *
	${X_3}^2$	2.58	1	2.58	2.72	0.1429
	Residual error	6.62	7	0.9462		
	Lack-of-fit	0.9668	3	0.3223	0.2279	0.8728
	Error	5.66	4	1.41		
	Sum	1771.74	16			
Miss-seeding index	Model	2088.30	9	232.03	190.83	<0.0001 **
	X_1	150.66	1	150.66	123.91	<0.0001 **
	X_2	5.91	1	5.91	4.86	<0.0001 **
	X_3	7.82	1	7.82	6.43	0.0044 **
	X_1X_2	9.80	1	9.80	8.06	0.0251 *
	X_1X_3	5.45	1	5.45	4.48	0.0720
	X_2X_3	0.0506	1	0.0506	0.0416	0.8441
	${X_1}^2$	51.62	1	51.62	42.46	0.0003 **
	${X_2}^2$	412.80	1	412.80	339.49	<0.0001 **
	${X_3}^2$	1.94	1	1.94	1.60	0.2469
	Residual error	8.51	7	1.22		
	Lack-of-fit	3.71	3	1.24	1.03	0.4685
	Error	4.80	4	232.03		
	Sum	2096.81	16	2.12		
Reseeding index	Model	19.05	9	5.19	11.99	0.0018 **
	X_1	5.19	1	0.0157	29.37	<0.0001 **
	X_2	0.0157	1	0.0018	0.0888	0.0014 **
	X_3	0.0018	1	1.35	0.0103	0.2281
	X_1X_2	1.35	1	0.0006	7.62	0.0281 *
	X_1X_3	0.0006	1	0.0289	0.0035	0.9542
	X_2X_3	0.0289	1	0.5663	0.1636	0.6979
	${X_1}^2$	0.5663	1	0.7578	3.21	0.1165
	${X_2}^2$	0.7578	1	0.0449	4.29	0.0770
	${X_3}^2$	0.0449	1	0.1766	0.2542	0.6296
	Residual error	1.24	7	0.2997		
	Lack-of-fit	0.8990	3	0.0843	3.55	0.1262
	Error	0.3373	4	2.12		
	Sum	20.29	16			

Note: ** means very significant ($p < 0.01$); * means significant ($0.01 < p < 0.05$).

The models for the qualification index, miss-seeding index, and reseeding index of the test parameters were all highly significant ($p < 0.01$), with some quadratic and interaction terms also showing significant effects. The regression equations for the three response indicators were not significantly misfitted ($p > 0.05$), indicating that they were well fitted with the experimental data. The quadratic regression equation fitted by the model was consistent with the numerical simulation test results, which thus correctly reflected the relationship between the three response indicators X_1, X_2, and X_3.

(1) Qualification index regression model

According to Table 7, under a significance level $\alpha = 0.05$, the quadratic regression model of the qualification index had a p of < 0.01, extremely significant. The lack-of-fit term was not significant ($p = 0.8728$), indicating that there was no other main factor affecting the qualification index, and the regression equation did not lose fit. The p values of X_2X_3 and X_3^2 were greater than 0.05, indicating no significant effect on the qualification index. The regression model equation after eliminating the insignificant factors in the interaction term was

$$Y_1 = 86.55 - 7.25X_1 + 10.38X_2 + 1.42X_3 + 2.15X_1X_2 + 1.15X_1X_3 - 3.13X_1^2 - 9.48X_2^2 \quad (15)$$

Table 7. Results of bench test.

Number	Qualification Index	Miss-Seeding Index	Reseeding Index
1	87.60%	7.60%	4.80%
2	92.40%	4.00%	3.60%
3	91.60%	5.20%	3.20%
Average	90.53%	5.60%	3.87%

Through the test of the regression coefficient in Equation (16), it was concluded that the factors affecting the qualification index in decreasing order of predominance were the angle of the lever X_2, the rotational speed of the rotating disc X_1, and the number of bearings X_3.

(2) Miss-seeding index regression model

According to Table 7, under a significance level α of 0.05, the quadratic regression model $p < 0.01$ for the miss-seeding index was extremely significant. The lack-of-fit term was not significant ($p = 0.4685$), indicating that there was no other main factor affecting the miss-seeding index, and the regression equation did not lose fitting. Among them, the p values of X_1X_3, X_2X_3, and X_3^2 were greater than 0.05, indicating no significant effect on the miss-seeding index. The regression model equation after eliminating the insignificant factors in the interaction term was

$$Y_2 = 9.76 + 8.43X_1 - 11.14X_2 - 1.61X_3 - 1.57X_1X_2 + 3.50X_1^2 + 9.90X_2^2 \quad (16)$$

Through the test of the regression coefficient using Equation (17), it was concluded that the factors affecting the miss-seeding index in descending order were the lever angle X_2, the rotational speed of the rotating disc X_1, and the number of bearings X_3.

(3) Reseeding index regression model

According to Table 7, under a significance level α of 0.05, the quadratic regression model of the reseeding index was $p < 0.01$, which is extremely significant. The loss-of-fit item was not significant ($p = 0.1262$), indicating that there were no other main factors affecting the reseeding index, and the regression equation was not lost. Among them, X_1 and X_2 had extremely significant effects, and X_2^2 had significant effects. The regression model equation after eliminating the insignificant factors in the interaction term was

$$Y_3 = 3.69 - 1.19X_1 + 0.7625X_2 - 0.1963X_3 - 0.5800X_1X_2 \quad (17)$$

Through the test of the regression coefficient using Equation (18), it was concluded that factors affecting the reseeding index in decreasing order were the rotating speed of the moving disc X_1, the lever angle X_2, and the number of bearings X_3.

4.1.2. Analysis of Model Interaction Terms

According to the regression model, a response surface diagram between the significant factors was made, where the shape of the response surface reflects the strength of the interaction factors [31]. This paper mainly analyzed the interaction items that had a significant impact ($p < 0.05$). We drew a response surface diagram of the speed of the hole seeder's moving disc and the angle of the moving hole seeder's lever to the three response indicators (Figure 16); we also drew a response surface diagram of the speed of the hole seeder's moving disc and the number of pendulum bearings to the eligibility index Y_1 (Figure 17).

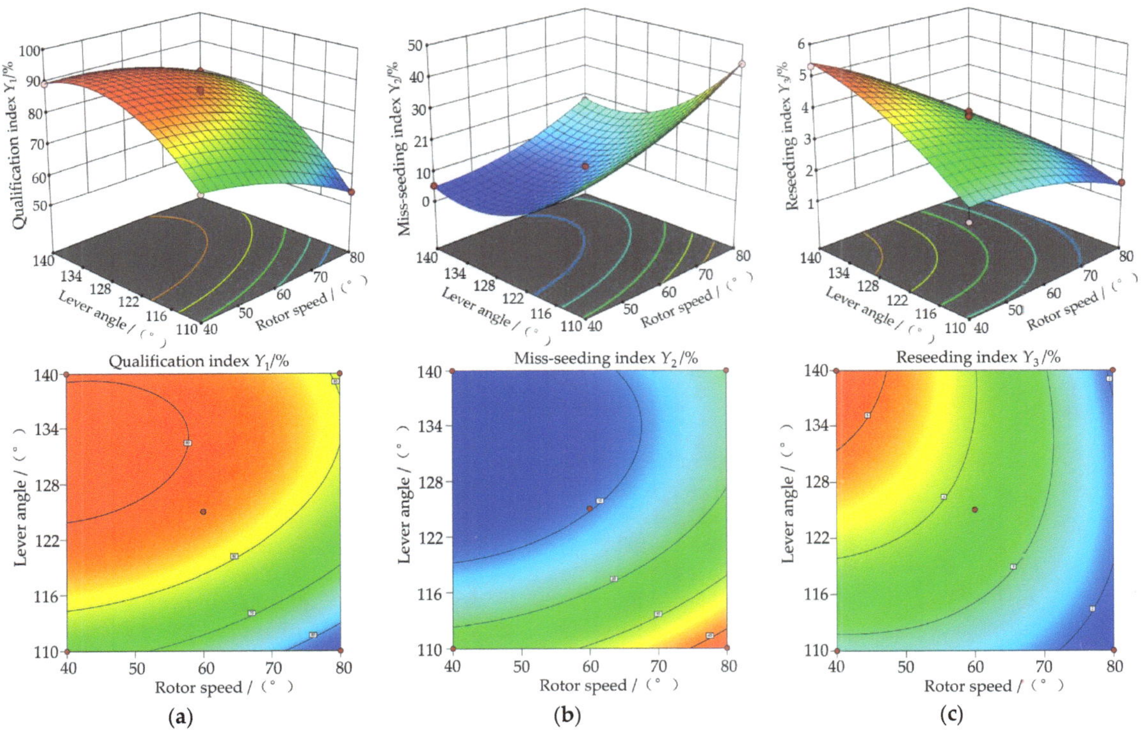

Figure 16. Effect of interaction of rotational speed of the rotating disc and angle of the lever on the response metrics. (**a**) Effect of the interaction of the rotational speed of the rotating disc and angle of the lever on the qualification index; (**b**) effect of the interaction of rotational speed of the rotating disc and angle of the lever on miss-seeding index; (**c**) effect of the interaction of rotational speed of the rotating disc and angle of the lever on the reseeding index.

(1) Interaction of rotational speed of the rotating disc and angle of the lever

Figure 16a illustrates the impact of the interaction between the rotating speed of the moving plate and the lever angle on the conformity index for four pendulum bearings. It can be observed that for a fixed rotation speed of the moving plate, the qualification index initially increases gradually with increasing lever angle, then gradually decreases. Similarly, at a constant lever angle, the qualification index slightly increases with a rising rotating speed of the moving plate before gradually decreasing after reaching its peak. The highest qualification index is achieved at a rotating speed of 40~60 r/min and a lever angle of 122°~134°.

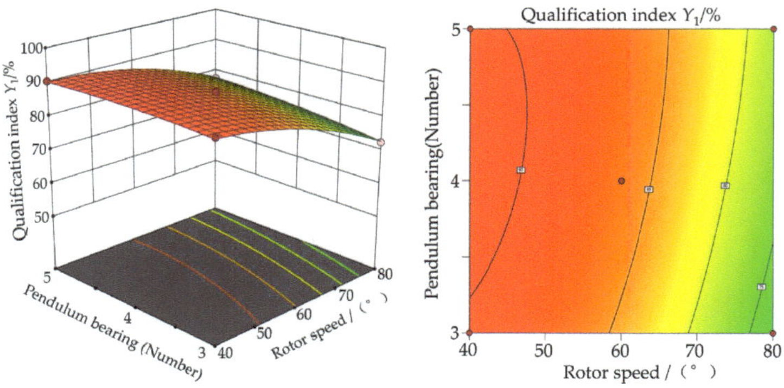

Figure 17. The influence of the interaction between the rotating speed of the moving disc and the number of bearings on the qualification index.

Figure 16b displays the influence of the interaction between the rotating speed of the moving plate and the lever angle on the miss-seeding index with four pendulum bearings. It is evident that at a constant rotation speed of the fixed moving plate, the miss-seeding index initially decreases gradually with increasing lever angle, then gradually increases. Conversely, at a constant lever angle, the miss-seeding index slightly decreases with increasing rotational speed of the moving plate before gradually increasing after reaching its peak. The lowest miss-seeding index was observed at a rotating speed of 40~60 r/min and a lever angle of 122°~134°.

Figure 16c depicts the impact of the interaction between the rotating speed of the moving plate and the lever angle on the reseeding index for four pendulum bearings. It is evident that, at a fixed speed of the moving plate, the reseeding index decreases with increasing lever angle. Similarly, at a constant lever angle, the reseeding index gradually decreases with increasing rotational speed of the moving plate. The lowest miss-seeding index was observed at a rotating speed of 80 r/min and a lever angle of 110 °.

The analysis revealed that when the lever angle remains constant, increasing the rotational speed of the moving plate enhances the rotation speed of the seeding wheel. This results in an accelerated seed delivery rate and an increased resultant force acting on the seeds. Consequently, some seeds fail to interact effectively and fill the designated holes, leading to a decline in both the qualification and reseeding indices, while the miss-seeding index notably escalates. Conversely, when the rotational speed of the moving plate remains constant and the lever angle decreases, the opening of the hole-forming device diminishes. Consequently, some seeds are unable to exit the hole-forming device, resulting in continuous decreases in the qualification and reseeding indices, accompanied by a significant increase in the miss-seeding index.

(2) Interaction between the rotational speed of the rotating disc and the number of pendulum bearings

Figure 17 shows the influence of the interaction between the rotating speed of the moving disc and the number of bearings on the qualification index when the angle of the lever is 125°. From the diagram, it can be seen that the rotating speed of the fixed moving plate is constant. With the increase in the number of bearings, the qualification index increases slightly first and then tends to be stable. The number of fixed bearings remains unchanged. With the increase in the rotating speed of the moving disc, the qualification index increases slightly and then decreases gradually after reaching the highest point. When the rotating speed of the moving disc is 40~60 r/min and the number of bearings is 4, the qualification index is the highest.

The analysis indicates that as the number of bearings increases, the openings of the hole-forming device gradually meet the requirements for discharging corn kernels. Initially, there is a slight increase in the qualification index, followed by stabilization. However, when the rotational speed of the rotating disc exceeds 60 r/min, the performance of the mechanical seed dispenser becomes a limiting factor. This results in an increase in empty holes on the seed-discharging wheel and a reduction in the seed-casting time of the hole-forming device, ultimately leading to a decrease in the qualification index.

4.1.3. Determination of Optimal Operating Parameters

To obtain the best combination of factors under the constraint conditions, the regression model was solved by multiobjective optimization to maximize the speed of the moving plate, minimize the angle of the lever, and minimize the number of bearings as the factor conditions, and to maximize the qualification index, minimize the reseeding index, and minimize the miss-seeding index as the evaluation indices. The regression equation and the constraint conditions were as follows:

$$\begin{cases} \max G(x) = Y_1(X_1, X_2, X_3) \\ \min G(x) = \begin{cases} Y_2 = (X_1, X_2, X_3) \\ Y_3 = (X_1, X_2, X_3) \end{cases} \end{cases} \tag{18}$$

$$\begin{cases} 40\text{r/min} \le \max X_1 \le 80\text{r/min} \\ 110^\circ \le \min X_2 \le 140^\circ \\ 3 \le \min X_3 \le 5, X_3 \in N^+ \end{cases} \tag{19}$$

The optimization solver in Design-Expert 12 software was used to optimize the regression equation models (12), (13) and (14) for objectives (18) and (19). Finally, the optimization test indices Y_1, Y_2, and Y_3 were obtained as the optimal working parameters: the speed of the moving plate was 49 r/min, the angle of the lever was 131°, and the number of the pendulum plate bearings was four. At this time, the seeding qualification index of the hole seeder was 91.70%, the miss-seeding seeding index was 4.57%, and the reseeding index was 3.73%.

4.2. Bench Test Results

To verify the reliability of the hole-seeding device and verify the seeding performance of the hole-seeding device, a bench test was carried out on 10 June 2023, at the Institute of Dry Farming Equipment Experts of Gansu Agricultural University, as shown in Figure 18.

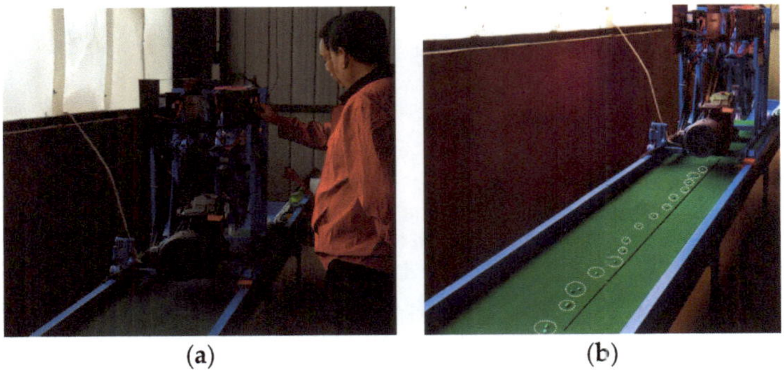

Figure 18. Bench test. (**a**) Diagram of the test bench operation and adjustment; (**b**) seedbed strip seeding maps.

The test was repeated three times, and the average value of 250 holes was counted continuously. The bench test results are presented in Table 7. The results indicate that when the speed of the hole seeder was 49 r/min, that is, the forward speed of the hole seeder was 5.76 km/h; the angle of the lever was 131 °; and the number of the pendulum bearing was four, the seeding qualification index was 90.53%, the miss-seeding seeding index was 5.60%, and the reseeding index was 3.87%.

4.3. Field Test Results

To investigate the field sowing performance of the pendulum-rod-type hole-sowing device, a field sowing test was carried out on 25 June 2023, in the experimental field of Taohe Tractor Manufacturing Co., Ltd., in Lintao County, Gansu Province. The experimental field was flat, north–south, and rectangular, and the soil was loessal soil. According to the agronomic technical requirements of whole-film mulching and double-ridge-furrow sowing, the surface was covered with film after ridging.

The corn planting variety was Jinsui No. 4, and the working power was 404 tractors with a power of 26.5 kW. The forward speed of the whole machine was stable at about 5.76 km/h, and the hole seeder was tested with the optimal working parameters. The unit forward was set to 50 m as the sampling length, referring to GB/T 6973-2005 "Test Methods for Single Grain (Precision) Seeders" [30] to determine the qualification index, miss-seeding seeding index, and reseeding index in the field test. The field sowing test results were averaged, and the field test results of the hole seeder were obtained. The qualification index was 89.13%, the miss-seeding seeding index was 5.46%, and the reseeding index was 6.41%. There was no obvious film tearing, film picking, or film hole dislocation during the test, as shown in Figure 19.

Figure 19. Field test.

5. Discussion

The simulation test results showed discrepancies compared to the indoor bench tests and field trials. The deviation in the qualification index in the simulation test from the bench test was 1.17% and that from the field test was 2.57%. The decrease in the qualification index, along with increases in the miss-seeding and reseeding indices, may be attributed to variations in the size and shape of corn kernels, leading to some holes in the seeding wheel accommodating two smaller kernels. Additionally, the vibrations from the field machinery may have caused larger kernels to fail to enter the seeding wheel holes, affecting the seeding process and slightly reducing the qualification index.

In an effort to enhance the performance of hole seeders, numerous scholars have conducted research on metering devices, proposing various structural designs and performance optimizations [32–34]. However, research on the seeding process of metering devices remains limited. The metering device selected for this study is a widely used mechanical planter with a simple structure [35]. Due to its limitations in high-speed operations, the qualification index decreased as the planter's performance diminished [36].

This was consistent with the significant increase in the miss-seeding index observed during simulation tests conducted at 80 r/min. Future research could focus on pneumatic seeders capable of accommodating higher speeds, aiming to mitigate the adverse effects caused by the seeding process of the hole-forming device during high-speed sowing.

6. Conclusions

1. We designed a pendulum plate and lever-type hole-seeding device, mainly composed of a fixed plate, a movable shaft, an electric actuator, a rotating disc, a connecting bracket, a fixed disc cover, a pendulum plate, a hole-former, and its structure. Bench testing of the hole seeder was conducted to meet different seed size requirements. By analyzing the working principle of the hole seeder, the structural parameters of the hole-forming device were optimized. A height-adjustment device suitable for hole seeders of different sizes was proposed and applied on an experimental bench. The device improved the operational performance of the hole seeder and reduced the rate of empty holes so as to achieve the purpose of improving operational efficiency and increasing production and income.
2. Utilizing Box–Behnken experimental design principles, we employed a three-factor, three-level response surface analysis method to conduct simulation experiments pf the seeding performance of the hole seeder under various combinations of operational parameters. With the aid of Design-Expert 12, we derived quadratic regression models for the qualification index, miss-seeding index, and reseeding index of the seeding process. We investigated the impact of the hole seeder's disc's rotational speed, the lever angle of the movable hole-former, and the number of bearings on the response values of these three indices. The optimal parameter combination identified was a rotating disc rotational speed of 49 r/min for the hole seeder (equating a forward speed of 5.76 km/h), a lever angle of 131° for the movable hole-former, and four bearings for the pendulum plate.
3. Under the optimal working parameters, field test results showed that the qualification index of the pendulum-lever-type hole seeder was 89.13%, the miss-seeding index was 5.46%, and the reseeding index was 6.41%. Compared to the simulation test, the relative errors of the qualification index, reseeding index, and miss-seeding index were 2.57%, 0.89%, and 2.68%, respectively. The small discrepancies between the actual results and the optimized values (<5%) demonstrate the reliability of the obtained working parameters of the pendulum-lever-type hole seeder.

Author Contributions: Conceptualization, F.D. and H.Z.; methodology, H.D. and H.P.; software, H.Z. and R.S.; validation, R.S., C.Z. and H.D.; formal analysis, H.Z. and H.P.; investigation, F.D.; resources, Q.Z. and H.D.; data curation, Q.Z. and C.Z.; writing—original draft preparation, F.D. and H.Z.; writing—review and editing, F.D. and R.S.; visualization, H.Z.; supervision, F.D.; project administration, C.Z.; funding acquisition, F.D. All authors have read and agreed to the published version of the manuscript.

Funding: This research was supported by Key Science and Technology Project of Gansu Province (23ZDNA008), National Natural Science Foundation of China (grant No. 52365029), Central Guide Local Science and Technology Development Fund Project (grant No. 23ZYQF305-1).

Institutional Review Board Statement: Not applicable.

Data Availability Statement: The data presented in this study are available on request from the authors.

Acknowledgments: The authors thank the editor for providing helpful suggestions for improving the quality of this manuscript.

Conflicts of Interest: The authors declare no conflicts of interest.

References

1. Dai, F.; Song, X.F.; Zhao, W.Y.; Sun, B.G.; Shi, R.J.; Zhang, Y. Numerical simulation and analysis of mechanized suppression process of seedbed with whole plastic film mulching on double ridges. *Int. J. Agric. Biol. Eng.* **2021**, *14*, 142–150. [CrossRef]
2. Dai, F.; Guo, W.; Song, X.; Shi, R.; Zhao, W.; Zhang, F. Design and field test of crosswise belt type whole plastic-film ridging-mulching corn seeder on double ridges. *Int. J. Agric. Biol. Eng.* **2019**, *12*, 88–96. [CrossRef]
3. Dai, F.; Zhao, W.Y.; Zhang, F.W.; Ma, H.J.; Xin, S.L.; Ma, M.Y. Research Progress Analysis of Furrow Sowing with Whole Plastic-film Mulching on Double Ridges Technology and Machine in Northwest Rainfed Area. *Trans. Chin. Soc. Agric. Mach.* **2019**, *50*, 1–16. [CrossRef]
4. Xie, J.H.; Zhang, R.Z.; Li, L.L.; Chai, Q.; Luo, Z.Z.; Cai, L.Q.; Qi, P. Effects of plastic film mulching patterns on maize grain yield, water use efficiency, and soil water balance in the farming system with one film used two years. *J. Appl. Ecol.* **2018**, *29*, 1935–1942 [CrossRef] [PubMed]
5. Zhao, W.Y.; Dai, F.; Yang, J.; Shi, Z.L.; Yang, Z.; Shi, L.R. Design and Experiment of direct insert precision hill-seeder with corn whole plastic-film mulching on double ridges. *Trans. Chin. Soc. Agric. Mach.* **2013**, *44*, 91–97. [CrossRef]
6. Dong, J.X.; Gao, X.J.; Zhang, S.L.; Liu, Y.; Chen, X.H.; Huang, Y.X. Design and Test of Maize Posture Control and Driving Precision Metering Device for High-speed Seeder. *Trans. Chin. Soc. Agric. Mach.* **2022**, *53*, 108–119. [CrossRef]
7. Gao, X.J.; Xie, G.F.; Li, J.; Shi, G.S.; Lai, Q.H.; Huang, Y.X. Design and validation of a centrifugal variable-diameter pneumatic high-speed precision seed-metering device for maize. *Biosyst. Eng.* **2023**, *227*, 161–181. [CrossRef]
8. Han, D. Optimization Simulation and Experimental Research of Inside Filling Air-Blowing Maize Precision Seed-Metering Device. Ph.D. Thesis, China Agricultural University, Beijing, China, 2018.
9. Zhang, C.; Xie, X.; Zheng, Z.; Wu, X.; Wang, W.; Chen, L. A Plant Unit Relates to Missing Seeding Detection and Reseeding for Maize Precision Seeding. *Agriculture* **2022**, *12*, 1634. [CrossRef]
10. Gao, X.J.; Cui, T.; Zhou, Z.Y.; Yu, Y.B.; Xu, Y.; Zhang, D.X.; Song, W. DEM study of particle motion in novel high-speed seed metering device. *Adv. Powder Technol.* **2021**, *32*, 1438–1449. [CrossRef]
11. Gao, X.J.; Zhao, P.F.; Li, J.; Xu, Y.; Huang, Y.X.; Wang, L. Design and Experiment of Quantitative Seed Feeding Wheel of Air-Assisted High-Speed Precision Seed Metering Device. *Agriculture* **2022**, *12*, 1951. [CrossRef]
12. Dong, J.X.; Gao, X.J.; Zhang, S.L.; Huang, Y.X.; Zhang, C.Q.; Shi, J.T. Design and Test of Guiding Seed Throwing Mechanism for Maize Posture Control and Driving Metering Device. *Trans. Chin. Soc. Agric. Mach.* **2023**, *54*, 25–34. [CrossRef]
13. Ma, J.; Sun, S.; Wang, J.; Hu, B.; Luo, X.; Xu, X. An Experimental Analysis of the Seed-Filling Mechanism of Maize-Precision Hole-Planter Clamping. *Agriculture* **2024**, *14*, 398. [CrossRef]
14. Wang, K.Y. Optimization Study of Self-Propelled Sunflower Coated Seeder Cavitator. Master's Thesis, Inner Mongolia Agricultural University, Hohhot, China, 2023.
15. Gu, L.L.; Zhang, W.Q.; Liu, M.J. The design and Experiment of the forced opening mechanism on the peanut dibbler. *J. Hunan Agric. Univ. (Nat. Sci.)* **2017**, *43*, 676–679. [CrossRef]
16. Adilet, S.; Zhao, J.; Sayakhat, N.; Chen, J.; Nikolay, Z.; Bu, L.; Sugirbayeva, Z.; Hu, G.; Marat, M.; Wang, Z. Calibration Strategy to Determine the Interaction Properties of Fertilizer Particles Using Two Laboratory Tests and DEM. *Agriculture* **2021**, *11*, 592. [CrossRef]
17. Gummer, A.; Sauer, B. Modeling planar slider-crank mechanisms with clearance joints in RecurDyn. *Multibody Syst. Dyn.* **2012**, *31*, 127–145. [CrossRef]
18. Liu, Q.F.; Chen, Y.; Liu, Y.; Wen, X.X.; Liao, Y.C. Coupling effects of plastic film mulching and urea types on water use efficiency and grain yield of maize in the Loess Plateau, China. *Soil Tillage Res.* **2016**, *157*, 1–10. [CrossRef]
19. Lu, H.D.; Xia, Z.Q.; Fu, Y.F.; Wang, Q.; Xue, J.Q.; Chu, J. Response of Soil Temperature, Moisture, and Spring Maize (*Zea mays* L.) Root/Shoot Growth to Different Mulching Materials in Semi-Arid Areas of Northwest China. *Agronomy* **2020**, *10*, 453. [CrossRef]
20. Wei, W.C.; Dai, F.; Zhang, F.W.; Zhang, S.L.; Shi, R.J.; Liu, Y.X. Numerical simulation on soil water-thermal effect under different film mulching methods of maize in the arid of North-west China. *Agric. Res. Arid Areas* **2020**, *38*, 13–21. [CrossRef]
21. Dai, F.; Gao, A.M.; Zhang, F.W.; Zhao, W.Y.; Han, Z.S.; Wang, S.L. Design and simulation of pneumatic direct insert hill-seeder with corn whole plastic film on double ridges. *Agric. Res. Arid Areas* **2018**, *36*, 284–289. [CrossRef]
22. Wang, G. Design and Test of Duckbill Seed-Metering Device for Rice Film Spreading and Hole Sowing Machine. Master's Thesis, Anhui Agricultural University, Hefei, China, 2023.
23. Shi, L.R.; Zhao, W.Y. Design and test of a rolling spoon type precision hole sower for caraway in northwest cold and arid agricultural region. *J. Jilin Univ. (Eng. Technol. Ed.)* **2023**, *53*, 2706–2717. [CrossRef]
24. Gu, L.L. Desing and Experimental Research on Peanut Dibbling Machine. Master's Thesis, Nanjing Agricultural University, Nanjing, China, 2019.
25. Li, H.; He, T.F.; Liu, H.; Shi, S.; Zhou, J.L.; Liu, X.C.; Wang, B.Q. Development of the profiling up-film transplanter for sweet potato in hilly and mountainous region. *Trans. Chin. Soc. Agric. Eng.* **2023**, *39*, 26–35. [CrossRef]
26. Qu, H.; Shi, L.R.; Xin, S.L.; Zhao, W.Y.; Guo, J.H.; Yang, T.; Liu, F.J. Design and experiment of flax precision hole-seeding combined operation machine with soil storage device. *J. Chin. Agric. Univ.* **2022**, *27*, 186–197. [CrossRef]
27. Shi, L.R.; Wu, J.M.; Sun, W.; Zhang, F.W.; Sun, B.G.; Liu, Q.W.; Zhao, W.Y. Simulation test for metering process of horizontal disc precision metering device based on discrete element method. *Trans. Chin. Soc. Agric. Eng.* **2014**, *30*, 40–48. [CrossRef]

Jin, X.N.; Zhang, J.C.; Xue, J.F.; Gou, C.C.; He, C.X.; Lu, L.Q. Calibration of Discrete Element Contact Parameters of Maize Seed and Rubber Belt. *J. Agric. Mach. Res.* **2022**, *44*, 39–43. [CrossRef]
Lu, B.; Ni, X.D.; Li, S.F.; Li, K.Z.; Qi, Q.Z. Simulation and Experimental Study of a Split High-Speed Precision Seeding System. *Agriculture* **2022**, *12*, 1037. [CrossRef]
GB/T 6973-2005; Testing Methods of Single Seed Drills. Standardization Administration of the P.R.C.: Beijing, China, 2005.
Zhang, B.; Chen, X.; Liang, R.; Wang, X.; Meng, H.; Kan, Z. Calibration and Test of Contact Parameters between Chopped Cotton Stalks Using Response Surface Methodology. *Agriculture* **2022**, *12*, 1851. [CrossRef]
Wang, J.; Yao, Z.; Xu, Y.; Guo, F.; Guan, R.; Li, H.; Tang, H.; Wang, Q. Mechanism Analysis and Experimental Verification of Side-Filled Rice Precision Hole Direct Seed-Metering Device Based on MBD-DEM Simulations. *Agriculture* **2024**, *14*, 184. [CrossRef]
33. Zhang, C.L.; Liu, T.; Zhang, Z.H.; Fang, J.; Xie, X.D.; Chen, L.Q. Design and test of the precision seeding dispenser with the staggered convex teeth for wheat sowing with wide seedling belt. *Trans. Chin. Soc. Agric. Eng.* **2024**, *40*, 47–59. [CrossRef]
34. Zhang, S.; He, H.L.; Yuan, Y.W.; Kuang, F.M.; Xiaog, W.; Li, Z.D.; Zhu, D.Q. Design and experiment of the orifice-groove combined hole of the oriented filling type precision hill-drop seed-metering device for rice. *Trans. Chin. Soc. Agric. Eng.* **2023**, *39*, 39–50. [CrossRef]
35. Yang, L.; Li, Z.M.; Zhang, D.X.; Li, C.; Cui, T.; He, X.T. Design and test of the T-shaped hole of centrifugal high-speed maize precision seed metering device. *Trans. Chin. Soc. Agric. Eng.* **2024**, *40*, 50–60. [CrossRef]
36. Gao, X.J.; Xu, Y.; Zhang, D.X.; Yang, L.; Lu, B.; Su, Y.; Xia, G.Y.; Cui, T. Design and experiment of air-assisted high speed precision maize seed metering device. *Trans. Chin. Soc. Agric. Eng.* **2019**, *35*, 9–20. [CrossRef]

MDPI AG
Grosspeteranlage 5
4052 Basel
Switzerland
Tel.: +41 61 683 77 34

Agriculture Editorial Office
E-mail: agriculture@mdpi.com
www.mdpi.com/journal/agriculture

www.ingramcontent.com/pod-product-compliance
Lightning Source LLC
Chambersburg PA
CBHW041402210926
51850CB00013B/3153

Response of Tree Rings to Climate Change and Climate Extremes

Response of Tree Rings to Climate Change and Climate Extremes

Guest Editors

Li Qin
Lushuang Gao
Vladimir V. Shishov
Ruibo Zhang

MDPI

Basel • Beijing • Wuhan • Barcelona • Belgrade • Novi Sad • Cluj • Manchester

Guest Editors

Li Qin
Institute of Desert Meteorology
China Meteorological Administration
Urumqi
China

Lushuang Gao
Research Center of Forest Management Engineering of State Forestry and Grassland Administration
Beijing Forestry University
Beijing
China

Vladimir V. Shishov
Department of Mathematical Methods and Information Technology
Siberian Federal University
Krasnoyarsk
Russia

Ruibo Zhang
Institute of Desert Meteorology
China Meteorological Administration
Urumqi
China

Editorial Office
MDPI AG
Grosspeteranlage 5
4052 Basel, Switzerland

This is a reprint of the Special Issue, published open access by the journal *Forests* (ISSN 1999-4907), freely accessible at: https://www.mdpi.com/journal/forests/special_issues/6N5VX6R165.

For citation purposes, cite each article independently as indicated on the article page online and as indicated below:

Lastname, A.A.; Lastname, B.B. Article Title. *Journal Name* **Year**, *Volume Number*, Page Range.

ISBN 978-3-7258-4749-5 (Hbk)
ISBN 978-3-7258-4750-1 (PDF)
https://doi.org/10.3390/books978-3-7258-4750-1

Cover image courtesy of Ruibo Zhang

Contents

forests

MDPI

Review

Vegetation Index Research on the Basis of Tree-Ring Data: Current Status and Prospects

Tongwen Zhang [1,*], Jinghui Song [2,3], Yuting Fan [3,*], Yan Liu [3], Shulong Yu [3], Dong Guo [3], Tianhao Hou [4] and Kailong Guo [4]

1 Xinjiang Uygur Autonomous Region Climate Center, Urumqi 830002, China
2 College of Geography and Remote Sensing Sciences, Xinjiang University, Urumqi 830046, China; songjh@stu.xju.edu.cn
3 Institute of Desert Meteorology, China Meteorological Administration, Urumqi 830002, China; liuyan@idm.cn (Y.L.); yushl@idm.cn (S.Y.); guodong@idm.cn (D.G.)
4 School of Geography and Tourism, Xinjiang Normal University, Urumqi 830054, China; houtianhao@stu.xjnu.edu.cn (T.H.); guokailong@stu.xjnu.edu.cn (K.G.)
* Correspondence: zhangtw@idm.cn (T.Z.); fanyt@idm.cn (Y.F.)

Abstract: The normalized difference vegetation index (NDVI) and tree-ring parameters are commonly used indicators in the research on forest ecology and responses to climate change. This paper compiles and analyzes the literature on vegetation index research on the basis of tree-ring information in the past 20 years and provides an overview of the relationship between tree-ring parameters and NDVI, as well as NDVI reconstruction. The research on the vegetation index based on tree-ring data is mainly concentrated in the middle and high latitudes, and relatively few studies are concentrated in the low latitudes. The tree-ring parameters have a strong correlation with the NDVI in the summer. In terms of tree-ring reconstruction NDVI, *Sabina przewalskii* is the tree with the longest reconstruction sequence so far, and the tree-ring width is the main proxy index. In addition, combining tree rings with the NDVI is useful for assessing forest decline, quantifying the forest response to drought, and monitoring forest productivity. In the future, it is necessary to consider a variety of environmental factors to find the optimal model construction parameters and carry out research on the climate response of forest tree growth and the reconstruction of the historical sequence of the vegetation index at large spatial scales.

Keywords: tree ring; vegetation index; tree-ring reconstruction; research progress

1. Introduction

Vegetation growth and distribution are closely related to climate change. Revealing the relationship between vegetation activity and climate change and its response pattern has become an important part of current global change research, providing a theoretical basis for coping with climate change and improving the sustainable development ability of ecosystems. Working Group I of the Sixth Assessment Report of the Intergovernmental Panel on Climate Change (IPCC, 2021) reported that the global average atmospheric temperatures and ocean temperatures are increasing, and this is leading to the widespread melting of snow and ice and global sea level rise [1]. Climate change is thought to be a major driver of vegetation growth and forest productivity. Tree-ring parameters and the normalized difference vegetation index (NDVI) are commonly used indicators in studies of forest ecology and responses to climate change [2–4]. Tree-ring parameters integrate abiotic (climatic and site conditions) and biotic (ecophysiological responses) factors to provide long-term historical information on the physiological processes of tree growth and responses to environmental changes at an annual time resolution [5–7]. Tree-ring parameters have been widely used in both small-scale climate response studies and large-scale reconstruction studies for their various advantages, including their ability to yield accurate date information, their high continuity and temporal resolution, their wide availability, and the ease with which several copies of tree rings can be obtained [8,9]. The main advantages of NDVI data on the basis of remote sensing satellite data are the ease with which these data can be acquired, their wide coverage, and their temporal resolution, which permits vegetation growth to be detected over larger areas [10–12]. However, the collection of remote sensing vegetation index data has a short history. The longest sequence of such data is a dataset, which has been provided by AVHRR since 1980. Given that obtaining a sequence of vegetation phenological parameters over the past hundred years is not possible [13,14], data that can serve as a proxy for vegetation change are important for the long-term monitoring of land surface characteristics and studies of climate and terrestrial ecosystem changes.

The growth process of trees can be divided into the primary growth of germination, flowering, branch, leaf, and root derivative growth and the secondary growth of the trunk, branch, and root [15,16]. During the growth process, trees are not only affected by their own factors, but also by external environmental factors [17,18]. As one of the important indicators of tree growth, tree-ring parameters such as the width, density, and stable isotope can continuously record the changes in the external climate and environment in the secondary growth process of trees and quantitatively display the changes in the climate and environment [19,20]. The remote sensing vegetation index can quantify the changes in the leaf area and greenness in the tree canopy. Leaves are the main organs of photosynthesis, and their growth conditions determine the material mass required for the radial growth of trees, reflecting the growth of trees from different levels [21]. In the past, scholars mainly used tree-ring data to reconstruct past climate changes or studied tree ring and vegetation dynamics separately. In recent years, domestic and foreign scholars have begun to combine tree-ring data with remote sensing data for research. The results show that when tree growth and regional vegetation are affected by common external environmental factors, and not affected by other non-climatic factors, the mathematical relationship between the tree-ring and NDVI in the same area and time period can be established. This can effectively deduce the change in the NDVI index in a historical period. With the increasing importance of tree-ring data in the analysis of long-term vegetation dynamic change, there have been studies that calculated the correlation between different vegetation index indicators and the tree-ring index. However, on a global scale, the response of forest growth to NDVI has a very high spatial and temporal diversity. Therefore, it is necessary to review the existing literature on tree rings and NDVI to find the relationship between tree-ring parameters and the vegetation index. Combining tree-ring data with remote sensing data can aid in our understanding of the factors affecting vegetation change; these data can also facilitate the identification of climatic events leading to vegetation change and improvements in the parameters of the global vegetation index reconstruction model. Here, we summarize the historical and recent vegetation index research on the basis of tree-ring data with a focus on tree-ring parameters, forest types, remote sensing data at different spatial resolutions, and future research directions to provide a reference for future research. Figure 1 shows the spatial distribution of sampling points for vegetation index research on the basis of tree-ring data.

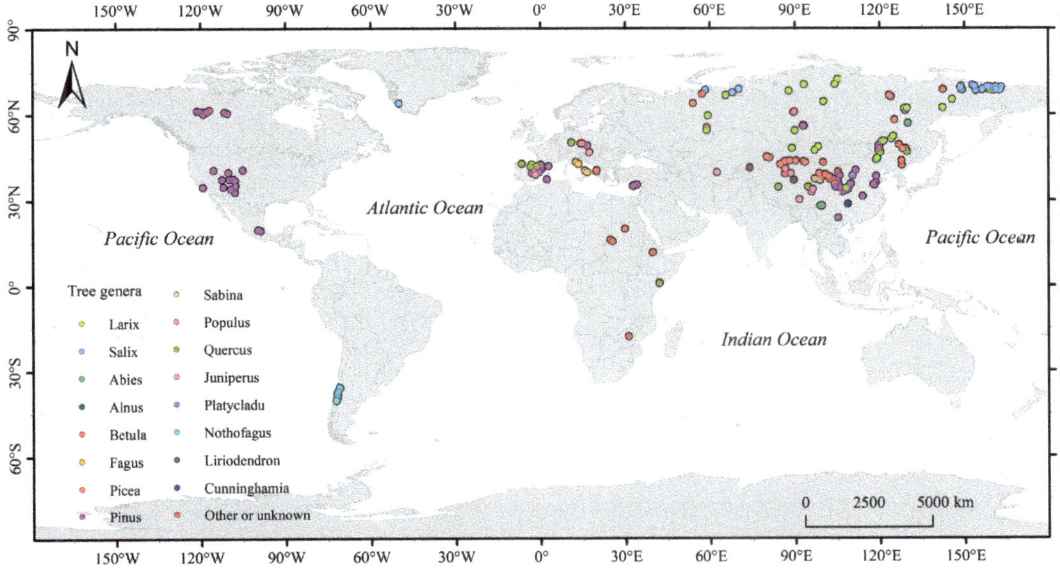

Figure 1. Distribution of sampling points of vegetation index studies on the basis of tree-ring data.

2. Study of the Relationships between Tree-Ring Parameters and Vegetation Index

An analysis of the relationships between tree-ring parameters and remote sensing data generally involves using tree-ring data to establish a long time series of vegetation change. Research employing tree-ring parameters combined with remote sensing data began in the late 1990s [22]. Tree-ring data have been used to test the reliability of models for estimating net primary productivity values; researchers initially explored the potential and limitations of tree-ring parameters in studies of vegetation productivity [23]. Some studies have detected strong correlations between tree-ring indexes and vegetation indexes (e.g., NDVI), and long-term sequences of vegetation change have been reconstructed on the basis of significant correlations [24–27]. Some studies have also shown that tree-ring parameters are decoupled from NDVI [28,29], and that many factors can affect the relationship between tree radial growth and the vegetation canopy, including the forest type, latitude, elevation, topography, and the spatial resolution of remote sensing data [30–38]. Therefore, the factors and processes affecting the relationship between tree-ring parameters and vegetation indexes are complex. All of the above results indicate that the relationship between tree-ring parameters and NDVI deserves further research.

2.1. Relationship between Tree-Ring width and Vegetation Index

Approximately 77.4% of NDVI and tree-ring width parameters are synchronized (i.e., positively correlated) during the growing season in arid regions of inland Asia. On the monthly scale, 69.4% of the maximum correlation coefficients between the NDVI and tree-ring width were concentrated in the period of May–July, especially in June [39]. This indicates that there is a strong synchronization between NDVI and the tree-ring width index from May to July [40]. This synchronous pattern has also been observed in many studies in the Northern Hemisphere. For example, the radial growth of trees in a semi-arid grassland in North China was found to be positively correlated with NDVI in the grassland from early May to early July [41]. Tree-ring width chronology was found to be positively correlated with the growing season NDVI and annual NDVI, and the correlation with the cumulative NDVI from June to October was the strongest [42]. There was a correlation between the tree-ring width and early-spring NDVI and late-autumn NDVI in the area south of 40° N. However, in the area north of 40° N, there was a correlation between the tree-ring width and NDVI in the summer, and the correlation between the tree-ring width and NDVI varied greatly among the months because of variation in the environmental conditions [43]. A low correlation was found between the tree-ring width chronology and NDVI from May to August in the Hejigong Mountain area of Nanzheng County, which mainly stemmed from differences in vegetation growing seasons caused by latitudinal differences [44]. The upper limit of the tree-ring width index was positively correlated with the NDVI in June, but no significant correlation was observed in July and August, when the vegetation growth was the most vigorous [45,46]. This might stem from the fact that cell division and expansion in several types of conifers ceased in June; although changes in the tree radial width were negligible, the vegetation canopy continued to grow, which resulted in a continuous increase in the regional NDVI value and in a weak correlation between the tree-ring width and NDVI from July to October. This indicates that accurate information on the tree radial growth and measurements of vegetation indexes at a high spatial resolution are necessary for clarifying the relationships between tree-ring parameters and the NDVI.

In areas dominated by snow that rarely melt in winter, snow can reflect large amounts of visible and red wavelengths and absorb near-infrared waves, which makes the calculated NDVI negative and reduces the NDVI value in the study area. However, winter precipitation provided sufficient water for tree-ring growth in the following year; consequently, the tree-ring width was negatively correlated with the NDVI in the non-growing season [47]. The relevant literature indicates that there is a weak or no correlation between the NDVI and tree-ring width in some areas [48]. For example, a weak correlation between the poplar tree-ring width and NDVI was found in the Tarim River, Xinjiang [16]. There was no common relationship found between the radial growth and NDVI of Norwegian spruce trees [49]. There was no significant correlation between the NDVI and tree-ring width at four Arctic tree system sites in North America [29]. The variation in the tree-ring width in Canada was not found to be consistent with the variation in the NDVI, and the reasons for this result remain unclear. This suggests that tree growth is not always correlated with the NDVI [50]. In addition, the scarcity of trees in the study area might explain why the NDVI data reflect both the canopy conditions of trees, as well as the canopy conditions of understory vegetation, thus leading to an increase in the NDVI and a decrease in the tree-ring width near the forest line in northern Russia in summer [51].

2.2. Relationship between Tree-Ring Density, Stable Isotope, and Vegetation Index

Because of the rapid development of modern science and technology and the use of new experimental equipment, the tree-ring width is no longer the only source of data for vegetation index research. The tree-ring density, tree-ring cell structure, and tree-ring stable isotopes have also begun to be used in studies of tree rings combined with remote sensing data. The tree-ring density index reflects the tree cell wall thickness, cell lumen area, and volume and provides indirect information on the tree hydraulic characteristics, carbon sequestration, and paleoclimate records [52,53]. Research on the tree-ring density combined with remote sensing data began in 2000. The chronology of the tree-ring maximum density was found to be negatively correlated with the NDVI in the summer and positively correlated with the NDVI in the late spring and early summer [23]. Current studies on the relationship between tree-ring data and the NDVI are mostly limited to analyses of certain tree-ring parameters. Researchers have analyzed the relationship between seven tree-ring parameters and NDVI and found that the maximum density of tree rings was highly correlated with NDVI. A significant variation was observed in the relationship between tree-ring data and remote sensing data [54]. The relationships of MODIS NDVI, MODIS EVI, and MODIS LAI with the minimum tree-ring density, average tree-ring density, and maximum tree-ring density were studied in two regions with similar climate, soil, and vegetation. The maximum density of tree rings in one region was strongly positively correlated with NDVI in March; the NDVI was negatively correlated with the minimum density of tree rings, and this was in contrast to the relationship observed between the NDVI and the average density of tree rings. This suggests that the tree-ring density and remote sensing information should be used to understand forest responses to climate change [30]. The relationship between the maximum tree-ring density and NDVI was found to be consistent with the relationship between the tree-ring width index and NDVI in temperature-restricted study areas [55]. There was a significant positive correlation between the tree-ring maximum density chronology and NDVI during the growing season. In the tundra–forest transition zone, the maximum tree-ring density and tree-ring width index were closely related to the NDVI in the growing season [56].

In terms of tree-ring carbon isotopes, to determine the NDVI and tree-ring parameters that are most suitable for forest carbon uptake research, the tree-ring width, tree-ring carbon isotopes, and NDVI were compared with the gross primary productivity (GPP) estimates using the vorticity covariance method. The tree-ring parameters were significantly positively correlated (tree-ring width) and negatively correlated (stable carbon isotope ratio ($\delta^{13}C$)) with the GIMMS NDVI in June and from June to August, and a 1-year lag effect was observed [57]. There were also strong correlations between the tree-ring width, tree-ring carbon isotope, and total forest primary productivity [58,59]. The forest carbon uptake can be more easily studied using tree-ring parameters than using satellite vegetation indexes [60,61]. In a previous study, the researchers explored the effects of volcanic activity on tree growth using stable ring isotopes from surviving trees that grew near the pre-eruption NDVI enhanced area of Mount Etna. The study found that additional soil water condensed by degassed water vapor promoted photosynthesis, leading to a local increase in the NDVI on Mount Etna's flanks prior to the 2002/2003 eruption. Other studies have shown that stable oxygen isotopes in tree rings can be used as indicators of past volcanic eruptions [62]. Insects can affect vegetation photosynthesis by depleting leaf epidermal cells and damaging stomatal regulation. Insect outbreaks were found to have a negative effect on poplar tree growth [63]. During insect outbreaks, the summer NDVI was positively correlated with radial growth and negatively correlated with $\delta^{13}C$. Insect dynamics are important for detecting tree growth at large scales. The ratio of carbon and oxygen isotopes was found to be mainly related to the NDVI at the beginning of the previous autumn and winter season and the beginning of the growing season. They also detected a positive correlation between tree-ring stable isotopes and the summer NDVI [31].

2.3. *Research on the Relationships between Tree-Ring Parameters and Vegetation Indexes at Different Spatial Resolutions*

In areas with complex vegetation types, the NDVI at coarse spatial resolutions tends to ignore heterogeneity in the vegetation landscape, stand density, and age structure, which results in a spatial disconnect between the NDVI and dendrochronology data, thus weakening the relationship between them. The responses of data at a medium spatial resolution (MODIS, 250 m) and coarse spatial resolution (GIMMS3g NDVI, 8 km) to tree growth were compared in Siberia using a random forest model. The MODIS NDVI growth signal in a conifer forest was stronger than that of the GIMMS3g NDVI. Among all the of the NDVI phenological indexes, the correlation between the summer NDVI and tree-ring width was the strongest, followed by the summer NDVI and spring NDVI of the previous year [64]. In contrast to previous studies, the MODIS NDVI (1 km) was used to predict the spatial distribution of tree-ring width from 2001 to 2017, revealed a positive correlation between the tree-ring width index and the period from June to August, and showed that the forest disturbance and restoration process in the study area were both related to hydrodynamics [65]. In addition, the NDVI values were calculated using high-resolution 30 m Landsat data to analyze and compare the climate signals of the leaf biomass and stem biomass under complex terrain. Compared with the NDVI data with a 250 m spatial resolution, the NDVI data with a 30 m spatial resolution did not enhance the correlation between the tree-ring width and NDVI, which indicates that the NDVI data with a 250 m spatial resolution may be sufficient to study the correlation between the tree-ring parameters and NDVI [66]. In addition to remote sensing observation data, phenological cameras and drones, as near-surface observation technologies, have also become new sources of vegetation data information [67]. Phenologists have found that the effectiveness of remote sensing to extract vegetation information in mountainous areas is much lower than that of phenological cameras. The resolution of vegetation information obtained using a phenological camera is much higher than that of remote sensing data and is less affected by weather, which can realize the observation of single vegetation, and thus improve the extraction and verification accuracy of remote sensing phenological data [68]. Its limitation is that phenological cameras need to be installed on specific fixed devices. The observation range is smaller than that of remote sensing data, and data collection in the region without signal is costly and susceptible to the influence of the observation angle [69]. Therefore, the establishment of operational phenological camera observation standards, and the combination of tree-ring data as well as vegetation data extracted using a phenological camera will be conducive to the reconstruction of a set of long-term stable vegetation data series.

2.4. Relationships between the Tree-Ring Parameters and Vegetation Indexes of Different Forest Types

Vegetation index studies on the basis of tree-ring data have been mostly conducted at middle and high latitudes, and most studies have been conducted on coniferous forests. In recent years, there has been increased interest in the study of the tree rings of broad-leaved trees and shrubs. The tree-ring growth of broad-leaved trees and shrubs in cold regions has a shorter response time to the cumulative NDVI compared with coniferous forests in temperate and arid regions. The correlations between tree-ring parameters and the NDVI of deciduous trees and conifers show virtually no temporal overlap. For deciduous tree species, the tree-ring parameters are mainly correlated with the NDVI in April, May, August, and October, whereas for conifers, the tree-ring parameters are mainly correlated with the NDVI in May, June, and July. This might stem from the different responses of different tree species to climate change. *Pinus sylvestris* var. *mongolica* Litv. was found to be more strongly correlated with the NDVI than *Picea abies*. The correlation between the two was the strongest in the plateau area, followed by the south slope, north slope, and valley. The tree-ring parameters are more closely related to climate factors than the NDVI. This might be affected by changes in biomass allocation during drought, the spectral distortion of remote sensing data, and the effects of different meteorological conditions on the stem and crown [66]. The relationship between the annual forest growth and NDVI were quantified in different forest types and regions. It showed that 67% of the global forest NDVI was significantly positively correlated with tree-ring growth. There were eight main models between the tree-ring growth and NDVI. The results indicated that the degree of correlation between the NDVI and tree-ring growth was similar in different forest types, and the spatial and temporal diversity of the relationship between forest secondary growth and the NDVI mainly stemmed from differences among tree species and environmental conditions [70]. In addition, high-resolution remote sensing data and the tree-ring parameters of different forest types were studied in Canada. The results showed that the relationships of the tree-ring parameters of cypress (*Thuja*), pine (*Pinus*), and spruce (*Picea asperata* Mast.) with the remote sensing satellite data were closer than the relationships of the tree-ring parameters of oak (*Quercus* L.) and poplar (*Populus* L.) with the remote sensing satellite data. The correlations between the tree-ring width chronology and satellite data were particularly weak for maple (*Acer palmatum* Thunb. in Murray). The negative response of the deciduous tree-ring width to summer air temperatures and the positive response of canopy to summer air temperatures might explain the weak correlation between the two [71]. A VS-oscilloscope model was used to study the relationship between vegetation canopy and trunk cambium phenology. The results showed that the forest canopy was significantly correlated with the end of season of the Chinese pine cambium. And the strength of this correlation was greater than that between grassland and cambium [72]. A significant positive correlation between the NDVI and shrub growth was found in July [73]. A relationship between the interannual variation in the NDVI with the radial growth of shrubs was detected. A significant positive correlation between the radial growth of shrubs and the summer NDVI were detected in relation to shrub expansion from multiple perspectives. Shrub expansion may be related to the rate of sustained warming in early summer and the frequency of extreme winter warming events [74]. In addition, shrub growth and interannual changes in the NDVI in the Patagonian steppe are regulated by winter precipitation, and tree radial growth explains 23%–62% of the seasonal NDVI changes. However, this study did not identify a clear relationship between the tree radial growth and NDVI in high-altitude mountain areas nor clarify the effect of intensive grazing on shrub growth [75].

2.5. Relationships between Tree-Ring Parameters and Vegetation Indexes at Different Altitudes

Altitude affects the relationships between the tree-ring parameters and NDVI mainly through temperature and precipitation. The tree-ring data were combined with remote sensing data to quantify the response of tree growth to climate change at different elevations in Cyprus. This showed that there was no common climate driver between tree-ring chronology and the NDVI in high-altitude forest areas. In the low-altitude arid areas dominated by grassland and shrubland, there was a close relationship between the standardized tree-ring chronology and NDVI, which indicates that trees are not representative of the primary productivity of grassland and shrubland [31,76]. In addition, the tree cores of *Picea schrenkiana* at three altitudes in four areas were collected on the north slope of Tianshan Mountain in China from the west to east, and forest NDVI values were extracted in each study area. The relationship between the tree-ring width index and NDVI were found to be gradually weakened from the west to east, and the positive correlation between them was found to be gradually decreased with altitude. Drought stress may be the main factor affecting the positive correlation between the tree-ring width and NDVI in arid mountainous evergreen coniferous forests [4]. The correlation between the canopy and trunk was found to be gradually decreased with altitude. The correlation coefficient was 0.371 in the alpine region, 0.413 in the valley region, and 0.583 in the desert region [77].

3. Other Research

3.1. Forest Decline

Forest decline has received much research attention, and is characterized by the loss of leaf biomass, the dieback of branches, and even tree death. These changes alter the structure, composition, and function of forests, as well as their response to climate change. Assessing the risk of forest decline is becoming increasingly important for sustainable forestry management as environmental change intensifies. However, given the complexity of the interactions between multiple processes, quantifying the various factors that affect forest decline remains a major challenge. Topographic patterns of the forest growth decline of *Pinus densiflora* in the Mengshan region of eastern China were studied and analyzed, and forest degradation events were identified from 2009 to 2014 using the NDVI and tree-ring width data. A decision tree model was established, and the topographic pattern of forest degradation was characterized. Serious forest degradation was found in this region in 2012, and the altitude factor indirectly determined the vulnerability of forest decline in the study area, indicating that it could be used to predict the risk of spatial decline [32]. Forest decline mainly occurs in five topographic backgrounds: upper slopes and high ridges at low altitudes, U-shaped valleys and upper slopes at high altitudes, and north and northwest directions at high altitudes [78–81]. In addition, repeated dieback and forest decline occurred in southwestern Hungary, and the summer water supply was the main factor limiting black pine growth [82–84]. During three consecutive droughts over the past few decades (1992–1993, 2000–2003, and 2011–2012), black pine was found to be a major contributor to growth. Black pine (*Pinus nigra* Arn.) is becoming less resilient, and the possibility that it might be a future alternative tree species in the Alps and Mediterranean should be reconsidered, given that other species that are better adapted to drought may exist [85].

3.2. Extreme Climate Events

Drought and climate warming will place increasing pressure on forests, and severe water scarcity will reduce the photosynthesis and growth rate of trees, thus harming the function of trees, limiting their radial growth, and leading to the formation of narrow or false rings. Linking historical drought to the tree-ring width can help to quantify the response of a forest to drought and predict its vulnerability to future extreme climate events [86]. The canopy height was used to characterize forest drought resistance, and forests with a medium canopy height (ca. 18 m) were found to have the strongest resistance to severe drought. Trunk growth and leaf growth were positively correlated with the canopy coverage of smaller trees (<18 m) and negatively correlated with the canopy coverage of taller trees (>18 m). Tree mortality was negatively correlated with the canopy coverage of smaller trees (<18 m) [87]. An analysis using the NDVI, cumulative NDVI, dendrochronology, water consumption, ecological flow regime, and groundwater depth data was conducted in the middle reaches of the Tarim River. There was a significant correlation between the tree-ring chronology and cumulative NDVI, and strategies for achieving potentially sustainable groundwater depths (-2.3 to -3.7 m) and discharge volumes (3.53 to 5.90×10^8 m^3) that could further promote vegetation growth were proposed [88].

In temperate forests, a late spring frost is a sudden and severe drop in temperature during mild weather, which can negatively affect tree productivity and growth. Large-scale remote sensing data were combined with tree-ring data to detect the magnitude of late spring frost and evaluate the recovery time. The EVI was found to be ta more sensitive indicator than he NDVI [89]. The NDVI was used to determine the locations of undamaged, weakly damaged, and severely damaged areas in the study area. Reliability tests were conducted on the tree damage classification using tree-ring data. The NDVI data could be used to remotely monitor the degree of damage to trees caused by insects in cedar forests [90]. In addition, the tree cytology, quantitative wood anatomy, tree-ring width, and remote sensing data were used to conduct the first large-scale reconstruction of the spatio-temporal dynamics of past insect outbreaks [91].

The resilience of forests to drought may vary among biomes. For example, biomes were selected from Mediterranean, temperate, and continental regions and on the basis of previously collected tree-ring and remote sensing data during and after four major droughts (1986, 1994–1995, 1999, and 2005). It was noted that the tree-ring parameters were positively correlated with the NDVI. The tree-ring width index is more sensitive to forest drought resistance than the NDVI data. Semi-arid Mediterranean forests dominated by evergreen gymnosperms had the lowest drought resistance, but a higher recovery rate than humid temperate forests dominated by deciduous angiosperms [92]. Deciduous trees are more susceptible to drought over short (1–3 months) and medium timescales than most evergreen conifers. In addition, to explore the sensitivity of Central Europe's major deciduous tree species, beech (*Fagus sylvatica*) and oak (*Quercus robur*), a principal component gradient analysis, the tree-ring parameters, and remote sensing data were used to quantify the drought sensitivity of tree growth. The results showed that there was a positive correlation between the NDVI and tree-ring width in the spring. Beech is more sensitive to drought events than oak [93].

3.3. Research on Forest Productivity on the Basis of Tree-Ring Data

A time series of field-measured tree rings along with remote sensing data can be used to generate a database for assessing spatial and temporal variability in forest productivity. Previous studies have demonstrated the utility of satellite remote sensing data and tree-ring series for monitoring tree productivity [23,42]. The NPP of trees increased significantly with tree age [94]. Tree radial growth, NDVI, and gross forest primary productivity under different environmental conditions in Spain were analyzed, which revealed a positive correlation between the annual NDVI and tree radial growth. The tree radial growth was most strongly correlated with the cumulative NDVI from June to October. And the tree radial growth and gross forest primary productivity were correlated. This study was the first attempt in Spain to study forest productivity on the basis of tree-ring data in various forest types under different environmental conditions [36]. To explore whether tree-ring stable isotopes can be used as alternatives to NPP, stable isotope ages at four sites were compared in three different hydroclimatic environments in the eastern United States with NPP products at three different temporal and spatial resolutions. Tree-ring stable isotopes can be useful for reconstructing NPP at large scales and over long periods [95]. The correlation between tree-ring stable isotopes and GPP was found to be stronger than that between the tree-ring width and GPP [96]. In addition, tree-ring data have been used to calibrate NPP values on the basis of model estimates. Previous studies have found that the tree-ring width can be used to calibrate models, and dominant tree species and old trees in the study area should be selected when using tree-ring data for model testing [97]. These results suggest that combining canopy and trunk growth studies is necessary for assessing forest productivity.

To sum up, studies on the vegetation index based on tree-ring data are mainly concentrated in the middle and high latitudes, and relatively few studies are conducted in the low latitudes. Tree-ring parameters have a strong correlation with the NDVI in the summer. The results showed that the NDVI of the previous year correlated better with the tree-ring parameters than the current year in forests in eastern North America and Tasmania, whereas this finding suggests that the accumulation of the NDVI from previous year to September in the current year showed a stronger predictive capacity of tree growth. This pattern was mainly found in the forests of Central and South Asia, southern Canada, the Northern USA and Siberia. The poor correlation between tree-ring parameters and the NDVI is mainly concentrated in cold biomes. Specifically, this pattern corresponded to the forests from cold biomes, such as those located in the Alps mountains or in the boreal forests in northern Canada and Scandinavia [70].

4. Research on the Reconstruction of Vegetation Change in the Historical Period on the Basis of Tree-Ring Data

4.1. Vegetation Index Reconstruction of Tree-Ring Data from Single Sample Sites

NDVI reconstruction on the basis of tree-ring data is achieved by leveraging the high temporal resolution of tree-ring data, establishing regression models according to tree physiological mechanisms, and conducting correlation analyses. These approaches permit the reconstruction of the NDVI over long periods and provide more complete datasets for future studies [25,98]. Most NDVI studies on the basis of tree-ring data have focused on the relationship between the tree-ring width and NDVI. However, few studies have examined NDVI reconstruction using tree-ring data [28,70]. Domestic research on NDVI reconstruction on the basis of tree-ring data began in 2005 [55]. In light of the increasing importance of tree-ring data in the analyses of long-term changes in vegetation, increasing numbers of researchers have used different tree species in different regions to conduct NDVI reconstruction research [36,99].

In most studies of NDVI reconstruction, the tree-ring width data of several sample points are used. The most well-researched areas include the Tianshan Mountains, Qilian Mountains, Helan Mountains, Qinghai–Tibet Plateau, Changbai Mountains, Xingan Mountains, and Qinling Mountains [100]. The results showed that the tree-ring width can reflect the change in the NDVI in the growing season in middle and high latitudes. For example, tree rings were used to reconstruct the NDVI (245 years) from May to July, the NDVI (206 years) from June to September, and the NDVI (163 years) from June to August in the southern mountains of the Badain Jaran Desert [101], Hasi Mountain [102], and Xinglong Mountain [103]. Both the tree-ring width and NDVI had a significant positive correlation with precipitation in the periods of July–August and May–June of last year, and a significant negative correlation with the temperature in the period of May–June of the same year. Due to the common climate influence, the tree-ring width had the strongest correlation with the period of July-August. Based on this relationship, the NDVI (442 years) from July to August was reconstructed in central Tibet using the standardized tree-ring chronology. The results show that in the past 20 years, with the increase in the NDVI index and the improvement in vegetation conditions, its coefficient of variation increased, and the frequency of extreme events increased, resulting in the increased risk of animal husbandry development in this region [104]. The NDVI (172 years and 162 years) in July was reconstructed in the Taibai Mountain Nature Reserve using the standardized tree-ring chronology of the northern and southern slopes, respectively [105]. The NDVI of the past 339 years was reconstructed by using the tree-ring width chronography of Spruce in Snow Ridge, and the single-particle Lagrange comprehensive track model mixed with the backward track model and wind field analysis showed that the NDVI anomaly was affected by precipitation brought on by westerly winds [106]. The reconstruction of the NDVI on the basis of tree-ring data is not only affected by tree-ring indexes but also by the latitude, altitude, and vegetation type. For example, the NDVI was considered to contain all vegetation information in a region, and the physiological processes of different vegetation types vary. Therefore, an analysis of the land cover type was conducted, and correlations of the forest NDVI and grassland NDVI with the tree-ring width index were determined. The correlation between the forest NDVI and tree-ring width was stronger than that between the grassland NDVI and tree-ring width, and the summer forest NDVI was reconstructed from 1920 to 2018 [107].

4.2. Vegetation Index Reconstruction on the Basis of Tree-Ring Data

In recent years, large-scale NDVI reconstructions on the basis of tree-ring data have been carried out. For example, the change trend of the NDVI in 193 years was reconstructed according to the tree-ring chronology of Pinus sylvestris. The variation in the NDVI in the western Greater Khingan Mountains was analyzed. After the 1950s, the scale and intensity of extreme changes in the NDVI increased significantly, and the effects of human activities on the NDVI were ranked in the order of importance as follows: livestock > afforestation > population > farmland [108]. It is found that there is a significant relationship between the winter and spring NDVI (December to March) and tree-ring index. Based on this significant relationship, the NDVI index of the past 151 years was reconstructed via linear regression. The results showed that the NDVI was lower than average during the 1869–1874, 1886–1889, 1956–1962, and 1975–1980 periods, and recently, the 2000–2004 period, which may be related to low growth and photosynthetic activity [76]. A total of 176 *Picea schrenkiana* were used to establish a regional tree-ring width chronology, and NDVI values over 167 years were reconstructed. And the reconstructed results were compared with historical disaster records and multiple regional climate reconstructions. The NDVI was found to be locally representative as well as reflect the large-scale circulation in Eurasia [26]. In addition, the vegetation changes over the past 377 years were reconstructed using a regression model. It was found that land reclamation policies and farmland conversion projects in historical periods affected the vegetation coverage by affecting the areas of forest land and farmland [98]. To explore whether climate change caused vegetation growth in Western Mongolia, the NDVI and the Siberian larch tree cycle chronology were modeled by using linear regression; meanwhile, the NDVI from 1940 to 2010 was reconstructed. The reconstructed NDVI index did not show a long-term growth trend, which rejects the hypothesis that climate change induced vegetation growth in Western Mongolia [109]. The relationship between the regional NDVI and climate variables and the NDVI and tree-ring width during the growing season (May–October) in the Hulunbuir Grassland were analyzed. The NDVI during the growing seasons in the past 116 years was reconstructed. The results show that the total number of sparse and dense NDVI anomalies in the past 116 years is 22 years. This may be related to Arctic and Pacific climate activity [99].

To improve the reconstruction accuracy, some researchers reconstructed the NDVI using the first principal component of the tree-ring width chronology, tree-ring width chronology of plots, or multiple tree-ring width indexes from tree-ring data. For example, the Ulan difference chronology and Tianjun autoregressive chronology were used to establish a binary regression equation with the grassland NDVI in August, and millennial changes in the grassland NDVI was reconstructed in the Buha River Basin [110]. It was found that the tree-ring width indexes and NDVI in the Delingha area were greatly affected by the precipitation in June. The tree-ring width indexes' chronology was significantly correlated with the grassland from June to September, and the correlation with NDV in August was the strongest. So, the first principal component of five tree-ring width index sequences was used to reconstruct millennial changes in the grassland NDVI in August in the Delingha area, which greatly enriches the tree-ring width and vegetation index databases [111]. Through a Pearson correlation analysis, it was found that the NDVI changes in the Qilian Mountains are highly correlated with the first principal component of the five width timelines. The NDVI from June to August over the past 160 years was reconstructed by using the first principal component of the tree-ring difference chronology of five sampling points [112]. The tree-ring width sequences of three sampling points in the study area were combined to establish a regional chronology. And both the NDVI changes in July and the NDVI changes in the grassland from May to July were reconstructed [113]. In addition, the tree-ring chronology and vegetation index data were smoothed to reduce the errors in the tree-ring data and remote sensing observations and increase the accuracy of the reconstructed cumulative NDVI values. It was found that the correlation between the tree-ring width chronology and the NDVI cumulative values of the growing season was significantly enhanced after the tree-ring width chronology and vegetation index data were smoothed [114].

By comparing the above research results, the NDVI reconstruction results were shown to be superior when the first principal component of two timelines was used compared with when the timelines of *Pinus tabuliformis* were used. Comprehensive timelines are superior to the timelines of single sampling sites for reconstructing the NDVI. In addition, when *Picea schrenkiana* was used for NDVI reconstruction, the reconstructed NDVI values were obtained over a longer period and the variance explained was greater when a combination of three sampling points was used compared with when a single sampling point was used. The robustness of the NDVI reconstruction results was higher in high-altitude areas than in low-altitude areas [106,113].

In sum, most studies of NDVI reconstruction have used tree-ring width index data; most studies have also considered multiple factors and used different types of proxy data for comparative verification (historical disaster records, atmospheric circulation, large-scale climate events, temperature reconstruction, precipitation reconstruction, volcanic eruption, and super storms). This approach verifies the reliability of NDVI reconstruction. An analysis of the reconstruction results of different tree species has revealed that most studies were conducted in areas with coniferous forests, and the robustness of the reconstruction results is relatively high. Few studies have reconstructed NDVI values on the basis of the tree-ring width data of broad-leaved forest and shrubs, and cedar is the tree species with the longest reconstructed NDVI time series. Within a certain range, the robustness of the reconstructed NDVI time series increases with the number of sampling points and sample size. However, the results of vegetation index reconstruction vary among regions when different tree species and tree-ring indexes are used. In the future, longer tree-ring chronologies and higher-resolution tree-ring networks should be used to reconstruct several types of vegetation indexes over longer periods and larger areas; this work will provide insights into the processes underlying vegetation changes over long historical periods.

5. Conclusions

An analysis of tree-ring parameters, such as the tree-ring width, density, cell, and stable isotopes of different tree species in different habitats, and the data mining of tree-ring samples are needed to enhance our understanding of the relationship between tree-ring data and remote sensing data [115]. The heterogeneity in the vegetation landscape pattern, stand density, age structure, and variation in growth rhythms among different vegetation types can be overlooked when monitoring changes in vegetation via remote sensing, thus weakening the relationship between the tree-ring parameters and vegetation index data. There is thus a need to consider the spatial matching between remote sensing data and tree-ring data in future research. Fine and accurate UAV technology could be combined with tree-ring data in future research to address this problem. In addition, more than 40 types of vegetation indexes have been defined to date. Vegetation indexes and remote sensing products other than NDVI could be combined with tree-ring data to increase the diversity of reconstruction indexes used. The relationship between tree ring and vegetation index data is affected by tree species, the spatial resolution of remote sensing data, terrain (altitude and slope direction), and other factors [68]. Therefore, multiple environmental factors should be considered to identify the optimal model construction parameters, and multiple sampling points and large sample sizes are ideal. Research on the responses of forest tree growth to climate change and the reconstruction of historical changes in vegetation indexes at large spatial scales will be a major focus of tree ring and vegetation index research in the future.

Author Contributions: Conceptualization, T.Z. and J.S.; resources, J.S. and T.H.; writing—original draft preparation, T.Z. and K.G.; writing—review and editing, Y.F.; visualization, Y.L. and D.G.; supervision, S.Y.; project administration, T.Z.; funding acquisition, S.Y. All authors have read and agreed to the published version of the manuscript.

Funding: This work was supported by the Desert Meteorological Science Research Foundation of China (Sqj2020002), the Natural Key Research and Development Program (2023YFE0102700), the Natural Science Foundation of Xinjiang Uigur Autonomous Region-Science Fund for Distinguished Young Scholars (2022D01E105), and the Natural Science Foundation of China (41975095).

Data Availability Statement: Data sharing is not applicable.

Conflicts of Interest: The authors declare no conflict of interest.

References

1. Masson-Delmotte, V.; Zhai, P.; Pirani, A.; Connors, S.L.; P´ean, C.; Berger, S.; Caud, N.; Chen, Y.; Goldfarb, L.; Gomis, M.I.; et al. *IPCC Summary for Policymakers. Climate Change 2021: The Physical Science Basis. Contribution of Working Group I to the Sixth Assessment Report of the Intergovernmental Panel on Climate Change*; Cambridge University Press: Cambridge, UK; New York, NY, USA, 2023.
2. Berner, L.T.; Massey, R.; Jantz, P.; Forbes, B.C.; Macias-Fauria, M.; Myers-Smith, I.; Kumpula, T.; Gauthier, G.; Andreu-Hayles, L.; Gaglioti, B.V.; et al. Summer warming explains widespread but not uniform greening in the Arctic tundra biome. *Nat. Commun.* **2020**, *11*, 4621. [CrossRef]
3. Jiang, P.; Liu, H.; Piao, S.; Ciais, P.; Wu, X.; Yin, Y.; Wang, H. Enhanced growth after extreme wetness compensates for post-drought carbon loss in dry forests. *Nat. Commun.* **2019**, *10*, 195. [CrossRef]
4. Wen, Y.; Jiang, Y.; Jiao, L.; Hou, C.; Xu, H. Inconsistent relationships between tree ring width and normalized difference vegetation index in montane evergreen coniferous forests in arid regions. *Trees* **2021**, *36*, 379–391. [CrossRef]
5. Granda, E.; Camarero, J.J.; Galván, J.D.; Sangüesa-Barreda, G.; Alla, A.Q.; Gutierrez, E.; Dorado-Liñán, I.; Andreu-Hayles, L.; Labuhn, I.; Grudd, H.; et al. Aged but withstanding: Maintenance of growth rates in old pines is not related to enhanced water-use efficiency. *Agric. For. Meteorol.* **2017**, *243*, 43–54. [CrossRef]
6. Levesque, M.; Andreu-Hayles, L.; Pederson, N. Water availability drives gas exchange and growth of trees in northeastern US, not elevated CO_2 and reduced acid deposition. *Sci. Rep.* **2017**, *7*, srep46158. [CrossRef]
7. Marvel, K.; Cook, B.I.; Bonfils, C.J.W.; Durack, P.J.; Smerdon, J.E.; Williams, A.P. Twentieth-century hydroclimate changes consistent with human influence. *Nature* **2019**, *569*, 59–65. [CrossRef]
8. Zhang, P.; Jeong, J.-H.; Yoon, J.-H.; Kim, H.; Wang, S.-Y.S.; Linderholm, H.W.; Fang, K.; Wu, X.; Chen, D. Abrupt shift to hotter and drier climate over inner East Asia beyond the tipping point. *Science* **2020**, *370*, 1095–1099. [CrossRef]
9. Büntgen, U.; Allen, K.; Anchukaitis, K.J.; Arseneault, D.; Boucher, É.; Bräuning, A.; Chatterjee, S.; Cherubini, P.; Churakova, O.V.; Corona, C.; et al. The influence of decision-making in tree ring-based climate reconstructions. *Nat. Commun.* **2021**, *12*, 3411.
10. Furusawa, T.; Koera, T.; Siburian, R.; Wicaksono, A.; Matsudaira, K.; Ishioka, Y. Time-series analysis of satellite imagery for detecting vegetation cover changes in Indonesia. *Sci. Rep.* **2023**, *13*, 8437. [CrossRef]
11. Pompa-García, M.; Vivar-Vivar, E.D.; Sigala-Rodríguez, J.A.; Padilla-Martinez, J.R. What Are Contemporary Mexican Conifers Telling Us? A Perspective Offered from Tree Rings Linked to Climate and the NDVI along a Spatial Gradient. *Remote Sens.* **2022**, *14*, 4506. [CrossRef]
12. Mao, D.; Wang, Z.; Luo, L.; Ren, C. Integrating AVHRR and MODIS data to monitor NDVI changes and their relationships with climatic parameters in Northeast China. *Int. J. Appl. Earth Obs. Geoinf.* **2012**, *18*, 528–536. [CrossRef]
13. Fensholt, R.; Proud, S.R. Evaluation of Earth Observation based global long term vegetation trends—Comparing GIMMS and MODIS global NDVI time series. *Remote Sens. Environ.* **2012**, *119*, 131–147. [CrossRef]

14. Huang, S.; Tang, L.; Hupy, J.P.; Wang, Y.; Shao, G.F. A commentary review on the use of normalized difference vegetation index (NDVI) in the era of popular remote sensing. *J. For. Res.* **2021**, *32*, 1–6. [CrossRef]
15. Fu, Y.H.; Zhao, H.; Piao, S.; Peaucelle, M.; Peng, S.; Zhou, G.; Ciais, P.; Huang, M.; Menzel, A.; Peñuelas, J.; et al. Declining global warming effects on the phenology of spring leaf unfolding. *Nature* **2015**, *526*, 104–107. [CrossRef]
16. Wang, L.L.; Gou, X.H.; Xia, J.Q.; Wang, F.; Zhang, F.; Zhang, J.Z. Research progress on cambial activity of trees and the influencing factors. *Chin. J. Appl. Ecol.* **2021**, *32*, 3761–3770.
17. Fang, K.Y.; Chen, Q.Y.; Liu, C.Z.; Cao, C.F.; Chen, Y.J.; Zhou, F.F. Research advances in dendrochronology. *Chin. J. Appl. Ecol.* **2014**, *25*, 1879–1888.
18. Peng, X.M.; Xiao, S.C.; Xiao, H.L. Advances in Methods of Building Tree-ring Width Chronolog *J. Desert Res.* **2013**, *33*, 857–865.
19. Qi, C.J.; Zhao, Y.L.; Yu, X.L.; Wu, Y. A comprehensive review on dendrochronology. *J. Cent. South Univ. For. Technol.* **2017**, *37*, 1–8.
20. Michelot, A.; Simard, S.; Rathgeber, C.B.K.; Dufrêne, E.; Damesin, C. Comparing the intra-annual wood formation of three European species (*Fagus sylvatica*, *Quercus petraea* and *Pinus sylvestris*) as related to leaf phenology and non-structural carbohydrate dynamics. *Tree Physiol.* **2012**, *32*, 1033–1045. [CrossRef]
21. Anyamba, A.; Tucker, C.J. Historical perspectives on AVHRR NDVI and vegetation drought monitoring. In *Remote Sensing of Drought. Innovative Monitoring Approaches*; Anderson, M.C., Verdin, J.P., Wardlow, B.D., Eds.; CRC Press: Boca Raton, FL, USA, 2012; pp. 23–49.
22. Malmström, C.M.; Thompson, M.V.; Juday, G.P.; Los, S.O.; Randerson, J.T.; Field, C.B. Interannual variation in global-scale net primary production: Testing model estimates. *Glob. Biogeochem. Cycles* **1997**, *11*, 367–392. [CrossRef]
23. D'Arrigo, R.D.; Malmström, C.M.; Jacoby, G.C.; Jacoby, G.C.; Los, S.O.; Bunker, D.E. Correlation between Maximum Latewood Density of Annual Tree Rings and NDVI Based Estimates of Forest Productivity. *Int. J. Remote Sens.* **2000**, *21*, 2329–2336 [CrossRef]
24. Gao, W.D.; Yuan, Y.J.; Zhang, R.B. Reconstruction of Normalized Difference Vegetation Index of Grassland in Hutobi River Basin Based on Tree-Ring. *J. North-East For. Univ.* **2012**, *40*, 26–30.
25. Wang, L.; Ye, M.; Gao, S.; Gou, X.; Xu, Q.; Zhang, T. Effects of hydrothermal factors on vegetation index and tree-ring index of Populus euphratica in the lower reaches of the Tarim River. *J. Nanjing For. Univ. Nat. Sci. Ed.* **2017**, *41*, 85–91.
26. Zhang, T.; Zhang, R.; Lu, B.; Mambetov, B.T.; Kelgenbayev, N.; Dosmanbetov, D.; Maisupova, B.; Chen, F.; Yu, S.; Shang, H.; et al. *Picea schrenkiana* tree-ring chronologies development and vegetation index reconstruction for the Alatau Mountains, Central Asia. *Geochronometria* **2018**, *45*, 107–118. [CrossRef]
27. Li, J.; Wang, S.; Qin, N.; Liu, X.; Jin, L. Vegetation Index Reconstruction and Linkage with Drought for the Source Region of the Yangtze River Based on Tree-ring Data. *Chin. Geogr. Sci.* **2021**, *31*, 684–695. [CrossRef]
28. Brehaut, L.; Danby, R.K. Inconsistent relationships between annual tree ring-widths and satellite-measured NDVI in a mountainous subarctic environment. *Ecol. Indic.* **2018**, *91*, 698–711. [CrossRef]
29. Beck, P.S.; Andreu-Hayles, L.; D'Arrigo, R.; Anchukaitis, K.J.; Tucker, C.J.; Pinzón, J.E.; Goetz, S.J. A large-scale coherent signal of canopy status in maximum latewood density of tree rings at arctic treeline in North America. *Glob. Planet. Chang.* **2013**, *100*, 109–118. [CrossRef]
30. Correa-Díaz, A.; Silva, L.C.R.; Horwath, W.R.; Gómez-Guerrero, A.; Vargas-Hernández, J.; Villanueva-Díaz, J.; Suárez-Espinoza, J.; Velázquez-Martínez, A. From Trees to Ecosystems: Spatiotemporal Scaling of Climatic Impacts on Montane Landscapes Using Dendrochronological, Isotopic, and Remotely Sensed Data. *Glob. Biogeochem. Cycles* **2020**, *34*, e2019GB006325. [CrossRef]
31. Coulthard, B.L.; Touchan, R.; Anchukaitis, K.J.; Meko, D.M.; Sivrikaya, F. Tree growth and vegetation activity at the ecosystem-scale in the eastern Mediterranean. *Environ. Res. Lett.* **2017**, *12*, 084008. [CrossRef]
32. Wang, Z.; Lyu, L.; Liu, W.; Liang, H.; Huang, J.; Zhang, Q.-B. Topographic patterns of forest decline as detected from tree rings and NDVI. *CATENA* **2020**, *198*, 105011. [CrossRef]
33. Montpellier, E.E.; Soule, P.T.; Knapp, P.A.; Shelly, J.S. Divergent growth rates of alpine larch trees (Larix lyallii Parl) in response to microenvironmental variability. *Arct. Antarct. Alp. Res.* **2018**, *50*, e1415626.
34. Oberhuber, W.; Kofler, W. Topographic influences on radial growth of Scots pine (*P. sylvestris* L.) at small spatial scales. *Plant Ecol.* **2000**, *146*, 231–240. [CrossRef]
35. Vicente-Serrano, S.M.; Gouveia, C.; Camarero, J.J.; Beguería, S.; Trigo, R.; Lopez-Moreno, J.I.; Azorin-Molina, C.; Pasho, E.; Lorenzo-Lacruz, J.; Revuelto, J.; et al. Response of vegetation to drought time-scales across global land biomes. *Proc. Natl. Acad. Sci. USA* **2013**, *110*, 52–57. [CrossRef] [PubMed]
36. Vicente-Serrano, S.M.; Martín-Hernández, N.; Camarero, J.J.; Gazol, A.; Sánchez-Salguero, R.; Peña-Gallardo, M.; El Kenawy, A.; Domínguez-Castro, F.; Tomas-Burguera, M.; Gutiérrez, E.; et al. Linking tree-ring growth and satellite-derived gross primary growth in multiple forest biomes. Temporal-scale matters. *Ecol. Indic.* **2019**, *108*, 105753. [CrossRef]
37. Zhou, Y.; Yi, Y.; Jia, W.; Cai, Y.; Yang, W.; Li, Z. Applying dendrochronology and remote sensing to explore climate-drive in montane forests over space and time. *Quat. Sci. Rev.* **2020**, *237*, 106292. [CrossRef]
38. Decuyper, M.; Chávez, R.O.; Čufar, K.; Estay, S.A.; Clevers, J.G.P.W.; Prislan, P.; Gircar, J.; Grepinsek, Z.; Merela, M.; de Luis, M.; et al. Spatio-temporal assessment of beech growth in relation to climate extreme in Slovenia—an integrated approach using remote sensing and tree-ring data. *Agric. For. Meteorol.* **2020**, *287*, 107925. [CrossRef]
39. Wu, Q.Q.; Zhang, X.; Xu, S.X.; Yang, X.H.; Liu, Y.S.; Li, H.Z.; Shi, Z.J. Climatic responses of NDVI and tree growth in the arid areas of inland Asia and their influencing factors. *J. Desert Res.* **2022**, *42*, 1–10.

40. Kaufmann, R.K.; D'arrigo, R.D.; Paletta, L.F.; Tian, H.Q.; Jolly, W.M.; Myneni, R.B. Identifying climatic controls on ring width: The timing of Correlations between the Timing Rings and NDVI. *Earth Interact.* **2008**, *12*, 114. [CrossRef]
41. Liang, E.Y.; Shao, X.M.; He, J.C. Relationships between tree growth and NDVI of grassland in the semi-arid grassland of north China. *Int. J. Remote Sens.* **2005**, *26*, 2901–2908. [CrossRef]
42. Wang, J.; Rich, P.M.; Price, K.P.; Kettle, W.D. Relations between NDVI and tree productivity in the central Great Plains. *Int. J. Remote Sens.* **2004**, *25*, 3127–3138. [CrossRef]
43. Leavitt, S.W.; Chase, T.N.; Rajagopalan, B.; Lee, E.; Lawrence, P.J. Southwestern U.S. tree-ring carbon isotope indices as a possible proxy for reconstruction of greenness of vegetation. *Geophys. Res. Lett.* **2008**, *35*, 33894. [CrossRef]
44. Wang, R.; Cheng, R.; Xiao, W.; Feng, X.; Liu, Z.; Wang, X. Relationship between Masson pine tree-ring width and NDVI in North Subtropical Region. *Acta Ecol. Sin.* **2011**, *31*, 5762–5770.
45. Guo, Y.F.; Gan, M.; Yang, M.L.; Ta, Z.; Yu, R. Relationship between tree radial growth and normalized difference vegetation index in eastern Tianshan Mountains. *J. Arid. Land Resour. Environ.* **2018**, *32*, 176–180.
46. Yan, P.; Xu, J.N.; Ju, H.; Taben, H.R.; Guo, F.Y.; Mu, X.W.; Ruan, C.Y. A Study on the Relation Between Tree Ring Width Index and NDVI in the Three Parallel Rivers Region, Yunnan Province. *For. Resour. Manag.* **2019**, *6*, 42–48.
47. Liu, J.; Wang, N.; Hou, Y.; Zhang, X.; Chang, J. Analysis on the Correlation between NDVI and Tree-ring Width Chronologies in the Upper Reaches of the Shiyang River. *Arid. Zone Res.* **2012**, *29*, 667–673.
48. Seftigen, K.; Frank, D.C.; Björklund, J.; Babst, F.; Poulter, B. The climatic drivers of normalized difference vegetation index and tree-ring-based estimates of forest productivity are spatially coherent but temporally decoupled in Northern Hemispheric forests. *Glob. Ecol. Biogeogr.* **2018**, *27*, 1352–1365. [CrossRef]
49. Lawrence, G.B.; Lapenis, A.G.; Berggren, D.; Aparin, B.F.; Smith, K.T.; Shortle, W.C.; Bailey, S.W.; Varlyguin, D.L.; Babikov, B. Climate Dependency of Tree Growth Suppressed by Acid Deposition Effects on Soils in Northwest Russia. *Environ. Sci. Technol.* **2005**, *39*, 2004–2010. [CrossRef]
50. Girardin, M.P.; Bouriaud, O.; Hogg, E.H.; Kurz, W.; Zimmermann, N.E.; Metsaranta, J.M.; de Jong, R.; Frank, D.C.; Esper, J.; Büntgen, U.; et al. No growth stimulation of Canada's boreal forest under half-century of combined warming and CO_2 fertilization. *Proc. Natl. Acad. Sci. USA* **2016**, *113*, E8406–E8414. [CrossRef]
51. Berner, L.T.; Beck, P.S.A.; Bunn, A.G.; Goetz, S.J. Plant response to climate change along the forest-tundra ecotone in north-eastern Siberia. *Glob. Change Biol.* **2013**, *19*, 3449–3462. [CrossRef]
52. Chave, J.; Coomes, D.; Jansen, S.; Lewis, S.L.; Swenson, N.G.; Zanne, A.E. Towards a worldwide wood economics spectrum. *Ecol. Lett.* **2009**, *12*, 351–366. [CrossRef]
53. Martinez-Meier, A.; Fernández, M.E.; Dalla-Salda, G.; Gyenge, J.; Licata, J.; Rozenberg, P. Ecophysiological basis of wood formation in ponderosa pine: Linking water flux patterns with wood microdensity variables. *For. Ecol. Manag.* **2015**, *346*, 31–40. [CrossRef]
54. He, J.C.; Wang, L.L.; Shao, X.M. The relationships between Mongolian Scotch pine tree ring indices and normalized difference vegetation index in Mohe, China. *Quat. Sci.* **2005**, *25*, 252–257.
55. Bunn, A.G.; Hughes, M.K.; Kirdyanov, A.V.; Losleben, M.; Shishov, V.V.; Berner, L.T.; Oltchev, A.; Vaganov, E.A. Comparing forest measurements from tree rings and a space-based index of vegetation activity in Siberia. *Environ. Res. Lett.* **2013**, *8*, 035034. [CrossRef]
56. Andreu-Hayles, L.; D'Arrigo, R.; Anchukaitis, K.J.; Beck, P.S.A.; Frank, D.; Goetz, S. Varying boreal forest response to Arctic environmental change at the Firth River, Alaska. *Environ. Res. Lett.* **2011**, *6*, 045503. [CrossRef]
57. Tei, S.; Sugimoto, A.; Kotani, A.; Ohta, T.; Morozumi, T.; Saito, S.; Hashiguchi, S.; Maximov, T. Strong and stable relationships between tree-ring parameters and forest-level carbon fluxes in a Siberian larch forest. *Polar Sci.* **2019**, *21*, 146–157. [CrossRef]
58. Babst, F.; Bouriaud, O.; Papale, D.; Gielen, B.; Janssens, I.A.; Nikinmaa, E.; Ibrom, A.; Wu, J.; Bernhofer, C.; Köstner, B.; et al. Above-ground woody carbon sequestration measured from tree rings is coherent with net ecosystem productivity at five eddy-covariance sites. *New Phytol.* **2013**, *201*, 1289–1303. [CrossRef]
59. Klesse, S.; Etzold, S.; Frank, D. Integrating tree-ring and inventory-based measurements of aboveground biomass growth: Research opportunities and carbon cycle consequences from a large snow breakage event in the Swiss Alps. *Eur. J. For. Res.* **2016**, *135*, 297–311. [CrossRef]
60. Dye, A.; Plotkin, A.B.; Bishop, D.; Pederson, N.; Poulter, B.; Hessl, A. Comparing tree-ring and permanent plot estimates of aboveground net primary production in three eastern U.S. forests. *Ecosphere* **2016**, *7*, e01454. [CrossRef]
61. Teets, A.; Fraver, S.; Hollinger, D.Y.; Weiskittel, A.R.; Seymour, R.S.; Richardson, A.D. Linking annual tree growth with eddy-flux measures of net ecosystem productivity across twenty years of observation in a mixed conifer forest. *Agric. For. Meteorol.* **2018**, *249*, 479–487. [CrossRef]
62. Seiler, R.; Hajdas, I.; Saurer, M.; Houlié, N.; D'Arrigo, R.; Kirchner, J.W.; Cherubini, P. Tree-ring stable isotopes and radiocarbon reveal pre- and post-eruption effects of volcanic processes on trees on Mt. Etna (Sicily, Italy). *Ecohydrology* **2021**, *14*, e2340. [CrossRef]
63. A Boyd, M.; Berner, L.T.; Doak, P.; Goetz, S.J.; Rogers, B.M.; Wagner, D.; Walker, X.J.; Mack, M.C. Impacts of climate and insect herbivory on productivity and physiology of trembling aspen (*Populus tremuloides*) in Alaskan boreal forests. *Environ. Res. Lett.* **2019**, *14*, 085010. [CrossRef]

64. Bhuyan, U.; Zang, C.; Vicente-Serrano, S.M.; Menzel, A. Exploring Relationships among Tree-Ring Growth, Climate Variability, and Seasonal Leaf Activity on Varying Timescales and Spatial Resolutions. *Remote Sens.* **2017**, *9*, 526. [CrossRef]
65. Xu, P.; Fang, W.; Zhou, T.; Zhao, X.; Luo, H.; Hendrey, G.; Yi, C. Spatial Upscaling of Tree-Ring-Based Forest Response to Drought with Satellite Data. *Remote Sens.* **2019**, *11*, 2344. [CrossRef]
66. Mašek, J.; Tumajer, J.; Lange, J.; Kaczka, R.; Fišer, P.; Treml, V. Variability in Tree-ring Width and NDVI Responses to Climate at a Landscape Level. *Ecosystems* **2023**, *26*, 1144–1157. [CrossRef]
67. Taylor, S.D.; Browning, D.M.; Baca, R.A.; Gao, F. Constraints and opportunities for detecting land surface phenology in dry lands. *J. Remote Sens.* **2021**, *2021*, 9859103.
68. Berra, E.F.; Gaulton, R.; Barr, S. Commercial off-the-shelf digital cameras on unmanned aerial vehicles for multitemporal monitoring of vegetation reflectance and NDVI. *IEEE Trans. Geosci. Remote Sens.* **2017**, *55*, 4878–4886. [CrossRef]
69. Zahawi, R.A.; Dandois, J.P.; Holl, K.D.; Nadwodny, D.; Reid, J.L.; Ellis, E.C. Using lightweight unmanned aerial vehicles to monitor tropical forest recovery. *Biol. Conserv.* **2015**, *186*, 287–295. [CrossRef]
70. Vicente-Serrano, S.M.; Camarero, J.J.; Olano, J.M.; Martín-Hernández, N.; Peña-Gallardo, M.; Tomás-Burguera, M.; Gazol, A.; Azorin-Molina, C.; Bhuyan, U.; El Kenawy, A. Diverse relationships between forest growth and the Normalized Difference Vegetation Index at a global scale. *Remote Sens. Environ.* **2016**, *187*, 14–29. [CrossRef]
71. Bonney, M.T.; He, Y. Temporal connections between long-term Landsat time-series and tree-rings in an urban–rural temperate forest. *Int. J. Appl. Earth Obs. Geoinf.* **2021**, *103*, 102523. [CrossRef]
72. Li, M.M.; Li, G. Effects of vegetation type on tree ring based NDVI reconstruction for Helan Mountains. *J. Gansu Agric. Univ.* **2020**, *55*, 152–161.
73. Macias-Fauria, M.; Forbes, B.C.; Zetterberg, P.; Kumpula, T. Eurasian Arctic greening reveals teleconnections and the potential for structurally novel ecosystems. *Nat. Clim. Chang.* **2012**, *2*, 613–618. [CrossRef]
74. Andreu-Hayles, L.; Gaglioti, B.V.; Berner, L.T.; Levesque, M.; Anchukaitis, K.J.; Goetz, S.J.; D'arrigo, R. A narrow window of summer temperatures associated with shrub growth in Arctic Alaska. *Environ. Res. Lett.* **2020**, *15*, 105012. [CrossRef]
75. Srur, A.M.; Villalba, R.; Baldi, G. Variations in *Anarthrophyllum rigidum* radial growth, NDVI and ecosystem productivity in the Patagonian shrubby steppes. *Plant Ecol.* **2011**, *212*, 1841–1854.
76. Correa-Díaz, A.; Silva LC, R.; Horwath, W.R.; Gómez-Guerrero, A.; Vargas-Hernández, J.; Villanueva-Díaz, J.; Velázquez-Martínez, A.; Suárez-Espinoza, J. Linking remote sensing and dendrochronology to quantify climate-induced shifts in high-elevation forests over space and time. *J. Geophys. Res. Biogeosci.* **2019**, *124*, 166–183. [CrossRef]
77. Yuan, K.; Xu, H.; Zhang, G. Is There Spatial and Temporal Variability in the Response of Plant Canopy and Trunk Growth to Climate Change in a Typical River Basin of Arid Areas. *Water* **2022**, *14*, 1573. [CrossRef]
78. Dyer, J.M. Assessing topographic patterns in moisture use and stress using a water balance approach. *Landsc. Ecol.* **2009**, *24*, 391–403. [CrossRef]
79. Silver, W.L.; Scatena, F.N.; Johnson, A.H.; Siccama, T.G.; Sanchez, M.J. Nutrient availability in a montane wet tropical forest: Spatial patterns and methodological considerations. *Plant Soil* **1994**, *164*, 129–145. [CrossRef]
80. Gaines, K.P.; Stanley, J.W.; Meinzer, F.C.; McCulloh, K.A.; Woodruff, D.R.; Chen, W.; Adams, T.S.; Lin, H.; Eissenstat, D.M. Reliance on shallow soil water in a mixed-hardwood forest in central Pennsylvania. *Tree Physiol.* **2015**, *36*, 444–458. [CrossRef]
81. Baldwin, D.C. *Catchment-Scale Soil Water Retention Characteristics and Delineation of Hydropedological Functional Units in the Shale Hills Catchment*; The Pennsylvania State University: State College, PA, USA, 2011; p. 126.
82. Eilmann, B.; Rigling, A. Tree-growth analyses to estimate tree species' drought tolerance. *Tree Physiol.* **2012**, *32*, 178–187. [CrossRef]
83. Martin-Benito, D.; Beeckman, H.; Cañellas, I. Influence of drought on tree rings and tracheid features of Pinus nigra and Pinus sylvestris in a mesic Mediterranean forest. *Eur. J. For. Res.* **2012**, *132*, 33–45. [CrossRef]
84. Herrero, A.; Zamora, R. Plant Responses to Extreme Climatic Events: A Field Test of Resilience Capacity at the Southern Range Edge. *PLoS ONE* **2014**, *9*, e87842. [CrossRef]
85. Marqués, L.; Camarero, J.J.; Gazol, A.; Zavala, M.A. Drought impacts on tree growth of two pine species along an altitudinal gradient and their use as early-warning signals of potential shifts in tree species distributions. *For. Ecol. Manag.* **2016**, *381*, 157–167. [CrossRef]
86. Urrutia-Jalabert, R.; Barichivich, J.; Rozas, V.; Lara, A.; Rojas, Y.; Bahamondez, C.; Rojas-Badilla, M.; Gipoulou-Zuñiga, T.; Cuq, E. Climate response and drought resilience of Nothofagus obliqua secondary forests across a latitudinal gradient in south-central Chile. *For. Ecol. Manag.* **2021**, *485*, 118962. [CrossRef]
87. Xu, P.; Zhou, T.; Yi, C.; Fang, W.; Hendrey, G.R.; Zhao, X. Forest drought resistance distinguished by canopy height. *Environ. Res. Lett.* **2018**, *13*, 075003. [CrossRef]
88. Ling, H.; Guo, B.; Yan, J.; Deng, X.; Xu, H.; Zhang, G. Enhancing the positive effects of ecological water conservancy engineering on desert riparian forest growth in an arid basin. *Ecol. Indic.* **2020**, *118*, 106797. [CrossRef]
89. Tonelli, E.; Vitali, A.; Malandra, F.; Camarero, J.J.; Colangelo, M.; Nolè, A.; Ripullone, F.; Carrer, M.; Urbinati, C. Tree-ring and remote sensing analyses uncover the role played by elevation on European beech sensitivity to late spring frost. *Sci. Total. Environ.* **2023**, *857*, 159239. [CrossRef] [PubMed]
90. Oguz, C.H.; Ramazan, ö.; Mustafa, A. Monitoring of damage from cedar shoot moth Dichelia cedricola Diakonoff (Lep.: Tortricidae) by multi-temporal Landsat imagery. *iForest Biogeosci. For.* **2014**, *7*, 126–131.

91. Prendin, A.L.; Carrer, M.; Karami, M.; Hollesen, J.; Pedersen, N.B.; Pividori, M.; Treier, U.A.; Westergaard-Nielsen, A.; Elberling, B.; Normand, S. Immediate and carry-over effects of insect outbreaks on vegetation growth in West Greenland assessed from cells to satellite. *J. Biogeogr.* **2019**, *47*, 87–100. [CrossRef]
92. Gazol, A.; Camarero, J.J.; Vicente-Serrano, S.M.; Sánchez-Salguero, R.; Gutiérrez, E.; de Luis, M.; Sangüesa-Barreda, G.; Novak, K.; Rozas, V.; Tíscar, P.A.; et al. Forest resilience to drought varies across biomes. *Glob. Chang. Biol.* **2018**, *24*, 2143–2158. [CrossRef]
93. Meyer, B.F.; Buras, A.; Rammig, A.; Zang, C.S. Higher susceptibility of beech to drought in comparison to oak. *Dendrochronologia* **2020**, *64*, 125780. [CrossRef]
94. Buras, A.; Thevs, N.; Zerbe, S.; Wilmking, M. Productivity and carbon sequestration of Populus euphratica at the Amu River, Turkmenistan. *For. Int. J. For. Res.* **2013**, *86*, 429–439. [CrossRef]
95. Levesque, M.; Andreu-Hayles, L.; Smith, W.K.; Williams, A.P.; Hobi, M.L.; Allred, B.W.; Pederson, N. Tree-ring isotopes capture interannual vegetation productivity dynamics at the biome scale. *Nat. Commun.* **2019**, *10*, 742. [CrossRef] [PubMed]
96. Diao, H.; Wang, A.; Gharun, M.; Saurer, M.; Yuan, F.; Guan, D.; Dai, G.; Wu, J. Tree-ring δ13C of Pinus koraiensis is a better tracer of gross primary productivity than tree-ring width index in an old-growth temperate forest. *Ecol. Indic.* **2023**, *153*, 110418. [CrossRef]
97. Erasmi, S.; Klinge, M.; Dulamsuren, C.; Schneider, F.; Hauck, M. Modelling the productivity of Siberian larch forests from Landsat NDVI time series in fragmented forest stands of the Mongolian forest-steppe. *Environ. Monit. Assess.* **2021**, *193*, 200. [CrossRef]
98. Sun, C.; Liu, Y.; Song, H.; Li, Q.; Cai, Q.; Wang, L.; Fang, C.; Liu, R. Tree-ring evidence of the impacts of climate change and agricultural cultivation on vegetation coverage in the upper reaches of the Weihe River, northwest China. *Sci. Total. Environ.* **2019**, *707*, 136160. [CrossRef]
99. Chen, Z.-J.; Li, J.-B.; Fang, K.-Y.; Davi, N.; He, X.-Y.; Cui, M.-X.; Zhang, X.-L.; Peng, J.-J. Seasonal dynamics of vegetation over the past 100 years inferred from tree rings and climate in Hulunbei'er steppe, northern China. *J. Arid. Environ.* **2012**, *83*, 86–93. [CrossRef]
100. Wang, H.; Chen, F.; Zhang, R.; Qin, L. Seasonal dynamics of vegetation of the central Loess Plateau (China) based on tree rings and their relationship to climatic warming. *Environ. Dev. Sustain.* **2016**, *19*, 2535–2546. [CrossRef]
101. Chen, F.; Martín, H.; Zhao, X.; Roig, F.; Zhang, H.; Wang, S.; Yue, W.; Chen, Y. Abnormally low precipitation-induced ecological imbalance contributed to the fall of the Ming Dynasty: New evidence from tree rings. *Clim. Chang.* **2022**, *173*, 13. [CrossRef]
102. Jia, F.F.; Lu, R.J.; Gao, S.Y. Variations of NDVI recorded by tree-ring in the HASI mountain, the southern margin of the tengger desert. *Quat. Sci.* **2018**, *38*, 327–335.
103. Wang, Y.J.; Ma, Y.Z.; Lu, R.J.; Gao, S.Y. Summer NDVI variability recorded by tree radial growth in the Xinglong Mountain, the eastward extension of the Qilian Mountains, since 1845 AD. *Geogr. Res.* **2016**, *35*, 653–663.
104. Shang, H.M.; Wei, W.S.; Yuan, Y.J.; Yu, S.L.; Zhang, R.B.; Hong, J.C.; Chen, F.; Zhang, T.W.; Fan, Z.A. Normalized vegetation variation index reconstruction based on the tree-ring width in central Tibet. *J. Lanzhou Univ. Nat. Sci.* **2016**, *52*, 18–23.
105. Zhu, X.L.; Li, S.H.; Bai, H.Y.; Hou, L.; Chen, L.; Qin, J. Reconstruction of July NDVI over 172 years based on tree-ring width of Larix chinensis in Taibai Mountain Nature Reserve, China. *J. Appl. Ecol.* **2018**, *29*, 2382–2390.
106. Zhang, Q.; Liu, Y.; Li, Q.; Sun, C.-F.; Li, T.; Li, P.; Ye, Y.-D. Reconstruction of summer NDVI over the past 339 years based on the tree-ring width of Picea schrenkiana in Bayinbuluke, Central Tianshan, China. *J. Appl. Ecol.* **2021**, *32*, 3671–3679. [CrossRef]
107. Li, M.-M.; Li, G. Relationship between phenology of vegetation canopy and phenology of tree cambium in Helan Mountains, China. *J. Appl. Ecol.* **2021**, *32*, 495–502.
108. Dong, J.; Yin, T.; Liu, H.; Sun, L.; Qin, S.; Zhang, Y.; Liu, X.; Fan, P.; Wang, H.; Zheng, P.; et al. Vegetation Greenness Dynamics in the Western Greater Khingan Range of Northeast China Based on Dendrochronology. *Biology* **2022**, *11*, 679. [CrossRef]
109. Hauck, M.; Klinge, M.; Erasmi, S.; Dulamsuren, C. No Signs of Long-term Greening Trend in Western Mongolian Grasslands. *Ecosystems* **2023**, *26*, 1125–1143. [CrossRef]
110. Wang, Q.C.; Shi, X.H.; Liu, Y.H.; Qin, N.S.; Shao, X.M. Relationship between tree ring indices and vegetation index (NDVI) in Buha river basin, Qinghai. *J. Glaciol. Geocryol.* **2012**, *34*, 1424–1432.
111. He, J.C.; Shao, X.M. Relationships between tree-ring width index and NDVI of grassland in Delingha. *Chin. Sci. Bull.* **2006**, *51*, 1083–1090.
112. Wang, W.Z.; Liu, X.H.; Chen, T.; An, W.L.; Xu, G.B. Reconstruction of regional NDVI using tree-ring width chronologies in the Qilian Mountains, northwestern China. *Chin. J. Plant Ecol.* **2010**, *34*, 1033–1044.
113. Ran, Y.-L.; Chen, Y.-P.; Chen, F.; Zhang, H.-L.; Jia, X.-B. May-July NDVI variation for the middle Qinling Mountains over the past 194 years indicated by tree rings of *Pinus tabuliformis*. *J. Appl. Ecol.* **2021**, *32*, 3661–3670. [CrossRef]

114. Zhao, X.Y.; Guo, Y.Y.; Zhu, L.K.; Tian, J.M.; Qu, X.Q.; Ren, J.X.; Li, W.J.; Bu, X.F.; Wang, S.X. Combining remote sensing and tree ring techniques to investigate forest vegetation dynamics and its response to climate change in Yimeng mountainous area from the early 20th century. *J. Earth Environ.* **2020**, *11*, 265–279.
115. Fajstavr, M.; Bednářová, E.; Nezval, O.; Giagli, K.; Gryc, V.; Vavrčík, H.; Horáček, P.; Urban, J. How needle phenology indicates the changes of xylem cell formation during drought stress in *Pinus sylvestris* L. *Dendrochronologia* **2019**, *56*, 125600. [CrossRef]

forests

MDPI

Article

Evaluating the Effect of Vegetation Index Based on Multiple Tree-Ring Parameters in the Central Tianshan Mountains

Jinghui Song [1,2,3], Tongwen Zhang [4,*], Yuting Fan [2,3,*], Yan Liu [3], Shulong Yu [2,3], Shengxia Jiang [2,3], Dong Guo [2,3], Tianhao Hou [2,3,5] and Kailong Guo [2,3,5]

1 College of Geography and Remote Sensing Sciences, Xinjiang University, Urumqi 830046, China; songjh@stu.xju.edu.cn
2 Key Laboratory of Tree-Ring Physical and Chemical Research of China Meteorological Administration, Urumqi 830002, China; yushl@idm.cn (S.Y.)
3 Institute of Desert Meteorology, China Meteorological Administration, Urumqi 830002, China; liuyan@idm.cn
4 Xinjiang Climate Center, Urumqi 830002, China
5 School of Geography and Tourism, Xinjiang Normal University, Urumqi 830054, China
* Correspondence: zhangtw@idm.cn (T.Z.); fanyt@idm.cn (Y.F.)

Abstract: Combining tree ring data with remote sensing data can help to gain a deeper understanding of the driving factors that influence vegetation change, identify climate events that lead to vegetation change, and improve the parameters of global vegetation index reconstruction models. However, it is currently not well understood how climate change at different elevations in the central Tianshan Mountains affects radial tree growth and the dynamics of forest canopy growth. We selected Schrenk spruce (*Picea schrenkiana*) tree core samples from different elevations in the central Tianshan Mountains. We analyzed the relationships of various tree-ring parameters, including tree-ring width, maximum latewood density (MXD), and minimum earlywood density (MID) chronologies, with 1982–2012 GIMMS (Global Inventory Modelling and Mapping Studies) NDVI (Normalized Difference Vegetation Index), 2001–2012 MODIS (moderate resolution imaging spectroradiometer) NDVI, and meteorological data. (1) There were strong correlations between tree-ring width chronologies and the lowest temperatures, especially in July. Tree-ring width chronologies at higher altitudes were positively correlated with temperature; the opposite pattern was observed at lower altitudes. MID chronologies were positively correlated with July temperature in high-altitude areas and mean temperature and highest temperature from May to September in low-altitude areas, and negatively correlated with precipitation during this period. MXD chronologies were mainly negatively correlated with precipitation. MXD chronologies were mainly positively correlated with temperature in April and May. (2) The correlations between MXD chronologies at each sampling point and NDVI in each month of the growing season were strong. Both MID and MXD chronologies were negatively correlated with GIMMS NDVI in July. The overall correlations between tree-ring parameters and MODIS NDVI were stronger than the correlations between tree-ring parameters and GIMMS NDVI in high-altitude areas; the opposite pattern was observed in low-altitude areas. Drought stress may be the main factor affecting tree ring parameters and NDVI. In the future, we should combine tree ring parameters with vegetation index to investigate a larger scale of forests.

Keywords: tree-ring parameters; climatic elements; vegetation index; climate response; NDVI

1. Introduction

The normalized difference vegetation index (NDVI) and tree-ring parameters are often used markers in research on the ecology of forests and their reactions to climate change [1–5]. Tree-ring characteristics give long-term historical information on tree growth processes and their reactions to environmental changes at an annual time resolution by integrating both biological (ecophysiological responses) and non-biological (climate and site circumstances) components [6–8]. However, obtaining tree-ring data for a larger forest area requires a significant amount of effort and resources [9].

As remote sensing technology has advanced, satellite remote sensing data inversion of the vegetation index has become an effective means to quantitatively present global and regional vegetation coverage and growth changes [10]. Currently, there are more than 40 types of remote sensing vegetation indices and some types of product sets, such as NDVI, Enhanced Vegetation Index (EVI), etc. [11–13]. Among them, NDVI is a popular vegetation index for remote sensing and has been widely applied in research on regional vegetation coverage, vegetation growth status, vegetation productivity estimation, and extreme climate prediction [14–17]. MODIS (Moderate Resolution Imaging Spectroradiometer) NDVI is thought to be an enhancement of the AVHRR NDVI, with increased spatial resolution and chlorophyll sensitivity, elimination of atmospheric water vapor interference, and adjustment of synthesis methods. It is an advancement and continuation of the AVHRR NDVI. The main advantages of NDVI data obtained from remote sensing satellite imagery include their accessibility, wide coverage, and temporal resolution [18–20]. However, the collection of NDVI data via remote sensing has a short history [21,22].

Photosynthesis in the vegetation canopy affects the production of plant carbohydrates and controls carbon storage in the woody parts of trees. This consequently influences the plants' radial growth. NDVI can accurately reflect vegetation greenness and photosynthetic activity [23,24]. The radial growth of trees reflects their growth status, and the width and density are the best indicators of radial growth [25,26]. The integration of tree-ring data with NDVI permits studies of tree-ring ecology from individual trees or forests to be conducted at a regional scale. This has ramifications for ecological research and forest management as well as improving our knowledge of how forests grow and adapt to climate change. Currently, many NDVI studies based on tree ring data have been carried out by researchers and yielded different results. The initial focus of the research was primarily on combining coarse-resolution NDVI with tree ring data. For instance, studies have shown a relationship in regions south of 40° latitude between tree ring width and the NDVI in the early spring and late autumn. Nonetheless, there is a relationship between summer NDVI and tree ring width in regions north of 40° latitude [27]. On a growing season scale in arid regions of Asia, approximately 77.4% of the variation in NDVI is found to be synchronous with changes in tree-ring width parameters [28]. However, coarse-resolution NDVI tends to overlook the heterogeneity of landscape patterns. It could cause the spatial scale of the NDVI and tree ring chronologies to diverge, thereby weakening the correlation between the two. Tree-ring growth and annual fluctuations in the high spatial resolution (1.21 km^2) NDVI were found to be positively correlated [29]. Bhuyan et al. compared the response of tree growth using a random forest model between MODIS NDVI and GIMMS3g NDVI [30]. The findings demonstrated that the MODIS NDVI growth signal outperformed the GIMMS3g NDVI in coniferous forest areas. Vicente-Serrano et al. showed that while there may be variations in the association between tree rings and NDVI at different spatial resolutions, there are similarities in the patterns of correlation between tree rings and NDVI over a range of NDVI periods and months. The spatial resolution of the NDVI sequence is mostly insensitive to the consistency of tree rings and NDVI [31]. Furthermore, Brehaut et al. discovered that tree rings and the GIMMS3g NDVI (8 km) and the Alaska Composite (1 km) have a poor association, suggesting that elevation may be the cause of their erratic correlation [32].

Elevation primarily influences the relationship between tree ring and NDVI through temperature and precipitation. Coulthard et al. found that in high-elevation forest areas, there are no common climatic driving factors between the NDVI and tree ring chronologies. However, standardized tree ring chronologies and NDVI are closely correlated with low-elevation dry regions that are dominated by grasslands and shrublands [33]. In addition, Wen et al. found that as the altitude increases, the positive correlation between tree ring width and NDVI gradually decreases. The primary component causing this association could be drought stress [34]. Yuan et al. discovered that as altitude rises, the relationship between the canopy and trunk progressively becomes less significant. In high mountain regions, the correlation coefficient is 0.371; in valley regions, it is 0.413; in desert regions, it is 0.583 [35]. In conclusion, a variety of factors can affect how radial tree growth and vegetation canopy interact. Additionally, the indicators that show a significant relationship between tree ring and NDVI index vary in different regions. Therefore, it is necessary to use multiple tree ring parameters and NDVI with different spatial resolutions for research purposes.

To investigate how the middle Tianshan Mountains' environment affects tree canopy and radial growth, we collected data from the Schrenk spruce (*Picea schrenkiana*) in Shawan Forest Farm. We established minimum earlywood density (MID), maximum latewood density (MXD), and tree-ring width using tree-ring samples gathered at eight distinct altitudes to explore how vegetation changes in response to climate change and how tree rings respond to it. Specifically, we aimed to (1) analyze the relationship between tree growth and the vegetation canopy, as well as the climate parameters that restrict tree growth, using climate variables related to tree-ring parameters; (2) evaluate differences in the relationships of vegetation index with different spatial resolutions with tree-ring width, MXD, and MID chronologies. Our findings provide new insights into ecological and environmental changes in the Tianshan Mountains' center region and will aid predictions of future changes in this region.

2. Materials and Methods

2.1. Study Area

The study location is situated in Shawan County's STG Forest location, in the middle of the Tianshan Mountains (Figure 1). In this region, the north has lower topography, while the south has higher terrain. The southern part is dominated by the Yilianhabierga Mountains, a branch of the Tianshan Mountains, and the northern part comprises the Gurban Tongut Desert. The area has a continental temperate arid climate, with cold winters, hot summers, short spring and autumn seasons, large temperature differences between winter and summer, abundant sunshine, low precipitation, and high evaporation. The primary forest is mainly concentrated on the windward slopes of the middle mountain belt at an elevation of 1700–2600 m. The vertical zonation of vegetation is quite distinct; grasses and shrubs are the main vegetation types in the forest. The soil is black-brown forest soil. *P. schrenkiana* forests are the predominant and extensively distributed type of forest on the Chinese northern slopes of the Tianshan Mountains.

2.2. Tree-Ring Data

We developed tree-ring width, MXD, and MID from *P. schrenkiana*. The sampling points were located in the STG Forest Area of Sha Wan City. The tree-ring samples were gathered during three separate periods from late August 2012 to August 2014. In 2012, one sample point was collected in the upper tree line area of the STG Forest Area. Samples were collected from five additional points at altitudes between 1740 and 2300 mm in the STG in late September 2013 considering the preliminary samples that had been collected at the top elevation of the STG in 2012 and the distribution of existing sampling points in Shaw Wan Forest Farm. The sampling points in STG had thin soil layers and steep slopes and lacked continuous slopes. The sampling points were mainly located on the steep slopes on the west side of the valley. High-altitude forests were covered in snow when samples were taken. The terrain in the STG Forest Area is steep; the thick snow, coupled with the slick roads, increases the difficulty of collecting tree samples between 2400 and 2500 m. Therefore, samples were taken from two further sites (stg 7 and stg 8) in the high-altitude forest area of STG in August 2014. The information on sampling points is shown in Table 1.

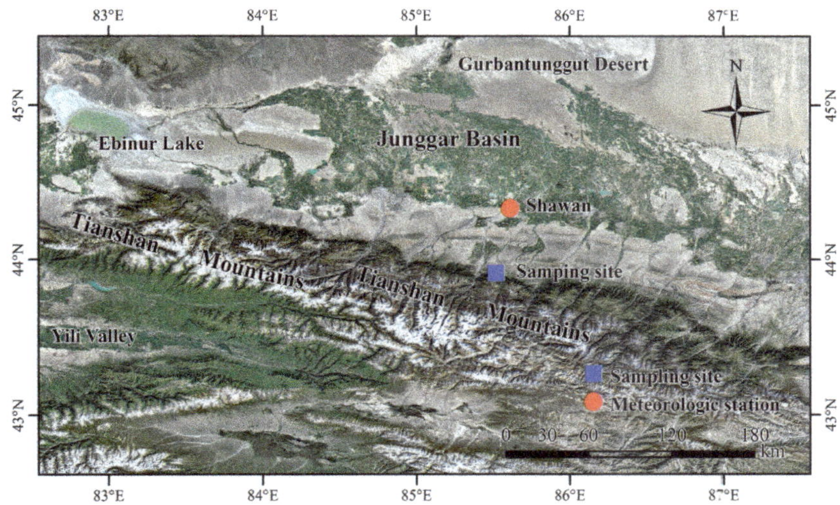

Figure 1. Map displaying the locations of meteorological stations and tree-ring collection sites in China's Tianshan Mountains.

Table 1. Tree-ring sampling site information.

Site	Latitude (N)	Longitude (E)	Elevation (m)	Aspect	Slope (°)	Canopy Density	No. of Trees/Cores
stg 1	43°52′59.5″	85°30′56.1″	2606	N	47	0.1	27/54
stg 7	43°53′02.7″	85°31′11.7″	2531	NW	40	0.3	20/40
stg 8	43°53′06.63″	85°31′09.46″	2400	NE	35	0.5	22/44
stg 2	43°53′11.5″	85°31′4.4″	2318	NE	35	0.3	22/44
stg 3	43°53′25.6″	85°31′4.8″	2206	NE	20	0.3	21/42
stg 4	43°53′54.7″	85°31′0.8″	2095	ENE	45	0.2	21/41
stg 5	43°54′10.3″	85°31′11.4″	1942	NNW	20	0.3	19/38
stg 6	43°54′42.2″	85°31′17.2″	1761	N	30	0.2	21/42

According to normal dendrochronological techniques, the core samples were transferred back to the laboratory for drying, fixing, grinding, and visual age determination [36]. With a Velmex Measuring System, the sanded cores were measured at a resolution of 0.001 mm while being viewed under a binocular microscope. We used the COFECHA application to ensure the quality of the cross-dating results, identify false rings and missing rings, eliminate errors in dating and measurement, remove unsuitable samples, ensure that the formation year of each tree ring is accurately determined, and keep the necessary documentation of this process [37].

The previously measured width samples were then desugared and defatted, and the samples were cut into trapezoidal blocks according to the angle of the wood fiber direction and fixed on the board. Using the DendroCut instrument, the core sample was cut into thin slices of 1 mm. Finally, X-ray transmission imaging was used to produce the film. The Dendro 2003 density measurement system was used to measure tree ring density values. When measuring the density of the overlapping sections, multiple tree rings were marked on the paper to indicate the overlapping part being measured and compared with the corresponding position on the test sample, in order to identify the overlapping tree rings and connect the segmented core samples. Finally, a data file is generated for each core sample [38,39].

Ultimately, the tree growth pattern was fitted and eliminated using the ARSTAN chronology development program by employing a negative exponential function in the fitting process. Ultimately, standardized chronologies (STD), differenced chronologies (RES), and autoregressive chronologies (ARS) for three tree ring parameters were established. This study is limited to a standardized tree-ring chronology of 8 sampling points. We further investigated using tree-ring width standardized chronologies, tree-ring maximum density standardized chronologies, and tree-ring minimum density standardized chronologies [40]. The sub-sample signal strength (SSS) greater than 0.85 was used to determine the reliable length of the chronology [41].

2.3. Meteorological Data

Shawan meteorological stations provided the meteorological data used in this study. (44°20′ N; 85°37′ E; altitude: 523 m a.s.l.; 48 km away from STG), which was close to the study area. We used precipitation (PCP), mean temperature (Tmean), mean maximum temperature (Tmax), and mean minimum temperature (Tmin) per month from 1961 to 2012. The above climatic data were from the Xinjiang Climate Center.

Analysis of changes in monthly PCP and average temperature at Shawan Meteorological Station (Figure 2) revealed that the average temperature in the region from 1961 to 2012 was approximately 7.69 °C. The average temperature of the hottest month (July) was 25.89 °C. The variation between the highest and lowest values for many years was only approximately 5 °C, indicating that the temperature was relatively stable. The average temperature in the coldest month (January) was −15.54 °C. The temperature difference was approximately 10 °C for many years, and changes in temperature over the years were significant. The average annual PCP was approximately 196 mm, and most PCP fell between April and July during the spring and summer seasons. Variation in PCP during the spring and summer seasons was much greater than that during the autumn and winter seasons.

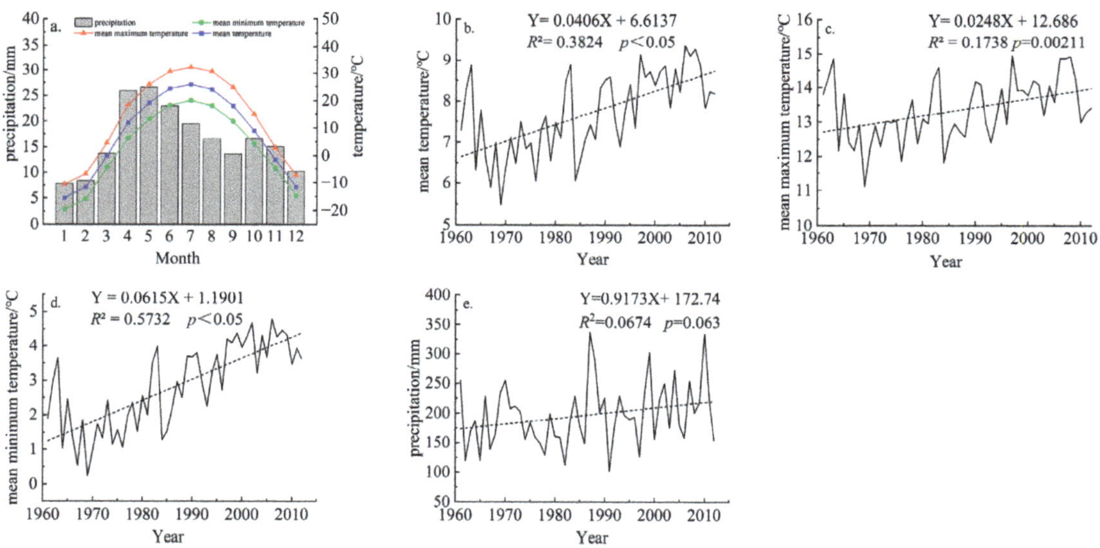

Figure 2. Climate data of the study area: (**a**) Precipitation, mean temperature, mean minimum temperature, and mean maximum temperature per month. Bars indicate precipitation (in mm), and curves with different colors indicate mean temperature (blue), mean minimum temperature (green), and mean maximum temperature (red) (in °C), (**b–d**) annual mean temperature, annual mean maximum temperature, and annual mean minimum temperature during 1961–2012, (**e**) annual precipitation during 1961–2012. Dashed lines indicate trends.

2.4. NDVI Data

We employed the Moderate Resolution Imaging Spectroradiometer (MODIS) NDVI and the Global Inventory Modeling and Mapping Studies 3g (GIMMS3g) NDVI, two NDVI datasets with varying spatial resolutions.

MODIS data were obtained from the MODIS products (MOD13Q1 product in HDF format, 16 days, 250 m, strip number: h24v04, http://glovis.usgs.gov/, accessed on 16 October 2021). Compared with AVHRR, MODIS has a narrower spectral channel and lacks the water vapor absorption band in the near-infrared band, which reduces the effects of water vapor and improves the ability to detect sparse vegetation. We analyzed data from 2001 (the first complete year of NDVI data) until 2012, the last year of our assembled tree-ring dataset.

With a temporal resolution of 15 days and a spatial resolution of 8 km, GIMMS3g is produced with the Advanced Very High Resolution Radiometer (AVHRR). This yields two maximum-value composites per month and 24 observations annually (http://ecocast.arc.nasa.gov/data/pub/gimms/3g.v0/, accessed on 16 October 2021). These data have undergone radiation correction and geometric coarse correction. Compared with GIMMS NDVI data, the improved GIMMS NDVI3g data have undergone satellite orbit drift correction, which eliminates common associations between the solar zenith angle and NDVI time series and improves the calibration procedure. The GIMMS3g NDVI data were first collected in July 1981. We analyzed data from 1982 (the first complete year of NDVI data) to the end of 2012 to maximize the overlap with the assembled tree-ring dataset. The latitude and longitude of each of the tree-ring sites were used to select corresponding pixels from the GIMMS3g and MODIS NDVI datasets.

Figure 3 shows that the values of GIMMS NDVI were highest from May to September, with a June peak. The annual average change was relatively stable. The MODIS NDVI values were higher at each sampling point than the GIMMS NDVI values; MODIS NDVI values were highest from May to September and peaked in July. As the elevation increased, the MODIS NDVI values gradually decreased (Figure 3).

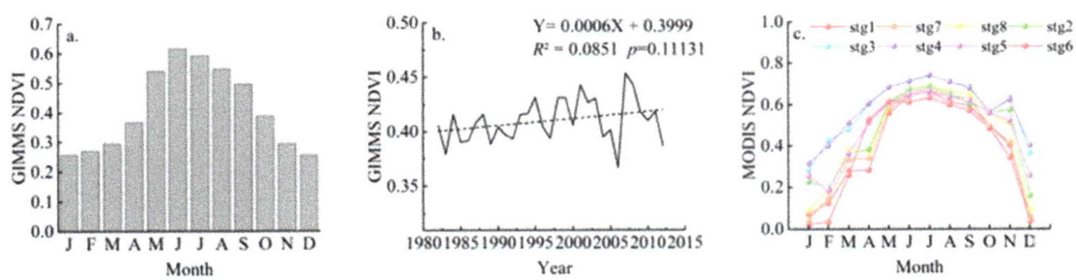

Figure 3. Two NDVI datasets with different spatial resolutions: (**a**) Monthly GIMMS3g NDVI (8 km, 1982–2012); (**b**) Annual mean NDVI during 1982–2012. Dashed lines indicate trends in GIMMS NDVI data. (**c**) Monthly MODIS NDVI (250 m, 2001–2012).

2.5. Statistical Analyses

The MOD13Q1 product first uses MRT V.4810 software to complete projection transformation and data format conversion and extract NDVI. The image output format is TIFF. Then, ArcGIS10.8 is used to batch clip the study area and exclude outliers, and finally, ArcGIS is used to synthesize monthly average NDVI values using the maximum synthesis method and extract the data to Excel v2019. The GIMMS NDVI3g product first uses Matlab R2022b to convert and project the downloaded NC data into a .tif format for image output. The study area is batch-cropped using ArcGIS, and the maximum composite method is used to extract the monthly average NDVI values of the study area. All data products use the Albers equal-area conic projection.

To analyze the relationships between connections between tree-ring at different altitudes and NDVI, using Pearson correlation analysis, the relationships between MXD, tree-ring width, and remote sensing vegetation index for all the sampling points were assessed. To determine whether climate had an impact on the relationship between tree-ring and NDVI, the relationships between climate and several tree growth indicators were examined. Given that the growth of trees might be affected by the climate conditions of the previous year, we conducted a monthly correlation analysis of the monthly Tmean, monthly mean Tmin, monthly mean Tmax, and PCP from April of the previous year to October of the current year, tree-ring width, MXD and MID, GIMMS NDVI, and MODIS NDVI.

3. Results

3.1. Statistical Characteristics of Chronologies

Tables 2–4 show that the tree-ring width normalized chronologies had mean sensitivity values ranging from 0.11 to 0.27. Below 2200 m, the average sensitivity is higher than the average sensitivity beyond this altitude. This indicates that tree-ring width chronologies in low-altitude regions are more sensitive to changes in climate factors. The mean sensitivity of MID was lower than that of tree-ring width chronologies. The MXD chronologies show a lower mean sensitivity and standard deviation in comparison to the MID chronologies. No significant changes in the mean sensitivity of the three chronologies with elevation were observed.

The first-order autocorrelation (AC1) between tree-ring width and MID chronologies was relatively high, and values were greater than 0.75 for all sampling points, with the exception of stg 4, stg 5, and stg 6. This suggests that the previous year's climate had a significant impact on tree growth, with high-altitude locations experiencing a longer-lasting effect on tree growth than low-altitude ones. However, the AC1 of MXD chronologies was lower compared with that of the other two chronologies, and values were relatively small in stg 4, stg 5, and stg 6. This indicates that the MID and tree-ring width of this year's tree rings are more influenced by changes in the previous year's climate characteristics than by the MXD, and the effect on tree-ring maximum density was greater in low-altitude regions than in high-altitude regions.

Table 2. Statistics of the STG tree-ring width standardized chronologies.

Statistic	stg 1	stg 7	stg 8	stg 2	stg 3	stg 4	stg 5	stg 6
Chronology length	513	350	242	213	146	142	132	172
Mean index (MI)	0.842	0.744	0.796	0.801	1.007	1.039	1.148	0.919
Mean sensitivity (MS)	0.159	0.166	0.121	0.152	0.119	0.268	0.236	0.262
Standard deviation (SD)	0.339	0.368	0.332	0.377	0.278	0.384	0.538	0.302
First-order autocorrelation (AC1)	0.878	0.911	0.931	0.911	0.849	0.564	0.818	0.556
Mean within-tree correlation	0.474	0.645	0.640	0.533	0.569	0.719	0.664	0.645
Signal-to-noise ratio (SNR)	11.357	10.694	19.295	12.305	19.921	14.796	7.721	17.032
Expressed population signal (EPS)	0.919	0.914	0.951	0.925	0.952	0.937	0.885	0.945
The first principal component (PC#1)	30.3	38.6	48.9	58.1	47.7	41.0	29.6	49.5
First year of SSS > 0.85	1646	1819	1851	1861	1875	1886	1932	1900

Table 3. Statistics of the STG MID standardized chronologies.

Statistic	stg 1	stg 7	stg 8	stg 2	stg 3	stg 4	stg 5	stg 6
Chronology length	513	350	242	213	146	142	132	172
Mean index (MI)	1.056	1.228	1.108	1.099	1.036	1.029	1.039	1.114
Mean sensitivity (MS)	0.064	0.105	0.149	0.102	0.097	0.136	0.101	0.141
Standard deviation (SD)	0.128	0.352	0.416	0.260	0.202	0.185	0.135	0.261
First-order autocorrelation (AC1)	0.755	0.889	0.883	0.817	0.743	0.416	0.434	0.669
Mean within-tree correlation	0.318	0.368	0.377	0.321	0.316	0.406	0.436	0.462
Signal-to-noise ratio (SNR)	8.891	5.679	5.140	4.742	5.879	16.250	6.235	5.645
Expressed population signal (EPS)	0.899	0.850	0.837	0.826	0.855	0.942	0.862	0.850
The first principal component (PC#1)	30.9	29.2	30.7	26.3	33.2	42.4	25.0	25.4
First year of SSS > 0.85	1651	1847	1872	1874	1882	1884	1933	1934

Table 4. Statistics of the STG MXD standardized chronologies.

Statistic	stg 1	stg 7	stg 8	stg 2	stg 3	stg 4	stg 5	stg 6
Chronology length	513	350	242	213	146	142	132	172
Mean index (MI)	47	40	35	35	20	45	20	30
Mean sensitivity (MS)	0.057	0.070	0.051	0.047	0.036	0.046	0.045	0.044
Standard deviation (SD)	0.070	0.083	0.066	0.053	0.049	0.065	0.072	0.064
First-order autocorrelation (AC1)	0.392	0.351	0.480	0.233	0.543	0.590	0.670	0.595
Mean within-tree correlation	0.430	0.398	0.490	0.412	0.396	0.405	0.381	0.335
Signal-to-noise ratio (SNR)	9.884	9.295	8.979	13.395	11.008	9.406	5.901	5.157
Expressed population signal (EPS)	0.908	0.903	0.900	0.931	0.917	0.904	0.855	0.838
The first principal component (PC#1)	29.4	30.5	37.3	38.5	37.1	33.8	23.9	25.0
First year of SSS > 0.85	1651	1819	1874	1858	1879	1888	1933	1934

The signal-to-noise ratio is a statistical measure of the amount of environmental information contained in a sample, and it is generally considered good if it is greater than 4 [41]. In this study, the signal-to-noise ratio of tree-ring width chronologies ranged from 7.8 to 20, the signal-to-noise ratio of MID ranged from 4.7 to 16.3, and the signal-to-noise ratio of MXD ranged from 5.1 to 13.4. These ranges indicate that the chronologies contain a significant amount of climate information and are suitable for dendroclimatology research.

Expressed population signals (EPSs) of all chronology samples were greater than 0.85, and the highest EPS value was observed at sampling point stg 1, indicating that the sampled cores provide an accurate representation of the study area.

3.2. Effects of Temperature and Precipitation on Forest Growth

Climate factors during the previous growing season have a major impact on tree-ring growth. Considering the physiological characteristics of trees and the fact that elevation results in a drop in temperature [42], we conducted a correlation analysis of Climate data from April of the previous year to October of the present one with the standardized tree-ring chronology (Figures 4–6).

The response of tree-ring width to Tmean, Tmin, and Tmax varied at different elevations. Overall, there were strong correlations between tree-ring width and Tmin. Reaching an extremely significant level of Tmin in July. Specifically, tree-ring width at higher elevations was positively linked with temperatures; the opposite pattern was observed at lower elevations (Figure 7a–c). Tree-ring width at upper elevations (stg 1, stg 7, and stg 8) responded positively to temperature factors in the previous spring and current year. Low-altitude tree-ring width (stg 5 and stg 6) responded negatively to temperature factors in the fall of the prior year and in the summer and fall of the current year.

However, the responses of tree-ring width in the forested zone to different temperature factors vary, especially the responses to Tmax and Tmin. The responses of stg 2 and stg 4 to the Tmin during the growing seasons, and their response to Tmax mostly shifted from positive to negative. This proves that at the same height, the impact of various temperature parameters on tree-ring width varies.

The responses of tree-ring to PCP differed at high and low elevations. There was a high correlation between the width of the tree rings at the upper forest boundary (stg 1, stg 7, and stg 8) and July PCP. By contrast, there was a significant positive relationship between May and September PCP in the previous year and September precipitation in the previous year, as well as tree-ring width at lower elevations.

From December to March of the previous year, MID was significantly negatively linked to temperature, indicating that low temperatures before the growing season favor the formation of higher MID. From April to May, there was an upward relationship between MID and temperature and an adverse connection with PCP. There was a large and robust association between the May temperature and the tree-ring width of stg 8.

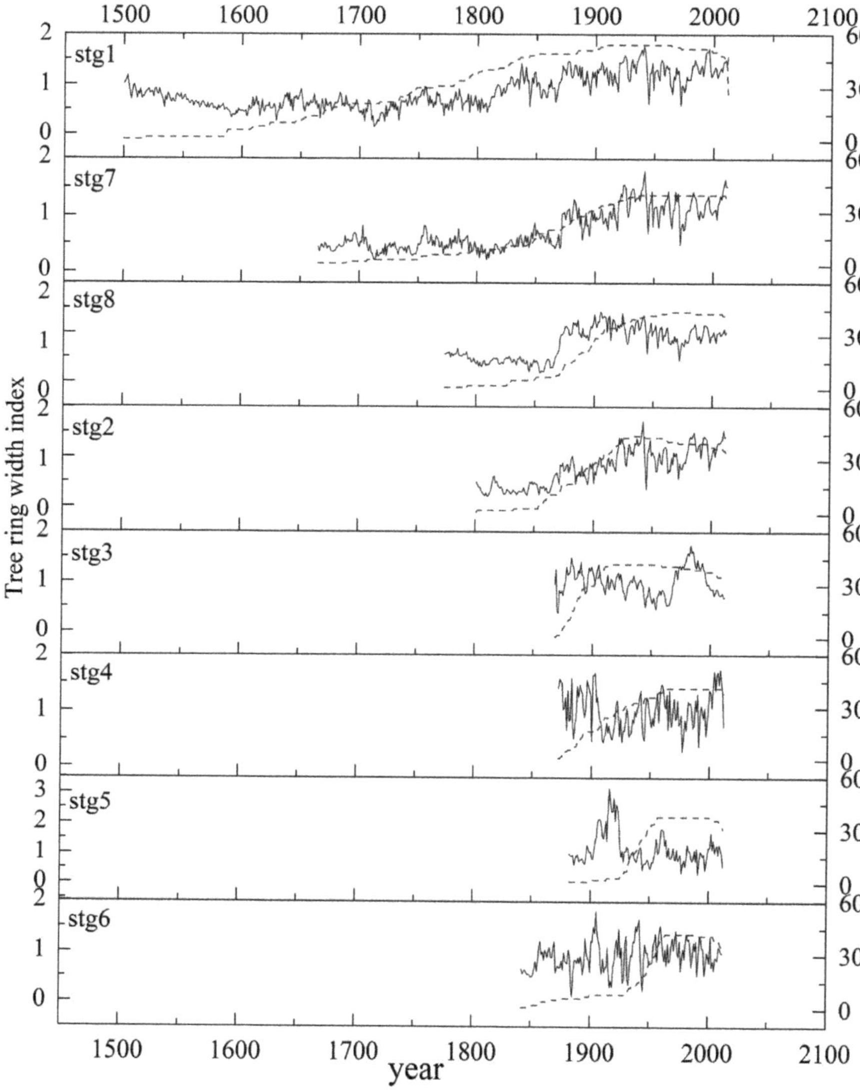

Figure 4. Tree-ring width standardized chronologies and sample depth. Solid line represents the tree ring width index; dashed line represents the sample size.

In high-altitude areas, MID chronologies had a negative correlation with PCP throughout this time and a positive correlation with July's Tmean and Tmax. In low-altitude areas, from May to September, there was a positive correlation between MID and temperature, and a negative correlation between MID and PCP (Figure 8). At every sampling location, MID showed a positive correlation with PCP in April and November of the preceding year and a negative correlation with PCP in July of that same year, as well as in May and July–September of the current year. Specifically, the correlations between MID in high-altitude areas (stg 1, stg 7, and stg 8) and Tmean in July were significant, and the correlations between MID chronologies and average temperature were weaker at other altitudes. Between April of the previous year and February of the present year, there was a negative correlation between the MID at stg 2 and the Tmin; there was a positive correlation with the minimum temperature from April to October of the current year, this correlation was particularly significant for Tmin s in May and June. The MID of all sampling sites was negatively correlated with the Tmax in December of the previous year and positively correlated with the Tmax in July. With the exception of stg 3, MID was positively correlated with the Tmax from March to May of the current year. Specifically, MID in high-altitude areas was positively correlated with Tmax from March to October, and this correlation was significant in July; the correlations between MID in low-altitude areas and Tmax were weak (Figure 8).

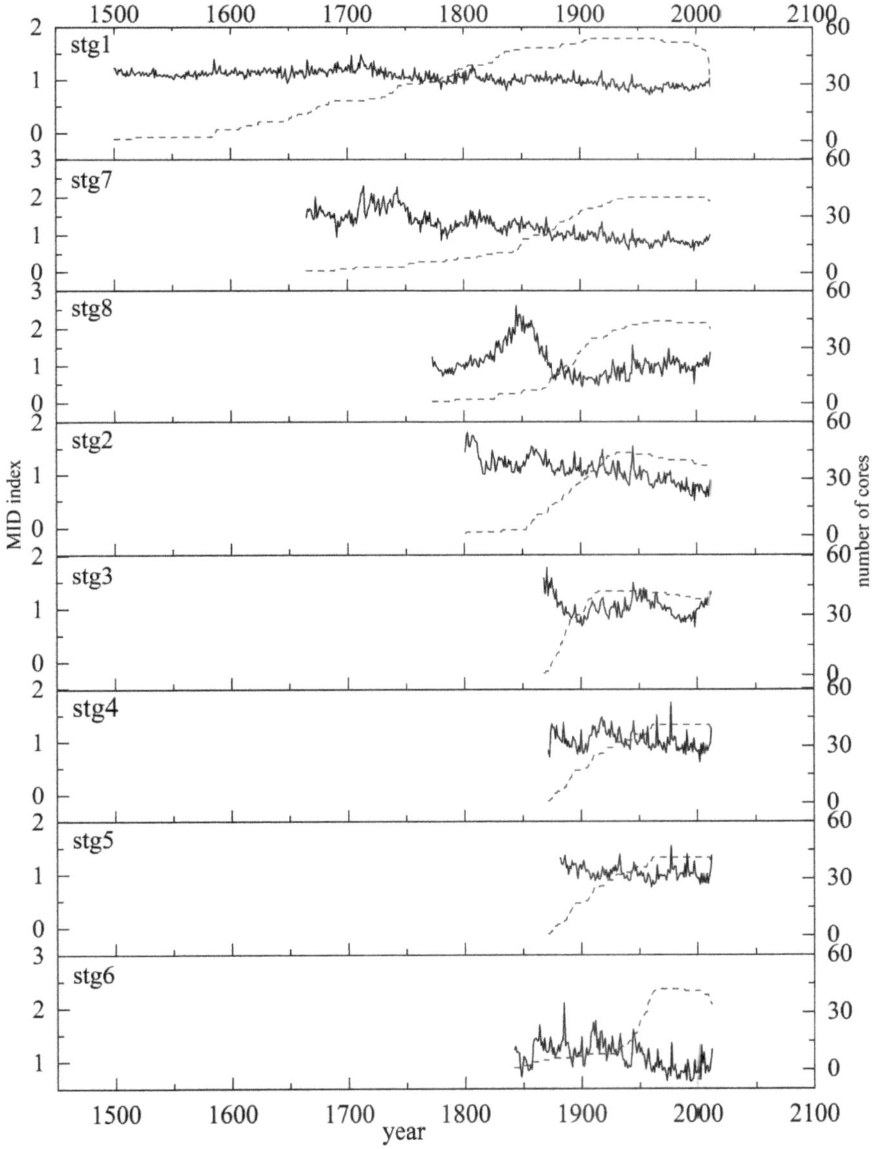

Figure 5. MID standardized chronologies and sample depth. Solid line represents the MID index; dashed line represents the sample size.

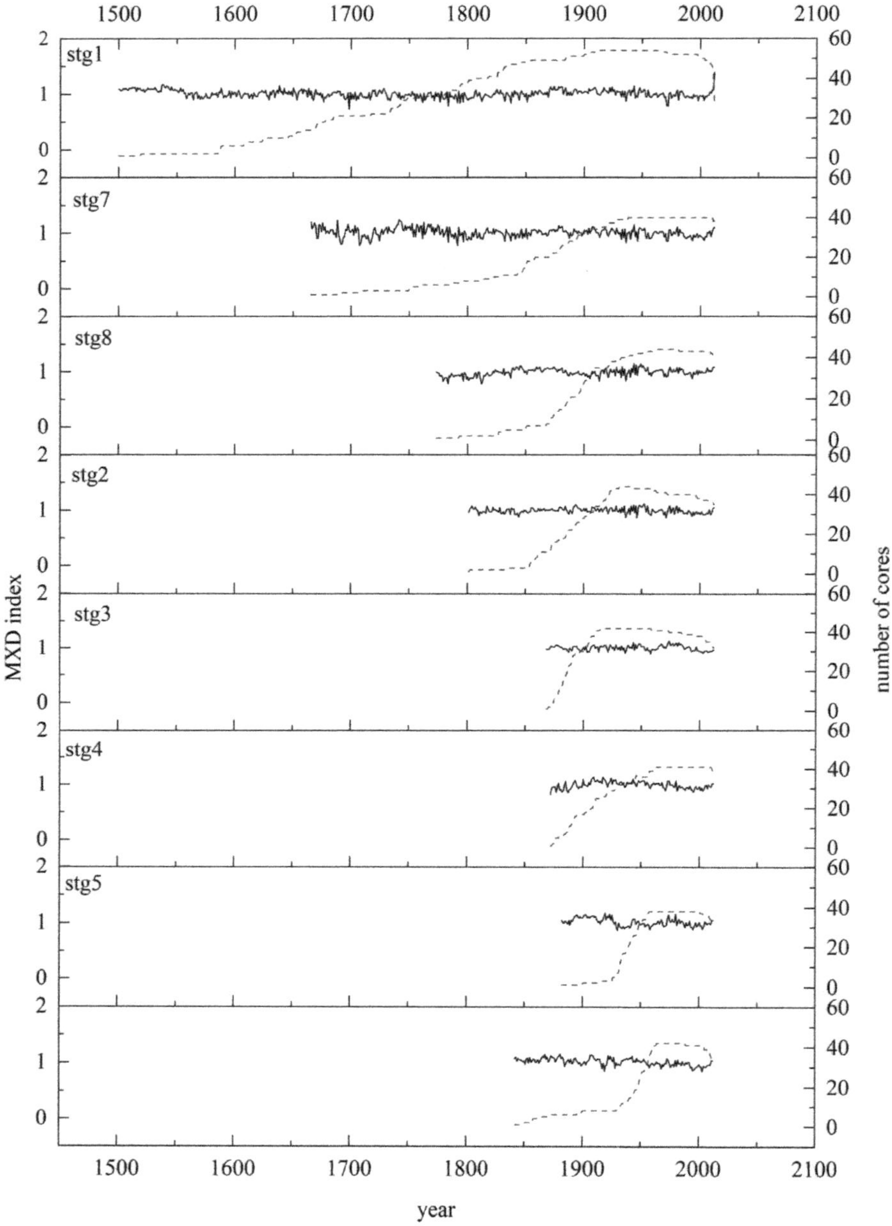

Figure 6. MXD standardized chronologies and sample depth. Solid line represents MXD index; dashed line represents the sample size.

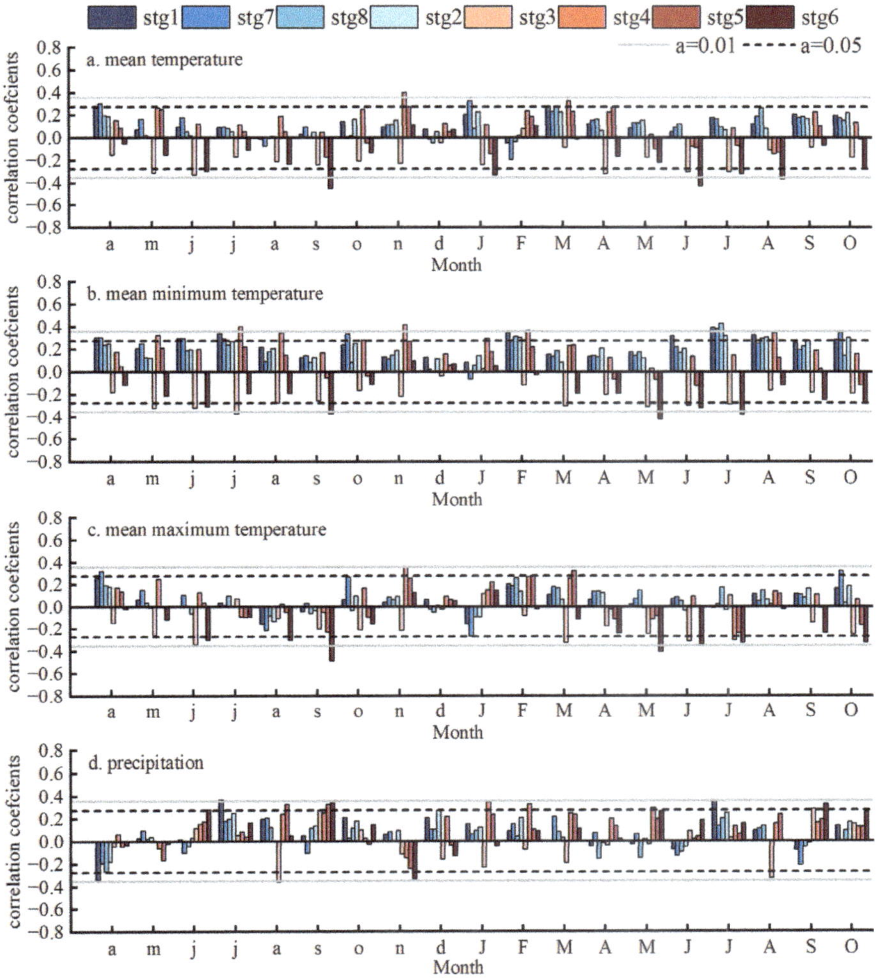

Figure 7. Pearson correlations between tree-ring width standardized chronologies and climatic data (monthly Tmean, monthly mean Tmin, monthly mean Tmax, and PCP). Lowercase and capital letters indicate months of the previous and current years, respectively.

Except for November, MXD was mainly negatively correlated with PCP (Figure 9a–c) and positively correlated with temperature factors in April and May of the current year, and negatively correlated with PCP during the same period. MXD chronologies at high altitudes were significantly positively correlated with the Tmean in April of the current year. MXD was negatively weakly correlated with the Tmean from November of the previous year to February of the current year and positively weakly correlated with the temperature from April to October of the current year (Figure 9).

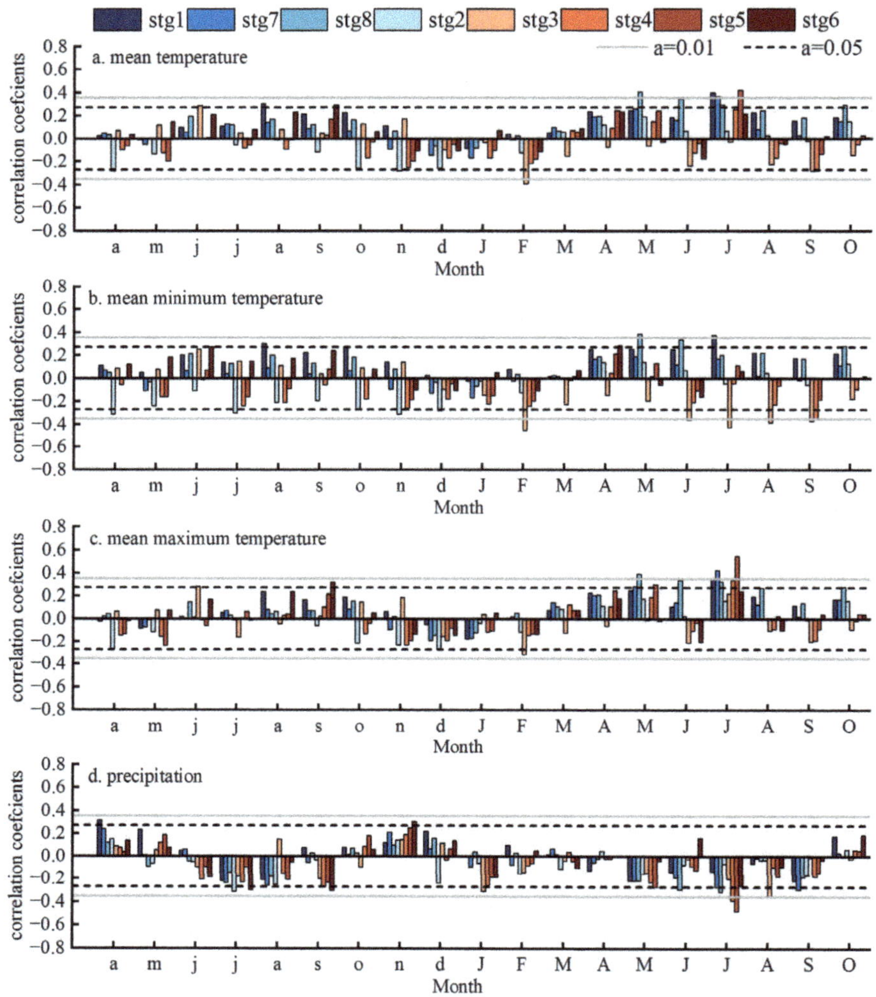

Figure 8. Pearson correlations between MID standardized chronologies and climatic data (monthly Tmean, monthly mean Tmin, monthly mean Tmax, and PCP). Lowercase and capital letters indicate months of the previous and current years, respectively.

Statistical analysis of NDVI and climate data (monthly Tmean, monthly Tmax, monthly Tmin, and monthly PCP) for the 12 months revealed variation in the correlations between NDVI and climate data (Figures 10 and 11). Overall, the effect of temperature and PCP from the previous year on the GIMMS NDVI from April to July was weak and not statistically significant. In April, the GIMMS NDVI and temperature showed a strong positive association, with the Tmax showing an especially significant connection. In May and June, the influence of temperature on the GIMMS NDVI was low. A remarkable positive connection was seen between the May GIMMS NDVI and the April–May PCP. Both the July GIMMS NDVI and Tmean and Tmax in May, as well as the July GIMMS NDVI and Tmin in May and June, showed a highly significant negative connection. A statistically significant positive connection was seen between the July GIMMS NDVI and the May PCP. The GIMMS NDVI for August of the previous year showed a strong positive association with the April temperature, and significant negative correlations of GIMMS NDVI with temperature in

April of the current year and PCP in November of the previous year. September GIMMS NDVI showed an important positive relationship with PCP of the current year in August and a significant negative correlation with the Tmin of the previous year in August. There were significant positive correlations between October NDVI and the temperature of the previous year from April to June and the temperature of the current year from February to March and August. Additionally, there was a significant opposite correlation between the October NDVI and the June and September PCP of the previous year. There were significant positive correlations between May to August GIMMS NDVI with the Tmean and Tmax of the previous year in August; there was also a highly significant positive correlation between May to August GIMMS NDVI and PCP in April.

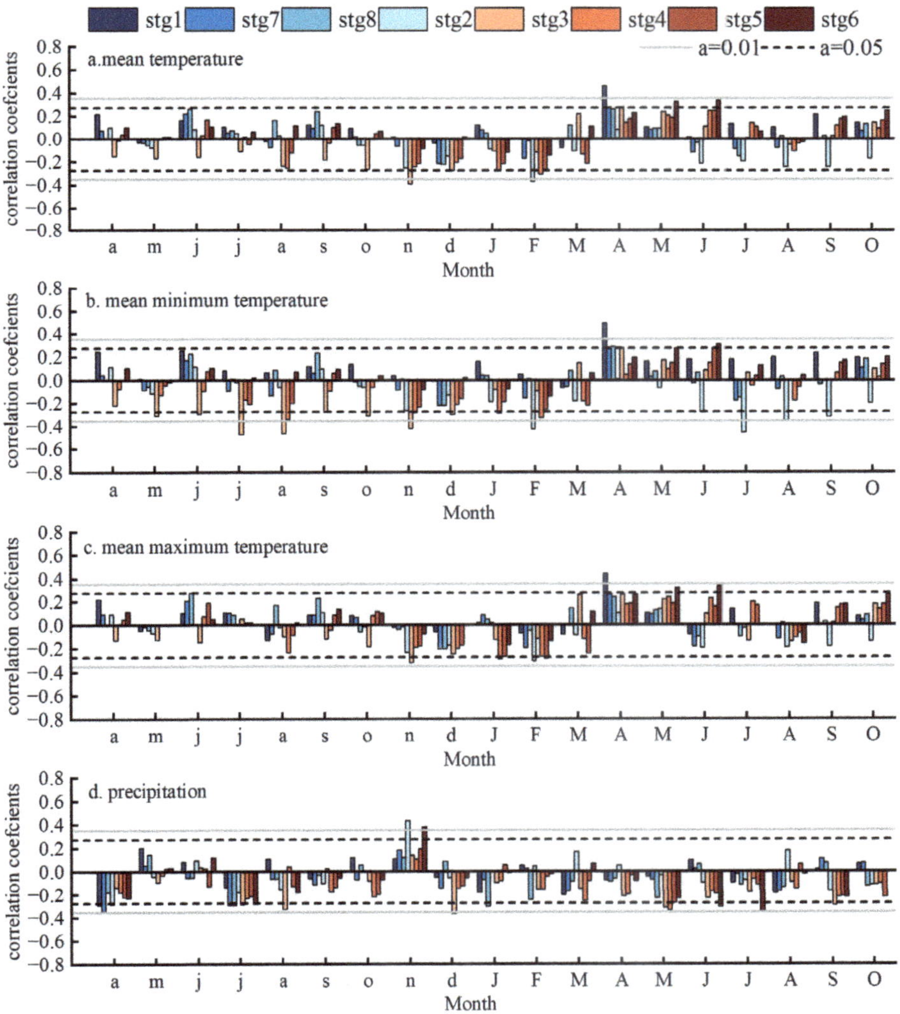

Figure 9. Pearson correlations between MXD standardized chronologies and climatic data (monthly Tmean, monthly mean Tmin, monthly mean Tmax, and PCP). Lowercase and capital letters indicate months of the previous and current years, respectively.

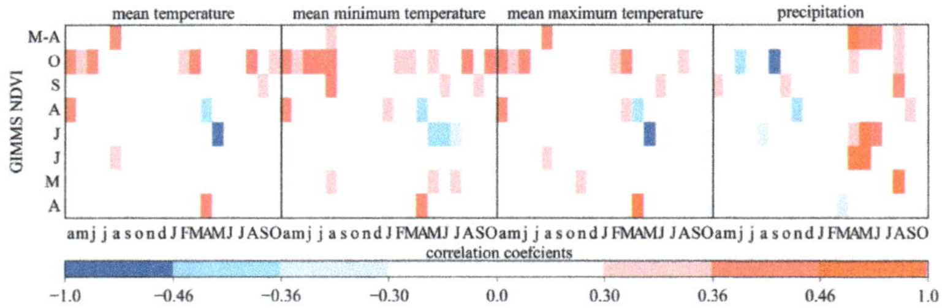

Figure 10. Correlations between GIMMS NDVI and climate factors (monthly Tmean, monthly mean Tmin, monthly mean Tmax, and PCP). The ordinate shows the monthly GIMMS NDVI from April to October, as well as the average NDVI from May to August. Lowercase and capital letters indicate months of the previous and current years, respectively. Deep red and deep blue indicate significance at $p < 0.01$; red and blue indicate significance at $p < 0.05$; light red and light blue indicate significance at $p < 0.1$.

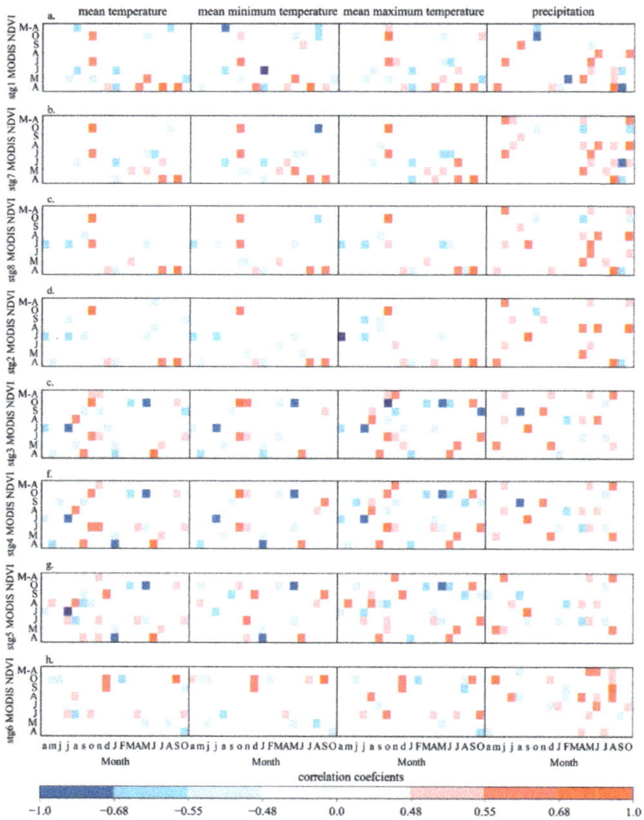

Figure 11. Correlations of MODIS NDVI with climate factors (monthly Tmean, monthly mean Tmin, monthly mean Tmax, and PCP). The ordinate shows the monthly MODIS NDVI from April to October, as well as the average NDVI from May to August. (**a–h**) shows the correlations between MODIS NDVI and climate factors at stg 1, stg 7, stg 8, stg 2, stg 3, stg 4, stg 5, and stg 6. Lowercase and capital letters indicate months of the previous and current years, respectively. Deep red and deep blue indicate significance at $p < 0.01$; red and blue indicate significance at $p < 0.05$; light red and light blue indicate significance at $p < 0.1$.

MODIS NDVI at high altitudes (stg 1, stg 7, stg 8, and stg 2) was correlated with temperature and PCP (Figure 11). May MODIS NDVI was significantly positively correlated with April PCP. July MODIS NDVI was significantly negatively correlated with April and July temperatures of the previous year. For stg1 and stg 7, July MODIS NDVI was significantly positively correlated with October temperatures of the previous year and significantly negatively correlated with the highest temperature in May of the same year; July MODIS NDVI was also positively correlated with May PCP of the same year. Among high-altitude sampling points, August MODIS NDVI was weakly correlated with temperature but significantly positively correlated with June PCP. The correlation between May to August MODIS NDVI and temperature was weak. Only MODIS NDVI at stg 1 was significantly negatively correlated with Tmean and Tmin of April in the current year. In the low-altitude region, April MODIS NDVI was significantly negatively correlated with January average temperature and Tmin. The correlation between April MODIS NDVI and PCP was weak and not significant. October MODIS NDVI was highly significantly negatively correlated with May temperatures of the current year, and its correlation with PCP was weak.

3.3. Relationships between Tree-Ring Parameters and NDVI

Figure 12 shows the correlations of tree-ring width, MXD, and MID chronologies with GIMMS NDVI and MODIS NDVI at different altitudes. Overall, the tree-ring width at all sampling sites was positively correlated with the GIMMS NDVI in July and December of the previous year and MODIS NDVI in February of the current year. The MID chronologies had a significant association with the MODIS NDVI of June of the past year and March of the current year, and a negative correlation with the GIMMS NDVI of October–December of the previous year and June to August of the current year. At each site, The GIMMS NDVI in January to February, July to August, and the growing season (May to August) showed negative correlations with the MXD chronologies, whereas the MODIS NDVI in June of both the prior and current year showed positive correlations.

There was a major positive association between tree-ring width at low altitudes (stg 4 and stg 5) and the GIMMS NDVI in October of that prior year, July of this present year, and during the growing period (May to August) of the current year. (Figure 12a). The response of tree-ring width to GIMMS NDVI at high altitudes (stg 1, stg 7, and stg 9) was not significant; however, in December of the prior year, there was a strong positive association between tree-ring width and GIMMS NDVI. Figure 12c illustrates the considerable negative correlation between GIMMS NDVI and MID chronologies at low altitudes in October of the preceding year and June to July of the current year. In January and February of this year, there was a substantial negative correlation between the MXD chronologies and the GIMMS NDVI (Figure 12e). MID and MXD chronologies at high altitudes responded weakly to GIMMS NDVI; highly significant correlations between MID chronology and GIMMS NDVI in April of the previous year and between MXD chronology and GIMMS in July of the current year were only observed at stg 1. Tree-ring parameters were more strongly correlated with MODIS NDVI than GIMMS NDVI at higher elevations. Tree-ring width at high altitudes (stg 1 and stg 7) was significantly positively correlated with MODIS NDVI in July (Figure 12b), and MID and MXD chronologies were significantly positively correlated with MODIS NDVI in June of the previous year (Figure 12d,f). However, the response of tree-ring parameters to MODIS NDVI and GIMMS NDVI differed in low-altitude areas. Tree-ring width and MID chronologies were significantly negatively correlated with MODIS NDVI in November of the previous year, and MXD chronologies were significantly positively correlated with MODIS NDVI in June of the current year. This indicates that the relationships between vegetation index with different spatial resolutions and tree-ring parameters varied with altitude. In high-altitude areas, the relationships between tree-ring parameters and MODIS NDVI were stronger than the relationships between tree-ring parameters and GIMMS NDVI; the opposite pattern was observed in low-altitude areas.

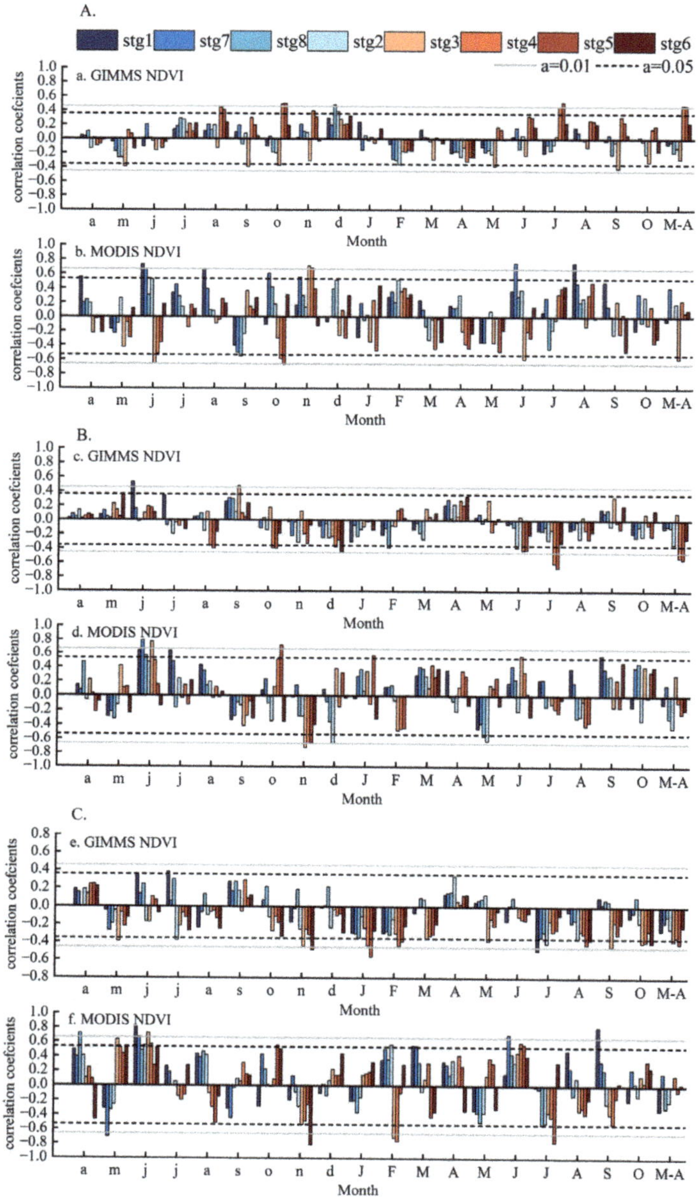

Figure 12. Relationships between tree-ring parameters and NDVI. (**A**) shows the correlations between tree-ring width standardized chronologies and NDVI. (**B**) shows the correlations between MID standardized chronologies and NDVI. (**C**) shows the correlations between MXD standardized chronologies and NDVI.

4. Discussion

4.1. Factors Affecting the Relationships between Tree-Ring Parameters and NDVI

This study revealed a correlation between monthly climatic parameters and the tree-ring width, MXD, and MID chronologies at various altitudes. The correlations between tree-ring width chronologies and Tmin were generally strong; the positive correlations between tree-ring width chronologies at high altitudes and Tmin in July were highly significant. The growth of tree rings requires suitable temperatures. In the research region, higher average low temperatures in July encourage photosynthesis in trees, which raises net nutrient accumulation, promotes cambial cell division, and ultimately causes the construction of larger tree rings. Higher temperatures during the growing season also accelerate the melting of ice and snow, which provides sufficient water for tree growth. Lower average temperatures in July can decrease the photosynthesis of trees, hinder the division of cambial cells, and even cause freezing or frost damage to trees, which results in narrower tree rings [43,44]. The tree-ring width chronology of stg 3 was negatively correlated with temperature. This might stem from the location of the sampling site in the PCP zone of the middle mountain belt on the northern slope of the Tianshan Mountains. The water and thermal conditions vary greatly in this region. Higher temperatures in this area lead to a decrease in soil moisture, an increase in the respiration of trees at night, a decrease in net photosynthesis, and ultimately the occurrence of narrower tree rings. MID and MXD chronologies were positively correlated with the temperature in the early growing season (April and May) and negatively correlated with PCP in April and May. April is a critical period for trees because it coincides with the transition from dormancy to an active state of growth. Moderate warming benefits trees by allowing them to break dormancy earlier, which stimulates the division of cells, improves photosynthetic efficiency, and promotes the buildup of carbohydrates. This gives plants enough energy to flourish. Additionally, higher temperatures in April can enhance snow melting, increase the soil moisture content, and thus improve the growth rate of plants during the growing season [45]. The MID shows a negative response to the temperature from December of the previous year to March of the current year. This negative response indicates that low temperatures have an adverse impact on tree growth and development. The higher temperatures during this period, from December to March, contribute to an accelerated photosynthesis process, which promotes the differentiation of cambial cells in trees and facilitates tree growth. As a result, the tree cells become larger and the cell walls thinner, leading to a smaller earlywood density value in the tree [46].

4.2. Relationships between Tree-Ring Parameters and NDVI with Different Spatial Resolutions

There is a mutual dependence between tree rings and tree crowns. Trees convert carbon dioxide, water, and mineral salts into organic compounds through photosynthesis in the canopy leaves, generating the nutrients necessary for their own growth and promoting radial tree growth. At various spatial resolutions, we detected inconsistencies in the connections between tree-ring characteristics and NDVI. This might be due to MODIS data's high spatial resolution, which is superior to GIMMS data's coarse spatial resolution (8 km) in capturing local climate and other information reflecting changes in vegetation or tree rings; at this resolution, the heterogeneity of vegetation landscape patterns, stand density, age structure, and other differences result in a spatial decoupling of NDVI and tree-ring chronology data, which weakens the relationship between the two [47,48]. Our research indicates that the correlations between tree-ring parameters and MODIS NDVI are generally stronger in high-altitude areas than the correlations between tree-ring parameters with GIMMS3g NDVI; however, the opposite pattern was observed in low-altitude areas. Bhuyan et al. [30] found similar results by comparing the response of tree growth to high spatial resolution (MODIS, 250 m) and coarse spatial resolution (GIMMS3g NDVI, 8 km) data using a random forest model. Specifically, they showed that the growth signal of MODIS NDVI was stronger than that of GIMMS3g NDVI in areas with coniferous forests. We analyzed the relationships between three tree-ring parameters and NDVI and found that the MXD chronologies of each sampling point were more strongly correlated with the NDVI of each month during the growing season than tree-ring width and MID chronologies. The MXD chronologies were negatively correlated with NDVI in July and August, which might stem from carbon allocation during the growing season to support bud growth (leaf area) and the initiation of cambial activity, given that cell division and expansion (resulting in wider tree rings) are prioritized over the thickening of the cell walls [49–51].

Although previous studies have demonstrated an increase in the strength of the correlation between NDVI and tree-ring width as the spatial resolution increases from 8 km to 250 m [32,52], our findings demonstrate the value of high-resolution dense time series for monitoring tree productivity and retrospective modeling. However, recent studies have observed correlation coefficients between tree-ring width and 30 m resolution NDVI ranging from −0.27 to 0.50. The use of 30 m resolution NDVI data did not enhance the correlations between tree-ring parameters and NDVI compared with 250 m resolution NDVI. Therefore, a 250 m spatial resolution might be sufficient for characterizing correlations between tree-ring and NDVI [53]. To improve our comprehension of the connection between the local canopy and cambial activity and evaluate the possible impacts of climate change on the development and productivity of forest trees, more research is required. Factors that might affect the NDVI signal such as soil moisture, topography, and vegetation type should be considered in future studies.

Compared with high elevations, there is a strong correlation between tree rings and GIMMS NDVI at low elevations; this is consistent with what Wen et al. found. According to research, it is possible to establish a mathematical relationship between tree rings and NDVI within the same region and period in order to effectively deduce historical changes in the NDVI when common external environmental factors influence the growth of trees and regional vegetation and when other non-climatic factors do not interfere [33]. In this study, the tree ring width chronology at higher altitudes is positively correlated with temperature, while at lower altitudes, it is the opposite. Due to its coarse spatial resolution, GIMMS NDVI fails to show the trend of altitude changes, but it is primarily positively correlated with temperature from April to August. The central part of the Tianshan Mountains may be a common environmental factor affecting tree ring parameters and NDVI under drought stress [54,55].

5. Conclusions

We analyzed the relationships between standardized tree-ring chronologies and NDVI in the central Tianshan Mountains. We found that the correlations between the two were generally weak, which might be explained by the lack of a consistent response of the radial growth of *P. schrenkiana* to climate factors. High-resolution tree-ring width data might not accurately reflect the changes in NDVI. Additional studies that use tree-ring parameters to reconstruct NDVI in this region are needed. Only eight sites were sampled in our study, and only climate factors, such as temperature and precipitation, were considered in our analysis. However, other factors, such as atmospheric CO_2 concentrations, drought, fire, forest pests and diseases, volcanic eruptions, earthquakes, glacier movement, and human-induced environmental pollution, can also affect tree growth. Therefore, additional research examining the effects of a wider range of factors on the relationships between tree-ring and NDVI at large scales and over a long time would be particularly valuable.

Author Contributions: T.Z., Y.F., S.Y. and S.J. conceived and designed the experiments; J.S., D.G., T.H. and K.G. performed the experiments; J.S. and Y.L. analyzed the data; J.S. wrote the manuscript. All authors have read and agreed to the published version of the manuscript.

Funding: This research was funded by the Natural Key Research and Development Program (Intergovernmental Key and Special Project, 2023YFE0102700), Natural Science Foundation of Xinjiang Uigur Autonomous Region-Science Fund for Distinguished Young Scholars (2022D01E105), Natural Science Foundation of China (41975095), Science and Technology Development Fund Project of Institute of Desert Meteorology, China Meteorological Administration (KJFZ202306).

Data Availability Statement: The data provided in this study are available from the authors upon request.

Conflicts of Interest: The authors declare that the research was conducted in the absence of any commercial or financial relationships that could be construed as a potential conflict of interest.

References

1. Berner, L.T.; Beck, P.S.A.; Bunn, A.G.; Goetz, S.J. Plant response to climate change along the forest-tundra ecotone in northeastern Siberia. *Glob. Chang. Biol.* **2013**, *19*, 3449–3462. [CrossRef] [PubMed]
2. Camarero, J.J.; Gutiérrez, E. Wood density of silver fir reflects drought and cold stress across climatic and biogeographic gradients. *Dendrochronologia* **2017**, *45*, 101–112. [CrossRef]
3. Babst, F.; Bouriaud, O.; Papale, D.; Gielen, B.; Janssens, I.A.; Nikinmaa, E.; Ibrom, A.; Wu, J.; Bernhofer, C.; Kostner, B.; et al. Above-ground woody carbon sequestration measured from tree rings is coherent with net ecosystem productivity at five eddy-covariance sites. *New Phytolodist.* **2014**, *201*, 1289–1303. [CrossRef] [PubMed]
4. Van Breugel, M.; Ransijn, J.; Craven, D.; Bongers, F.; Hall, J.S. Estimating carbon stock in secondary forests: Decisions and uncertainties associated with allometric biomass models. *For. Ecol. Manag.* **2011**, *262*, 1648–1657. [CrossRef]
5. Miyeni, R.B.; Hall, F.G.; Sellers, P.J.; Marshak, A.L. The interpretation of spectral vegetation indexes. *IEEE Trans. Geosci. Remote Sens.* **1995**, *33*, 481–486.
6. Erasmi, S.; Klinge, M.; Dulamsuren, C.; Schneider, F.; Hauck, M. Modelling the productivity of Siberian larch forests from Landsat NDVI time series in fragmented forest stands of the Mongolian forest-steppe. *Environ. Monit. Assess.* **2021**, *193*, 200. [CrossRef]
7. Levesque, M.; Andreu-Hayles, L.; Smith, W.K.; Williams, A.P.; Hobi, M.L.; Allred, B.W.; Pederson, N. Tree-ring isotopes capture interannual vegetation productivity dynamics at the biome scale. *Nat. Commun.* **2019**, *10*, 742. [CrossRef]
8. Zhang, T.; Zhang, R.; Lu, B.; Mambetov, B.T.; Kelgenbayev, N.; Dosmanbetov, D.; Maisupova, B.; Chen, F.; Yu, S.; Shang, H.; et al. *Picea schrenkiana* tree-ring chronologies development and vegetation index reconstruction for the Alatau Mountains, Central Asia. *Geochronometria* **2018**, *45*, 107–118. [CrossRef]
9. Camarero, J.J.; Franquesa, M.; Sangüesa-Barreda, G. Timing of drought triggers distinct growth responses in holm oak: Implications to predict warming-induced forest defoliation and growth decline. *Forests* **2015**, *6*, 1576–1597. [CrossRef]
10. Wang, M.Y.; Luo, Y.; Zhang, Z.Y. Recent advances in remote sensing of vegetation phenology: Retrieval algorithm and validation strategy. *Natl. Remote Sens. Bull.* **2020**, *26*, 431–455. [CrossRef]
11. Rouse, J.W.; Haas, R.H.; Schell, J.A. Monitoring vegetation systems in the Great Plains with ERTS. In *Third Earth Resources Technology Satellite-1 Symposium, Proceedings of a Symposium, Goddard Space Flight Center, Washington, DC, USA, 10–14 December 1973*; NASA: Washington, DC, USA, 1974; Volume 351, pp. 309–317.
12. Huete, A.; Didan, K.; Miura, T. Overview of the radiometric and biophysical performance of the MODIS vegetation indices. *Remote Sens. Environ.* **2002**, *83*, 195–213. [CrossRef]
13. Kauth, R.J.; Thomas, G.S. The tasselled cap a graphic description of the spectra temporal development of agriculture crops as seen by Landsat. In Proceedings of the 2nd International Symposium on Machine Processing of Remotely Sensed Data, West Lafayette, IN, USA, 29 June–1 July 1976.
14. Lopatin, E.; Kolström, T.; Spiecker, H. Determination of forest growth trends in Komi Republic (northwestern Russia): Combination of tree-ring analysis and remote sensing data. *Boreal Environ. Res.* **2006**, *11*, 341–353.
15. Liang, E.Y.; Eckstein, D. Dendrochronological Potential of the Alpine Shrub Rhododendron nivale on the South-Eastern Tibetan Plateau. *Ann. Bot.* **2009**, *104*, 665–670. [CrossRef] [PubMed]
16. Pettorelli, N.; Vik, J.O.; Mysterud, A.; Gaillard, J. Using the satellite-derived NDVI to assess ecological responses to environmental change. *Trends Ecol. Evol.* **2005**, *20*, 503–510. [CrossRef] [PubMed]
17. Gamon, J.A.; Kovalchuck, O.; Wong, C.Y.S. Monitoring seasonal and diurnal changes in photosynthetic pigments with automated PRI and NDVI sensors. *Biogeosciences* **2015**, *12*, 4149–4159. [CrossRef]
18. Furusawa, T.; Koera, T.; Siburian, R.; Wicaksono, A.; Matsudaira, K.; Yoshinori, I. Time-series analysis of satellite imagery for detecting vegetation cover changes in Indonesia. *Sci. Rep.* **2023**, *13*, 8437. [CrossRef]
19. Mao, D.; Wang, Z.M.; Luo, L.; Ren, C.Y. Integrating AVHRR and MODIS data to monitor NDVI changes and their relationships with climatic parameters in Northeast China. *Int. J. Appl. Earth Obs. Geoinf.* **2012**, *18*, 528–536. [CrossRef]
20. Vivar-Vivar, E.D.; Pompa-García, M.; Rodríguez-Trejo, D.A.; Leyva-Ovalle, A.; Wehenkel, C.; Carrillo-Parra, A.; García-Montiel, E.; Moreno-Anguiano, O. Drought responsiveness in two Mexican conifer species forming young stands at high elevations. *For. Syst.* **2021**, *30*, e012. [CrossRef]
21. Fensholt, R.; Proud, S.R. Evaluation of Earth Observation based global long term vegetation trends Comparing GIMMS and MODIS global NDVI time series. *Remote Sens. Environ.* **2012**, *199*, 131–147. [CrossRef]
22. Huang, S.; Tang, L.N.; Hupy, J.P.; Wang, Y.; Shao, G.F. A commentary review on the use of normalized difference vegetation index (NDVI) in the era of popular remote sensing. *J. For. Res.* **2020**, *32*, 1–6. [CrossRef]
23. Decuyper, M.; Chávez, R.O.; Cufar, K.; Estay, S.A.; Clevers, J.G.P.W.; Prislan, P.; Gricar, J.; Crepinsek, Z.; Merela, M.; de Luis, M. Spatio-temporal assessment of beech growth in relation to climate extremes in Slovenia—An integrated approach using remote sensing and tree-ring data. *Agric. For. Meteorol.* **2020**, *287*, 107925. [CrossRef]
24. Andreu-Hayles, L.; Gaglioti, B.V.; Berner, L.T.; Levesque, M.; Anchukaitis, K.J.; Goetz, S.; D'Arrigo, R. A narrow window of summer temperatures associated with shrub growth in Arctic Alaska. *Environ. Res. Lett.* **2020**, *15*, 105012. [CrossRef]
25. Bonney, M.T.; He, Y.H. Temporal connections between long-term Landsat time-series and tree-rings in an urban–rural temperate forest. *Int. J. Appl. Earth Obs. Geoinf.* **2021**, *103*, 102523. [CrossRef]

26. Tonelli, E.; Vitali, A.; Malandra, F.; Camarero, J.J.; Colangelo, M.; Nole, A.; Ripullone, F.; Carrer, M.; Urbinati, C. Tree-ring and remote sensing analyses uncover the role played by elevation on European beech sensitivity to late spring frost. *Sci. Total Environ.* **2023**, *857*, 159239. [CrossRef]
27. Kaufmann, R.K.; D'arrigo, R.D.; Paletta, L.F.; Tian, H.Q.; Jolly, W.M.; Myneni, R.B. Identifying climatic controls on ring width: The timing of Correlations between The Timing Rings and NDVI. *Earth Interact.* **2008**, *12*, 114. [CrossRef]
28. Wu, Q.Q.; Zhang, X.; Xu, S.X.; Yang, X.H.; Liu, Y.S.; Li, H.Z.; Shi, Z.J. Climatic responses of NDVI and tree growth in the arid areas of inland Asia and their influencing factors. *J. Desert Res.* **2022**, *42*, 1–10.
29. Vicente-Serrano, S.M.; Martin-Hernandez, N.; Camarero, J.J.; Gazol, A.; Sanchez-Salguero, R.; Pena-Gallardo, M.; El Kenawy, A.; Dominguez-Castro, F.; Tomas-Burguera, M.; Gutierrez, E.; et al. Linking tree-ring growth and satellite-derived gross primary growth in multiple forest biomes. Tempor-Scale Matters. *Ecol. Indic.* **2020**, *108*, 105753. [CrossRef]
30. Bhuyan, U.; Zang, C.; Vicente-Serrano, S.M.; Menzel, A. Exploring relationships among tree-ring growth, climate variability, and seasonal leaf activity on varying timescales and spatial resolutions. *Remote Sens.* **2017**, *9*, 526. [CrossRef]
31. Vicente-Serrano, S.M.; Camarero, J.J.; Olano, J.M.; Martin-Hernandez, N.; Pena-Gallardo, M.; Tomas-Burguera, M.; Gazol, A.; Azorin-Molina, C.; Bhuyan, U.; El Kenawy, A.A. Diverse relationships between forest growth and the Normalized Difference Vegetation Index at a global scale. *Remote Sens. Environ.* **2016**, *187*, 14–29. [CrossRef]
32. Brehaut, L.; Danby, R.K. Inconsistent relationships between annual tree ring-widths and satellite-measured NDVI in a mountainous subarctic environment. *Ecol. Indic.* **2018**, *91*, 698–711. [CrossRef]
33. Coulthard, B.L.; Touchan, R.; Anchukaitis, K.J.; Meko, D.M.; Sivrikaya, F. Tree growth and vegetation activity at the ecosystem-scale in the eastern Mediterranean. *Environ. Res. Lett.* **2017**, *12*, 084008. [CrossRef]
34. Wen, Y.; Jiang, Y.; Jiao, L.; Hou, C.X.; Xu, H. Inconsistent relationships between tree ring width and normalized difference vegetation index in montane evergreen coniferous forests in arid regions. *Trees* **2022**, *36*, 379–391. [CrossRef]
35. Yuan, K.Y.; Xu, H.L.; Zhang, G.P. Is There Spatial and Temporal Variability in the Response of Plant Canopy and Trunk Growth to Climate Change in a Typical River Basin of Arid Areas. *Water* **2022**, *14*, 1573. [CrossRef]
36. Speer, J.H. *Fundamentals of Tree-Ring Research*; The University of Arizona Press: Tucson, AZ, USA, 2010.
37. Holmes, R.L. Computer-assisted quality control in tree-ring dating and measurement. *Tree Ring Bull.* **1983**, *43*, 68–78.
38. Cook, E.R. *A Time Series Analysis Approach to Tree-Ring Standardization*; The University of Arizona: Tucson, AZ, USA, 1985.
39. Fritts, H.C. *Tree Rings and Climate*; Academic Press: London, UK, 1976; p. 567.
40. Cook, E.R.; Krusic, P.J. *Program ARSTAN: A Tree-Ring Standardization Program Based on Detrending and Autoregressive Time Series Modeling, with Interactive Graphics*; Lamont-Doherty Earth Observatory, Columbia University: New York, NY, USA, 2005; p. 5.
41. Wigley, T.M.L.; Briffa, K.R.; Jones, P.D. On the average value of correlated time series, with applications in dendroclimatology and hydrometeorology. *J. Clim. Appl. Meteorol.* **1984**, *23*, 201–213. [CrossRef]
42. Wang, Z. *Plant Physiology*; China Agriculture Press: Beijing, China, 2010; pp. 314–368.
43. Jiao, L.; Ma, L.; Zhang, T.W.; Wang, S.J. Changes of mean minimum temperature in June–July since 1798 in central Altay Mountain recorded by tree rings. *Acta Ecol. Sin.* **2021**, *41*, 1944–1958.
44. Shang, J.X.; Shi, Z.J.; Gao, J.X.; Xu, L.H.; Lü, S.H.; Feng, C.Y.; Wang, L.X. Response of tree-ring width of Pinus sylvestris var. mongolica to climate change in Hulunbuir sand land, China. *Acta Ecol. Sin.* **2012**, *32*, 1077–1084. [CrossRef]
45. Zhang, Q.; Yu, R.D.; Zheng, H.W. Response Analysis of Larix sibirica to Climate Warming at Different Elevations in the Eastern Tianshan Mountains. *Bull. Bot. Res.* **2018**, *38*, 14–25.
46. Chen, F.; Yuan, Y.; Wei, W. Chronology development and climate response analysis of Schrenk Spruce (*Picea schrenkiana*) tree-ring parameters in the Urumqi River basin, China. *Geochronometria* **2010**, *36*, 17–22. [CrossRef]
47. Hauck, M.; Klinge, M.; Erasmi, S.; Dulamsuren, C. No Signs of Long-term Greening Trend in Western Mongolian Grasslands. *Ecosystems* **2023**. [CrossRef]
48. Xu, P.P.; Fang, W.; Zhou, T.; Zhao, X.; Luo, H.; Hendrey, G.; Yi, C.X. Spatial Upscaling of Tree-Ring-Based Forest Response to Drought with Satellite Data. *Remote Sens.* **2019**, *11*, 2344. [CrossRef]
49. Correa-Díaz, A.; Gómez-Guerrero, A.; Vargas-Hernández, J.J.; Rozenberg, P.; Horwath, W.R. Long-term wood micro-density variation in alpine forests at Central México and their spatial links with remotely sensed information. *Forests* **2020**, *11*, 452. [CrossRef]
50. Andreu-Hayles, L.; D'Arrigo, R.; Anchukaitis, K.J.; Beck, P.S.A.; Frank, D.; Goetz, S. Varying boreal forest response to Arctic environmental change at the Firth River, Alaska. *Environ. Res. Lett.* **2011**, *6*, 045503. [CrossRef]
51. D'Arrigo, R.D.; Malmstrom, C.M.; Jacoby, G.C.; Jacoby, G.C.; Los, S.O.; Bunker, D.E. Correlation between Maximum Latewood Density of Annual Tree Rings and NDVI Based Estimates of Forest Productivity. *Int. J. Remote Sens.* **2000**, *21*, 2329–2336. [CrossRef]
52. Bunn, A.G.; Hughes, M.K.; Kirdyanov, A.V.; Losleben, M.; Shishov, V.V.; Berner, L.; Oltchev, A.; Vaganov, E.A. Comparing forest measurements from tree rings and a space-based index of vegetation activity in Siberia. *Environ. Res. Lett.* **2013**, *8*, 035034. [CrossRef]
53. Mašek, J.; Tumajer, J.; Lange, J.; Kaczka, R.; Fiser, P.; Treml, V. Variability in Tree-ring Width and NDVI Responses to Climate at a Landscape Level. *Ecosystems* **2023**. [CrossRef]

54. Huo, Y.X.; Gou, X.H.; Liu, W.H. Climate–growth relationships of Schrenk spruce (*Picea schrenkiana*) along an altitudinal gradient in the western Tianshan mountains, northwest China. *Trees* **2017**, *31*, 429–439. [CrossRef]
55. Jiao, L.; Jiang, Y.; Wang, M.C. Responses to climate change in radial growth of *Picea schrenkiana* along elevations of the eastern Tianshan Mountains, northwest China. *Dendrochronologia* **2016**, *40*, 117–127. [CrossRef]

forests

MDPI

Article

Short-Term Tree-Ring Series of *Pinus hartwegii* Lindl. Taken at Ground Level Correlate to Normalized Difference Vegetation Index Series

Luis R. Montoya-Jiménez [1], Armando Gómez-Guerrero [1,*], Felipe J. A. Pedraza-Oropeza [2], Teresa M. González-Martínez [1] and Arian Correa-Díaz [3]

[1] Posgrado en Ciencias Forestales, Colegio de Postgraduados, Texcoco 56230, Mexico; montoya.luis@colpos.mx (L.R.M.-J.); teresa.gonzalez@colpos.mx (T.M.G.-M.)
[2] Posgrado en Hidrociencias, Colegio de Postgraduados, Texcoco 56230, Mexico; fpedraza@colpos.mx
[3] Centro Nacional de Investigación Disciplinaria en Conservación y Mejoramiento de Ecosistemas Forestales, Instituto Nacional de Investigaciones Forestales, Agrícolas y Pecuarias, Av. Progreso 5, Barrio de Santa Catarina, Delegacion Coyoacán, Mexico City 04010, Mexico; correa.arian@inifap.gob.mx
* Correspondence: agomezg@colpos.mx

Abstract: In this work, we propose that small wood core samples ($\leq$10 cm length), taken from 0.3 m above the soil, represent an accurate means of correlating forest growth with remote sensing data. Short-term correlations between the Normalized Difference Vegetation Index (NDVI) and tree-ring components of the last three to four decades were tested, using 210 *Pinus hartwegii* wood cores collected at two locations, Tláloc (TLA) and Jocotitlán (JOCO) in central Mexico. The NDVI time series were generated with the Google Earth Engine (GEE) using Landsat 8 images. Also, seasonal trends in NDVI (e.g., spring, summer, autumn, winter) were analyzed through longitudinal analysis. The results showed more statistically significant dendrochronological indices in TLA than in JOCO, but both locations consistently showed an NDVI decrease in 2018 and 2020, indicating a reduction in vegetation vigor. At the two locations, the minimum and maximum NDVI occurred in April and October, respectively. Seasonal NDVI changes for spring were mainly seen at TLA with a decreasing trend, which may be related to a less defined dry season. The significant correlations ($p < 0.05$) between tree-ring components and the NDVI occurred in the dry season, indicating that the productivity of a given year is defined by the tree vigor shown in April and May, in the case of TLA, and between January and March, for JOCO. Although the NDVI values of JOCO were higher than those of TLA, tree growth, expressed by tree-ring indices, was lower. Our proposed field method to correlate tree-ring information and the NDVI is reliable and can be used in other coniferous forests.

Keywords: highland forest; radial growth; dendrochronology; vegetation index; *Pinus hartwegii*; NDVI; tree-ring correlation methods; remote sensing

1. Introduction

Pinus hartwegii Lindl. inhabits elevated areas of Mexico and Guatemala [1]. In Mexico, it is found mainly from elevations of 3500 m along the Trans-Mexican Volcanic Belt, defining the tree-zone limit in the study area [2]. It is a species adapted to the hostile climate and soil conditions that prevail at high elevations, such as high temperature oscillations of more than 30 °C in a single day and sandy shallow soils of low fertility [3]. According to vegetation prediction models, the natural distribution area of *Pinus hartwegii* will decrease because of climate change impacts [4]. The conditions preventing the migration of this species to higher elevations include shallow and low-fertility soils with slow rates of formation. Specifically, phosphorus is an element of low availability in *Pinus hartwegii* forests [3]. In addition to the reduction in habitable area, changes in the chemical composition of plant components and nutrient dynamics are likely to occur in these forests as climate change progresses [5].

One strategy to investigate how tree growth is influenced by recent climate variation is to correlate tree-ring width measurements with remotely sensed data. Recently, dendrochronological methods have been used to explain the response of forests to climate change and whether their resistance and resilience are changing [6–8]. The historical information recorded in tree rings helps us to understand other changes occurring in the ecosystem, such as droughts [9].

The advantage of using information from remote sensing is that the vegetation vigor can be inferred from red and near-infrared bands, allowing for spatial evaluation. Remote sensing captures forest canopy reflectance and stores the information of different bands (wavelengths), allowing for the calculation of vegetation indices, which can provide information about vegetation vigor and tree growth [10].

The series of annual tree rings provide information on radial growth rates, and their correlation with vegetation indices is possible because, if the vegetation shows good vigor, it is likely that there are favorable conditions for tree growth and therefore tree rings are wider [11,12]. However, topographic variation in sites and their climatic conditions also influence the degree of correlation between rings and vegetation indices [13]. The Normalized Difference Vegetation Index (NDVI) is a well-known vegetation index that was developed by Rouse et al. (1973) and that has been shown to be reliable in representing vegetation vigor variation [12,14,15].

The correlation analysis between vegetation indices and tree-ring series is a reliable strategy to understand forest growth. Studies such as Beck, et al. [16] confirmed the link between the NDVI, ring width index (RWI), and maximum latewood density (MXD) in Alaska and western Canada. Correa-Díaz et al. [12] also confirmed significant correlations in *Pinus hartwegii* forests using MODIS sensor data.

This work aimed to find a short-term correlation between standardized tree-ring measurements and the NDVI, but, in contrast to previous studies, incorporated a high-spatial-resolution data set (using series from LANDSAT8 from 2013 to 2021) and collected short-term samples at 0.3 m above the ground (at stump height), where tree rings are wider. Two levels of elevation and aspect were considered to better capture the variation in the landscape and test our method. The NDVI variation is related to the growing season and can be useful in detecting whether extended periods of tree growth are occurring [17]. The hypotheses proposed were: (i) there are statistically significant correlations between standardized tree-ring measurements and the NDVI, and (ii) annual and seasonal variations in NDVI are different at the two studied locations, TLA and JOCO.

2. Materials and Methods

2.1. Study Area

The study sites were two mountains in central Mexico: Tláloc (TLA), located at coordinates 19°24′44.86″ N and 98°42′45.34″ W, with 4125 m of maximum elevation, and Jocotitlán (JOCO), with coordinates 19°42′26″ N 99°47′12″ W and 3910 m of maximum elevation. The vegetation at the two locations is represented by natural monospecific stands of *Pinus hartwegii* that grow from 3500 and 3700 m of elevation for TLA and JOCO, respectively (Figure 1). The predominant climate for both elevations is C(E)(W_2) (W)b(i), which is described as a semi-cold, sub-humid climate with summer rainfall [18]. A reconstruction of the climatic variables from 1980 to 2021 for the two study areas is shown in Figure S1 in the Supplementary Information. The database procedure used for this reconstruction is as indicated by Thornton et al. (2022) [19]. Daily temperatures recorded with HOBO (®) dataloggers, every four hours, from 2017 to 2023, for the two study locations, are shown in Figure S2 of the Supplementary Information.

Mean tree density for the sites is 156 and 180 tree ha^{-1}, with tree volume of 220 and 183 m^3 ha^{-1}, and mean tree age ranging from 40 to 80 years and from 52 to 89 years for TLA and JOCO, respectively [20,21].

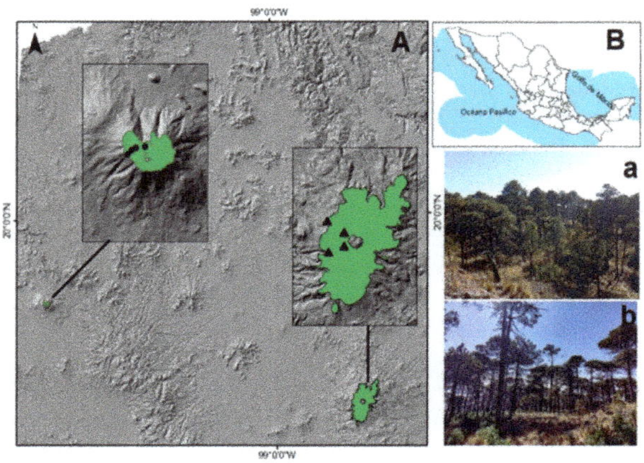

Figure 1. Location of the two study areas. (**A**) From left to right, the Jocotitlán and Tláloc mountains, highlighted with green polygons, respectively. Black triangles and circles represent sampling sites within each mountain area. (**B**) Macro location of the study areas. General views of the forests studied (**a**,**b**).

The soils at both sites are Andosols, developed from volcanic material, and are characterized by their dark color, high water-retention capacity and richness in organic matter [3]. To capture the higher variability of the landscape, sites were distributed along two elevations and two aspects. For TLA, the elevations were 3500 and 3900 m and the aspects were northwest (NW) and southwest (SW). For JOCO, the elevations were 3700 and 3800 m, with NW and SW aspects.

2.2. Dendrochronological Sampling

At each site, 30 wood cores of *P. hartwegii* were extracted at 0.30 m above ground level, ensuring that information was obtained from at least the last three or four decades (30 or 40 tree rings). The tree-ring segments were sampled in 2022, because we had to wait for the complete formation of the 2021 tree ring. It is important to note that tree-ring samples are usually taken at 1.3 m from the ground, to correlate the information with regular forest inventories, but in this study, we took samples at 0.3 m above ground level because at this height, the tree rings are wider, ensuring more accurate measurements. Nevertheless, a preliminary sampling indicated high linear correlation ($R^2 = 0.96$, $p < 0.001$) between measurements at 1.3 m and 0.3 m. Diameter at stump height (0.3) is not an unknown variable in forestry; its use can improve volume estimations, and it is useful in estimating volume removals after disturbances [22]. The diameter at stump height of the sampled trees ranged from 45 to 50 cm (9.7 sd) in TLA, and from 32 to 36 cm (6.5 sd) in JOCO.

We used a 10 cm wood core sampler (Haglöf) because it is faster to obtain tree-ring samples with wider rings that also perfectly correspond to the historical period of LANDSAT 8 images. Sampled trees were dominant and/or codominant, healthy, and without apparent mechanical damage.

2.3. Tree-Ring Sample Dating and Measurement

Wood core samples were air-dried then mounted and glued onto pieces of wood to prevent twisting. Although we planned to collect 240 tree-ring samples from the two mountains, only 210 were collected due to a recent forest fire (JOCO, one site) that allowed for dense undergrowth, creating risks to sampling due to the rocky soil surface and sloping terrain. However, historical NDVI was calculated for all the sites at the two locations.

Wood core samples were sanded with sandpaper of different grain size (80, 120, 240, 240, 280, 360, 600, and 1200) to clarify the earlywood and latewood bands. After that, samples were scanned in an EPSON scanner (HP MODEL J131B) at a resolution of 4200 dpi. Tree-ring measurements were performed with the WinDENDRO® software 2008a (Regent Instruments, Quebec, QC, Canada), following standard dendrochronological procedures [23,24].

2.4. Tree-Ring Standardization and Short-Term Chronologies

Tree-ring measurements were analyzed with the dendrochronology library dplR in R version 4.1.1 [25,26]. For detrending purposes, we used a spline function with the criteria of 50% response frequency and 67% of the length of the series [27,28]. Standardized series for ring width (TRW), earlywood (EW) and latewood (LW) were generated for both locations. As we measured the diameter of the trunk at 0.3 m and the bark width of every tree, we used the bai.out instruction of the dplR library to calculate the basal area increment (BAI), assuming concentric tree rings.

2.5. NDVI Data Processing through Google Earth Engine (GEE)

To construct the NDVI time series in GEE, we used a JavaScript code to access the Landsat 8 OLI images from 2013 to 2021 and to identify pixels contaminated with clouds, snow, or shadows. The OLI sensor contains 10 multispectral bands, with pixel sizes of 15 or 30 m, depending on the band. The NDVI is the ratio between red and near-infrared light reflected by the canopy and is an indirect measurement of photosynthesis and forest productivity. NDVI series were calculated for the two locations, and for the four sites within every location. The trends in NDVI series were analyzed for the 2013–2021 period, using average monthly values, and conducting the analysis according to the season of the year. We also found that in this study, the NDVI was statistically correlated to other indices, such as the soil-adjusted vegetation index (SAVI) and the enhanced vegetation index (EVI) (see Supplementary Information). However, for the purposes of this work, we only considered the most common vegetation index, the NDVI.

2.6. Tree Ring and NDVI Correlation

To investigate the correlation of NDVI values with the standardized measurements of TRW, EW, and LW, we used the treeclim R package [29]. Although treeclim is normally used with temperature and precipitation data, in this study, NDVI values were used to temporally correlate tree-ring components using a significant level of 5%.

2.7. Statistical Analysis of the NDVI Time Series

The NDVI time series were analyzed in two ways. First, a smoothed curve regression model (spline) was fitted with the PROC TRANSREG SAS procedure, in order to maximize correlation and to represent monthly NDVI fluctuations for the period from 2013 to 2021. The SAS output of the spline was plotted with the R package. Second, the monthly mean values for the twelve months of the year were analyzed and a cubic regression model was fitted with longitudinal analysis procedures, as explained by Fitzmaurice et al. [30]. The longitudinal analysis was conducted with the SAS software (V. 9.4), using the PROC MIXED module and an autoregressive variance–covariance matrix. The output of predicted values was plotted with an Excel spreadsheet. To investigate the seasonal changes in NDVI at each elevation, the effect of time and the time × season interaction were tested at 5% statistical significance, using longitudinal analysis [30].

3. Results and Discussion

3.1. Dendrochronological Analysis

When the data were analyzed for each location, inter-correlation series equal or greater than 0.32 were not obtained for any tree-ring component (Table 1). The best dendrochronological indices were seen for the total tree ring (TRW), and for this reason, dendroecological details for the EW and LW components are not shown. The low inter-correlation at each mountain is due to the high variability between trees and environmental conditions at each site. However, when the series were analyzed by site, and grouped by combinations of elevation and aspect, the results improved significantly. In the case of TLA, all four sites, except site S-3900, reached the minimum inter-correlation of 0.32 with only 30 trees. In the case of JOCO, the correlations did not improve when separating the series by site (Table 1).

Table 1. Dendrochronological analysis for the sample trees at the two mountain sites.

Mountain	Site	Trees	Expressed Population Signal (EPS)	Mean Sensitivity	Inter-Correlation Series	Chronology Extension	Period (Year)
TLA	3900-NO	30	0.754	0.36	0.33	37	1985–2021
	3900-SO	30	0.703	0.38	0.17	37	1985–2021
	3500-NO	30	0.348	0.31	0.40	37	1985–2021
	3500-SO	30	0.428	0.37	0.42	37	1985–2021
	(four sites)	120	0.826	0.37	0.23	37	1985–2021
JOCO	3700-NO	30	0.547	0.42	0.15	37	1985–2021
	3800-NO	30	0.65	0.38	0.21	37	1985–2021
	3800-SO	30	0.594	0.38	0.19	37	1985–2021
	(three sites)	90	0.768	0.40	0.20	37	1985–2021

The probable explanation for not finding correlations higher than 0.32 in JOCO is that on this mountain, other abiotic factors, such as soil variability and microclimate conditions, were more important in explaining the variation in tree-ring growth. Another possible explanation is that we did not collect full wood core samples down to the center or pith of the tree; we sampled the last three to four decades, when the global climate has been more variable [31]. Despite these results, it is worth highlighting the good inter-correlation for most TLA sites with a sample of 30 trees with values higher than 0.32.

The TRW series at the mountain level and for the period from 1985 to 2021 are shown in Figure 2. Despite the low inter-correlations at most sites, for both mountains, a slight increase in radial growth is observed around 2017, but this episode of good growth has decreased considerably by the year 2021. It is noticeable that, from 1985 to date, neither mountain has returned to growth levels close to the historical mean corresponding to the value of 1.0 in Figure 2, which could be evidence of the recent loss of forest resilience indicated by other authors [6].

We found an expressed population signal (EPS) lower than the threshold of 0.85 suggested by other authors [32]. This result indicates that a larger sample size is required to find a more defined trend for the tree-ring series. However, it is evident that when pooling all the ring series by mountain, the EPS reached values of 0.83 and 0.77 for TLA and JOCO, respectively. Other studies for the same species, carried out at the Trans-Mexican Volcanic Belt, have also reported EPS values of the same 0.80 to 0.88, which are close to those we found in TLA [33,34]. The mean sensitivity values at all sites were low, ranging from 0.31 to 0.42, indicating the low influence of the previous year's measurements on the actual year values. A possible explanation for this result is the high yearly climate variability in recent decades [31], which prevents the sites from reaching sensitivities that are 0.5 greater than those reported in other studies [34].

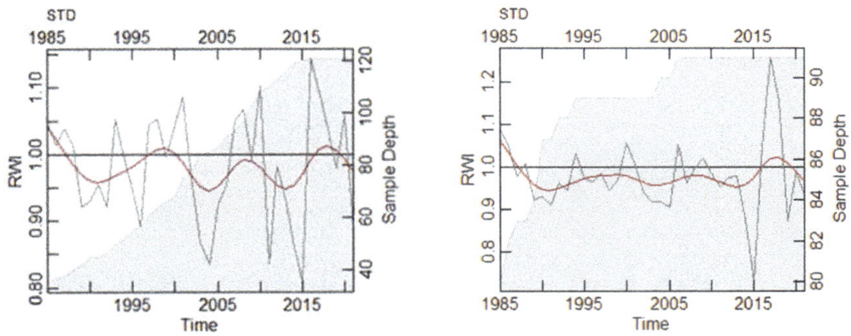

(A) (B)

Figure 2. Standardized measurements of total ring width (RWI) in the period from 1985 to 2021 for *Pinus hartwegii* on Mount Tláloc (**A**) and Jocotitlán (**B**).

3.2. Temporal Analysis of the NDVI

The monthly NDVI mean values at the two mountain sites show a seasonal change corresponding to the expected vegetation vigor variation over the year. According to the fitted longitudinal model, the extreme low and high NDVI values occur in April and October, respectively (Figure 3a). This period corresponds to the observed minimum and maximum values for *Pinus hartwegii* forests across the Trans-Mexican Volcanic Belt, which defines the growing season [17]. The historical NDVI in TLA fluctuated between 0.5 and 0.7, depending on the season of the year; however, in the dry seasons of 2013, 2018, and 2021, there was a considerable decrease in NDVI that could denote periods of stress in the vegetation (Figure 3b). The NDVI variation in JOCO is similar to that of TLA, but two points should be highlighted. First, the NDVI values for JOCO were statistically higher than those for TLA ($p < 0.05$), despite its lower tree growth, as indicated by the tree-ring index (Figures 2 and 3c, Table 2). This is extremely important since NDVI values cannot be assumed to be indicators of forest growth, as other studies have suggested [12]. Direct measurements from the tree stand are required to infer the productivity of each site. The second point is that the decrease in NDVI in JOCO is also seen in the years 2013, 2018, and 2021 (Figure 3d).

Table 2. Type III Tests of Fixed Effects for the monthly NDVI.

Effect	DF	Den DF [a]	F Value	Pr > F
TIME	1	178	90.93	<0.0001
$TIME^2$	1	178	97.69	<0.0001
$TIME^3$	1	178	83.63	<0.0001
MOUNTAIN [b]	1	178	4.23	0.0411
TIME × MOUNTAIN	1	178	1.41	0.2358

[a] Degrees of freedom associated to the model errors. [b] Mountains are Tláloc or Jocotiltán.

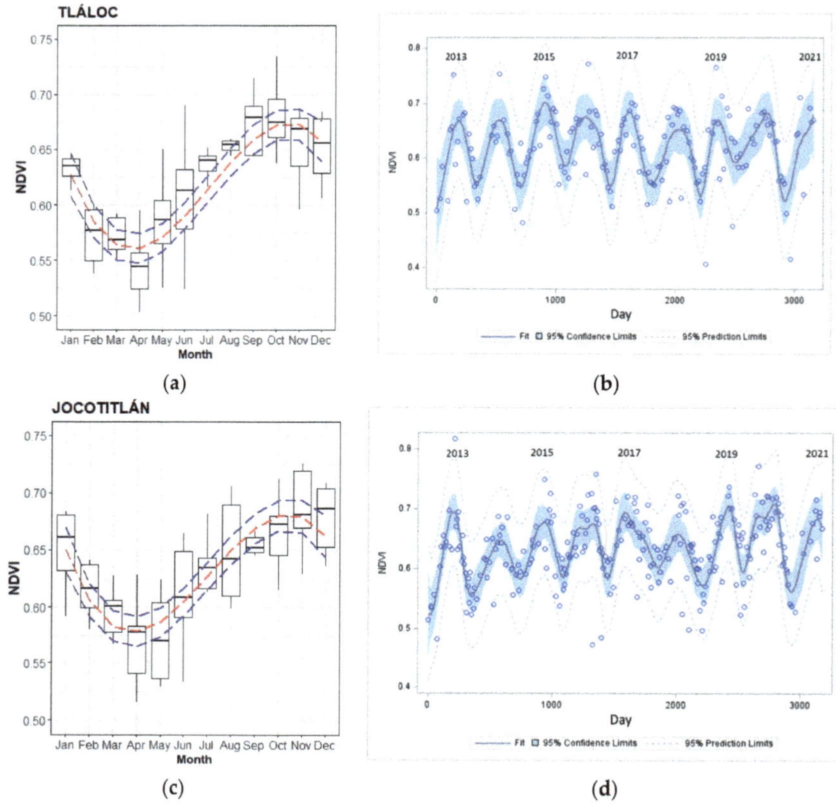

Figure 3. Box plot and cubic model fitted via longitudinal analysis (**a**,**c**). The blue lines represent the 95% confidence intervals. Fitted spline for the NDVI values at the Tláloc (**b**) and Jocotitlán (**d**) mountains for the period 2013 to 2021. The R^2 for the spline is 0.53 ($p < 0.05$) and 0.52 ($p < 0.05$) for Tláloc and Jocotitlán, respectively.

3.3. Tree-Ring and NDVI Correlation

We found significant and positive correlations ($p < 0.01$, $p < 0.05$) between the monthly NDVI and ring components at the mountain level (Table 3). As we found a highly significant correlation ($p < 0.001$) between the standardized tree-ring measurements and the BAI, we assumed these two variables as proxies of tree growth (see supplementary information). In the case of TLA, there were significant correlations from the previous year ("december") to spring of the actual year (APRIL–MAY). For JOCO, correlations were observed from winter (JANUARY) of the current year to early spring (MARCH). The correlations found were observed in all wood components (TRW, EW, LW). Our results agree with those found by Kaufmann et al. [35], who reported a statistically significant correlation between ring width and the NDVI in the Northern Hemisphere, but the authors pointed out that the correlation depends on the season. Tree species located south of 40° exhibited the best correlations in early spring, while for species located north of 40°, summer was the best season to find significant correlations. Other authors have also found significant relationships with the NDVI from the previous year "december" to March of the actual year, but using the MODIS sensor at Tlaloc Mountain [12]. Likewise, Bunn et al. [36] reported significant correlations between the NDVI and ring width for species of three genera (Pinus, Picea, and Larix) in the Siberian Taiga. This confirms the reliability of tree-ring measurements in supporting remote sensing information to better understand forest growth and related processes, like carbon sequestration [37].

Table 3. Main correlations between the tree-ring components of *Pinus hartwegii* and the NDVI at sites on two mountains, Tláloc and Jocotitlán.

Mountain	Variable	Tree-Ring Component	Correlation to the NDVI
Tláloc	$NDVI_{APRIL}$	TRW	0.664 *
	$NDVI_{MAY}$	TRW	0.866 **
	$NDVI_{MAY}$	EW	0.851 **
	$NDVI_{MAY}$	LW	0.772 *
	$NDVI_{\text{"december"}}$	LW	0.727 *
Jocotitlán	$NDVI_{MARCH}$	TRW	0.646 *
	$NDVI_{MARCH}$	EW	0.570 *
	$NDVI_{MARCH}$	LW	0.798 *
	$NDVI_{JANUARY}$	LW	0.687 *

Note: Months of the previous year are in lowercase letters, while those of the current year are in capital letters. Significance levels ($p < 0.01$ **, $p < 0.05$ *).

Some authors have reported weak and divergent correlations between the NDVI and tree-ring measurements of coniferous and deciduous species [38–40], but they used data from sensors of lower spatial resolution, which may explain this lack of significance. Mašek et al. [13] reported that tree rings and the NDVI were decoupled for *Pinus sylvetris* and *Abies picea*; however, they suggested that topography is very important in explaining NDVI variation and tree growth. This is in agreement with our results, and explains why, when our data were grouped by elevation and aspect, better dendrochronological parameters could be identified. It is also known that NDVI variation is influenced by a synergistic effect of temperature and precipitation [41]. Although, in general, the relationship between the NDVI and precipitation and temperature is positive, this result is not always consistent, especially for temperature [42]. Our results show higher historical maximum temperatures (Tmax) in JOCO, which is consistent with the positive effect on the NDVI (Figure S1). However, in both locations, the variation in Tmax has increased since 2019 (Figures S1 and S2).

The correlations between the NDVI and tree-ring components indicate that the NDVI values of the dry season (March–May) are critical in explaining the expected tree growth in a given year. The use of a standardized index for tree rings allows us to compare the NDVI and tree growth relationships for both locations. All correlations are positive, indicating that the higher the NDVI of the dry season, the higher the tree growth expected, which applies to both mountains. Kaufmann et al. [43] found a positive correlation between tree rings and the NDVI from April to July and in October in coniferous forests in North America and Eurasia, concluding that summer changes may have important implications for Northern Hemisphere forests.

However, it was not expected that the "december" NDVI of the previous year would correlate significantly with the latewood width in the subsequent year (0.73). This memory effect between the wood formed during the previous winter and the autumn of the following year is not easy to explain but can be attributed in part to the marginal memory effect of the previous and subsequent year measurements, as shown in Table 1 (mean sensitivity: 0.31 to 0.42). In addition, it is possible that the level of vigor with which forests finish the previous year determines the overall tree growth during the following year.

3.4. Seasonal Trends in NDVI

The longitudinal analysis indicated statistically significant interaction of time × season; therefore, seasonal NDVI trends have different patterns at each mountain (Figure 4). However, the most evident seasonal change was seen in TLA, as the NDVI of spring decreased in the years 2019 to 2021. The NDVI trends for JOCO are more uniform, as NDVI changes for spring and summer behave similarly, and for fall and summer, no significant changes are seen over time. Other studies have suggested that seasonal changes in NDVI in the Trans-Mexican Volcanic Belt are due to a shift in maximum NDVI values later in the year for several mountains [17]. In contrast, our data for TLA showed a trend of reduced vegetation vigor in summer. Beck et al. [16] used MODIS and NDVI data to spatially represent early spring and late autumn in northern Europe and found a trend of earlier autumn onset in southernmost Sweden. The results of this study show that autumn has been the most stable season for both mountains. The distribution of precipitation over time in TLA is greater, even compared to JOCO (Figure S1). This fact could explain why TLA is more sensitive to seasonal changes in NDVI that depend on precipitation. The JOCO location shows two sharp dry seasons, before and after the growing season, and that may be related to the higher NDVI. Additionally, TLA is a site more affected by changes in temperature and precipitation in the most recent years (Figures S1 and S2).

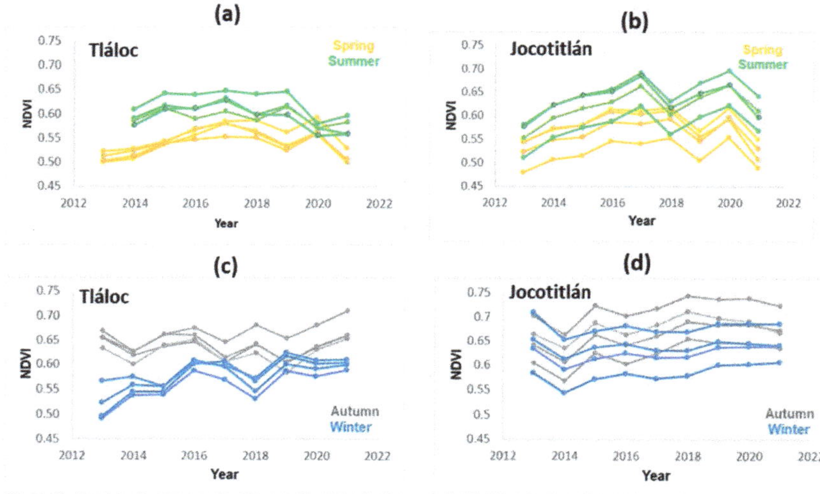

Figure 4. Predicted values of NDVI trends estimated via longitudinal analysis.

4. Conclusions

Short-term series of radial growth taken at the base of the tree (0.3 m from the ground) with a small wood core sample (10 cm length) had acceptable dendrochronological parameters and allowed us to determine a statistical correlation to remote sensing data. Standardized short-term tree-ring series are statistically correlated to the NDVI and reliable for understanding tree growth. The annual and seasonal NDVI were different between the mountains, but the NDVI of the dry season is what defines the annual tree growth for each site. The location with more defined dry periods, before and after the growing season, showed a higher NDVI. The previous year's "december" NDVI correlated with the latewood width of the actual year. Although this result was not fully explained, it may be related to a natural memory effect that occurs in coniferous forests. Most importantly, the NDVI cannot always be used as a proxy of forest growth without terrain-based data, like tree-ring measurements. In this study, the Jocotitlán mountain had statistically higher NDVI values than the Tláloc mountain, but its tree growth was lower. Therefore, for this type of study, it is essential to calibrate the information from remote sensing with the direct measurement of tree parameters, such as tree-ring width.

Supplementary Materials: The following supporting information can be downloaded at: https://www.mdpi.com/article/10.3390/f15020324/s1, Figure S1: Temporal distribution of climatic variables for the two study sites, from 1980 to 2021. (a) Daily precipitation, (b) Minimum temperature and (c) Maximum temperature. Estimations were performed according to Thornton et al. (2022) [19]. Note that daily precipitation is more even distributed at TLÁLOC site, while at JOCOTILÁN, two sharp dry periods are defined before and after the growing season. Minimum temperatures in the growing season are higher at JOCOTILTÁN. Also note that at TLÁLOC maximum temperatures, before the growing season have increased in recent years; Figure S2: Daily temperatures recorded with HOBO (®) dataloggers every four hours for three sites in the TLÁLOC location. Note the increase in temperatures in the year 2020 and 2021, specially in plots (a,c); Figure S3: Daily temperatures recorded with HOBO (®) dataloggers every four hours for three sites in the JOCOTILÁN location. Note the higher contrast in temperature variation between the 3800-SW site (a) and the 3700-NO site (c); Figure S4: The standardized values of ring width (RWI) are highly correlated to Basal area increments (BAI) indicating that RWI is a reasonable a proxy of tree growth. BAI area in cm^2 $year^{-1}$; Table S1: Correlation Matrix for the vegetation indices, NDVI, EVI and SAVI for the two studied mountains. Note that in all cases the correlation is highly significant ($p < 0.05$), implying that all indices would generate similar results and conclusions. However, we emphasized on one of the most common vegetation indices like the NDVI.

Author Contributions: Conceptualization, A.G.-G.; Data Curation, A.G.-G. and L.R.M.-J.; Investigation, A.G.-G. and L.R.M.-J.; Original draft, A.G.-G. and L.R.M.-J.; Writing—review and editing, A.G.-G., A.C.-D., T.M.G.-M. and F.J.A.P.-O.; Methodology, A.G.-G. and L.R.M.-J.; Visualization, A.G.-G. and A.C.-D. All authors have read and agreed to the published version of the manuscript.

Funding: This research was partially funded by the National Council of Humanities, Science and Technology (CONAHCYT) with a graduate studies scholarship granted to the first author.

Data Availability Statement: The data presented in this study are available upon request from the corresponding author.

Conflicts of Interest: The authors declare no conflicts of interest.

References

1. García-Amorena, I.; Moreno-Amat, E.; Aulló-Maestro, M.E.; Mateo-Sánchez, M.C.; Merino-De-Miguel, S.; Ribalaygua, J.; Marchant, R. Combining Remote Sensing and Species Distribution Modelling to Assess Pinus hartwegii Response to Climate Change and Land Use from Izta-Popo National Park, Mexico. *Land* **2021**, *10*, 1037. [CrossRef]
2. Farjon, A.; de la Rosa, J.A.P.; Styles, B.T. *A Field Guide to the Pines of Mexico and Central America*; Royal Botanic Gardens: Kew, UK, 1997.
3. Marín, L.E.; Escolero-Fuentes, O.; Trinidad-Santos, A. Physical geography, hydrogeology, and forest soils of the basin of Mexico. In *Urban Air Pollution and Forests: Resources at Risk in the Mexico City Air Basin*; Springer: Berlin/Heidelberg, Germany, 2002; pp. 44–67.
4. Guitérrez, E.; Trejo, I. Efecto del cambio climático en la distribución potencial de cinco especies arbóreas de bosque templado en México. *Rev. Mex. Biodivers.* **2014**, *85*, 179–188. [CrossRef]
5. Gómez-Guerrero, A.; Doane, T. The response of forest ecosystems to climate change. In *Developments in Soil Science*; Elsevier: Amsterdam, The Netherlands, 2018; Volume 35, pp. 185–206.
6. Correa-Díaz, A.; Villanueva-Díaz, J.; Gómez-Guerrero, A.; Martínez-Bautista, H.; Castruita-Esparza, L.; Horwath, W.; Silva, L. A comprehensive resilience assessment of Mexican tree species and their relationship with drought events over the last century. *Glob. Chang. Biol.* **2023**, *29*, 3652–3666. [CrossRef]
7. Gómez-Guerrero, A.; Correa-Díaz, A.; Castruita-Esparza, L. Cambio Climático y Dinámica de Los Ecosistemas Forestales. *Rev. Fitotec. Mex.* **2021**, *44*, 673–682. [CrossRef]
8. Lloret, F.; Keeling, E.G.; Sala, A. Components of tree resilience: Effects of successive low-growth episodes in old ponderosa pine forests. *Oikos* **2011**, *120*, 1909–1920. [CrossRef]
9. Gartner, B.L.; Aloni, R.; Funada, R.; Lichtfuss-Gautier, A.N.; Roig, F.A. Clues for dendrochronology from studies of wood structure and function. *Dendrochronologia* **2002**, *20*, 53–61. [CrossRef]
10. Pandey, P.C.; Sharma, L.K. *Advances in Remote Sensing for Natural Resource Monitoring*; John Wiley & Sons: Hoboken, NJ, USA, 2021.
11. Vicente-Serrano, S.M.; Camarero, J.J.; Olano, J.M.; Martín-Hernández, N.; Peña-Gallardo, M.; Tomás-Burguera, M.; Gazol, A.; Azorin-Molina, C.; Bhuyan, U.; El Kenawy, A. Diverse relationships between forest growth and the Normalized Difference Vegetation Index at a global scale. *Remote Sens. Environ.* **2016**, *187*, 14–29. [CrossRef]
12. Correa-Díaz, A.; Silva, L.C.R.; Horwath, W.R.; Gómez-Guerrero, A.; Vargas-Hernández, J.; Villanueva-Díaz, J.; Velázquez-Martínez, A.; Suárez-Espinoza, J. Linking remote sensing and dendrochronology to quantify climate-induced shifts in high-elevation forests over space and time. *J. Geophys. Res. Biogeosci.* **2019**, *124*, 166–183. [CrossRef]
13. Mašek, J.; Tumajer, J.; Lange, J.; Kaczka, R.; Fišer, P.; Treml, V. Variability in Tree-ring Width and NDVI Responses to Climate at a Landscape Level. *Ecosystems* **2023**, *26*, 1144–1157. [CrossRef]

14. Gessesse, A.A.; Melesse, A.M. Temporal relationships between time series CHIRPS-rainfall estimation and eMODIS-NDVI satellite images in Amhara Region, Ethiopia. In *Extreme Hydrology and Climate Variability*; Elsevier: Amsterdam, The Netherlands, 2019; pp. 81–92.
15. Ma, Y.; Guan, Q.; Sun, Y.; Zhang, J.; Yang, L.; Yang, E.; Li, H.; Du, Q. Three-dimensional dynamic characteristics of vegetation and its response to climatic factors in the Qilian Mountains. *Catena* **2022**, *208*, 105694. [CrossRef]
16. Beck, P.S.; Andreu-Hayles, L.; D'Arrigo, R.; Anchukaitis, K.J.; Tucker, C.J.; Pinzón, J.E.; Goetz, S.J. A large-scale coherent signal of canopy status in maximum latewood density of tree rings at arctic treeline in North America. *Glob. Planet. Chang.* **2013**, *100*, 109–118. [CrossRef]
17. Correa-Díaz, A.; Romero-Sánchez, M.; Villanueva-Díaz, J. The greening effect characterized by the Normalized Difference Vegetation Index was not coupled with phenological trends and tree growth rates in eight protected mountains of central Mexico. *For. Ecol. Manag.* **2021**, *496*, 119402. [CrossRef]
18. García, E. *Modificaciones al Sistema de Clasificación Climática de Köpen (Para Adaptarlo a las Condiciones de la República Mexicana)*, 5th ed.; Instituto de Geografía UNAM: Mexico City, Mexico, 2004; p. 98.
19. Thornton, M.; Shrestha, R.; Wei, Y.; Thornton, P.; Kao, S.; Wilson, B. *Daymet: Daily Surface Weather Data on a 1-Km Grid for North America*; Version 4; ORNL DAAC: Oak Ridge, TN, USA, 2022. [CrossRef]
20. Acosta Mireles, M.; Carrillo Anzures, F.; Delgado, D.; Velasco Bautista, E. Establecimiento de parcelas permanentes para evaluar impactos del cambio climático en el Parque Nacional Izta-Popo. *Rev. Mex. Cienc. For.* **2014**, *5*, 6–29.
21. Nuñez-Garcia, A.; Gomez-Guerrero, A.; Terrazas-Salgado, T.M.; Vargas-Hernández, J.J.; Villanueva-Diaz, J. Analysis of basal area increment of Pinus hartwegii Lindl. at different elevations and aspects on Jocotitlan Mountain, State of Mexico. *Rev. Chapingo Ser. Ciencias For. Ambient* **2020**, *27*, 77–88.
22. Pond, N.C.; Froese, R.E. Evaluating published approaches for modelling diameter at breast height from stump dimensions. *For. Int. J. For. Res.* **2014**, *87*, 683–696. [CrossRef]
23. Fritts, H. *Rings and Climate*; Academic Press: Cambridge, MA, USA, 1976.
24. Stokes, M.; Smiley, T. *An Introduction to Tree-Ring Dating*; University of Arizona Press: Tucson, AZ, USA, 1968.
25. Bunn, A.G. A dendrochronology program library in R (dplR). *Dendrochronologia* **2008**, *26*, 115–124. [CrossRef]
26. Bunn, A.G. Statistical and visual crossdating in R using the dplR library. *Dendrochronologia* **2010**, *28*, 251–258. [CrossRef]
27. Cook, E.R. The decomposition of tree-ring series for environmental studies. *Tree-Ring Bull.* **1987**, *47*, 37–59.
28. Cook, E.R.; Peters, K. The smoothing spline: A new approach to standardizing forest interior tree-ring width series for dendroclimatic studies. *Tree-Ring Bull.* **1981**, *41*, 45–53.
29. Zang, C.; Biondi, F. treeclim: An R package for the numerical calibration of proxy-climate relationships. *Ecography* **2015**, *38*, 431–436. [CrossRef]
30. Fitzmaurice, G.M.; Laird, N.M.; Ware, J.H. *Applied Longitudinal Analysis*; Wiley: Hoboken, NJ, USA, 2004.
31. Williams, A.P.; Allen, C.D.; Macalady, A.K.; Griffin, D.; Woodhouse, C.A.; Meko, D.M.; Swetnam, T.W.; Rauscher, S.A.; Seager, R.; Grissino-Mayer, H.D. Temperature as a potent driver of regional forest drought stress and tree mortality. *Nat. Clim. Chang.* **2013**, *3*, 292–297. [CrossRef]
32. Briffa, K.R.; Jones, P.D. *Basic Chronology Statistics and Assessment in Methods of Dendrochronology: Applications in the Environmental Sciences*; Cook, E.R., Kairiukstis, L.A., Eds.; Kluwer Academic Publishers: Boston, MA, USA, 1990; pp. 137–152.
33. Manzanilla-Quinones, U.; Aguirre-Calderón, Ó.A.; Jimenez-Perez, J.; Treviño-Garza, E.J.; Yerena-Yamallel, J.I. Current and future distribution of the Pinus hartwegii Lindl subalpine forest in the Tranverse Neovolcanic Belt. *Madera Bosques* **2019**, *25*, e2521804.
34. Manzanilla-Quiñones, U.; Aguirre-Calderón, Ó.A.; Jiménez-Pérez, J.; Villanueva-Díaz, J. Sensibilidad climática en anchuras de anillos de crecimiento de Pinus hartwegii: Una especie alpina mexicana con potencial dendroclimático. *Rev. Mex. Biodivers.* **2020**, *91*, abril-junio. [CrossRef]
35. Kaufmann, R.; D'arrigo, R.; Paletta, L.; Tian, H.; Jolly, W.M.; Myneni, R. Identifying climatic controls on ring width: The timing of correlations between tree rings and NDVI. *Earth Interact.* **2008**, *12*, 1–14. [CrossRef]
36. Bunn, A.G.; Hughes, M.K.; Kirdyanov, A.V.; Losleben, M.; Shishov, V.V.; Berner, L.T.; Oltchev, A.; Vaganov, E.A. Comparing forest measurements from tree rings and a space-based index of vegetation activity in Siberia. *Environ. Res. Lett.* **2013**, *8*, 035034. [CrossRef]
37. Evans, M.E.; DeRose, R.J.; Klesse, S.; Girardin, M.P.; Heilman, K.A.; Alexander, M.R.; Arsenault, A.; Babst, F.; Bouchard, M.; Cahoon, S.M. Adding tree rings to North America's national forest inventories: An essential tool to guide drawdown of atmospheric CO_2. *Bioscience* **2022**, *72*, 233–246. [CrossRef]
38. Brehaut, L.; Danby, R.K. Inconsistent relationships between annual tree ring-widths and satellite-measured NDVI in a mountainous subarctic environment. *Ecol. Indic.* **2018**, *91*, 698–711. [CrossRef]
39. D'Andrea, G.; Šimůnek, V.; Castellaneta, M.; Vacek, Z.; Vacek, S.; Pericolo, O.; Zito, R.G.; Ripullone, F. Mismatch between annual tree-ring width growth and NDVI index in Norway spruce stands of Central Europe. *Forests* **2022**, *13*, 1417. [CrossRef]
40. Wen, Y.; Jiang, Y.; Jiao, L.; Hou, C.; Xu, H. Inconsistent relationships between tree ring width and normalized difference vegetation index in montane evergreen coniferous forests in arid regions. *Trees* **2022**, *36*, 379–391. [CrossRef]
41. Chen, Y.; Ma, L.; Liu, T.; Huang, X.; Sun, G. The Synergistic Effect between Precipitation and Temperature for the NDVI in Northern China from 2000 to 2018. *Appl. Sci.* **2023**, *13*, 8425. [CrossRef]

42. Garai, S.; Khatun, M.; Singh, R.; Sharma, J.; Pradhan, M.; Ranjan, A.; Rahaman, S.M.; Khan, M.L.; Tiwari, S. Assessing correlation between Rainfall, normalized difference Vegetation Index (NDVI) and land surface temperature (LST) in Eastern India. *Saf. Extrem. Environ.* **2022**, *4*, 119–127. [CrossRef]
43. Kaufmann, R.K.; D'Arrigo, R.D.; Laskowski, C.; Myneni, R.B.; Zhou, L.; Davi, N. The effect of growing season and summer greenness on northern forests. *Geophys. Res. Lett.* **2004**, *31*, L09205. [CrossRef]

forests

MDPI

Article

Daily Climatic Data Better Explain the Radial Growth of Swiss Stone Pine (*Pinus cembra* L.) in High-Elevation Cliffs in the Carpathians

Katarzyna Izworska [1,2,*], Tomasz Zielonka [1], Paweł Matulewski [3] and Elżbieta Muter [4]

1 Institute of Biology and Earth Sciences, Pedagogical University of Cracow, Podchorążych 2, 30-084 Kraków, Poland; tomasz.zielonka@up.krakow.pl
2 W. Szafer Institute of Botany, Polish Academy of Sciences, Lubicz 46, 31-512 Kraków, Poland
3 Institute of Geoecology and Geoinformation, Faculty of Geographical and Geological Sciences, Adam Mickiewicz University, B. Krygowskiego 10, 61-680 Poznań, Poland; pawel.matulewski@amu.edu.pl
4 Department of Forest Biodiversity, Faculty of Forestry, University of Agriculture in Krakow, 29 Listopada 46, 31-425 Kraków, Poland; elzbieta.muter@urk.edu.pl
* Correspondence: k.izworska@botany.pl

Abstract: Information about climate–growth relationships is crucial for predicting the potential climatic impact on tree species, especially those growing on the edges of their distribution range, for instance, in high-elevation forests. This study aimed to determine changes in the relationships between tree-ring widths and daily climatic data in high-elevation forests in the Western Carpathians over time. Climate–growth relationships were calculated to obtain the TRWI (tree-ring-width index) chronology (based on 104 trees) and day-wise aggregated data for temperature (mean, minimum, and maximum) and sums of precipitation. The radial growth of stone pine was mostly determined by the mean temperature in the period between mid-June (21st) and the beginning of July (4th) for the critical 14-day window width ($r = 0.44$). The negative influence of precipitation on the radial growth occurred in summer ($r = -0.35$) and overlapped with the period of the positive influence of temperature. Dendroclimatic studies based on daily data may define the exact periods (expressed in calendar days) that influence the radial growth of trees much better than the commonly used monthly means. This is particularly important in analysing the growth of trees at high elevations, where the climatic factor strongly limits radial growth.

Keywords: climate change; cliff forest; daily climatic data; dendrochronology; *Pinus cembra*; tree rings

1. Introduction

Climate change is a phenomenon observed worldwide in recent times. It is related to increases in annual temperature and the distribution of precipitation [1–3]. Although we are still far from the Holocene climatic optimum, recent models predict further progress towards higher temperature and lower precipitation [4–6].

Thus, it is predicted that the distribution range of tree species will be changing parallel to climate change. The possible delays in tree responses will be minor or the range changes will be unable to keep up with climate change; therefore, the adaptation of species in accordance with environmental conditions will significantly worsen [7–11]. The frequency and intensity of extreme climatically induced events are also expected to change [4,12,13]. Climate change is widespread, affecting forests within a variety of geographical regions where warmth (e.g., in mountain or boreal ranges) or water availability (e.g., in arid and Mediterranean areas) are limiting environmental factors [13]. Trees on the edge of their distribution range are subjected to environmental stress, and their growth response to climate can reveal ecological thresholds beyond which they may not be able to persist. This information is critical for predicting the potential impacts of climate change on the distribution of tree species and the vitality of species populations.

Moreover, climate impacts on forests range from single extreme events with a difficult-to-predict time and location, to more obvious gradual changes. Therefore, it is necessary to conduct research on different spatial scales, from local to global, with the use of data of varying resolutions and species. Retrospective tree-ring-based studies are a useful tool for investigating how trees may adapt to forecasted changes [5,14–16]. A long-living organisms exposed to environmental variability, trees seem to be the best indicators of the reactions of living organisms to climate change [14,15]. This is especially important for trees growing on the edge of their distribution range, such as the north and upper timberlines.

High-elevation forests are the sentinels of change due to their location at the edges of their distribution range. They react quickly and strongly to environmental changes, including climate [17]. Extreme weather conditions are common in high-elevation ecosystems, due to the large amplitude of day–night and seasonal temperatures [17–19]. A low nutrient supply, as well as significant disruptions produced by gravity—rock and avalanche, snowfall and snow accumulation, and winds—contribute to the harsh environment of cliffs [17]. The Swiss stone pine (*Pinus cembra* L.) is a taxon with scattered populations (the Alps and Carpathian Mountains), exhibiting the finest adaptations for colonisation and growth in high-elevation cliffs [9,20–24]. Previous studies from the Alps [9,24–26] and Carpathian Mountains [20,27,28] have shown that the response of this species to climate changes is not stable over time.

The analysis of the relationship between radial growth and climate has typically been carried out using monthly meteorological data, e.g., [25,26,29,30]. Such data can provide a broader perspective on the long-term relationship between growth and climate, as it captures seasonal and annual fluctuations in environmental conditions. It is useful to identify general patterns and trends in growth responses to climate variables over time. However, it is important to consider the limitations of monthly data and its disadvantage to miss short-term climatic effects. Nevertheless, most of these studies have used the mean monthly climatic data when the months are artificially time-compressed [31–33]. Methodological improvements, access to gridded climate datasets and computing power to manage big databases have allowed for conducting more comprehensive studies with high-resolution data sets [16]. Daily climatic data offer a more detailed view of environmental conditions, which can help identify subtle effects of climate on tree growth that may be missed with monthly data. The daily resolution of climatic data detects the exact days of the year when temperature and precipitation significantly influence the radial growth of trees [32,33]. Overall, the use of daily climatic data can be valuable in analysing climate–growth relationships, but it requires careful consideration of advantages and disadvantages and the research question addressed.

This study aimed to enhance the precision and temporal resolution of the climate–growth correlation for Swiss stone pine in an extreme environment of highly elevated cliff forests. Specifically, we aimed to (1) examine climate–growth relationships using a daily response function, (2) determine the exact days of the year in which climatic parameters significantly affect growth, and (3) explore how the growth response to climate changes over time (throughout the last 90 years). It was hypothesised that the strength of the climate–growth relationship decreased in the warming climate over the last decades.

2. Materials and Methods

2.1. Study Area and Climate

The study site was located in the upper treeline ecotone in the Tatras (the Western Carpathian Mountains) (49°12′54.48″ N; 20°4′34.68″ E) (Figure 1). The Tatra Mountains, having classic alpine terrain and elevations of over 2600 m a.s.l., represent the highest and most rugged Carpathian mountain range [22,34]. The studied stone pines grow in high-elevation cliff forests on steep, granite slopes (ca. 70–80°) between the subalpine zone and dwarf pine scrub (*Pinus mugo* Turra) (ca. 1480 m a.s.l.). The study area was described in detail in the previous paper analysing the same tree samples [20]. It is a protected area of the Tatra National Park under a temperate-continental climate zone with significant differences between the daily maximum and minimum temperature and large amounts of precipitation [34]. The annual mean temperature of the sample site was 3.32 °C covering the period of 1921–2009, and the sum of annual precipitation was 1212 mm (Figure 2a,b).

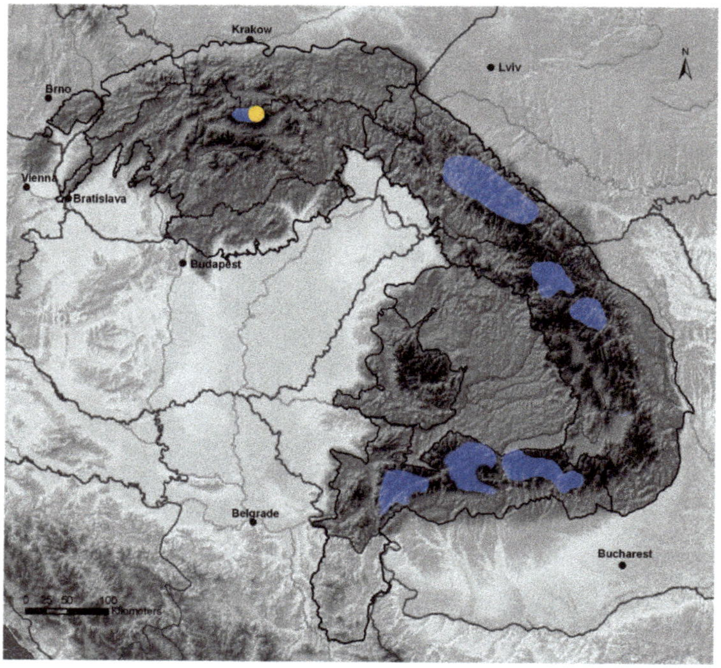

Figure 1. Map showing study site location (yellow dot) and the distribution of *Pinus cembra* in the Carpathian Mountains (blue areas) [35].

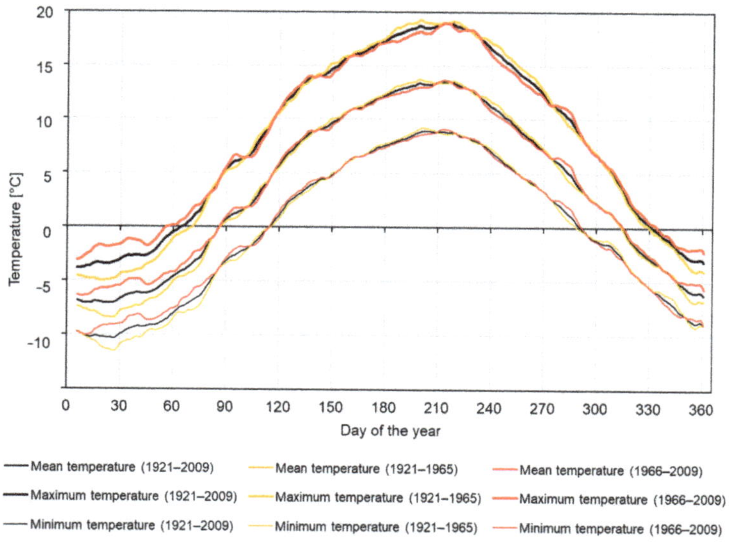

(**a**)

Figure 2. *Cont.*

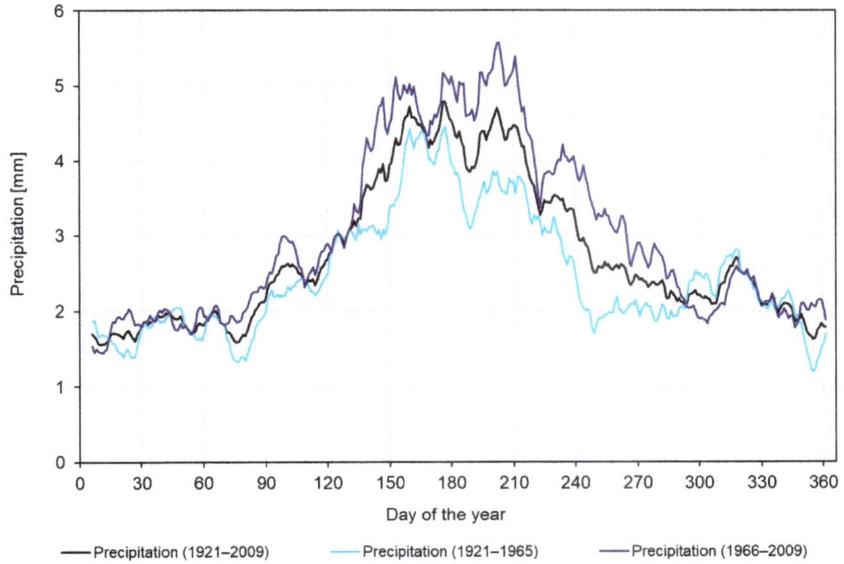

(b)

Figure 2. Climatic diagrams for the analysed period (1921–2009) and two subperiods (1921–1965 and 1966–2009) based on daily data for the study area: (**a**) Eleven-day running means of the daily mean, maximum, and minimum temperature; and (**b**) eleven-day running means of the daily sums of precipitation.

2.2. Tree-Ring and Climatic Data

The increment cores were extracted at breast height (ca. 1.3 m) from stone pine trees growing in the study area with a Pressler borer. The cores were labelled, sanded, and then scanned at a 2400 DPI resolution. The ring widths were measured using the WinDendro software (https://www.regentinstruments.com/assets/windendro_about.html (accessed on 21 April 2023)). The accuracy of cross-dating was checked with COFECHA [36,37]. Time series that did not correlate with others were excluded to avoid potential errors in cross-dating. The diameter at breast height (DBH), height, and age of the studied trees as well as the tree-ring width (TRW) chronology characteristics were comprehensively described by Izworska et al. [20].

Daily climatic data: Temperature (mean, maximum, and minimum) and a sum of precipitation were obtained from the closest grid point for the study site (49°12′54.48″ N; 20°4′34.68″ E) from the E-OBS gridded climate dataset [38]. E-OBS version 25.0e (0.1-degree regular grid) covers the time from 1 January 1920 to now.

2.3. Analysis of Daily Climate–Growth Relationships

Climate–growth relationships were calculated using residual chronology (TRWI—tree-ring-width index) for the 1921–2009 period. To remove the age-related growth trends and competition effect, detrending was performed using the spline function (frequency response of 0.50 and cut off at 0.67 series length) in the 'dplR' (in R version 4.0.0) [39,40]. Statistics of the detrended chronology for the climate analysis (1921–2009), including interseries correlation (Rbar), expressed population signal (EPS), subsample signal strength (SSS) [41], and mean sensitivity (MS) [42], were calculated using the 'dplR' with the 'rwi.stats' functions [43].

Climate–growth relationships for the TRWI chronology and day-wise data for temperature (mean, maximum, and minimum) and the sum of precipitation during the 1921–2009 period, and the two subperiods of 1921–1965 and 1966–2009, were analysed. These subperiods were chosen as a comparable length (45 years and 44 years, respectively) series out of the 89-year climatic data to indicate possible changes in the growth reaction due to climate change over the last decades. We used the 'daily_response()' function from the 'dendroTools' [31,40]. The primary objective of the 'daily_response()' function was to examine changes over time in the relationships between tree rings and daily climatic data. The function used a moving window through daily data and aggregated data within each window by calculating its averages and then calculating statistical metrics (i.e., correlation coefficient). The moving window was based on the window width (number of days) and daily data within the position of the matrix [31]. The daily correlations were computed by taking into account all window widths ranging from 7 to 210 consecutive days from July of the previous year to September of the year of growth. To calculate partial correlations, a bootstrap procedure was applied within 1000 replicates.

3. Results

3.1. Site-Specific Chronology

The TRWI chronology for the 1921–2009 period covered the available daily climatic data (Figure 3); the statistical parameters are provided in Table 1. The chronology shows the EPS value, as well as SSS exceeding the threshold of 0.85, indicating the strong climate signal in the site chronology and allowing us to conduct further climatic analyses [44,45]. The relatively low value of Rbar might be because our sampled trees were not a homogenous group with a large number of samples (104 trees). Alongside that, the trees were of different ages and growing in various microhabitats, reflecting individual growth conditions in the cliffs. Rbar values are usually higher in more homogenous groups/areas.

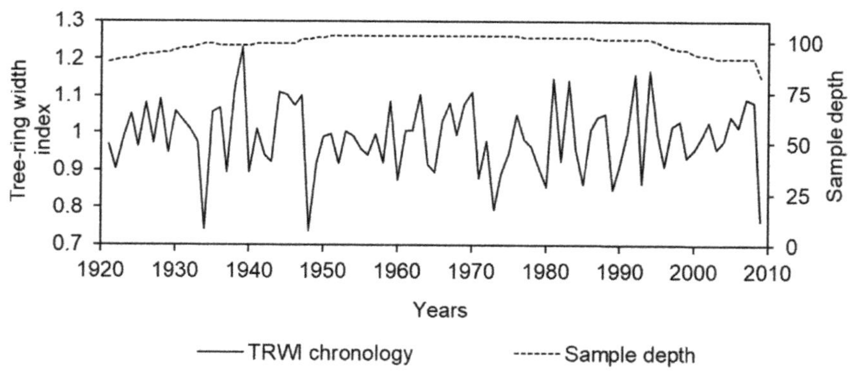

Figure 3. The TRWI chronology (solid line) and sample depth (dotted line) for 1921–2009 period.

Table 1. Statistics of the detrended chronology for the analysed period (1921–2009) and two subperiods (1921–1965 and 1966–2009).

Parameter	Number of Years	Number of Trees	Interseries Correlation (Rbar)	Expressed Population Signal (EPS)	Subsample Signal Strength (SSS)	Mean Sensitivity (MS)
1921–2009	89	104	0.16	0.951	0.998	0.196
1921–1965	45	104	0.17	0.953	0.997	0.187
1966–2009	44	104	0.14	0.942	0.998	0.188

3.2. Climate–Growth Relationships

3.2.1. Mean Temperature

The highest positive significant correlation coefficient (r = 0.44) between the TRWI chronology and the mean temperature was found for the 14-day window width between 21 June and 4 July (DOY 172–185) during the year of growth (Figure 4). However, an important period with a high correlation (r > 0.40) was found for the window width range between 14 and 50 days, spanning mid-June to the beginning of August (DOY 168–222). Additionally, tree growth was positively influenced by the mean temperature of autumn the previous year (0.20 < r < 0.30), and an important period with relatively high correlation coefficients for this period spanned mid-September to the end of October (DOY 257–306; 39–46-day window width) (Figure 4).

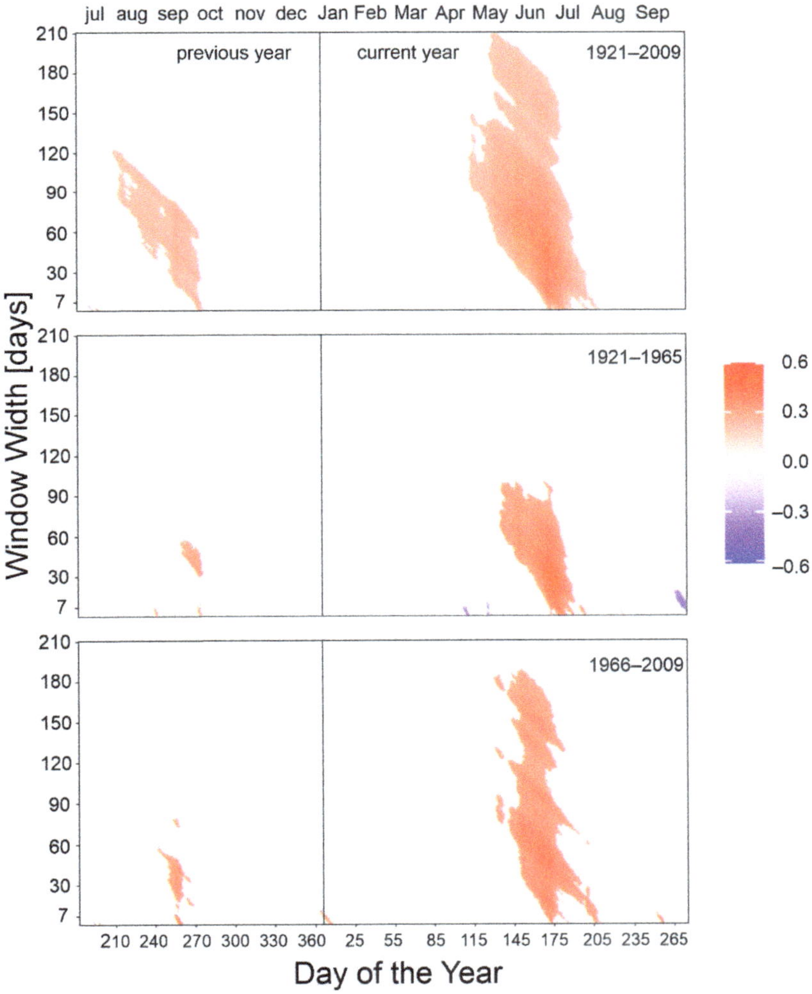

Figure 4. Correlations between the mean daily temperature and the TRWI chronology for the analysed periods. The colours show a significant ($p < 0.05$) correlation coefficient for consecutive window widths. The given values show the beginning of the specific time window.

In the first subperiod (1921–1965), the highest positive significant correlation coefficient (r = 0.51) between the TRWI chronology and the mean temperature was found for the 29-day window width between 21 June and 19 July (DOY 172–200) during the year of growth (Figure 4). The period with a high correlation (r > 0.40) was found for the window width range between 10 and 33 days and spanned the end of June to the end of July (DOY 172–211). Tree growth was positively influenced by the mean temperature of autumn the previous year (0.30 < r < 0.40), and an important period with relatively high correlation coefficients was found for the period spanning the end of September to mid-November (DOY 272–316; 35–43-day window width) (Figure 4).

In the second subperiod (1966–2009), the highest positive significant correlation coefficient (r = 0.44) between the TRWI chronology and the mean temperature was found for the 14-day window width between 20 June and 3 July (DOY 171–184) during the year of growth (Figure 4). The period with a high correlation (r > 0.40) was found for the window width range between 14 and 47 days and spanned mid-June to the beginning of August (DOY 168–218). Tree growth was positively influenced by the mean temperature of autumn the previous year (0.30 < r < 0.40), and an important period with relatively high correlation coefficients was found for the period spanning mid-September to the end of October (DOY 256–300; 23–43-day window width) (Figure 4).

3.2.2. Minimum Temperature

The highest positive significant correlation coefficient (r = 0.37) between the TRWI chronology and the minimum temperature was found for the 14-day window width between 21 June and 4 July (DOY 172–185) during the year of growth (Figure 5). However, an important period with a high correlation (r > 0.35) was found for the window width range between 14 and 38 days, spanning mid-June to the end of July (173–192). Additionally, tree growth was positively influenced by the minimum temperature at the end of summer and autumn the previous year (0.25 < r < 0.30). There were two important periods with relatively high correlation coefficients (r > 0.25), spanning the end of September to the end of October (DOY 267–302, 30–35-day window width) and mid-August to the first days of November (DOY 232–307; 64–70-day window width) (Figure 5).

In the first subperiod (1921–1965), the highest positive significant correlation coefficient (r = 0.49) between the TRWI chronology and the minimum temperature was found for the 13-day window width between 22 June and 4 July (DOY 173–185) during the year of growth (Figure 5). However, an important period with a high correlation (r > 0.45) was found for the window width range between 11 and 23 days, spanning the end of June to the end of July (DOY 171–202). There were two important periods with relatively high correlation coefficients (r > 0.33), spanning the end of September to mid-November (DOY 272–324, 7–50-day window width) and mid-August to mid-November (DOY 231–316; 73–83-day window width) (Figure 5).

No distinct impact of minimum temperature was observed in the second subperiod (1966–2009) during the year of growth and the previous year (Figure 5).

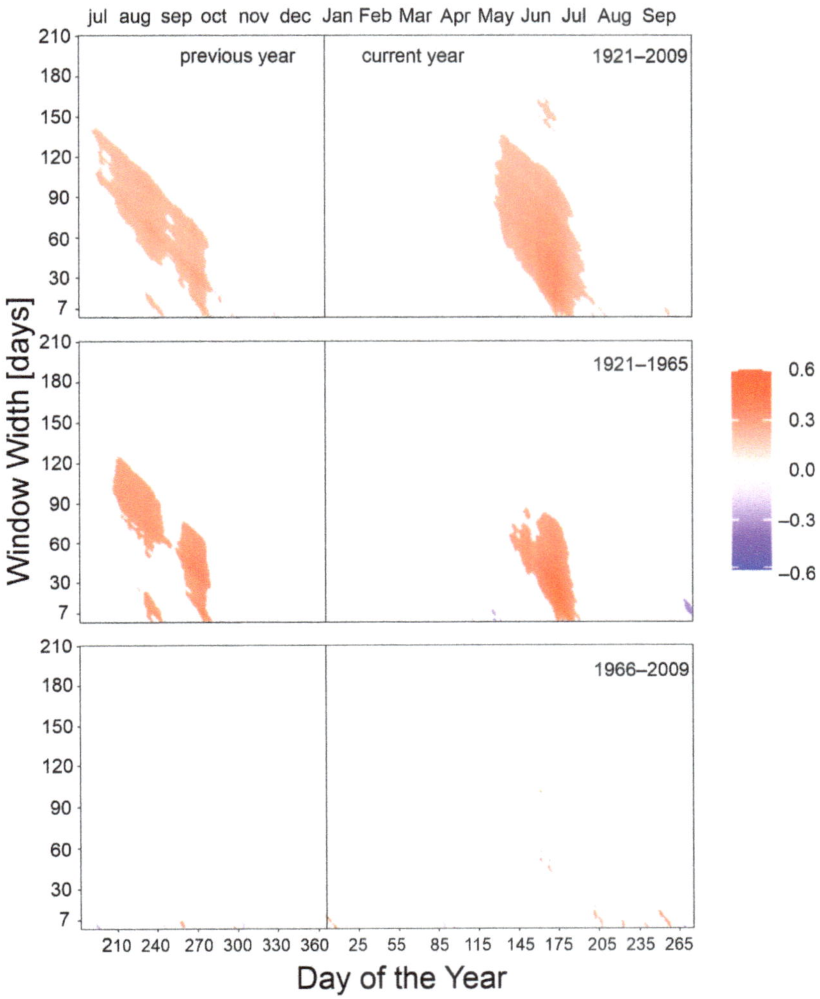

Figure 5. Correlations between the minimum daily temperature and the TRWI chronology for the analysed periods. The colours show a significant ($p < 0.05$) correlation coefficient for consecutive window widths. The given values show the beginning of the specific time window.

3.2.3. Maximum Temperature

The highest positive significant correlation coefficient (r = 0.48) between the TRWI chronology and the maximum temperature was found for the 38-day window width between 21 June and 28 July (DOY 172–209) during the year of growth (Figure 6). However, an important period with a high correlation (r > 0.44) was found for the window width range between 14 and 55 days, spanning mid-June to the beginning of September (DOY 160–250). Additionally, tree growth was positively influenced by the maximum temperature of autumn the previous year (0.20 < r < 0.26), and an important period with relatively high correlation coefficients was found for the period spanning mid-September and mid-November (DOY 256–315; 23–58-day window width) (Figure 6).

In the first subperiod (1921–1965), the highest positive significant correlation coefficient (r = 0.56) between the TRWI chronology and the maximum temperature was found for the 39-day window width between 21 June and 29 July (DOY 172–210) during the year of growth (Figure 6). However, an important period with a high correlation (r > 0.45) was found for the window width range between 15 and 45 days, spanning mid-June to the beginning of August (DOY 169–222). No significant impact of maximum temperature was observed in the first period of the previous year (Figure 6).

In the second subperiod (1966–2009), the highest positive significant correlation coefficient (r = 0.45) between the TRWI chronology and the maximum temperature was found for the 14-day window width between 20 June and 3 July (DOY 171–184) during the year of growth (Figure 6). However, an important period with a high correlation (r > 0.33) was found for the window width range between 11 and 55 days, spanning the end of June to mid-August (DOY 171–225). Additionally, tree growth was positively influenced by the maximum temperature of autumn the previous year (0.37 < r < 0.46), and an important period with relatively high correlation coefficients was found for the period spanning between mid-September and mid-October (DOY 255–292; 12–36-day window width) (Figure 6).

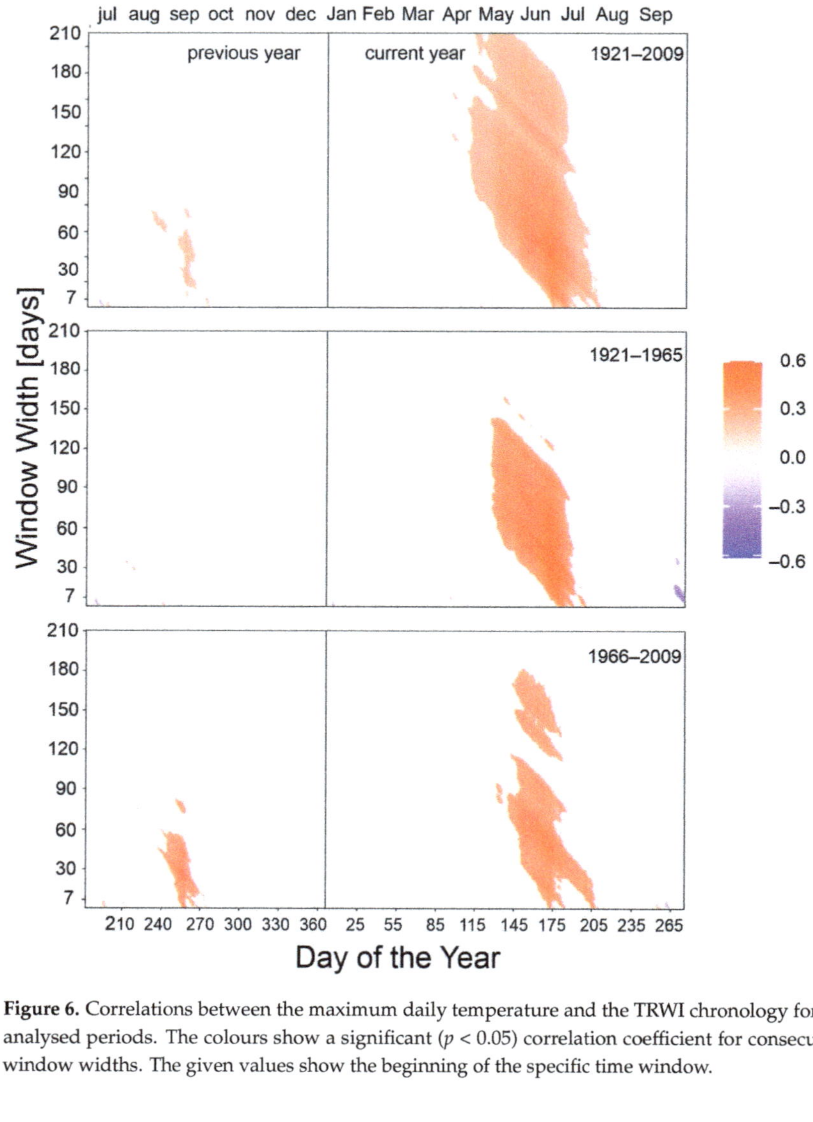

Figure 6. Correlations between the maximum daily temperature and the TRWI chronology for the analysed periods. The colours show a significant ($p < 0.05$) correlation coefficient for consecutive window widths. The given values show the beginning of the specific time window.

3.2.4. Precipitation

The strongest negative significant correlation coefficient (r = −0.35) between the TRWI chronology and the sum of precipitation was found for the 15-day window width between 19 June and 3 July (DOY 170–184) during the year of growth (Figure 7). An important period with a strong correlation coefficient (r < −0.33) was found for the window width range between 13 and 21 days, spanning mid-June to mid-July (DOY 169–192). No distinct impact of precipitation was observed in the previous year (Figure 7).

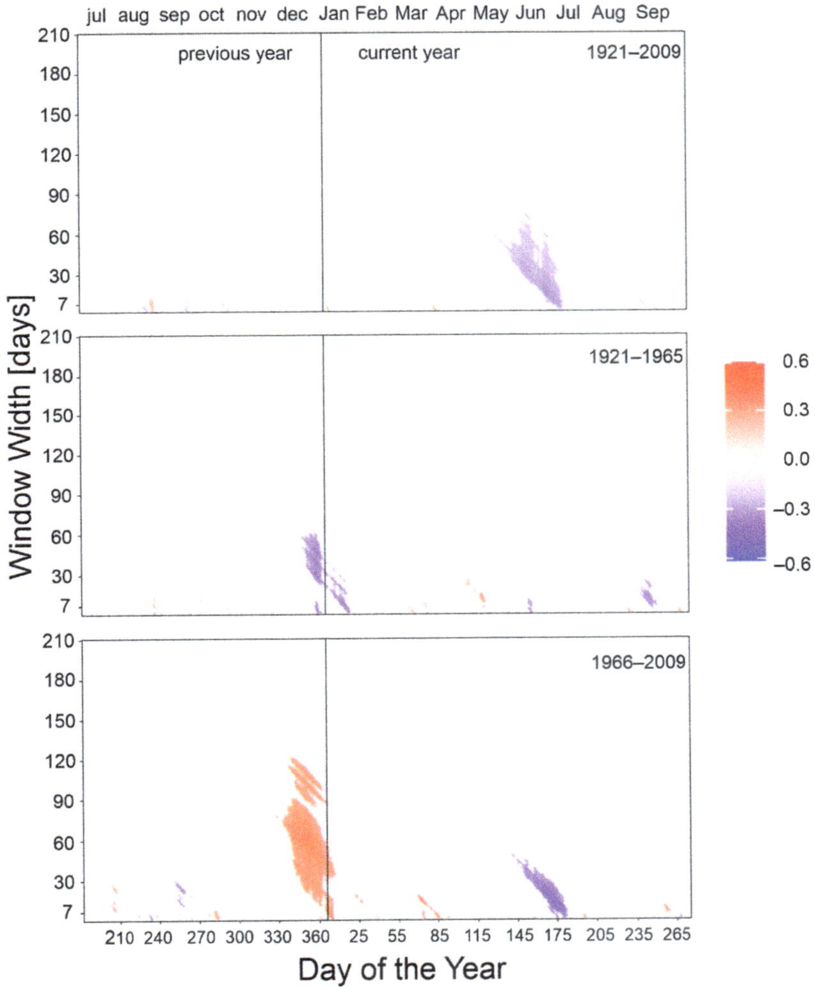

Figure 7. Correlations between daily precipitation and the TRWI chronology for the analysed periods. The colours show a significant ($p < 0.05$) correlation coefficient for consecutive window widths. The given values show the beginning of the specific time window.

In the first subperiod (1921–1965), the strongest negative significant correlation coefficient (r = −0.38) between the TRWI chronology and the sum of precipitation was found for the 39-day window width at the turn of the year from 25 December to 2 February (DOY 360–33) (Figure 7). An important period with a strong correlation coefficient (r < −0.30) was found for the window width range between 27 and 54 days, spanning the end of December to mid-February (DOY 358–49) (Figure 7).

In the second subperiod (1966–2009), the strongest negative significant correlation coefficient (r = −0.53) between the TRWI chronology and the sum of precipitation was found for the 15-day window width in the period between 19 June and 3 July (DOY 170–184) during the year of growth (Figure 7). An important period with strong correlation coefficients (r < −0.40) was found for the window width range between 11 and 20 days, spanning mid-June to mid-July (DOY 168–194). Additionally, the period with significant correlation coefficients (r > 0.40) was found for the window width range between 52 and 63 days, spanning mid-December to the end of February (DOY 348–54) (Figure 7).

4. Discussion

This is the first study that analyses the potential of applying daily climatic variables to enhance the precision and temporal resolution of the climate–growth correlation for stone pine in the Western Carpathians. Most previous dendroclimatic studies about the growth of *Pinus cembra* were based on monthly climatic data. Papers about the radial growth of stone pine discuss higher altitudes of the Alps and Carpathians [20,24–28,46,47]. Only a few authors used daily data for the dendroclimatic study and mostly with reference to wood anatomy. Day-wise aggregated climatic data were used by Știrbu et al. [48] for the South-Western Carpathians and Carrer et al. [23] for the Eastern Alps to analyse xylem anatomical traits, maximum density (MXD), and the ring widths of stone pine.

From another perespective, summer temperature remains a key limiting factor of stone pine growth at higher elevations. Our analyses clearly show that the most important period for radial growth of stone pine is limited to two–three weeks in the second half of June and the first half of July, when the correlations between the ring widths and temperature are the strongest. By expanding the number of analysed days (the window width), significant reactions were found for the period between mid-June and the beginning of August (with a lower correlation coefficient). A similar pattern was found in the Alps and Carpathians, where the temperature from mid-June to the beginning of July was the most important factor for radial growth and explained by favourable conditions for the cell number and tangential cell-wall thickness. The radial cell-wall thickness was positively correlated with the temperature of a later period (from mid-July to the beginning of September) [23,48]. The lumen area was negatively correlated with the temperature from the end of June to the end of August [48]. The TRW analysed in this study was a parameter, which comprised all anatomical features; thus, the relationship between TRW and daily temperature data is not so strong. The changes in the growth–temperature relationship over time were more obvious when analysing the maximum and minimum daily temperatures. The minimum daily temperature, known as the night-time low, is probably the closest indicator limiting tree growth at high elevations [42]. The warmer day-time temperature cannot compensate for the cold nights, and what is more, the low minimum temperature in spring can delay the growing season [49]. The biggest difference between the subperiods of 1921–1965 and 1966–2009 was obvious for minimum temperatures. In the first period the minimum temperature of summer positively and significantly influenced the growth; while in the second subperiod, the minimum temperature lost its relevance. It is possible that the minimum temperature was greater than the threshold for affecting tree growth in the second subperiod.

The association between growth and the temperature of autumn the previous year is also well documented in the literature [25,46,50–52]. Our study allowed us to determine the period of importance more precisely. The most positive temperature occurred during the time between mid-September and mid-October of the previous year, but this interval expanded from the beginning of August to the beginning of November once the minimum temperature is taken into consideration. The crucial role of the autumn temperature results from the timing period was the bud set and the accumulation of reserves for building effective protection against winter frost, drought, or wind [18,53–55]. However, this pattern changed over time when considering shorter subperiods and the minimum and maximum temperatures. For example, this relationship disappeared for maximum temperatures in the years 1921–1965, which can be related to the cooler autumn during this period; however, it becomes clear in the second half of the 20th century when the maximum temperatures of autumn were higher. Our analysis did not reflect the temperature of the soil, which is crucial for the physiological activity of trees within their distribution limits (upper timberline), as soil retains heat for several days during unfavourable weather conditions, especially in autumn when trees enter dormancy [49,56].

Our research confirmed that summer precipitation had a negative influence on tree growth at high elevations [20,24,26]. In our study, this period of negative correlations almost completely overlaps with the days with the highest positive influence of the temperature and lasts between mid-June and the beginning of July. This relationship was not clear in the first half of the 20th century (1921–1965), which can be attributed to the lower precipitation during this time (Figure 2b). It is noteworthy that the radial growth of stone pine was influenced by winter precipitation from the preceding vegetation season during the shorter analysed periods. Additionally, the negative impact of winter precipitation on radial growth observed in the first subperiod (1921–1965) became positive in the second subperiod (1966–2009). The same trend was observed in the Alps [26,50]. This shift may be due to changes in precipitation patterns, possibly because of the amount and timing of precipitation having shifted. It is possible that warmer temperature during the dormant season with higher precipitation results in substantial snow depth, which can help prevent deep soil frost and mitigate the risk of damage caused by frost drought [17].

The general outcome of our study was that temperature became less important as a growth factor at high elevations in the Carpathians, which confirmed our hypothesis. The change in the temperature may potentially promote an upward shift of this species in the Tatra Mountains. Our study essentially validated the conclusions of Briffa et al. [57], Paulsen et al. [58], and Körner and Hoch [49]: The climate pattern fluctuates and reflects a threshold response. As the climate continues to warm, the treeline is also shifting upslope. One of the biggest challenges is to determine the rate and range at which local tree limits align with the warming climate [49]. As a result, trees that once grew within the treeline are now exposed to conditions that no longer resemble their boundary from the past. It means that stone pine in the Tatras has been growing in better climatic conditions over the last decades due to the shifting climatic treeline [59]. The warming climate opens space for the colonisation of higher elevations, and stone pine may benefit from these changes. As a zoochoric species dispersed by nutcracker (*Nucifraga caryocatactes*), stone pine can migrate upslope. This is especially important for the species due to their relatively small and fragmented, but stable, population with a significant degree of genetic variability in the Western Carpathians [22,60–62].

5. Conclusions

Our study showed that the temperature between mid-June and the beginning of July was a key factor for the growth Swiss stone pine in cliffs. The reaction of trees to climate was not stable over time. The changes in the growth–temperature relationship are the clearest with the minimum daily temperature. The negative influence of precipitation on growth almost completely overlaps in time with the highest positive influence of temperature (mid-June to beginning of July). Therefore, the potential migration of *Pinus cembra* to higher elevations is promoted by climate warming.

Dendroclimatic studies based on daily data define the periods (exact calendar days) that influence the radial growth of trees more precisely than monthly data. This is especially important in analysing the growth of trees at high elevations, where the vegetation season is short, and the climate factor strongly limits radial growth. Finally, the forest stands near the treeline are prone to climate change, and daily climatic data allow us to track these changes with a higher resolution.

Author Contributions: Conceptualization, K.I., T.Z. and E.M.; Data curation, K.I. and P.M.; Formal analysis, K.I.; Investigation, K.I., T.Z. and E.M.; Methodology, K.I. and E.M.; Resources, T.Z.; Visualization, K.I. and P.M.; Writing—original draft, K.I., T.Z. and E.M.; Writing—review and editing, K.I., T.Z. and E.M. All authors have read and agreed to the published version of the manuscript.

Funding: This research received no external funding.

Data Availability Statement: Data will be made available on request.

Acknowledgments: We are grateful to Natalia Dubaj for help with fieldwork and laboratory assistance. We would like to thank the reviewers whose comments improved the paper.

Conflicts of Interest: The authors declare no conflict of interest.

References

1. Adler, C.; Wester, P.; Bhatt, I.D.; Huggel, C.; Insarov, G.; Morecroft, M.; Muccione, V.; Prakash, A.; Alcántara-Ayala, I.; Allen, S.K.; et al. Cross-Chapter Paper 5: Mountains. In *Climate Change 2022: Impacts, Adaptation and Vulnerability. Contribution of Working Group II to the Sixth Assessment Report of the Intergovernmental Panel on Climate Change;* Pörtner, H.O., Roberts, D.C., Tignor, M., Poloczanska, E.S., Mintenbeck, K., Alegría, A., Craig, M., Langsdorf, S., Löschke, S., Möller, V., et al., Eds.; Cambridge University Press: Cambridge, NY, USA, 2022; pp. 2273–2318. [CrossRef]
2. Esper, J.; Klippel, L.; Krusic, P.J.; Konter, O.; Raible, C.C.; Xoplaki, E.; Luterbacher, J.; Büntgen, U. Eastern Mediterranean summer temperatures since 730 CE from Mt. Smolikas tree-ring densities. *Clim. Dyn.* **2020**, *54*, 1367–1382. [CrossRef]
3. Chen, D.; Rojas, M.; Samset, B.H.; Cobb, K.; Diongue Niang, A.; Edwards, P.; Emori, S.; Faria, S.H.; Hawkins, E.; Hope, P.; et al. Framing, Context, and Methods. In *Climate Change 2021: The Physical Science Basis. Contribution of Working Group I to the Sixth Assessment Report of the Intergovernmental Panel on Climate Change;* Masson-Delmotte, V., Zhai, P., Pirani, A., Connors, S.L., Péan, C., Berger, S., Caud, N., Chen, Y., Goldfarb, L., Gomis, M.I., et al., Eds.; Cambridge University Press: Cambridge, NY, USA, 2021; pp. 147–286. [CrossRef]
4. Beniston, M.; Keller, F.; Koffi, B.; Goyette, S. Estimates of snow accumulation and volume in the Swiss Alps under changing climatic conditions. *Theor. Appl. Climatol.* **2003**, *76*, 125–140. [CrossRef]
5. Frank, D.; Esper, J. Characterization and climate response patterns of a high-elevation, multi-species tree-ring network in the European Alps. *Dendrochronologia* **2005**, *22*, 107–121. [CrossRef]
6. Zscheischler, J.; Westra, S.; van den Hurk, B.J.; Seneviratne, S.I.; Ward, P.J.; Pitman, A.; AghaKouchak, A.; Bresch, D.N.; Leonard, M.; Wahl, T.; et al. Future climate risk from compound events. *Nat. Clim. Chang.* **2018**, *8*, 469–477. [CrossRef]
7. Babst, F.; Bouriaud, O.; Poulter, B.; Trouet, V.; Girardin, M.P.; Frank, D.C. Twentieth century redistribution in climatic drivers of global tree growth. *Sci. Adv.* **2019**, *5*, eaat4313. [CrossRef] [PubMed]
8. Camarero, J.J.; Linares, J.C.; Sangüesa-Barreda, G.; Sánchez-Salguero, R.; Gazol, A.; Navarro-Cerrillo, R.M.; Carreira, J.A. The multiple causes of forest decline in Spain: Drought, historical logging, competition and biotic stressors. In *Dendroecology;* Amoroso, M., Daniels, L., Baker, P., Camarero, J.J., Eds.; Springer: New York, NY, USA, 2017; pp. 307–323. [CrossRef]
9. Casalegno, S.; Amatulli, G.; Camia, A.; Nelson, A.; Pekkarinen, A. Vulnerability of *Pinus cembra* L. in the Alps and the Carpathian Mountains under present and future climates. *For. Ecol. Manag.* **2010**, *259*, 750–761. [CrossRef]
10. Dyderski, M.K.; Paź, S.; Frelich, L.E.; Jagodziński, A.M. How much does climate change threaten European forest tree species distributions? *Glob. Chang. Biol.* **2018**, *24*, 1150–1163. [CrossRef]
11. Lindner, M.; Maroschek, M.; Netherer, S.; Kremer, A.; Barbati, A.; Garcia-Gonzalo, J.; Marchetti, M. Climate change impacts, adaptive capacity, and vulnerability of European forest ecosystems. *For. Ecol. Manag.* **2010**, *259*, 698–709. [CrossRef]
12. Allen, C.D.; Macalady, A.K.; Chenchouni, H.; Bachelet, D.; McDowell, N.; Vennetier, M.; Kitzberger, T.; Rigling, A.; Breshears, D.D.; Hogg, E.H. (Ted); et al. A global overview of drought and heat-induced tree mortality reveals emerging climate change risks for forests. *For. Ecol. Manag.* **2010**, *259*, 660–684. [CrossRef]
13. Spinoni, J.; Vogt, J.V.; Naumann, G.; Barbosa, P.; Dosio, A. Will drought events become more frequent and severe in Europe? *Int. J. Climatol.* **2018**, *38*, 1718–1736. [CrossRef]
14. Schweingruber, F.H. *Tree Rings and Environment: Dendroecology;* Haupt-Verlag: Bern, Switzerland, 1996; p. 609.
15. Speer, J.H. *Fundamentals of Tree-Ring Research;* University of Arizona Press: Tucson, AZ, USA, 2010; p. 368.
16. Pearl, J.; Keck, J.; Tintor, W.; Siekacz, L.; Herrick, H.; Meko, D.; Pearson, C. New frontiers in tree-ring research. *Holocene* **2020**, *30*, 923–941. [CrossRef]
17. Körner, C. *Alpine Treelines: Functional Ecology of the Global High Elevation Tree Limits;* Springer: Berlin/Heidelberg, Germany, 2012; p. 220.
18. Tranquillini, W. Physiological ecology of the alpine timberline. In *Ecological Study;* Springer Nature: Berlin/Heidelberg, Germany; New York, NY, USA, 1979; Volume 31, pp. 1–137.
19. Körner, C.; Paulsen, J.A. World-wide study of high altitude treeline temperatures. *J. Biogeogr.* **2004**, *31*, 713–732. [CrossRef]
20. Izworska, K.; Muter, E.; Matulewski, P.; Zielonka, T. Tree rings as an ecological indicator of the reaction of Swiss stone pine (*Pinus cembra* L.) to climate change and disturbance regime in the extreme environment of cliff forests. *Ecol. Indic.* **2023**, *148*, 110102. [CrossRef]

21. Tóth, E.G.; Tremblay, F.; Housset, J.M.; Bergeron, Y.; Carcaillet, C. Geographic isolation and climatic variability contribute to genetic differentiation in fragmented populations of the long-lived subalpine conifer *Pinus cembra* L. in the western Alps. *BMC Evol. Biol.* **2019**, *19*, 190. [CrossRef]
22. Zięba, A.; Różański, W.; Bukowski, M.; Pałka, B.; Szwagrzyk, J. Distribution and habitat conditions of *Pinus cembra* forests in the Tatra Mountains. *Dendrobiology* **2019**, *81*, 86–96. [CrossRef]
23. Carrer, M.; Unterholzner, L.; Castagneri, D. Wood anatomical traits highlight complex temperature influence on *Pinus cembra* at high elevation in the Eastern Alps. *Int. J. Biometeorol.* **2018**, *62*, 1745–1753. [CrossRef] [PubMed]
24. Carrer, M.; Urbinati, C. Age-dependent tree-ring growth responses to climate in *Larix decidua* and *Pinus cembra*. *Ecology* **2004**, *85*, 730–740. [CrossRef]
25. Oberhuber, W.; Kofler, W.; Pfeifer, K.; Seeber, A.; Gruber, A.; Wieser, G. Long-term changes in tree-ring–climate relationships at Mt. Patscherkofel (Tyrol, Austria) since the mid-1980s. *Trees* **2008**, *22*, 31–40. [CrossRef]
26. Saulnier, M.; Edouard, J.L.; Corona, C.; Guibal, F. Climate/growth relationships in a *Pinus cembra* high-elevation network in the Southern French Alps. *Ann. For. Sci.* **2011**, *68*, 189–200. [CrossRef]
27. Popa, I.; Nechita, C.; Hofgaard, A. Stand structure, recruitment and growth dynamics in mixed subalpine spruce and Swiss stone pine forests in the Eastern Carpathians. *Sci. Total Environ.* **2017**, *598*, 1050–1057. [CrossRef]
28. Popa, I.; Kern, Z. Long-term summer temperature reconstruction inferred from tree-ring records from the Eastern Carpathians. *Clim. Dyn.* **2009**, *32*, 1107–1117. [CrossRef]
29. Roibu, C.C.; Popa, I.; Kirchhefer, A.J.; Palaghianu, C. Growth responses to climate in a tree-ring network of European beech (*Fagus sylvatica* L.) from the eastern limit of its natural distribution area. *Dendrochronologia* **2017**, *42*, 104–116. [CrossRef]
30. Rozas, V. Dendrochronology of pedunculate oak (*Quercus robur* L.) in an old-growth pollarded woodland in northern Spain: Tree-ring growth responses to climate. *Ann. For. Sci.* **2005**, *62*, 209–218. [CrossRef]
31. Jevšenak, J.; Levanič, T. dendroTools: R package for studying linear and nonlinear responses between tree-rings and daily environmental data. *Dendrochronologia* **2018**, *48*, 32–39. [CrossRef]
32. Jevšenak, J. Daily climate data reveal stronger climate-growth relationships for an extended European tree-ring network. *Quat. Sci. Rev.* **2019**, *221*, 105868. [CrossRef]
33. Vaganov, E.A.; Hughes, M.K.; Shashkin, A.V. Growth dynamics of conifer tree rings: Images of past and future environments. In *Ecological Study*; Springer Nature: Berlin/Heidelberg, Germany; New York, NY, USA, 2006; Volume 183, p. 350.
34. Niedzwiedz, T. Climate of the Tatra Mountains. *Mt. Res. Dev.* **1992**, *12*, 131–146. [CrossRef]
35. Caudullo, G.; Welk, E.; San-Miguel-Ayanz, J. Chorological maps for the main European woody species. *Data Brief* **2017**, *12*, 662–666. [CrossRef] [PubMed]
36. Grissino-Mayer, H.D. Evaluating crossdating accuracy: A manual and tutorial for the computer program COFECHA. *Tree-Ring Res.* **2001**, *57*, 5–21. Available online: http://hdl.handle.net/10150/251654 (accessed on 21 April 2023).
37. Holmes, R.L. Computer-assisted quality control in tree-ring dating and measurement. *Tree-Ring Bull.* **1983**, *43*, 69–75.
38. Cornes, R.C.; van der Schrier, G.; van den Besselaar, E.J.; Jones, P.D. An ensemble version of the E-OBS temperature and precipitation data sets. *J. Geophys. Res. Atmos.* **2018**, *123*, 9391–9409. [CrossRef]
39. Bunn, A.G. Statistical and visual crossdating in R using the dplR library. *Dendrochronologia* **2010**, *28*, 251–258. [CrossRef]
40. R Development Core Team. R: A Language and Environment for Statistical Computing. Available online: https://www.r-project.org (accessed on 21 April 2023).
41. McCarroll, D.; Loader, N.J. Stable isotopes in tree rings. *Quat. Sci. Rev.* **2004**, *23*, 771–801. [CrossRef]
42. Fritts, H.C. *Tree Rings and Climate*; Academic Press: London, UK, 1976; p. 582.
43. Bunn, A.G. A dendrochronology program library in R (dplR). *Dendrochronologia* **2008**, *26*, 115–124. [CrossRef]
44. Buras, A. A comment on the expressed population signal. *Dendrochronologia* **2017**, *44*, 130–132. [CrossRef]
45. Wigley, T.M.L.; Briffa, K.R.; Jones, P.D. On the average value of correlated time series, with applications in dendroclimatology and hydrometeorology. *J. Appl. Meteorol. Climatol.* **1984**, *23*, 201–213. [CrossRef]
46. Bednarz, Z. *Relationship of Tree-Ring Widths in the Tatra Mountains to Variations in Monthly Temperature and Precipitation*; Mitteilungen der Forstlichen Bundes-Versuchsanstalt: Wien, Austria, 1981; Volume 142, pp. 367–389.
47. Büntgen, U.; Trnka, M.; Krusic, P.J.; Kyncl, T.; Kyncl, J.; Luterbacher, J.; Zorita, E.; Ljungqvist, F.C.; Auer, I.; Konter, O.; et al. Tree-ring amplification of the early nineteenth-century summer cooling in Central Europe. *J. Clim.* **2015**, *28*, 5272–5288. [CrossRef]
48. Știrbu, M.I.; Roibu, C.C.; Carrer, M.; Mursa, A.; Unterholzner, L.; Prendin, A.L. Contrasting climate sensitivity of *Pinus cembra* tree-ring traits in the Carpathians. *Front. Plant Sci.* **2022**, *13*, 855003. [CrossRef]
49. Körner, C.; Hoch, G. Not every high-latitude or high-elevation forest edge is a treeline. *J. Biogeogr.* **2023**, *50*, 838–845. [CrossRef]
50. Carrer, M.; Nola, P.; Eduard, J.L.; Motta, R.; Urbinati, C. Regional variability of climate–growth relationships in *Pinus cembra* high elevation forests in the Alps. *J. Ecol.* **2007**, *95*, 1072–1083. [CrossRef]
51. Housset, J.M.; Tóth, E.G.; Girardin, M.P.; Tremblay, F.; Motta, R.; Bergeron, Y.; Carcaillet, C. Tree-rings, genetics and the environment: Complex interactions at the rear edge of species distribution range. *Dendrochronologia* **2021**, *69*, 125863. [CrossRef]
52. Vittoz, P.; Rulence, B.; Largey, T.; Freléchoux, F. Effects of climate and land-use change on the establishment and growth of cembran pine (*Pinus cembra* L.) over the altitudinal treeline ecotone in the Central Swiss Alps. *Arct. Antarct. Alp. Res.* **2008**, *40*, 225–232. [CrossRef]

53. Mayr, S.; Hacke, U.; Schmid, P.; Schwienbacher, F.; Gruber, A. Frost drought in conifers at the alpine timberline: Xylem dysfunction and adaptations. *Ecology* **2006**, *87*, 3175–3185. [CrossRef]
54. Oberhuber, W. Influence of climate on radial growth of Pinus cembra within the alpine timberline ecotone. *Tree Physiol.* **2004**, *24*, 291–301. [CrossRef]
55. Simard, S.; Giovannelli, A.; Treydte, K.; Traversi, M.L.; King, G.M.; Frank, D.; Fonti, P. Intra-annual dynamics of non-structural carbohydrates in the cambium of mature conifer trees reflects radial growth demands. *Tree Physiol.* **2013**, *33*, 913–923. [CrossRef] [PubMed]
56. Körner, C.; Hoch, G. A test of treeline theory on a montane permafrost island. *Arct. Antarct. Alp. Res.* **2006**, *38*, 113–119. [CrossRef]
57. Briffa, K.R.; Schweingruber, F.H.; Jones, P.D.; Osborne, T.J.; Shiyatov, S.G.; Vaganov, E.A. Reduced sensitivity of recent tree-growth to temperature at high northern latitudes. *Nature* **1998**, *391*, 678–682. [CrossRef]
58. Paulsen, J.; Weber, U.M.; Körner, C. Tree growth near treeline: Abrupt or gradual reduction with altitude? *Arct. Antarct. Alp. Res.* **2000**, *32*, 14–20. [CrossRef]
59. Körner, C. The cold range limit of trees. *Trends Ecol. Evol.* **2021**, *36*, 979–989. [CrossRef]
60. Dzialuk, A.; Chybicki, I.; Gout, R.; Mączka, T.; Fleischer, P.; Konrad, H.; Curtu, L.; Sofletea, N.; Valadon, A. No reduction in genetic diversity of Swiss stone pine (*Pinus cembra* L.) in Tatra Mountains despite high fragmentation and small population size. *Conserv. Genet.* **2014**, *15*, 1433–1445. [CrossRef]
61. Gugerli, F.; Brodbeck, S.; Lendvay, B.; Dauphin, B.; Bagnoli, F.; van der Knaap, W.O.; Tinner, W.; Höhn, M.; Vendramin, G.G.; Morales-Molino, C.; et al. A range-wide postglacial history of Swiss stone pine based on molecular markers and palaeoecological evidence. *J. Biogeogr.* **2023**, *50*, 1049–1062. [CrossRef]
62. Zwijacz-Kozica, T.; Żywiec, M. Fifty-year changes in a strictly protected stone pine population in the Tatra National Park. *Nat. Conserv.* **2007**, *64*, 73–82.

forests

MDPI

Article

Ecological Adaptation of Two Dominant Conifer Species to Extreme Climate in the Tianshan Mountains

Xuan Wu [1,2], Liang Jiao [1,2,*], Xiaoping Liu [1,2], Ruhong Xue [1,2], Changliang Qi [1,2] and Dashi Du [1,2]

1 College of Geography and Environment Science, Northwest Normal University, No.967, Anning East Road, Lanzhou 730070, China; 2020212684@nwnu.edu.cn (X.W.); 2018212281@nwnu.edu.cn (X.L.); 2022120260@nwnu.edu.cn (R.X.); 2019212378@nwnu.edu.cn (C.Q.); 2020212664@nwnu.edu.cn (D.D.)

2 Key Laboratory of Resource Environment and Sustainable Development, Oasis Northwest Normal University, Lanzhou 730070, China

* Correspondence: jiaoliang@nwnu.edu.cn

Abstract: With global warming, the frequency, intensity, and period of extreme climates in more areas will probably increase in the twenty first century. However, the impact of climate extremes on forest vulnerability and the mechanisms by which forests adapt to climate extremes are not clear. The eastern Tianshan Mountains, set within the arid and dry region of Central Asia, is very sensitive to climate change. In this paper, the response of *Picea schrenkiana* and *Larix sibirica* to climate fluctuations and their stability were analyzed by Pearson's correlation based on the observation of interannual change rates of climate indexes in different periods. Additionally, their ecological adaptability to future climate change was explored by regression analysis of climate factors and a selection of master control factors using the Lasso model. We found that the climate has undergone significant changes, especially the temperature, from 1958 to 2012. Around 1985, various extreme climate indexes had obvious abrupt changes. The research results suggested that: (1) the responses of the two tree species to extreme climate changed significantly after the change in temperature; (2) Schrenk spruce was more sensitive than Siberian larch to extreme climate change; and (3) the resistance of Siberian larch was higher than that of Schrenk spruce when faced with climate disturbance events. These results indicate that extreme climate changes will significantly interfere with the trees radial growth. At the same time, scientific management and maintenance measures are taken for different extreme weather events and different tree species.

Keywords: ecological adaptability; resistance indexes; lasso model; future prediction; tree-ring; climate change

1. Introduction

Forest ecosystems are an important part of terrestrial ecosystems. Plants form a long-term, relatively stable and dynamic balance with the regional climate during their growth [1]. The occurrence of extreme climatic events has strongly disturbed the inherent balance between trees and environmental factors, causing significant changes in the phenological activities, productivity levels, and community succession statuses of forest ecosystems and thereby increasing community survival pressure and the potential risk of local extinction [2–4].

In the context of worldwide warming, the frequency, intensity, and period of maximum weather events still increase, and the impacts of climate on tree growth are growing stronger [5,6]. Therefore, it is particularly important to study the response relationship between extreme climate indicators and tree growth in order to explore tree growth strategies and formulate management and protection measures under future climate change. At present, in order to further strengthen forest management and protection and scientific planning, climate change modeling tools such as the Forest Landscape Model (RCMs), which includes the effects of climate and management on forest dynamics, have been used to predict the layout of forest growth under future climate impacts, especially in forest areas with complex terrain [7]. At low latitudes in Bangladesh, the radial growth of *Chukrasia tabularis* was reduced by 54% and 48.7% by the two extreme drought events in 1999 and 2006, respectively [8]. At an identical time, extreme precipitation events are more likely to flood plants in tropical regions due to abundant water and heat resources, which will lead to hypoxia of plant roots, reduced respiration, and eventually plant death [9]. At middle and high latitudes in the virgin forests of southern Europe, the sensitivity of *Picea abies* and *Abies alba* to climatic and non-climatic disturbances increased due to atmospheric warming and increased drought [10]. Compared to those in low-latitude regions, trees in mid- and high-latitude regions were much more at risk of extreme drought and freezing damage throughout the growing season [11]. Although the impact of extreme precipitation on tree growth is not as direct and significant as the impacts of extreme high-temperature and drought events, studies have shown that extreme precipitation events affect forest carbon sinks and ecohydrological processes, thereby affecting the function and structure of entire forest ecosystems [12–14].

Tree ring width is an important indicator for recording climate change, and its change can truly and objectively reflect the impact of climate change on trees and the ecological response of trees to climate change due to its high resolution, easy preservation, and long time series [15,16].

Tree rings will appear with narrower, false, and missing rings when there are sudden and dramatic variations in climate, such as extreme high temperature, high-intensity precipitation, and frost [17,18]. Specifically, these growth–climate relationships will significantly change with a continuous decrease in the number of cold nights and continuous increases in the frequency and intensity of drought [19,20]. For example, the development of unstable responses of tree growth to temperature change has been documented worldwide [21–24], and therefore, the responses of various tree species to temperature change are clearly varied thanks to variations in physiological and ecological thresholds [9,25,26].

Ecological adaptability refers to the adjustment and recovery abilities of forest ecosystems in the face of extreme climatic conditions. Resistance indexes have been widely used to assess the dynamic changes of forests in response to extreme environmental disturbance events in recent years based on tree rings [27,28]. In addition, a great number of studies have found that differences in the resistance of trees to extreme weather events are affected by many factors, such as age, tree species, and forest stand [23,29–31]. *Fagus sylvatica* in southern Germany exhibited low resistance but high resilience to extreme frost events [32]. In addition, the resistance and resilience of trees are not only related to tree species but are also related to the types of extreme weather events, according to an analysis of the resistance differences of the five main European tree species to spring frost and summer drought [33].

The increase in temperature at the mid-high latitudes of the Northern Hemisphere is considerably beyond the worldwide average attributable to the magnification of the Arctic [34,35]. The forest ecosystems within the arid and dry regions of Central Asia are particularly sensitive and at risk of global climate change thanks to the tough environmental conditions in these regions [3].

As the largest mountain range in Central Asia, the Tianshan Mountains are the origin of most inland rivers and support oasis development in Central Asia. The forest ecosystems in the Tianshan Mountains, with water conservation functions, are of great significance to the natural ecological environment and socioeconomic development in Central Asia [36]. To date, research on tree growth and climate relationships in the Tianshan Mountains has mainly focused on climate reconstruction and conventional climate responses [37,38]. In these studies, the reconstruction of climate indicators using accurate long-sequence tree-ring data is conducive to revealing long-term climate dynamics in different regions of the world, and exploring the response relationship between tree growth and climate is conducive to predicting forest growth patterns under future climate change. However, the effect of extreme climate on forest vulnerability and the mechanisms of forest response and resistance to extreme climate remain unclear. Therefore, we constructed models of the relationship between radial growth of the dominant tree species (Schrenk spruce and Siberian larch) and extreme climate factors in the eastern Tianshan Mountains, an area with more arid conditions. We aimed to compare the responses and resistances of the two tree species to extreme climate.

2. Materials and Methods

2.1. Site Description

The study area is found on Barkol Mountain in the eastern Tianshan Mountains (Figure 1). The Barkol region incorporates a typical temperate continental climate with an outsized annual temperature variation. The annual mean temperature is 2.02 °C and presents a significant increasing trend of 0.611 °C/10a ($p < 0.001$), and the annual total precipitation is 220.33 mm, showing an increasing trend of 9.6 mm/10a ($p < 0.05$). The increasing trend of temperature is more obvious than that of total precipitation (Figure 2). The results of the Pettitt test show that the annual mean temperature in the Barkol area underwent a significant abrupt change in 1985 ($p < 0.001$) (Figure 2).

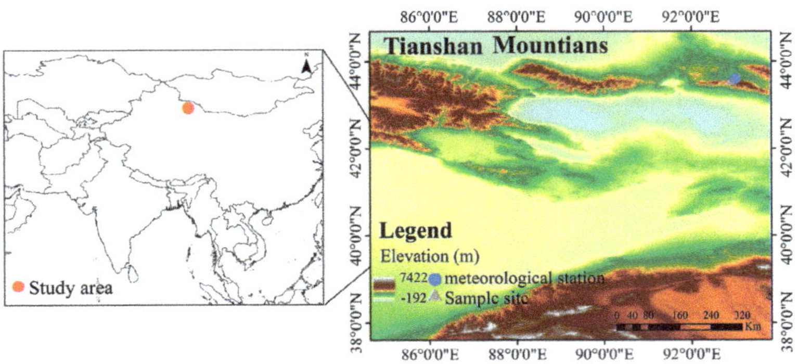

Figure 1. Locations of the sampling sites and the nearest meteorological station.

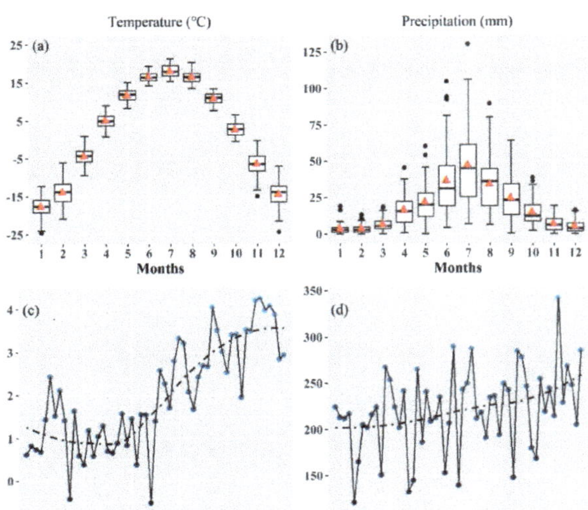

Figure 2. Monthly (**a**,**b**) and interannual (**c**,**d**) variation trends of mean temperature and annual total precipitation at the Barkol meteorological station in the eastern Tianshan Mountains from 1958 to 2012.

In panels a and b, the black dots are outliers; red triangles are average values; lower dash is the minimum (excluding outliers); the bottom of the box demarcates 25% of the data; the middle line of the box demarcates 50% of the data (median); the top of the box demarcates 75% of the data; and the upper dash is the maximum (excluding outliers). The dotted line in panels (c) and (d) indicates the interannual change trend of climate elements according to local weighted regression.

2.2. Tree Ring Sample Collection and Processing

The sampling sites are settled within the original coniferous mixed forest close to the upper forest line on the northeastern slope of Barkol Mountain with an altitude of roughly 2550–2600 m (Table 1). The sampling sites were less affected by human activity than surrounding areas, with relatively low forest canopy closure, large spacing between trees, and less influence of density. In August 2013, a total of 120 core samples were collected. After damaged and unidentifiable samples were excluded, there remained 60 and 58 sample cores of two species of trees.

Table 1. Sampling information of two species of trees.

Species	Schrenk Spruce	Siberian Larch
Altitude (m)	2552	2590
Latitude	43°32.100′ N	43°32.085′ N
Longitude	92°56.329′ E	92°56.662′ E
Slope	27°	33°
Aspect	Northeast (40° from North)	North
Canopy closure	0.3	0.2
Average tree spacing (m)	3	4.5
Average breast diameter (cm)	32.4	33.8
Average Tree height (m)	12.3	9.7
Average Crown width (m)	2.7	3.7

In the laboratory, the core samples of the two tree species were air-dried naturally, fixed in a wooden sample tank with latex, and sanded with mesh sandpapers until the tree rings were visible and clear. First, we used the LINTAB measurement system to measure the ring width of each core with a resolution of 0.001 mm, and then we used the COFECHA program to determine the quality of the cross-dating [39]. Finally, the standard chronology of the two species was obtained by ARSTAN procedure [40].

2.3. Chronological Statistical Parameter Calculation

The following statistical parameters of the chronologies, which could be used to assess the reliability of the chronologies, were calculated through the ARSTAN program (Table 2). The SD represents the fluctuation of the tree rings, MS reflects the richness of climate information contained in the chronological sequences, AC1 indicates the influence degree of the previous year's climate on tree growth, R represents the similarity of the chronological sequences, PC1 represents the common information in the tree core samples, the SNR can indicate the amount of climate and environmental information shared by the chronologies, and the EPS reveals the representativeness of the subsample for the whole sample.

Table 2. Statistical parameters of the STD of Schrenk spruce and Siberian larch (1958–2012).

Statistical Parameters	Schrenk Spruce	Siberian Larch
Sample depth (core/tree number)	60/30	58/29
Sequence length	1735–2012 (288)	1761–2012 (252)
Standard deviation (SD)	0.313	0.312
Mean sensitivity (MS)	0.196	0.214
AC1	0.681	0.712
Correlation coefficient ®	0.483	0.382
Mean correlation among trees (R1)	0.771	0.723
Mean correlation between trees (R2)	0.475	0.366
First principal comment (PC1)	0.507	0.435
Signal-to-noise ratio (SNR)	36.366	14.213
Expressed population signal (EPS)	0.973	0.934

2.4. Meteorological Data and Preprocessing

The meteorological data were obtained from the Barkol National Alpine Meteorological Station in Xinjiang (93°05′ E, 43°06′ N, 1677.2 m). The data set of main climatic factors during 1958–2012 can be obtained from the China Meteorological Data Network (http://data.cma.cn/, accessed on 18 June 2020).

Twenty-seven climate indexes were obtained through the RClimDex program (Table 3). These indexes were developed by the Expert Group on Climate Change Detection, Monitoring, and Indexes (ETCCDI) and the World Meteorological Organization (WMO) cooperative proposal and formulation [41,42]. We further divided the extreme temperature indicators into extreme warmth indicators and extreme cold indicators for better observation and analysis [43]. Among the 27 indexes, 12 included both monthly and annual statistics, and 15 included only annual statistics.

Table 3. Extreme climate indexes.

Type	Class	ID	Indicator Name	Definitions	Unit
Extreme temperature indexes	Extreme temperature warm indexes	TXx	Max Tmax	Monthly maximum value of daily maximum temp	°C
		TXn	Min Tmax	Monthly minimum value of daily maximum temp	°C
		TX90p	Warm days	Percentage of days when TX > 90th percentile	d
		TN90p	Warm nights	Percentage of days when TN > 90th percentile	d
		WSDI	Warm spell duration indicator	Annual count of days with at least 6 consecutive days when TX > 90th percentile	d
		SU25	Summer days	Annual count when TX (daily maximum) > 25 °C	d
		TR20	Tropical nights	Annual count when TN (daily minimum) > 20 °C	d
	Extreme temperature cold indexes	TNx	Max Tmin	Monthly maximum value of daily minimum temp	°C
		TNn	Min Tmin	Monthly minimum value of daily minimum temp	°C
		TX10p	Cool days	Percentage of days when TX < 10th percentile	d
		TN10p	Cool nights	Percentage of days when TN < 10th percentile	d
		CSDI	Cold spell duration indicator	Annual count of days with at least 6 consecutive days when TN < 10th percentile	d
		FD0	Frost days	Annual count when TN (daily minimum) < 0 °C	d
		ID0	Ice days	Annual count when TX (daily maximum) < 0 °C	d
	Other temperature indexes	GSL	Growing season length	Annual (1st Jan to 31st Dec in NH, 1st July to 30th June in SH) count between first span of at least 6 days with TG > 5 °C and first span after July 1 (January 1 in SH) of 6 days with TG < 5 °C	d
		DTR	Diurnal temperature range	Monthly mean difference between TX and TN	°C

Table 3. *Cont.*

Type	Class	ID	Indicator Name	Definitions	Unit
Extreme precipitation indexes	Precipitation frequency indexes	R10	Number of heavy precipitation days	Annual count of days when PRCP ≥ 10 mm	d
		R20	Number of very heavy precipitation days	Annual count of days when PRCP ≥ 20 mm	d
		Rnn	Number of days above nn mm	Annual count of days when PRCP ≥ nn mm, nn is user defined threshold	d
		CDD	Consecutive dry days	Maximum number of consecutive days with RR < 1 mm	d
		CWD	Consecutive wet days	Maximum number of consecutive days with RR ≥ 1 mm	d
	Precipitation magnitude indexes	RX1day	Max 1-day precipitation amount	Monthly maximum 1-day precipitation	mm
		Rx5day	Max 5-day precipitation amount	Monthly maximum consecutive 5-day precipitation	mm
		R95p	Very wet days	Annual total PRCP when RR > 95th percentile	mm
		R99p	Extremely wet days	Annual total PRCP when RR > 99th percentile	mm
		PRCPtot	Annual total wet-day precipitation	Annual total PRCP in wet days (RR ≥ 1 mm)	mm
	Precipitation intensity index	SDII	Simple daily intensity index	Annual total precipitation divided by the number of wet days (defined as PRCP ≥ 1.0 mm) in the year	mm/day

2.5. Data Analysis

2.5.1. Climate Data Analysis

First, one-variable linear regression was used to observe the interannual change rates of the 28 climate indexes (annual mean temperature and 27 extreme climate indexes), and then local polynomial regression was used to fit the climate indexes to further observe local changes in the data. The Pettitt test was then used to test for changes in each climate index sequence. Among all the climate indexes (18) that passed the significance test, 12 indexes displayed change years in the mid-1980s, and 6 displayed change years in the mid-1990s. Observing these six indicators, we found that although the change times detected by the test were in the mid-1990s, the increases (or decreases) detected by the test began in the mid-1980s (Figure 3). Therefore, we set the year of abrupt change to 1985 in all cases.

2.5.2. Analysis of the Relationships between Tree Rings and Climate Factors

The correlation coefficients between each climate index and the STDs of the two tree species before climate change (1958–1985) and after climate change (1986–2012) were analyzed by Pearson's correlation. The stability and climate responses of the two tree species were observed when the climatic conditions of the study area change significantly. The climate data from September of the previous year to October of the current year were selected because the radial growth of trees was restricted not only by the climatic factors of the current year but also affected by those of the previous year [15].

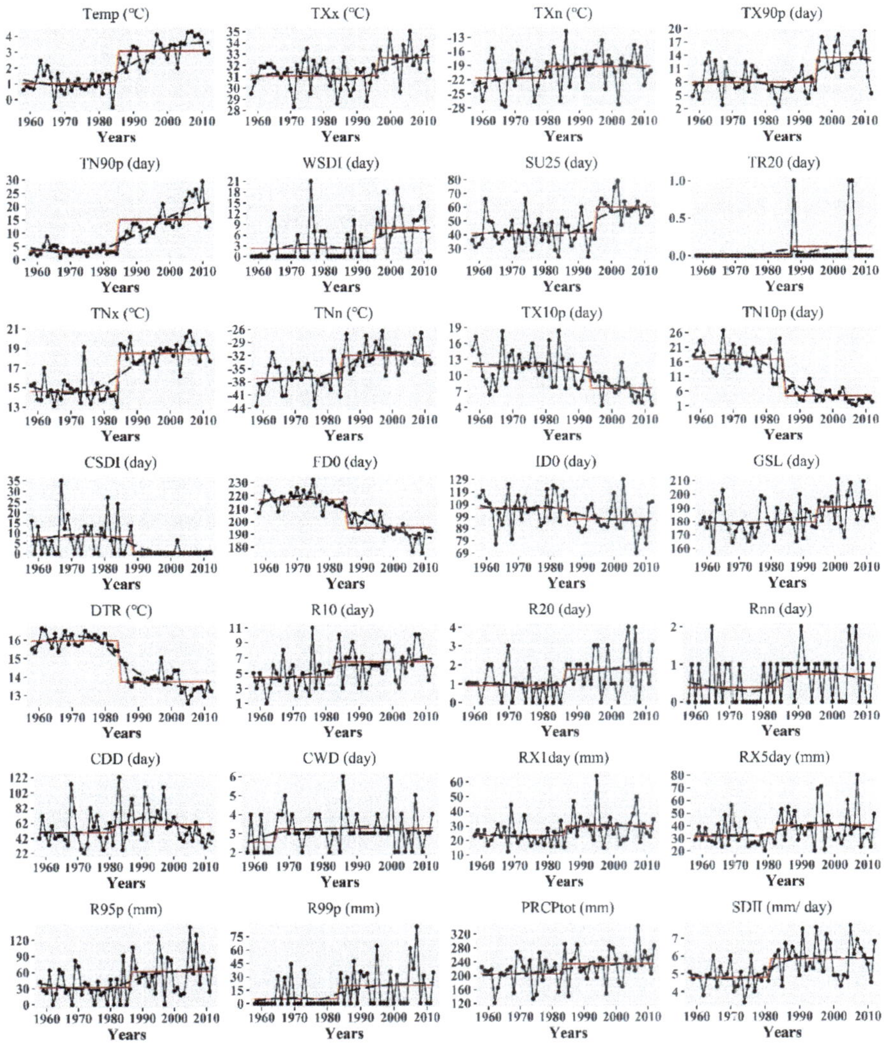

Figure 3. Interannual variability and abrupt change of 28 climate indexes in 1958–2012.

2.5.3. Pointer Year Selection

We used the STDs to screen the narrow-ring years of the two tree species before and after climate change and used the basal area increment (BAI, $cm^2/year$) for verification. BAI was calculated based on non-standardized raw measurement ring width data as the following equation (Monserud and Sterba, 1996):

$$BAI_t = \pi\left(r_t^2 - r_{t-1}^2\right)$$

where r_t is a given annual ring corresponding to radial radius at t year and r_{t-1} is a given annual ring corresponding to radial radius at $t - 1$ year.

The two tree species generally returned to the growth state observed before the environmental disturbance two years after the appearance of the narrow ring. We chose two years as a window to calculate the 2-year moving average of the chronological sequence to avoid overlap with other disturbance events. In addition, we set two narrow-ring levels at 80% and 75% of the average of the previous two years and then integrated the screening results of the two chronologies for the tree species. Finally, the years in which both tree species had narrow rings without other narrow rings in the previous two years were selected as pointer years.

2.5.4. Resistance Index Calculation

The resistances of the two tree species to interference events were reflected by their resistance, recovery, and resilience. The calculation formula is as follows (Lloret et al., 2011):

$$RT = TRW_i/TRW_{i-2};$$

$$RC = TRW_{i+2}/TRW_i;$$

$$RS = TRW_{i+2}/TRW_{i-2};$$

where TRW_i represents TRW in the year of environmental stress, TRW_{i-2} represents the average TRW 2 years before the environmental stress, and TRW_{i+2} represents the average TRW 2 years after the environmental stress. RT characterizes the ability of a tree to resist disturbance. RC indicates the degree of recovery of a tree after being disturbed by the environment. RS can measure the growth of a tree after being disturbed and after a period of recovery. Instances of two consecutive drought years were regarded as drought events (for example, 1984 and 1985 in the following text). Afterwards, one-way analysis of variance was used to evaluate the significance of the differences in resistance indexes (RT, RC, and RS) of the two tree species in pointer years before and after the climatic conditions changed.

The above data analyses were all performed in R using the packages "statas", "trend", "car", and "ggplot2".

3. Results

3.1. Climate Change Characteristics in the Study Area

One-variable linear regression of the 27 climatic factors revealed a very obvious trend of increasing temperature (Figure 3). According to the interannual variations of various extreme climate indicators, the growing season length and extreme temperature warmth index except for tropical night showed a significant increasing trend ($p < 0.05$). The annual diurnal temperature range and the rest of the extreme climate cold indexes showed significant decreasing trends ($p < 0.05$) expect for Max Tmin and Min Tmin. All extreme precipitation indexes showed increasing trends; specifically, the number of heavy precipitation days, number of very heavy precipitation days, very wet days, extremely wet days, and simple daily intensity index showed significant increasing trends ($p < 0.05$).

Pettitt tests of the 27 extreme climate indicators revealed obvious points of abrupt change in the extreme temperature indexes, except for Min Tmax and TR20 ($p < 0.05$) (Table 4). The only extreme precipitation indexes with abrupt change points were R10, R95P, and SDII ($p < 0.05$). The abrupt climate change mainly manifested as a significant increase in temperature, providing evidence that the magnitude and significance of the variation in each precipitation index were smaller than those of the temperature indexes.

Table 4. Unary linear regression and Pettit test of 28 climate indexes.

ID	Unary Linear Regression Equation	R2	*p* Value of Equation	Mutation Year	*p* Value of Pettitt Test
Temp	y = 0.3100 + 0.0611x	0.6200	1.3×10^{-12} **	1985	9.3×10^{-9} **
TXx	y = 30.700 + 0.0308x	0.1100	0.011 *	1996	0.01202 *
TXn	y = −22.100 + 0.0707x	0.0940	0.023 *	1979	0.0843
TX90p	y = 6.5800 + 0.11x	0.1900	0.001 **	1995	0.0004654 **
TN90p	y = −1.3800 + 0.382x	0.6900	2.9×10^{-15} **	1984	4.415×10^{-9} **
WSDI	y = 0.5050 + 0.127x	0.1200	0.0085 **	1995	0.03846 *
SU25	y = 35.600 + 0.412x	0.2600	6.9×10^{-5} **	1995	4.995×10^{-5} **
TR20	y = −0.0034 + 0.0032x	0.0490	0.1000	1987	1.501
TNx	y = 13.6000 + 0.1060x	0.5400	1.6×10^{-10} **	1984	3.522×10^{-8} **
TNn	y = −38.6000 + 0.1450x	0.3200	5.5×10^{-6} **	1984	1.931×10^{-5} **
TX10p	y = 13.50–0 − 0.1170x	0.2800	3.2×10^{-5} **	1993	0.0007737 **
TN10p	y = 21.20–0 − 0.3750x	0.7000	2.4×10^{-15} **	1985	4.002×10^{-9} **
CSDI	y = 10.40–0 − 0.2040x	0.1700	0.0016 **	1988	0.002104 **
FD0	y = 2–6 − 0.7110x	0.7200	3.4×10^{-16} **	1985	7.16×10^{-9} **
ID0	y = 1–8 − 0.2310x	0.0770	0.0400 *	1985	0.03497 *
GSL	y = 175 + 0.2910x	0.1400	0.005 **	1994	0.02881 *
DTR	y = 16.70–0 − 0.0662x	0.7500	1×10^{-17} **	1984	3.232×10^{-9} **
R10	y = 3.8900 + 0.0597x	0.1700	0.0016 **	1982	0.006723 **
R20	y = 0.7330 + 0.0212x	0.1000	0.016 *	1986	0.05441
Rnn	y = 0.3660 + 0.0071x	0.0320	0.1900	1985	0.3658
CDD	y = 53.7000 + 0.0521x	0.0013	0.7900	1981	0.4919
CWD	y = 3.0100 + 0.00491x	0.0066	0.5600	1956	0.6511
RX1day	y = 22.5000 + 0.1420x	0.0560	0.0820	1985	0.05505
RX5day	y = 31.8000 + 0.1710x	0.0420	0.1300	1982	0.1928
R95p	y = 24 + 0.769x	0.1200	0.0080 **	1986	0.0333 *
R99p	y = 1.8800 + 0.4070x	0.0960	0.0210 *	1983	0.1893
PRCPtot	y = 194 + 0.9600x	0.1200	0.0100 *	1983	0.07576
SDII	y = 4.6100 + 0.0279x	0.2000	0.0006 **	1981	0.001395 *

* means significant, ** means extremely significant.

3.2. Statistical Parameters of Tree-Ring Width Chronologies for the Two Tree Species

First, the SNR and EPS values of chronologies for Schrenk spruce and Siberian larch were relatively high, indicating that the two chronologies were reliable and suitable for studying the response of tree radial growth to climate change (Table 2). Moreover, the high values of SD, MS, and AC1 also indicated that the chronologies contained much climate information for the two tree species. However, Schrenk spruce exhibited higher consistency between core sequences and was more sensitive to climate change than Siberian larch due to its higher R and PC1 values.

3.3. Relationships between Radial Growth and Climatic Factors for the Two Species

The responses of radial growth of the two tree species to climatic factors significantly differed before and after the abrupt temperature change, based on the Pearson's correlation results between the STDs of the two tree species and the annual indexes of 28 climatic factors (Figures 4 and 5). Before the abrupt temperature change, the correlations between Schrenk spruce indexes and the annual climatic factors were poor. However, the STD of Schrenk spruce had a significant negative correlation with summer days (SU25) ($p < 0.05$) and significant positive correlations with cool days (TX10p) and frost days (FD0) ($p < 0.05$) after the abrupt temperature change. The STD of Siberian larch had significant positive correlations with annual mean temperature (Temp), TXn, and TNn ($p < 0.01$) and significant negative correlations with TX10p ($p < 0.01$) before the abrupt temperature change. However, the STD of Siberian larch had a significant positive correlation with FD0 ($p < 0.05$) and a significant negative correlation with R10 ($p < 0.01$) after the abrupt temperature change.

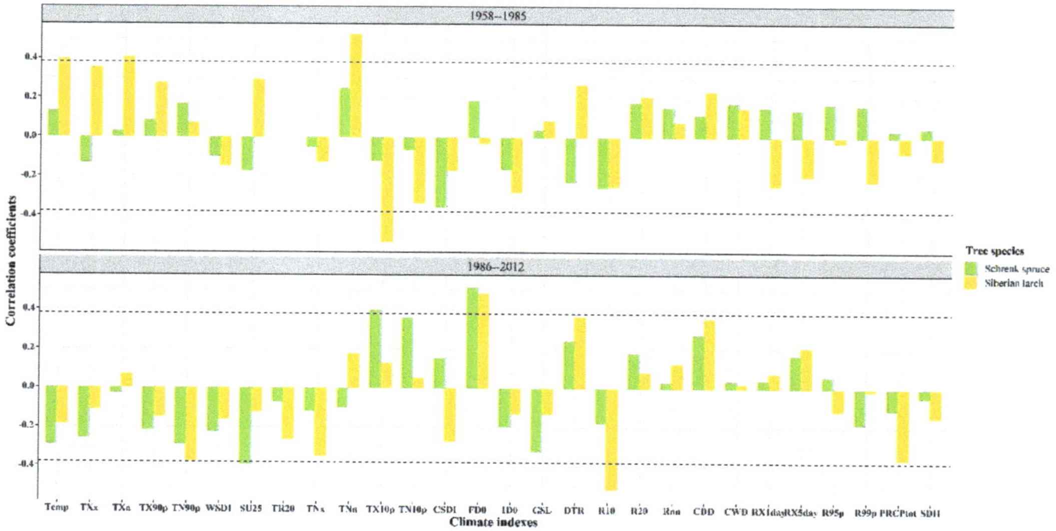

Figure 4. Correlation coefficients between the standard chronologies of the two species and annual values of 28 climate indexes during 1958–2012. Dotted line: critical value of significance level $p < 0.05$.

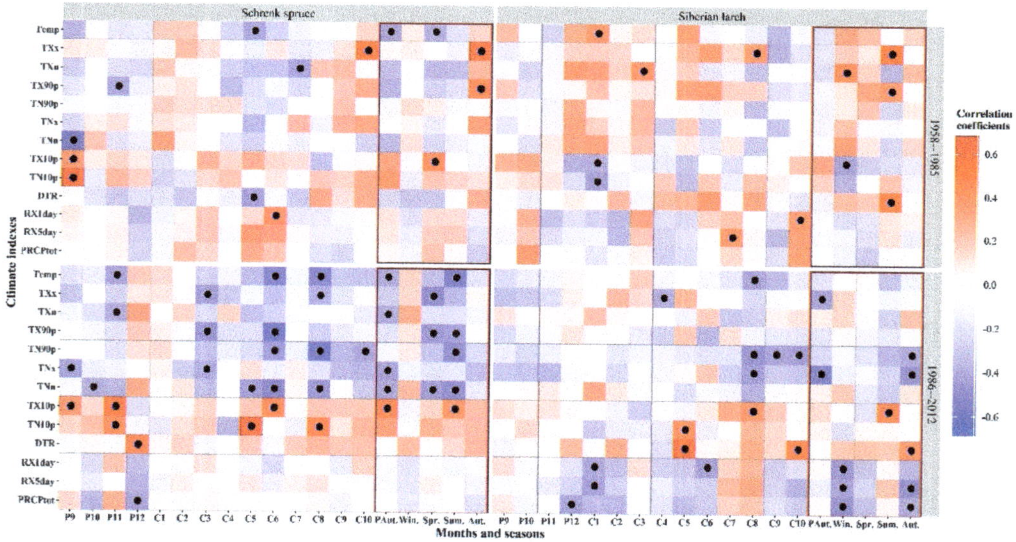

Figure 5. Correlation coefficients between the standard chronologies of the two species and monthly values of 13 climate indexes during 1958–2012. Blue: negative correlation, red: positive correlation; black dots: significance level < 0.05, red rectangular box: the correlation coefficients between the standard chronologies of the two species and seasonal climate indicators. P: previous year, C: current year (for example, P9: September of the previous year, C1: January of current year).

The correlation results between the indexes of the two tree species and seasonal factors showed that the STD of Schrenk spruce had a significant correlation with previous autumn temp, spring temp, TX10p, autumn TXx, and TX90p before climate change (1958–1985). That of Schrenk spruce had significant relationships with temp, TXN, TNX, TNN, and TX10P in the autumn of the previous year; TXx, TX90p, and TNn in the spring; and Temp,

TX90p, TN90p, TNn, and TX10p in the summer of the current year after abrupt climate change (1986–2012). The STD of Siberian larch was significantly correlated with only TXn, TX10p in winter, TXx, TX90p, and DTR in summer during the period of 1958–1985. That of Siberian larch had significant correlation with TXX; TNX in the autumn; RX1day, RX5day, and PRCPtot in the winter of the previous year; TX10p in the summer; and TN90p, TNx, DTR, RX5day, and PRCPtot in the autumn of the current year during the period of 1958–1985.

The correlation results between the radial growth of the two tree species and 13 monthly climate indexes from September of the previous year to October of the current year are shown in Figure 5. The STD of Schrenk spruce had significant correlations with P9-TNn (month-climatic factors), P9-TX10p, P9-TN10p, P11-TX90p, C5-Temp, C5-TNx, C5-DTR, C6-RX1day, C7-TXn, and C10-Txx ($p < 0.05$) during the period of 1958–1985. The STD of Schrenk spruce had significant correlations with 25 climatic factors during the period of 1985–2012. The STD of Siberian larch had significant correlations with P10-PRCPtot, P12-TX10p, C1-Temp, C1-TX10p, C1-TN10p, C3-TXn, C7-RX5day, C8-TXx, C10-RX1day, and C10-PRCPtot during the period of 1958–1985. The STD of Siberian larch had significant correlations with 14 climatic factors during the period of 1985–2012. In addition, the negative correlations between the radial growth of Siberian larch and the three precipitation indexes (RX1day, RX5day, and PRCPtot) were significantly enhanced. Schrenk spruce was strongly restricted by the temperatures at the end of the growing season of the previous year and in the early and middle growing seasons of the current year after 1985. However, Siberian larch was mainly affected by the temperatures in the middle and late growing seasons of the current year.

3.4. Comparison of the Resistance Indexes of the Two Tree Species to Climate Change

Both Schrenk spruce and Siberian larch were detected in 9 low-value years (Schrenk spruce: 1958, 1974, 1978, 1981, 1995, 1998, 2003, 2008, and 2012; Siberian larch: 1966, 1976, 1981, 1984–1985, 1998–1999, 2003, and 2008) defined as narrow-ring years with a value 20% lower than the average value of the previous two years based on the STDs and BAI, (Figure 6). Schrenk spruce showed only 4 low-value years (STDs: 1974, 2003, 2008, and 2012; BAI: 1998, 2003, 2008, and 2012) and Siberian larch showed 7 low-value years (STDs: 1966, 1976, 1984–1985, 1998–1999, and 2003; BAI: 1966, 1984–1985, 1998–1999, 2003, and 2008) defined as narrow-ring years with a value 25% lower than the average value of the previous two years. The results suggested three years (1981, 1998, and 2003) that were narrow-ring years for both tree species. Since the Siberian larch showed narrow rings in both 1998 and 1999, we chose 1981 and 2003 as pointer years to compare resistance to environmental disturbances between the two tree species.

Figure 7 shows the results of one-way analysis of variance (ANOVA) for the three resistance indicators of the two tree species in 1981 and 2003. In 1981, the resistance indicators of Siberian larch were significantly higher than those of Schrenk spruce (Schrenk spruce: RT = 1.063, RC = 1.06, and RS = 1.128; Siberian larch: RT = 1.1354, RC = 1.1354, and RS = 1.3046, $p < 0.05$). In 2003, the three resistance indicators of Siberian larch were still higher than those of Schrenk spruce (Schrenk spruce: RT = 1.028, RC = 1.033, and RS = 1.063; Siberian larch: RT = 1.05, RC = 1.041, and RS = 1.094, $p < 0.05$). However, the difference between the two tree species in terms of resistance to environmental disturbances gradually decreased after the temperature changed. The resistance indexes of both tree species declined in 2003 compared with 1981, especially those of Siberian larch.

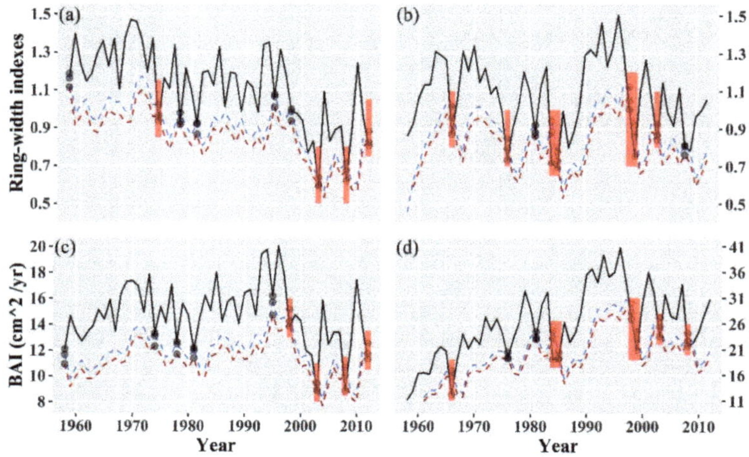

Figure 6. Variability of the standard chronologies for the two tree species and two narrow-ring horizontal lines. (**a**): Schrenk spruce, (**b**): Siberian larch. (**c**): BAI value of Schrenk spruce, (**d**): BAI value of Siberian larch. Solid black line: tree ring width of the standard chronologies, blue dotted line: 80% of the two-year moving average of the standard chronology, red dotted line: 75% of the two-year moving average of the standard chronology, gray dots: narrow years in the standard chronology with a value 20% less than the average of the previous two years, red rectangular box: narrow years in the standard chronology with a value 25% less than the average of the previous two years.

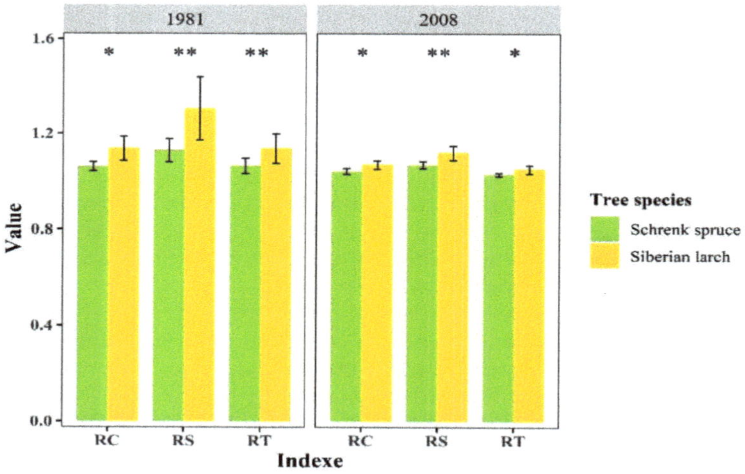

Figure 7. Histogram of the resistance indexes (RT: resistance, RC: recovery and RS: resilience) of the two species in 1981 and 2003. Asterisks indicate significance differences in the resistance index between Schrenk spruce and Siberian larch according to ANOVA, with *: significance level < 0.05, **: significance level < 0.01.

4. Discussion

4.1. Evaluation of the Responses of the Two Tree Species to Extreme Climate

The processes of tree leaf emergence and development, carbon accumulation, and xylem growth are all affected by climate, especially unstable extreme climate, which will have a more significant effect on tree-ring growth [44]. TRW includes abundant environmental change signals, which are direct and clear data that represent the responses of tree radial growth to climate change [18]. Therefore, TRW can be used to accurately analyze the impact of climate change on forest ecosystems and evaluate the ecological response of tree growth to climate change.

Since the end of the last century, the trend of global warming has been obvious. The eastern Tianshan Mountains, located in the middle latitude of the Northern Hemisphere and central Asia, has a particularly significant interannual variation, especially the trend of the temperature rising and the abrupt change of the most extreme climate index in the 1980s (Figure 3; Table 4). After the temperature changed, the extreme warmth indexes (TXx, TX90P, TX10P, and SU25) and the minimum temperature indexes reflecting extreme cold (TNx and TNn) showed clear increasing trends, and the negative responses of Schrenk spruce and Siberian larch to those climate factors significantly strengthened (Figures 3 and 4). Under natural conditions, extreme high-temperature events are often accompanied by decreasing precipitation and increasing vapor pressure deficit (VPD), which further induce drought and aggravate drought stress on tree growth [1,45,46]. Extreme high temperature and drought damage the hydraulic structure of trees by cavitating the xylem and increasing duct embolism and can also cause tree "carbon starvation" by stimulating the closing of the stomata and inhibiting photosynthesis [47,48]. Decreasing trends could extend the growing season of plants (Figure 3). However, the probability of trees suffering from freezing damage has also increased with the extension of the growing season [33,49]. Specifically, trees in high-latitude areas have a narrower safety threshold for frost and greater risks than those in low-latitude areas [11]. Moreover, increasing temperatures before the growing season would increase the melting of snow and evaporation, which would aggravate the restriction of subsequent tree radial growth by limited water availability [19, 50]. In addition, an increase in minimum temperature in the winter could reduce the overwintering mortality of pathogens and pests, thereby increasing the probability of pests and diseases in the coming year and ultimately affecting the radial growth of trees [48].

We also detected large differences in the times and types of responses of Schrenk spruce and Siberian larch to extreme climatic factors. After the temperature change during 1986–2012, the responses of Schrenk spruce to the extreme temperature indicators in the autumn of the previous year and the spring and summer of the current year became more significant (Figure 5). However, Siberian larch was more restricted by the extreme cold temperature indexes and extreme precipitation in the winter of the previous year and climatic factors in the autumn of the current year (Figure 5). Compared with Siberian larch, which is drought-tolerant and low-nutrient-tolerant, Schrenk spruce prefers a more humid and richer growth environment [51,52]. Therefore, Schrenk spruce is more susceptible to the obvious impact of climate change under significant increases in temperature and the frequency of extreme weather events (high temperature, drought, and frost) [53]. Moreover, Schrenk spruce, with shallow roots, is more susceptible to drought restrictions during the growing season due to its limited capacity for water storage and access to deep soil water resources [54]. In addition, evergreen Schrenk spruce was more significantly affected by climatic factors at the end of the growing season of the previous year than was deciduous Siberian larch because evergreen tree species continue to photosynthesize at the end of the growing season in order to maintain aboveground biomass [55,56]. However, the radial growth of Siberian larch is very sensitive to extreme precipitation variation in winter (Figure 5). The precipitation in winter is mostly in the form of snow in high-altitude areas at mid-to-high latitudes. Evergreen Schrenk spruce still retains a large number of branches and leaves in winter, and the amount of snow under the forest is correspondingly reduced since its canopy can intercept most of the snow. However, the amount of snow under larch forest is larger, and it takes longer for the snow to melt, which can delay the initiation of tree growth [57–59].

4.2. Comparison of the Resistances of the Two Tree Species to Extreme Climate

The frequency of narrow rings in the STDs and the resistance, recovery, and resilience of Schrenk spruce and Siberian larch were significantly different when the trees faced environmental disturbances, according to the occurrence of narrow rings and the resistance indicators reflecting radial growth (Figure 7). The narrow-ring years appearing in the STD of Schrenk spruce were all low-value years (1974, 1978, 2003, 2008, 2012, etc.), while Siberian larch was prone to consecutive low-value years. Schrenk spruce was more susceptible to occasional environmental disturbances, but the impact was relatively weak, while Siberian larch was more susceptible to sustained strong environmental stress, and the impact was relatively strong (Figure 6). High-frequency environmental interference might cause Schrenk spruce to develop strong tolerance in the face of subsequent interference [44]. Therefore, evergreen tree species and deciduous tree species may adopt different biomass allocation strategies under environmental stress conditions, with evergreen trees potentially choosing to preserve more aboveground biomass in order to maintain growth but larch choosing to abandon its leaves and branches and increase carbohydrate reserves in the rhizomes for subsequent recovery [60,61]. In addition, many studies have found that the growth rate of deciduous tree species is generally higher than that of evergreen tree species, and extreme climate events have more sudden and profound effects on deciduous tree species [62–64].

The radial growth of Schrenk spruce and Siberian larch was significantly lower in 1981 and 2003 than in other years (Figure 6). We first checked the historical data and ruled out the occurrence of biological interference in the two years [65]. For 1981, the warm autumn of the previous year may have allowed the trees to remain more active at the end of the growing season, consume more photosynthetic nutrients, and reduce the resource reserves for growth in the coming year. Therefore, the spring drought in 1981 (C3–C7) further aggravated the effect of water stress on tree growth during the growing season [19]. Before the abrupt climate change, Schrenk spruce was obviously restricted by the temperatures of the previous autumn and the spring of the current year, while Siberian larch was less restricted by the climate during these periods (Figures 4 and 5). Therefore, Schrenk spruce was more constrained by drought caused by high temperature and thus exhibited lower resistance and resilience in 1981. In 2003, the temperature was low throughout the year, especially in spring, and the precipitation in the previous winter increased significantly (Figure 2). The large amount of snow in winter and the low temperature at the beginning of the growing season slowed the increase in soil temperature, affected the emergence of new leaves, delayed growth, and shortened the growth period, thereby affecting the radial growth of trees [59,66,67]. After the abrupt temperature change, Siberian larch exhibited a significant negative correlation with the precipitation of the previous winter (Figure 5). Evergreen coniferous species usually use nonstructural carbohydrates from old leaves for bud burst, and it takes longer for new evergreen leaves to emerge [68,69]. However, the bud burst period of deciduous trees is earlier and is more affected by the low temperatures and frost in spring because these trees rely on the NSC from stems and branches for leaf development [70,71]. Therefore, the low temperature in the spring of 2003 had a significant impact on the early budding and radial growth of Siberian larch. In addition, the fluctuations in climate in 2003 were more obvious than those in 1981. Studies have shown that the resistance of trees to climate change is closely related to the intensity of extreme climate events [28]. This might be one of the important reasons why the resistance difference between Schrenk spruce and Siberian larch decreased. On two occasional climate disturbance events, Siberian larch showed better resistance and resilience, so we believe that Siberian larch may be better at adapting to extreme environmental events in the context of climate warming.

The frequency and intensity of extreme climate events have increased significantly since the beginning of the 21st century. The occurrence of extreme climate events has significantly disrupted the growth, death, and regeneration of trees, thereby affecting the structure and function of entire forest ecosystems [72]. Especially in arid and semiarid regions, trees have shown significant declines in growth rates and increased mortality due to drought and climate warming [73]. Evergreen Schrenk spruce and deciduous Siberian larch are dominant species in the eastern Tianshan Mountains. The adaptations of these two tree species to climate change exhibit obvious variation due to significant differences in genetic physiology and ecological thresholds between the species [74]. According to our research results, Schrenk spruce is more sensitive to climate and more affected by extreme climatic changes, while Siberian larch exhibits better resistance and a greater recovery ability when faced with the same climate disturbances. Moreover, extreme climate events do not occur in isolation. For example, extreme high temperature and heat wave events are often accompanied by drought, extreme precipitation events are often accompanied by hurricanes or storms, and extreme low-temperature events may also be accompanied by cold waves and hail [75]. The response of trees to extreme climate events is also affected by factors such as tree species and the type and intensity of the extreme event [76]. Therefore, we should pay more attention to the effects of extreme climate on tree growth and the adaptation of tree growth to climate change in ecologically fragile and sensitive areas. We also need to conduct more research to assess the adaptability of different tree species to extreme climatic events under future climate change.

5. Conclusions

The growth of trees and the structure and stability of forest ecosystems have been disturbed by the more frequent occurrence of extreme climatic events with global warming. Schrenk spruce and Siberian larch are the most dominant mountain conifers in the arid and semiarid regions of central Asia. Therefore, studying their growth patterns and responses to extreme climate in this region can provide importance references. Our research results showed that Schrenk spruce and Siberian larch had strong responses to 27 extreme climatic indexes. Specifically, the correlations between characteristics of the two tree species and the extreme climatic indexes increased after the temperature changed. This confirmed that the restrictive effect of extreme climate on forest ecosystems gradually increased with the intensification of climate change. Therefore, it is necessary to further strengthen the dynamic monitoring of extreme climate and forest ecosystems and research on their impact mechanisms. In addition, our study also revealed that the evergreen spruce species was more susceptible to extreme climate, while the deciduous species (Siberian larch) had stronger resistance and a greater recovery ability after being disturbed by the extreme environment. Therefore, forest ecosystems should be protected and managed more specifically according to differences in the responses and resistances of tree species to extreme climatic events under future climate change.

Author Contributions: Methodology: L.J. and X.W.; software: X.W.; validation: L.J. and X.W.; formal analysis: L.J. and X.W.; investigation: L.J., X.W., X.L., C.Q. and R.X.; resources: L.J. and X.L.; data curation: L.J. and X.W.; writing—original draft: L.J. and X.W.; writing—review and editing: L.J., X.L. and X.W.; visualisation: X.L., C.Q., R.X. and D.D.; supervision: L.J., X.L., D.D., C.Q. and R.X.; project administration: L.J. and X.W.; funding acquisition: L.J. All authors have read and agreed to the published version of the manuscript.

Funding: This research was supported by Natural Science Foundation of Gansu (No. 21JR7RA111), CAS "Light of West China" Program (2020XBZG-XBQNXZ-A), and the 2022 Major Scientific Research Project Cultivation Plan of Northwest Normal University (WNU-LKZD2022-04).

Data Availability Statement: The data set analyzed during the current research period is provided in China Meteorological Data Network. Other data sets analyzed during the current research period may be obtained from the corresponding author upon reasonable request.

Acknowledgments: We also thank the anonymous referees for helpful comments on the manuscript.

Conflicts of Interest: The authors declare that they have no known competing financial interests or personal relationships that could have appeared to influence the work reported in this paper.

References

1. Piao, S.; Zhang, X.; Chen, A.; Liu, Q.; Lian, X.; Wang, X.; Peng, S.; Wu, X. The impacts of climate extremes on the terrestrial carbon cycle: A review. *Sci. China Earth Sci.* **2019**, *62*, 1551–1563. [CrossRef]
2. Evans, P.; Brown, C.D. The Boreal-Temperate Forest Ecotone Response to Climate Change. *Environ. Rev.* **2017**, *25*, 423–431. [CrossRef]
3. He, Z.B.; Du, J.; Chen, L.F.; Zhu, X.; Lin, P.F.; Zhao, M.M.; Fang, S. Impacts of recent climate extremes on spring phenology in arid-mountain ecosystems in China. *Agric. For. Meteorol.* **2018**, *260–261*, 31–40. [CrossRef]
4. Siegmund, J.F.; Wiedermann, M.; Donges, J.F.; Donner, R.V. Impact of temperature and precipitation extremes on the flowering dates of four german wildlife shrub species. *Biogeosciences* **2016**, *13*, 5541–5555. [CrossRef]
5. Depardieu, C.; Girardin, M.P.; Nadeau, S.; Lenz, P.; Bousquet, J.; Isabel, N. Adaptive genetic variation to drought in a widely distributed conifer suggests a potential for increasing forest resilience in a drying climate. *New Phytol.* **2020**, *227*, 427–439. [CrossRef]
6. Field, C.B.; Barros, V.; Stocker, T.F.; Dahe, Q. *Managing the Risks of Extreme Events and Disasters to Advance Climate Change Adaptation: Special Report of the Intergovernmental Panel on Climate Change*; Cambridge University Press: Cambridge, UK, 2012.
7. Suárez-Muñoz, M.; Bonet-García, F.J.; Navarro-Cerrillo, R. Forest management scenarios drive future dynamics of Mediterranean planted pine forests under climate change. *Landsc. Ecol.* **2023**, 1–6. [CrossRef]
8. Islam, M.; Rahman, M.; Bräuning, A. Impact of extreme drought on tree-ring width and vessel anatomical features of Chukrasia tabularis. *Dendrochronologia* **2019**, *53*, 63–72. [CrossRef]
9. Kramer, P.; Kozlowski, T.T. *Physiology of Woody Plants*; Elsevier: Amsterdam, The Netherlands, 2012.
10. Begović, K.; Rydval, M.; Mikac, S.; Čupić, S.; Svobodova, K.; Mikoláš, M.; Kozák, D.; Kameniar, O.; Frankovič, M.; Pavlin, J.; et al. Climate-growth relationships of Norway Spruce and silver fir in primary forests of the Croatian Dinaric mountains. *Agric. For. Meteorol.* **2020**, *12*, 288–289. [CrossRef]
11. Zohner, C.M.; Mo, L.; Sebald, V.; Renner, S.S. Leaf-out in northern ecotypes of wide-ranging trees requires less spring warming, enhancing the risk of spring frost damage at cold range limits. *Glob. Ecol. Biogeogr.* **2020**, *29*, 1065–1072. [CrossRef]
12. Chang, C.T.; Wang, L.J.; Huang, J.C.; Liu, C.P.; Wang, C.P.; Lin, N.H.; Wang, L.X.; Lin, T.C. Precipitation controls on nutrient budgets in subtropical and tropical forests and the implications under changing climate. *Adv. Water Resour.* **2017**, *103*, 44–50. [CrossRef]
13. Huang, J.C.; Lee, T.Y.; Lin, T.C.; Hein, T.; Lee, L.C.; Shih, Y.T.; Kao, S.J.; Shiah, F.K.; Lin, N.H. Effects of different N sources on riverine DIN export and retention in a subtropical high-standing island Taiwan. *Biogeosciences* **2016**, *13*, 1787–1800. [CrossRef] [PubMed]
14. Jia, L.L.; Fang, Y.C.; Munger, W.; Jiang, P.; Whitby, T.G.; Chen, S.Y.; Wei, W.J.; Xiu, L.M. Precipitation extremes influence patterns and partitioning of evapotranspiration and transpiration in a deciduous boreal larch forest. *For. Ecol. Manag.* **2020**, *287*, 107936. [CrossRef]
15. Fritts, H.C. *Tree Rings and Climate*; Academic: London, UK, 1976; p. 582.
16. Shao, X.M. Progress of tree-ring chronology. *Quat. Sci.* **1997**, *17*, 265–271.
17. Li, C.J.; Chen, T.; Wang, B.; Xu, G.B.; Zhang, X.W.; Wu, G.J. Advances in research on the abnormal structure of tree-rings. *Acta Ecol. Sin.* **2019**, *38*, 1538–1550.
18. Wu, X.D. *Tree Rings and Climate Change*; Meteorological: Beijing, China, 1990; pp. 65–75.
19. Gao, S.; Liu, R.S.; Zhou, T.; Fang, W.; Yi, C.X.; Lu, R.J.; Zhao, X.; Luo, H. Dynamic responses of tree-ring growth to multiple dimensions of drought. *Glob. Chang. Biol.* **2018**, *24*, 5380–5390. [CrossRef]
20. Parobeková, Z.; Sedmáková, D.; Kucbel, S.; Pittner, J.; Jaloviar, P.; Saniga, M.; Balanda, M.; Vencurik, J. Influence of disturbances and climate on high-mountain Norway spruce forests in the Low Tatra Mts. Slovakia. *For. Ecol Manag.* **2016**, *380*, 128–138. [CrossRef]
21. DeSoto, L.; Camarero, J.J.; Olano, J.M. Geographically structured and temporally unstable growth responses of Juniperus thurifera to recent climate variability in the Iberian Peninsula. *Eur. J. For. Res.* **2012**, *131*, 905–917. [CrossRef]
22. Hofgaard, A.; Ols, C.; Drobyshev, I.; Kirchhefer, A.J.; Söderström, L. Non-stationary Response of Tree Growth to Climate Trends Along the Arctic Margin. *Ecosystems* **2019**, *22*, 434–451. [CrossRef]
23. Jiao, L.; Chen, K.; Wang, S.J.; Liu, X.P. Stability evaluation of radial growth of Picea schrenkiana in different age groups in response to climate change in the eastern Tianshan Mountains. *J. Mt. Sci.-Engl.* **2020**, *17*, 1735–1748. [CrossRef]
24. Zhang, Y.; Yin, D.C.; Tian, K.; Zhang, W.G.; He, R.H.; He, W.Q.; Sun, J.M.; Liu, Z.Y. Radial growth responses of Picea likiangensis to climate variabilities at different altitudes in Yulong Snow Mountain, southwest China. *Chin. J. Plant Ecol.* **2018**, *042*, 629–639. [CrossRef]
25. Altman, J.; Treydte, K.; Pejcha, V.; Cerny, T.; Petrik, P.; Srutek, M.; Song, J.S.; Trouet, V.; Dolezal, J. Tree growth response to recent warming of two endemic species in Northeast Asia. *Clim. Chang.* **2020**, *162*, 1345–1364. [CrossRef]
26. Jia, F.F.; Sun, C.Y.; Sun, H.Y.; Li, X. Responses of radial growth of two dominant tree species to climate change on Changling Mountain. *Acta Ecol. Sin.* **2019**, *39*, 6332–6340. [CrossRef]
27. Lloret, F.; Keeling, E.G.; Sala, A. Components of tree resilience: Effects of successive low-growth episodes in old ponderosa pine forests. *Oikos* **2011**, *120*, 1909–1920. [CrossRef]

28. Wang, X.F.; Yang, B.; Li, G. Drought-induced tree growth decline in the desert margins of Northwestern China. *Dendrochronologia* **2020**, *60*, 125685. [CrossRef]
29. Andivia, E.; Ruiz-Benito, P.; Díaz-Martínez, P.; Carro-Martínez, N.; Madrigal-González, J. Inter-specific tolerance to recurrent droughts of pine species revealed in saplings rather than adult trees. *For. Ecol. Manag.* **2020**, *459*, 117848. [CrossRef]
30. Bottero, A.; D'Amato, A.W.; Palik, B.J.; Bradford, J.B.; Fraver, S.; Battaglia, M.A.; Asherin, L.A. Density-dependent vulnerability of forest ecosystems to drought. *J. Appl. Ecol.* **2017**, *54*, 1605–1614. [CrossRef]
31. Pretzsch, H.; Schütze, G.; Uhl, E. Resistance of European tree species to drought stress in mixed versus pure forests: Evidence of stress release by inter-specific facilitation. *Plant Biol.* **2013**, *15*, 483–496. [CrossRef]
32. Príncipe, A.; van der Maaten, E.; van der Maaten-Theunissen, M.; Struwe, T.; Wilmking, M.; Kreyling, J. Low resistance but high resilience in growth of a major deciduous forest tree (*Fagus sylvatica* L.) in response to late spring frost in southern Germany. *Trees* **2017**, *31*, 743–751. [CrossRef]
33. Vitasse, Y.; Bottero, A.; Cailleret, M.; Bigler, C.; Fonti, P.; Gessler, A.; Lévesque, M.; Rohner, B.; Weber, P.; Rigling, A.; et al. Contrasting resistance and resilience to extreme drought and late spring frost in five major European tree species. *Glob. Chang. Biol.* **2019**, *25*, 3781–3792. [CrossRef]
34. Cohen, J.; Screen, J.A.; Furtado, J.C.; Barlow, M.; Whittleston, D.; Coumou, D.; Francis, J.; Dethloff, K.; Entekhabi, D.; Overland, J.; et al. Recent arctic amplification and extreme mid-latitude weather. *Nat. Geosci.* **2014**, *7*, 627–637. [CrossRef]
35. IPCC. *Climate Change 2013: The Physical Science Basis: Contribution of Working Group I to the Fifth Assessment Report of the Intergovernmental Panel on Climate Change*; Cambridge University: Cambridge, UK, 2013.
36. Qin, L.; Yuan, Y.J.; Shang, H.M.; Yu, S.L.; Liu, W.P.; Zhang, R.B. Impacts of Global Warming on the Radial Growth and Long-Term Intrinsic Water-Use Efficiency (iWUE) of Schrenk Spruce (*Picea schrenkiana* Fisch. et Mey) in the Sayram Lake Basin, Northwest China. *Forests* **2020**, *11*, 380. [CrossRef]
37. Shi, B.B. Responses of the upper and lower tree-ring width of the Schrenk spruce forest in the central Tianshan Mountains to climate change. *Chin. J. Appl. Environ. Biol.* **2017**, *5*, 1–12. (In Chinese)
38. Zhang, R.B.; Yuan, Y.J.; Gou, X.H.; Qing, H. Tree-ring-based moisture variability in western Tianshan Mountains since AD 1882 and its possible driving mechanism. *Agric. For. Meteorol.* **2016**, *218*, 267–276. [CrossRef]
39. Holmes, R.L. Computer-assisted quality control in tree-ring dating and measurement. *Tree-Ring Bull.* **1983**, *43*, 69–95.
40. Cook, E.R. A Time Series Analysis Approach to Tree Ring Standardization. Doctoral Dissertation, University of Arizona, Tucson, AZ, USA, 1985.
41. Ali, S.; Eum, H.I.; Cho, J.; Li, D.; Khan, F.; Firdos, K.; Dairaku, K.; Lall, S.M.; Syewoon, H.; Wajid, N.; et al. Assessment of climate extremes in future projections downscaled by multiple statistical downscaling methods over Pakistan. *Atmos. Res.* **2019**, *222*, 114–133. [CrossRef]
42. Bonsal, B.R.; Zhang, X.; Vincent, L.A.; Hogg, W.D. Characteristics of daily and extreme temperatures over Canada. *J. Clim.* **2001**, *14*, 1959–1976. [CrossRef]
43. Zhao, N.; Jiao, Y.M.; Ma, T.; Zhao, M.M.; Yue, T.X. Estimating the effect of urbanization on extreme climate events in the Beijing-Tianjin-Hebei region, China. *Sci. Total Environ.* **2019**, *688*, 1005–1015. [CrossRef]
44. Pallardy, S.G. *Physiology of Woody Plants*; Academic: London, UK, 2010.
45. Grossiord, C.; Buckley, T.N.; Cernusak, L.A.; Novick, K.A.; Poulter, B.; Siegwolf, R.T.W.; Sperry, J.S.; McDowell, N.G. Plant responses to rising vapor pressure deficit. *New Phytol.* **2020**, *226*, 1550–1566. [CrossRef]
46. Will, R.E.; Wilson, S.M.; Zou, C.B.; Hennessey, H.C. Increased vapor pressure deficit due to higher temperature leads to greater transpiration and faster mortality during drought for tree seedlings common to the forest-grassland ecotone. *New Phytol.* **2013**, *200*, 366–374. [CrossRef]
47. Bréda, N.; Huc, R.; Granier, A.; Dreyer, E. Temperate forest trees and stands under severe drought: A review of ecophysiological responses, adaptation processes and long-term consequences. *Ann. For. Sci.* **2006**, *63*, 625–644. [CrossRef]
48. McDowell, N.; Pockman, W.T.; Allen, C.D.; Breshears, D.D.; Cobb, N.; Kolb, T.; Plaut, J.; Sperry, J.; West, A.; Williams, D.G.; et al. Mechanisms of plant survival and mortality during drought: Why dosome plants survive while others succumb to drought? *New Phytol.* **2008**, *178*, 719–739. [CrossRef] [PubMed]
49. Wang, Z.; Xu, J.L.; Wei, H.X. The occurrence of "back spring cold" weather and its influence on the spring growth of trees in China. *North. Hortic.* **2015**, *16*, 195–201. [CrossRef]
50. Wang, X.C.; Pederson, N.; Chen, Z.J.; Lawton, K.; Zhu, C.; Han, S.J. Recent rising temperatures drive younger and southern Korean pine growth decline. *Sci. Total Environ.* **2019**, *649*, 1105–1116. [CrossRef] [PubMed]
51. Abaimov, A.P. Geographical distribution and genetics of Siberian larch species. In *Permafrost Ecosystems: Siberian Larch Forests*; Osawa, A., Zyryanova, O.A., Matsuura, Y., Kajimoto, T., Wein, R.W., Eds.; Springer: Dordrecht, The Netherlands, 2010; pp. 41–58.
52. Peng, J.F.; Gou, X.H.; Chen, F.H.; Li, J.B. Climatic records of tree-ring width in Picea schrenkiana Fisch and Larix sibirica Ledb. *Ecol. Environ.* **2005**, *14*, 460–465. (In Chinese)
53. Reich, P.B.; Walters, M.B.; Ellsworth, D.S. Leaf life-span in relation to leaf, plant, and stand characteristics among diverse ecosystems. *Ecol. Monogr.* **1992**, *62*, 365–392. [CrossRef]
54. King, G.; Fonti, P.; Nievergelt, D.; Buentgen, U.; Frank, D. Climatic drivers of hourly to yearly tree radius variations along a 6 degrees C natural warming gradient. *Agric. For. Meteorol.* **2013**, *168*, 36–46. [CrossRef]

55. Mooney, H.A.; Dunn, E.L. Convergent evolution of Mediterranean-climate evergreen sclerophyll shrubs. *Evolution* **1970**, *24*, 292–303. [CrossRef]
56. Westoby, M.; Falster, D.S.; Moles, A.T.; Vesk, P.A.; Wright, I.J. Plant ecological strategies: Some leading dimensions of variation between species. *Annu. Rev. Ecol. Syst.* **2002**, *33*, 125–159. [CrossRef]
57. Schneider, E.E.; Affleck, D.L.R.; Larson, A.J. Tree spatial patterns modulate peak snow accumulation and snow disappearance. *For. Ecol. Manag.* **2019**, *441*, 9–19. [CrossRef]
58. Storck, P.; Lettenmaier, D.P.; Bolton, S.M. Measurement of snow interception and canopy effects on snow accumulation and melt in a mountainous maritime climate, Oregon, United States. *Water Resour. Res.* **2002**, *38*, 5-1–5-16. [CrossRef]
59. Watson, E.; Luckman, B.H. An investigation of the snowpack signal in moisture-sensitive trees from the Southern Canadian Cordillera. *Dendrochronologia* **2016**, *38*, 118–130. [CrossRef]
60. Tomlinson, K.W.; Lourens, P.; Frans, B.; Borghetti, F.; Jacobs, L.; van Langevelde, F. Relative growth rate variation of evergreen and deciduous savanna tree species is driven by different traits. *Ann. Bot.-Lond.* **2014**, *114*, 315–324. [CrossRef] [PubMed]
61. Tomlinson, K.W.; van Langevelde, F.; Ward, D.; Bongers, F.; da Silva, D.A.; Prins, H.H.T.; de Bie, S.; Sterck, F.J. Deciduous and evergreen trees differ in juvenile biomass allometries because of differences in allocation to root storage. *Ann. Bot.-Lond.* **2013**, *112*, 575–587. [CrossRef]
62. Castellano, P.L.; Srur, A.M.; Lucas, O.; Bianchi, L.O. Climate-growth relationships of deciduous and evergreen Nothofagus species in Southern Patagonia, Argentina. *Dendrochronologia* **2019**, *58*, 125646. [CrossRef]
63. Martínez-Vilalta, J.; López, B.C.; Loepfe, L.; Lloret, F. Stand-and tree-level determinants of the drought response of Scots pine radial growth. *Oecologia* **2012**, *168*, 877–888. [CrossRef]
64. Zhou, B.; Fan, Z.X.; Qi, J.H. Inter-annual radial growth of evergreen and deciduous tree species and their response to climatic factors in a montane moist evergreen broad-leaved forest in the Ailao Mountains, Southwest China. *Acta Ecol. Sincia* **2020**, *40*, 1699–1708. [CrossRef]
65. Wen, K.G.; Shi, Y.G. *China Meteorological Disaster Dictionary, Xinjiang Volume*; Meteorological: Beijing, China, 2006.
66. Liu, B. The weather process in northern Xinjiang is frequent, and the low temperature repeatedly breaks the extreme value; the precipitation in southern Xinjiang is unevenly distributed and the temperature is close to normal. *Desert Oasis Meteorol.* **2003**, *26*, 33–34. (In Chinese)
67. Vaganov, E.A.; Hughes, M.K.; Kirdyanov, A.V.; Schweingruber, F.H.; Silkin, P.P. Influence of snowfall and melt timing on tree growth in subarctic Eurasia. *Nature* **1999**, *400*, 149–151. [CrossRef]
68. Egger, B.; Einig, W.; Sclereth, A.; Wallenda, T.; Magel, E.; Loewe, A.; Hampp, R. Carbohydrate metabolism in one- and two-year-old spruce needles, and stem carbohydrates from three months before until three months after bud break. *Physiol. Plant.* **1996**, *96*, 91–100. [CrossRef]
69. Vitasse, Y.; Schneider, L.; Rixen, C.; Christen, D.; Rebetez, M. Increase in the risk of exposure of forest and fruit trees to spring frosts at higher elevations in Switzerland over the last four decades. *Agric. For. Meteorol.* **2018**, *248*, 60–69. [CrossRef]
70. Taschler, D.; Neuner, G. Summer frost resistance and freezing patterns measured in situ in leaves of major alpine plant growth forms in relation to their upper distribution boundary. *Plant Cell Environ.* **2004**, *27*, 737–746. [CrossRef]
71. Bigler, C.; Bugmann, H. Climate-induced shifts in leaf unfolding and frost risk of European trees and shrubs. *Sci. Rep.* **2018**, *8*, 9865. [CrossRef]
72. Allen, C.D.; Macalady, A.K.; Chenchouni, H.; Bachelet, D.; McDowell, N.; Vennetier, M.; Kitzberger, T.; Rigling, A.; Breshears, D.D.; Hogg, E.H.; et al. A global overview of drought and heat-induced tree mortality reveals emerging climate change risks for forests. *For. Ecol. Manag.* **2020**, *259*, 660–684. [CrossRef]
73. Poulter, B.; Pederson, N.; Liu, H.; Zhu, Z.; D'Arrigo, R.; Ciais, P.; Davi, N.; Frank, D.; Leland, D.; Myneni, R.; et al. Recent trends in Inner Asian forest dynamics to temperature and precipitation indicate high sensitivity to climate change. *Agric. For. Meteorol.* **2013**, *178–179*, 31–45. [CrossRef]
74. Jiao, L.; Jiang, Y.J.; Zhang, W.T.; Wang, M.C.; Wang, S.J.; Liu, X.R. Assessing the stability of radial growth responses to climate change by two dominant conifer trees species in the tianshan mountains, northwest China. *For. Ecol. Manag.* **2019**, *466*, 667–677. [CrossRef]
75. Sillmann, J.; Roeckner, E. Indices for extreme events in projections of anthropogenic climate change. *Clim. Chang.* **2008**, *86*, 83–104. [CrossRef]
76. Xu, C.Y.; Liu, H.Y.; Anenkhonov, O.A.; Korolyuk, A.Y.; Sandanov, D.V.; Balsanova, L.D.; Naidanov, B.B.; Wu, X.C. Long-term forest resilience to climate change indicated by mortality, regeneration, and growth in semiarid southern Siberia. *Glob. Chang. Biol.* **2016**, *23*, 2370–2382. [CrossRef] [PubMed]

forests

MDPI

Article

Rapid Warming Exacerbates Winter Drought Stress in Trees at High-Altitude Areas in Northeast China

Rui Zhang [1], Xueping Bai [1,*], Xun Tian [1], Zhenju Chen [2], Haiyue Zhang [1] and Haiting Liu [1]

1 College of Life Sciences and Food Engineering, Inner Mongolia Minzu University, Tongliao 028000, China; ruizhang990123@163.com (R.Z.); tianxun@imun.edu.cn (X.T.); 15114710254@163.com (H.Z.); 16604992147@163.com (H.L.)

2 College of Forestry, Shenyang Agricultural University, Shenyang 110866, China; chenzhenju@syau.edu.cn

* Correspondence: baixueping@imun.edu.cn or xuepingbai90@hotmail.com

Abstract: Against the background of global warming, trees in high-latitude and high-altitude areas are more sensitive to rapid warming. Revealing the response patterns of trees at different altitudes to rapid warming in typical alpine mountain environments can help to predict the future distribution of forests in the region and the associated changes in the timberline. This study used tree-ring width data (band sampling) from Dahurian larch (*Larix gmelinii* Rupr.) along an altitudinal gradient (970–1409 m) on Oakley Mountain to establish 10 chronologies and to analyze the growth–climate response of larch to warming and altitudinal changes along a gradient. The results showed that before rapid warming, larch growth was strongly influenced mainly by precipitation in July–August, while after warming, the growth was controlled mainly by precipitation (snowfall) in winter (October–April) and showed a significant positive correlation with the SPEI in winter and a negative correlation with temperature in February–April (early spring). This indicates that the rapid warming event led to a drastic change in the water heat balance during the pre-growth period of the trees in the study area, which shifted the tree growth from being restricted by drought in the growing season before warming to being restricted by winter drought (i.e., the lagged effect of snowfall in the pre-growing season) after warming and that the trees at low altitudes suffered from more severe winter drought. In the future, as the global climate warms further, the growth of mountain larch at low altitudes (below 1200 m) will continue to decrease, and the coupling of winter snowfall and seasonal (February–April) warming will increase the upper altitude limit at which winter drought occurs for larch in the study area; additionally, larch at higher altitudes will be able to cope with this ecological process better.

Keywords: tree-ring; rapid warming; winter drought; altitude gradient; *Larix gmelinii* Rupr.

1. Introduction

The sixth report of the IPCC states that human activities have led to a sustained and rapid increase in global temperatures and a high frequency of multiple extreme events, with irreversible impacts on the development of human societies and the evolution of terrestrial ecosystems [1,2]. Forest ecosystems are important components of terrestrial ecosystems and provide resources and ecological services for human survival and development [3]. Rapid climate change will likely put enormous pressure on forest ecosystems in this century [4].

This rapid warming phenomenon is already having major impacts on forest ecosystems, including changes in tree growth and development and in forest composition and function, especially in some alpine and high-altitude areas [5–8]. Altitude is an important factor that indirectly affects tree growth in mountain habitats, and several studies have reported that rapid warming may negatively affect trees at lower altitudes but favor tree growth at higher altitudes [9,10]. Reduced water availability due to rapid warming in arid and semiarid zones inhibits tree growth and leads to tree mortality and forest decline [11,12]. In addition, several studies have emphasized the important role of seasonal temperatures at high altitudes and latitudes in determining tree growth [10,13].

Located in the northernmost part of China, the Great Xing'an Mountains is the largest and best-preserved original cold-temperate needle-leaved forest area in China and is also one of the regions experiencing the most drastic climate warming in China [8,13,14]. Numerous studies have shown that the forests of the Great Xing'an Mountains have been experiencing rapid warming and localized extreme drought events over the past half century [14–16], and associated dendroecological studies have shown that warming has resulted in significant differences in larch growth at different altitudes [17].

Mountain forests account for a large proportion of the total forest area in the study area, and the high-altitude areas referred to in previous studies on Dahurian larch along the altitudinal gradient are mainly flatlands in the high-altitude zone rather than mountains. This topographic difference greatly affects tree growth in the permafrost zone. For example, during summer in high-altitude flatlands, compared to mountainous terrain, where trees are well watered due to seasonal freezing and thawing of perennial permafrost, trees in mountainous environments are more susceptible to preseason drought constraints [18]. In this study, in order to explore the effect of winter drought on the radial growth of Dahurian larch at different altitudes in Oakley Mountain, a typical peak of the Great Xing'an Range against the background of rapid warming, we collected tree ring samples from 86 Dahurian larch species along an altitudinal gradient (970–1409 m), explored the relationship between growth and climatic factors by means of dendrochronology, and revealed the spatial and temporal variations of the distribution pattern of winter climate-driven tree growth under sudden temperature change. Our objectives were to (1) analyze the variability in the effects of dominant climate factors on the growth of Dahurian larch before and after abrupt temperature changes [19] in the context of global warming and (2) assess the spatial and temporal heterogeneity in the growth response of Dahurian larch to winter drought along an altitudinal gradient. We hypothesize that warming-induced winter drought stress is already a major limiting factor for the radial growth of low-elevation trees in alpine regions and that this stress is exacerbated by rapid warming.

2. Materials and Methods

2.1. Study Area

The study area is located on Oakley Mountain (51°50′ N, 122°02′ E, 1523 m above sea level) in the Great Xing'an Mountains in northeast China, which, as the highest peak in the northern section of the Great Xing'an Mountains, can better reflect the characteristics of the vertical distribution of vegetation in the forested area (Figure 1a). According to the actual survey results and literature records, the distribution of mountain vegetation exhibits significant vertical zonation along the altitudinal gradient [20,21]. The main coniferous or coniferous and broadleaved mixed forest belts (750–1350 m) and subalpine forest belts (1350–1500 m) are included, and the specific tree species from low to high are *Larix gmelinii* Rupr.—*Betula platyphylla* Sukaczev—*Rhododendron simsii* Planch., *Larix gmelinii* Rupr.—*Betula platyphylla* Sukaczev—*Pinus pumila* Pall., *Larix gmelinii* Rupr.—*Betula ermanii* Cham.—*Pinus pumila* Pall., *Betula ermanii* Cham.—*Pinus pumila* Pall.—*Sabina davurica* Pall. and *Pinus pumila* Pall. pure forests [22]. A variety of extremely drought-tolerant plants (*Pinus pumila* Pall., *Sabina davurica* Pall.) occur in the upper elevation coniferous forest belt, which is the main environmental characteristic of the region's typical high-latitude, high-altitude, alpine, and arid environment. Dahurian larch is the main tree species of zonal vegetation in the northern part of the Great Xing'an Mountains, with a vertical distribution range of approximately 400–1350 m above sea level, and it is the forest species with the largest distribution area and widest range in the region [23,24]. In this study, only Dahurian larch within a continuous altitudinal gradient (970–1409 m) on Oakley Mountain were selected for growth–climate analysis to avoid the influence of microtopography and other factors.

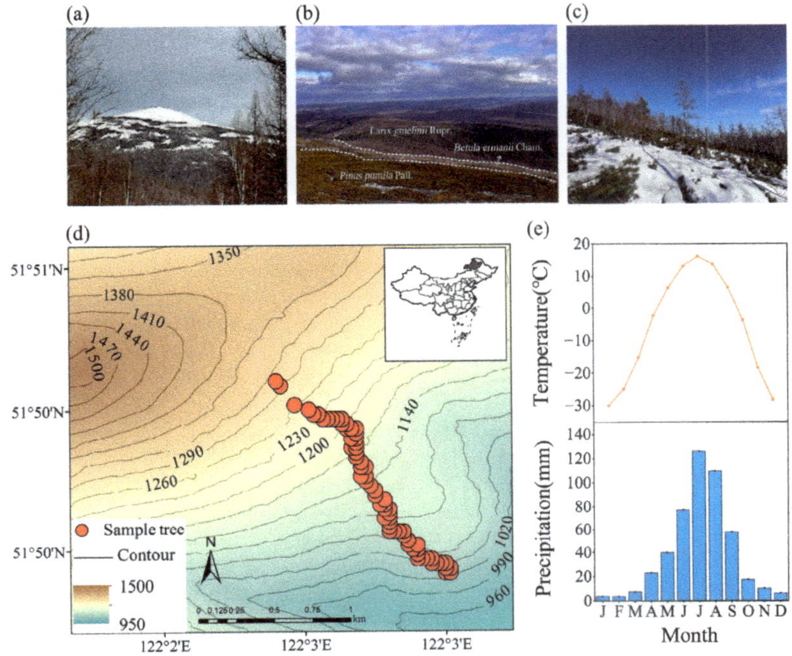

Figure 1. Geographical locations of the samples in the study area (**a**–**d**) and climatic characteristics of the monthly values for 1950–2020 (**e**).

Oakley Mountain has a cold-temperate continental monsoon climate that is cold and dry in winter with a short, hot, and humid summer. The temperature difference between day and night is large, the average annual snow storage period on the summit of the mountain is 240 days, and the frost-free period is short. The mean annual temperature is −5.6 °C (Figure 1e), the average precipitation is 470 mm, and the highest average temperature occurs in July (16.2 °C), while the lowest average temperature occurs in January (−30.5 °C); the precipitation is mainly concentrated in the months of June–September, which accounts for 77% of the total amount of the year. There has been a significant warming trend in the study area since 1950 (Figure 2). The mean annual temperature showed a significant upward trend (0.29 °C·$(10a)^{-1}$), with a stronger warming trend in winter (October–April) (0.34 °C·$(10a)^{-1}$), and there was no significant change in annual precipitation.

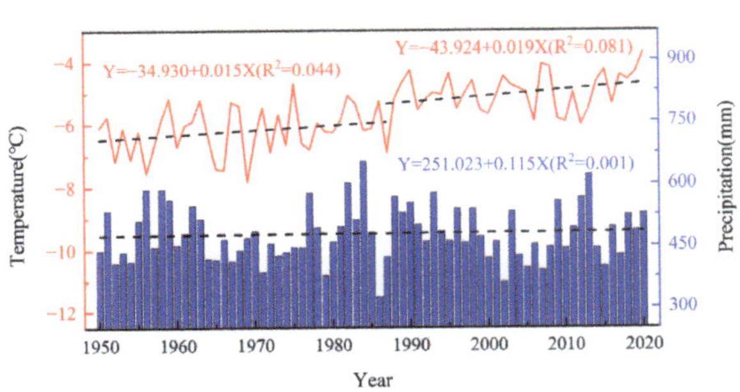

Figure 2. Interannual variation of the mean annual temperature (line) and annual precipitation (bar) in the study area from 1950 to 2020. Dotted lines represent the trend of the corresponding series.

2.2. Sample Collection and Chronology Development

In this study, we selected Dahurian larch trees, which are characterized by little anthropogenic disturbance and good growth (except near the tree line), for tree sampling along a continuously increasing elevational gradient (970–1409 m) on the southeastern slope of Oakley Mountain. Using an increment borer with an inner diameter of 5.15 mm, 1–4 cores were taken in the vertical direction from a height of approximately 1.3 m per tree. Samples within every 40 m along the elevational gradient were divided into one group during the sampling process, and 10 groups of tree ring samples were collected, totaling 174 sample cores from 86 trees. In this study, to ensure the consistency of the growth of the same group of trees within a certain elevation or slope direction (within 40 m), the sampling did not cross a large vertical or horizontal space. Therefore, the sample size of each chronology group was smaller than that of a conventional study. In addition, at an elevation of 1330 m above the Dahurian larch tree line height, there were fewer tree samples, so the sample data above the tree line height (1330–1409 m) were merged into one group (Table 1). The obtained cores were sealed in custom-made plastic tubes, and the number and sample information of each tree were recorded to facilitate the subsequent processing of the samples. The samples were brought back to the laboratory and fixed in a special wooden trough with white latex. After the samples were completely dried, they were sanded and polished with sandpaper of different meshes from coarse to fine, until the boundaries of the tree ring were clear. Cross-dating was carried out using a microscope, and the width of the tree ring was subsequently measured using the Velmex Tree Ring Analysis System (accuracy of 0.001 mm). The measurement results were verified using the COFECHA program to ensure the accuracy of the dating results [25]. The ARSTAN program was used to address the missing values (gaps) generated by the loss of sample breaks; the negative exponential function and the two-thirds step length spline function were used to detrend for the tree-ring width series and establish a standard chronology of 10 groups of Dahurian larch. We chose the STD chronology as a representative of the tree growth changes along the altitudinal gradient to analyze their climatic responses [26,27].

Table 1. Statistical characteristics of the tree ring width chronologies.

Altitudinal Classes	Site ID	Time Span	Trees (Cores)	MS	SD	EPS	SNR
970–1009	KLD1	1880–2021	7 (14)	0.267	0.507	0.970	31.907
1010–1049	KLD2	1849–2021	13 (26)	0.181	0.312	0.934	14.205
1050–1089	KLD3	1849–2021	8 (16)	0.231	0.416	0.859	6.112
1090–1129	KLD4	1845–2021	13 (26)	0.194	0.366	0.920	11.491
1130–1169	KLD5	1873–2021	8 (16)	0.343	0.223	0.851	5.695
1170–1209	KLD6	1880–2021	5 (11)	0.252	0.377	0.914	10.644
1210–1249	KLD7	1794–2021	10 (20)	0.258	0.439	0.925	12.346
1250–1289	KLD8	1745–2021	12 (24)	0.220	0.282	0.928	12.795
1290–1329	KLD9	1811–2021	8 (16)	0.251	0.314	0.920	11.484
1330–1409	KLD10	1769–2021	3 (5)	0.292	0.270	0.675	2.073

MS: mean sensitivity; SD: standard deviation; EPS: expressed population signal; SNR: signal-to-noise ratio.

2.3. Climate Data

Climatic information was obtained from gridded meteorological data (CRU TS 4.06 gridded dataset, resolution 0.5° × 0.5°) using the KNMI Climate Explorer data sharing website (http://climexp.knmi.nl), with selected gridded coordinates ranging from 51.5° to 52.5° N and 121.5° to 122.5° E. The selected climatic data span the period of 1950–2020 and include the mean monthly temperature, mean monthly precipitation, actual vapor pressure, and standardized precipitation evapotranspiration index (SPEI). The relative humidity (RH) is used to analyze the effects of changing hydrothermal conditions on tree growth. In this study, RH data were calculated from actual and saturated vapor pressure data [28], and saturated water vapor pressure was calculated from grid point temperature data [29]. The relevant formulae are as follows:

$$\mathrm{RH} = \left(\frac{e}{E}\right) \times 100\%$$

$$E = 6.112 \exp\left(\frac{17.67T}{T + 243.5}\right)$$

where RH is the relative humidity, E is the saturated water vapor pressure, e is the actual water vapor pressure, and T is the air temperature.

2.4. Data Analysis

Mutation points in the time series of major climate factors were tested using the Mann–Kendall method [30]. Due to the strong low-frequency variation and autocorrelation of meteorological data series (e.g., temperature), to reduce the influence of trends in chronological and climatic variables on the results of the correlation analyses, Pearson's correlation coefficients were calculated on first-order differences for both the chronology and the meteorological data. Indeed, this correlation coefficient represents the relationship between original tree growth and climate data in our study. The selected calculation periods are 1950–1986 and 1987–2020, respectively. The single-month meteorological factor was selected from the previous October to the current September and the winter meteorological factor was selected for the combination month of October–April (pOA). The regression analysis was conducted between the correlations of growth and winter climate and altitude, and validated by covariance analysis. The correlation analyses and charting were performed using the packages 'stats', 'ggbiplot' and 'ggplot2' of R4.3.0 software [31], the charting and study area mapping were performed using Origin 2022 and Arcmap v10.2 software.

3. Results

3.1. Winter Climatic Characteristics

Since 1950, the study area has experienced a significant winter warming trend with a rate of 0.34 °C·(10a)$^{-1}$ and no significant trend in winter precipitation (snow) (Figure 3). The mutation point test analysis of the mean annual temperatures for the years of 1950–2020 showed that there was an abrupt change in the temperature before and after 1987, and the t test results also indicated that there was a significant difference of the temperature data in approximately 1987 ($p < 0.05$). After the sudden temperature change in 1987, the mean winter temperatures (October–April) for the two periods (1950–1986 and 1987–2020) were −18.27 °C and −16.88 °C, respectively, while the warming during this period was 1.39 °C. The values of the extreme low temperatures around the time of warming were −20.85 °C in 1966 and −19.47 °C in 2001. The mean winter precipitation in the two periods was basically the same (71.59 mm and 72.77 mm), but the precipitation extremes were lower and more frequent in the last 20 years, with the minimum precipitation decreasing from 48.57 mm in 1953 to 40.95 mm in 2008. The overall trend of the winter SPEI is consistent with that of winter precipitation, which has fluctuated greatly over the last 20 years, indicating that rapid warming has been followed by more drastic winter hydrothermal changes in the study area.

3.2. Radial Growth Characteristics

The chronology spans from 141 to 276 years, with the longest dating back to 1745 (Table 1). The main purpose of this study was to explore the processes and patterns of Dahurian larch growth in response to climatic factors under rapid warming; therefore, chronological data from 1950 to 2020 matched with valid meteorological record data were used for the relevant analyses. The mean sensitivity (MS) of the chronology varied between

0.181 and 0.343, the standard deviation varied between 0.223 and 0.507, and the mean signal-to-noise ratio (SNR) of the chronologies was 11.876, indicating that the Dahurian larch chronology contained more climate signals. Samples of KLD10 were located near the Dahurian larch tree line, with smaller sample sizes and younger trees not reaching the public interval of chronological statistics; thus, they had lower overall representative (EPS) value. In the remaining nine chronologies, EPS values were above 0.85. Overall, the chronologies used in this paper met the criteria for conducting dendrochronological statistical analyses of trees.

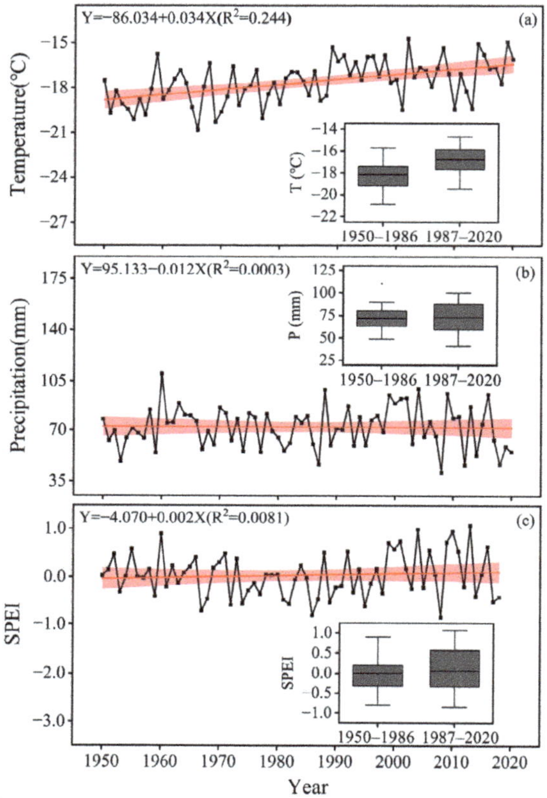

Figure 3. Winter climate trends in the study area (1950–2020, red line). The boxplots show the distributional differences in the values of the corresponding climate factors before and after rapid warming. (**a**) Temperature; (**b**) precipitation; (**c**) standardized precipitation evapotranspiration index (SPEI). The red shaded areas indicate 95% confidence interval.

In this study, we analyzed the radial growth characteristics and segmentation trends of Dahurian larch before and after rapid warming (Figure 4). Eight chronologies along the altitudinal gradient exhibited a slow upward trend with large interannual fluctuations throughout the study period (1950–2020). After suffering a sudden temperature change in the 1980s, the growth of Dahurian larch at low altitudes (below 1200 m) increased rapidly for a short period of time and then showed a decreasing trend (Figure 4), whereas the growth of trees at high altitudes showed no significant changes during this time period. In addition, before the rapid warming in 1987, for the two low altitude areas, the chronologies showed an increasing trend, and for the middle altitude and most high-altitude areas, the chronologies showed a slow decreasing trend; after 1987, approximately 60% of all the chronologies showed a decreasing trend, with the decreasing trend in the low altitude areas being more pronounced and the interannual fluctuations more drastic.

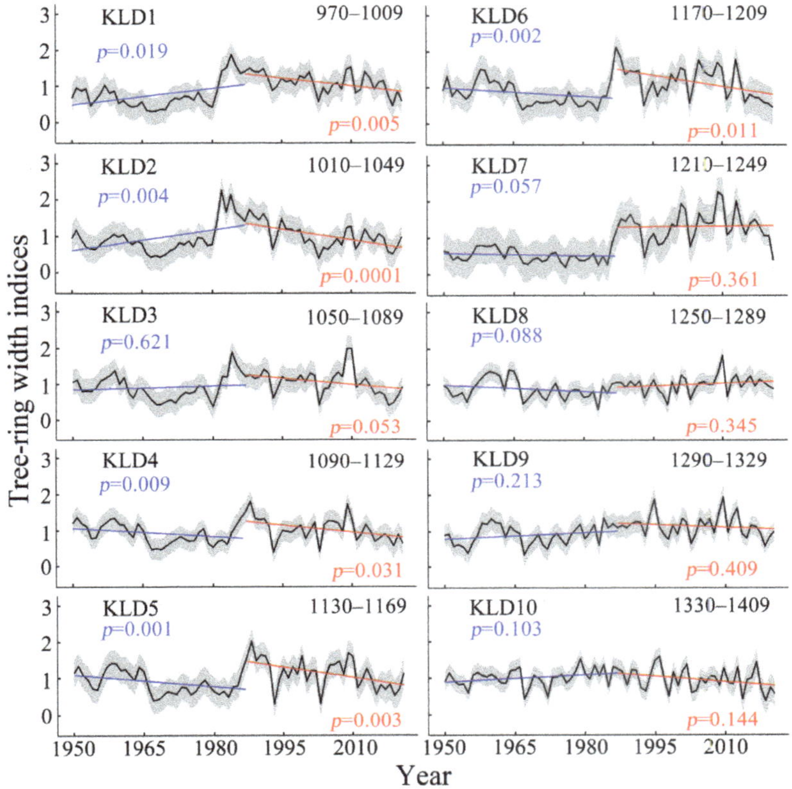

Figure 4. Changes in the tree ring width index along the altitudinal gradient from 1950 to 2020. The blue line represents the growth trend from 1950 to 1986, the red line represents the growth trend from 1987 to 2020, and gray represents the standard deviation. The altitude of the sampling site (m above sea level) is given at the top right of each subplot.

3.3. Growth–Climate Relationships

The results of correlation analysis revealed significant differences in the effects of climate factors on the radial growth of Dahurian larch in the study area before and after rapid warming. Before rapid warming, the growth was mainly positively correlated with the mean temperature in April–June, the correlation with temperature in April changed from positive to negative with increasing elevation, and the negative correlation was dominant in the other months (Figure 5a). After rapid warming, the growth was significantly positively correlated with the mean temperature in May and significantly negatively correlated with the mean temperature from February to April, while the correlation with temperature in April changed from negative to significantly negative with increasing elevation (Figure 5e). Before rapid warming (Figure 5b), the growth was mainly positively correlated with summer (July–August) precipitation (especially at low altitudes) and weakly correlated with winter precipitation (snow). After rapid warming, the growth was mainly positively correlated with winter (October–April) precipitation and reached a significant positive correlation with precipitation in January and April. On the contrary, the higher precipitation in summer was detrimental to growth compared with that in the prewarming period (Figure 5f).

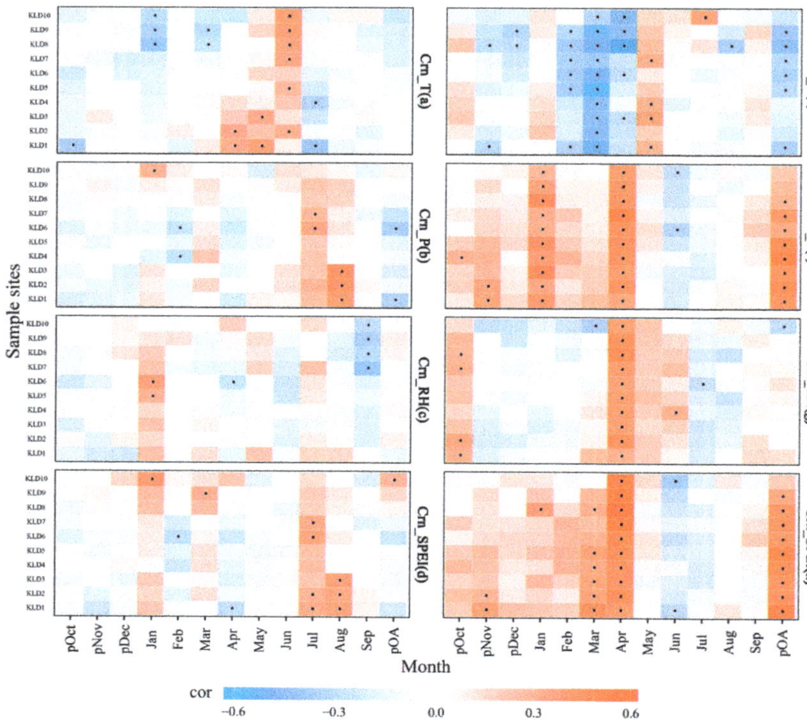

Figure 5. Correlations between chronologies and climatic factors at different altitudes for the periods of 1950–1986 (**a–d**) and 1987–2020 (**e–h**). t, p, RH, and SPEI represent the temperature, precipitation, relative humidity, and standardized evapotranspiration index (SPEI), respectively; lowercase p represents the previous year; pOA represents the period from October of the previous year to April of the current year; blue indicates a negative correlation, and red indicates a positive correlation; black dots represent correlation coefficients reaching the 95% confidence level; and vertical coordinate sampling points are arranged from bottom to top according to the increase in elevation.

Before rapid warming, the RH had no significant effect on the growth during the growing season (Figure 5c). After rapid warming, the growth was mainly negatively correlated with the RH in November–February and reached a significant negative correlation at high altitudes. Additionally, the correlation between growth and RH in March changed from a positive correlation to a negative correlation as the altitude increased. The growth is mainly positively correlated with RH from April to June and reached a significant positive correlation in April (Figure 5g). Before rapid warming, the correlation between growth and SPEI in winter (October–April) shifted from negative to positive with increasing altitude (Figure 5d); after rapid warming, the correlation between growth and SPEI in winter (October–April) was positive (Figure 5h) and became significantly positive in March–April.

Before rapid warming, the winter climate had no significant effect on Dahurian larch growth, and water deficit was the main factor limiting the growth of the trees in summer (July–August), whereas higher temperatures and less snowfall in winter after warming resulted in Dahurian larch growth being significantly affected by winter drought.

3.4. Altitudinal Divergence of Growth–Winter Drought Associations

We conducted covariance analysis and the results showed significant differences in the relationship between growth and climate along the altitudinal gradients before and after rapid warming; the *p*-values are all less than 0.05. Before rapid warming, the growth of Dahurian larch was predominantly insignificantly negatively correlated with winter (October–April) temperatures and the negative correlation increased with altitude ($R^2 = 0.930$, $p < 0.001$; Figure 6a). Also before rapid warming, growth and winter precipitation were predominantly negatively correlated and tended to weaken significantly with altitude ($R^2 = 0.261$, $p < 0.05$); after rapid warming, growth and winter precipitation shifted to a positive correlation across the entire altitudinal gradient, and the correlation tended to increase highly significantly with decreasing altitude ($R^2 = 0.572$, $p < 0.0001$; Figure 6b). Before and after rapid warming, the correlations between winter RH and Dahurian larch growth showed opposite results across the altitudinal gradient; neither reached significance, but there was an altitudinal trend in these correlations (Figure 6c). Before rapid warming, the correlation between growth and the SPEI shifted from a negative correlation to a positive correlation with increasing altitude ($R^2 = 0.453$, $p < 0.0001$; Figure 6d); after warming, the correlation became positive across the entire altitudinal gradient, and there was a highly significant altitudinal trend ($R^2 = 0.763$, $p < 0.0001$). Taken together, these results suggest that under the influence of rapid warming, the radial growth of Dahurian larch is increasingly affected by winter drought in the study area, which was particularly obvious before and after warming at low altitudes.

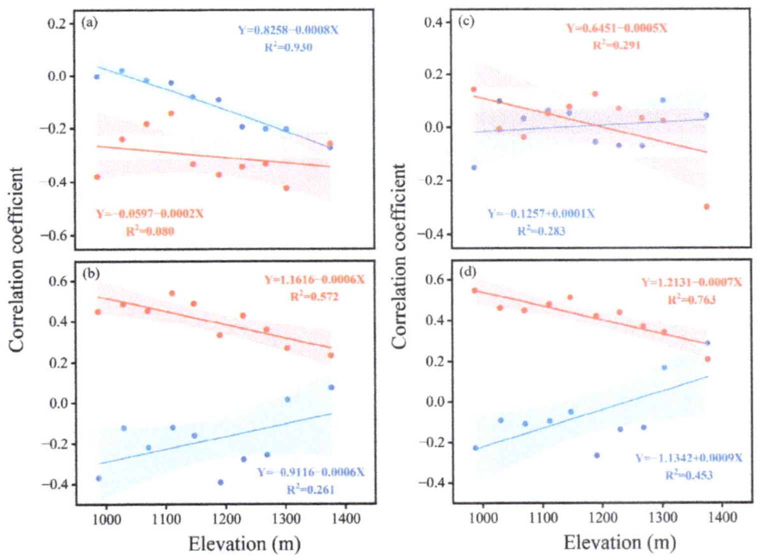

Figure 6. Change in correlations, between growth and winter T (**a**), P (**b**), RH (**c**), and SPEI (**d**), with altitude for the periods of 1950–1986 (blue) and 1987–2020 (red), respectively. The shaded areas indicate 95% confidence interval.

4. Discussion

4.1. Temporal–Spatial Differences in Radial Growth

In the context of rapid warming, frequent and extreme climatic events can directly or indirectly trigger forest decline and lead to changes in its distribution area [10,32]. Numerous studies have shown that at high latitudes, which are sensitive to warming, there is a general increase in tree growth, while tree growth at low altitudes tends to decrease due to drought limitations [11,17,33]. This finding does not contradict our hypothesis that tree growth in high-altitude mountain habitats is subject to drought stress because the high and low altitudes, as referred to in this study, are relative, and in fact those areas greater than 900 m in the Great Xing'an Mountains are considered high-altitude environments. Since the 1980s, the trees at different altitudes in the study area have exhibited spatial and temporal growth trends (Figure 4), with the growth occurring at low altitudes showing a trend of increasing and then decreasing, while the growth at high altitudes (above 1200 m) has shown a slight decreasing trend. This suggests that low-altitude trees benefited from adequate water heating in the short term during the rapid warming process, whereas at high altitudes, Dahurian larch approached the tree line where low temperatures acted as a major limiting factor, and this rapid warming did not completely alter their long-term acclimatization to a state of water heat equilibrium. In conclusion, the sample sizes at each altitude in this study, although small, still represented the significant variability in tree growth across the altitudinal gradient.

4.2. Temporal–Spatial Differences in Growth–Climate Associations

Before and after rapid warming, the correlations between Dahurian larch growth and major climatic factors changed significantly, especially between precipitation and temperature and between the SPEI in winter (October–April) (Figure 7). This indicates that the hydrothermal conditions of the tree habitats in the study area were strongly affected by rapid warming, which caused a shift in the previously stable growth–climate relationship. Before warming, the winter climate had no significant effect on the growth of Dahurian larch, while summer drought was the main limiting factor for their growth at low altitudes, which decreased with increasing altitude. After warming, Dahurian larch growth was negatively correlated with winter temperature (Figure 7e) and significantly positively correlated with winter precipitation (especially at low altitudes; Figure 7f), indicating that abrupt warming gradually changed Dahurian larch growth at mountainous low altitudes from drought limitation in the growing season in the past to drought limitation in the pre-growing season. This shift is primarily driven by the amount and timing of winter snow storage. For example, less snowpack in winter and earlier snowmelt in spring can lead to insufficient soil available moisture as trees enter the pre-growth period, which is indirectly reflected in the impact of winter drought stress on the radial growth of Dahurian larch in the following year [34,35]. The average winter temperature in the study area reaches −25 °C, and trees are dormant [36]. Thus, the lag effect of winter moisture (snowpack) on trees is primarily related to changes in temperature the following spring.

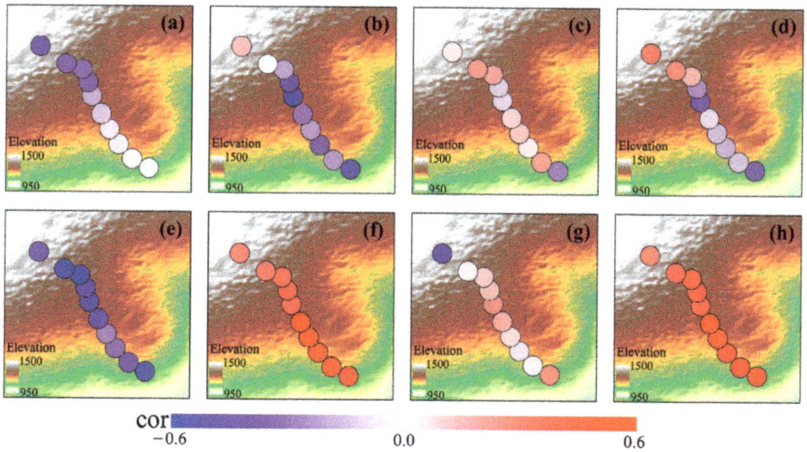

Figure 7. The correlation-scales between growth and climate along the altitudinal gradient. (**a**,**e**), (**b**,**f**), (**c**,**g**), (**d**,**h**) represent the winter T, P, RH, and SPEI, respectively. (**a**–**d**) (1950–1986) and (**e**–**h**) (1987–2020) represent before and after warming, respectively. Correlations range from −0.6 (blue) to 0.6 (red).

Before warming, temperature promoted the growth of Dahurian larch in April–June (Figure 5a), which was the period during which the physiological activities of the trees gradually recovered (leaf spread in early June) in the study area. The relatively high temperatures and slower rate of snowmelt in April–May replenish effective soil moisture, and it has been documented that larch has a high percentage of fine roots in shallow organic soils [37,38]. It mainly absorbs shallow soil moisture early in the growing season, and snowmelt replenishes the surface soil water content. [39]. Together, the water retention of the humus layer and the higher temperatures in June triggered rapid activity in the cambium (especially at high altitudes, Figure 5a), promoting tree growth in the current year [40]. After rapid warming, the February–April temperatures significantly inhibited Dahurian larch growth (Figure 5e). February–April is the most significant warming period in the study area, and excessive temperatures and early snowmelt depletion during this period can result in insufficient available soil moisture during the early part of the growing season, causing tree growth to suffer from early spring physiological drought stress [6, 16,41]. In particular, the inhibitory effect of temperature on Dahurian larch growth in April gradually diminished with decreasing altitude, which was attributed to the gradual increase in temperature in April, when the soil warmed more quickly at low altitudes than at high altitudes, and when soil conditions unfavorable to the tree root system lessened.

The heterogeneity of increasing warming and decreasing humidity leads to severe water deficit and inhibits tree growth. Before warming, the growth was positively correlated with precipitation and the SPEI in July and August (Figure 5b,d) and was more significant at low altitudes. After warming, the growth was positively correlated with winter precipitation and the winter SPEI (Figure 5f,h), while it was positively correlated with relative humidity only in April–June (Figure 5g). This suggests that increased evapotranspiration of water during the pre-growing season further exacerbates winter drought stress caused by soil moisture deficit. In addition, the location of the sampling environment on the southeastern leeward slopes, coupled with the rapid warming process during the winter season, exacerbated the effects of winter (early spring) drought on the growth of trees at low altitudes [42,43]. However, this drought effect gradually weakened with altitude (Figure 7f,h) because water evapotranspiration decreases rapidly with the increase in altitude, and trees are less susceptible to water stress at high altitudes than at low altitudes [9]. This variability was also confirmed by the negative correlation between Dahurian larch growth at high altitudes and relative humidity from November to March.

4.3. Future Winter Warming-Drying Trends and Dahurian larch Distribution

Major changes in temperature and precipitation patterns occur under global warming [2]. The study area is located at a high latitude and high altitude and its forest ecosystems are highly sensitive to climate change [24,44,45]. During the past 70 years, dramatic climate change in the study area has led to a shift in the correlation between tree growth and climate, with winter precipitation (snowfall) becoming the main limiting factor for Dahurian larch growth. The drought effect of rapid climate warming on tree growth in the study area increased across the altitudinal gradient, but the magnitude of the effect tended to decrease significantly with altitude (Figure 6); i.e., trees in high-altitude environments were more hydrothermally stable in response to this rapid warming (Figure 7). Compared to those at low altitudes, the chronologies at high altitudes experienced no significant change in radial growth and were less affected by warming, which differed from the results of other studies that showed an increase in the growth of larch at high altitudes [14,16,17]. This is, first and foremost, because the present study site is a high-altitude mountainous environment (unlike the high-altitude flatland tundra) where tree growth is limited by a combination of heat and water; it is also because some of the samples in the high-altitude area are close to the timberline (tree line), and their microhabitats have a greater impact on the trees (e.g., infertile soils and greater wind speeds, etc.). In summary, the sensitivity of tree growth to drought will increase under continued warming in the future, with trees at low altitudes facing more severe challenges and trees at high altitudes being better able to balance the relationship between water and heat and cope with short-term warming events.

5. Conclusions

Our study was designed to explore the growth trends of Dahurian larch and climate response patterns with altitude under rapid warming conditions. In the 1980s, rapid warming led to an increase and then a decrease in the growth of Dahurian larch at low altitudes but had a smaller effect on growth at high altitudes (above 1200 m). The response of larch forests to climate has shifted from drought limitations in the growing season before warming to winter drought (the lag effects of snowfall before the next growing season) limitations after warming, especially for low-altitude trees. As the climate continues to warm, the coupled effects of winter snowpack and rapid warming will drive an increase in the upper altitude limit at which winter drought occurs in larch on the studied altitudinal gradient, which are currently better able to cope with this ecological change in higher-altitude environments. In conclusion, rapid warming increases the risk of forest decline at the lower altitude, and in the future forest management and strategy formulation, we should focus on the growth dynamics of dominant conifer species at low and medium altitudes.

Author Contributions: R.Z. and X.B. conceived and designed the study; R.Z. contributed to data measurement and wrote the manuscript; Field work was completed by X.B., Z.C. and R.Z.; X.T., H.Z. and H.L. contributed to investigation, writing and data analyses. All authors have read and agreed to the published version of the manuscript.

Funding: This study was funded by the National Natural Science Foundation of China (grant number 32001136), and partly supported by the Doctoral Research Start-up Funds for the Inner Mongolia Minzu University (KYQD19072).

Data Availability Statement: The raw/processed data required to reproduce these findings cannot be shared at this time as the data also form part of an ongoing study.

Acknowledgments: We would like to thank the Laboratory of Dendrochronology, College of Forestry, Shenyang Agricultural University, for providing the experimental space, Zhenju Chen, Junxia Li, and Yuting Jin for helping with my experiments.

Conflicts of Interest: The authors declare that they have no known competing financial interests or personal relationships that could have appeared to influence the work reported in this paper.

References

1. IPCC. *Climate Change 2021: The Physical Science Basis. Contribution of Working Group I to the Sixth Assessment Report of the Intergovernmental Panel on Climate Change*; Cambridge Press: Cambridge, UK, 2021.
2. IPCC. *Climate Change 2022: Impacts, Adaptation, and Vulnerability. Contribution of Working Group II to the Sixth Assessment Report of the Intergovernmental Panel on Climate Change*; Cambridge Press: Cambridge, UK, 2022.
3. Anderegg, W.R.L.; Trugman, A.T.; Badgley, G.; Anderson, C.M.; Bartuska, A.; Ciais, P.; Cullenward, D.; Field, C.B.; Freeman, J.; Goetz, S.J.; et al. Climate-driven risks to the climate mitigation potential of forests. *Science* **2020**, *368*, eaaz7005. [CrossRef]
4. Anderegg, W.R.L.; Wu, C.; Acil, N.; Carvalhais, N.; Pugh, T.A.M.; Sadler, J.P.; Seidl, R. A climate risk analysis of Earth's forests in the 21st century. *Science* **2022**, *377*, 1099–1103. [CrossRef]
5. Du, H.; Liu, J.; Li, M.H.; Buntgen, U.; Yang, Y.; Wang, L.; Wu, Z.; He, H.S. Warming-induced upward migration of the alpine treeline in the Changbai Mountains, northeast China. *Glob. Chang. Biol.* **2018**, *24*, 1256–1266. [CrossRef] [PubMed]
6. Gaire, N.P.; Fan, Z.X.; Bräuning, A.; Panthi, S.; Rana, P.; Shrestha, A.; Bhuju, D.R. Abies spectabilis shows stable growth relations to temperature, but changing response to moisture conditions along an elevation gradient in the central Himalaya. *Dendrochronologia* **2020**, *60*, 125675. [CrossRef]
7. Shi, H.; Zhou, Q.; Xie, F.; He, N.; He, R.; Zhang, K.; Zhang, Q.; Dang, H. Disparity in elevational shifts of upper species limits in response to recent climate warming in the Qinling Mountains, North-central China. *Sci. Total Environ.* **2020**, *706*, 135718. [CrossRef]
8. Zheng, L.; Gaire, N.P.; Shi, P.; Schöb, C. High-altitude tree growth responses to climate change across the Hindu Kush Himalaya. *J. Plant Ecol.* **2021**, *14*, 829–842. [CrossRef]
9. Yuan, S.; Jiang, Y.; Zhao, Z.; Cui, M.; Shi, D.; Wang, S.; Kang, M. Different trends and divergent responses to climate factors in the radial growth of Abies georgei along elevations in the central Hengduan Mountains. *Dendrochronologia* **2023**, *80*, 126114. [CrossRef]
10. Panthi, S.; Fan, Z.X.; van der Sleen, P.; Zuidema, P.A. Long-term physiological and growth responses of Himalayan fir to environmental change are mediated by mean climate. *Glob. Chang. Biol.* **2020**, *26*, 1778–1794. [CrossRef]

11. Jiao, L.; Jiang, Y.; Zhang, W.T.; Wang, M.C.; Zhang, L.N.; Zhao, S.D. Divergent responses to climate factors in the radial growth of Larix sibirica in the eastern Tianshan Mountains, northwest China. *Trees* **2015**, *29*, 1673–1686. [CrossRef]
12. Lyu, L.; Zhang, Q.B.; Pellatt, M.G.; Büntgen, U.; Li, M.H.; Cherubini, P. Drought limitation on tree growth at the Northern Hemisphere's highest tree line. *Dendrochronologia* **2019**, *53*, 40–47. [CrossRef]
13. Gao, J.; Liu, Y. Climate warming and land use change in Heilongjiang Province, Northeast China. *Appl. Geogr.* **2011**, *31*, 476–482. [CrossRef]
14. Yasmeen, S.; Wang, X.; Zhao, H.; Zhu, L.; Yuan, D.; Li, Z.; Zhang, Y.; Ahmad, S.; Han, S. Contrasting climate-growth relationship between *Larix gmelinii* and *Pinus sylvestris* var. *mongolica* along a latitudinal gradient in Daxing'an Mountains, China. *Dendrochronologia* **2019**, *58*, 125645. [CrossRef]
15. Zhang, X.; Liu, X.; Zhang, Q.; Zeng, X.; Xu, G.; Wu, G.; Wang, W. Species-specific tree growth and intrinsic water-use efficiency of Dahurian larch (*Larix gmelinii*) and Mongolian pine (*Pinus sylvestris* var. *mongolica*) growing in a boreal permafrost region of the Greater Hinggan Mountains, Northeastern China. *Agric. For. Meteorol.* **2018**, *248*, 145–155. [CrossRef]
16. Zhang, X.; Bai, X.; Hou, M.; Chen, Z.; Manzanedo, R.D. Warmer Winter Ground Temperatures Trigger Rapid Growth of Dahurian Larch in the Permafrost Forests of Northeast China. *J. Geophys. Res. Biogeosci.* **2019**, *124*, 1088–1097. [CrossRef]
17. Bai, X.P.; Zhang, X.L.; Li, J.X.; Duan, X.Y.; Jin, Y.T.; Chen, Z.J. Altitudinal disparity in growth of Dahurian larch (*Larix gmelinii* Rupr.) in response to recent climate change in northeast China. *Sci. Total Environ.* **2019**, *670*, 466–477. [CrossRef]
18. Li, Y.; Liu, H.; Zhu, X.; Yue, Y.; Xue, J.; Shi, L. How permafrost degradation threatens boreal forest growth on its southern margin? *Sci. Total Environ.* **2021**, *762*, 143154. [CrossRef]
19. Pithan, F.; Mauritsen, T. Arctic amplification dominated by temperature feedbacks in contemporary climate models. *Nat. Geosci.* **2014**, *7*, 181–184. [CrossRef]
20. Gao, M.; Meng, F. On the delineation of forest vegetation zones in the Oakleigh Pile Mountains. *Inn. Mong. For. Investig. Des.* **1999**, *4*, 141–143.
21. Zhang, L. The Research on the East Slope of Middle Region of Daxing'an Mountains Vertical Zone. Master's Thesis, Northeast Normal University, Changchun, China, 2007.
22. Huang, X.; Zhang, X.; Yuan, L. Vegetation Types and Vertical Distribution Patterns of Oakley Mountain. *High. Educ. Forum.* 2009. Available online: https://www.docin.com/p-236712927.html (accessed on 20 February 2024).
23. Yang, Z.; Zhou, G.; Yin, X.; Jia, B. Geographic distribution of *Larix gmelinii* natural forest in China and its climatic suitability. *Chin. J. Ecol.* **2014**, *33*, 1429–1436. [CrossRef]
24. Zhang, H.; Zhang, S.; Song, W.; Tigabu, M.; Fu, M.; Xue, H.; Sun, A.; Zhao, M.; Cai, K.; Li, Y.; et al. Climate response of radial growth and early selection of *Larix olgensis* at four trials in northeast China. *Dendrochronologia* **2022**, *73*, 125955. [CrossRef]
25. Holmes, R.L. Computer-Assisted Quality Control in Tree-Ring Dating and Measurement. *Tree-Ring Bull.* **1983**, *43*, 51–67.
26. Cook, E.R. A Time Series Approach to Tree-Ring Standardization. Ph.D. Thesis, The University of Arizona, Tucson, AZ, USA, 1985.
27. Holmes, R.L.; Cook, E.R. *Guide for Computer Program ARSTAN*; Laboratory of Tree-Ring Research, University of Arizona: Tucson, AZ, USA, 1994.
28. Peixoto, J.; Oort, A.H. The Climatology of Relative Humidity in the Atmosphere. *J. Clim.* **1996**, *9*, 3443–3463. [CrossRef]
29. Liu, J.; Guo, H.; Li, Y.; Liu, H.; Wu, B. *Fundamentals of Physical Quantity Calculation in Weather Analysis and Forecasting*; China Meteorological Press: Beijing, China, 2005.
30. Kendall, M.G. Rank Correlation Methods. *Biometrika* **1957**, *44*, 146–163. [CrossRef]
31. Team R Core. R: A language and environment for statistical computing. *MSOR Connect.* **2014**, *1*, 12–21.
32. Hepting, G.H. Climate and Forest diseases. *Annu. Rev. Phytopathol.* **1963**, *60*, 133–149. [CrossRef]
33. Huang, R.; Zhu, H.; Liu, X.; Liang, E.; Grießinger, J.; Wu, G.; Li, X.; Bräuning, A. Does increasing intrinsic water use efficiency (iWUE) stimulate tree growth at natural alpine timberline on the southeastern Tibetan Plateau? *Glob. Planet. Chang.* **2017**, *148*, 217–226. [CrossRef]
34. Su, J.; Gou, X.; Deng, Y.; Zhang, R.; Liu, W.; Zhang, F.; Lu, M.; Chen, Y.; Zheng, W. Tree growth response of Fokienia hodginsii to recent climate warming and drought in southwest China. *Int. J. Biometeorol.* **2017**, *61*, 2085–2096. [CrossRef] [PubMed]
35. Bogachev, M.I.; Gafurov, A.M.; Iskandirov, P.Y.; Kaplun, D.I.; Kayumov, A.R.; Lyanova, A.I.; Pyko, N.S.; Pyko, S.A.; Safonova, A.N.; Sinitca, A.M.; et al. Reversal in the drought stress response of the Scots pine forest ecosystem: Local soil water regime as a key to improving climate change resilience. *Heliyon* **2023**, *9*, e21574. [CrossRef]
36. Freeland, R.O. Apparent Photosynthesis in Some Conifers During Winter. *Plant Physiol.* **1944**, *19*, 179–185. [CrossRef]
37. Iversen, C.M.; Childs, J.; Norby, R.J.; Ontl, T.A.; Kolka, R.K.; Brice, D.J.; McFarlane, K.J.; Hanson, P.J. Fine-root growth in a forested bog is seasonally dynamic, but shallowly distributed in nutrient-poor peat. *Plant Soil* **2017**, *424*, 123–143. [CrossRef]
38. Zhang, Y.P.; Jiang, Y.; Wang, B.; Jiao, L.; Wang, M.c. Seasonal water use by Larix principis-rupprechtii in an alpine habitat. *For. Ecol. Manag.* **2018**, *409*, 47–55. [CrossRef]
39. Sugimoto, A.; Naito, D.; Yanagisawa, N.; Ichiyanagi, K.; Kurita, N.; Kubota, J.; Kotake, T.; Ohata, T.; Maximov, T.C.; Fedorov, A.N. Characteristics of soil moisture in permafrost observed in East Siberian taiga with stable isotopes of water. *Hydrol. Process.* **2002**, *17*, 1073–1092. [CrossRef]
40. Pompa García, M.; Cerano Paredes, J.; Fulé, P.Z. Variation in radial growth of *Pinus cooperi* in response to climatic signals across an elevational gradient. *Dendrochronologia* **2013**, *31*, 198–204. [CrossRef]

41. Ren, P.; Rossi, S.; Camarero, J.J.; Ellison, A.M.; Liang, E.; Penuelas, J. Critical temperature and precipitation thresholds for the onset of xylogenesis of Juniperus przewalskii in a semi-arid area of the north-eastern Tibetan Plateau. *Ann. Bot.* **2018**, *121*, 617–624. [CrossRef] [PubMed]
42. Qiao, J.; Sun, Y. Effects of altitude and slope on the climate–radial growth relationships of *Larix olgensis* A. Henry in the southern Lesser Khingan Mountains, Northeast China. *Ecol. Process.* **2022**, *11*, 46. [CrossRef]
43. Li, J.; Li, S.; Liu, C.; Guo, D.; Zhang, Q. Response of Chinese pine regeneration density to forest gap and slope aspect in northern China: A meta-analysis. *Sci. Total Environ.* **2023**, *873*, 162428. [CrossRef] [PubMed]
44. Gaire, N.P.; Zaw, Z.; Bräuning, A.; Grießinger, J.; Sharma, B.; Rana, P.; Bhandari, S.; Basnet, S.; Fan, Z.-X. The impact of warming climate on Himalayan silver fir growth along an elevation gradient in the Mt. Everest region. *Agric. For. Meteorol.* **2023**, *339*, 109575. [CrossRef]
45. Zhang, W.; Gou, X.; Zhang, F.; Liu, W.; Zhang, Y.; Gao, L. Divergent responses of Qinghai spruce (*Picea crassifolia*) to recent warming along elevational gradients in the central Qilian Mountains, Northwest China. *J. Geogr. Sci.* **2023**, *33*, 151–168. [CrossRef]

forests

MDPI

Article

Distinct Impact of Drought on Radial Growth at Different Heights and Parts of *Populus euphratica* in the Oasis at the Hinterland of the Taklimakan Desert

Anwar Abdureyim [1,2], Yue Dai [2,3], Qingdong Shi [1,2,*], Feng Zhang [2,3], Yanbo Wan [1,2], Haobo Shi [1,2] and Lei Peng [1,2]

[1] College of Ecology and Environment, Xinjiang University, Urumqi 830017, China; anwar_jan@stu.xju.edu.cn (A.A.); wanyb1995@163.com (Y.W.); shi_haobo@163.com (H.S.); penglei1028@stu.xju.edu.cn (L.P.)
[2] Key Laboratory of Oasis Ecology, Xinjiang University, Urumqi 830017, China; daiyue@xju.edu.cn (Y.D.); zhangfeng@xju.edu.cn (F.Z.)
[3] College of Geography and Remote Sensing Science, Xinjiang University, Urumqi 830017, China
* Correspondence: shiqd@xju.edu.cn

Abstract: Warming and persistent droughts may exacerbate drought stress in water-scarce areas, thereby negatively affecting tree growth. When riparian plants in arid regions experience severe drought stress, they sacrifice non-dominant branches with less competitive sap flow to improve the sap flow of dominant branches and thus ensure strong plant growth. *Populus euphratica* is one of the most dominant tree species in the riparian forest ecosystems of inland river basins in arid zones and is a reliable indicator of ecological change because of its diversity in growth and environment. To understand the adaptability of *P. euphratica* to the environment, the relationship between radial growth and meteorological factors, the growth decline and resistance of different heights and components of *P. euphratica*, as well as the resilience and resilience after recession were investigated. The results indicated that tree-ring width decreased gradually with increasing height and branching class of *P. euphratica*. Growth decreased at the bottom of the stem earlier than at the middle and top. Temperature, precipitation, and the Palmer drought index contributed to the growth at the bottom of *P. euphratica*, while precipitation contributed to growth at the top. The decline in the *P. euphratica* growth change rate was highly synchronized across heights and parts, with relatively high declines at the bottom and top. There were no significant differences in the recovery values for different heights and parts of *P. euphratica*, but the resistance, resilience, and relative resilience for the bottom and top were significantly lower than those for the other components, indicating vulnerability in the bottom and top of *P. euphratica* to drought. The relative resilience gradually decreased with the increase in branching class, and that of the secondary lateral branches at different heights was the lowest. In conclusion, the sensitivity of the top and lateral branches of *P. euphratica* to drought would cause the phenomenon of "breaking its arm" under drought disturbance in the future.

Keywords: tree-ring width; basal area increment; resistance; recovery; resilience

1. Introduction

The dramatic increase in greenhouse gas emissions since the beginning of the industrial revolution has led to global warming. Global temperatures are likely to continue to increase throughout the century, followed by more severe, widespread, and irreversible global impacts [1,2]. An IPCC (Intergovernmental Panel on Climate Change) report confirmed that the average global temperature rose by 0.85 °C from 1880 to 2012. Global warming has rapidly accelerated in recent decades, especially at high altitudes [3]. Increasing temperatures are accompanied by significant changes in global precipitation patterns, leading to extreme climate events such as severe droughts, with potential impacts on plant growth, phenology, distribution, and ecosystem services [4,5]. The arid zone of northwest China is located within the mid-latitudes and consists of the Gobi Desert and its surrounding arid zone. These areas represent fragile ecosystems and are sensitive to climate change, making the region a key site for studying the environmental impacts of global climate change [6]. Warming and persistent drought may exacerbate drought stress in water-scarce areas, thereby negatively affecting tree growth [7]. Irreversible damage, such as defoliation and dieback, occurs when a tree's physiological damage exceeds a certain threshold [8]. In response to harmful environmental factors, such as drought and heat, trees may reduce their vigor or growth rate, experience a decline in growth, and die.

Tree growth serves as an indicator of tree vitality and capacity to adapt to environmental pressures [9]. The recovery of trees following stress is a protracted process that may entail time lags or legacy effects [10,11]. Tree-ring width (TRW) provides a continuous and high-resolution record of tree growth that accurately reflects environmental changes during growth [12]. Tree chronologies have been widely utilized in studies investigating forest resilience to drought [13]. Prolonged drought may reduce the resilience of trees owing to hydraulic failure [14,15] or carbon starvation [16]. If accumulated stress surpasses a tree's threshold, it may result in irreversible damage and ultimately lead to the tree's decline or mortality [17,18]. Moreover, the chronology of trees can be used to retrospectively detail their growth for each preceding year, ascertain their age, and comprehend the temporal manifestations of processes such as growth, branching, and flowering [19].

Most dendrochronological studies utilize core and disc measurements of ring width at breast height (1.3 m above ground) because this height is characterized by a relatively stable radial growth pattern and many rings [20,21]. Radial growth at breast height, which is influenced by various climatic factors, is often used as an indicator of the overall stem growth pattern [22]. Factors such as tree condition, natural environment, and physiological conditions can affect the stem, resulting in differences in radial growth at different stem heights [23]. In addition, the relationship between the radial growth response and climatic elements varies among trees with different stem heights. Chhin [24] studied the relationship between the radial growth of *P. thunbergii* Parl. at different stem heights and climatic factors and demonstrated that the growth of the lower part of the stem was mainly influenced by temperature and moisture conditions during the growing season in the year of tree-ring formation; by contrast, the growth of the upper part was more influenced by environmental conditions during the year of growing. Maaten [22] analyzed differences in radial growth responses to climate change in different stems of *A. alba* Mill. and *P. abies* (L.) Karst. at different elevations. They found that high temperatures in early summer reduced the growth of high-elevation fir in the upper stem portion, whereas high temperatures in summer limited the growth of high-elevation *A. alba* and *P. abies* growth, especially at breast height. For trees at lower elevations, both high summer temperatures and low precipitation reduced stem growth, with a greater impact on growth at breast height. Zhang [21] discovered that the sensitivity of *P. schrenkiana* tree-ring width to climatic elements was stronger at a height of 10 m on the stem than at other heights.

Moreover, a plant's ability to withstand drought stress is heavily dependent on its water transport system, which encompasses the root, stem, and leaf structures. Studies have demonstrated that in favorable water conditions, plants compete for light, heat, and nutrients by augmenting their hydraulic conductivity [25]. The hypothesis of branch connection limitation has been confirmed in studies on the hydraulic structure of woody plants. This hypothesis suggests that dominant apical branches have stronger hydraulic conductivity than lateral branches because they control the sap flow of branch connections [26]. To maintain their viability, riparian plants actively reduce the water transport capacity of twigs to compete for the limited water delivered by the root system. They prioritize supplying water to the top branches to ensure a sufficient supply [27]. When riparian plants in arid regions experience severe drought stress, they tend to sacrifice non-dominant branches with less competitive sap flow to enhance the sap flow of the dominant branches and ensure optimal plant growth [28]. Therefore, it is imperative to evaluate the lateral branch growth of *P. euphratica* and its response to climatic factors to comprehensively understand plant growth adaptations. Although tree-ring widths can detail growth distribution patterns across different periods and components, there remains a relative dearth of research about radial growth within plant branching systems. In general, analyses of climatic responses to tree growth based on different stem heights are limited and have mostly focused on coniferous forests. Other tree species, such as temperate broadleaf forests, have received less attention, and it remains unclear whether they exhibit differential responses to temperature changes.

As one of the most dominant tree species in riparian forest ecosystems of inland river basins in arid zones, *P. euphratica* is a reliable indicator of ecological change owing to its diverse growth rates and environmental adaptability [29]. The growth of *P. euphratica* is influenced by various environmental factors, including groundwater, surface water, and temperature. Groundwater is pivotal in the growth of *P. euphratica* [30,31]. Previous studies have primarily focused on reconstructing past groundwater depth data or examining the relationship between groundwater and radial growth using tree chronology. However, a comprehensive investigation of the factors influencing *P. euphratica* growth under sufficient groundwater resources and the response of *P. euphratica* growth at different stem heights and lateral branches to climate change has not been conducted. Therefore, this study examined the response of *P. euphratica* radial growth to climate change across different stem heights and components using tree chronology analysis.

The Daliyaboyi Oasis is situated in the Tarim Basin of China and is an alluvial oasis formed by the deep penetration of the Keriya River into the hinterland of the Taklamakan Desert [31]. The oases remain in a 'pristine' state with minimal anthropogenic disturbance in this region, which distinguishes them from most of the extant modern oases in Central Asia [32]. *P. euphratica* is a dominant species in the oasis and is indispensable to protecting it and maintaining the stability of riparian ecosystems. Moreover, *P. euphratica* can provide valuable information on long-term environmental changes. However, most studies have focused on the impact of hydrological factors on *P. euphratica* growth, leaving a gap in the literature regarding individual *P. euphratica* growth responses at different stem heights. Therefore, we hypothesized that *P. euphratica* growth and growth response to drought can vary at different heights and in different parts of the plant, with the bottom and top of *P. euphratica* being relatively sensitive to drought. The objectives of this study were to evaluate differences in growth, rates of growth change, and recovery from drought across heights and parts of *P. euphratica* and to explore the vulnerability of *P. euphratica* to drought.

2. Materials and Methods

2.1. Study Area

The Daliyaboyi Oasis is located in the Taklimakan Desert of China's hinterland at 38°16′ N/38°37′ N, 81°41′ E/82°20′ E, marking the last oasis of the Keriya River, the second largest river on the southern margin of the Tarim Basin [31]. The oasis has an area of approximately 342 km^2 and elevations of 1100–1300 m, with a typical warm temperate arid climate [32] and significant diurnal temperature differences. The region experiences dust storms and highly variable weather [31], with an annual precipitation of less than 10 mm and a potential evaporation rate of 2480 mm [33]. The interior of the oasis is characterized by reticular fluvial deposition resulting from the scouring of the Keriya River [31]. The vegetation of the oasis is primarily dominated by *P. euphratica*, *Tamarix chinensis*, and *Phragmites australis*, exhibiting simple structures and types with limited distributions, which are typical characteristics of desert ecosystems [33]. Sandy soils constitute the predominant soil type in the oases.

2.2. Experimental Design and Sample Collection

The experiment was conducted in June 2021 within a *P. euphratica* community located adjacent to the main perennial river (the river has water year round), situated in the southern region of the Daliyaboyi Oasis at a distance of 150 m from the riverbank. A

50 × 50 m sample square was established, which was dominated by *P. euphratica*, with a few *T. chinensis* instances. The number of individual *P. euphratica* in the sample was 51, the average tree height was 17.72 m, the average diameter at breast height was 55.84 cm, the average crown width was 7.43 × 5.92 m, and the cover was 97.5%. Three *P. euphratica* that were healthy, structurally characterized, and unshaded at 5–10 m were selected in the sample site (Figure 1).

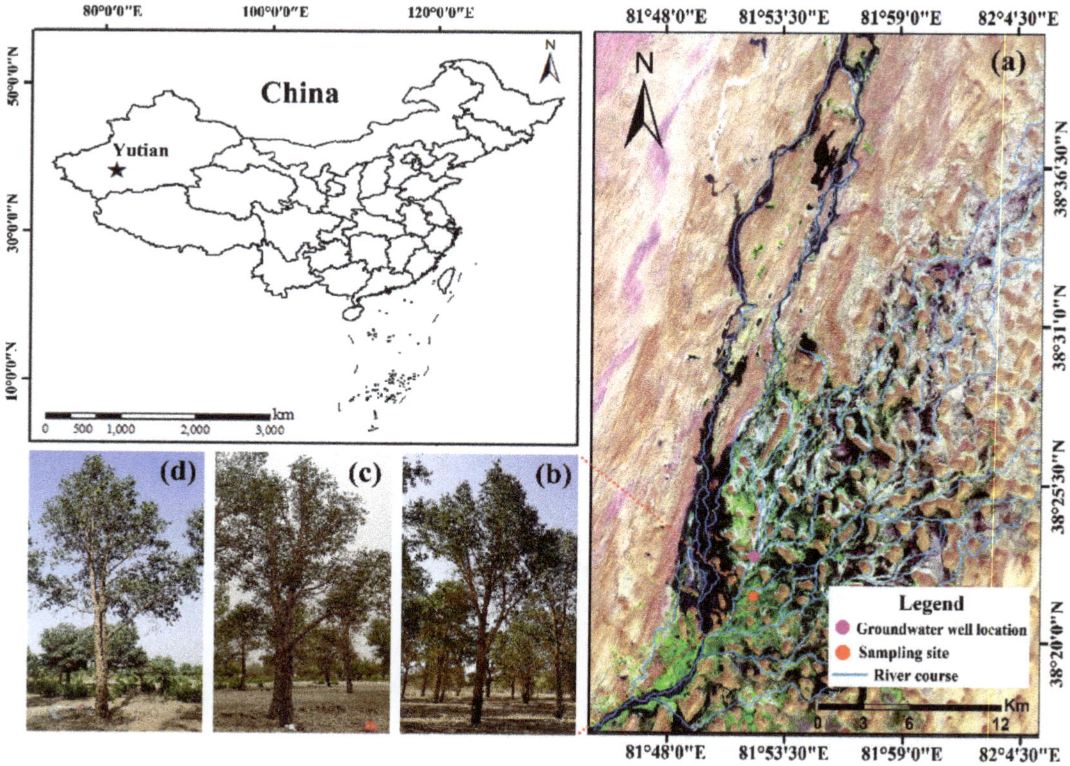

Figure 1. Location of the study sites. (**a**) Daliyaboyi Oasis, (**b**) first tree, (**c**) second tree, (**d**) third tree. In Figure 1a, the purple circle is a nearby groundwater level observation well; the red circle is a sampling site; and the light blue horizontal line is the stream channel.

The three plants were sampled from the bottom 1 m to the top 8 m of the stem using a 5 mm increment borer (Haglöf increment borer, Torsång, Sweden), and two cores were collected at each height. Based on the determined branch order, primary and secondary lateral branches of different heights were collected from the cores and discs at the bases of the tree samples (Table 1). The *P. euphratica* plants were divided into three parts according to the distance of the primary lateral branches from the ground, as follows: bottom—from 0–4 m, containing the stem bottom (STB), primary lateral branches at the bottom of the stem (PLB), and secondary lateral branches of primary lateral branches at the bottom of the stem (SLB); middle—between 4–6.5 m, including the stem middle area (STM), primary lateral branches at the middle of the stem (PLM), and secondary lateral branches of primary lateral branches at the middle of the stem (SLM). Top: between 6.5–8 m, containing the stem's top area (STT) and primary lateral branches at the top of the stem (PLT). The grouping basis is shown in Figure 2.

Table 1. Information on sampling points. Note: sample size refers to the number of cores collected. For example, ① indicates 16 cores collected at 8 different heights; ② indicates 16 cores collected at 8 different primary lateral branches. Two cores were collected per height, per lateral branch.

Sample Trees	Longitude and Latitude	Height (m)	Diameter at Breast Height (cm)	Canopy (m)	Age (a)	
First tree	38°21′23.19″ N 81°51′36.32″ E	11.78	39.3	6.47 × 6.52	50	
Second tree		11.63	41.4	7.09 × 5.87	60	
Third tree		11.51	48.6	7.43 × 6.02	56	
Different parts	**Stem**		**Primary lateral branch**		**Secondary lateral branch**	
Sample trees	**Sample size**	**Diameter range (cm)**	**Sample size**	**Diameter range (cm)**	**Sample size**	**Diameter range (cm)**
First tree	16 ①	10.2~39.3	16 ②	5~14	14	4~10
Second tree	16	15.5~41.4	16	6~19	12	4~7
Third tree	16	9.4~48.6	18	5~22	22	4~13

Category	Full Title	Abbreviation
Characteristics of ring chronology	Tree-ring width	TRW
	Ring width index	RWI
	Basal area increment	BAI
	Standardized chronology	STD
	Standard deviation	SD
Different parts of *Populus euphratica*	Bottom of stem	STB
	Primary lateral branches at the bottom of the stem	PLB
	Secondary lateral branches of primary lateral branches at the bottom of the stem	SLB
	Middle of stem	STM
	Primary lateral branches at the middle of the stem	PLM
	Secondary lateral branches of primary lateral branches at the middle of the stem	SLM
	Top of stem	STT
	Primary lateral branches at the top of the stem	PLT
Soil	Soil water content	SWC
Drought index	Palmer drought severity index	PDSI
Indicators of drought response	Resistance	Rt
	Recovery	Rc
	Resilience	Rs
	Relative resilience	RRs

2.3. Gravimetric Soil Water Content, Groundwater Depth, and Meteorological Data

In June 2021, soil samples were collected by soil drilling method near the plant samples, stratified every 20 cm, 3 replicates per layer until the water was produced. The soil sample was placed in an aluminum box and brought back to the laboratory to measure the gravimetric soil water content (SWC) by oven drying method (hereafter referred to as SWC). Depth to groundwater data were obtained from the groundwater level monitoring well closest to the sampling site (2.7 km from the sampling site) and included data from 2013–2021.

Meteorological data for the study area were obtained from grid-point meteorological data from the Royal Netherlands Meteorological Institute data-sharing website (http://climexp.knmi.nl (accessed on 20 March 2023). Monthly temperature, precipitation (CRUTS4.06 grid point dataset with 0.5° × 0.5° resolution), and Monthly Palmer Drought Index (PDSI) data (CRU self-calibrating PDSI grid point dataset with 0.5° × 0.5° resolution) for 1961–2021 were acquired. Based on monthly men precipitation, temperature, and PDSI data from

1961–2021, annual (average of January through December of each year) and multi-year monthly mean (average of each month in the 1961–2021 period) precipitation and temperature data, as well as the average of PDSI from April of the previous year through October of the current year, were calculated.

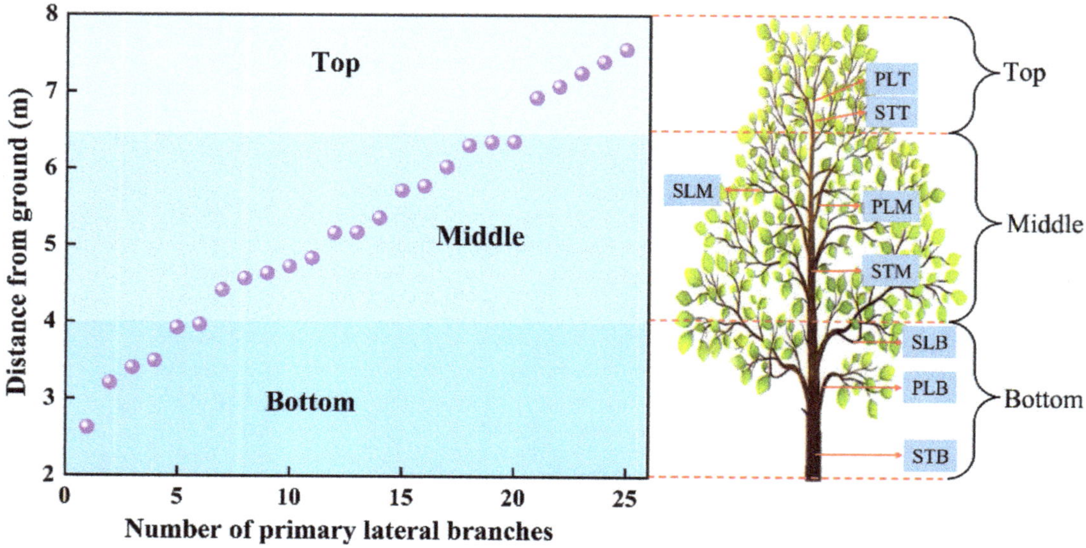

Figure 2. Stem and lateral branches are grouped according to the distance from the ground of the primary lateral branches. The number of primary lateral branches refers to the sum of the number of primary lateral branches of the three trees. The purple color indicates the distance of the primary lateral branches from the ground. STB, STM, and STT represent the bottom, middle, and top of the stem, respectively; PLB, PLM, and PLT represent the primary lateral branches at the bottom, middle, and top of the stem, respectively; and SLB and SLM represent the secondary lateral branches at the bottom and middle primary lateral branches of the stem, respectively.

The PDSI serves as a measure of drought severity as well as a measure of hydrologic drought [34,35]. It can "establish drought" and has greater persistence [36]. In addition, Alley [37] analyzed the relationship between the PDSI and regional runoff and groundwater indices and showed that the groundwater index is the most conservative indicator of the end of a drought. Together with that, the groundwater depth at the sampling sites of this study is within the range of suitable burial depth for *P. euphratica* growth. Therefore, drought severity and duration were not represented by the groundwater index. However, the PDSI is used as the main indicator to reflect the severity of drought and measure hydrological drought in this paper.

To analyze the adaptability of tree growth to extreme drought, the year with an average PDSI ≤ -1 from October of the previous year to April of the current year was set as the drought year [38] concerning the national standard of Meteorological Drought Rating (GB/T20481–2017). To understand the variation pattern of adaptive capacity over time, years with $-2 <$ PDSI ≤ -1 were set as light drought years, years with $-3.0 <$ PDSI ≤ -2 were set as medium drought, years with $-4.0 <$ PDSI $\leq$ than -3.0 were set as severe drought and years with PDSI ≤ -4 were set as exceptional drought (Figure 3).

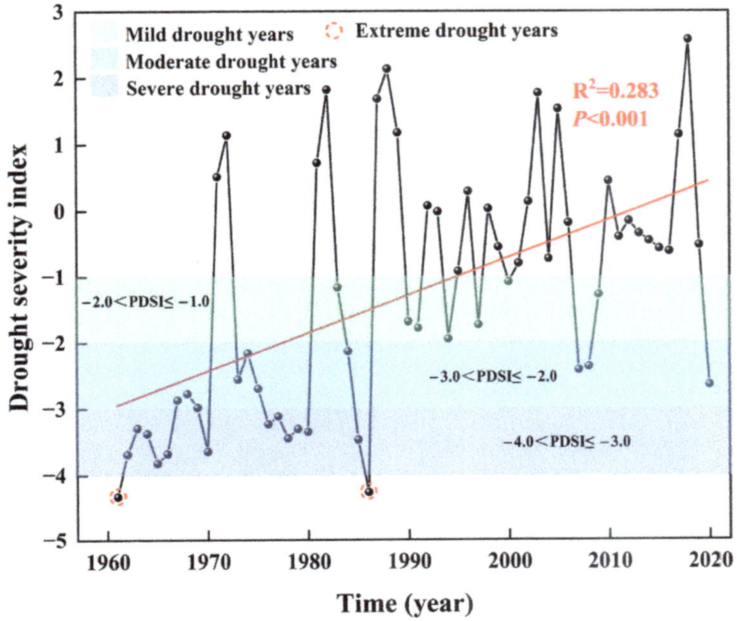

Figure 3. Mean PDSI from April of the previous year to October of the current year. Red circles indicate extreme drought years.

2.4. Tree-Ring Width, Chronology, and Basal Area Increment

The collected cores were processed in the laboratory according to common tree chronology treatment methods and subsequently fixed, dried, sanded, polished, and visually dated under a microscope [7]. TRW was measured (accuracy 0.001 mm) using a LINTABTM6 (RINNTECH, Heidelberg, Germany) measuring system. The dating and measurement results were verified using the COFECHA program. The negative exponential function in the ARSTAN program was used to remove tree growth trends and obtain the ring width index (RWI) for the three studied trees at different heights and different parts of the stem [39,40]. The basal area increment (BAI) was calculated using Equation (1) [41]:

$$BAI_t = \pi\left(R_t^2 - R_{t-1}^2\right) \quad (1)$$

where R is the tree radius and n is the year of tree-ring formation.

2.5. Vulnerability to Drought

The resistance (R_t), recovery (R_c), resilience (R_s), and relative resilience (RR_s) of tree growth to extreme drought can reflect a tree's drought adaptability [7]. The response of *P. euphratica* to drought was quantified using the BAI approach. The following four metrics have been described for the responses of trees to drought [7,42].

R_t is the ratio in tree growth between a drought event and before the drought event and represents the ability of trees to maintain growth under drought stress. R_c is the ratio in tree growth between a drought event and after a drought event, characterizing the ability of trees to recover their normal growth after being subjected to drought stress. R_s is the ratio in tree growth before and after a drought event, characterizing the ability of trees to recover to previous growth levels after a drought event. RR_s are the weighted amount of damage to R_s during a drought [7]. The indicators were calculated as follows:

$$R_t = BAI_D / BAI_{Pre} \quad (2)$$

$$R_c = BAI_{Post}/BAI_D \quad (3)$$

$$R_S = BAI_{Post}/BAI_{Pre} \quad (4)$$

$$RR_S = (BAI_{Post} - BAI_D)/BAI_{Pre} \quad (5)$$

where BAI_D is the average BAI during the drought event, and BAI_{pre} and BAI_{post} are the average BAI during the 3-year period before and after the drought event, respectively. In this study, the adaptation of growth at different heights and in different parts of *P. euphratica* to drought events was analyzed using BAI at the base, middle, and top of the stem.

The radial growth change rate of trees was used to determine whether growth depression occurred in *P. euphratica* and was calculated as follows [43]:

$$GC_i = (M_2 - M_1)/M_1 \quad (6)$$

where GC_i is the radial growth change rate for each of the 5 years before and after the tree was evaluated in the year i. To minimize the effect of growth changes in individual years on the detection of growth inhibition events, the radial growth change rate of *P. euphratica* was calculated as a sliding average over 5 years, and M_1 and M_2 represent the average of the tree-ring width indices for the initial 5 years (including the current year) and the final 5 years (excluding the current year), respectively. A radial growth change of less than –25% indicates growth inhibition and decline, whereas a change of more than 75% indicates growth release [44].

2.6. Data Analysis

The growing season of *P. euphratica* is from April to October, and studies have shown that the response of tree growth to post-climatic changes has a "hysteresis effect" [30]. Therefore, in this study, climate factors and tree-ring chronology from April of the previous year to October of the current year were selected for a correlation analysis. Pearson correlation coefficients were calculated using the SPSS 25 statistical software (IBM, Inc., Armonk, NY, USA). The plotting process was executed using the Origin 2021 Plotting and Statistics software (OriginLab Corp., Northampton, MA, USA) for data visualization and analysis.

3. Results

3.1. Soil Water Content and Groundwater Depth at Sampling Points

The soil water content (SWC) at the sampling sites increased with increasing depth. The variation in SWC ranged from 1.95% to 34.19%, with a mean value of 20.85% (Figure 4a). Groundwater depth at the sampling site was 2 m. Data from the nearest groundwater bathymetry observation point at the sampling site showed that the groundwater bathymetry became progressively shallower (3.03–2.45 m) from 2013 to 2020, and then declined in 2021, to 3.49 m, a long period of time at a level suitable for the growth of Euonymus (2–4 m) (Figure 4b).

3.2. Changes in Meteorological Factors

From 1961 to 2021, the total annual precipitation of the Daliyaboy Oasis was 141.8 mm, with an average annual precipitation of 2.3 mm and large fluctuations in precipitation, and recorded wet and dry years. The total annual precipitation showed a significant increasing trend at a rate of 0.015 mm/a (R^2 = 0.07, $p < 0.05$) (Figure 5a). The mean annual temperature varied between 11.3 °C and 13.9 °C with a significant upward trend (R^2 = 0.550, $p < 0.01$), with a rate of increase of 0.03 °C/a (Figure 5a). According to the national standards of the Meteorological Drought Rating (GB/T20481–2017), 1984, 1990–1991, 1994–1995, and 1997 were classified as light drought years; 1967–1969, 1973–1975, 2007–2008, and 2020 were classified as medium drought years; 1963–1966, 1970, 1976–1980 and 1985 were classified as heavy drought years; and 1961 and 1986 were classified as exceptional drought years (Figure 3). Meteorological data from 1961–2021 indicated local warming and humidification.

Increases in PDSI values ranged from −4.70 to 2.95, and there were no severe or exceptional drought events after 1986 (Figure 3).

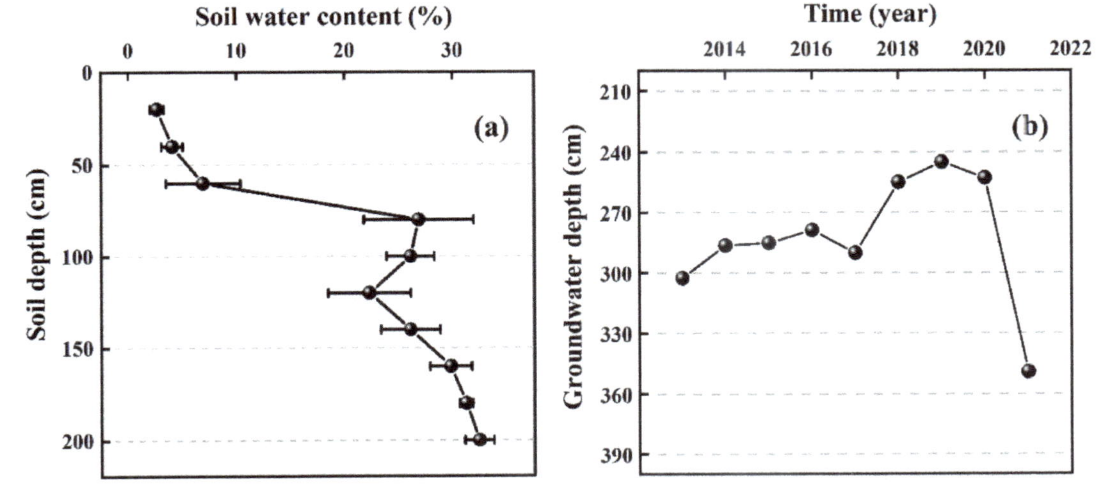

Figure 4. Variation in the vertical profiles of soil water content, $n = 3$, mean ± SD (**a**) and groundwater depth (**b**) at sampling sites. SWC collected in June 2021. Groundwater depth data for 2013–2021.

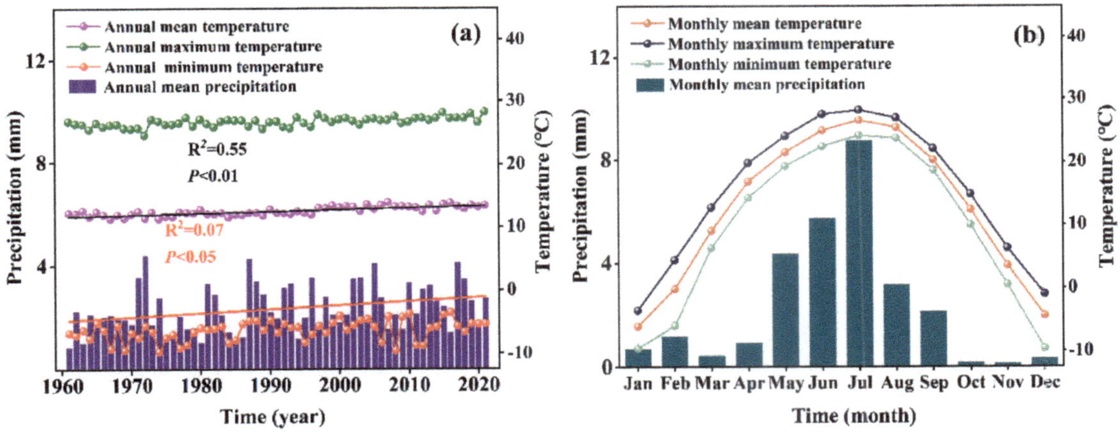

Figure 5. Variation in annual mean temperature, annual maximum and minimum temperatures, and annual mean precipitation (**a**); monthly mean temperature, monthly maximum and minimum temperature, and monthly mean precipitation (**b**) in Daliyaboyi Oasis. The black line is the mean temperature trend; the red line is a precipitation trend.

3.3. Tree-Ring Width and Chronology

The TRW was narrowest in the years when growth began and became progressively wider with age. The variation in TRW of the different parts and different heights was generally consistent, with the widest ring width occurring in 2005. The TRW decreased gradually as the branching class of different parts of *P. euphratica* increased and as the tree height increased (Figure 6a, mean ring width). Interannual fluctuations in the TRW were more variable in different parts of the bottom, and flatter in the middle and top (Figure 6a). The cumulative value of TRW of the stem and primary lateral branches of *P. euphratica* decreased with increasing height, with the cumulative value of TRW of the stem decreasing from 90.945 mm at the bottom to 66.245 mm at the top. The cumulative value of TRW of the primary lateral branches decreased from 56.987 mm at the bottom to 27.768 mm at the top, and the cumulative value of TRW of the secondary lateral branches increased from 28.795 mm at the bottom to 29.828 mm in the middle (Figure 6b).

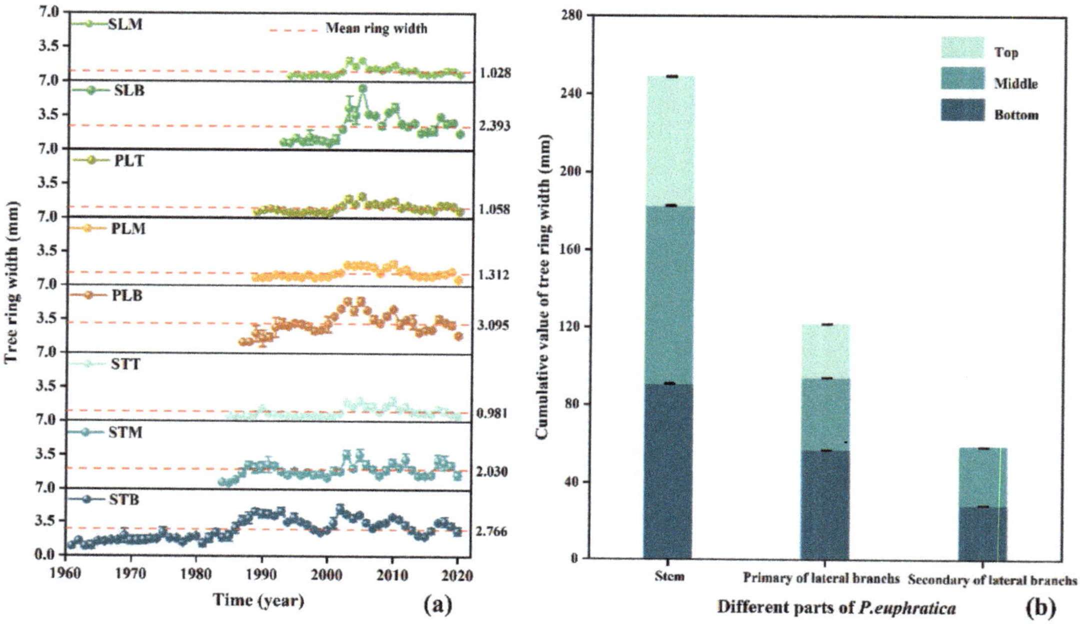

Figure 6. Variation in the tree-ring width of different parts of *P. euphratica* (**a**) and cumulative tree-ring width from 1994 to 2020 (**b**). (**a**) n varies from 61 at the bottom to 27 at the top, mean ± SE, (**b**) n = 27, mean ± SE.

The variations in RWI were highly consistent among different heights of the stem (STB, STM, and STT, Figure 7a), primary lateral branches (PLB, PLM, and PLT, Figure 7b), and secondary lateral branches (SLB and SLM, Figure 7c). Growth trends were consistent across different heights and parts, with the smallest RWI occurring in 2000 (Figure 7).

3.4. Radial Growth of P. euphratica in Relation to Climatic Variables

The TRW of STB was significantly correlated with PDSI, temperature, and precipitation ($p < 0.01$). The TRW of STT was significantly correlated with precipitation ($p < 0.01$) (Figure 8a). The TRW of PLB was significantly correlated with PDSI and precipitation ($p < 0.01$), and the TRW of PLT was significantly correlated with precipitation ($p < 0.05$) (Figure 8b). The TRW of SLM was positively correlated with the mean annual precipitation ($p < 0.05$) (Figure 8c).

The RWI of SLB, STM, PLM, and SLM was positively correlated with the temperature in October of that year ($p < 0.05$), the RWI of STM, SLM, and STT were positively correlated with the temperature in June of the previous year ($p < 0.05$), and the RWI of SLB and STM were positively correlated with the temperature of the previous November (Figure 9, left).

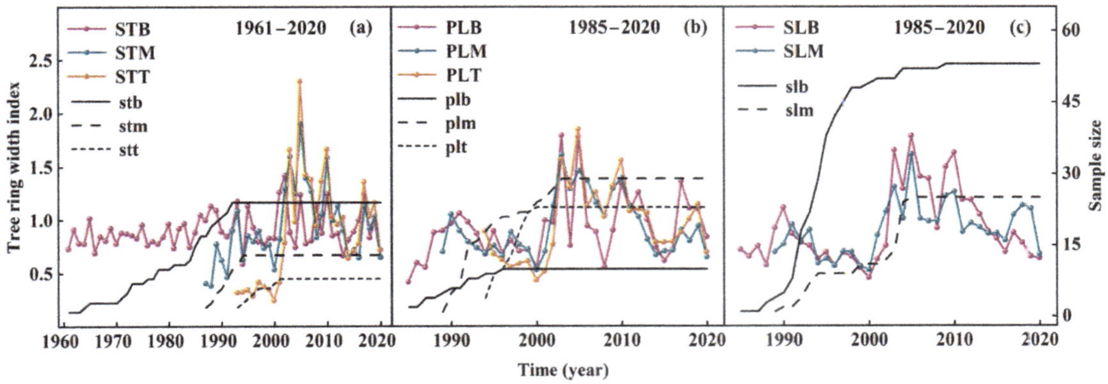

Figure 7. Tree-ring width indices of different components of *P. euphratica*. (**a**) Stem: STB, STM, and STT; (**b**) primary lateral branches: PLB, PLM, and PLT; (**c**) secondary lateral branches: SLB and SLM. Lowercase letters indicate the corresponding sample size.

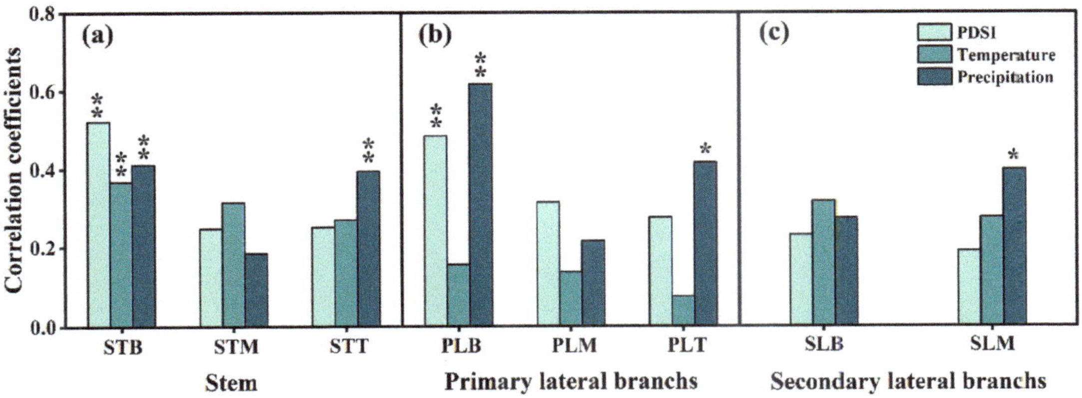

Figure 8. Correlation of tree-ring width and mean annual climatic factors (* and ** denote $p < 0.05$ and $p < 0.01$, respectively); (**a**) different heights of the stem; (**b**) primary lateral branches of different heights; (**c**) secondary lateral branches of different heights.

There was a positive correlation between the RWI of SLB and the temperature in October of the previous year, and the RWI of SLM was positively correlated with the temperature in March of that year. The RWI of the PLB and SLB were positively correlated with precipitation in January of that year ($p < 0.05$) (Figure 9, middle).

The RWI of STB was positively correlated with PDSI in July–October of that year ($p < 0.01$). Positive correlations were observed between PDSI and RWI of PLB in November–December of the previous year and in January–February, April–June, and August–October of the current year ($p < 0.01$). The RWI of SLM was positively correlated with PDSI in September–October of that year ($p < 0.05$ and $p < 0.01$) (Figure 9, right).

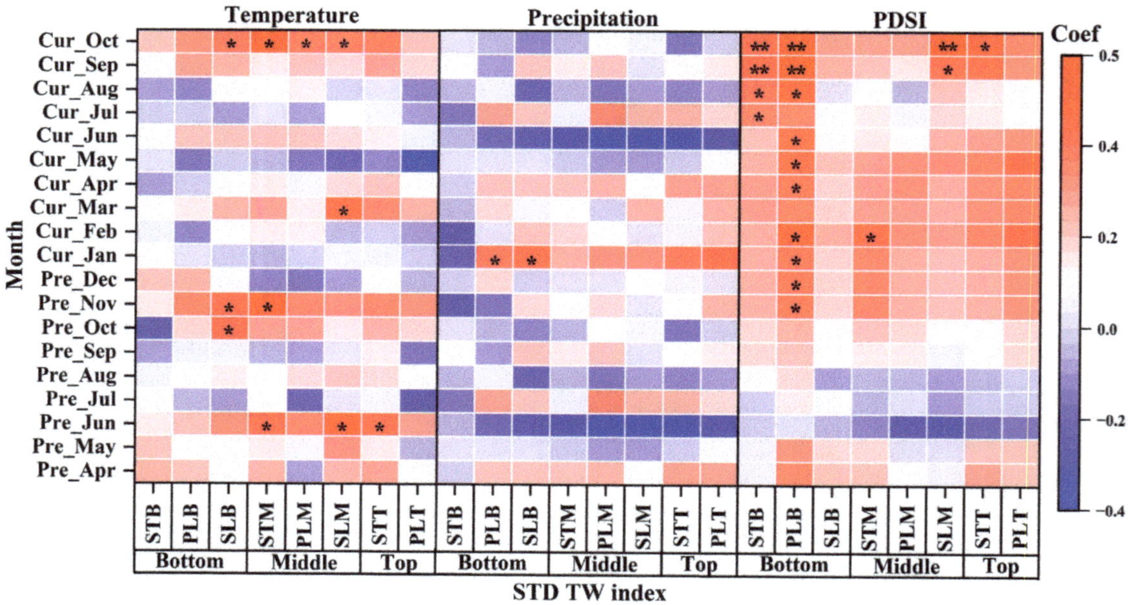

Figure 9. Correlation between the tree ring index of *P. euphratica* and monthly climatic factors. Cur_month represents the current year of ring formation and Pre_month represents the climate variables during the year preceding ring formation (* and ** denote $p < 0.05$ and $p < 0.01$, respectively).

3.5. Growth Decline of P. euphratica at Different Heights and Parts

Between 1990–1997 and 2005–2019, plants of *P. euphratica* at different heights experienced varying degrees of growth decline. Growth increased in 1998–2004. The increase and decrease in the growth of plants of different heights and different parts were essentially synchronized. The radial growth change rates of different parts of the bottom ranged from −25.26% to −83.79%, with STB and SLB being the most pronounced. The growth recession in the middle ranged from −27.56% to −86.37%, with STM and PLM being the most pronounced. The growth recession in the top ranged from −24.27% to −85.71%, with STT being more pronounced than PLT. Primary/secondary lateral branches at the bottom and top of the *P. euphratica* stem showed a higher degree of growth decline than in the middle (Figure 10).

3.6. Adaptation of Different Parts of P. euphratica to Drought

The BAI data were analyzed for R_t, R_c, R_s, and RRs during the drought event. As tree height and branching class increased, R_t first increased and then decreased, and $R_t > 1$. The R_t differed significantly between the STB and STT ($p < 0.05$)(Figure 11a). R_c, R_s, and RR_s increased with tree height and decreased with branching class and $R_c > 1$ (Figure 11b–d). There were no significant differences between the different heights of *P. euphratica* and different parts in terms of R_c ($p < 0.05$) (Figure 11b). Significant differences were found in R_s between different parts at the bottom and middle of *P. euphratica* ($p < 0.05$) (Figure 11c).

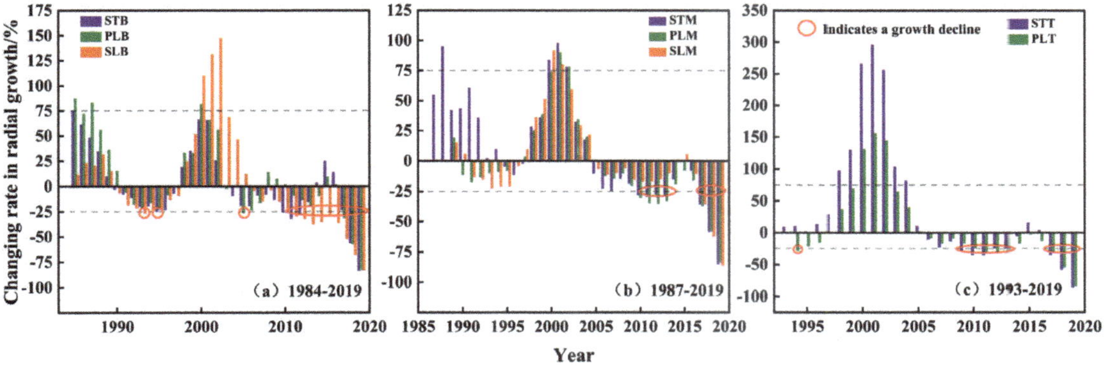

Figure 10. Change rate in radial growth of *P. euphratica* at different heights. (**a**) Change in radial growth rate of bottom of stem, primary lateral branches at the bottom of the stem, and secondary lateral branches of primary lateral branches at the bottom of the stem. (**b**) Change in the radial growth rate of the middle of the stem, primary lateral branches at the middle of the stem, and secondary lateral branches of primary lateral branches at the middle of the stem. (**c**) Change in the radial growth rate of the top of stem and primary lateral branches at the top of the stem. The dashed lines represent the 75% and −25% control lines.

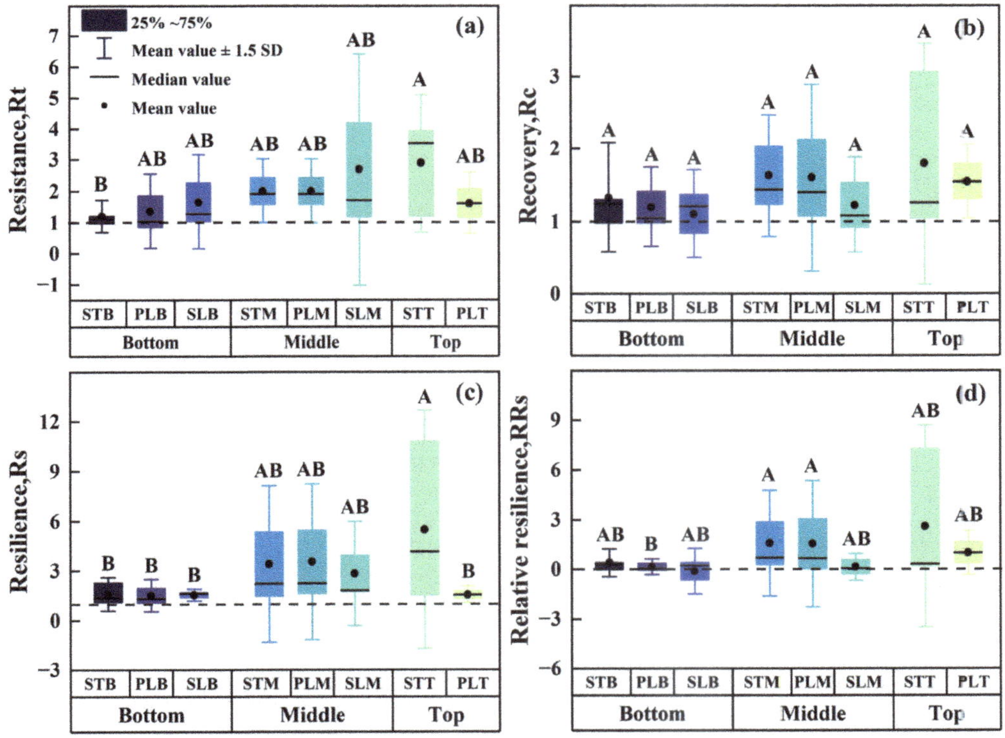

Figure 11. Mean values of resistance (**a**), recovery (**b**), resilience (**c**), and relative resilience (**d**) of *P. euphratica* in different drought events. Different letters in the figure indicate the differences among Rt, Rc, Rs, and RRs of different parts of *P. euphratica*, with different letters indicating significant differences ($p < 0.05$) and the same letters indicating no significant differences ($p < 0.05$).

4. Discussion

4.1. Meteorological Factors and Radial Growth

At the beginning of the 21st century, the climate of northwest China was changing from warm and dry to warm and wet [6]. Zhang [6] analyzed the process and scale characteristics of climate humidification in the northwest arid zone of China using satellite remote sensing, field observations, and coupled models and found that the temperature and precipitation increased significantly from 1961 to 2018. Temperature and precipitation in our study area showed a significant increasing trend, at rates of 0.03 °C/a and 0.015 mm/a, respectively, from 1961 to 2020 (Figure 5). The PDSI trend in the study area was severe drought → moderate drought → light drought, and there were no exceptional or severe drought years after 1986 (Figure 3). The climate change observed in the study area is consistent with the trend of climate change in the Northwest Arid Zone, and there is a clear phenomenon of "warming and humidification". Although there was an increasing trend in precipitation in the study area, an increase in temperature increased evaporation and offset the increase in precipitation. Therefore, the climate in the study area is significantly "warmer" than "wetter".

Factors such as a tree's condition, natural environment, and physiological conditions can affect the stem, resulting in differences in radial growth at different stem heights [23]. In the present study, the TRW at different stem heights increased and then decreased with time. The TRW of the primary and secondary lateral branches of *P. euphratica* varied consistently at different heights (Figure 6a). We argue that diameter growth during tree growth is a result of the processes of new sapwood replenishment and conversion of old sapwood into core wood, and therefore trees as a complete system have consistent growth trends at different heights of the stem and lateral branches of TRW. Gartner [45] showed that the TRW of torch pines was positively correlated with stem height. In the present, the TRW of *P. euphratica* increased from the STB to the STM and then decreased at the STT (Figure 6b). Yu [20] showed an increase followed by a decrease in the TRW of torch pines with stem height, which is consistent with the results of the present study. The results of Zhang [46] on the radial growth characteristics of red pines at different tree heights on Changbai Mountain showed that the trends of TRW variation based on height were mostly the same, and the TRW variation was most obvious at the base and top (0.3, 1.3, and 20 m). In the present study, the TRW at the STB, STM, and STT of the stem of *P. euphratica* were highly undulating (Figure 6a). The results of the analysis of the variation in TRW of different parts of *P. euphratica* at different heights showed that the variation patterns of TRW of the stem, primary, and secondary lateral branches of *P. euphratica* at different heights were relatively consistent (Figure 6a). Variations in the PLM, SLM, and PLT annual cycle widths became increasingly stable over time (Figure 6a). The TRW of the primary and secondary lateral branches of *P. euphratica* were smaller than those of their stems (Figure 6b). The TRW of the primary lateral branches of *P. euphratica* gradually decreased with increasing height. Previous studies have shown that in trees with physiologically active branches, the TRW of the branch is greater at the base of the stem; as the tree matures, the lower branches become less physiologically active or even begin to die, and the maximum growth point moves upward [47]. In the present study, the TRW of the secondary lateral branches increased with height (Figure 6b). Farrar [48] showed that TRW is relatively narrow at the apical internode, reaches its first maximum at the branch with the highest number of leaves, and then decreases at the lower part of the canopy. The variation in TRW with stem height is controlled by growth hormone production, distribution, and nutrient flow [49]. In a simulation study of meristematic growth in Pinus sylvestris, Ford [50] showed that as branches elongate, leaves move farther away from the stem. Additional leaves require additional branches to support them, and the potential biomass exported from branches to the stem first increases with the number of leaves and then decreases with the amount of support required. An increase in the number of branches and leaves reduces the number of photosynthetic products exported to the main stem, resulting in less stem growth than lateral branch growth [50]. This is inconsistent with the results of the present study, in which the stem growth was greater than that of the lateral branches. The differences in tree species and the areas where the trees grow may lead to different resource allocation (exchange) patterns between the stem and lateral branches in the aboveground parts, resulting in differences in growth allocation. However, differences in biomass allocation can be explained by the typical pipe model theory of the existence of a linear relationship between the part of the tree stem cross-section with conductive capacity and the corresponding number of supporting leaves [51]. The part of the tree stem cross-section with conduction capacity is the sapwood, which implies that the relationship between the sapwood area and the number of responsive leaves, as well as the difference between the sapwood area of the stem and the lateral branches, may be the key factors contributing to the differences in the growth of different parts of the tree.

4.2. Relationship between Radial Growth and Meteorological Factors

Analysis of the correlation between different heights and different parts of *P. euphratica* and meteorological factors showed that radial growth rates at varying heights and distinct parts of the stem responded to different degrees of climate change. The TRW at the STB of *P. euphratica* was highly significantly and positively correlated with annual mean PDSI, temperature, and precipitation (Figure 8). This correlation with temperature is consistent with the results of a previous study [30]. There was no significant correlation between the TRW at breast height of the stem of *P. euphratica* and the mean annual precipitation in a previous study [30], which is inconsistent with the results of this study. In this study, the TRW at the STT and the PLT of *P. euphratica* were significantly affected by precipitation (Figure 8a,b). Although *P. euphratica* grows in a warm temperate continental climate zone with sparse precipitation, precipitation has an effect on the growth of the STB, PLB, STT, PLT, and SLM.

Correlations between different heights and different parts RWI of *P. euphratica* and climatic factors from April of the previous year to October of the current year showed no significant correlation between the STB of *P. euphratica* and temperature (Figure 9, left). Previous studies have shown that the RWI at the breast height of the stem of *P. euphratica* was negatively correlated with the temperature in September of the previous year. This indicates that growth at the bottom of the stem of *P. euphratica* is not limited by temperature [52]. The correlation with temperature showed that the lateral branches in the middle and top of the stem were more sensitive to temperature than the bottom of the stem, and the temperature in June of the previous year had a certain "hysteresis effect" on the growth of STM, SLM, and STT in the current year. Correlation with precipitation showed that the RWI of primary/secondary lateral branches at the bottom of the stem of *P. euphratica* was significantly and positively correlated with precipitation in January of that year (Figure 9, middle). This may have resulted from low precipitation in the study area, which does not contribute much to tree growth [52]. The RWI at the STB was significantly correlated with PDSI in spring and summer (Figure 9, right). When temperature is no longer a limiting factor for tree growth, the limiting role of moisture as well as the correlation between it and PDSI becomes increasingly prominent [44]. Climate change affects the growth of the top of the stem of trees differently from that at breast height [24] and the top of the stem is less sensitive to climate change [22].

4.3. Growth Decline and Response to Drought

Tree growth decline events can be determined from a tree's radial growth history. Studies have shown that the occurrence of growth depression in trees is a sign of growth decline, and prolonged growth depression indicates that trees are definitively experiencing growth decline [53]. Different degrees of growth decline occurred in the different components of *P. euphratica* growth from 1984 to 2019 (Figure 10). The timing of growth decline in the stem, primary lateral branches, and secondary lateral branches occurred with some degree of synchronization throughout the growth period. Growth declined at the base of *P. euphratica* in 1993, 1995, 2005, and 2010–2019. Figure 5a demonstrates that the average temperature in the study area consistently increased from 1961 to 2019. High temperatures and low precipitation increase the intensity of drought and enhance the effects of water stress on physiological processes that are detrimental to the formation and storage of photosynthetic products in trees [54,55]. SLB underwent growth decline more frequently and was more sensitive to drought than STB and PLB (Figure 10a). Figure 10b indicates that a growth decline occurred in the central portions of *P. euphratica* from 2010 to 2013 and from 2017 to 2019. Growth decline occurred most frequently in the STM and PLM, with synchronous growth decline occurring in different parts in response to drought from 2017 to 2019 (Figure 10b). Growth decline occurred in different parts of the top of *P. euphratica* in 1994, 2009–2013, and 2017–2019, and growth decline occurred most frequently in the PLT, indicating a greater sensitivity to drought and a strong synchronization in drought responses (Figure 10c). Growth decline occurred simultaneously at the bottom, middle, and top of *P. euphratica* from 2010 to 2013 and from 2017 to 2019. Studies on forest decline on the Tibetan Plateau have found that recent forest decline is associated with increased drought frequency owing to climate warming [56]. In the present study, more than 30 drought events occurred between 1961 and 2020 (Figure 3), and a growth decline occurred in different parts of *P. euphratica* corresponding to the occurrence of drought events. Thus, different parts of *P. euphratica* undergo growth decline owing to drought events. Sensitivity to drought was higher at the middle and top of *P. euphratica* than at the bottom.

Global tree degradation and mortality affect forest health and ecological services [57]. Decreases or increases in forest mortality may be related to thresholds that affect the specific components of a tree's ability to recover its growth performance before stress [7]. Under drought stress, some trees adapt through internal regulatory mechanisms, and growth resumes once the stress disappears [58]. In the present study, there were no significant differences between the R_c values of the different parts of *P. euphratica* (Figure 11b). There was a significant difference between the R_t values in the STB and STT *P. euphratica* (Figure 11a). The R_t at the bottom of *P. euphratica* was lower than those at the middle and top (Figure 11a), which shows that growth in the middle and top of the stem of *P. euphratica* was more sensitive to drought events, and therefore showed a higher resistance to drought events (Figure 11a). Zhou [30] studied the relationship between the radial growth of *P. euphratica* at breast height and groundwater and temperature and showed that temperature was the main limiting factor for *P. euphratica* growth when the groundwater burial depth was less than 6 m. Additionally, the frequency of young/moderate drought events was high (10 occurrences) in the study area from 1984 to 2019 (Figure 3). Thus, repeated mild to moderate drought events increased tree resistance to extreme drought stress [59]. In addition to the R_t of PLT, that of lateral branches was higher than that of the stems (Figure 11a). Young trees exhibit a higher R_t than mature wood [60], including that the R_t of trees may be age dependent. Tree resistance is related to water use efficiency, and trees in areas with inadequate water availability have reduced resistance to drought [61]. The groundwater burial depth in the *P. euphratica* growing area in this study was shallower than that suitable for *P. euphratica* growth (Figure 4b), with no water stress. Thus, the R_t of *P. euphratica* showed moisture-independent variation at different heights and parts. In this study, the resilience of the top of *P. euphratica* was higher than that of the bottom and middle. As the height of the stem increased, R_c increased; the R_c of the primary lateral branches was largest at the top and smallest at the bottom, and the R_c of the middle of the secondary lateral branches was larger than that of the bottom (Figure 11b). Studies have found that physiological damage caused by prolonged drought, such as xylem embolism, reduced reserve or preventative capacity, pest attack, and infection, may reduce the resilience of trees to subsequent drought [62]. The results of Zhou [27] on the xylem hydraulic conductivity and embolism of *P. euphratica* and its response to drought stress showed that riparian plants adapt to mild drought stress by limiting the xylem hydraulic conductivity of branches to coordinate plant growth. In this study, the R_c at the top of the stem (STT) was the highest. Studies have shown that plants compete for additional light, heat, and nutrient resources by increasing their hydraulic conductivity in well-watered environments [25]. *P. euphratica* improves the chances of plant survival by increasing the competitiveness of dominant branches and sacrificing non-dominant branches, thereby optimizing the resilience of the whole plant [27].

Tree growth resilience is the ability to restore an original growth rate after being disturbed by extreme events [63]. The values of $R_s < 1$ reflect the persistent impact of extreme events. In this study, the Rs values of different parts of *P. euphratica* increased with height and were consistently greater than 1, except for PLT (Figure 11c). This indicated that all parts of *P. euphratica*, except for the top lateral branches, recovered their pre-disturbance growth rates following an extreme event disturbance.

RR_s are defined as R_s weighted by the damage that occurred during a drought. RR_s are <0 if the growth rate after a drought event is lower than that during the drought event [7]. In the present study, the RR_s decreased in different parts of *P. euphratica*, indicating that the trees were subjected to drought stress, accumulating from the stem to the secondary lateral branches and from the bottom to the middle. The RR_s of *P. euphratica* SLB was <1 (Figure 11d), indicating that the growth rate of *P. euphratica* lateral branches was lower than the growth rate during drought. An $RR_s < 0$ indicates that trees experience long-term stress, which leads to loss of vigor and growth decline [64]. The gradual decrease in RR_s, which eventually became <0 in the secondary lateral branches, indicated the vulnerability of *P. euphratica* lateral branches to drought. The combined results proved that the hypothesis that the resistance of the lateral branches of *P. euphratica* would be higher than that of the stem was valid, but the hypothesis that the recovery force of the lateral branches and the recovery elasticity would be higher than those of the stem was not valid.

Drought is one of the most serious challenges facing species and ecosystems. Drought tolerance is related to water relations in plants [65]. Water relations in plants maintain healthy branches, and prolonged drought conditions for plant growth may affect water relations [66]. Stomatal conductance, hydraulic properties, and water status determine water relations in branches [67]. Past studies have shown that hydraulic properties are related to plant drought resistance [68]. The results of this study showed that growth decline occurred in the top and top branches of the stem of *P. euphratica* (Figure 10). And the lateral branches of *P. euphratica* were weaker than the stem in restoring pre-drought growth after drought (Figure 11). This may be related to the sensitivity of water status and water transport efficiency, as well as xylem hydraulic conductivity to drought at different heights and in different parts of *P. euphratica* [66], or since the physiological activity of apical branches relative to lateral branches can be explained by the lower photosynthesis of last ones because of shading. In conclusion, the effect of drought on the dynamic balance between water supply, water status, and water transport efficiency of *P. euphratica* is one of the main reasons for the "broke its arm" of *P. euphratica* under future drought stress.

5. Conclusions

TRW decreased gradually with increasing height and branching class of *P. euphratica*. Growth declines at the bottom of the stem earlier than at the middle and top. Temperature, precipitation, and PDSI contributed to the growth of the bottom of *P. euphratica*, while precipitation contributed to the growth of the top. *P. euphratica* growth change rate decline was highly synchronized across heights and parts, with relatively high declines at the bottom and top. There were no significant differences in the R_c values for different heights and parts of *P. euphratica*, but the R_t, R_s, and RR_s values for the bottom and top were significantly lower than those for the other components, indicating the vulnerability of the bottom and top of *P. euphratica* to drought. The RR_s gradually decreased with the increase in branching class, and the RR_s values of the secondary lateral branches at different heights were the lowest. The sensitivity of the bottom and top lateral branches of *P. euphratica* to drought ultimately leads to their growth decline or even death. Therefore, when drought stress occurs *P. euphratica* preferentially abandons bottom and top lateral branches and increases resource utilization efficiency to offset the effects of drought on *P. euphratica* growth.

Author Contributions: Conceptualization, Q.S., Y.D. and F.Z.; investigation, A.A., Y.W., L.P. and H.S.; writing—original draft preparation, A.A.; writing—review and editing, A.A. and Y.D. All authors have read and agreed to the published version of the manuscript.

Funding: This work was funded by the National Natural Science Foundation of China (Nos. U1703237; 32160260 and 31800613).

Data Availability Statement: Data are contained within the article.

Acknowledgments: Special thanks go to the Urumqi Desert Meteorological Research Institute of the China Meteorological Administration for providing us with experimental support. We are also grateful to the government of Daliyaboyi Township, Xinjiang, China, for its strong support of scientific research and to our driver, Ali, who drove us through the Taklamakan Desert.

Conflicts of Interest: The authors declare no conflict of interest.

References

1. Allen, C.; Macalady, A.; Chenchouni, H.; Bachelet, D.; McDowell, N.; Vennetier, M.; Kitzberger, T.; Rigling, A.; Breshears, D.; Hogg, E.; et al. A global overview of drought and heat-induced tree mortality reveals emerging climate change risks for forests. *For. Ecol. Manag.* **2010**, *4*, 660–684. [CrossRef]
2. Wang, J.; Xu, C.; Hu, M.; Li, Q.; Yan, Z.; Jones, P. Global land surface air temperature dynamics since 1880. *Int. J. Climatol.* **2018**, *38*, e466–e474. [CrossRef]
3. IPCC. Climate change: Synthesis report. In *Contribution of Working Groups, I, II and III to the Fifth Assessment Report of the Intergovernmental Panel on Climate Change*; Core Writing Team; Pachauri, R., Meyer, L., Eds.; IPCC: Geneva, Switzerland, 2014; p. 151.
4. Liu, H.; Williams, A.; Allen, C.; Guo, D.; Wu, X.; Anenkhonov, O.; Badmaeva, N. Rapid warming accelerates tree growth decline in semi–arid forests of Inner Asia. *Glob. Chang. Biol.* **2013**, *19*, 2500–2510. [CrossRef] [PubMed]
5. Thackeray, S.; Henrys, P.; Hemming, D.; Bell, J.; Botham, M.; Burthe, S.; Helaouet, P.; Johns, D.; Jones, I.; Leech, D.; et al. Phenological sensitivity to climate across taxa and trophic levels. *Nature* **2016**, *535*, 241–245. [CrossRef]
6. Zhang, Q.; Yang, J.; Wang, W.; Ma, P.; Lu, G.; Liu, X.; Fang, F. Climatic Warming and Humidification in the Arid Region of Northwest China: Multi–Scale Characteristics and Impacts on Ecological Vegetation. *J. Meteorol. Res.* **2021**, *35*, 113–127. [CrossRef]
7. Sun, S.; Lei, S.; Jia, H.; Li, C.; Zhang, J.; Meng, P. Tree–Ring Analysis Reveals Density–Dependent Vulnerability to Drought in Planted Mongolian Pines. *Forests* **2020**, *11*, 98. [CrossRef]
8. Jump, A.; Ruiz-Benito, P.; Greenwood, S.; Allen, C.; Kitzberger, T.; Fensham, R.; Martínez–Vilalta, J.; Lloret, F. Structural overshoot of tree growth with climate variability and the global spectrum of drought–induced forest dieback. *Glob. Chang. Biol.* **2017**, *23*, 3742–3757. [CrossRef]
9. Dobbertin, M. Tree growth as indicator of tree vitality and of tree reaction to environmental stress: A review. *Eur. J. For. Res.* **2005**, *124*, 319–333. [CrossRef]
10. Castagneri, D.; Battipaglia, G.; von Arx, G.; Pacheco, A.; Carrer, M. Tree–ring anatomy and carbon isotope ratio show both direct and legacy effects of climate on bimodal xylem formation in *Pinus pinea*. *Tree Physiol.* **2018**, *38*, 1098–1109. [CrossRef]
11. Huang, M.; Wang, X.; Keenan, T.; Piao, S. Drought timing influences the legacy of tree growth recovery. *Glob. Chang. Biol.* **2018**, *24*, 3546–3559. [CrossRef]
12. Gao, S.; Liu, R.; Zhou, T.; Fang, W.; Yi, C.; Lu, R.; Zhao, X.; Luo, H. Dynamic responses of tree–ring growth to multiple dimensions of drought. *Glob. Chang. Biol.* **2018**, *24*, 5380–5390. [CrossRef] [PubMed]
13. Bottero, A.; D'Amato, A.; Palik, B.; Bradford, J.; Fraver, S.; Battaglia, M.; Asherin, L. Density–dependent vulnerability of forest ecosystems to drought. *J. Appl. Ecol.* **2017**, *4*, 1605–1614. [CrossRef]
14. Anderegg, W.R.L.; Kane, J.M.; Anderegg, L.D.L. Consequences of widespread tree mortality triggered by drought and temperature stress. *Nat. Clim. Chang.* **2012**, *3*, 30. [CrossRef]
15. Liu, Y.; Wang, A.; An, Y.; Lian, P.; Wu, D.; Zhu, J.; Meinzer, F.; Hao, G. Hydraulics play an important role in causing low growth rate and dieback of aging *Pinus sylvestris* var. Mongolia trees in plantations of Northeast China. *Plant Cell Environ.* **2018**, *41*, 1500–1511. [CrossRef]
16. Salmon, Y.; Torres-Ruiz, J.; Poyatos, R.; Martinez-Vilalta, J.; Meir, P.; Cochard, H.; Mencuccini, M. Balancing the risks of hydraulic failure and carbon starvation: A twig scale analysis in declining Scots pine. *Plant. Cell. Environ.* **2015**, *38*, 2575–2588. [CrossRef]
17. Guada, G.; Camarero, J.; Sánchez, S.; Cerrillo, R. Limited growth recovery after drought-induced forest dieback in very defoliated trees of two *pine* species. *Front. Plant Sci.* **2016**, *7*, 418. [CrossRef] [PubMed]
18. Colangelo, M.; Camarero, J.; Borghetti, M.; Gentilesca, T.; Oliva, J.; Redondo, M.; Ripullone, F. Drought and phytophthora are associated with the decline of *Oak* Species in Southern Italy. *Front. Plant Sci.* **2018**, *9*, 1595. [CrossRef]
19. Heuret, P.; Meredieu, C.; Coudurier, T.; Courdier, F.; Barthélémy, D. Ontogenetic trends in the morphological features of main stem annual shoots of *Pinus pinater* Ait (Pinaceae). *Am. J. Bot.* **2006**, *93*, 1577–1587. [CrossRef]
20. Yu, M.; Cheng, X.; He, Z.; Wu, T.; Yin, Z. Longitudinal Variation of Ring Width, Wood Density and Basal Area Increment in 26–Year–Old Loblolly Pine (*Pinus taeda*) Trees. *Tree-Ring Res.* **2014**, *70*, 137–144. [CrossRef]

21. Zhang, T.; Huang, L.; Zhang, R.; Gao, Y.; Hu, D.; Yu, S.; Jiang, S. The impacts of climatic factors on radial growth patterns at different stem heights in Schrenk spruce (*Picea schrenkiana*). *Trees* **2020**, *34*, 163–175. [CrossRef]
22. Maaten, M.; Bouriaud, O. Climate–growth relationships at different stem heights in silver fir and Norway *spruce*. *Can. J. For. Res.* **2012**, *42*, 958–969.
23. Meng, S.; Lieffers, V.; Reid, D. Reducing stem bending increases the height growth of tall pines. *J. Exp. Bot.* **2006**, *57*, 3175–3182. [CrossRef] [PubMed]
24. Chhin, S.; Hogg, E.H.; Lieffers, V.J.; Huang, S. Growth–climate relationships vary with height along the stem in lodgepole pine. *Tree Physiol.* **2010**, *30*, 335–345. [CrossRef] [PubMed]
25. Nardini, A.; Pedà, G.; Rocca, N. Trade–offs between leaf hydraulic capacity and drought vulnerability: Morpho–anatomical bases, carbon costs and ecological consequences. *New Phytolog.* **2012**, *196*, 788–798. [CrossRef] [PubMed]
26. Simcha, L.; Roni, A. Vascular differentiation in branch junctions of circular patterns and functional significance. *Trees* **1990**, *4*, 49–54.
27. Zhou, H.; Chen, Y.; Li, W.; Ayup, M. Xylem hydraulic conductivity and embolism in riparian plants and their responses to drought stress in desert of Northwest China. *Ecohydrology* **2013**, *6*, 984–993. [CrossRef]
28. Rice, K.; Matzner, S.; Byer, W.; Brown, J. Patterns of tree dieback in Queensland, Australia: The importance of drought stress and the role of resistance to cavitation. *Oecologia* **2004**, *139*, 190–198. [CrossRef]
29. Li, Q.; Feng, Q.; Zhai, L. Study of the height growth dynamic based on tree-ring data in *Populus euphratica* from the lower reach of the Heihe River, China. *Dendrochronologia* **2010**, *28*, 49–64.
30. Zhou, H.; Chen, Y.; Zhu, C.; Li, Z.; Fu, A. Climate change may accelerate the decline of desert riparian forest in the lower Tarim River, Northwestern China: Evidence from tree–rings of *Populus euphratica*. *Ecol. Indicat.* **2020**, *111*, 105997. [CrossRef]
31. Shi, H.; Zhang, F.; Shi, Q.; Li, M.; Dai, Y.; Zhang, Z.; Zhu, C. Responses of arid plant species diversity and composition to environmental factors. *J. For. Res.* **2023**, *34*, 1723–1734. [CrossRef]
32. Imin, B.; Dai, Y.; Shi, Q.; Guo, Y.; Li, H.; Nijat, M. Responses of two dominant desert plant species to the changes in groundwater depth in hinterland natural oasis, Tarim Basin. *Ecol. Evol.* **2021**, *11*, 9460–9471. [CrossRef] [PubMed]
33. Shi, H.; Shi, Q.; Zhou, X.; Bilal, E.; Li, H.; Zhang, W.; Yasin, K. Effect of the competition mechanism of between co–dominant species on the ecological characteristics of *Populus euphratica* under a water gradient in a desert oasis. *Glob. Ecol. Conserv.* **2021**, *27*, e01611. [CrossRef]
34. Havens, A.V.; Snow, W.B.; Horowitz, J.L.; Liu, C.S. *Drought Frequency, Intensity, and Duration: Its Correlation to Streamflow and Its Impact Upon Synthetic Hydrology*; New Jersey Water Resources Research Institute: New Brunswick, NJ, USA, 1968; p. 49.
35. Draper, S.H.; Palmer, R.N.; Lettenmaier, D.P.; Burges, S.J. *Water Resource System Reliability under Drought Conditions: The Seattle Water Supply System as a Case Study*; Charles W. Harris Hydraulics Laboratory Technical Report No. 72; University of Washington: Seattle, DC, USA, 1981; p. 72.
36. Alley, W.M. The Palmer Drought Severity Index: Limitations and Assumptions. *J. Clim. Appl. Meteor.* **1984**, *23*, 1100–1109. [CrossRef]
37. Alley, W.M. The palmer drought severity index as a measure of hydrologic drought. *J. Am. Water Resour. Assoc.* **1985**, *21*, 105–114. [CrossRef]
38. *GB/T 20481–2017*; Classification of Meteorological Drought. China Meteorological Administration: Beijing, China. Standards Press of China: Beijing, China, 2017.
39. Cook, E.R.; Kairiukstis, L.A. *Methods of Dendrochronology: Applications in the Environmental Sciences*; Kluwer Academic Publishers: Alphen aan den Rijn, The Netherlands, 1990; p. 394.
40. Stokes, M.A.; Smiley, T.L. *An Introduction to Tree–Ring Dating*; University of Arizona Press: Tucson, AZ, USA, 1968.
41. Silva, L.C.; Anand, M.; Leithead, M.D. Recent widespread tree growth decline despite increasing atmospheric CO_2. *PLoS ONE* **2010**, *5*, e11543. [CrossRef]
42. Lloret, F.; Keeling, E.; Sala, A. Components of tree resilience: Effects of successive low-growth episodes in old ponderosa pine forests. *Oikos* **2011**, *120*, 1909–1920. [CrossRef]
43. Nowacki, G.; Abrams, M. Radial–growth averaging criteria for reconstructing disturbance histories from presettlement–origin oaks. *Ecol. Monogr.* **1997**, *67*, 225–249. [CrossRef]
44. Shen, J.; Li, S.; Huang, X.; Wang, S.; Su, J. Ecological resilience and growth degradation of *Pinus Yunnanensis* at different altitudes in Jinsha River Basin. *Sci. Silvae Sin.* **2019**, *56*, 1–11.
45. Gartner, B.; North, E.; Johonson, G.; Singleton, R. Effects of live crown on vertical patterns of wood density and growth in Douglas-fir. *Can. J. For. Res.* **2002**, *32*, 439–447. [CrossRef]
46. Zhang, X.; Gao, L.; Qiu, Y.; Guo, J. Characteristics of Korean pine (*Pinus koraiensis*) radial growth at different heights and its response to climate change on Changbai Mountain. *Acta Ecol. Sinic.* **2015**, *5*, 2978–2984.
47. Peterson, D.; Arbaugh, M.; Robinson, L. Effects of ozone and climate on ponderosa pine (*Pinus ponderosa*) growth in the Colorado Rocky Mountains. *Can. J. For. Res.* **1993**, *23*, 1750–1759. [CrossRef]
48. Farrar, J. Longitudinal variation in the thickness of the annual ring. *For. Chron.* **1961**, *37*, 323–349. [CrossRef]
49. Larson, P. Stem Form Development of Forest Trees. *For. Sci.* **1963**, *9* (Suppl. S2), a0001-42. [CrossRef]
50. Ford, E.; Avery, A.; Ford, R. Simulation of branch growth in the Pinaceae: Interactions of morphology, phenology, foliage productivity, and the requirement for structural support, on the export of carbon. *J. Theor. Biol.* **1990**, *146*, 15–36. [CrossRef]

51. Lehnebach, R.; Beyer, R.; Letort, V.; Heuret, P. The pipe model theory half a century on: A review. *Ann. Bot.* **2018**, *121*, 773–795. [CrossRef] [PubMed]
52. Peng, X.; Xiao, S.; Cheng, G.; Xiao, H.; Tian, Q.; Zhang, Q. Human activity impacts on the stem radial growth of *Populus euphratica* riparian forests in China's Ejina Oasis, using tree-ring analysis. *Trees* **2017**, *31*, 379–392. [CrossRef]
53. Amoroso, M.; Daniels, L.; Larson, B. Temporal patterns of radial growth in declining *Austrocedrus chilensis* forests in Northern Patagonia: The use of tree-rings as an indicator of forest decline. *For. Ecol. Manag.* **2012**, *265*, 62–70. [CrossRef]
54. Liang, E.; Leuschner, C.; Dulamsuren, C.; Wagner, B.; Hauck, M. Global warming-related tree growth decline and mortality on the north-eastern Tibetan plateau. *Climat. Chang.* **2015**, *134*, 163–176. [CrossRef]
55. Ren, P.; Rossi, S.; Gricar, J.; Liang, E.; Cufar, K. Is precipitation a trigger for the onset of Xylo genesis in Juniperus przewalskii on the north-eastern Tibetan Plateau? *Ann. Bot.* **2015**, *115*, 629–639. [CrossRef]
56. Fang, O.; Alfaro, R.; Zhang, Q. Tree rings reveal a major episode of forest mortality in the late 18th century on the Tibetan Plateau. *Glob. Planet Chang.* **2018**, *163*, 44–50. [CrossRef]
57. Gauthier, S.; Bernier, P.; Kuuluvainen, T.; Shvidenko, A.; Schepaschenko, D. Boreal Forest health and global change. *Science* **2015**, *349*, 819–822. [CrossRef] [PubMed]
58. Arend, M.; Sever, K.; Pflug, E.; Gessler, A.; Schaub, M. Seasonal photosynthetic response of European beech to severe summer drought: Limitation, recovery and post–drought stimulation. *Agric. For. Meteorol.* **2016**, *220*, 83–89. [CrossRef]
59. Backhaus, S.; Kreyling, J.; Grant, K.; Beierkuhnlein, C.; Walter, J.; Jentsch, A. Recurrent mild drought events increase resistance toward extreme drought stress. *Ecosystems* **2014**, *17*, 1068–1081. [CrossRef]
60. Rubio–Cuadrado, Á.; Camarero, J.; Aspizua, R.; Sánchez–González, M.; Gil, L.; Montes, F. Abiotic factors modulate post-drought growth resilience of Scots pine plantations and rear-edge Scots pine and oak forests. *Dendrochronologia* **2018**, *51*, 54–65. [CrossRef]
61. Sánchez, S.R.; Camarero, J.J.; Rozas, V.; Génova, M.O.; José, M.; Arzac, A. Resist, recover or both? growth plasticity in response to drought is geographically structured and linked to intraspecific variability in pinus pinaster. *J. Biogeogr.* **2018**, *45*, 1126–1139. [CrossRef]
62. Brodribb, T.; Powers, J.; Cochard, H.; Choat, B. Hanging by a thread? Forests and drought. *Science* **2020**, *368*, 261–266. [CrossRef]
63. Folke, C.; Carpenter, S.; Walker, B.; Scheffer, M.; Elmqvist, T.; Gunderson, L.; Holling, C. Regime Shifts, Resilience, and Biodiversity in Ecosystem Management. *Annu. Rev. Ecol. Evol. Syst.* **2004**, *35*, 557–581. [CrossRef]
64. De Grandpré, L.; Kneeshaw, D.; Perigon, S.; Boucher, D.; Marchand, M.; Pureswaran, D.; Girardin, M. Adverse climatic periods precede and amplify defoliator-induced tree mortality in eastern boreal North America. *J. Ecol.* **2019**, *107*, 452–467. [CrossRef]
65. Engelbrecht, B.M.; Comita, L.S.; Condit, R.; Kursar, T.A.; Tyree, M.T.; Turner, B.L.; Hubbell, S.P. Drought sensitivity shapes species distribution patterns in tropical forests. *Nature* **2007**, *447*, 80. [CrossRef]
66. Li, D.; Si, J.; Zhang, X. Comparison of Branch Water Relations in Two Riparian Species: *Populus euphratica* and *Tamarix ramosissima*. *Sustainability* **2019**, *11*, 5461. [CrossRef]
67. Hetherington, A.M.; Woodward, F.I. The role of stomata in sensing and driving environmental change. *Nature* **2003**, *424*, 901. [CrossRef] [PubMed]
68. Cochard, H.; Herbette, S.; Hernández, E.; Hölttä, T.; Mencuccini, M. The effects of sap ionic composition on xylem vulnerability to cavitation. *J. Exp. Bot.* **2009**, *61*, 275–285. [CrossRef] [PubMed]

forests

MDPI

Article

Tree-Ring Chronologies from the Upper Treeline in the Russian Altai Mountains Reveal Strong and Stable Summer Temperature Signals

Alexander V. Kirdyanov [1,2,3,*], Alberto Arzac [3], Alina A. Kirdyanova [4], Tito Arosio [1], Dmitriy V. Ovchinnikov [2], Dmitry A. Ganyushkin [4], Paul N. Katjutin [4,5], Vladimir S. Myglan [3], Andrey N. Nazarov [3], Igor Y. Slyusarenko [6], Tatiana Bebchuk [1] and Ulf Büntgen [1,7,8]

1 Department of Geography, University of Cambridge, Cambridge CB2 3EN, UK; ta530@cam.ac.uk (T.A.); tb649@cam.ac.uk (T.B.); ub223@cam.ac.uk (U.B.)
2 Sukachev Institute of Forest SB RAS, Federal Research Center 'Krasnoyarsk Science Center SB RAS', Akademgorodok, Krasnoyarsk 660036, Russia; dovch@mail.ru
3 Institute of Ecology and Geography, Siberian Federal University, 79 Svobodnii, Krasnoyarsk 660041, Russia; aarzac@gmail.com (A.A.); v.myglan@gmail.com (V.S.M.); nazar_69@mail.ru (A.N.N.)
4 Institute of Earth Sciences, St. Petersburg State University, St. Petersburg 199034, Russia; akird2002@mail.ru (A.A.K.); ganushkinspbgu@mail.ru (D.A.G.); paurussia@binran.ru (P.N.K.)
5 Botanical Institute of RAS, ul. Professor Popov 2, St. Petersburg 197022, Russia
6 Institute of Archaeology and Ethnography SB RAS, Pr. Akademika Lavrentieva 17, Novosibirsk 630090, Russia; slig1963@yandex.ru
7 Department of Geography, Masaryk University, 61137 Brno, Czech Republic
8 Global Change Research Centre, 61300 Brno, Czech Republic
* Correspondence: ak2118@cam.ac.uk

Abstract: Radial tree growth at high-elevation and high-latitude sites is predominantly controlled by changes in summer temperature. This relationship is, however, expected to weaken under projected global warming, which questions the reliability of tree-ring chronologies for climate reconstructions. Here, we examined the growth–climate response patterns of five tree-ring width (TRW) and maximum latewood density (MXD) chronologies of larch (*Larix sibirica*) from upper-treeline ecotones in the Altai Mountains, which is a key region for developing millennial-long dendroclimatic records in inner Eurasia. The TRW and MXD chronologies exhibited significant year-to-year coherency within and between the two parameters ($p < 0.001$). While TRW is mostly influenced by temperature changes during the first half of the growing season from June to July ($r = 0.66$), MXD is most strongly correlated with May–August temperatures ($r = 0.73$). All seasonal temperature signals are statistically significant at the 99% confidence level, temporally stable back to 1940 CE, the period with reliable instrumental measurements, and spatially representative for a vast area of inner Eurasia between northeastern Kazakhstan in the west, northern Mongolia in the east, southern Russia in the north and northwestern China in the south. Our findings demonstrate the paleoclimatic potential of TRW and especially MXD chronologies and reject any sign of the ´divergence problem´ at these high-elevation, mid-latitude larch sites.

Keywords: climate reconstruction potential; dendroclimatology; divergence problem (DP); global warming; spatiotemporal homogeneity

1. Introduction

Tree growth in most of the high-elevation and high-latitude regions in Eurasia is currently temperature-sensitive and predominantly controlled by summer temperature variability [1]. Tree-ring data from such regions are considered a valuable climate proxy and are widely used in reconstructions of local to global temperatures [2]. However, the dependence of tree growth on temperature is expected to weaken under projected global warming and even decouple from temperature during the 21st century [3].

The loss of sensitivity of tree growth in ecosystems presumably limited by temperature to rising summer temperatures in high- or low-frequency domains or both, the so-called 'divergence problem' (DP) [4], was first reported for white spruce (*Picea glauca*) in Alaska [5]. The study demonstrated weakened signals of the tree-ring width and maximum latewood density at high-latitude and high-elevation sites in response to increasing temperatures since the late 1970s. Lately, the DP has been described for a variety of tree species in different regions, mostly in high-latitude and high-elevation ecosystems (see [4] for a review). Briffa et al. [6] and Wilson et al. [7] provided evidence of the DP as a widespread phenomenon in circum-polar high-latitude forests. The DP was also found in several high-elevation forests in lower latitudes [8–10]. However, some recent studies have shown that the DP is not observed in all temperature-limited sites, implying that the DP is a spatially heterogeneous phenomenon [2,11–13]. A better understanding of the scale and regionality of the DP is required as this phenomenon may affect the ability of tree-ring data to serve as a proxy for temperature during past warm periods [2]. Special attention has to be paid to regions with high paleo-dendroclimatic potential, like the Altai Mountains [14,15].

The Altai Mountains are one of the largest mountain ranges in Asia, which extends between 47–53° N and 84–92° E and lies in the cross-border region extending into Russia in the north, China in the south, Kazakhstan in the south-west, and Mongolia in the south-east. Altai serves as a crucial climatic boundary, where North Atlantic climatic systems from the west interact with Pacific climatic systems from the east. This interaction has led to recent temperature increases in both North and South Altai [16]. The area is an origin of long-living trees [17] and a source of relict wood, which is used to develop multi-millennia-long tree-ring width (TRW) chronologies [14,18–20] and shorter, but extremely valuable, chronologies of other tree-ring parameters, including the maximum latewood density (MXD), tree-ring stable isotope composition, cell structure, and blue intensity (BI) [21–23]. These data provide information not only on forest dynamics in the region [24–27] but also allow an understanding of tree growth's dependence on the climate and serve as proxies for different climate variables, such as temperature [15,19,28,29], precipitation [30,31], droughts [32,33], air relative humidity [34], and frost events [35].

Tree-ring-based reconstructions from the Altai Mountains and neighboring mountain ridges serve as a basis for understanding climate variability over a vast region in Central Asia [15,36]. These data can be matched with climate dynamics reconstructed from ice cores [37], elemental contents [38], pollen data from lake sediments [39,40], pollen data from soil profiles and geomorphological data [41], and lipids from peat cores [42]. In this context, the further development and updating of the existing tree-ring network in the Altai Mountains and understanding the recent changes in tree rings' response to the climate under current warming are urgent. Special attention has to be paid to TRW and MXD measurements from the upper treeline because these parameters are usually closely connected to summer temperature changes and may, consequently, be prone to the DP.

Here, we explored the radial growth and climate sensitivity of five upper treeline larch sites across Central Altai. We analyzed the temporal stability and spatial extent of the climatic response of recently developed tree-ring width and maximum latewood density chronologies to test their exposure to the 'divergence problem' and assess their reconstruction potential. We hypothesized that tree growth in the study region is directly affected by climate warming, and tree-ring parameters retain their sensitivity to temperature changes.

2. Materials and Methods

Wood samples were collected at upper treeline sites in the Altai Mountains (Figure 1). According to the instrumental records from the high-elevation meteorological station Kara-Tyurek (50°02′ N, 86°27′ E; WMO 36442; 1940–2022), the climate in the study area is characterized as extremely continental, with low annual mean temperatures of −5.4 °C. The warmest and coldest months are July and January, with monthly temperatures of 7.1 °C and −16.4 °C, respectively. The annual precipitation totals are around 620 mm, of which 46% fall during summer months (from June to August).

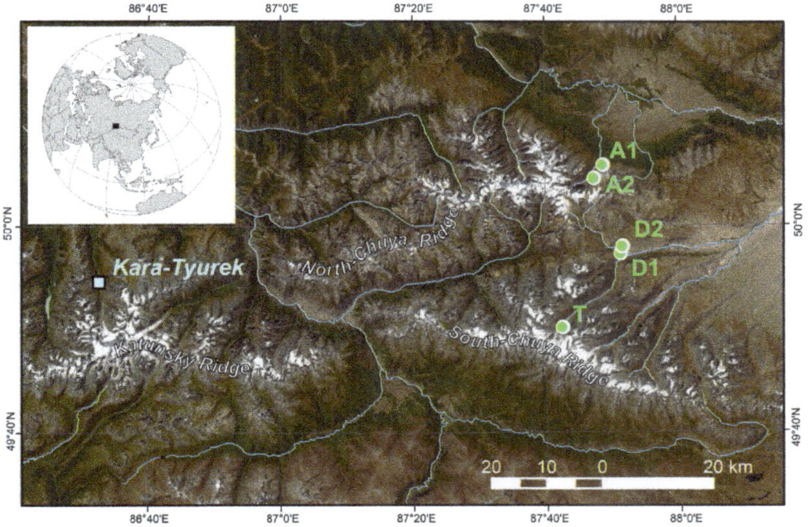

Figure 1. Location of the tree-ring sampling sites (light-green circles) and the meteorological station (light-blue square).

The annual and summer (June–August) temperature means were relatively stable till the beginning of the 1980s (Figure 2a). Since 1983, temperatures have been increasing by 0.38 and 1.33 °C/decade annually and in summer, respectively ($p < 0.001$). The annual and summer precipitation totals have not shown statistically significant changes ($p > 0.05$) (Figure 2b).

Siberian larch (*Larix sibirica* Ledeb.), which was shown to be sensitive to summer temperature variability [14,15,22,23], is one of the dominant tree species at the upper treeline in the study region. During several field campaigns between 2007 and 2023, we sampled five different stands in the upper treeline ecotone in three valleys in Central Altai associated with the glaciers Ak-Tru (sites A1 and A2), Djelo (D1 and D2), and Taldura (T) (Figures 1 and S1–S4; Tables 1 and S1). We collected wood from healthy dominant and co-dominant living larch trees at elevations from 2100–2150 m a.s.l. at site A2 to 2400–2500 at site T. At all the sites, except A2, larch was the only tree species. Although we aimed to core the old-growth trees, the stand at the highest site T was formed of larch trees of <100 years old because it has developed in an area recently covered by a glacier [43]. A minimum of 16 living trees were cored, with at least two cores per tree, of which one core was used for further TRW and MXD measurements. At three sites, A2, D2, and T, the TRW was measured at an accuracy of 0.01 mm using a LINTAB measurement device and the TSAPWin 4.68c software (RINNTECH e.K., Heidelberg, Germany). For the cores from sites A1 and D1, we measured the TRW and MXD using a DENDRO-2003 X-ray microdensitometer (WALESCH Electronic GmbH, Effretikon, Switzerland) according to [44]. All the series were visually cross-dated and statistically checked for missing rings and dating errors using the program COFECHA 6.06 [45].

The individual TRW and MXD series were standardized to remove non-climatic, tree geometry, and age-related trends [46]. Due to the different average ages of the trees at the sites, the individual TRW and MXD series were standardized with an age-dependent cubic spline (ARSTAN software version v49, https://www.geog.cam.ac.uk/research/projects/dendrosoftware/, last accessed on 25 June 2024). In addition, prior to growth trend removal, adaptive power transformation (PT; [47]) was applied to the MXD series. The bi-weight robust means of the individual measurement series were calculated to produce dimensionless index chronologies. The standard versions of the chronologies were chosen in further analyses to trace the trends in the climatic data.

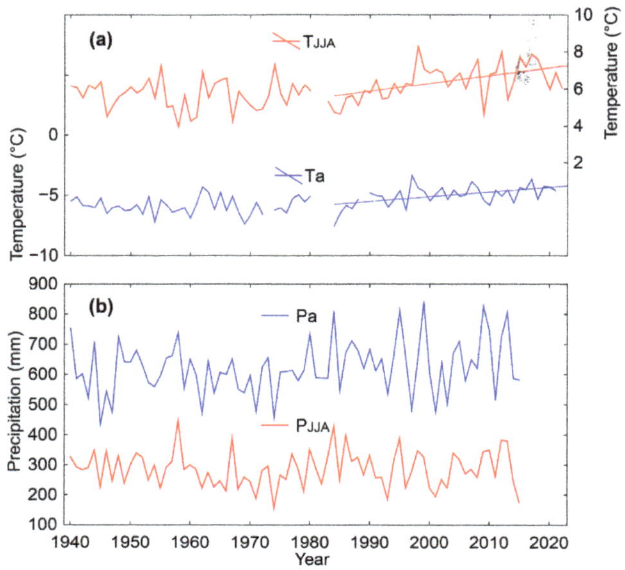

Figure 2. Summer (red) and annual mean (blue) temperatures (**a**) and precipitation (**b**) recorded at the high-elevation Kara-Tyurek meteorological station near the tree sites. Lines indicate statistically significant ($p < 0.05$) trends over time since 1984.

Table 1. Statistics of tree-ring width (TRW) and maximum latewood density (MXD) local chronologies (MSL = mean segment length (mean ± SD = mean TRW or MXD value ± standard deviation); CS = coefficient of sensitivity; Rbar = inter-series correlation; EPS = expressed population signal).

Site	Elevation, m a.s.l.	Tree-Ring Parameter	N of Series	Period CE	MSL	Mean ± SD	Mean CS	Mean Rbar	Mean EPS
A1	2200–2300	TRW	18	1653–2014	262	0.67 ± 0.32 mm	0.202	0.472	0.916
A2	2100–2150	TRW	20	1554–2021	269	0.79 ± 0.35 mm	0.225	0.508	0.915
D1	2200–2300	TRW	23	1478–2006	325	0.42 ± 0.22 mm	0.271	0.503	0.925
D2	2360–2400	TRW	16	1715–2021	198	0.62 ± 0.30 mm	0.310	0.520	0.899
T	2400–2500	TRW	20	1936–2022	63	1.17 ± 0.44 mm	0.234	0.456	0.925
A1	2200–2300	MXD	18	1653–2014	262	0.86 ± 0.10 g/cm^3	0.07	0.526	0.934
D1	2200–2300	MXD	23	1478–2006	325	0.86 ± 0.10 g/cm^3	0.08	0.531	0.934

In total, we developed five local (site) TRW and two MXD chronologies (Table 1). The lengths of the TRW chronologies varied from 87 (site T) to 529 (D1) years, with the lowest and highest mean segment length (MSL) for the same sites, respectively. The average growth rate was generally lower in older trees. The TRW chronologies were sensitive to an environmental signal (coefficient of sensitivity > 0.200), which synchronized the radial growth of individual trees and explained the high inter-series correlation (Rbar > 0.456). The multi-century MXD chronologies were also characterized by a high Rbar but a much lower sensitivity compared with the TRW. The expressed population signal (EPS), which measures how well chronologies built on a limited number of samples represent a theoretically infinite population, was higher than the commonly accepted threshold of 0.85 [48] for all TRW and MXD chronologies. The replication of 16 to 23 series and the high Rbar and EPS values confirmed the suitability of the TRW and MXD index chronologies for dendroclimatic analysis during the period with the available temperature and precipitation instrumental records.

For further analyses, we also used regional chronologies, which were calculated as the averages of the local index chronologies in ARSTAN. The regional TRW chronology was built from the five local TRW index chronologies, and the two local MXD index chronologies were used to build the regional MXD chronology. This procedure allowed us to equalize the weights of the data from sites with different local conditions, sample replications, and tree ages.

To evaluate the climate sensitivity of the TRW and MXD records, Pearson's correlation coefficients were calculated against the monthly temperature means and precipitation totals from the nearest meteorological station Kara-Tyurek from the previous year September–September of the current year. The June–July (JJ), summer (JJA), and May–August (MJJA) temperatures, as well as the total precipitation during summer, were used to assess the seasonal climatic influences. To estimate the temporal stability of the relations between the regional tree-ring chronologies and climate records, we used the running correlations calculated for a 25-year window with a 1-year step with growing seasonal temperature means. The spatial correlations between the regional TRW and MXD chronologies and gridded seasonal temperature means (CRU TS4.07, [49]) were calculated with the KNMI Climate Explorer (https://climexp.knmi.nl/start.cgi, last accessed on 25 June 2024) for the period with the available instrumental temperature records at the nearest high-elevation station Kara-Tyurek from 1940.

3. Results

The local TRW index chronologies demonstrated high synchronicity in the radial growth indices in different sites (Table 2; Figure 3). High statistically significant correlations ($p < 0.000001$) of up to $r = 0.79$ (N = 362) were observed between the four longer chronologies A1, A2, D1, and D2. For the shortest chronology T, the correlations were lower but statistically significant at $p < 0.002$. The correlation between the two available MXD index chronologies was higher than that for TRW, at $r = 0.62$ and 0.49 (N = 354), respectively. There were also notable correlations between the TRW and MXD records from the same sites, which varied from $r = 0.50$ (N = 362) for A1 to $r = 0.71$ (N = 529) for site D1.

Table 2. Correlation coefficients and number of cases (N) between the local index chronologies.

Site	A2	D1	D2	T
A1	**0.79** N = 362	**0.49** N = 354	**0.48** N = 300	0.36 N = 79
A2		**0.61** N = 453	**0.52** N = 307	0.50 N = 86
D1			**0.63** N = 292	0.50 N = 71
D2				0.33 N = 86

Note: Correlation coefficients are significant at $p < 0.002$, and values significant at $p < 0.000001$ are marked in **bold**.

The regional TRW and MXD index chronologies demonstrated a common pattern of high- and medium-frequency variability (Figure 3). Remarkably low index values in both records were observed around the year 1700 CE and during the first decades of the 19th century. Later, the indices were characterized by a generally increasing trend till the last decade of the 20th century. Interestingly, both the TRW and MXD indices generally stabilized or even decreased in the 21st century. The correlation coefficient between the regional chronologies was 0.63 (N = 537; $p < 0.000001$), which increased to 0.73 for the records smoothed with a 30-year cubic spline.

The monthly climate response analysis showed that the TRW of larch in the upper treeline was mostly dependent on temperature from May to July (Figure 4a). The highest correlations were found for the local A1 ($r = 0.58$), A2 ($r = 0.55$), and T ($r = 0.54$) and regional

($r = 0.60$) index chronologies with the June temperature means. The correlations with May temperatures were generally lower and statistically significant ($p < 0.05$) only for A2, D1, and the regional chronologies. July temperatures were important for secondary tree growth at all the sites, except D2. The correlations of the TRW standard chronologies with the seasonal mean temperatures were generally higher than those with the monthly means, especially for JJ (up to $r = 0.66$ with a regional chronology; $p < 0.01$).

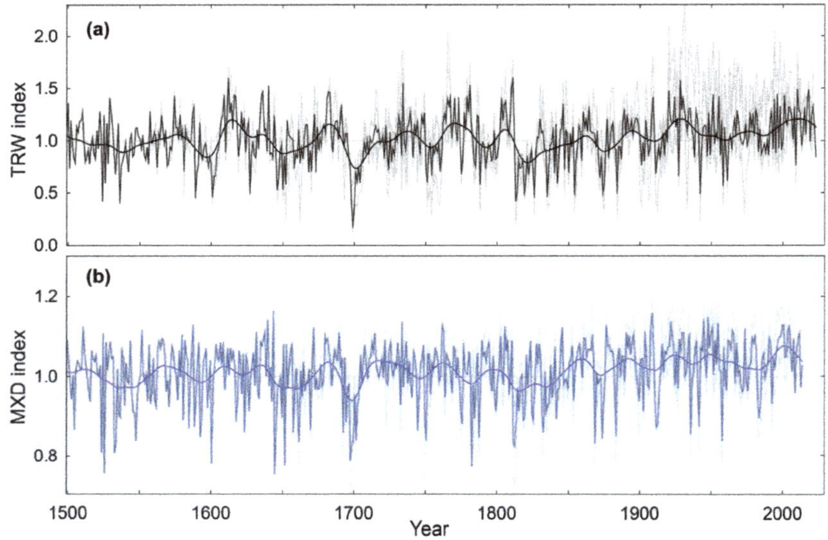

Figure 3. Regional tree-ring width (TRW) (**a**) and maximum latewood density (MXD) (**b**) chronologies with local index chronologies shown in light color. The chronologies were smoothed with a 30-year cubic spline.

The MXD was influenced by temperature variability over a longer interval of the growing season than the TRW (Figure 4b). The correlations of the MXD index chronologies with the monthly temperature means were statistically significant ($p < 0.05$) from May to August and showed the lowest and the highest values in August (up to $r = 0.30$ for D1) and July (up to $r = 0.61$ for D1), respectively. For the seasonal temperature means, the correlations further increased and reached the highest values with the temperature of the extended May–August period ($r = 0.69$, 0.74, and 0.73 for A1, D1, and the regional chronologies, respectively; $p < 0.01$).

The correlations of the tree-ring parameters with the monthly precipitation totals were lower than those with the temperature (Table S2). Statistically significant ($p < 0.05$) negative values were observed for the TRW mainly in the June instrumental record for sites A2 ($r = -0.23$), D2 (-0.26), and T (-0.28) and the regional chronology (-0.24). The MXD negatively correlated with the July precipitation totals at A1 ($r = -0.35$) and A2 (-0.45) and regionally (-0.41).

The 25-year window running correlations between the TRW regional chronology and seasonal temperature means showed that the dependence of radial growth on temperature was positive and mostly statistically significant at $p < 0.05$ over the period from 1940 (Figure 5a). All three sequences of running correlation coefficients were characterized by slightly decreasing trends toward the present. Generally higher correlations were observed over time for JJ temperatures (mean $r = 0.63$), which reached $r > 0.7$ during the first part of the analyzed period and, later, fell to around 0.55 ($p < 0.01$). The correlation of the MXD with the seasonal temperature means was stronger than that for TRW, with the correlation coefficients often being higher than 0.8 (Figure 5b). The dependence of the MXD on temperature was stable over time, and the mean correlations were slightly higher for MJJA ($r = 0.78$) than for JJA (0.77) and JJ (0.76).

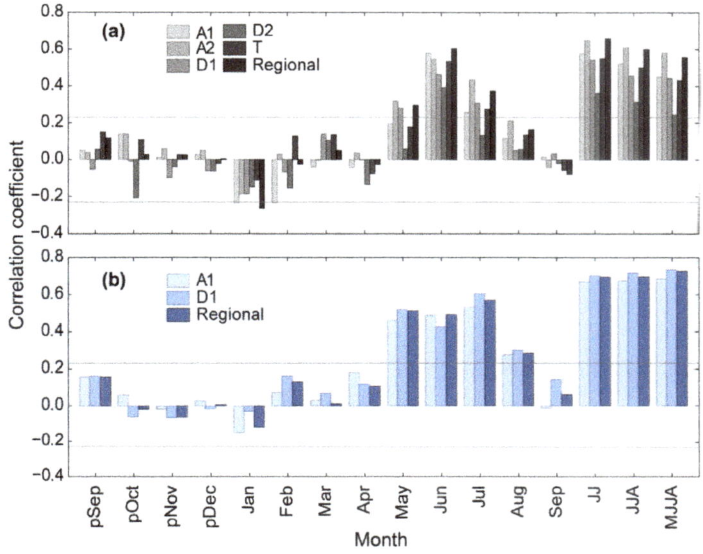

Figure 4. Correlation coefficients between standard tree-ring width (**a**) and maximum latewood density (**b**) local and regional chronologies and monthly temperature means from the previous year September–September of ring formation and seasonal temperature means for June–July (JJ), summer (JJA), and from May to August (MJJA). Grey horizontal lines indicate the significance level $p < 0.05$.

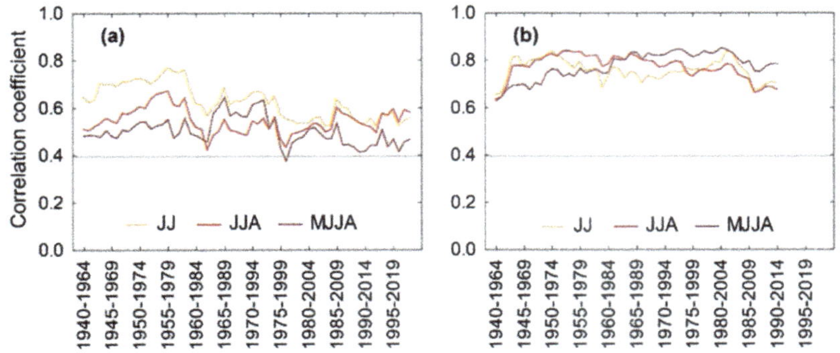

Figure 5. Twenty-five-year window running correlations between the regional tree-ring width (**a**) and maximum latewood density (**b**) index chronologies and seasonal temperature means of June–July (JJ), summer (JJA), and May–August (MJJA). Grey horizontal lines indicate the significance level $p < 0.05$.

Both the TRW and MXD regional chronologies well captured the inter-annual and multi-decadal changes in the JJ and MJJA instrumental temperature records, respectively (Figure 6). Following the tendencies in the temperature changes, the regional tree-ring chronologies were characterized by a slightly decreasing trend in the 21st century, which followed the increase from the 1980s and relatively stable decadal temperatures before that. The high correlations obtained between the annually resolved tree-ring and climate records (see Figure 4) further increased to 0.97 (N = 83) between the TRW and JJ and 0.85 (N = 75) between the MXD and MJJA temperatures for the records smoothed with a 30-year spline (Figure 6).

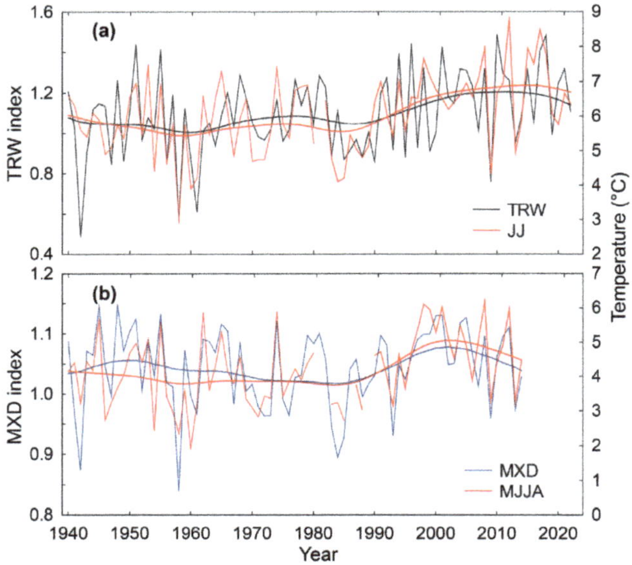

Figure 6. Tree-ring width (TRW) scaled against June–July (JJ) temperature means (**a**) and maximum latewood density (MXD) versus May–August (MJJA) temperature means (**b**). The chronologies were smoothed with a 30-year cubic spline.

The spatial temperature signature of the MXD regional chronology (Figure 7b) was larger than that of the TRW (Figure 7a). However, the high field correlations of the tree-ring records with the gridded summer temperature means covered a substantial fraction of southern Siberia and northern Inner Asia.

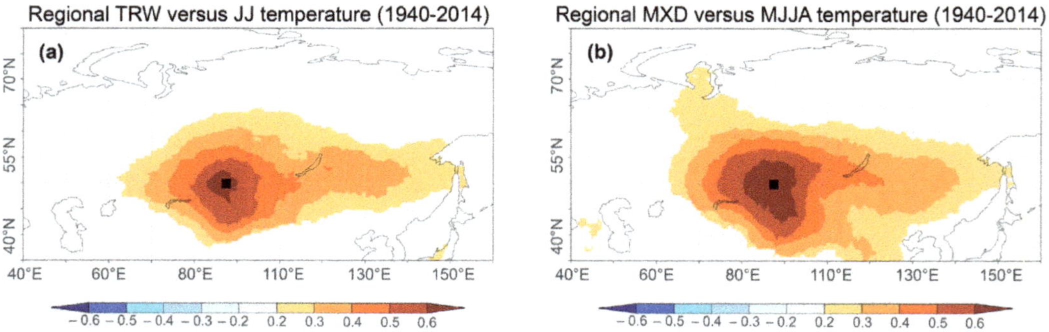

Figure 7. Correlation fields of tree-ring width (TRW) (**a**) and maximum latewood density (MXD) (**b**) regional chronologies against gridded June–July (JJ) and May–August (MJJA) averaged temperature means (TS4.07, [49]) for the period of 1940–2014. Black rectangles indicate the study region. Built with KNMI Climate Explorer (https://climexp.knmi.nl/start.cgi, last accessed on 25 June 2024).

4. Discussion

Our results show that larch trees at the upper treeline in the central Altai Mountains exhibit highly coherent growth in different environments. The altitudinal range of 350 m of the study sites located from 2100 to 2450 m a.s.l. corresponds to an approximately 2 °C difference in the mean annual temperature between the habitats [50]. Such a temperature difference may lead to considerable site-specific changes in the intra- and inter-annual variability of tree seasonal activity in larch [51,52]. Local conditions, including slope, aspect, stand density, ground vegetation, exposure to wind, permafrost conditions, etc., can further significantly modulate tree radial growth dynamics [53–57]. However, despite the differences in local conditions, the TRW and MXD were highly synchronized within and between our study sites, indicating that there is a common climate factor strongly limiting tree growth within the region.

Previous studies in Altai and neighboring mountain regions have indicated summer temperature as the main factor that defines tree radial growth at the upper treeline [15,29,57–59], and our results confirm these earlier findings (Figure 3). Temperature signatures were earlier found not only in the TRW and MXD but also in other tree-ring parameters, including the latewood cell wall thickness, stable isotope composition of tree-ring cellulose, and BI [21,22,60]. However, the MXD typically has greater accuracy in capturing meteorological information [12,61–63], and it is usually considered the highest-quality tree-ring proxy variable for temperature reconstructions [64,65]. Cell structure data can sometimes perform better than the MXD in representing growing seasonal temperatures [22,66]; however, the process of wood anatomy measurements is even more labor- and time-consuming than that for MXD. In this context, the approaches that rely on a binary (black–white) representation of anatomical structures on the surfaces of tree-ring samples and combine the advances of the BI, wood anatomy, and MXD techniques are promising [61,67,68].

In this study, we showed that the MXD is highly correlated with temperature variables and accumulates climate information over a longer period of the growing season (May–August) compared with the TRW (June–July). This phenomenon is well-known in tree-ring science [44,69,70] and is explained by the physiology and phenology of tree-ring and latewood formation [23,71,72]. These findings confirm the superiority of larch MXD from the upper treeline in Altai as a proxy for growing seasonal temperatures. The wide spatial coverage of the statistically significant correlations demonstrates the ability of our tree-ring records to capture temperature signals over a vast area of inner Eurasia, which includes north-eastern Kazakhstan in the west, north-western Mongolia in the east, southern Siberia (Russia) in the north, and north-eastern China in the south.

We found that both tree-ring parameters and especially MXD demonstrated a highly temporally stable sensitivity to the growing seasonal temperature. Moreover, the chronologies followed the recent warming and temperature stabilization in the last few years (Figure 6). These findings testify to the absence of the "divergence problem" in the study area. Although the DP is a region- and site-specific phenomenon [11,13], we assume that the upper treeline in the Altai Mountains is less prone to the weakening of the growth–temperature coupling because the DP is usually less pronounced in mid-latitudes [6,73], and Altai is remote from industrial pollution sources [74]. Moreover, the summer temperature in the upper treeline at the study sites in the Altai Mountains is still low, not reaching the threshold above which the influence of temperature on tree-ring formation drastically decreases [72]. In addition, the studied high-elevation sites are located close to glaciers and are hardly accessible, receiving no direct impact from human activity.

Currently, Altai is one of the few regions in the world where (i) tree rings provide reliable information on temperature changes, (ii) old trees are growing [17], and (iii) relict wood is available. A number of millennia-long chronologies have already been produced in the region [14,19]. The current efforts of our team are directed toward the construction of continuous absolutely dated well-replicated regional TRW and MXD chronologies, which cover the past two–three millennia. In this context, our findings of the wide spatial signature and temporal stability of temperature signals in the TRW, and especially MXD, are of great importance for future paleoclimatic reconstructions. When developed, an MXD-based regional temperature reconstruction will not only provide data on the local and regional late Holocene climate but may also serve to better understand the land use and human history in the Altai Mountains [15,75,76].

5. Conclusions

The results of our study demonstrate that temperature signals in the local and regional tree-ring width and maximum latewood density chronologies of Siberian larch from the upper treeline in Central Altai are strong and temporally stable. This testifies to the absence of the 'divergence problem' in tree-ring data and indicates the high paleoclimatic potential of upper treeline larch tree-ring chronologies. Current efforts directed at developing a network of well-replicated, absolutely dated millennia-long tree-ring chronologies will be the basis for collaborative multiparameter dendroclimatology that will provide the opportunity for a better understanding of paleoclimate and human history in the Altai and surrounding regions.

Supplementary Materials: The following supporting information can be downloaded at: https://www.mdpi.com/article/10.3390/f15081402/s1. Figure S1: Site A1. Photo by D.V. Ovchinnikov; Figure S2: Site A2. Photos by A.V. Kirdyanov; Figure S3: Site D1. Photos by V.S. Myglan; Figure S4: Site T. Photos by D.A. Ganyushkin; Table S1: Site location; Table S2: Correlation coefficients between standard tree-ring width (TRW) and maximum latewood density (MXD) local and regional chronologies and monthly precipitation totals from the previous year September to September of ring formation.

Author Contributions: Conceptualization, A.V.K.; methodology, A.V.K. and U.B.; investigation, A.V.K. and T.A.; data curation A.V.K., A.A., A.A.K., D.V.O., P.N.K., V.S.M., A.N.N. and I.Y.S.; writing—original draft preparation, A.V.K. and U.B.; writing—review and editing, A.A.K., A.A., A.A.K., T.A., P.N.K., I.Y.S., T.B. and U.B.; visualization, A.V.K. and D.A.G.; funding acquisition, A.A., D.V.O., D.A.G. and U.B. All authors have read and agreed to the published version of the manuscript.

Funding: This work was supported by the projects of the Ministry of Science and Higher Education of the Russian Federation (FSRZ-2020-0014 and FSRZ-2023-0007) (which provided the equipment). The data for one of the sites were collected and analyzed under the Russian Science Foundation (project no. 22-67-00020). U.B. received funding from the Czech Science Foundation (#23-08049S; Hydro8), the ERC Advanced Grant (#882727; Monostar), and the ERC Synergy Grant (#101118880; Synergy-Plague).

Data Availability Statement: The data presented in this study are available upon request from the corresponding author.

Conflicts of Interest: The authors declare no conflicts of interest.

References

1. George, S.S. An overview of tree-ring width records across the Northern Hemisphere. *Quat. Sci. Rev.* **2014**, *95*, 132–150. [CrossRef]
2. Büntgen, U.; Allen, K.; Anchukaitis, K.; Arseneault, D.; Boucher, É.; Bräuning, A.; Chatterjee, S.; Cherubini, P.; Churakova, O.V.; Corona, C.; et al. The influence of decision-making in tree ring-based climate reconstructions. *Nat. Commun.* **2021**, *12*, 3411. [CrossRef] [PubMed]
3. Camarero, J.J.; Gazol, A.; Sánchez-Salguero, R.; Fajardo, A.; McIntire, E.J.B.; Gutiérrez, E.; Batllori, E.; Boudreau, S.; Carrer, M.; Diez, J.; et al. Global fading of the temperature-growth coupling at alpine and polar treelines. *Glob. Chang. Biol.* **2021**, *27*, 1879–1889. [CrossRef] [PubMed]
4. D'Arrigo, R.; Wilson, R.; Liepert, B.; Cherubini, P. On the 'Divergence Problem' in Northern Forests: A review of the tree-ring evidence and possible causes. *Glob. Planet. Chang.* **2008**, *60*, 289–305. [CrossRef]
5. Jacoby, G.C.; D'Arrigo, R. Tree-ring width and density evidence of climatic and potential forest change in Alaska. *Glob. Biogeochem. Cycles* **1995**, *9*, 227–234. [CrossRef]
6. Briffa, K.R.; Schweingruber, F.H.; Jones, P.D.; Osborn, T.J.; Shiyatov, S.G.; Vaganov, E.A. Reduced sensitivity of recent tree-growth to temperature at high northern latitudes. *Nature* **1998**, *391*, 678–682. [CrossRef]
7. Wilson, R.; D'Arrigo, R.D.; Buckley, B.M.; Büntgen, U.; Esper, J.; Frank, D.; Luckman, B.; Payette, S.; Vose, R.; Youngblut, D. A matter of divergence: Tracking recent warming at hemispheric scales using tree ring data. *J. Geophys. Res.* **2007**, *112*, D17103. [CrossRef]
8. Zhang, Y.; Wilmking, M.; Gou, X. Changing relationships between tree growth and climate in Northwest China. *Plant. Ecol.* **2009**, *201*, 39–50. [CrossRef]
9. Jiao, L.; Jiang, Y.; Zhang, W.T.; Wang, M.-C.; Zhang, L.-N.; Zhao, S.-D. Divergent responses to climate factors in the radial growth of *Larix sibirica* in the eastern Tianshan Mountains, northwest China. *Trees* **2015**, *29*, 1673–1686. [CrossRef]

10. Li, J.X.; Li, J.B.; Li, T.; Au, T.F. Tree growth divergence from winter temperature in the Gongga Mountains, southeastern Tibetan Plateau. *Asian Geogr.* **2020**, *37*, 1–15. [CrossRef]
11. Anchukaitis, K.J.; D'Arrigo, R.D.; Andreu-Hayles, L.; Frank, F.; Verstege, A.; Curtis, A.; Buckley, B.M.; Jacoby, G.C.; Cook, E.R. Tree-Ring-Reconstructed Summer Temperatures from Northwestern North America during the Last Nine Centuries. *J. Clim.* **2013**, *26*, 3001–3012. [CrossRef]
12. Büntgen, U.; Reinig, F.; Verstege, A.; Piermattei, A.; Kunz, M.; Krusic, P.; Slavin, P.; Štěpánek, P.; Torbenson, M.; del Castillo, E.M.; et al. Recent summer warming over the western Mediterranean region is unprecedented since medieval times. *Glob. Planet. Chang.* **2024**, *232*, 104336. [CrossRef]
13. Yin, H.; Li, M.-Y.; Huang, L. Summer mean temperature reconstruction based on tree-ring density over the past 440 years on the eastern Tibetan Plateau. *Quat. Int.* **2021**, *571*, 81–88. [CrossRef]
14. Myglan, V.S.; Oidupaa, O.C.; Vaganov, E.A. A 2367-Year Tree-Ring Chronology for the Altai–Sayan Region (Mongun-Taiga Mountain Massif). *Archaeol. Ethnol. Anthropol. Eurasia* **2012**, *40*, 76–83. [CrossRef]
15. Büntgen, U.; Myglan, V.S.; Ljungqvist, F.C.; McCormick, M.; Di Cosmo, N.; Sigl, M.; Jungclaus, J.; Wagner, S.; Krusic, P.J.; Esper, J.; et al. Cooling and societal change during the Late Antique Little Ice Age from 536 to around 660 AD. *Nat. Geosci.* **2016**, *9*, 231–236. [CrossRef]
16. Zhang, T.; Yuan, Y.; Chen, F.; Yu, C.; Zhang, R.; Qin, L.; Jiang, S. Reconstruction of hydrological changes based on tree-ring data of the Haba River, northwestern China. *J. Arid Land* **2018**, *10*, 53–67. [CrossRef]
17. Taynik, A.V.; Myglan, V.S.; Barinov, V.V.; Oidupaa, O.C.; Churakova, O.V. Ancient larch trees in the Tuva Republic, land of the oldest trees in Russia. *J. Prot. Mt. Areas Res. Manag.* **2023**, *15*, 13–19. [CrossRef]
18. Panyushkina, I.; Sljusarenko, I.; Bikov, N.; Bogdanov, E. Floating larch tree-ring chronologies from Archaeological timbers in the Russian Altai between about 800 BC and AD 800 E. *Radiocarbon* **2007**, *49*, 693–702. [CrossRef]
19. Myglan, V.S.; Oidupaa, O.C.; Kirdyanov, A.V.; Vaganov, E.A. 1929-year tree-ring chronology for the Altai-Sayan. Region (Western Tuva). *Archaeol. Ethnol. Anthropol. Eurasia* **2008**, *36*, 25–31. [CrossRef]
20. Nazarov, A.N.; Solomina, O.N.; Myglan, V.S. Absolute and relative age of moraines of the Aktru and Historical stages of glaciers of Central Altai based on lichenometry and dendrochronology. *Led i Sneg. Ice Snow* **2022**, *62*, 387–409. (In Russian)
21. Loader, N.J.; Helle, G.; Los, S.O.; Lehmkuhl, F.; Schleser, G.H. Twentieth-century summer temperature variability in the southern Altai Mountains: A carbon and oxygen isotope study of tree-rings. *Holocene* **2010**, *20*, 1149–1156. [CrossRef]
22. Sidorova, O.V.; Saurer, M.; Myglan, V.S.; Eichler, A.; Schwikowski, M.; Kirdyanov, A.V.; Bryukhanova, M.V.; Gerasimova, O.V.; Kalugin, I.A.; Daryin, A.V.; et al. A multi-proxy approach for revealing recent climatic changes in the Russian Altai. *Clim. Dyn.* **2012**, *38*, 175–188. [CrossRef]
23. Chen, F.; Yuan, Y.; Wei, W.; Fan, Z.; Zhang, T.; Shang, H.; Zhang, R.; Yu, S.; Ji, C.; Qin, L. Climatic response of ring width and maximum latewood density of *Larix sibirica* in the Altay Mountains, reveals recent warming trends. *Ann. For. Sci.* **2012**, *69*, 723–733. [CrossRef]
24. Dulamsuren, C.; Khishigjargal, M.; Leuschner, C.; Hauck, M. Response of tree-ring width to climate warming and selective logging in larch forests of the Mongolian Altai. *J. Plant Ecol.* **2014**, *7*, 24–38. [CrossRef]
25. Wu, X.; Liu, H.; He, L.; Qi, Z.; Anenkhonov, O.A.; Korolyuk, A.Y.; Yu, Y.; Guo, D. Stand-total tree-ring measurements and forest inventory documented climate-induced forest dynamics in the semi-arid Altai Mountains. *Ecol. Indic.* **2014**, *36*, 231–241. [CrossRef]
26. Cazzolla Gatti, R.; Callaghan, T.; Velichevskaya, A.; Dudko, A.; Fabbio, L.; Battipaglia, G.; Liang, J. Accelerating upward treeline shift in the Altai Mountains under last-century climate change. *Sci. Rep.* **2019**, *9*, 7678. [CrossRef] [PubMed]
27. Savchuk, D.A.; Timoshok, E.E.; Filimonova, E.O.; Nikolaeva, S.A. The dynamics of the upper forest line on the Katunsky Range (the Altai Mountains) over the last 120 years. *Russ. J. Ecol.* **2023**, *54*, 482–487. [CrossRef]
28. Adamenko, M.F. Dynamics of larch tree growth as indicator of thermical regimes of summer in the Altai Mountain. In *Regional Geographical Investigations in the Western Siberia*; Nauka: Novosibirsk, Russia, 1978; pp. 20–23. (In Russian)
29. Zhang, T.; Yuan, Y.; Hu, Y.; Wei, W.; Shang, H.; Huang, L.; Zhang, R.; Chen, F.; Yu, S.; Fan, Z.; et al. Early summer temperature changes in the southern Altai Mountains of Central Asia during the past 300 years. *Quat. Int.* **2015**, *358*, 68–76. [CrossRef]
30. Chen, F.; Yuan, Y.J.; Wei, W.S.; Zhang, T.W.; Shang, H.M.; Zhang, R.B. Precipitation reconstruction for the southern Altay Mountains (China) from tree rings of Siberian spruce, reveals recent wetting trend. *Dendrochronologia* **2014**, *32*, 266–272. [CrossRef]
31. Pang, W.; Li, Q.; Liu, Y.; Song, H.; Sun, C.; Wang, J.; Yan, Y.; Cai, Q.; Ren, M. Precipitation variations in China's Altay Mountains detected from tree rings dating back to AD 1615. *Forests* **2023**, *14*, 1496. [CrossRef]
32. Jiang, S.; Zhang, T.; Yuan, Y.; Yu, S.; Shang, H.; Zhang, R. Drought reconstruction based on tree-ring earlywood of *Picea obovata* Ledeb. for the southern Altay Mountains. *Geogr. Ann. A* **2020**, *102*, 267–286. [CrossRef]
33. Davi, N.; Jacoby, G.; Fang, K.; Li, J.; D'Arrigo, R.; Baatarbileg, N.; Robinson, D. Reconstructing drought variability for Mongolia based on a large-scale tree ring network: 1520–1993. *J. Geophys. Res.* **2010**, *115*, D22103. [CrossRef]
34. Xu, G.; Liu, X.; Qin, D.; Chen, T.; Wang, W.; Wu, G.; Sun, W.; An, W.; Zeng, X. Relative humidity reconstruction for northwestern China's Altay Mountains using tree-ring δ^{18}O. *Chin. Sci. Bull.* **2014**, *59*, 190–200. [CrossRef]
35. Barinov, V.V.; Myglan, V.S.; Taynik, A.V.; Oydupaa, O.C.; Vaganov, E.A. Extreme climatic events in the Republic of Tuva according to tree-ring analysis. *Contemp. Probl. Ecol.* **2015**, *8*, 414–422. [CrossRef]

36. Oyunmunkh, B.; Weijers, S.; Loeffler, J.; Byambagerel, S.; Soninkhishig, N.; Buerkert, A.; Goenster-Jordan, S.; Simme, C. Climate variations over the southern Altai Mountains and Dzungarian Basin region, central Asia, since 1580 CE. *Int. J. Climatol.* **2019**, *39*, 4543–4558. [CrossRef]
37. Eichler, A.; Olivier, S.; Henderson, K.; Laube, A.; Beer, J.; Papina, T.; Gäggeler, H.W.; Schwikowski, M. Temperature response in the Altai region lags solar forcing. *Geophys. Res. Lett.* **2009**, *36*, L01808. [CrossRef]
38. Babich, V.V.; Daryin, A.V.; Rudaya, N.A.; Markovich, T.I. Two Millennia of Climate History for the Russian Altai: Integrated Reconstruction from Lake Sediment Data. *Russ. Geol. Geophys.* **2023**, *64*, 1217–1226. [CrossRef]
39. Blyakharchuk, T.A.; Wright, H.E.; Borodavko, P.S.; van der Knaap, W.O.; Ammann, B. Late Glacial and Holocene vegetational changes on the Ulagan high-mountain plateau, Altai Mountains, southern Siberia. *Palaeogeogr. Palaeoclimatol. Palaeoecol.* **2004**, *209*, 259–279. [CrossRef]
40. Huang, X.; Peng, W.; Rudaya, N.; Grimm, E.C.; Chen, X.; Cao, X.; Zhang, J.; Pan, X.; Liu, S.; Chen, C.; et al. Holocene vegetation and climate dynamics in the Altai Mountains and surrounding areas. *Geophys. Res. Lett.* **2018**, *45*, 6628–6636. [CrossRef]
41. Schlütz, F.; Lehmkuhl, F. Climatic change in the Russian Altai, southern Siberia, based on palynological and geomorphological results, with implications for climatic teleconnections and human history since the middle Holocene. *Veget. Hist. Archaeobot.* **2007**, *16*, 101–118. [CrossRef]
42. Rao, Z.; Guo, H.; Cao, J.; Shi, F.; Jia, G.; Li, Y.; Chen, F. Consistent long-term Holocene warming trend at different elevations in the Altai Mountains in arid central Asia. *J. Quat. Sci.* **2020**, *35*, 1036–1045. [CrossRef]
43. Ostanin, O.V.; Mikhailov, N.N. *Current Changes in High-Mountain Geosystems (Case Study of Central and South-Eastern Altai)*; Altai State University: Barnaul, Russia, 2013; 171p.
44. Schweingruber, F.H. *Tree Rings: Basics and Applications of Dendrochronology*; Kluwer Academic Publishers: Dordrecht, The Netherlands; Boston, MA, USA, 1988; 276p.
45. Holmes, R.L. Computer-assisted quality control in tree-ring datingand measurement. *Tree-Ring Bull.* **1983**, *44*, 69–75.
46. Cook, E.R.; Kairiukstis, L.A. *Methods of Dendrochronology: Applications in the Environmental Sciences*; Kluwer: Dordrecht, The Netherlands, 1990; pp. 104–122.
47. Cook, E.R.; Peters, K. Calculating unbiased tree-ring indices for the study of climatic and environmental change. *Holocene* **1997**, *7*, 361–370. [CrossRef]
48. Wigley, T.M.L.; Briffa, K.R.; Jones, P.D. On the average value of correlated times series, with applications in dendroclimatology and hydrometeorology. *J. Clim. Appl. Meteorol.* **1984**, *23*, 201–213. [CrossRef]
49. Harris, I.; Osborn, T.J.; Jones, P.; Lister, D. Version 4 of the CRU TS monthly high-resolution gridded multivariate climate dataset. *Sci. Data* **2020**, *7*, 109. [CrossRef] [PubMed]
50. Galakhov, V.P.; Mukhametov, R.M. *Glaciers of the Altai*; Nauka: Novosibirsk, Russia, 1999; 136p. (In Russian)
51. King, G.M.; Gugerli, F.; Fonti, P.; Frank, D. Tree growth response along an elevational gradient: Climate or genetics? *Oecologia* **2013**, *173*, 1587–1600. [CrossRef] [PubMed]
52. Moser, L.; Fonti, P.; Büntgen, U.; Esper, J.; Luterbacher, J.; Franzen, J.; Frank, D. Timing and duration of European larch growing season along altitudinal gradients in the Swiss Alps. *Tree Physiol.* **2010**, *30*, 225–233. [CrossRef]
53. Sugimoto, A.; Yanagisawa, N.; Naito, D.; Fujita, N.; Maximov, T.C. Importance of permafrost as a source of water for plants in east Siberian taiga. *Ecol. Res.* **2002**, *17*, 493–503. [CrossRef]
54. Vaganov, E.A.; Shiatov, S.G.; Mazepa, V.S. *Dendroclimatic Research in the Ural-Siberian Subarctic Zone*; Nauka: Novosibirsk, Russia, 1996; 246p. (In Russian)
55. Kirdyanov, A.V.; Prokushkin, A.S.; Tabakova, M.A. Tree-ring growth of Gmelin larch under contrasting local conditions in the north of Central Siberia. *Dendrochronologia* **2013**, *31*, 114–119. [CrossRef]
56. Kirdyanov, A.V.; Saurer, M.; Arzac, A.; Knorre, A.A.; Prokushkin, A.S.; Churakova, O.V.; Arosio, T.; Bebchuk, T.; Siegwolf, R.; Büntgen, U. Thawing permafrost can mitigate warming-induced drought stress in boreal forest trees. *Sci. Total Environ.* **2024**, *912*, 168858. [CrossRef]
57. Petrov, I.A.; Kharuk, V.I.; Golyukov, A.S.; Im, S.T.; Ondar, S.O.; Shushpanov, A.S. Siberian pine and larch response to warming-drying climate in the southern boundary of their range. *Forests* **2024**, *15*, 1054. [CrossRef]
58. Ovchinnikov, D.V.; Panjushkina, I.P.; Adamenko, M.F. The millennial tree-ring chronology of the Altai upland larch-tree and its use for the reconstruction of summer temperature. *Geogr. Nat. Resour.* **2002**, *1*, 102–108. (In Russian)
59. Taynik, A.V.; Barinov, V.V.; Oidupaa, O.C.; Myglan, V.S.; Reinig, F.; Büntgen, U. Growth coherency and climate sensitivity of *Larix sibirica* at the upper treeline in the Russian Altai-Sayan Mountains. *Dendrochronologia* **2016**, *39*, 10–16. [CrossRef]
60. Davi, N.K.; Rao, M.P.; Wilson, R.; Andreu-Hayles, L.; Oelkers, R.; D'Arrigo, R.; Nachin, B.; Buckley, B.; Pederson, N.; Leland, C.; et al. Accelerated recent warming and temperature variability over the past eight centuries in the Central Asian Altai from Blue Intensity in tree rings. *Geophys. Res. Lett.* **2021**, *48*, e2021GL092933. [CrossRef]
61. Park, W.K.; Telewski, F. Measuring maximum latewood density by image-analysis at the cellular level. *Wood Fiber Sci.* **1993**, *25*, 326–332.
62. Schweingruber, F.H.; Briffa, K.R.; Nogler, P. A tree-ring densitometric transect from Alaska to Labrador: Comparison of ring-width and maximum-latewood-density chronologies in the conifer belt of northern North America. *Int. J. Biometeorol.* **1993**, *37*, 151–169. [CrossRef]

63. Wilson, R.J.S.; Luckman, B.H. Dendroclimatic reconstruction of maximum summer temperatures from upper tree-line sites in interior British Columbia. *Holocene* **2003**, *13*, 853–863. [CrossRef]
64. Esper, J.; Krusic, P.; Ljungqvist, F.; Luterbacher, J.; Carrer, M.; Cook, E.; Davi, N.K.; Hartl-Meier, C.; Kirdyanov, A.; Konter, O.; et al. Ranking of tree-ring based temperature reconstructions of the past millennium. *Quat. Sci. Rev.* **2016**, *145*, 134–151. [CrossRef]
65. Björklund, J.; von Arx, G.; Nievergelt, D.; Wilson, R.; Van den Bulcke, J.; Günther, B.; Loader, N.J.; Rydval, M.; Fonti, P.; Scharnweber, T.; et al. Scientific Merits and Analytical Challenges of Tree-Ring Densitometry. *Rev. Geophys.* **2019**, *57*, 1224–1264. [CrossRef]
66. Björklund, J.; Seftigen, K.; Stoffel, M.; Fonti, M.V.; Kottlow, S.; Frank, D.C.; Esper, J.; Fonti, P.; Goosse, H.; Grudd, H.; et al. Fennoscandian tree-ring anatomy shows a warmer modern than medieval climate. *Nature* **2023**, *620*, 97–103. [CrossRef]
67. Silkin, P.P.; Kirdyanov, A.V.; Krusic, P.J.; Ekimov, M.V.; Barinov, V.V.; Büntgen, U. A new approach to measuring tree-ring density parameters. *J. Sib. Fed. Univ. Biol.* **2022**, *15*, 441–455.
68. Rydval, M.; Björklund, J.; von Arx, G.; Begović, K.; Lexa, M.; Nogueira, J.; Schurman, J.S.; Jiang, Y. Ultra-high-resolution reflected-light imaging for dendrochronology. *Dendrochronologia* **2024**, *83*, 126160. [CrossRef]
69. Schweingruber, F.H. *Tree Rings and Environment. Dendroecology*; Haupt: Berne, Switzerland, 1996; 609p.
70. Kirdyanov, A.V.; Vaganov, E.A.; Hughes, M.K. Separating the climatic signal from tree-ring width and maximum latewood density records. *Trees* **2007**, *21*, 37–44. [CrossRef]
71. Kramer, P.J.; Kozlowski, T.T. *Physiology of Woody Plants*; Academic Press: New York, NY, USA; San Francisco, CA, USA; London, UK, 1979; 811p.
72. Vaganov, E.A.; Hughes, M.K.; Shashkin, A.V. *Growth Dynamics of Conifer Tree Rings: Images of Past and Future Environments (Ecological Studies, Vol. 183)*; Springer Science & Business Media: Berlin, Germany, 2006; 356p.
73. Cook, E.R.; Esper, J.; D'Arrigo, R. Extra-tropical Northern Hemisphere land temperature variability over the past 1000 years. *Quat. Sci. Rev.* **2004**, *23*, 2063–2074. [CrossRef]
74. Kirdyanov, A.V.; Krusic, P.J.; Shishov, V.V.; Vaganov, E.A.; Fertikov, A.I.; Myglan, V.S.; Barinov, V.V.; Browse, J.; Esper, J.; Ilyin, V.A.; et al. Ecological and conceptual consequences of Arctic pollution. *Ecol. Lett.* **2020**, *23*, 1827–1837. [CrossRef]
75. Unkelbach, J.; Dulamsuren, C.; Behling, H. Late Holocene climate and land-use history in the Mongolian Altai Mountains: Combined evidence from palynological, macro-charcoal and tree-ring analyses. *Trees For. People* **2021**, *4*, 100073. [CrossRef]
76. Xiang, L.; Huang, X.; Sun, M.; Panizzo, V.N.; Huang, C.; Zheng, M.; Chen, X.; Chen, F. Prehistoric population expansion in Central Asia promoted by the Altai Holocene Climatic Optimum. *Nat. Commun.* **2023**, *14*, 3102. [CrossRef]

forests

MDPI

Article

The More the Merrier or the Fewer the Better Fare? Effects of Stand Density on Tree Growth and Climatic Response in a Scots Pine Plantation

Yulia A. Kholdaenko [1,2], Elena A. Babushkina [1,2], Liliana V. Belokopytova [1,2,*], Dina F. Zhirnova [1,2], Nataly N. Koshurnikova [2], Bao Yang [3,4] and Eugene A. Vaganov [2,5]

[1] Khakass Technical Institute, Siberian Federal University, 655017 Abakan, Russia; kropacheva_yulechka@mail.ru (Y.A.K.); babushkina70@mail.ru (E.A.B.); dina-zhirnova@mail.ru (D.F.Z.)
[2] Institute of Ecology and Geography, Siberian Federal University, 660036 Krasnoyarsk, Russia; nkoshurnikova@sfu-kras.ru (N.N.K.); eavaganov@hotmail.com (E.A.V.)
[3] Key Laboratory of Desert and Desertification, Northwest Institute of Eco-Environment and Resources, Chinese Academy of Sciences, Lanzhou 730000, China; yangbao@lzb.ac.cn
[4] School of Geographic and Oceanographic Sciences, Nanjing University, Nanjing 210093, China
[5] Department of Dendroecology, V.N. Sukachev Institute of Forest, Siberian Branch of the Russian Academy of Science, 660036 Krasnoyarsk, Russia
* Correspondence: white_lili@mail.ru

Abstract: In forests, the growth and productivity of individual trees and stands as a whole are regulated by stand density among other factors, because access to vital resources is limited by competition between trees. On 18 experimental plots of Scots pine (*Pinus sylvestris* L.) planted with a density of 500–128,000 trees/ha in the south taiga (Middle Siberia), interactions between stand density, tree- and stand-scale productivity, and tree-ring parameters were investigated. Tree-scale productivity variables, tree-ring width, and latewood width had stable negative allometric relationships with stand density ($R^2 > 0.75$), except for tree height (insignificant for inventory surveys at ages of 20 and 25 years; $R^2 > 0.4$ at the age of 35 years), while positive allometry was registered for stand productivity variables ($R^2 > 0.7$) and the all-time average latewood ratio ($R^2 = 0.5$ with planting density). Tree-ring parameters aside from the age trends correlate ($p < 0.05$) between the plots and demonstrate common responses to moderate moisture deficit. Although, its seasonality apparently depends on the resource base and intensity changes with stand density. February–June precipitation is more important for pine growth in dense stands, July–August conditions affect the latewood ratio stronger in sparse stands, and medium-density stands are more resistant to winter frosts.

Keywords: stand density; productivity; tree-ring width; latewood width; latewood ratio; Scots pine; growth–climate relationship

1. Introduction

Forest ecosystems play an indispensable role in the mitigation of climate change and in the protection of the environment [1,2]. However, shifts in the phenology and spatial distribution of tree species due to observed climate warming are leading to changes in forest productivity, biodiversity, structure, and, ultimately, the functioning of forest ecosystems as such [3–6]. Therefore, understanding the mechanism of forests' response to climate change is of great importance for sustainable forest management [7].

Tree growth is modulated by a large number of abiotic and biotic factors: from climate, landscape, and catastrophic natural phenomena to interrelations within the biocenosis, the impact of phytophages, etc. [8–14]. At the same time, there is a significant contribution of the individual tree characteristics, including the genotype, phenotype, age, etc., to tree growth and its response to external impacts [15–17].

Due to complex tree–environment interactions, the intensity of competition plays a vital role within a homogeneous habitat limited by common conditions, such as the soil substrate, topography, established biocenosis structure, and so on [18–24]. Access to vital resources can be limited by interactions between trees, mainly competition dependent on stand density; this leads to the gradual differentiation of the stand structure with age [25–27]. With a rapidly changing environment, including a warming climate, competition can intensify dramatically, with effects ranging from increased water stress and altered pest and disease resistance to altered availability of basic nutrients [9,28–31]. As a result, the competition for limited resources can fundamentally determine the ability of individual trees to allocate biomass, particularly, radial growth and its response to abiotic factors including droughts [31–36].

The result of the intra-specific competition for each individual tree depends on the morphophysiological characteristics and individual heredity, as well as on development conditions such as the micro-landscape, an excess or lack of moisture, cover from frost and sun, etc. [37–39]. In this regard, the phytosocial structure of stands is a key factor that significantly regulates the availability of life-sustaining resources such as light, water, and soil nutrients for each tree [35,40,41]. Additionally, competition significantly affects the allometric ratios and the distribution of biomass along the stem, branches, and foliage [42–44]. Social status may be crucial in the process of differentiating the biomass accumulation by parts of a tree [27]. The result of such competitive relationships is clearly visible in the external appearance of woody plants, since in almost any forest (mono- or mixed-species and even- or mixed-aged), we can visually observe the differentiation of trees in terms of growth and development. However, there is no clear answer to the question of whether dominant trees always gain a competitive and survival advantage [45–47].

As stand density changes, tree species vary greatly in their ability to alter their radial growth [48] due to the interaction between competition and environmental conditions (e.g., [32]). Scots pine as a subject for dendroclimatic research is unique in its versatility due to the wide ecological range from extremely xeric sites in semiarid climates to humid wetland habitats [49,50]. The radial growth of this species, as a good indicator of water stress [51], can be successfully used to assess forest management options designed to mitigate climate change in regions with moisture deficiency. In addition, the regulation of stand density will be an option here which assists forest stands in adaptation to the future climate [52].

This study was carried out in unique experimental plantations in the southern taiga of Middle Siberia, where the conifers were planted with a wide gradient of stand density [20,24,53,54]. Evaluation of the contribution of the initial planting density (and its dynamics due to self-thinning) to the dynamics of productivity, tree radial growth, the ratio of earlywood and latewood, and their response to external stresses provides data useful in sylviculture. For example, it assists in determining a forest management regime that provides a compromise between the tree growth stability in a changing climate, the stock of organic matter in the forest stand (carbon deposition), and the quality of the developed wood. In practical terms, the results of this experiment are applicable for optimizing the phytosocial structure of artificial plantations and natural forest stands on a fairly large spatial scale due to the investigation of widespread species that are dominant trees in many Siberian forests.

In addition to the relationship between the competition intensity (stand density) and the productivity and growth indicators in the Scots pine stand, the study addressed the question as to how this factor impacts: (i) the reactions of pine growth to climatic variability and (ii) the dynamics of the tree ring width, latewood width, and latewood ratio.

2. Materials and Methods

2.1. Study Area and Experimental Plantation

The study was carried out in Middle Siberia (central part of the Krasnoyarsk Krai), in the subzone of the southern taiga represented by conifer forests with the addition of broad-leaf species. The landscape of the study area is flat; the climate is continental, with large amplitudes of daily and seasonal temperature fluctuations. This study object is the experimental plantations of conifers planted by the Sukachev Institute of Forest, Siberian Branch of the USSR Academy of Sciences (Krasnoyarsk) in the vicinity of Talovka village (57°12′18″ N 93°08′36″ E, 240 m a.s.l.) on agricultural lands. These territories were historically covered with pine–larch forests before the Russians settled there, so the soil on the plantation is a loamy alfisol. For the experiment in 1982, two-year-old seedlings of Siberian spruce (*Picea obovata* Ledeb.), Scots pine (*Pinus sylvestris* L.), and Siberian larch (*Larix sibirica* Ledeb.) were planted on the flat plain in a mostly rectangular pattern using 18 planting densities ($DENS_0$ = 500–128,000 trees/ha; Figure 1a and Table 1). At least 500 individuals were planted for each planting density.

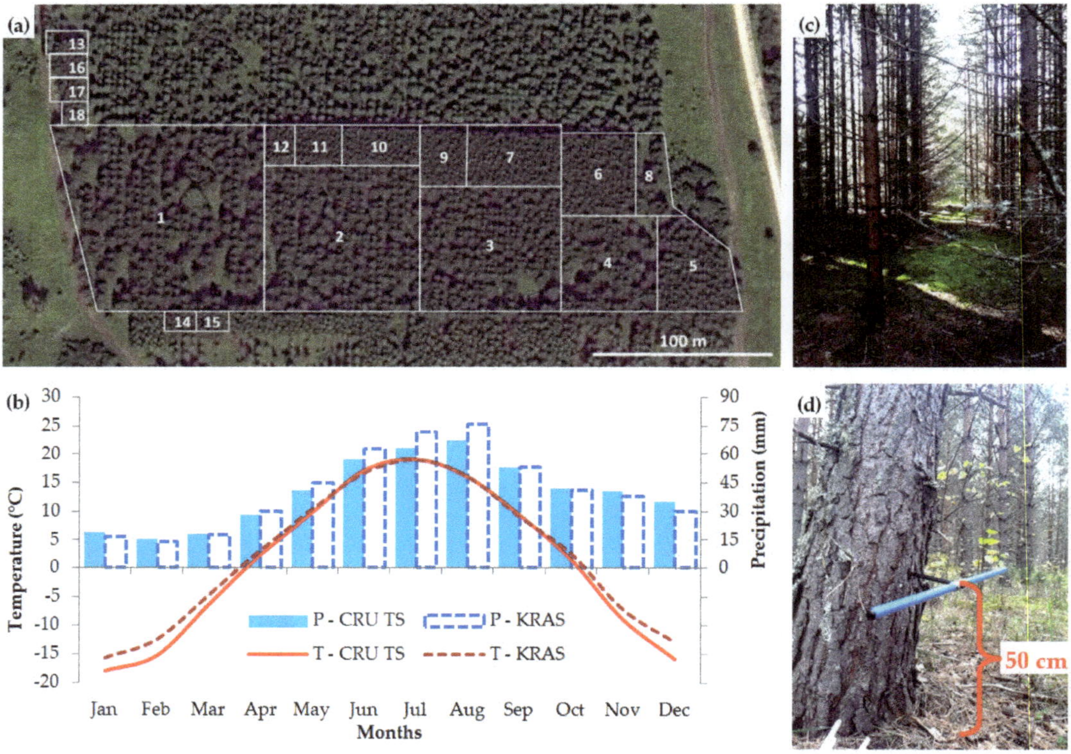

Figure 1. The study area: (**a**) aerial map of experimental plantation; the plots where the Scots pine was planted are marked and numbered according to Table 1; plot #8 was omitted from the study due to the failure of many seedlings soon after planting; (**b**) monthly temperature T and precipitation P (1988–2019) from the Krasnoyarsk climatic station (KRAS; dashed lines) and from the CRU TS field (solid lines) for the grid cell corresponding to the plantation's location; (**c**) plots of Siberian larch (left) and Scots pine (right) divided by clearing; (**d**) collecting of a wood sample (core) by an increment borer from a pine tree in the low-density plot.

In this study, a monthly series of mean temperature T and precipitation P from the spatially distributed field CRU TS (https://climexp.knmi.nl/selectfield_obs2.cgi; accessed on 20 January 2023) were used for a geographic grid cell including the plantation location (57–57.5° N 93–93.5° E), as well as monthly and daily series of the Krasnoyarsk weather station 120 km south of the study area (56°02′ N 92°45′ E, 277 m a.s.l.; http://meteo.ru/data; accessed on 20 January 2023; Figure 1b). For the period 1988–2020 used for the analysis of the climatic response in the studied plantations, the average annual temperatures are

+0.9 °C in the study area and +2.0 °C in Krasnoyarsk; the average annual precipitation is 480 and 496 mm, respectively. The warmest month is July (+19.0 and +19.1 °C); the coldest month is January (−18.0 and −15.8 °C, respectively). Temperatures exceed +5 °C from May to September on average, during which time about 60% of the annual precipitation falls. The maximum monthly precipitation is observed in August (67 and 76 mm, respectively); the minimum precipitation is observed in February (14–15 mm).

Table 1. Characteristics of the tree-ring width data.

Plot	**Raw TRW Measurements**					**Standard TRW Chronology**			
Code	**DENS$_0$ (Trees/ha)**	**Trees/ Cores**	**Period (Years)**	**Length (Years)**	*r-bar*	*SD*	*sens*	*r-bar*	ar_{-1}
1PS	500	10/20	1987–2020	33	0.68	0.11	0.13	0.41	0.08
2PS	750	10/20	1986–2020	34	0.78	0.10	0.12	0.45	0.01
3PS	1000	10/19	1986–2020	34	0.91	0.10	0.11	0.48	0.07
4PS	1500	9/18	1987–2020	33	0.88	0.09	0.11	0.37	0.03
5PS	2000	10/20	1987–2020	33	0.91	0.11	0.12	0.40	0.13
6PS	3000	10/20	1986–2020	34	0.92	0.11	0.14	0.49	0.01
7PS	4000	11/22	1988–2020	32	0.92	0.12	0.13	0.49	0.06
9PS	8000	10/20	1986–2020	34	0.88	0.15	0.15	0.54	0.17
10PS	10,000	10/19	1983–2020	37	0.83	0.20	0.16	0.47	0.07
11PS	12,000	10/20	1987–2020	33	0.88	0.13	0.15	0.52	0.03
12PS	16,000	10/19	1985–2020	35	0.85	0.17	0.15	0.53	0.31
13PS	24,000	10/20	1985–2020	35	0.84	0.15	0.15	0.42	0.07
14PS	36,000	10/20	1985–2020	35	0.81	0.13	0.15	0.50	−0.02
15PS	48,000	10/20	1986–2020	34	0.78	0.14	0.14	0.45	0.13
16PS	64,000	10/20	1985–2020	35	0.63	0.10	0.12	0.37	−0.04
17PS	96,000	10/20	1984–2020	36	0.71	0.16	0.18	0.33	−0.06
18PS	128,000	10/20	1984–2020	36	0.71	0.14	0.16	0.27	−0.07

DENS$_0$, initial stand density (planting density); TRW, tree-ring width; *SD*, standard deviation; *sens*, coefficient of sensitivity; *r-bar*, inter-series correlation coefficient; ar_{-1}, first-order autocorrelation coefficient.

2.2. Data on Tree Morphometry and Radial Growth

Inventory measurements were carried out in 2000, 2005, and 2015 according to methods generally accepted in Russia [55–58]. For each plot, the stand density (DENS, number of trees per hectare), the diameter at 1.3 m from the soil surface (DBH), and the height (H) of trees were measured. As inventory indicators of tree productivity for each plot, we used root-mean-square DBH values based on measurements for all living trees except the outer rows and arithmetic mean H based on measurements for 25–30 model trees of different sizes. The outer rows were omitted due to differing stand densities on the boundary of the plot (edge effect). The basal area of the trunk was calculated based on the root mean square diameter (BA = $\pi \cdot DBH^2/4$). Trunk volume V was calculated using the formula $V = BA \cdot H \cdot f$ [59], where $f = Q_{0.5}^2$ is the form factor (ratio of tree trunk volume to the volume of the cylinder with diameter DBH and height H) and $Q_{0.5}$ is the form quotient (ratio of the tree diameter at the half height to DBH). We implemented the average value of the form quotient for the Scots pine species, $Q_{0.5} = 0.67$ [60]. Stand productivity per area unit was estimated in each plot using the total basal area and timber stock, calculated with multiplying the BA and V values, respectively, by the stand density DENS.

To collect wood samples (cores) in 2020, ten dominant (the largest and most well developed) trees per plot were selected, also with the exception of the outer rows. The choice of dominant individuals allows one to focus on the growth dynamics of trees that are more likely to survive in the future since both self-thinning and thinning tend to remove depressed trees [61,62].

Cores were collected and processed using standard dendrochronology techniques [63], with two cores per tree. Individual tree ring width (TRW) series were measured on a LINTAB measuring platform using the TSAP program [64] and cross-dated using the COFECHA program [65]. Then, in the ARSTAN program [66], age trends described by a

67% cubic smoothing spline were removed from the original series, using the division of the measured values by the trend values. Based on the raw and indexed series of individual cores, generalized chronologies for the plots were obtained as the bi-weight mean. In the same way as TRW, generalized chronologies of latewood width (LWW) were obtained. The boundary between earlywood and latewood within the ring was determined visually [67]. In addition, for each site, the dynamics of the latewood ratio (LWR = LWW/TRW) was calculated based on the generalized raw chronologies of TRW and LWW. For the period 1988–2020, within which this parameter was steadily increasing for all plots, linear trends were calculated and standardization (indexing) of the LWR series was carried out as a division of the actual values by the trend values.

For the raw and standardized series of TRW, the inter-series correlation coefficients (*r-bar*) were calculated to estimate the common external signal. The variability of local standardized TRW chronologies was assessed by the standard deviation (*SD*), sensitivity coefficient ($sens = |TRW_t - TRW_{t-1}| / mean(TRW_t, TRW_{t-1})]$), and first-order autocorrelation coefficient (ar_{-1}). Due to the sampling of two cores per tree, all TRW chronologies were found to have an expressed population signal (*EPS*) above 0.85 [68].

2.3. Statistical Analysis

Relationships among variables describing the growth of individual trees and total forest stand were assessed using simple linear regression models, where both variables or one of them were usually transformed into a logarithmic scale by a natural logarithmic function. In the first case, the linear function in log–log scale $\ln(y) = a \cdot \ln(x) + b$ (where y is the dependent variable, x is the independent variable, and a and b are the numerical coefficients) corresponds to the simple allometric power function $y = b' \cdot x^a$, where $b' = \exp(b)$. In the second case, the linear function for log-scale dependent variable $\ln(y) = a \cdot x + b$ corresponds to the exponential function $y = b' \cdot \exp(a \cdot x)$. For each relationship, we tested both functions and used the one better fitting the actual data. Fitting of numerical coefficients with the least squares method and calculation of the determination coefficient R^2 were performed in the linear forms of the regression models.

Dendroclimatic analysis was carried out by calculating Pearson's pairwise correlation coefficients between generalized indexed chronologies of TRW, LWW, and LWR and climatic series for the period 1988–2020 (common period of all of the time series). Correlations were calculated for the period from August of the year prior to tree ring formation (previous year) to August of the year of tree ring formation (current year). Comparison of monthly climate data series showed that over the period of dendroclimatic analysis, the correlations between the CRU TS series for the study area and the Krasnoyarsk station series for the same months were 0.89–0.98 (0.96 on average) for temperatures and 0.42–0.91 (0.75 on average) for precipitation. Such spatial synchronicity confirms the suitability for using dendroclimatic analysis not only for the local monthly CRU TS 4.04 series, but also, with some caution, for the daily data from the remote Krasnoyarsk station. In the latter case, to assess the detailed response of tree-ring parameters to climatic factors, a moving series with a window of 21 days and a step of 1 day (for example, 1–21 May, 2–22 May, etc.) were calculated on the basis of daily data and used in the dendroclimatic analysis.

The significance level p of all relationships was estimated with a two-tailed t-test.

3. Results

3.1. Dynamics of Pine Stand Density, Productivity, and Radial Growth Parameters

When establishing experimental stands, planting densities in the range of 500–128,000 trees per hectare were selected with approximately uniform steps on a logarithmic scale (Figure 2). Negative trends in the stand density with age in the absence of logging describe the process of self-thinning, i.e., die-off of the most oppressed trees. Regression models of the stand density versus age, using a negative exponential function $DENS = DENS_0 \cdot \exp[-a(A - A_0)]$ (where $A_0 = 2$ years is the absolute age of seedlings when planted), describe 80 to 99% of its variation (Table A1). In this function, the numerical coefficient a

(regulating the tree mortality rate) slightly decreases with planting density at first, reaches a minimum in plots with planting densities of 3000–8000 trees/ha, and then significantly increases depending on the logarithm of $DENS_0$. This relationship can be described as a linear (R^2 = 0.566, $p < 0.05$) or quadratic regression (R^2 = 0.922, $p < 0.05$). Indeed, at the time of the last inventory (2015), the actual mortality was 43%–61% of trees in stands with $DENS_0$ = 500–2000 trees/ha, 33%–49% in stands with $DENS_0$ = 3000–10,000 trees/ha, and then there was a gradual increase in mortality from 48% to 91% in the densest stands. As a result, the range of stand density decreased to 228–11,750 trees/ha by the age of 35 years.

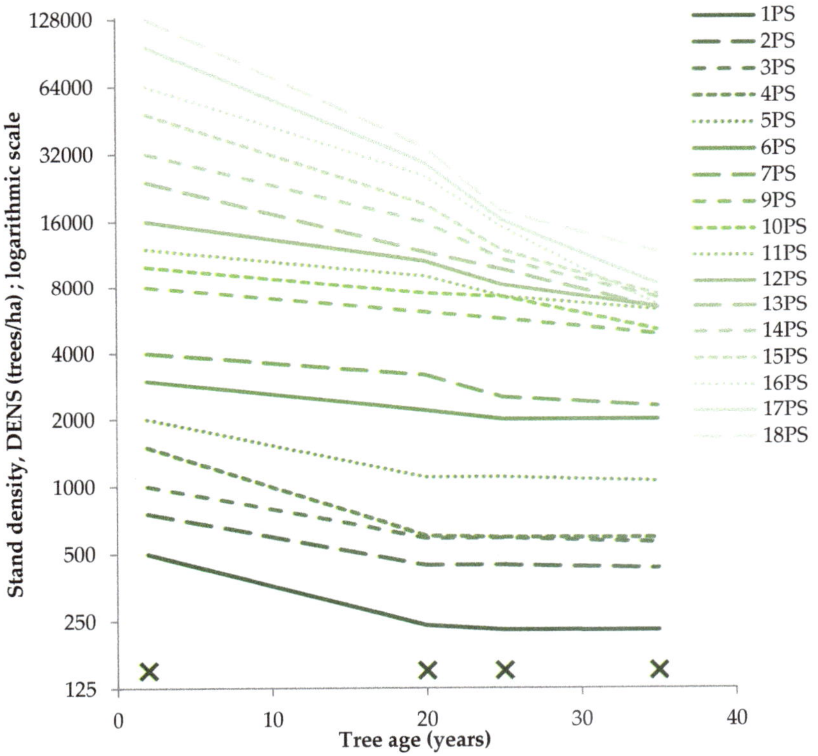

Figure 2. Dynamics of stand density DENS versus the absolute age of the trees in each experimental plot. The planting of 2-year-old seedlings in 1982 and inventories in 2000, 2005, and 2015 are marked with crosses.

The average tree diameter during each inventory significantly ($p < 0.05$) negatively depended on both the planting density and stand density at the moment of inventory (Table 2), and the dependence is closest to linear for the logarithmic scales of both variables. The fitness of the relationship is consistently high, but with age, an increase in the slope of the linear function was observed for functions of contemporary density DENS, both due to the accumulation of differences between the plots in diameter and due to a decrease in the range of the stand density. At the same time, the negative dependence of tree height on stand density became statistically significant ($p < 0.05$) only at the last inventory.

For derivative indicators of tree productivity, average basal area, and trunk volume, the dependence on stand density indicators is negative and consistently high, as it is for DBH, but the slope angle changes little with tree age in functions of both $DENS_0$ and DENS. All indicators of growth and productivity of individual trees, except for height, depend to a greater extent on the stand density at the same moment than on the planting density. On the other hand, their dependences on planting density have slopes more stable with age.

Table 2. Dependences of tree diameter, height, and derivative productivity indicators on stand density variables in experimental plots for various tree ages.

Age (Years)	Function of Planting Density $DENS_0$ (Trees/ha)			Function of Contemporary Stand Density DENS (Trees/ha)		
	Equation	R	R^2	Equation	R	R^2
Tree diameter at the breast height, DBH (cm)						
20	$\ln(DBH) = -0.246 \cdot \ln(DENS_0) + 4.377$	**0.981**	0.962	$\ln(DBH) = -0.268 \cdot \ln(DENS) + 4.404$	**0.982**	0.965
25	$\ln(DBH) = -0.248 \cdot \ln(DENS_0) + 4.597$	**0.981**	0.962	$\ln(DBH) = -0.306 \cdot \ln(DENS) + 4.858$	**0.988**	0.976
35	$\ln(DBH) = -0.221 \cdot \ln(DENS_0) + 4.688$	**0.954**	0.910	$\ln(DBH) = -0.325 \cdot \ln(DENS) + 5.262$	**0.987**	0.974
Tree height, H (m)						
20	$\ln(H) = -0.010 \cdot \ln(DENS_0) + 2.153$	0.306	0.094	$\ln(H) = -0.010 \cdot \ln(DENS) + 2.148$	0.285	0.081
25	$\ln(H) = -0.007 \cdot \ln(DENS_0) + 2.486$	0.247	0.061	$\ln(H) = -0.008 \cdot \ln(DENS) + 2.489$	0.232	0.054
35	$\ln(H) = -0.030 \cdot \ln(DENS_0) + 2.966$	**0.737**	0.543	$\ln(H) = -0.036 \cdot \ln(DENS) + 2.987$	**0.637**	0.406
Tree trunk volume, V (m^3)						
20	$\ln(V) = -0.501 \cdot \ln(DENS_0) + 0.655$	**0.972**	0.944	$\ln(V) = -0.545 \cdot \ln(DENS) + 0.702$	**0.972**	0.945
25	$\ln(V) = -0.503 \cdot \ln(DENS_0) + 1.428$	**0.976**	0.952	$\ln(V) = -0.620 \cdot \ln(DENS) + 1.953$	**0.982**	0.964
35	$\ln(V) = -0.472 \cdot \ln(DENS_0) + 2.090$	**0.956**	0.914	$\ln(V) = -0.687 \cdot \ln(DENS) + 3.258$	**0.979**	0.958
Tree basal area, BA (m^2)						
20	$\ln(BA) = -0.491 \cdot \ln(DENS_0) - 0.697$	**0.981**	0.961	$\ln(BA) = -0.535 \cdot \ln(DENS) - 0.645$	**0.982**	0.965
25	$\ln(BA) = -0.496 \cdot \ln(DENS_0) - 0.257$	**0.981**	0.962	$\ln(BA) = -0.612 \cdot \ln(DENS) + 0.265$	**0.988**	0.976
35	$\ln(BA) = -0.443 \cdot \ln(DENS_0) - 0.076$	**0.954**	0.910	$\ln(BA) = -0.650 \cdot \ln(DENS) + 1.073$	**0.987**	0.974
Total wood volume, i.e., timber stock, ΣV (m^3/ha)						
20	$\ln(\Sigma V) = 0.404 \cdot \ln(DENS_0) + 0.874$	**0.927**	0.859	$\ln(\Sigma V) = 0.455 \cdot \ln(DENS) + 0.702$	**0.960**	0.923
25	$\ln(\Sigma V) = 0.287 \cdot \ln(DENS_0) + 2.463$	**0.883**	0.780	$\ln(\Sigma V) = 0.380 \cdot \ln(DENS) + 1.953$	**0.953**	0.909
35	$\ln(\Sigma V) = 0.193 \cdot \ln(DENS_0) + 3.998$	**0.794**	0.631	$\ln(\Sigma V) = 0.313 \cdot \ln(DENS) + 3.258$	**0.909**	0.827
Total basal area, ΣBA (m^2/ha)						
20	$\ln(\Sigma BA) = 0.413 \cdot \ln(DENS_0) - 0.478$	**0.945**	0.892	$\ln(\Sigma BA) = 0.465 \cdot \ln(DENS) - 0.645$	**0.977**	0.954
25	$\ln(\Sigma BA) = 0.294 \cdot \ln(DENS_0) + 0.778$	**0.902**	0.813	$\ln(\Sigma BA) = 0.388 \cdot \ln(DENS) + 0.265$	**0.970**	0.941
35	$\ln(\Sigma BA) = 0.222 \cdot \ln(DENS_0) + 1.833$	**0.864**	0.747	$\ln(\Sigma BA) = 0.350 \cdot \ln(DENS) + 1.073$	**0.957**	0.917

R is the correlation between the actual values and the model; significant correlations ($p < 0.05$) are in bold; R^2 is the determination coefficient.

Stand productivity indicators (total basal area and timber stock per area unit), on the contrary, are positively related to both density variables. The fitness of these relationships is consistently high for all considered tree ages ($R^2 > 0.9$ for most of functions). On the contrary, a decrease in the slope of the function was observed with age, more pronounced for planting density.

The long-term mean value of the tree ring width was strongly negatively correlated with the planting density on a log–log scale (Table 3). Comparing the data of further inventories, its dependence on the stand density gradually decreased with age, although the slope increased. It is also noticeable in the graphs of the raw TRW values (Figure 3a) that the maximum radial increment gradually shifts from 1995 in the 1PS–5PS plots to 1987 in the 17PS–18PS plots. Until the occurrence of this maximum, the tree-ring width dynamics is similar in all plots. The synchronous nature of age trends within each plot is confirmed by the very high inter-series correlations of the raw TRW series (0.63–0.92) compared to the ones of the standardized series (0.27–0.54). At the same time, the standardized TRW chronologies (Figure 3b) show a high similarity of interannual growth variation between all plots. For the common period 1988–2020, the correlations between the standardized chronologies are all significant (0.36–0.96, $p < 0.05$) (Figure 4). The highest correlations were observed between chronologies from plots of similar densities, but even correlations between chronologies of the most and the least dense stands became significant after standardization. The variability of standard TRW chronologies is low, both in general (SD = 0.09–0.20) and year-to-year ($sens$ = 0.12–0.18).

Table 3. Dependencies of tree-ring parameters (average values since respective inventories) on stand density in experimental plots for various tree ages.

Age (Years)	Function of Stand Density DENS (Trees/ha)		
	Equation	R	R^2
Tree-ring width, TRW (mm)			
2	$\ln(TRW) = -0.222 \cdot \ln(DENS_0) + 2.911$	**0.961**	0.923
20	$\ln(TRW) = -0.278 \cdot \ln(DENS) + 2.918$	**0.947**	0.897
25	$\ln(TRW) = -0.295 \cdot \ln(DENS) + 2.881$	**0.925**	0.856
35	$\ln(TRW) = -0.333 \cdot \ln(DENS) + 2.934$	**0.883**	0.780
Latewood width, LWW (mm)			
2	$\ln(LWW) = -0.198 \cdot \ln(DENS_0) + 1.233$	**0.946**	0.894
20	$\ln(LWW) = -0.237 \cdot \ln(DENS) + 1.339$	**0.935**	0.874
25	$\ln(LWW) = -0.255 \cdot \ln(DENS) + 1.388$	**0.916**	0.838
35	$\ln(LWW) = -0.321 \cdot \ln(DENS) + 1.712$	**0.867**	0.751
Latewood ratio, LWR (dimensionless)			
2	$\ln(LWR) = 0.031 \cdot \ln(DENS_0) - 1.642$	**0.708**	0.502
20	$\ln(LWR) = 0.038 \cdot \ln(DENS) - 1.522$	**0.735**	0.540
25	$\ln(LWR) = 0.034 \cdot \ln(DENS) - 1.434$	**0.632**	0.399
35	$\ln(LWR) = 0.012 \cdot \ln(DENS) - 1.215$	0.209	0.044

R is the correlation between the actual values and the model; significant correlations ($p < 0.05$) are in bold; R^2 is the determination coefficient.

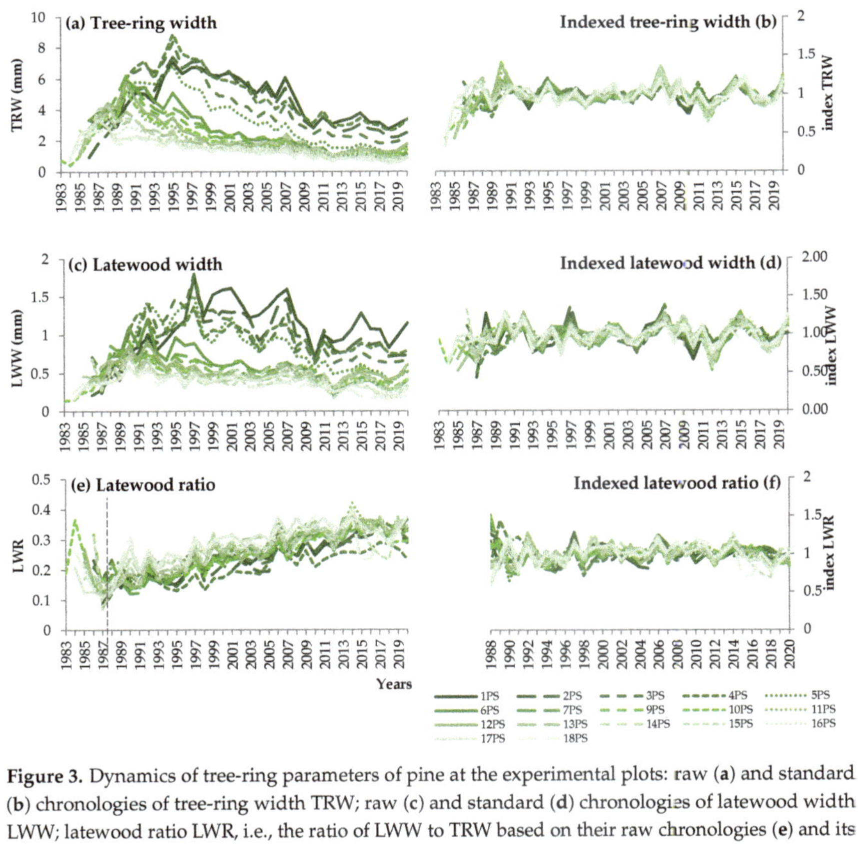

Figure 3. Dynamics of tree-ring parameters of pine at the experimental plots: raw (**a**) and standard (**b**) chronologies of tree-ring width TRW; raw (**c**) and standard (**d**) chronologies of latewood width LWW; latewood ratio LWR, i.e., the ratio of LWW to TRW based on their raw chronologies (**e**) and its index after elimination of linear trend calculated for the period 1988–2020 (**f**).

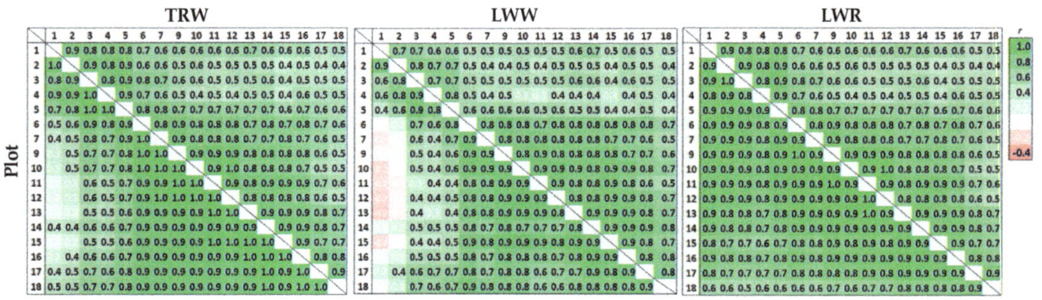

Figure 4. Between-plot correlations of raw (below diagonal) and indexed (above diagonal) chronologies (1988–2020): tree-ring width (TRW), latewood width (LWW), and latewood ratio (LWR). The color gradient marks the correlation values (legend). Black values are significant at $p < 0.05$.

Latewood width LWW also has a significant ($p < 0.05$) negative allometric dependence on planting density (Table 3) and, to a lesser extent, on the contemporary values of the stand density. Its raw series and standard chronologies (Figure 3c,d) are slightly less synchronous with each other compared to TRW (raw: $r = -0.22$–0.95, $p < 0.05$ for 82% of correlations; standard: $r = 0.28$–0.95, 98%; Figure 4). Nevertheless, there is a common pattern in the correlations of the TRW and LWW chronologies: the presence of two groups, 1PS–5PS and 6PS–18PS, where within-group correlations are higher than between-group ones (for both raw and standard chronologies). The latewood ratio in the ring LWR (Figure 3e), in contrast, increases with the planting density. The dependence of the latewood ratio in subsequent rings on the stand density is the maximum for the age of 20 years and then weakens, losing significance by the age of 35 years (Table 3). For the raw LWR series over 1988–2020, the correlations between the plots are the most stable in comparison to other tree-ring parameters under consideration, being in the range $r = 0.52$–0.97. After removing long-term trends (Figure 3f), the correlations decrease to $r = 0.36$–0.96 but still remain significant at $p < 0.05$.

Note that the raw LWW and TRW values are closely positively related to each other for all plots: $r = 0.67$–0.95 (Table 4). The negative relationship between LWR and TRW is somewhat less pronounced but stably significant; the correlations range from −0.48 to −0.93. The width and proportion of latewood also correlate negatively in most stands (from −0.80 to 0.31), but this relationship is not always significant. After standardization of the LWW and TRW chronologies, the relationship between them somewhat weakens: $r = 0.46$–0.86. The LWW and LWR indices, in contrast to their raw values, are positively related to each other ($r = 0.14$–0.71), while the correlations between the TRW and LWR indexed chronologies range from −0.43 to 0.08 and are mostly not significant. Along the stand density gradient, the correlations between the raw tree-ring parameter series and between the TRW and LWW indices increase from 1PS to 5PS–I7PS, are stable up to 13PS–14PS, and then weaken again in the densest stands. Calculating the above relationships for the full sample from all plots together, all of them are significant at the $p < 0.05$ level except for the correlation between the LWW and LWR indices.

3.2. Climatic Response of Tree-Ring Parameters at Different Planting Densities

Due to the short period of dendroclimatic analysis, the threshold of significance at level $p < 0.05$ is about $|r| = 0.34$; thus, most of the correlations between the tree-ring chronologies and monthly or seasonal climatic variables are not significant (Figure 5). However, if a significant reaction of the tree rings to climatic factors is observed in some of the plots, insignificant correlations of the same sign to the same factors in other plots are probably associated with similar but weaker reactions.

Table 4. Between-parameter correlations: within-plot (for generalized chronologies) and in total.

Plot	Sample Depth	Correlations between Two Parameters					
		Raw Chronologies			Indexed Chronologies		
		TRW LWW	LWW LWR	TRW LWR	TRW LWW	LWW LWR	TRW LWR
1PS	20	**0.67**	0.31	**−0.48**	0.67	**0.71**	0.08
2PS	20	**0.72**	−0.04	**−0.69**	0.55	**0.57**	−0.32
3PS	19	**0.87**	**−0.41**	**−0.78**	0.70	**0.55**	−0.13
4PS	18	**0.76**	−0.30	**−0.82**	0.68	0.14	**−0.41**
5PS	20	**0.87**	**−0.54**	**−0.86**	0.62	**0.44**	**−0.36**
6PS	20	**0.95**	**−0.76**	**−0.89**	0.84	**0.52**	0.10
7PS	22	**0.90**	**−0.80**	**−0.93**	0.86	0.20	−0.23
9PS	20	**0.86**	**−0.59**	**−0.87**	0.84	**0.45**	−0.01
10PS	19	**0.93**	**−0.74**	**−0.89**	0.75	0.21	**−0.35**
11PS	20	**0.80**	**−0.47**	**−0.86**	0.78	**0.39**	−0.16
12PS	19	**0.84**	**−0.54**	**−0.86**	0.77	**0.36**	−0.22
13PS	20	**0.87**	**−0.63**	**−0.89**	0.80	0.28	−0.21
14PS	20	**0.78**	**−0.39**	**−0.85**	0.80	**0.38**	−0.18
15PS	20	**0.84**	**−0.39**	**−0.76**	0.67	**0.56**	−0.09
16PS	20	**0.72**	−0.30	**−0.83**	0.71	**0.52**	−0.11
17PS	20	**0.72**	−0.25	**−0.81**	0.67	**0.48**	−0.12
18PS	20	**0.84**	−0.28	**−0.71**	0.46	**0.40**	**−0.43**
Total	337	**0.89**	**−0.38**	**−0.71**	0.72	**0.42**	−0.19

Bold values are significant at $p < 0.05$.

The climatic response of pine TRW is similar for all plots, but its intensity and seasonality depend on the planting density (Figure 5). In August of the previous year, the temperature slightly negatively affects the growth of the pine plantations (significant at $p < 0.05$ only at 3PS). The positive response of the tree-ring width to the temperature of the cold season for all plots includes the period from January to April (for the temperature averaged over these months, $r = 0.24$–0.49) and is less pronounced in plots with middle and high planting densities (insignificant at 9PS–11PS, 13PS, 15PS, and 18PS). In low-density plantations, the period of positive reaction also includes May (up to 0.51 for January–May temperatures at the plot 2PS), and it includes December in denser stands (7PS–16PS). In June, the sign of the reaction of pine growth to temperature changes to negative but not significant. Furthermore, in July–August, a positive, mostly insignificant reaction is observed again.

Precipitation contributes to an increase in pine growth closer to the end of the cold season (the only significant responses are observed at six of the denser plots in February). In April, precipitation tends to slow down the subsequent radial growth (significant only at 10PS), but its effect is again positive in May–June ($r = 0.11$–0.56, significant at 7PS–18PS for two-month sum). The overall response of TRW to precipitation is more pronounced for denser stands.

For the standard LWW, a slightly negative response to temperature is observed from August to October of the previous year ($r = -0.14$–-0.46, significant at 1PS, 2PS, and 13PS). Then, the positive reaction to temperature lasts from January to May ($r = 0.16$–0.49, significant at 10 plots of various densities). In June, the LWW reacts slightly negatively to temperature in stands of high and especially medium density (significant at 7PS and 10PS–11PS), stronger than TRW reacts. The response to winter precipitation is weakly positive for the period from December to March r = 0.19–0.41 (significant at 1PS, 3PS, 4PS, 13PS, and 17PS) but negative in April for all stands except for the sparsest ones 1PS-5PS (significant at 6PS and 15PS). From May or June to August, the influence of precipitation is again positive; it is more pronounced and begins earlier for the plots of medium and high density 6PS–17PS ($r = 0.31$–0.59, significant at 7PS–9PS and 11PS–17PS).

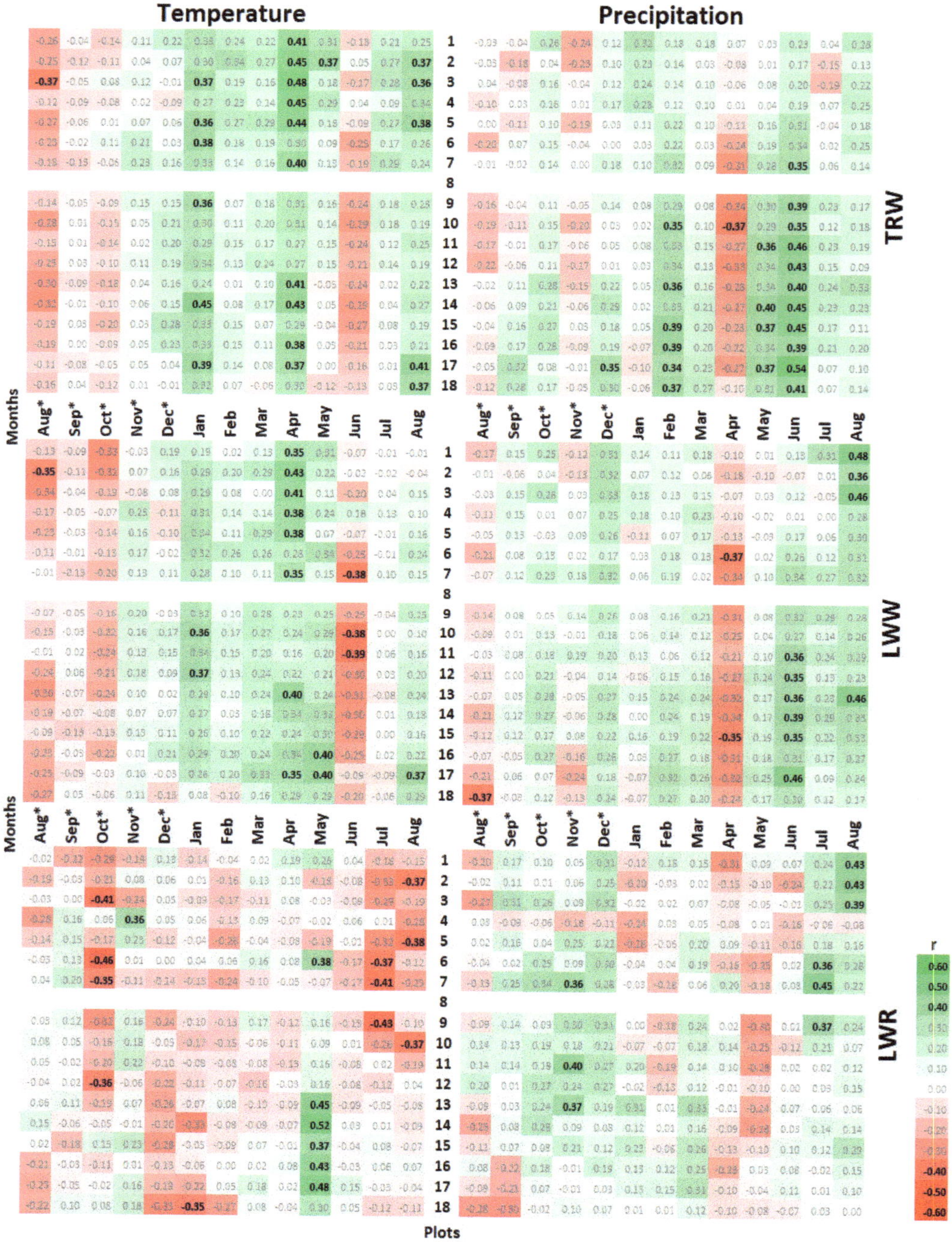

Figure 5. Correlations of tree-ring width (TRW), latewood width (LWW), and latewood ratio (LWR) indexed chronologies with monthly temperature and precipitation from the CRU TS 4.04 grid series (previous August–current August, 1988–2020). Asterisks mark the months of the previous year. The color gradient marks the correlation values (see legend); significant correlations ($p < 0.05$) are marked with black bold font.

For the LWR, the climatic response is less stable and differs sharply from the above parameters of pine tree rings. A positive response to temperature is observed for dense stands (13PS–17PS) in May (r = 0.37–0.52), while for the sparser stands (1PS–12PS), the latewood ratio is weakly negatively related to the temperatures of the previous October (up to −0.46, significant at four plots), and the current July–August (up to −0.54, significant at 2PS–3PS and 5PS–9PS). The response to precipitation is also positive only for sparser stands from September or October to December of the previous year (up to 0.52, significant at 3PS, 6PS–9PS, and 11PS–13PS) and in July to August of the current year (up to 0.46, significant at 1PS–3PS and 6PS–9PS). There are no pronounced shifts in the timing of the climatic response for this wood parameter.

An obvious disadvantage of the analysis performed above is the low temporal resolution being accurate to a calendar month, which makes it difficult to identify the smaller-scale differences in the seasonality of the climatic signal. Therefore, the response was detailed on the basis of the moving climatic series of the Krasnoyarsk station (Figure A1). Note that the short duration of the considered climatic series here means that each correlation registers only a small part on the total climatic signal registered in the integral tree-ring parameters. Thus, insignificant but consistent dendroclimatic correlations during some intra-seasonal interval can be transformed into one significant response to climatic factor generalized over this interval by the sum or average.

For this type of analysis, it becomes noticeable that the positive effect of temperature on pine growth begins in early November, reaches significant values (in some plots) in the first half of January, and ends around 20 May, regardless of the planting density. Only the intensity of the response differs between the plots. On the other hand, a weak negative reaction to the June temperature lasts longer by about 10 days in denser stands. For the positive effect of winter precipitation, the planting density affects only the intensity of the response. In May, the positive reaction to precipitation is not only stronger but also begins earlier for high-density plantations; the difference in the start date of the response reaches two weeks. The end of this response is less abrupt, fading away by about the 10th of August.

For the latewood width, the beginning of the negative effect of the temperatures in the August of the previous season depends on the planting density (occurring earlier in denser stands; the difference reaches 2 weeks), but the end of this response in the last days of October is synchronous for all plots. For the negative impact of June temperatures, a non-linear dependence of intensity on planting density is observed, but the beginning and end of the response are stable. In response to precipitation, a dependence on planting density is observed for the transition from a negative response in April to a positive response in May. This transition shifts up to two weeks earlier in dense stands. The intensity of the positive response to precipitation in the second half of July and early August also increases with the density.

4. Discussion

The conditions of the ongoing long-term experiment, such as the homogeneity of the local soil and the landscape conditions and the even-age structure of forest stands within the plantation, made it possible to simplify the interpretation of the results of this study. The exclusion of species-specific and habitat differences and the synchronicity of ontogenetic development allowed us to focus on the variation in growth and productivity of Scots pine due to the planting density gradient.

4.1. Stand Density Dynamics and Tree and Stand Productivity

The density of the even-aged forest stands at the early ages (before reproduction) changes only due to the death of trees. Therefore, for the experimental plots in this study, in the absence of logging, the dynamics of the stand density describes only the process of natural mortality, i.e., self-thinning. During the first years after planting, seedlings practically do not compete with each other even in the densest plots. This means that mortality during this period is independent of planting density. However, over forty years, the gradual closure of crowns and root systems as trees increase in size is leading to the emergence and gradual increase of competition for limited resources [69–72], primarily sunlight and water. The nonlinear increase in size with age causes the observed negative exponent in the dynamics of stand density. Competition stratifies the population according to social roles, leading to growth inhibition and the subsequent death of some individuals under environmental stresses [73–75]. A higher stand density reduces the living space and resources available to each individual tree, which increases the intensity of competition. In addition, the "race for height" (an allocation of a larger share of resources for growth in height) in dense stands leads to the formation of vertically elongated trees with thin trunks and underdeveloped crowns [35,69,76–79]. As a result, depressed trees in such areas, in addition to climatically determined stresses, also become more vulnerable to mechanical loads of wind and snow. On the contrary, the beneficial effect of stand density on the growth of young trees has been previously recorded within the range from tens to several thousand trees per hectare [80–82]. Since the mentioned observations were made under conditions of moisture deficiency of various severities, it could be possible that lower moisture losses in the soil during shading of its surface contribute to the growth and survival of trees, slowing down self-thinning. These phenomena can explain the observed non-linear dependence of the mortality rate on the planting density. It is worth noting that the presence of an "optimal" range of planting density, which ensures the maximum survival of Scots pine seedlings, may be of interest for forest management in the conditions of the region.

Allometric relationships of the morphometric variables of individual plants among themselves and with the density of their growth can be described by power functions [83, 84], which are linearized by a logarithmic transformation of variables. It is also known that specific quantitative ratios depend on habitat conditions and plant species, and their long-term dynamics can be disturbed by uneven self-thinning [85–87]. It is logical that the size, shape, and allometric ratios of different parts of a tree also change with age, as well as their relationship with stand density [71,87,88]. This includes the dynamics of the influence of density variables on growth and productivity observed in this study. In this study, well-known patterns are observed. With the increasing density of the forest stand, (1) the total productivity indicators (the sum of productivity variable values per area unit) also increase, but (2) the size and radial growth of individual trees decrease (for example [89,90]). However, the particular numerical terms of these relationships change with age.

Tree height is less dependent on stand density than other growth indicators, as shown in previous studies [25,80,91–93]. Indeed, the race for height due to competition for light compensates to some extent for the lesser availability of resources in dense stands. However, the cumulative impact of resource constraints has resulted in 40%–50% of the average stand height variability at the time of the last inventory being explained by density variables. For the remaining indicators, the effect of planting density or stand density during further inventories follows allometric functions more closely and from an early age. It should be noted that, unlike the indicators of individual trees, the indicators of the total productivity of the forest stand are regulated to a greater extent by the contemporary density of the forest stand than by the initial one, i.e., having a higher potential for regulation by partial logging. This can be explained by the fact that the morphometric tree structure, e.g., the proportions of trunk diameter with tree height or crown size, becomes less flexible with age [94,95] (cf. lesser reaction of radial growth to thinning for old trees in [29]). At the same time, the utilization of resources (light and water) by the forest stand as a total constantly tends to the maximum, retreating from it for a short time only after the massive die-off of trees during catastrophic events (strong winds, droughts, or snowfall) or cutting down.

Based on the TRW maximum and its subsequent suppression, an analysis of the dynamics of radial growth made it possible to identify the moment when significant intra-specific competition appeared in the forest stands. Interestingly, the latewood ratio in the ring starts to increase at approximately the same moment. It is possible that the beginning of the race for height causes an increase in the requirements for the mechanical strength of the trunk wood, which is provided, among other things, by a greater proportion of latewood. This is indicated by the reasoning of Zeide [33] about the limitation of tree morphology (the dependence of crown size on tree diameter and height) in terms of the mechanical load on the trunk. For conifers, in particular *Pinus* species, a pith-to-bark increase in wood density is typical [96–98], which may also be associated with an increase in the proportion of denser latewood [98] and is driven by a combination of mechanical strength, hydraulic efficiency, and safety requirements [97,99].

An increase in correlations after standardization between the chronologies of pine radial growth between stands of different density indicates that competition in the forest stand affects the long-term dynamics of growth much more than its inter-annual variability. Nevertheless, the experimental stands are clearly divided by density into two groups, the differences in the growth dynamics between which, as shown below, are due to shifts in the intensity and seasonality of the response to climatic factors.

The very close relationships between measurements of latewood width and the total tree-ring width show that cell production in latewood is largely "pre-programmed", presumably through the state of the tree at the beginning of the growing season. Both positive correlations between the standardized chronologies of these tree-ring parameters [67,100] and their similar climatic response [101] have already been noted not only for conifers but also for deciduous trees [102]. The very close relationships observed in this study both before and after standardization of chronologies show that their common component includes both the impact of external factors, including climatic variables (significant relationships between the standard chronologies), and common long-term trends (high correlations between the raw series). To confirm this hypothesis and to analyze the stability of these ratios over the course of time, further studies are required using data from mature forest stands. The counter-intuitive negative relationship between the LWW and LWR are obviously associated with opposite long-term trends—a decrease in the latewood growth clearly connected to a decrease with age in the total radial growth within the considered time interval—while the latewood ratio in the ring increases steadily with the exception of the very first rings. However, it should be noted that the almost completely insignificant relationships between the indexed chronologies of the LWR and TRW demonstrate the independence of the external signal in these variables, i.e., the fundamentally different effect of climatic fluctuations on the quantity (radial growth, i.e., TRW) and quality (density and mechanical strength, positively related to LWR) of developing wood.

4.2. Climatic Response of Pine Radial Growth Parameters and its Dependence on the Stand Density

An analysis of the influence of climatic dynamics on the tree-ring width, latewood width, and ratio of Scots pine in artificial plantations in the southern taiga zone showed the presence of similar patterns for all planting densities, due to the common natural and climatic conditions of the study area and the ecophysiological characteristics of the species.

For TRW, there is a positive response to winter–spring temperatures, which has been previously recorded in other areas within the southern and middle taiga zones [103–106] from Germany to Central Siberia. On the contrary, in the mountain taiga of Southern Siberia and Mongolia, despite the similar seasonal dynamics of temperature and precipitation, such a response does not occur in pine (for example [107,108]). We assume that the positive correlations with the temperatures in the cold season may be associated with damage to the root system during severe frosts due to the freezing of soil saturated with moisture after late autumn precipitation in the plains [109], in contrast to soil quickly losing moisture and being frozen while dry on the mountain slopes.

During the growing season, the limitation of pine growth by the moisture deficit can be observed, expressed as a negative relationship with temperatures and a positive one with precipitation [110–113]. The intensity of this response is low, which corresponds to the location of the plantation within the inner part of the species distribution range and, consequently, a relatively weak moisture deficit compared to forest-steppe stands and band forests located on the southern semiarid borders of the Scots pine distribution range [106–108,114–116].

The climatic signal in latewood ratio chronologies, if looked at closely, more or less corresponds to those parts of the LWW response in that are independent of the TRW. For example, a positive reaction to precipitation in July–August and to the temperature in May, negative to the temperature of the previous October is common for the LWW and LWR but is not reflected in the TRW. This is to be expected given the lack of statistically significant relationships or weak relationships between the TRW and LWR indexed chronologies. However, this regularity means that the latewood ratio, apparently, reflects not so much the influence of current climatic fluctuations on cambial activity directly during the formation of earlywood and latewood, but rather a climate-dependent trigger for the transition to differentiation of cells into the latewood-type tracheids (small and thick-walled) during the subsequent stages of xylogenesis. However, the dynamics of the switch in the sign of the LWR climatic response within the growth season indicates that in May, cool and humid conditions contribute to an increase in the earlywood ratio in the ring and in July–August they drive an increase in the latewood ratio. The fact that according to the more detailed 21-day dendroclimatic analysis, the effect of temperature and precipitation on the latewood ratio appears decoupled in June, may be due to the year-to-year instability of the transition to latewood within this month (see other examples of variation in the phenology of xylogenesis in [117–121]).

The revealed differences in the climatic response between stands of different densities can be related directly to the degree of competition, as well as to the size of the trees and the living space they occupy. E.g., a decrease in competition and a corresponding increase in the resources available to an individual tree occurs after the artificial thinning of forest stands. Dendroclimatic studies show that this changes both the overall climate response and vulnerability to adverse events, including droughts [36,48,122–125]. In less dense forest stands, the soil contains more nutrients and available moisture in general [16,126] and even more so per individual trees, which can regulate their response to droughts [127]. On the other hand, differences in sensitivity to climate fluctuations have been recorded within one stand, associated with such morphometric characteristics of individual trees such as crown class, DBH, and root depth [46,61,62,128], which are obviously also modulated by the density of the forest stand [35]. It should be noted that the components of the water and the thermal balance of the habitat also depend on the stand density through components such as the interception of precipitation by tree crowns, the soil moisture losses through transpiration and evaporation from the soil surface, and the micro-scale regulation of surface and air temperature by the ratio of shaded and sunlit surfaces [129–131]. The complex nature of all the listed factors and interactions ensures uncertainty in the direction of change in the climatic response depending on the planting density and the degree of competition. On the one hand, there is evidence of the "complacency" of growth, i.e., a weakened climatic response, for the most dominant, large, or fast-growing trees in the stand [106,132,133]. On the other hand, extreme degrees of competition, i.e., restrictions on living space and resource availability, can make it impossible for a tree to take full advantage of favorable conditions and increase growth, thus also reducing tree growth variability and climatic signals [46,134–136]. As our previous study showed, the dependence of the climatic response on the degree of competition can even be nonlinear within the same habitat [137].

We can suggest that in the study area, the competition of pine trees for water increases with the stand density, which leads to an increase in their response to summer precipitation. As for the uneven effect of winter frosts on growth, among the possible contributing factors are different depths of fine roots damaged by frost, different degrees of growth stability after these injuries, and the presence of dense grass cover in sparser plantations additionally isolating the soil in winter [138]. In addition, a greater degree of intercept of the snow cover by canopies is possible in denser stands [139,140]. This was indirectly confirmed by the shift to earlier dates of the climatic response in April–May, possibly associated with an earlier end of snowmelt and the thaw of soil [141,142].

Temporal shifts in the climatic response, which are more pronounced for parameters of latewood, obviously indicate an increase in the duration of cambial activity and a shift in the timing of the transition to formation of latewood toward later periods in sparse stands. Previously, a similar hypothesis was put forward for a change in the climatic response of trees in a thinned Scots pine stand, which was presumably associated with an extension of the period of latewood cell division due to more favorable moisture conditions in the second half of the growing season with less competition [126]. Finally, the stable end dates for the climatic response during the previous autumn in all of the plots clearly indicate that air temperature regulates the transition from nutrient storage (active photosynthesis in evergreen pine needles) to dormancy regardless of stand density.

5. Conclusions

An analysis of tree growth in Scots pine plantations of different planting densities with the natural dynamics of tree stand self-thinning demonstrated that the expected opposite gradients, depending on the stand density, are observed for the productivity indicators of individual trees and the total stand (per unit area). The productivity shifts from the maximum on the tree scale and the minimum on the stand scale in the sparsest plots toward the opposite pattern in the densest plots. As for the radial growth, the tree-ring width and latewood width direction depend negatively on stand the density, but the latewood ratio in the ring increases with the stand density. For most of the considered timeframe, the first two tree-ring variables have negative age trends, while the latewood ratio increases; thus, we can assume that the similar influence of age dynamics and stand density on tree-ring development is based on the intensity of competition. All of the observed dependencies are well described by allometric power functions (i.e., linear relationships in a log–log scale).

The contribution of planting density to the climate response is more ambiguous. In the seasonality of the response and, accordingly, the phenology of xylogenesis, there is a direct dependence of the duration of active growth on the available resource base, whereas the intensity of the influence of climatic fluctuations on the production and quality of wood is ambiguous, and for some seasons, it even nonlinearly depends on the planting density. The amount of precipitation in February–June is more important for pine growth in dense stands, the conditions in July–August have a stronger effect on the latewood ratio in sparse stands, and trees from medium-density stands show the greatest resistance to winter frosts.

Author Contributions: Conceptualization, E.A.V.; methodology, L.V.B., N.N.K. and E.A.V.; validation, B.Y. and E.A.V.; formal analysis, Y.A.K. and L.V.B.; investigation, Y.A.K., E.A.B., L.V.B., D.F.Z. and N.N.K.; resources, E.A.B. and D.F.Z.; data curation, L.V.B.; writing—original draft preparation, Y.A.K., L.V.B. and D.F.Z.; writing—review and editing, all authors; visualization, Y.A.K. and L.V.B.; supervision, E.A.B. and E.A.V.; project administration, D.F.Z.; funding acquisition, D.F.Z., B.Y. and E.A.V. All authors have read and agreed to the published version of the manuscript.

Funding: This research was funded by the Russian Science Foundation grant no. 23-44-00067 and the National Natural Science Foundation of China grant no. 42261134537 in the framework of a joint Russian–Chinese project (information on pine phenology and data analysis); Russian Science Foundation grant no. 22-06-00232 (processing of dendrochronological samples).

Data Availability Statement: The data presented in this study are available from the corresponding author upon reasonable request.

Conflicts of Interest: The authors declare no conflict of interest. The funders had no role in the design of the study; in the collection, analysis, or interpretation of the data; in the writing of the manuscript; or in the decision to publish the results.

Appendix A

Table A1. Dependences of stand density DENS on absolute tree age A.

Plot	Stand Density DENS (Trees/ha) at Ages A (Years) *				Function ** DENS = $DENS_0 \cdot \exp[-a(A - A_0)]$		
	$A_0 = 2$	20	25	35	a ***	R	R^2
1PS	500	239	228	228	−0.0274	0.911	0.830
2PS	750	445	445	431	−0.0190	0.915	0.837
3PS	1000	567	567	560	−0.0196	0.924	0.854
4PS	1500	599	590	588	−0.0330	0.899	0.808
5PS	2000	1103	1103	1057	−0.0218	0.919	0.844
6PS	3000	2196	1997	1997	−0.0136	0.947	0.897
7PS	4000	3198	2523	2300	−0.0157	**0.968**	0.938
9PS	8000	6108	5683	4867	−0.0139	**0.999**	0.999
10PS	10,000	7524	7227	5085	−0.0166	**0.968**	0.937
11PS	12,000	9070	7236	6268	−0.0182	**0.983**	0.966
12PS	16,000	10,639	8236	6487	−0.0250	**0.991**	0.983
13PS	24,000	11,668	9742	6388	−0.0370	**0.999**	0.999
14PS	36,000	15,951	10,826	7181	−0.0414	**0.994**	0.988
15PS	48,000	19,207	11,890	7449	−0.0525	**0.995**	0.990
16PS	64,000	25,662	15,062	6627	−0.0594	**0.987**	0.974
17PS	96,000	29,243	16,115	8367	−0.0682	**0.996**	0.992
18PS	128,000	34,336	17,795	11,750	−0.0706	**0.990**	0.980

R is the correlation between the actual values and the model; significant correlations ($p < 0.05$) are in bold; R^2 is the determination coefficient.* Planting density $DENS_0$ was measured at the tree age of $A_0 = 2$ years because plantation was established with 2-year-old seedlings.** Dependence of stand density DENS on the absolute tree age A was initially described as a log-linear function in relative units ($DENS/DENS_0$) because within the plot, $DENS_0$ can be considered as constant. Taking into account that DENS = $DENS_0$ at the age A_0 in all plots, it can be written as $\ln(DENS/DENS_0) = a(A–A_0)$. Then, the equation was transformed to the exponential form to use DENS as the dependent variable. *** Numerical term a in the equation DENS = f(A) changes with the planting density. Its dependence can be described with function $_a = -0.00433 \cdot \ln(DENS_0)2 + 0.0697 \cdot \ln(DENS_0) - 0.297$, which explains $R^2 = 0.922$ of the actual variance in a.

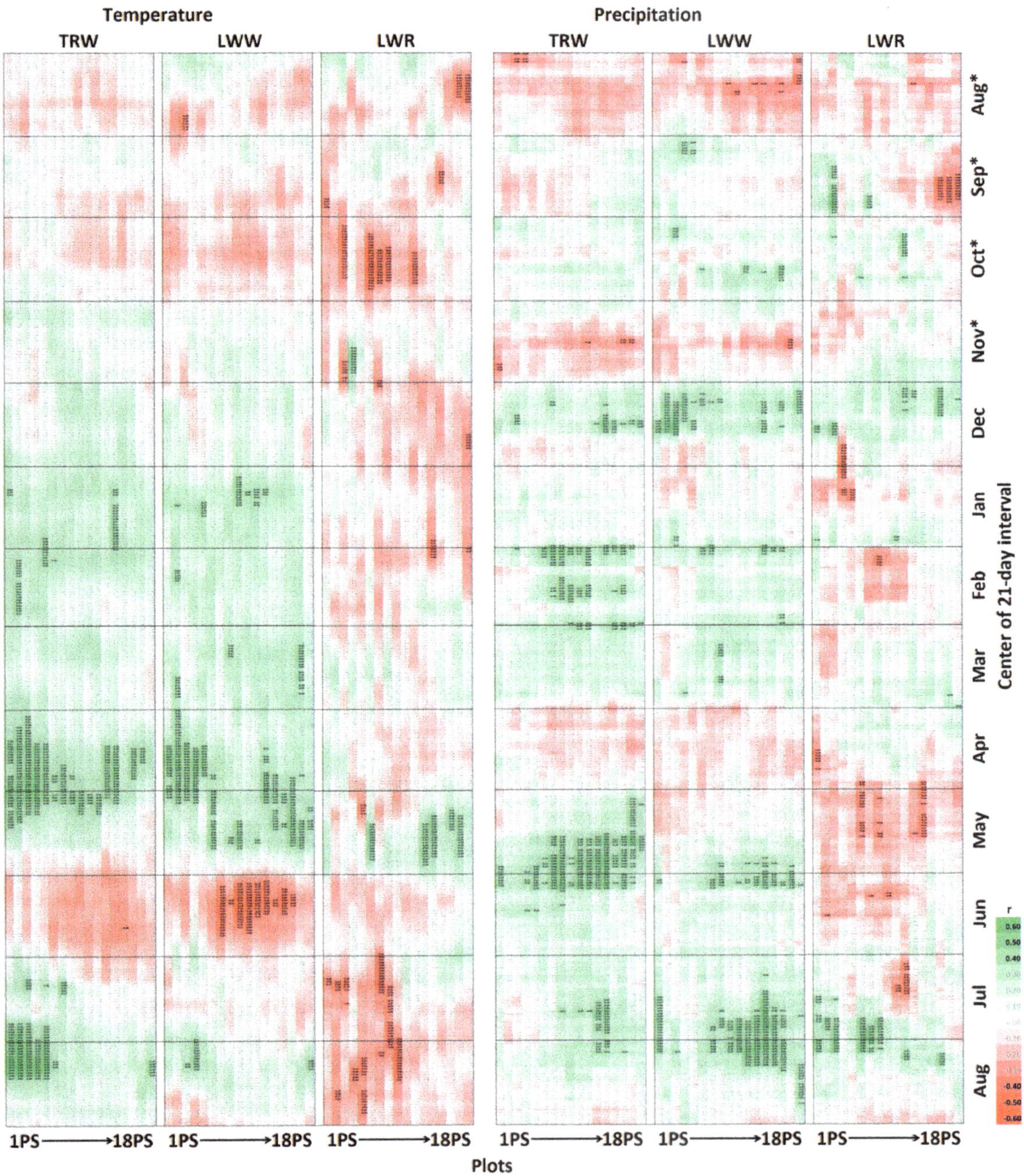

Figure A1. Correlations of the indexed chronologies of tree-ring width (TRW), latewood width (LWW), and latewood ratio (LWR) with the 21-day temperature and precipitation from Krasnoyarsk station, 1988–2019. The color gradient marks the correlation values (see legend); black marks show significant correlations ($p < 0.05$); asterisks "*" mark months of the previous year.

References

1. Edwards, D.P.; Tobias, J.A.; Sheil, D.; Meijaard, E.; Laurance, W.F. Maintaining ecosystem function and services in logged tropical forests. *Trends Ecol. Evol.* **2014**, *29*, 511–520. [CrossRef]
2. Brockerhoff, E.G.; Barbaro, L.; Castagneyrol, B.; Forrester, D.I.; Gardiner, B.; Gonzalez-Olabarria, J.R.; Lyver, P.O.B.; Meurisse, N.; Oxbrough, A.; Taki, H.; et al. Forest biodiversity, ecosystem functioning and the provision of ecosystem services. *Biodivers. Conserv.* **2017**, *26*, 3005–3035. [CrossRef]
3. Babst, F.; Carrer, M.; Poulter, B.; Urbinati, C.; Neuwirth, B.; Frank, D. 500 years of regional forest growth variability and links to climatic extreme events in Europe. *Environ. Res. Lett.* **2012**, *7*, 045705. [CrossRef]
4. Pederson, N.; D'Amato, A.W.; Dyer, J.M.; Foster, D.R.; Goldblum, D.; Hart, J.L.; Hessl, A.E.; Iverson, L.R.; Jackson, S.T.; Martin-Benito, D.; et al. Climate remains an important driver of post-European vegetation change in the eastern United States. *Glob. Chang. Biol.* **2015**, *21*, 2105–2110. [CrossRef]
5. Clark, J.S.; Iverson, L.; Woodall, C.W.; Allen, C.D.; Bell, D.M.; Bragg, D.C.; D'Amato, A.W.; Davis, F.W.; Hersh, M.H.; Ibanez, I.; et al. The impacts of increasing drought on forest dynamics, structure, and biodiversity in the United States. *Glob. Chang. Biol.* **2016**, *22*, 2329–2352. [CrossRef]
6. Dyderski, M.K.; Paz, S.; Frelich, L.E.; Jagodzinski, A.M. How much does climate change threaten European forest tree species distributions? *Glob. Change Biol.* **2018**, *24*, 1150–1163. [CrossRef]
7. Keenan, R.J.; Reams, G.A.; Achard, F.; de Freitas, J.V.; Grainger, A.; Lindquist, E. Dynamics of global forest area: Results from the FAO Global Forest Resources Assessment 2015. *For. Ecol. Manag.* **2015**, *352*, 9–20. [CrossRef]
8. Rigling, A.; Waldner, P.O.; Forster, T.; Bräker, O.U.; Pouttu, A. Ecological interpretation of tree-ring width and intraannual density fluctuations in *Pinus sylvestris* on dry sites in the central Alps and Siberia. *Can. J. For. Res.* **2001**, *31*, 18–31. [CrossRef]
9. Zhang, J.; Ritchie, M.W.; Maguire, D.A.; Oliver, W.W. Thinning ponderosa pine (*Pinus ponderosa*) stands reduces mortality while maintaining stand productivity. *Can. J. For. Res.* **2013**, *43*, 311–320. [CrossRef]
10. Altman, J.; Fibich, P.; Santruckova, H.; Dolezal, J.; Stepanek, P.; Kopacek, J.; Hunova, I.; Oulehle, F.; Tumajer, J.; Cienciala, E. Environmental factors exert strong control over the climate-growth relationships of *Picea abies* in Central Europe. *Sci. Total Environ.* **2017**, *609*, 506–516. [CrossRef] [PubMed]
11. Malanson, G.P. Mixed signals in trends of variance in high-elevation tree ring chronologies. *J. Mt. Sci.* **2017**, *14*, 1961–1968. [CrossRef]
12. Morozov, G.; Guman, V. (Eds.) *The Teaching of the Types of Plantings*; Yurayt: Moscow, Russia, 2019; 371p. (In Russian)
13. Huang, X.; Dai, D.; Xiang, Y.; Yan, Z.; Teng, M.; Wang, P.; Zhou, Z.; Zeng, L.; Xiao, W. Radial growth of *Pinus massoniana* is influenced by temperature, precipitation, and site conditions on the regional scale: A meta-analysis based on tree-ring width index. *Ecol. Indic.* **2021**, *126*, 107659. [CrossRef]
14. Anderson-Teixeira, K.J.; Herrmann, V.; Rollinson, C.R.; Gonzalez, B.; Gonzalez-Akre, E.B.; Pederson, N.; Alexander, M.R.; Allen, C.D.; Alfaro-Sanchez, R.; Awada, T.; et al. Joint effects of climate, tree size, and year on annual tree growth derived from tree-ring records of ten globally distributed forests. *Glob. Chang. Biol.* **2022**, *28*, 245–266. [CrossRef]
15. Vaganov, E.A.; Hughes, M.K.; Shashkin, A.V. *Growth Dynamics of Conifer Tree Rings: Images of Past and Future Environments*; Springer: Berlin/Heidelberg, Germany, 2006; 358p. [CrossRef]
16. Chen, Y.; Cao, Y. Response of tree regeneration and understory plant species diversity to stand density in mature *Pinus tabulaeformis* plantations in the hilly area of the Loess Plateau, China. *Ecol. Eng.* **2014**, *73*, 238–245. [CrossRef]
17. Jiao, L.; Jiang, Y.; Wang, M.; Zhang, W.; Zhang, Y. Age-effect radial growth responses of *Picea schrenkiana* to climate change in the eastern Tianshan Mountains, Northwest China. *Forests* **2017**, *8*, 294. [CrossRef]
18. Perry, D.A. The competition process in forest stands. In *Attributes of Trees as Crop Plants*; Cannell, M.G.R., Jackson, J.E., Eds.; Institute of Terrestial Ecology: Abbots Ripton, UK, 1985; pp. 481–506.
19. Sheriff, D.W. Responses of carbon gain and growth of *Pinus radiata* stands to thinning and fertilizing. *Tree Physiol.* **1996**, *16*, 527–536. [CrossRef] [PubMed]
20. Pshenichnikova, L.S. Pine growth in the experimental planting of different density. *Lesn. Z. For. J.* **2001**, *1*, 25–31. (In Russian with English abstract).
21. Kazmierczak, K. Selected measures of the growth space of a single tree in maturing pine stand. *Sylwan* **2009**, *153*, 298–303. (In Polish with English abstract).
22. Lundqvist, L.; Elfving, B. Influence of biomechanics and growing space on tree growth in young *Pinus sylvestris* stands. *For. Ecol. Manag.* **2010**, *260*, 2143–2147. [CrossRef]
23. Gómez-Aparicio, L.; García-Valdes, R.; Ruiz-Benito, P.; Zavala, M.A. Disentangling the relative importance of climate, size and competition on tree growth in Iberian forests: Implications for management under global change. *Glob. Change Biol.* **2011**, *17*, 2400–2414. [CrossRef]
24. Kuzmichev, V.V.; Pshenichnikova, L.S. Growth of Scots pine stands planted in a range of density in southern taiga of Krasnoyarsk region. *Hvojnye Borealnoj Zony* **2014**, *32*, 83–88. (In Russian with English abstract).
25. Burkhart, H.E. Comparison of maximum size–density relationships based on alternate stand attributes for predicting tree numbers and stand growth. *For. Ecol. Manag.* **2013**, *289*, 404–408. [CrossRef]
26. Martín-Benito, D.; del Río, M.; Heinrich, I.; Helle, G.; Cañellas, I. Response of climate-growth relationships and water use efficiency to thinning in a *Pinus nigra* afforestation. *For. Ecol. Manag.* **2010**, *259*, 967–975. [CrossRef]

27. Wertz, B.; Bembenek, M.; Karaszewski, Z.; Ochal, W.; Skorupski, M.; Strzelinski, P.; Wegiel, A.; Mederski, P.S. Impact of stand density and tree social status on aboveground biomass allocation of Scots pine *Pinus sylvestris* L. *Forests* **2020**, *11*, 765. [CrossRef]
28. Cochran, P.H. *Examples of Mortality and Reduced Annual Increments of White Fir Induced by Drought, Insects, and Disease at Different Stand Densities*; USDA Forest Service, Pacific Northwest Experiment Station: Portland, OR, USA, 1998; 19p. [CrossRef]
29. Latham, P.; Tappeiner, J. Response of old-growth conifers to reduction in stand density in western Oregon forests. *Tree Physiol.* **2002**, *22*, 137–146. [CrossRef] [PubMed]
30. D'Amato, A.W.; Bradford, J.B.; Fraver, S.; Palik, B.J. Effects of thinning on drought vulnerability and climate response in north temperate forest ecosystems. *Ecol. Appl.* **2013**, *23*, 1735–1742. [CrossRef]
31. Bello, J.; Vallet, P.; Perot, T.; Balandier, P.; Seigner, V.; Perret, S.; Couteau, C.; Korboulewsky, N. How do mixing tree species and stand density affect seasonal radial growth during drought events? *For. Ecol. Manag.* **2019**, *432*, 436–445. [CrossRef]
32. Piutti, E.; Cescatti, A. A quantitative analysis of the interactions between climatic response and intraspecific competition in European beech. *Can. J. For. Res.* **1997**, *27*, 277–284. [CrossRef]
33. Zeide, B. How to measure stand density. *Trees* **2005**, *19*, 1–14. [CrossRef]
34. Buzykin, A.I.; Pshenichnikova, L.S. Influence of Scots pine stands of different planting densities on radial growth. *Hvojnye Borealnoj Zony* **2011**, *28*, 188–192. (In Russian with English abstract).
35. Trouve, R.; Bontemps, J.D.; Seynave, I.; Collet, C.; Lebourgeois, F. Stand density, tree social status and water stress influence allocation in height and diameter growth of *Quercus petraea* (Liebl.). *Tree Physiol.* **2015**, *35*, 1035–1046. [CrossRef] [PubMed]
36. Steckel, M.; Moser, W.K.; del Río, M.; Pretzsch, H. Implications of reduced stand density on tree growth and drought susceptibility: A study of three species under varying climate. *Forests* **2020**, *11*, 627. [CrossRef]
37. Weiner, J. Size hierarchies in experimental populations of annual plants. *Ecology* **1985**, *66*, 743–752. [CrossRef]
38. Silvertown, J.; Charlesworth, D. *Introduction to Plant Population Biology*, 4th ed.; Wiley-Blackwell: Hoboken, NJ, USA, 2009; 368p.
39. Chhetri, P.K.; Bista, R.; Shrestha, K.B. How does the stand structure of treeline-forming species shape the treeline ecotone in different regions of the Nepal Himalayas? *J. Mt. Sci.* **2020**, *17*, 2354–2368. [CrossRef]
40. Chapin, F.S.; Matson, P.A.; Mooney, H.A. *Principles of Terrestrial Ecosystem Ecology*; Springer: Berlin/Heidelberg, Germany, 2011; 529p. [CrossRef]
41. Kerhoulas, L.P.; Kolb, T.E.; Koch, G.W. Tree size, stand density, and the source of water used across seasons by ponderosa pine in northern Arizona. *For. Ecol. Manag.* **2013**, *289*, 425–433. [CrossRef]
42. Nilsson, U.; Albrektson, A. Productivity of needles and allocation of growth in young Scots pine trees of different competitive status. *For. Ecol. Manag.* **1993**, *62*, 173–187. [CrossRef]
43. Mäkinen, H. Effect of intertree competition on biomass production of *Pinus sylvestris* (L.) half-sib families. *For. Ecol. Manag.* **1996**, *86*, 105–112. [CrossRef]
44. Ochal, W.; Grabczynski, S.; Orzel, S.; Wertz, B.; Socha, J. Aboveground biomass allocation in Scots pines of different biosocial positions in the stand. *Sylwan* **2013**, *157*, 737–746. (In Polish with English abstract).
45. Zobel, B.J.; Van Buijtenen, J.P. *Wood Variation: Its Causes and Control*; Springer: Berlin/Heidelberg, Germany, 1989; 363p.
46. Martin-Benito, D.; Cherubini, P.; del Rio, M.; Canellas, I. Growth response to climate and drought in *Pinus nigra* Arn. trees of different crown classes. *Trees* **2008**, *22*, 363–373. [CrossRef]
47. Martinez-Vilalta, J.; Lopez, B.C.; Loepfe, L.; Lloret, F. Stand- and tree-level determinants of the drought response of Scots pine radial growth. *Oecologia* **2012**, *168*, 877–888. [CrossRef] [PubMed]
48. Plauborg, K.U. Analysis of radial growth responses to changes in stand density for four tree species. *For. Ecol. Manag.* **2004**, *188*, 65–75. [CrossRef]
49. Leibundgut, H. *Unsere Waldbäume. Eigenschaften und Leben [Our Forest Trees. Properties and Life]*; Verlag Huber: Frauenfeld, Switzerland, 1984; 168p. (In German)
50. Schweingruber, F.H. *Trees and Wood in Dendrochronology*; Springer: Berlin/Heidelberg, Germany, 1993; 402p. [CrossRef]
51. Locosselli, G.M.; Buckeridge, M.S.; Moreira, M.Z.; Ceccantini, G. A multi-proxy dendroecological analysis of two tropical species (*Hymenaea* spp., Leguminosae) growing in a vegetation mosaic. *Trees* **2013**, *27*, 25–36. [CrossRef]
52. Puettmann, K.J. Silvicultural challenges and options in the context of global change: "Simple" fixes and opportunities for new management approaches. *J. For.* **2011**, *109*, 321–331. [CrossRef]
53. Buzykin, A.I.; Pshenichnikova, L.S.; Sobachkin, D.S.; Sobachkin, R.S. Natural thinning of young trees in experimental pine plantings of different density. *Hvojnye Borealnoj Zony* **2008**, *25*, 244–249. (In Russian with English abstract).
54. Pshenichnikova, L.S. Productivity of young pine forests of different density. In *Factors of Forest Productivity: Collected Works*; Nauka: Novosibirsk, Russia, 1978; pp. 36–52. (In Russian)
55. Pobedinsky, A.V. *Study of Reforestation Processes*; Nauka: Moscow, USSR, 1966; 60p. (In Russian)
56. Moiseev, V.S. *Inventory of Saplings*; Forest Engineering Academy: Leningrad, USSR, 1971; 344p. (In Russian)
57. Buzykin, A.I. On the productivity of forests and the levels of its regulation. In *Problems of Forest Science in Siberia*; Nauka: Moscow, USSR, 1977; pp. 7–24. (In Russian)
58. Buzykin, A.I.; Pshenichnikova, L.S. Influence of density on the morphostructure and productivity of pine plantations. *Lesovedenie [For. Sci.]* **1999**, *3*, 38–43. (In Russian)
59. Peng, P.H.; Kuo, C.H.; Wei, C.H.; Hsieh, Y.T.; Chen, J.C. The relationship between breast height form factor and form quotient of *Liquidambar formosana* in the Eastern part of Taiwan. *Forests* **2022**, *13*, 1111. [CrossRef]

60. Tishin, D.V. *Assessment of Tree Stand Productivity*; Educational-Methodical Manual; Kazan University: Kazan, Russia, 2011; 31p. (In Russian)
61. Coomes, D.A.; Allen, R.B. Mortality and tree-size distributions in natural mixed-age forests. *J. Ecol.* **2007**, *95*, 27–40. [CrossRef]
62. Chen, H.Y.H.; Fu, S.; Monserud, R.A.; Gillies, I. Relative size and stand age determine *Pinus banksiana* mortality. *For. Ecol. Manag.* **2008**, *255*, 3980–3984. [CrossRef]
63. Cook, E.R.; Kairiukstis, L.A. (Eds.) *Methods of Dendrochronology. Application in Environmental Sciences*; Kluwer Academic Publishers: Dordrecht, Germany, 1990; 394p. [CrossRef]
64. Rinn, F. *TSAP-Win: Time Series Analysis and Presentation for Dendrochronology and Related Applications: User Reference*; RINNTECH: Heidelberg, Germany, 2003; 91p.
65. Holmes, R.L. Computer-assisted quality control in tree-ring dating and measurement. *Tree-Ring Bull.* **1983**, *43*, 69–78.
66. Cook, E.R.; Krusic, P.J. *Program ARSTAN: A Tree-Ring Standardization Program Based on Detrending and Autoregressive Time Series Modeling, with Interactive Graphics*; Lamont-Doherty Earth Observatory, Columbia University: Palisades, NY, USA, 2005; 14p.
67. Meko, D.M.; Baisan, C.H. Pilot study of latewood-width of conifers as an indicator of variability of summer rainfall in the North American monsoon region. *Int. J. Climatol.* **2001**, *21*, 697–708. [CrossRef]
68. Wigley, T.M.L.; Briffa, K.R.; Jones, P.D. On the average value of correlated time series, with applications in dendroclimatology and hydrometeorology. *J. Clim. Appl. Met.* **1984**, *23*, 201–213. [CrossRef]
69. Ford, E.D. Branching, crown structure and the control of timber production. In *Attributes of Trees as Crop Plants*; Cannell, M.G.R., Jackson, J.E., Eds.; Institute of Terrestrial Ecology: Abbotts Ripton, UK, 1985; pp. 228–252.
70. Oliver, C.D.; Larson, B.C. *Forest Stand Dynamics*; McGraw-Hill, Inc.: New York, NY, USA, 1996; 467p.
71. Xue, L.; Chen, F.X.; Feng, H.F. Time-trajectory of mean component weight and density in self-thinning *Pinus densiflora* stands. *Eur. J. For. Res.* **2010**, *129*, 1027–1035. [CrossRef]
72. Sola, G.; Marchelli, P.; Gallo, L.; Chauchard, L.; El Mujtar, V. Stand development stages and recruitment patterns influence fine-scale spatial genetic structure in two Patagonian *Nothofagus* species. *Ann. For. Sci.* **2022**, *79*, 21. [CrossRef]
73. Lane, B.; Prusinkiewicz, P. Generating spatial distributions for multilevel models of plant communities. In Proceedings of the Graphics Interface 2002, Calgary, AL, Canada, 27–29 May 2002; pp. 69–80.
74. Morris, E. How does fertility of the substrate affect intraspecific competition? Evidence and synthesis from self-thinning. *Ecol. Res.* **2003**, *18*, 287–305. [CrossRef]
75. Kamara, M.; Kamruzzaman, M. Self-thinning process, dynamics of aboveground biomass, and stand structure in overcrowded mangrove *Kandelia obovata* stand. *Reg. Studies Mar. Sci.* **2020**, *38*, 101375. [CrossRef]
76. Waring, R.H. Characteristics of trees predisposed to die: Stress causes distinctive changes in photosynthate allocation. *BioScience* **1987**, *37*, 569–574. [CrossRef]
77. Baldwin, V.C., Jr.; Peterson, K.D.; Clark, A., III; Ferguson, R.B.; Strub, M.R.; Bower, D.R. The effects of spacing and thinning on stand and tree characteristics of 38-year-old loblolly pine. *For. Ecol. Manag.* **2000**, *137*, 91–102. [CrossRef]
78. Zhang, S.Y.; Lei, Y.C.; Bowling, C. Quantifying stem quality characteristics in relation to initial spacing and modeling their relationship with tree characteristics in black spruce (*Picea mariana*). *North. J. Appl. For.* **2005**, *22*, 85–93. [CrossRef]
79. del Río, M.; Bravo-Oviedo, A.; Ruiz-Peinado, R.; Condes, S. Tree allometry variation in response to intra-and inter-specific competitions. *Trees* **2019**, *33*, 121–138. [CrossRef]
80. Woodruff, D.R.; Bond, B.J.; Ritchie, G.A.; Scott, W. Effects of stand density on the growth of young Douglas-fir trees. *Can. J. For. Res.* **2002**, *32*, 420–427. [CrossRef]
81. Tepley, A.J.; Swanson, F.J.; Spies, T.A. Post-fire tree establishment and early cohort development in conifer forests of the western Cascades of Oregon, USA. *Ecosphere* **2014**, *5*, 1–23. [CrossRef]
82. Abbas, A.M.; Al-Kahtani, M.; Novak, S.J.; Soliman, W.S. Abundance, distribution, and growth characteristics of three keystone *Vachellia* trees in Gebel Elba National Park, south-eastern Egypt. *Sci. Rep.* **2021**, *11*, 1284. [CrossRef]
83. Reineke, L.H. Perfecting a stand-density index for even-aged forests. *J. Agric. Res.* **1933**, *46*, 627–638.
84. Yoda, K.; Kira, T.; Ogawa, H.; Hozumi, K. Self-thinning in overcrowded pure stands under cultivated and natural conditions (Intraspecific competition among higher plants XI). *J. Inst. Polytech. Osaka City Univ. Ser. D* **1963**, *14*, 107–129.
85. Lonsdale, W.M. The self-thinning rule: Dead or alive? *Ecology* **1990**, *71*, 1373–1388. [CrossRef]
86. Pretzsch, H.; Biber, P. A re-evaluation of Reineke's rule and stand density index. *For. Sci.* **2005**, *51*, 304–320. [CrossRef]
87. Zhang, W.P.; Zhao, L.; Larjavaara, M.; Morris, E.C.; Sterck, F.J.; Wang, G.X. Height-diameter allometric relationships for seedlings and trees across China. *Acta Oecol.* **2020**, *108*, 103621. [CrossRef]
88. Zhang, W.P.; Jia, X.; Morris, E.C.; Bai, Y.Y.; Wang, G.X. Stem, branch and leaf biomass-density relationships in forest communities. *Ecol. Res.* **2012**, *27*, 819–825. [CrossRef]
89. Zhang, J.; Oliver, W.W.; Ritchie, M.W. Effect of stand densities on stand dynamics in white fir (*Abies concolor*) forests in northeast California, USA. *For. Ecol. Manag.* **2007**, *244*, 50–59. [CrossRef]
90. Isaac-Renton, M.; Stoehr, M.; Statland, C.B.; Woods, J. Tree breeding and silviculture: Douglas-fir volume gains with minimal wood quality loss under variable planting densities. *For. Ecol. Manag.* **2020**, *465*, 118094. [CrossRef]
91. Skovsgaard, J.A.; Vanclay, J.K. Forest site productivity: A review of the evolution of dendrometric concepts for even-aged stands. *For. Int. J. For. Res.* **2008**, *81*, 13–31. [CrossRef]
92. Burkhart, H.E.; Tome, M. *Modeling Forest Trees and Stands*; Springer: Dordrecht, Germany, 2012; 458p. [CrossRef]

93. Kholdaenko, Y.A.; Belokopytova, L.V.; Zhirnova, D.F.; Upadhyay, K.K.; Tripathi, S.K.; Koshurnikova, N.N.; Sobachkin, R.S.; Babushkina, E.A.; Vaganov, E.A. Stand density effects on tree growth and climatic response in *Picea obovata* Ledeb plantations. *For. Ecol. Manag.* **2022**, *519*, 120349. [CrossRef]
94. Mäkinen, H.; Isomäki, A. Thinning intensity and long-term changes in increment and stem form of Norway spruce trees. *For. Ecol. Manag.* **2004**, *201*, 295–309. [CrossRef]
95. Weiskittel, A.R.; Maguire, D.A.; Monserud, R.A. Modeling crown structural responses to competing vegetation control, thinning, fertilization, and Swiss needle cast in coastal Douglas-fir of the Pacific Northwest, USA. *For. Ecol. Manag.* **2007**, *245*, 96–109. [CrossRef]
96. Kofman, G.B. *Growth and Form of Trees*; Nauka: Novosibirsk, Russia, 1986; 211p. (In Russian)
97. Gartner, B.L. Patterns of xylem variation within a tree and their hydraulic and mechanical consequences. In *Plant Stems: Physiology and Functional Morphology*; Academic Press: Cambridge, MA, USA, 1995; pp. 125–149. [CrossRef]
98. Sarkhad, M.; Ishiguri, F.; Nezu, I.; Aiso, H.; Ngadianto, A.; Tumenjargal, B.; Baasan, B.; Chultem, G.; Ohshima, J.; Yokota, S. Preliminary evaluation of anatomical characteristics of four common Mongolian softwoods. *For. Sci. Technol.* **2022**, *18*, 87–97. [CrossRef]
99. Domec, J.C.; Gartner, B.L. Age-and position-related changes in hydraulic versus mechanical dysfunction of xylem: Inferring the design criteria for Douglas-fir wood structure. *Tree Physiol.* **2002**, *22*, 91–104. [CrossRef] [PubMed]
100. Liu, Y.; Shi, J.; Shishov, V.; Vaganov, E.; Yang, Y.; Cai, Q.; Sun, J.; Wang, L.; Djanseitov, I. Reconstruction of May–July precipitation in the north Helan Mountain, Inner Mongolia since AD 1726 from tree-ring late-wood widths. *Chin. Sci. Bull.* **2004**, *49*, 405–409. [CrossRef]
101. Lebourgeois, F. Climatic signals in earlywood, latewood and total ring width of Corsican pine from western France. *Ann. For. Sci.* **2000**, *57*, 155–164. [CrossRef]
102. Zhang, S.Y. Variations and correlations of various ring width and ring density features in European oak: Implications in dendroclimatology. *Wood Sci. Technol.* **1997**, *31*, 63–72. [CrossRef]
103. Savva, Y.V.; Schweingruber, F.H.; Vaganov, E.A.; Milyutin, L.I. Influence of climate changes on tree-ring characteristics of Scots pine provenances in southern Siberia (forest-steppe). *IAWA J.* **2003**, *24*, 371–383. [CrossRef]
104. Bauwe, A.; Koch, M.; Kallweit, R.; Konopatzky, A.; Strohbach, B.; Lennartz, B. Tree-ring growth response of Scots pine (*Pinus sylvestris* L.) to climate and soil water availability in the lowlands of North-Eastern Germany. *Balt. For.* **2013**, *19*, 212–225.
105. Voltas, J.; Aguilera, M.; Gutierrez, E.; Shestakova, T.A. Shared drought responses among conifer species in the middle Siberian taiga are uncoupled from their contrasting water-use efficiency trajectories. *Sci. Total Environ.* **2020**, *720*, 137590. [CrossRef] [PubMed]
106. Arzac, A.; Tabakova, M.A.; Khotcinskaia, K.; Koteneva, A.; Kirdyanov, A.V.; Olano, J.M. Linking tree growth and intra-annual density fluctuations to climate in suppressed and dominant *Pinus sylvestris* L. trees in the forest-steppe of Southern Siberia. *Dendrochronologia* **2021**, *67*, 125842. [CrossRef]
107. Kolar, T.; Kusbach, A.; Cermak, P.; Sterba, T.; Batkhuu, E.; Rybnicek, M. Climate and wildfire effects on radial growth of *Pinus sylvestris* in the Khan Khentii Mountains, north-central Mongolia. *J. Arid Environ.* **2020**, *182*, 104223. [CrossRef]
108. Zhirnova, D.F.; Belokopytova, L.V.; Meko, D.M.; Babushkina, E.A.; Vaganov, E.A. Climate change and tree growth in the Khakass-Minusinsk Depression (South Siberia) impacted by large water reservoirs. *Sci. Rep.* **2021**, *11*, 14266. [CrossRef]
109. Kargas, G.; Kerkides, P.; Poulovassilis, A. Infiltration of rain water in semi-arid areas under three land surface treatments. *Soil Tillage Res.* **2012**, *120*, 15–24. [CrossRef]
110. Fritts, H.C. Growth-rings of trees: Their correlation with climate: Patterns of ring widths in trees in semiarid sites depend on climate-controlled physiological factors. *Science* **1966**, *154*, 973–979. [CrossRef]
111. Allen, C.D.; Macalady, A.K.; Chenchouni, H.; Bachelet, D.; McDowell, N.; Vennetier, M.; Kitzberger, T.; Rigling, A.; Breshears, D.D.; Hogg, E.H.; et al. A global overview of drought and heat-induced tree mortality reveals emerging climate change risks for forests. *For. Ecol. Manag.* **2010**, *259*, 660–684. [CrossRef]
112. Liu, H.; Park Williams, A.; Allen, C.D.; Guo, D.; Wu, X.; Anenkhonov, O.A.; Liang, E.; Sandanov, D.V.; Yin, Y.; Qi, Z.; et al. Rapid warming accelerates tree growth decline in semi-arid forests of Inner Asia. *Glob. Change Biol.* **2013**, *19*, 2500–2510. [CrossRef] [PubMed]
113. Long, A.L. Drought. In *Disturbance and Sustainability in Forests of the Western United States*; Barrett, T.M., Robertson, G.C., Eds.; USDA Forest Service, Pacific Northwest Research Station: Portland, OR, USA, 2021; pp. 67–79.
114. Yasmeen, S.; Wang, X.; Zhao, H.; Zhu, L.; Yuan, D.; Li, Z.; Zhang, Y.; Ahmad, S.; Han, S. Contrasting climate-growth relationship between *Larix gmelinii* and *Pinus sylvestris* var. *mongolica* along a latitudinal gradient in Daxing'an Mountains, China. *Dendrochronologia* **2019**, *58*, 125645. [CrossRef]
115. Agafonov, L.I.; Gurskaya, M.A.; Kukarskih, V.V.; Bubnov, M.O.; Devi, N.M.; Galimova, A.A. Insular pine forests of the Southern Urals and ribbon pine forests of the Altai as objects of dendroclimatic research. *Russ. J. Ecol.* **2021**, *52*, 349–357. [CrossRef]
116. Arzac, A.; Tychkov, I.; Rubtsov, A.; Tabakova, M.A.; Brezhnev, R.; Koshurnikova, N.; Knorre, A.; Büntgen, U. Phenological shifts compensate warming-induced drought stress in southern Siberian Scots pines. *Eur. J. For. Res.* **2021**, *140*, 1487–1498. [CrossRef]
117. Jayawickrama, K.J.; McKeand, S.E.; Jett, J.B.; Wheeler, E.A. Date of earlywood-latewood transition in provenances and families of loblolly pine, and its relationship to growth phenology and juvenile wood specific gravity. *Can. J. For. Res.* **1997**, *27*, 1245–1253. [CrossRef]

118. Yang, B.; He, M.; Shishov, V.; Tychkov, I.; Vaganov, E.; Rossi, S.; Ljungqvist, F.C.; Bräuning, A.; Grießinger, J. New perspective on spring vegetation phenology and global climate change based on Tibetan Plateau tree-ring data. *PNAS* **2017**, *114*, 6966–6971. [CrossRef] [PubMed]
119. Fonti, M.V.; Babushkina, E.A.; Zhirnova, D.F.; Vaganov, E.A. Xylogenesis of Scots pine in an uneven-aged stand of the Minusinsk Depression (Southern Siberia). *J. Sib. Fed. Univ. Biol.* **2020**, *13*, 197–207. [CrossRef]
120. Gricar, J.; Cufar, K.; Eler, K.; Gryc, V.; Vavrcík, H.; De Luis, M.; Prislan, P. Transition dates from earlywood to latewood and early phloem to late phloem in Norway spruce. *Forests* **2021**, *12*, 331. [CrossRef]
121. Babushkina, E.A.; Sitnikov, G.A.; Upadhyay, K.K.; Zhirnova, D.F.; Zelenov, G.K.; Vaganov, E.A.; Belokopytova, L.V. Seasonal growth of pine tree rings: Comparison of direct observations and simulation. *Forests* **2022**, *13*, 1978. [CrossRef]
122. Deleuze, C.; Herve, J.C.; Colin, F.; Ribeyrolles, L. Modelling crown shape of *Picea abies*: Spacing effects. *Can. J. For. Res.* **1996**, *26*, 1957–1966. [CrossRef]
123. Misson, L.; Nicault, A.; Guiot, J. Effects of different thinning intensities on drought response in Norway spruce (*Picea abies* (L.) Karst.). *For. Ecol. Manag.* **2003**, *183*, 47–60. [CrossRef]
124. Sohn, J.A.; Gebhardt, T.; Ammer, C.; Bauhus, J.; Haberle, K.H.; Matyssek, R.; Grams, T.E.E. Mitigation of drought by thinning: Short-term and long-term effects on growth and physiological performance of Norway spruce (*Picea abies*). *For. Ecol. Manag.* **2013**, *308*, 188–197. [CrossRef]
125. Olivar, J.; Bogino, S.; Rathgeber, C.; Bonnesoeur, V.; Bravo, F. Thinning has a positive effect on growth dynamics and growth–climate relationships in Aleppo pine (*Pinus halepensis*) trees of different crown classes. *Ann. For. Sci.* **2014**, *71*, 395–404. [CrossRef]
126. Peltola, H.; Kilpeläinen, A.; Sauvala, K.; Räisänen, T.; Ikonen, V.P. Effects of early thinning regime and tree status on the radial growth and wood density of Scots pine. *Silva Fenn.* **2007**, *41*, 489–505. [CrossRef]
127. Perevalova, E.A. Drought effect on the cambial activity of Scots pine in stands of different planting density. In *Forestry Universities in the Implementation of the Concept of the Engineering Education Revival: Socio-Economic and Environmental Problems of the Forestry Complex*; Ural State Forest Engineering University: Ekaterinburg, Russia, 2019; pp. 219–221, (In Russian with English abstract).
128. Coomes, D.A.; Allen, R.B. Effects of size, competition and altitude on tree growth. *J. Ecol.* **2007**, *95*, 1084–1097. [CrossRef]
129. Aussenac, G. Interactions between forest stands and microclimate: Ecophysiological aspects and consequences for silviculture. *Ann. For. Sci.* **2000**, *57*, 287–301. [CrossRef]
130. Liu, J.; Chen, J.M.; Cihlar, J. Mapping evapotranspiration based on remote sensing: An application to Canada's landmass. *Water Resour. Res.* **2003**, *39*, 1189. [CrossRef]
131. Moreno-Gutierrez, C.; Battipaglia, G.; Cherubini, P.; Saurer, M.; Nicolas, E.; Contreras, S.; Querejeta, J.I. Stand structure modulates the long-term vulnerability of *Pinus halepensis* to climatic drought in a semiarid Mediterranean ecosystem. *Plant Cell Environ.* **2012**, *35*, 1026–1039. [CrossRef] [PubMed]
132. Van Den Brakel, J.A.; Visser, H. The influence of environmental conditions on tree-ring series of Norway spruce for different canopy and vitality classes. *For. Sci.* **1996**, *42*, 206–219. [CrossRef]
133. Merian, P.; Lebourgeois, F. Size-mediated climate–growth relationships in temperate forests: A multi-species analysis. *For. Ecol. Manag.* **2011**, *261*, 1382–1391. [CrossRef]
134. Campelo, F.; Vieira, J.; Nabais, C. Tree-ring growth and intra-annual density fluctuations of *Pinus pinaster* responses to climate: Does size matter? *Trees* **2013**, *27*, 763–772. [CrossRef]
135. Ford, K.R.; Breckheimer, I.K.; Franklin, J.F.; Freund, J.A.; Kroiss, S.J.; Larson, A.J.; Theobald, E.J.; HilleRisLambers, J. Competition alters tree growth responses to climate at individual and stand scales. *Can. J. For. Res.* **2017**, *47*, 53–62. [CrossRef]
136. Campbell, E.M.; Magnussen, S.; Antos, J.A.; Parish, R. Size-, species-, and site-specific tree growth responses to climate variability in old-growth subalpine forests. *Ecosphere* **2021**, *12*, e03529. [CrossRef]
137. Babushkina, E.A.; Vaganov, E.A.; Belokopytova, L.V.; Shishov, V.V.; Grachev, A.M. Competitive strength effect in the climate response of Scots pine radial growth in south-central Siberia forest-steppe. *Tree-Ring Res.* **2015**, *71*, 106–117. [CrossRef]
138. Matyshak, G.V.; Goncharova, O.Y.; Bogatyrev, L.G.; Riazantseva, M.I. Influence of plant cover on hydrothermal conditions in soils of large lysimeters of the MSU soil station: Results of a 60-year experiment. *Mosc. Univ. Soil Sci. Bull.* **2021**, *76*, 134–139. [CrossRef]
139. Duyar, A. The effects of forest canopy cover and altitude on snow accumulation and melting in the upper watersheds. *Fresenius Environ. Bull.* **2018**, *27*, 9642–9649.
140. Chen, L.; Chen, Z.; Jia, G.; Zhou, J.; Zhao, J.; Zhang, Z. Influences of forest cover on soil freeze-thaw dynamics and greenhouse gas emissions through the regulation of snow regimes: A comparison study of the farmland and forest plantation. *Sci. Total Environ.* **2020**, *726*, 138403. [CrossRef] [PubMed]
141. Fu, Q.; Hou, R.; Li, T.; Yan, P.; Ma, Z. The critical depth of freeze-thaw soil under different types of snow cover. *Water* **2017**, *9*, 370. [CrossRef]
142. Li, Z.; Qi, Z.; Smith, W.; Pattey, E.; Qian, B. Long-term simulation of snow cover and its potential impacts on seasonal frost dynamics in croplands across Southern Canada. *Water Resour. Res.* **2022**, *58*, e2021WR031674. [CrossRef]

forests

MDPI

Article

Impact of Hydrocarbon Emissions from Oil and Gas Deposits on $\delta^{13}C$ Variability in Pine Tree Rings from the Tatarstan Republic

Olga V. Churakova (Sidorova) [1,2,3,*], Georgii Batalin [3], Bulat Gareev [3], Gazinur Mingazov [3], Andrey Terekhin [3], Denis Tishin [4], Dilyara Kuzina [3] and Danis Nurgaliev [3]

1 Laboratory of Biogeochemistry Ecosystems, Institute of Ecology and Geography, Siberian Federal University, Svobodny 79, Krasnoyarsk 660041, Russia
2 Department of Forest Dynamics, Swiss Federal Institute for Forest, Snow and Landscape Research WSL, Zürcherstrasse 111, CH-8903 Birmensdorf, Switzerland
3 Institute of Geology and Petroleum Technologies, Kazan Federal University, Kremlyovskaya Str. 18, Kazan 420008, Russia; batalinga@yandex.ru (G.B.); bulat@gareev.net (B.G.); gazikovsky@gmail.com (G.M.); andrey.tereokhin@kpfu.ru (A.T.); di.kuzina@gmail.com (D.K.); dnk@kpfu.ru (D.N.)
4 Institute of Ecology and Environmental Management, Kazan Federal University, Kremlyovskaya Str. 18, Kazan 420008, Russia; denis.tishin@kpfu.ru
* Correspondence: ochurakova@sfu-kras.ru

Abstract: Human-caused anthropogenic greenhouse emissions impact the climate globally. In this pilot study, we aim to reveal the influence of hydrocarbon emissions on pine forests by applying a stable carbon isotope analysis in pine tree rings ($\delta^{13}C_{ptrw}$). Our study was conducted in an industrial giant oil field reservoir (UVRT) and natural reserve (Raifa) sites, the Tatarstan Republic, Russia. Our results show a decreasing $\delta^{13}C_{ptrw}$ at the UVRT site in 1943, when oil extraction started, and in 1970, when it reached maximum production. We found that the $\delta^{13}C_{ptrw}$ from UVRT indicates developing unfavourable drier conditions and a suppressed tree growth caused by both human-induced oil and deposit infrastructures and natural processes compared to the undisturbed Raifa site. A 5-year running correlation analysis showed a significant difference between the sites in 1965 over the period of 1930 to 2021. The $\delta^{13}C_{ptrw}$ values from Raifa are more negative compared to UVRT, which can be explained by a higher forest sensitivity to human-induced impacts. From an eco-physiological point of view, the decreasing of intercellular (c_i)-to-ambient (c_a) CO_2 concentration ratios at the leaf level and the increasing of intrinsic water-use efficiency (iWUE) along with a decreasing of tree-ring widths at the UVRT site (1970–2021) indicate the development of drought conditions.

Keywords: CO_2; CH_4; $\delta^{13}C$ in tree-ring wood; soil; iWUE; climate

1. Introduction

The accelerated development of energy resources around the world has significantly increased forest changes associated with oil and gas activities, leading to both carbon dioxide and methane emissions. The impacts of these anthropogenic indirect greenhouse gases (GHGs) play a significant role on forest ecosystems at the regional and global scales [1]. GHGs attributed up to 65% to human activities [2], including hydrocarbon emissions from the oil and gas infrastructures [3], causing a global average air temperature increase of 1.4 °C [4]. Moreover, the extraction and refining of oil produces about 48% of hydrocarbons and 44% of carbon monoxide [5], which can significantly impact the environment. In the case of accidents or during intensive oil and gas deposit developments, the toxic contamination of soil can lead to the reduction of air exchange in the soil, hindering the water flow into the soil compared to clean natural soils. Water shortage in the soil can lead to soil fertility and the reduction of microbial activity impacting the forest ecosystem negatively. Additionally, local oil and gas production delivers up to 90% of hydrocarbons, which contribute to global GHG emissions [6]. Due to the intensive development of oil and gas depositions, oil and oil products refine the problem of their emissions locally and globally, impacting the environment and humans' health.

Recent climate change and drought-induced tree mortality affects forest ecosystems worldwide [7]. It is well known that forest soils and permafrost are an important sink for atmospheric methane emissions, where a GHG contributes roughly 20% to global warming [8,9]. Soil microorganisms remove about 30 million tons of GHGs from the atmosphere annually, which is 6–10% of its annual flow [10]. Living and dead trees have the potential to be CH_4 sources or sinks or both [9]. Different tree species have different effects on the activity of CH_4 oxidation in the soil—the highest is noted under *P. cembra* L. (27.68 mg kg^{-1}) in Europe, in *Larix sibirica* Led. (7.98 mg kg^{-1}) and *P. sylvestris* L. (4.96 mg kg^{-1}), and the lowest in spruce (4.62 mg kg^{-1}) forests [10].

The carbon atoms fixed in tree-ring widths originate from the atmospheric carbon dioxide to which the tree's canopy is exposed [9]. During photosynthesis, several fractionation steps take place: first, when CO_2 from the atmosphere (c_a) diffuses into the leaf (needle) intercellular spaces (c_i) and second, during CO_2 fixation by the enzyme Rubisco [11]. The opening and closure of the stomata (g_s) determines the water control. Under warm and dry conditions, trees respond to limited water resources by reducing stomatal conductance (g_s), resulting in a decreasing CO_2 uptake and biomass production, in a reduced intercellular CO_2 concentration (c_i). Changes in the assimilation rate of the needles will therefore influence the intercellular CO_2 concentration (c_i) through changes in the rate at which the CO_2 is utilized to form sugars, and an increase or decrease in stomatal conductance will affect the rate at which this internal CO_2 (c_i) can be replenished. Trees discriminate more strongly against ^{13}C under conditions of high (c_i), when stomata are relatively wide open, or when photosynthesis is low [11]. Under an increasing CO_2 concentration, the water vapor exchange between the needles and the ambient air (c_a) is reduced and stomatal conductance decreases [12–14]. The carbon isotopic ratio ($^{13}C/^{12}C$ or $\delta^{13}C$) in tree rings reflects signals of water availability and air humidity as a result of the climate impact on photosynthesis. Atmospheric CO_2 is well-mixed, but the sub-canopy air space can become depleted in ^{13}C due to inputs from soil and plant interactions. Tree growth is influenced by many factors such as solar irradiance, ambient air temperature, precipitation, air humidity, soil and ground water and nutrient availability. The heavier stable carbon isotope (^{13}C) in tree rings is modulated by environmental parameters like temperature and moisture regime changes due to fractionation processes during CO_2 uptake as well as those related to land management and disturbances like insects [7,11,12,14–16]. Photosynthetic limitations of intrinsic water use efficiency (iWUE) can be an indicator of oil refinery in the case of an increase, which corresponds with NO_x pollution [17] and moisture changes, reducing climate sensitivity [18]. If trees uptake a significant amount of hydrocarbon emitted from gas and oil deposits, then we suggest that the tree-ring $\delta^{13}C$ should also receive more negative values.

Data on hydrocarbon (HC) production are available for the study area; however, it is not possible to estimate how much hydrocarbons are released from gas and oil deposits into the air, surface water, groundwater, and soil. The HC emissions most likely occurred from the subsurface prior to oil and gas deposit developments. This had a relatively constant flow, for hundreds of thousands and even millions of years, which could change only in the case of events like earthquakes [19]. However, after the beginning of field developments, the equilibrium established during a long geological time was broken which could have led to an increase in the flow of hydrocarbons both from natural fractures and faults and due to losses during production and transportation. In the first stage of extraction, oil is extracted through natural processes. As a result, it is replaced by water. If the pressure in the reservoir does not allow oil to come to the surface, then special pumps are used to extract it. Later, secondary methods are used to extract oil. This is carried out by introducing liquids and gases into oil-bearing formations to provide the necessary amount of energy to extract oil from the earth's bowels [20]. At this stage of production, the natural state of the reservoirs is disturbed and a significant emission of hydrocarbons into the air, water and soil is possible. When using the hydraulic fracturing method, HC emissions can be very large, comparable to other global sources of greenhouse gases [21]. A large increase in U.S. methane emissions over the past decade was inferred from satellite data and surface observations [22,23].

In the Romashkinskoye UVRT field, the hydraulic fracturing method was not used, but there could also have been noticeable HC emissions during production. Methods for assessing hidden HC emissions during oil field developments were not developed. Usually, only accidents in which a large volume of HC is released into the soil or water are evaluated.

In this study, we hypothesized that during oil and gas deposit developments and active oil production, additional emissions of hydrocarbons can be emitted into the atmosphere, which can be recorded in trees growing close to the oil and gas deposit site.

To test this hypothesis, we developed new tree-ring $\delta^{13}C$ in pine wood chronologies from the oil and gas deposits (UVRTs) and for the natural reserve Raifa (Raifa) site to reveal the impact of the Romashkinskoye oil and gas deposit site on pine forests.

2. Materials and Methods

2.1. Study Sites

The study sites are located in the Republic of Tatarstan, within 54–55° N and 49–52° E, the Russian Federation. We investigated two sites: UVRT—the oil and gas deposit infrastructure (UVRT) site near the Romashkinskoye deposit industrial site—and Raifa—a natural reserve site, which is located 250 km away from UVRT (Figure 1a). The territory is located on the southern border of the taiga natural zone in the Volga–Vyatka high-plain complex of dark coniferous broad-leaved forests [22].

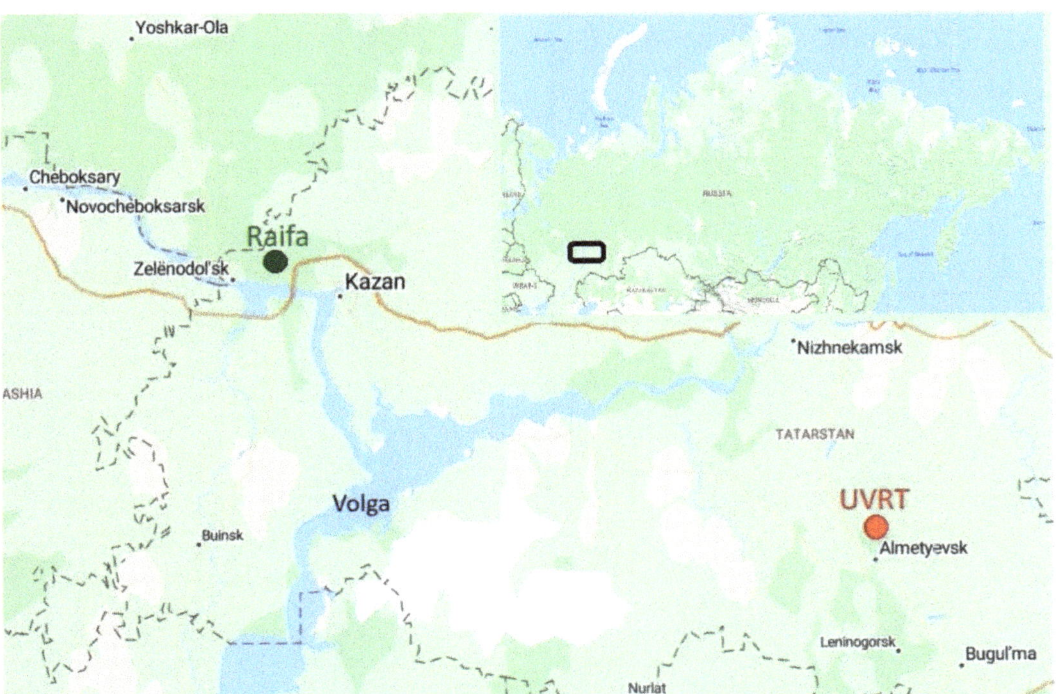

(a)

Figure 1. *Cont.*

(b)

(c)

Figure 1. Map location with the study sites (black rectangle) (**a**) and photos from (**b**) the oil and gas deposit infrastructure UVRT (red circle) and (**c**) the natural reserve Raifa (green circle) sites. Average numbers of the δ^{13}C atmospheric CO_2 [24] and $\delta^{13}C_{ptrw}$ are presented for the common period from 1930 to 2021. The δ^{13}C CO_2 and δ^{13}C CH_4 are presented for the available year of 2023 with a schematic interaction of water evaporation and CH_4 and CO_2 soil–atmosphere–plant exchanges.

The Romashkinskoye UVRT oil and gas deposit site is located 54° N, 52° E, which is 15 km away from the Almetyevsk city in the northwestern part of the Bugulma plateau, confined to the South Tatar (Almetyevsk) arch at 200–210 m asl (Figure 1a). Geological oil reserves are 5 billion tons, while proved and recoverable reserves are3 billion tons. Exploratory drilling started in 1943 [19,20], while in 1948, a thick Devonian formation was discovered. Intensive developments and exploitations started in 1953. At present, the Romashkinskoye field is one of the largest oil fields in the world. More than 400 deposits have been identified. The main ones, in terms of size are the oil deposits of the Devonian terrigenous complex in the Kynovsko-Pashy sediments, which is about 70% of the explored reserves. In terms of reserve replenishments, commercial oil reserves are growing, although cumulative production is also growing and already exceeds previously approved reserves. The depletion of sediments of the Pashinsky horizon of the Romashkinskoye field is regularly estimated to be more than 85%. It was announced that residual oil reserves are concentrated in undeveloped low-permeability reservoirs. Then, petrophysical works, hydrodynamic studies and seismic exploration are carried out, and the reserves are replenished. The composition of the Kynovsko-Pashy sediments indicates that anomalous zones of light oil from the Zhivet complex flowed into the wells. The Zhivetsky complex is represented by sandstones, siltstones and clays with thin interlayers of marls and limestones and is divided into the Afoninsky and Starooskolsky horizons [19,21].

The relief of the study site is characterised as a flat with small hills and a dry forest dominated by pine (*P. sylvestris* L.) (Figure 1b). The herbage is dominated by ground reedgrass, sedges and various steppe plant species. The soils are dry sod-podzol soils on sandy and sandy loam sediments.

The Raifa natural reserve site 55° N, 49° E (Figure 1a) is represented by pine trees as the main dominant species (*P. sylvestris* L.) (Figure 1c), with rare spruces and birches (green-moss pine forest). Pine trees (*P. sylvestris* L.) can reach up to 150 years. The forest stand encounters single species of *Betula pendula* Roth., *P. fennica* (Regel) Kom. The understory is sparse, represented by *Euonymus verrucosus* Scop., *Tillia cordata* Mill. Dominants of the herbaceous layer are *Calamagrostis epigejos* (L.) Roth and *Convallaria majalis* L. The soil is soddy-podzolic on sandy and sandy loam deposits of the third floodplain terrace of the Volga River [22].

2.2. Local Weather Station Observations

The climate of the region is continental with cold winters and warm summers. According to the local Kazan weather station (55° N, 49° E), the average annual temperature is 5.2 °C. The January average temperature reached −10.4 °C, and the July average temperature was 20.8 °C for the period from 1970 to 2021. The average snow depth in the winter reached up to 40 cm. The sum of precipitation was 584 mm per year. Maximum precipitation fell in July with up to 70 mm and the minimum was recorded in March with up to 33 mm, according to the average monthly data from the Russian Research Institute of Hydrometeorological Information—World Data Center [23]—for the period from 1970 to 2021.

2.3. Sampling

Tree-core sampling from four dominant pine trees (*P. sylvestris* L.) was carried out in July 2022 at each study location (UVRT and Raifa), respectively. Pine tree cores were collected at a height of 1.3 m at the breast height on the south and north sides of the tree from visually healthy, without a double crown and not-damaged, living *P. sylvestris* L. trees using the Pressler increment borer (Figure 1b,c) according to the standard methodology [25]. All tree cores were packaged in a metal foil and placed in a hard tube for transportation to a stable isotope laboratory.

Soil samples were collected from February–March 2023 using a soil drill borer. The depth of sampling ranged from 1.0 to 1.5 m [25]. Soil samples were placed in a glass container immediately after extraction from the pits, filled with a salt solution and hermetically sealed.

2.4. Tree-Ring Width and Stable Isotope Analyses

Tree-ring cores were prepared for both tree-ring widths (TRWs) [25] and stable carbon isotope analyses [15]. Wood cores were marked with a needle for calendar dating.

The TRWs were measured for the UVRT site for the period from 1925 to 2022 and for Raifa for the period from 1887 to 2021 using Lintab-6 with the TSAPWin software package (Version 6.02, USA) [26]. The quality of the cross-dating chronologies was accessed by using the Cofecha software (Version 4.70a, Germany) [27,28]. Only the common period (1930–2021) for both study sites was used. After that, each annual tree ring was split using a sharp BA-170P NT blade under a Leica M50 microscope (Leica, Berlin, Germany).

For the stable carbon isotope analysis, each annual tree-ring sample for each year separately was packed ca. 200 µg into a tin capsule. All measurements for each sample were repeated three times. International Atomic Energy Agency (IAEA) standards were used as a reference sample for the control: USGS-40 and IAEA-CH-7 with known carbon isotopic ratios.

Stable carbon isotope measurements in wood samples were performed for each tree and each year separately using a Delta V Plus isotope mass spectrometer (Thermo Fisher Scientific, Bremen, Germany) via a Flash HT Plus in the constant flow mode at the Kazan

Federal University, the Russian Federation. The tree-ring $\delta^{13}C$ in wood chronologies was corrected according to the $\delta^{13}C$ atmospheric CO_2 for both study sites.

The $\delta^{13}C$ CO_2 and $\delta^{13}C$ CH_4 in the soil samples were analysed for both sites using Delta V Plus (Thermo Fisher Scientific, Germany, Bremen, SN08893D). Gas samples were separated and prepared using a TRACE 1310 gas chromatography-mass spectrometer (ThermoFisher Scientific, Bremen, Germany, 713101387) coupled with an ISQ quadrupole mass detector (Thermo Fisher Scientific, Bremen, Germany) connected to the isotope mass spectrometer via a GC Isolink interface unit. The gas sample was injected automatically by means of the TriPlus RSH autosampler (CTC Analytica AG, Zwingen, Switzerland) from a vial with gas with a gas-tight syringe and was introduced into the chromatograph evaporator. The syringe was purged with pure helium before sampling. The volume of the injected sample varied from 0.1 mL to 2.5 mL, depending on the content of the target gases in the sample. Two chromatographic columns were used for the complete separation of methane, carbon dioxide and air group gases (which interfere with the determination of the carbon isotopic ratio due to their interference with nitrogen oxides). The results were processed using the "Isodat" data processing program (Version 3.0.94.12, Germany) [29].

The stable carbon isotopic ratio ($^{13}C/^{12}C$) is typically expressed in a delta notation ($\delta^{13}C$) (Equation (1)), which is the relative deviation of the ratio in the organic sample (R_{sample}) from that of an internationally accepted standard (R_{standard}), Vienna Pee Dee Belemnite (VPDB).

$$\delta_{\text{sample}} = (R_{\text{sample}} - R_{\text{standard}} - 1) * 1000 \quad (1)$$

2.5. CO_2 and Greenhouse Gas Emissions (GHGs)

The correction of the raw tree-ring $\delta^{13}C$ was necessary because the combustion of fossil fuels and biomass and land-use changes have resulted in a decrease in $\delta^{13}C$ of the atmospheric CO_2 over the last 150 years. Changes in the isotopic ratio of atmospheric CO_2 are directly reflected in the isotopic ratios of the products of photosynthesis. Calculating the differences for each year to the pre-industrial value (1850) for $\delta^{13}C$ of the atmospheric CO_2 obtained from ice cores and direct atmospheric measurements at the Mauna Loa Observatory, Hawaii [24]; https://gml.noaa.gov/ccgg/trends/ (accessed on 22 March 2023), we subtracted these differences from the raw isotope series for the $\delta^{13}C$ values for each year. Because isotope fractionation is additive, this completely removes the trend due to a decreasing atmospheric $\delta^{13}C$ from fossil fuel emissions and land-use changes.

Local data from the oil extraction and cumulative oil production from the oil and gas UVRT site are available for the period from 1943 to 2021 [19] for comparison with our newly developed $\delta^{13}C_{ptrw}$ from the UVRT and natural reserve Raifa sites. Global oil and gas emission data from Russia and globally [6] were used for a regression analysis with $\delta^{13}C_{ptrw}$ from both the UVRT and Raifa sites.

2.6. Relation to Intrinsic Water-Use Efficiency (iWUE)

The calculation of intrinsic water-use efficiency (iWUE) was based on Equations (2) and (3) and is described in detail by Saurer and Voelker [30,31].

$$\text{iWUE} = (c_a - c_i)/1.6 \quad (2)$$

$$c_i = CO_2 * ((\delta^{13}C_{atm} - \delta^{13}C_{plant} - 4.4)/(27 - 4.4)) \quad (3)$$

where c_a is the ambient CO_2 concentration, c_i is the intercellular CO_2 concentration and 1.6 is the ratio of the water diffusivities and CO_2 in the air. Diffusion through the stomata is described as (a = 4.4‰) and b is the discrimination associated with carboxylation (b = 27‰) [13,31].

The $\delta^{13}C_{plant}$ is the uncorrected tree-ring $\delta^{13}C$ measurement in pine wood relative to the δ^{13}Catm atmospheric CO_2. The $\delta^{13}C_{atm}$ data is available from https://www.esrl.noaa.gov/gmd/outreach/isotopes/c13tellsus.html (accessed on 22 March 2023).

2.7. Statistical Analyses

The statistical analysis was performed in the STATISTICA Ultimate Academic 13 RUS/EN Rus for Windows). The mean, minimum and maximum values, the standard deviation (SD) and the Pearson correlation coefficients were calculated for stable carbon isotope chronologies from both study sites. A 5-year running correlation analysis was performed between $\delta^{13}C_{ptrw}$ from the UVRT and Raifa sites to reveal a breakpoint and differences between the natural reserve Raifa and gas and oil deposit UVRT sites. Stable carbon isotope chronologies for both study sites were detrended using a time series forecasting analysis with the "no trend" function. A multiple regression analysis was performed between local and global gas and oil emissions versus $\delta^{13}C_{ptrw}$. The dependent variable was associated with $\delta^{13}C_{ptrw}$, while the independent variables were local and global oil gas emissions. Bivariant regression plots were computed. A Pearson correlation matrix was computed and a correlation analysis between the non-detrended and the detrended tree-ring $\delta^{13}C_{ptrw}$ from both study sites (UVRT and Raifa) with the climatic parameters from the local Kazan weather station was performed. The monthly averaged temperature (*t*) and precipitation (*p*) data from September of the previous year (*t*, *p_9*) to August of the current year (*p*, *p_8*), the annual air temperature and precipitation data (*p_1_12; p_1_12*), averaged spring–summer (*t,p_4567*) and summer (*t,p_678*), were used for calculations.

To reduce the impact of trends, we used detrended stable isotope chronologies. The level of statistical significance is expressed as a p-value ≤ 0.05.

The iWUE data were smoothed by an 11-year Hamming window [32] to reveal low-frequency variations and to compare long-term trends between the study sites.

2.8. Spatial Climate Patterns

Spatial correlation gridded maps were computed within the gridded net $0.5 \times 0.5°$ within the coordinates 50–60° N, 40–55° E between June–August. Evapotranspiration, differences between the maximal and minimal air temperatures ($T_{max\text{-}min}$), precipitation and cloud fractions are available at the KNMI Climate Explorer website (https://climexp.knmi.nl/) (accessed on 29 March 2023) as well as the detrended tree-ring $\delta^{13}C_{ptrw}$ from the UVRT site for the period 1980–2020. The level of significance is a p-value of ≤ 0.05.

3. Results

3.1. Stable Carbon Isotopes in Wood vs. Oil and Gas Emissions

Stable carbon isotope chronologies were developed for the UVRT and Raifa sites based on four individual trees for each study site for the common period from 1930 to 2021 (Figure 2a).

The mean $\delta^{13}C$ value is more negative for Raifa (−23.6‰) compared to UVRT (−22.3‰). The standard deviation (SD) is higher for the UVRT site (1.2) compared to Raifa (0.8). The maximum and minimum values are more negative for Raifa (max −21.5‰ and min −25.4%) compared to UVRT (max −19.7‰ and min −24.3‰). The $\delta^{13}C$ from Raifa showed a significant correlation with UVRT (r = 0.89 and $p < 0.001$). Moreover, we found more negative values for both $\delta^{13}C$ CO_2 and $\delta^{13}C$ CH_4 in the soil samples from the Raifa site (−19.3 ‰ and −64.4 ‰) compared to UVRT (−12.4 and −51.1 ‰), respectively.

A 5-year running correlation analysis between the $\delta^{13}C_{ptrw}$ from UVRT and Raifa (Figure 2b) showed a significant difference in the year of 1965, showing a negative correlation (r = −0.38 and $p < 0.05$) over the studied period 1930–2020. During other subperiods, positive correlations were revealed.

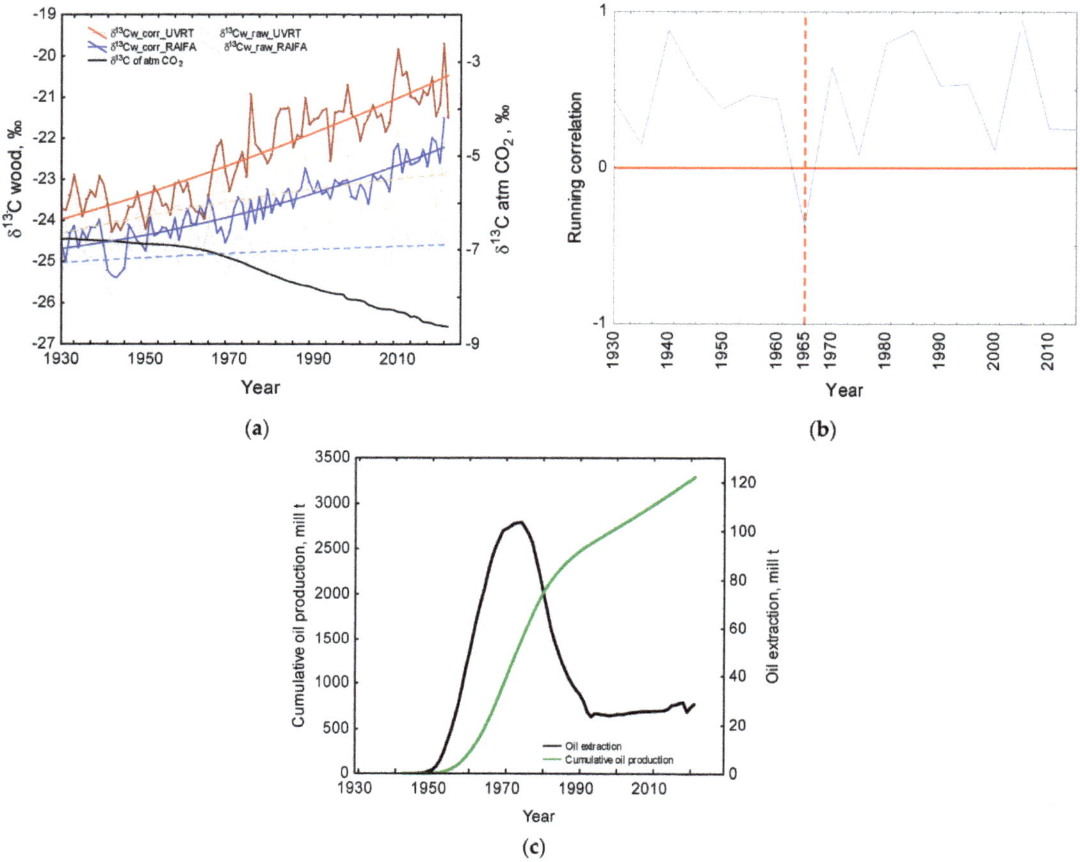

Figure 2. Raw (light-blue and light-red lines) and corrected (bold-blue and bold-red lines) according to the $\delta^{13}C$ of atmospheric CO_2 [24] tree-ring $\delta^{13}C$ in pine wood chronologies ($\delta^{13}C_{ptrw}$) from UVRT (red-coloured lines) and Raifa (blue-coloured lines) sites versus $\delta^{13}C$ atmospheric CO_2 (bold black) from 1930 to 2021 (**a**). A 5-year window running correlations between $\delta^{13}C_{ptrw}$ from UVRT and Raifa sites. A running correlation is presented as a blue line and the break-point year is represented as the dotted line in 1965 according to the normalized reference line (relative to the zero, red line) (**b**). Oil extractions (black line) and cumulative oil production (green line) from the Romashkinskoye oil and gas deposit UVRT site [19] for the period from 1943 to 2022 in comparison with $\delta^{13}C_{ptrw}$ from UVRT and Raifa (**c**).

Oil and gas extractions from the deposits started in 1943 and reached their maximum in 1970 (Figure 2c), which corresponds with decreasing $\delta^{13}C$ values from both study sites, with the most pronounced decrease for the Raifa natural reserve site (Figure 2a).

The regression analysis showed significant relationships between the local Romashkinskoye oil extraction and detrended stable carbon isotope chronologies for the UVRT and Raifa sites (r^2 = 0.19 and r^2 = 0.04), respectively (Figure 3a,b), with a higher significant correlation for the gas and oil deposit UVRT site only.

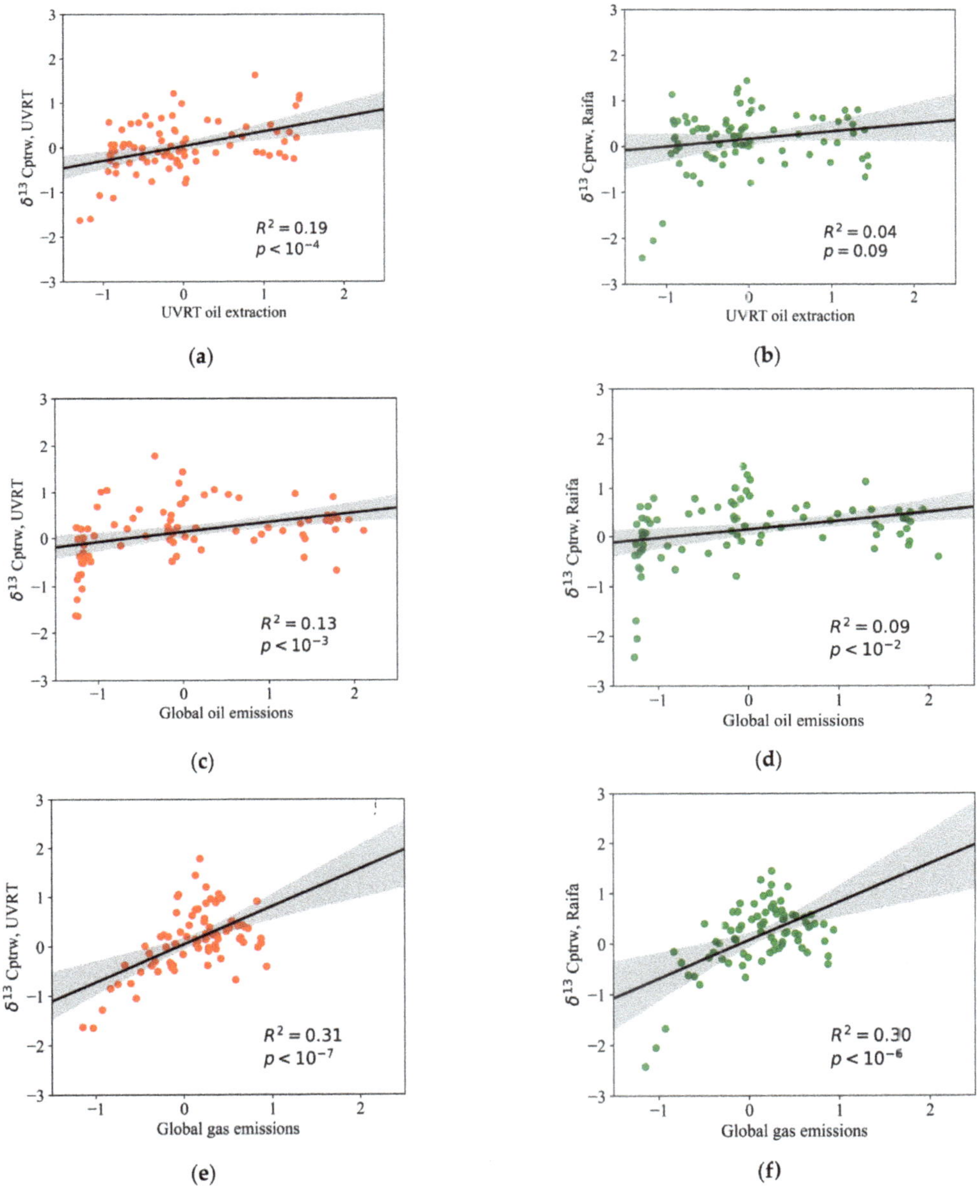

Figure 3. Detrended tree-ring $\delta^{13}C_{ptrw}$ from the UVRT (left panel, red dots) and Raifa (right panel, green dots) sites versus detrended local Romashkinskoe oil extraction [19] (**a**,**b**), global oil (**c**,**d**) and gas (**e**,**f**) [6] emissions for the period 1943 to 2021. Confidential interval is 0.95 and $p < 0.05$.

The global oil and gas emissions correlate significantly with the tree-ring $\delta^{13}C_{ptrw}$ for UVRT (r^2 = 0.13 and r^2 = 0.31) (Figure 3c,e) and for Raifa (r^2 = 0.09 and r^2 = 0.30) (Figure 3d,f), with $p < 0.05$, respectively.

3.2. Intrinsic Water-Use Efficiency (iWUE) versus Tree-Ring Width and Tree-Ring $\delta^{13}C$

The earlier 1930s averaged tree-ring width (TRW) chronology from UVRT shows lower mean values (1.42 mm) and SDs (1.21) compared to the Raifa site for TRWs (1.55 mm) and SDs (1.44), respectively. However, starting from the 1960s, the tree-ring widths from Raifa showed higher values, compared to the suppressed UVRT site (Figure 4a).

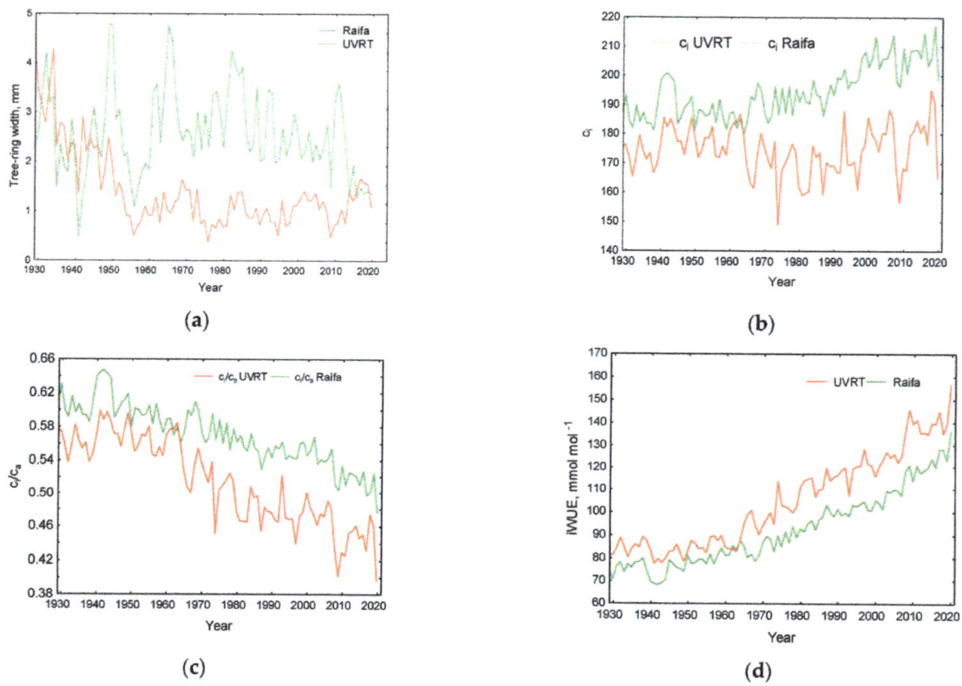

Figure 4. Averaged tree-ring width chronologies (**a**), intercellular CO_2 concentration (c_i) (**b**) and intercellular CO_2 concentration versus ambient CO_2 concentration (c_i/c_a) from Raifa versus the oil and gas reservoir UVRT site (**c**). Intrinsic water-use efficiency (iWUE) was calculated for both the Raifa and UVRT study sites (**d**).

The intercellular CO_2 concentration (c_i) showed a higher offset between the two sites starting from 1965, when oil and gas production from Romashkinskoye was doubly increased (Figure 4b) with a higher c_i for the Raifa site. The lower c_i values found for the UVRT site can be explained by stomata closure and the incorporation of a higher CO_2 concentration leading to drier conditions.

A strong decreasing c_i/c_a trend with lower values (Figure 4c) and a continuous increase of the iWUE (Figure 4d) were revealed for the UVRT site. Both UVRT and Raifa showed similar increasing trends from 1930 towards 2020 (Figure 4d).

3.3. The Impact of Local Environmental Parameters on Tree-Ring $\delta^{13}C$ Variability

The Pearson correlation analysis between the non-detrended and detrended tree-ring $\delta^{13}C_{ptrw}$ from both study sites showed positive significant ($p < 0.05$) correlations with spring–summer air temperatures (April–July) (r = 0.65 and 0.52) for the Raifa and (r = 0.61 and 0.45) UVRT sites, respectively.

The highest correlation was found between the annual air temperature between the non-detrended $\delta^{13}C_{ptrw}$ for Raifa (r = 0.74) and UVRT (r = 0.68) (Figure 5). The detrended

$\delta^{13}C_{ptrw}$ showed lower values but still had significant correlations (r = 0.44) for Raifa and for UVRT (r = 0.31).

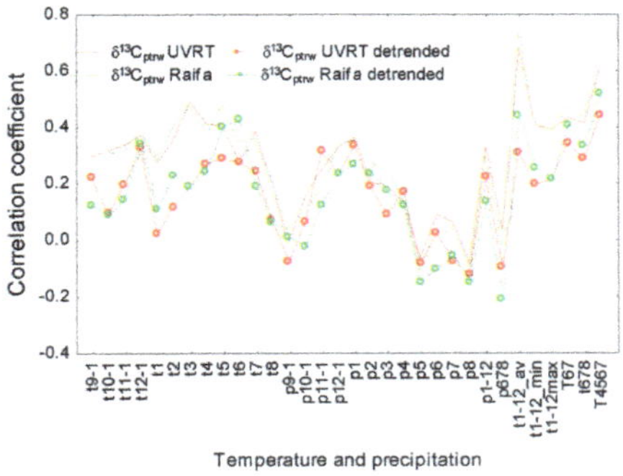

Figure 5. Pearson correlation analysis between the $\delta^{13}C_{ptrw}$ chronology from UVRT (red line), the $\delta^{13}C_{ptrw}$ chronology from Raifa (green line) as well as the detrended $\delta^{13}C_{ptrw}$ from UVRT (red line with circle) and the $\delta^{13}C_{ptrw}$ from Raifa (green line with green circle) versus climatic data from the local Kazan weather station. The temperature and precipitation from September of the previous year are marked as *t9-1* and *p9-1*, respectively. The monthly averaged temperature and precipitation are indicated in numbers.

A negative significant correlation was found between the tree-ring $\delta^{13}C$ from the Raifa site and the summer (June–August) precipitation (r = −0.21 and $p < 0.05$). Positive significant correlations were found between the detrended $\delta^{13}C_{ptrw}$ from both sites, with a spring air temperature and precipitation of the previous year and the current one.

3.4. Spatial Correlation Patterns

The detrended tree-ring $\delta^{13}C$ chronology from the UVRT site showed positively significant ($p < 0.05$) spatial correlations within the gridded net 0.5 × 0.5° for the coordinates within 50–60° N and 40–55° E with an averaged June–August evapotranspiration (r = 0.5) (Figure 6a) and a June–August $T_{max-min}$ (r = 0.5) (Figure 5b), while negative ones were shown with a June–August precipitation (r = −0.5) (Figure 6c) and a June–August cloud fraction (r = −0.6) (Figure 6d).

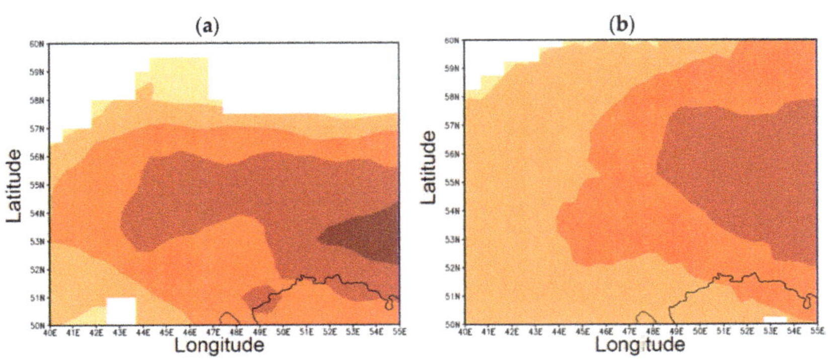

Figure 6. *Cont.*

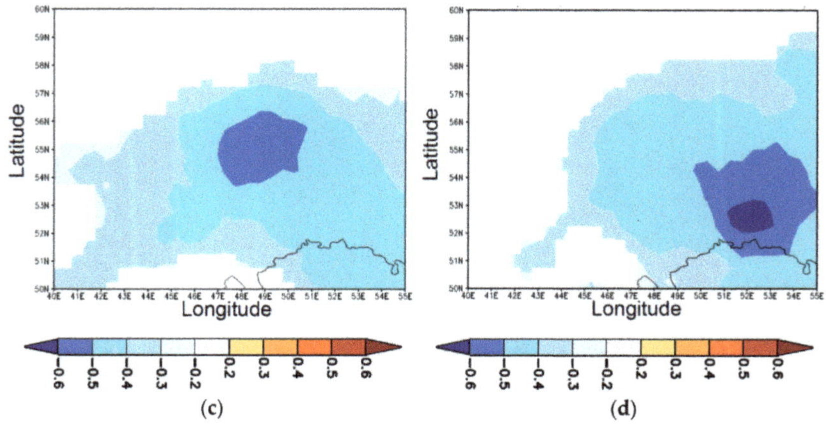

Figure 6. Spatial distribution of correlation coefficients within the gridded 0.5 × 0.5° net coordinates 50–60° N (latitude) and 40–55° E (longitude) between the detrended tree-ring $\delta^{13}C_{ptrw}$ from UVRT and the averaged June–August reference evapotranspiration (**a**), the differences between the maximal and minimal air temperature ($T_{max\text{-}min}$) (**b**), precipitation (**c**) and cloud fractions (**d**) computed for the period from 1980–2020. Right vertical scales represent correlation coefficients within the range from −0.6 (dark blue, negative correlations) to +0.6 (dark red, positive correlations), with $p < 0.05$.

4. Discussion

The oil and gas industry started to be developed in Romashkinskoye, the Tatarstan Republic, the Russian Federation, in 1943 [19], and in the 1970s, it reached the maximum in oil extractions, which was recorded in the tree-ring pine $\delta^{13}C_{ptrw}$ chronologies from both the UVRT and Raifa sites. Recently, oil and gas production were increased by 3.5% compared to the past. A further increase of oil and gas production is expected in the near future [20]. The accelerating rate of GHG emissions will undouble and will change the environment, biodiversity and humans' health. Only accidents in which a large volume of hydrocarbons released into the soil or water have been evaluated so far. Therefore, our pioneering study highlights how trees could respond to oil and gas development infrastructures near to the deposit and in the remote site locations of the region.

Our results, at the tree-ring level, show a higher tree growth suppression during oil extractions for the UVRT site, which is also reflected in the tree-ring $\delta^{13}C$ values. This can be explained by the negative impact of the developed oil and gas infrastructure on the pine forest. Our finding is in line with Pickell et al. [33], who showed an anthropogenic disturbance in developed oil and gas activities on forest landscapes in the USA [34]. The heavier stable carbon isotope (^{13}C) in tree rings is modulated by hydrocarbon emissions from the infrastructure site and to environmental parameters like the temperature, precipitation, sunshine duration and evapotranspiration. Moreover, anthropogenic CO_2s increase due to the Suess effect and the global GHG emissions accumulated in tree rings, which is reflected in the drastic $\delta^{13}C$ increase towards the recent decades.

Interestingly, pine trees from the natural reserve Raifa site, which is undisturbed by direct human impacts, show a more positive tree ring variability compared to the human-induced UVRT site. However, the tree-ring $\delta^{13}C$, CH_4 and CO_2 in the soil showed more negative values compared to the oil and gas deposit UVRT site. One of the explanations for this could be that pine trees from natural reserve sites are more sensitive to the impact of greenhouse emissions, while pine trees from human-induced sites are already stressed by local and global impacts. A significant impact of local oil extractions and global HC emissions was revealed for the UVRT site, which supports our hypothesis that both local and global hydrocarbon emissions affect local pine forests. An increasing intrinsic water-use efficiency (iWUE) along with a decreasing tree-ring width chronology from the oil and gas deposits indicate the development of drought conditions (Figure 4) during the vegetation period, which is confirmed by the local and spatial climatic correlation analyses. Despite site-specific and species-specific differences, the study by Guerrieri et al. [17] showed that NO_x pollutions alter the iWUE by confirming the impact of anthropogenic factors on oak trees. It is well known that the variation of c_i depends on various environmental factors and on species and site conditions. Our study demonstrates the site-specific response of trees to oil and gas productions after 1965, showing an offset between two sites. The c_i/c_a ratio has decreased in recent decades, indicating the closure of the stomata, which is also reflected in increased iWUE trends for both sites. However, a higher iWUE is observed for the natural reserve Raifa site.

An increasing evapotranspiration and extreme changes in air temperatures, less clouds and a decreasing of precipitation lead to developing drier environmental conditions. Pine trees from the Raifa site record more pronounced drought conditions during July over the period from 1970 to 2021, which is also recorded in the $\delta^{13}C_{ptrw}$ values. Early spring temperatures can impact tree growth significantly, by shifting the vegetation period to earlier dates. Such an early spring temperature shift was observed in the southern part of Siberia due to early snowmelt under recent anthropogenic warming [35,36]. However, it is challenging to separate the impacts from global and local anthropogenic emissions here. Therefore, further studies are needed for complex and multi-parametric approaches.

5. Conclusions

We conclude that pine trees from the UVRT site indicate developing unfavourable drier conditions and a suppressed tree growth caused by both human-induced oil and deposit infrastructures and natural eco-physiological processes compared to the undisturbed natural reserve Raifa site. Understanding the mechanisms behind local hydrocarbon emissions from oil and gas deposits and their contribution to global GHG emissions is of great importance and should be further investigated among transects. A comparative analysis may provide a comprehensive description of anthropogenic impacts among available oil and gas deposit infrastructures in the region, locally and globally.

Based on our pilot study, we would recommend performing seasonal and annual multi-proxy parameter approaches by analysing the $\delta^{13}C$ in needles and the $\delta^{13}C$ and CH_4 in the soil at different depths and by analysing tree-ring cellulose instead of wood, which is a more sensitive parameter to both eco-physiological and climatological changes. Further local oil and gas developments and their impacts on forest ecosystems should be better tracked and analysed along the transect for long-term monitoring.

Author Contributions: Conceptualization, all authors; methodology, G.B., B.G., G.M., A.T., D.T., D.K. and D.N.; software, O.V.C. and G.M.; validation, O.V.C. and G.M.; formal analysis, O.V.C., G.M., A.T., D.T., G.B. and B.G.; resources, D.N.; data curation, O.V.C.; writing—original draft preparation, all authors; writing, all authors; visualization, O.V.C.; supervision, D.N.; project administration, D.K.; funding acquisition, D.N. All authors have read and agreed to the published version of the manuscript.

Funding: This research was funded by the Kazan Federal University Strategic Academic Leadership Program (PRIORITY-2030).

Data Availability Statement: Research data will be uploaded to the Zenodo repository upon acceptance of the manuscript.

Conflicts of Interest: The authors declare no conflict of interest.

References

1. Jones, M.W.; Peters, G.P.; Gasser, T.; Andrew, R.M.; Schwingshackl, C.; Gütschow, J.; Houghton, R.A.; Friedlingstein, P.; Pongratz, J.; Le Quéré, C. National contributions to climate change due to historical emissions of carbon dioxide, methane, and nitrous oxide since 1850. *Sci. Data* **2023**, *10*, 155. [CrossRef]
2. Saunois, M.; Bousquet, P.; Poulter, B.; Peregon, A.; Ciais, P.; Canadell, J.G.; Dlugokencky, E.J.; Etiope, G.; Bastviken, D.; Houweling, S.; et al. The global methane budget 2000–2012. *Earth Syst. Sci. Data* **2016**, *8*, 697–751. [CrossRef]

3. Bruhwiler, L.M.; Basu, S.; Bergamaschi, P.; Bousquet, P.; Dlugokencky, E.; Houweling, S.; Ishizawa, M.; Kim, H.S.; Locatelli, R.; Maksyutov, S.; et al. CH_4 emissions from oil and gas production: Have recent large increases been detected? *J. Geophys. Res.-Atmos.* **2017**, *122*, 4070–4083. [CrossRef]
4. IPCC. Summary for Policymakers. In *Climate Change 2021: The Physical Science Basis. Contribution of Working Group I to the Sixth Assessment Report of the Intergovernmental Panel on Climate Change*; Masson-Delmotte, V., Zhai, P., Pirani, A., Connors, S.L., Péan, C., Berger, S., Caud, N., Chen, Y., Goldfarb, L., Gomis, M.I., et al., Eds.; Cambridge University Press: Cambridge, UK; New York, NY, USA, 2021; pp. 3–32. [CrossRef]
5. Privalova, N.M.; Dvadnenko, M.V.; Nekrasova, A.A.; Privalov, D.M.; Popova, O.S. Impact of oil and oil products on the environment. *Tech. Sci.* **2017**, *125*, 309–318. [CrossRef]
6. Ritchie, H.; Roser, M.; Rosado, P. Greenhouse Gas Emissions. 2020. Databased Platform. Available online: https://ourworldindata.org/co2-and-greenhouse-gas-emissions (accessed on 22 March 2023).
7. Allen, C.D.; Macalady, A.K.; Chenchouni, H.; Bachelet, D.; McDowell, N.; Vennetier, M.; Kitzberger, T.; Rigling, A.; Breshears, D.D.; Hogg (Ted), E.H.; et al. A global overview of drought and heat-induced tree mortality reveals emerging climate change risks for forests. *For. Ecol. Manag.* **2010**, *259*, 660–684. [CrossRef]
8. IPCC. *Climate Change 2007: Synthesis Report. Contribution of Working Groups I, II and III to the Fourth Assessment Report of the Intergovernmental Panel on Climate Change*; Core Writing Team, Pachauri, R.K., Reisinger, A., Eds.; IPCC: Geneva, Switzerland, 2017; p. 104.
9. Covey, K.R.; Megonigal, J.P. Methane production and emissions in trees and forests. *Transley Rev. New Phytol.* **2019**, *222*, 35–51. [CrossRef]
10. Menyailo, O.V.; Abraham, W.-R.; Conrad, R. Tree species affect atmospheric CH_4 oxidation without altering community composition of soil methanotrophs. *Soil Biol. Biochem.* **2010**, *42*, 101–107. [CrossRef]
11. Farquhar, G.D.; Ehleringer, J.R.; Hubick, K.T. Carbon isotope discrimination and photosynthesis. *Annu. Rev. Plant Phisiol. Plant Mol. Biol.* **1989**, *40*, 503–537. [CrossRef]
12. Siegwolf, R.T.W.; Lehmann, M.M.; Goldsmith, G.R.; Churakova (Sidorova), O.V.; Mirande-Ney, C.; Timofeeva, G.; Weigt, R.B.; Saurer, M. Updating dual C and O isotope—Gas exchange model: A concept to understand plant responses to the environment and its implications for tree rings. *Plant Cell Environ.* **2023**, *46*, 2606–2627. [CrossRef] [PubMed]
13. Farquhar, G.D.; Hubick, K.T.; Condon, A.G.; Richards, R.A. Carbon isotope fractionation and plant water-use efficiency. In *Stable Isotopes in Ecological Research. Ecological Studies*; Rundel, P.W., Ehleringer, J.R., Nagy, K.A., Eds.; Springer: New York, NY, USA, 1989; Volume 68. [CrossRef]
14. Siegwolf, R.T.W.; Brooks, J.R.; Roden, J. *Stable Isotopes in Tree Rings Inferring Physiological, Climatic and Environmental Responses Stable Isotope in Tree Rings*; Springer: Berlin/Heidelberg, Germany, 2022; p. 773. [CrossRef]
15. McCarroll, D.; Loader, N.J. Stable isotopes in tree rings. *Quat. Sci. Rev.* **2004**, *23*, 771–801. [CrossRef]
16. Quinby, B.M.; Creighton, J.C.; Flacherty, E.A. Stable isotope ecology in insects: A review. *Ecol. Entomol.* **2020**, *45*, 1231–1246. [CrossRef]
17. Guerrieri, R.; Siegwolf, R.T.W.; Saurer, M.; Ripullone, F.; Mencuccini, M.; Borghetti, M. Anthropogenic NOx emissions alter the intrinsic water-use efficiency (WUEi) for *Quercus cerris* stands under Mediterranean climate conditions. *Environ. Pollut.* **2010**, *158*, 2841–2847. [CrossRef]
18. Leonelli, G.; Battipaglia, G.; Siegwolf, R.T.W.; Saurer, M.; Morra di Cella, U.; Cherubini, P.; Pelfini, M. Climatic isotope signals in tree rings masked by air pollution: A case study conductance along the Mont Blanc Tunnel access road (Western Alps, Italy). *Atmos. Environ.* **2012**, *61*, 169–1719. [CrossRef]
19. Muslimov, R.K. (Ed.) *Romashkino Oil Field*; VNIIOENG: Moscow, Russia, 1995; Volume 1, p. 492.
20. Khasanova, A.; Bakhtina, O. Tatneft Estimates Reserves of the Romashkinskoye Field. 2023. Available online: https://neftegaz.ru/news/gas/774735-tatneft-otsenila-zapasy-romashkinskogo-mestorozhdeniya/ (accessed on 22 March 2023).
21. Selley, R.; Sonnenberg, S.A. *Elements of Petroleum Geology*, 3rd ed.; Elsevier: Amsterdam, The Netherlands, 2014. [CrossRef]
22. Bakin, O.V.; Rogova, T.V.; Sitnikov, A.P. *Vascular Plants of Tatarstan*; University Press Kazan: Kazan, Russia, 2000; p. 496.
23. RIHMI-WDC. Baseline Climatological Data Sets. Obninsk, Russia. 2022. Available online: http://meteo.ru/english/data (accessed on 3 December 2022).
24. Francey, R.J.; Allison, C.E.; Etheridge, D.M.; Trudinger, C.M.; Enting, I.G.; Leuenberger, M.; Langenfelds, R.L.; Michel, E.; Steele, E. A 1000-year high precision record of $\delta^{13}C$ in atmospheric CO_2. *Tellus B* **1999**, *51*, 170–193. [CrossRef]
25. Cook, E.R.; Kairiukstis, L.A. Methods of Dendrochronology. In *Applications in the Environmental Sciences*; Springer: Berlin/Heidelberg, Germany, 1990; p. 394. [CrossRef]
26. Rinn, F. *TSAPWin—Time Series Analysis and Presentation for Dendrochronology and Related Applications*; V. 0.53, User Reference; Heidelberg, Germany, 2005.
27. Holmes, R.L. Computer-assisted quality control in tree-ring dating and measurement. *Tree-Ring Bull.* **1983**, *43*, 51–67.
28. Grissino-Mayer, H.D. Evaluating cross-dating accuracy: A manual and tutorial for the computer program COFECHA. *Tree-Ring Res.* **2001**, *57*, 205–221.27.
29. Thermo Electron (Bremen) GmbH. Isodat 2.0 Operating Manual. In *Finnigan Advanced Mass Spectrometry*; Thermo Electron (Bremen) GmbH: Bremen, Germany, 2004.

30. Saurer, M.; Voelker, S. Intrinsic water-use efficiency derived from stable carbon isotopes of tree rings. In *Stable Isotopes in Tree Rings*; Springer: Berlin/Heidelberg, Germany, 2023; pp. 481–498.
31. Dawson, T.E.; Siegwolf, R. *Book Stable Isotopes as Indicators of Ecological Change*; Todd, E., Dawson, T.E., Rolf, T.W., Siegwolf, R., Eds.; Academic Press: Cambridge, MA, USA, 2007; ISBN 978-0-12-373627-7.
32. Blackman, R.; Tukey, J.W. The Measurement of Power Spectra from the Point of View of Communications Engineering—Part I. *Bell Syst. Tech. J.* **1958**, *37*, 185–282. [CrossRef]
33. Pickell, P.D.; Gergel, S.E.; Coops, N.C.; Andison, D.W. Monitoring Forest Change in Landscapes Under-Going Rapid Energy Development: Challenges and New Perspectives. *Land* **2014**, *3*, 617–638. [CrossRef]
34. Turner, A.J.; Jacon, D.J.; Benmergui, S.C.; Wofsy, J.D.; Maasakkers, A.; Butz, O.; Hasekamp, S.C.; Biraud, A. large increase in U.S. methane emissions over the past decade inferred from satellite data and surface observations. *Geophys. Res. Lett.* **2016**, *43*, 2218–2224. [CrossRef]
35. Vaganov, E.A.; Hughes, M.K.; Kirdyanov, A.V.; Schweingruber, F.H.; Silkin, P.P. Influence of snowfall and melt timing on tree growth in subarctic Eurasia. *Nature* **1999**, *400*, 149–151. [CrossRef]
36. Knorre, A.A.; Siegwolf, R.T.W.; Saurer, M.; Sidorova, O.V.; Vaganov, E.A.; Kirdyanov, A.V. Twentieth century trends in tree ring stable isotopes ($\delta^{13}C$ and $\delta^{18}O$) of *Larix sibirica* under dry conditions in the forest steppe in Siberia. *J. Geophys. Res.* **2010**, *115*, G03002. [CrossRef]

forests

MDPI

Article

Differences in the Responses of Tree-Ring Stable Carbon Isotopes of *L. sibirica* and *P. schrenkiana* to Climate in the Eastern Tianshan Mountains

Yi Jia [1,2], Guanghui Lv [1], Hailili Guligena [3], Li Qin [2], Zhengbing Peng [4], Ruxianguli Abudureheman [2] and Ruibo Zhang [2,*]

1 Key Laboratory of Oasis Ecology of Education Ministry, College of Ecology and Environment, Xinjiang University, Urumqi 830017, China; 17837158917@163.com (Y.J.)
2 Xinjiang Key Laboratory of Tree Ring Ecology, Key Laboratory of Tree-Ring Physical and Chemical Research (CMA), Institute of Desert Meteorology, China Meteorological Administration, Urumqi 830002, China
3 Center of Xinjiang Meteorological Information, Urumqi 830002, China
4 State Key Laboratory of Marine Geology, Tongji University, Shanghai 200092, China; 2110887@tongji.edu.cn
* Correspondence: river0511@163.com

Abstract: The eastern Tianshan Mountains are located in the arid interior of Asia, where tree growth is especially sensitive to climate. The ratio of stable carbon isotopes ($\delta^{13}C$) in the tree rings can provide information on changes in atmospheric CO_2 concentrations, water availability, and physiological processes within the tree. In particular, the use of tree-ring $\delta^{13}C$ values as a proxy for past atmospheric CO_2 concentrations has gained widespread acceptance. In this study, detrended stable carbon isotope chronologies ($^{13}C_{corr}$) of *Larix sibirica* Ledeb. and *Picea schrenkiana* Fisch. et Mey was established using tree-ring samples from high elevations in the eastern Tianshan Mountains of Xinjiang, China. The relationships between the tree-ring $^{13}C_{corr}$ and different climatic factors were explored using the correlation function and collinearity analysis. Our results demonstrate that the tree-ring $\delta^{13}C_{corr}$ of *L. sibirica* is significantly and negatively correlated with precipitation and relative humidity during the growing season. The main climate factor affecting the stable carbon isotope fractionation of *L. sibirica* during the growing season is relative humidity during the growing season. The tree-ring $\delta^{13}C_{corr}$ of *P. schrenkiana* is significantly and negatively correlated with the mean temperature, mean minimum temperature, precipitation, and vapor pressure deficit from the end of the previous growing season and throughout the current growing season, especially in summer. However, it is significantly and positively correlated with relative humidity, indicating that the relationship between the climate factors and the tree-ring stable carbon isotope fractionation of *P. schrenkiana* is more complex. Further analysis showed that summer temperature and summer precipitation jointly controlled the tree-ring stable carbon isotope fractionation of *P. schrenkiana* at a high elevation. This research has important implications for our understanding of past and future climate change, as well as for the development of effective strategies to mitigate and adapt to these changes. This study also contributed to the development of a more in-depth understanding of the effects of climate change on tree growth in extremely arid environments and provided evidence to support effective forest management in arid regions.

Keywords: *Larix sibirica* Ledeb. and *Picea schrenkiana* Fisch. et Mey; tree-ring; stable carbon isotope; climate response; eastern Tianshan Mountains

1. Introduction

Due to their high accuracy, continuity, and strong sensitivity to environmental fluctuations, stable carbon isotopes in tree rings have become a crucial tool in dendroclimatological research [1,2]. The ratios of stable isotopes preserved in xylem cells contain information regarding past environmental conditions and are an important proxy in paleoclimate studies. They are crucial to researching both stage-specific historical environmental changes and the global carbon cycle [3]. The response of stable carbon isotopes in tree rings to climate factors varies from region to region and from species to species [4]. In many parts of the world, the response of tree-ring ^{13}C to the climate in various coniferous and broadleaf tree species has been studied and used to accurately reconstruct climate data over hundreds and even thousands of years [5–14].

Northwestern China and the Qinghai-Tibet Plateau are ideal for dendroclimatology studies because tree growth in both regions is more sensitive to climate change, but there are significant regional differences in the responses of tree growth to climate [15–20]. Two tree species—*Larix sibirica* Ledeb. and *Picea schrenkiana* Fisch. et Mey—predominate in the Tianshan Mountains: the largest mountain system in the inland arid region of Asia. *L. sibirica* and *P. schrenkiana*, which are both endemic to the mountains and capable of withstanding cold and drought, are well-suited to dendroclimatological studies. In arid inland Asia, most dendroclimatological studies are based on tree-ring width parameter reconstruction, and tree-ring stable isotope studies are scarce [21–24]. Zhang et al. [25,26] reconstructed temperature changes in southeast Kazakhstan based on tree-ring stable carbon isotopes and combined the tree-ring width and stable carbon isotopes to reconstruct glacier mass balance changes. Qin et al. [27,28] analyzed differences in the tree-ring stable carbon isotope fractionation of *P. schrenkiana* in response to climate on the northern and southern slopes of the Tianshan Mountains. They also reconstructed changes in relative humidity in the eastern Tianshan Mountains over the past 443 years using tree-ring widths from *L. sibirica*. Nevertheless, tree-ring width can only reflect moisture information, and it is difficult to capture temperature information in the extremely arid eastern Tianshan Mountains. Therefore, other tree-ring parameters should be extracted for further study. Shang et al. [29] analyzed the stable carbon isotope composition of *P. schrenkiana* on the northern slope of the eastern Tianshan Mountains and explored its climatic and environmental significance. Zhang et al. [30] revealed the relationship between tree-ring stable carbon isotopes of *L. sibirica* and climate elements and discussed the climate response mechanism.

However, the responses of *L. sibirica* and *P. schrenkiana* to the climate in the same region remain unclear. Further studies are needed to understand how climate affects the tree-ring stable carbon isotope fractionation of different tree species in extremely arid inland Asia. In this study, the tree-ring stable carbon isotope chronologies of *L. sibirica* and *P. schrenkiana* were established, and the relationship between the tree-ring stable carbon isotopes of these two species and multiple climatic factors was investigated. The response relationships between tree-ring stable carbon isotope fractionation processes in response to climate were further explored based on the physiological mechanisms and the characteristics of the tree species itself.

2. Data and Methods

2.1. Overview of the Study Area and Sample Collection

The Tianshan Mountains are located in the middle of Asia and stretch from west to east across central Xinjiang, China, Kyrgyzstan, and Kazakhstan. This massive mountain ecosystem is characterized by low precipitation, prolonged sunshine hours, significant diurnal temperature differences, and an uneven distribution of precipitation that is primarily influenced by the westerly wind belt [31].

In July 2017, we collected 60 tree core samples by using growth cones measuring 10 mm in diameter from 25 *L. sibirica* individuals at breast height in the Banfang Gorge (BFL). We also collected 48 core samples from 24 *P. schrenkiana* individuals growing at the Tianshan Temple (TSS) and located in the eastern Tianshan Mountains of Hami (Figure 1. Details regarding the sampling sites are shown in Table 1.

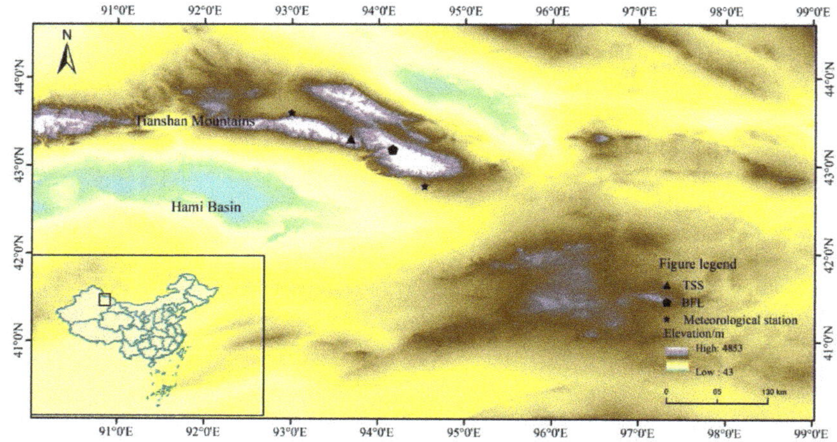

Figure 1. Locations of sampling sites and meteorological stations.

Table 1. Sampling site information.

	Sampling Sites	Code Name	Latitude/N	Longitude/E	Elevation/m	Aspect	Slope (°)	Canopy Closure
L. sibirica	Banfang Gorge	BFL	43°11′	94°10′	2900–2950	N	10–15	0.3
P. schrenkiana	Tianshan Temple	TSS	43°18′	93°41′	2620	NE	30	0.5

2.2. Stable Carbon Isotope Chronologies Development

In accordance with fundamental dendrochronological principles and procedures, samples were air-dried and fixed, sanded, and visually dated in the lab. Then, ring-width measurements were made using a ring-width meter with an accuracy of 0.001 mm and the MeasureJ2X program. Cross-dating tests were carried out using the COFECHA program [32]. Outlier measurements, as well as chronologies that were poorly correlated with the master chronology, were removed in order to increase the accuracy of the tests. Tree-ring chronology was standardized using ARSTAN [33].

We selected 14 cores from 7 trees of *L. sibirica* and *P. schrenkiana* to develop the tree-ring stable isotope chronologies, respectively. Tree cores with no missing rings, no obvious damage, a high correlation with primary chronology, and distinct tree-ring boundaries were selected for further study based on cross-dating. The first thirty years of each tree core were removed to prevent bias due to juvenile effects [34,35]. Using a scalpel, the remaining cores were segmented year by year under a microscope. Segmentation was carried out by mixing core samples from the same calendar year with cross-dated ring-width data. Samples from each year were thoroughly crushed and mixed using a mixing ball mill (MM400, Retsch GmbH, Haan-Germany). Afterward, all samples were subjected to year-by-year α-cellulose extraction using the acetate and nitrate mixing method of Brendel et al. [36]. A 17% NaOH treatment was added to remove lignin and non-cellulosic polysaccharides [37], which is a method that is widely used in tree-ring stable isotope studies [25,26,38].

Year-by-year cellulose samples of 70–100 µg were wrapped in tin cups in the form of cubes or spheres, and stable carbon isotopes were measured using an Elemental Analyzer (Flash EA 1112; Thermo Fisher Scientific, Waltham, MA, USA) and a Stable Isotope Mass Spectrometer (MAT253, Thermo Fisher Scientific, Bremen GmbH, Bremen, Germany) at the Key Laboratory of the Ministry of the Western Environment, Lanzhou University. A laboratory-known graphite standard (−16.0‰) was measured simultaneously for every 7 samples measured. The analytical error (standard deviation) of the isotopic measurements was less than 0.05‰. The stable carbon isotope expression was defined as the relative abundance compared to the South Carolina Cretaceous Pee Dee Formation Stratigraphic American aragonite fossil (Vienna Pee Dee Belemnite VPDB) standard [39] and was calculated as follows:

$$\delta^{13}C = \left[\frac{(^{13}C/^{12}C)_{sample}}{(^{13}C/^{12}C)_{VPDB}} - 1 \right] \times 10^3 ‰ \quad (1)$$

Studies have shown that the concentration of CO_2 in the atmosphere has continued to rise since the industrial revolution due to the extensive use of fossil fuels [40]. Trees continuously absorb atmospheric CO_2 through photosynthesis as they develop; as a result, variations in the amount of CO_2 in the atmosphere inevitably affect the values of ^{13}C in tree rings. Carbon isotopes in tree rings are affected by the balance between plant photosynthesis and respiration, which can be influenced by environmental factors such as temperature, precipitation, and solar radiation. However, the isotopic fractionation of carbon can also be influenced by the concentration of atmospheric CO_2, as plants tend to discriminate against the heavier isotope of carbon when they fix CO_2. This results in a decrease in the isotopic ratio of carbon in tree rings as the CO_2 concentration increases. Since this change is not directly related to climate, it must be disregarded when using tree-ring ^{13}C to study past climate change [4].

We used the best-estimate value of $\delta^{13}C_{corr}$ in 1850 since the industrial revolution, which is based on ice core measurements and is about −6.7‰ [41]. We used the measured values of tree-ring stable carbon isotopes plus the difference in ice core $\delta^{13}C_{corr}$ relative to the standard value as the corrected stable carbon isotope value. The corrected value was then defined as the detrended chronology of the $\delta^{13}C_{corr}$ tree-ring [42] (Figure 2). Based on this chronology, we analyzed the responses to the climate of the stable carbon isotopes of *L. sibirica* and *P. schrenkiana* in the eastern Tianshan Mountains.

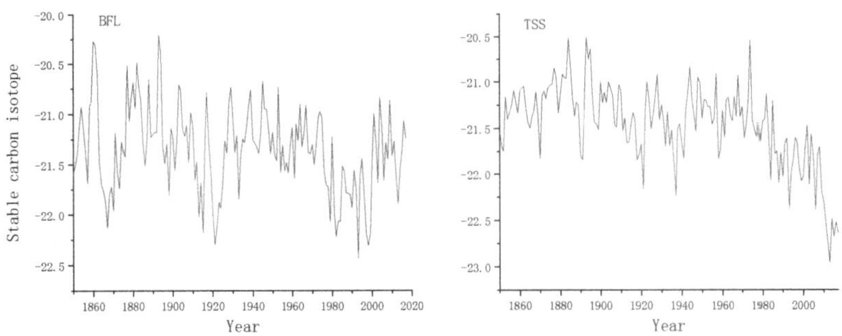

Figure 2. Two tree-ring $\delta^{13}C_{corr}$ chronologies. BFL represents the *L. sibirica* chronology and TSS represents the *P. schrenkiana* chronology.

2.3. Meteorological Data

The meteorological data used in this study were obtained from the China Meteorological Data Network and the Xinjiang Uygur Autonomous Region Meteorological Information Center. Monthly meteorological data were obtained from the Balikun meteorological station (1957–2015; 43°36′ N, 93°00′ E) and from the Qincheng meteorological station (1959–1996; 42°46′ N, 94°32′ E). A total of six meteorological elements were obtained from each station: mean temperature, mean maximum temperature, mean minimum temperature, precipitation, relative humidity, and a saturated water vapor pressure deficit. The saturated water vapor pressure deficit was calculated using the following equation [43]:

$$VPD = (1 - RH) \times 0.6108 \times e^{(17.27T/(T+273.3))} \quad (2)$$

where VPD is the month-by-month saturated water vapor pressure deficit, RH is the monthly mean relative humidity, and T is the monthly mean temperature.

This analysis showed that the annual mean temperature at the Qincheng meteorological station, which is located on the southern slope of the Tianshan Mountains, was 5.44 °C. Moreover, annual precipitation was 112 mm, and rain and heat occurred during the same period, with precipitation occurring primarily during the summer months (June to August). The annual mean temperature at the Balikun meteorological station, which is located on the northern slope of the Tianshan Mountains, was 2.08 °C, and the annual precipitation was 185.15 mm. Compared with Qincheng, Balikun had a lower mean temperature and more precipitation, mostly concentrated in the middle and end of the growing season from July to September. Precipitation at Balikun exhibited a significant increasing trend (Figures 3 and 4).

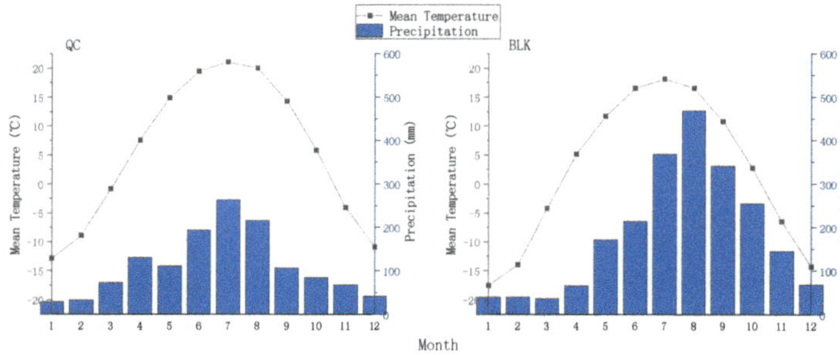

Figure 3. Variations in mean monthly temperature and precipitation at the Qincheng and Balikun meteorological stations.

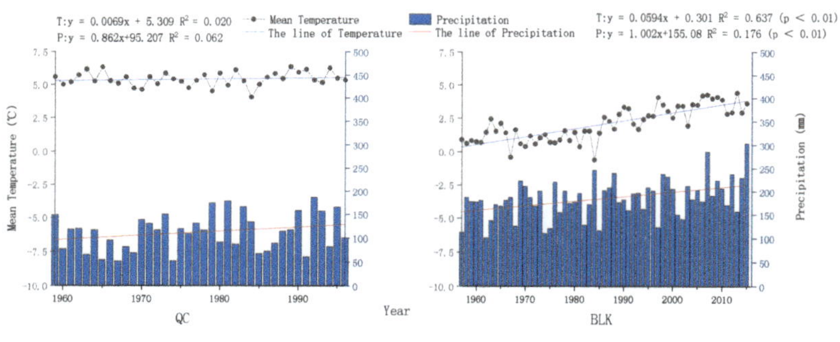

Figure 4. Variations in mean annual temperature and precipitation variation at the Qincheng and Balikun meteorological stations.

2.4. Analysis Methods

Data were analyzed using mathematical and statistical methods that are common in traditional tree-ring climate studies [44]. Using the Pearson correlation, we conducted an analysis of the relationship between climate and tree-ring parameters. In order to extract seasonal-scale climate signals, we analyzed the correlation between the tree-ring carbon isotope chronology and monthly average climate variables from October of the previous year to September of the current year. To further investigate the relationship between stable carbon isotope chronologies and monthly climatological data, we performed a combined correlation analysis. We examined six meteorological variables from each of the two meteorological stations and then selected the meteorological data that best correlated with each of the chronologies. We then analyzed these correlations on a month-by-month basis. The combination of precipitation was calculated by summing the precipitation values across all months, and the other meteorological elements were calculated by averaging their values. In addition, we used collinearity analysis to examine the contribution of each climatic factor to the stable carbon isotope fractionation of the tree rings.

3. Results

3.1. Statistical Analysis of the Tree-Ring Stable Carbon Isotope Chronologies

The stable carbon isotope values of the two species were similar, though *L. sibirica* exhibited a slightly lower variance and a lower standard deviation than *P. schrenkiana*. *L. sibirica* tree rings, which varied from −20.287‰ to −23.619‰, with a mean value of −21.822‰ and a coefficient of variation of −0.0337. The stable carbon isotope values of *P. schrenkiana* varied from −20.581‰ to −24.603‰, with a mean value of −21.912‰ and a coefficient of variation of −0.0382. The interannual variation in stable carbon isotopes in *L. sibirica* tree rings was relatively more stable than that of *P. schrenkiana* tree rings (Table 2 and Figure 5).

Table 2. Statistics of the tree-ring stable carbon isotope chronologies.

Codename	Year	Maximum	Minimum	Mean	Variance	Standard Deviation	Coefficient of Variation
BFL	1850–2017	−20.287‰	−23.619‰	−21.822‰	0.541	0.735	−0.0337
TSS	1850–2017	−20.581‰	−24.603‰	−21.912‰	0.701	0.837	−0.0382

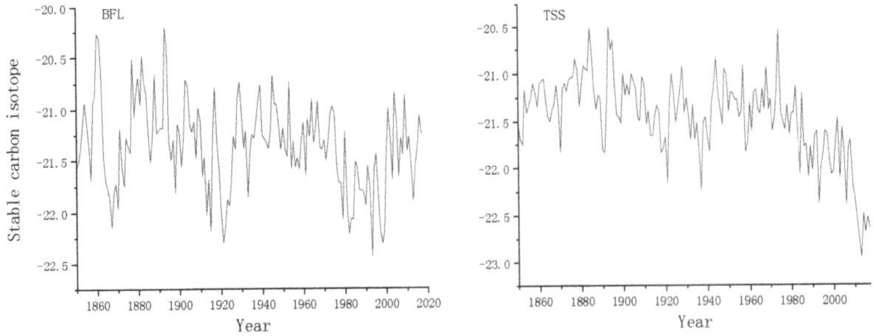

Figure 5. Two tree-ring $\delta^{13}C_{corr}$ chronologies. BFL represents the *L. sibirica* chronology and TSS represents the *P. schrenkiana* chronology.

3.2. Response of Tree-Ring Stable Carbon Isotopes to Climate

We used correlation analysis to examine the relationships between the $\delta^{13}C_{corr}$ chronologies and the meteorological parameters from both the Balikun and the Qincheng meteorological stations. Specifically, we used monthly values from October of the previous year to September of the current year for the mean temperature, mean maximum temperature, mean minimum temperature, precipitation, relative humidity, and saturated saturation water vapor pressure deficit.

The $\delta^{13}C_{corr}$ chronology of *P. schrenkiana* showed a significantly negative correlation with the mean growing season temperature and passed the 99% significance test, but that of *L. sibirica* did not correlate well with the mean temperature. Neither chronology was significantly correlated with the mean maximum temperature except in July. However, the *P. schrenkiana* chronology correlated well with the mean minimum temperature. In all months other than January and March, the correlations between the *P. schrenkiana* $\delta^{13}C_{corr}$ chronology and mean minimum temperature passed the 99% significance test. In July, this correlation was as high as −0.66 (n = 59, $p < 0.01$). The *P. schrenkiana* $\delta^{13}C_{corr}$ chronology showed a significant negative correlation with growing season precipitation, as well as a significant positive correlation with relative humidity in both the previous year's growing season and the current growing season. The highest correlation with relative humidity occurred in August, with a correlation coefficient of −0.409 (n = 59, $p < 0.01$). This chronology had a significantly negative correlation with the saturated water vapor pressure deficit (Figure 6).

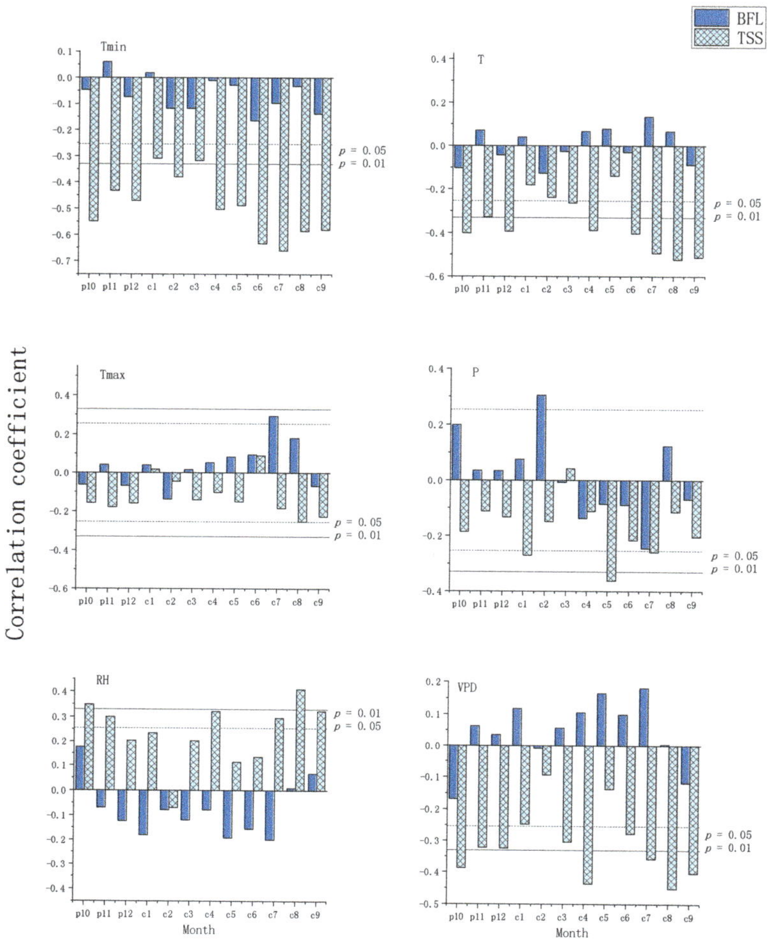

Figure 6. Relationships between the tree-ring stable carbon isotope chronologies and key climatic factors as observed at the Balikun meteorological station. T: mean temperature; Tmin: mean minimum temperature; Tmax: mean maximum temperature; P: precipitation; RH: relative humidity; VPD: saturated water vapor pressure. p10 represents October of the previous year; c9 represents January of the current year.

The analysis of climatological variables observed at the Qincheng meteorological station showed that the mean and mean maximum growing season temperatures were significantly positively correlated with the tree-ring chronologies of both species. With a high correlation coefficient of −0.644 (n = 37, $p < 0.01$) in July of the current year, the $\delta^{13}C_{corr}$ chronology of *L. sibirica* expressed a significantly negative correlation with relative humidity in the current growing season and in October of the previous year, with both passing the 99% significance test. The *P. schrenkiana* $\delta^{13}C_{corr}$ chronology had a significant negative correlation with the relative humidity of the previous winter and from May to July of the current year. The maximum correlation coefficient was −0.683 in July of the current year (n = 37, $p < 0.01$). Compared to the $\delta^{13}C_{corr}$ chronology of *L. sibirica*, that of *P. schrenkiana* was more strongly correlated with climatic elements and was somewhat more influenced by precipitation during the previous winter and by current-year spring temperatures (Figure 7).

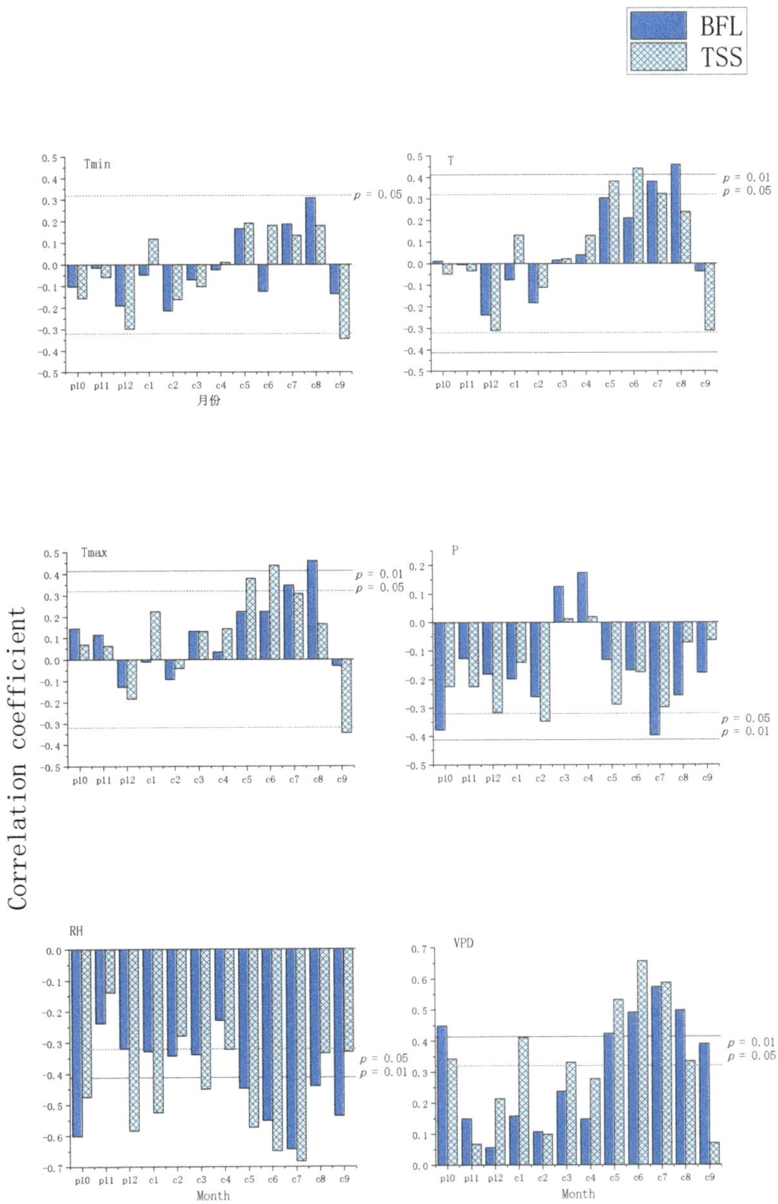

Figure 7. Relationships between the tree-ring stable carbon isotope chronologies and key climatic factors as observed at the Qincheng meteorological station. T: mean temperature; Tmin: mean minimum temperature; Tmax: mean maximum temperature; P: precipitation; RH: relative humidity; VPD: saturated water vapor pressure. p10 represents October of the previous year; c9 represents January of the current year.

Figures 6 and 7 show the correlations between $\delta^{13}C_{corr}$ chronologies of both the two tree species and the meteorological variables.

Further analysis revealed that the $\delta^{13}C_{corr}$ chronologies of *L. sibirica* showed significant positive correlations with monthly mean temperatures, monthly mean maximum temperatures, and various monthly combinations during the growing season of the current year. There were also significant negative correlations with relative humidity: the highest correlations between relative humidity and $\delta^{13}C_{corr}$ chronologies appeared during the growing season (Table 3) and were as high as −0.759.

Table 3. Correlations between various meteorological variables observed at Qingcheng meteorological station and the stable carbon isotope chronology of BFL during the growing season. p7c6 represents the combination from July to June of the previous year; c6c10 represents the combination from June to October of the current year.

	p7c6	p8c7	p9c8	p10c9	c5c8	c5c9	c6c7	c6c8	c6c9	c6c10	c7c8
T	−0.035	0.021	0.061	0.250	0.521 **	0.493 **	0.411 *	0.529 **	0.418 *	0.298	0.497 **
Tmax	0.140	0.198	0.247	0.302	0.49 **	0.477 **	0.383 *	0.526 **	0.439 **	0.36 *	0.495 **
P	−0.435 **	−0.581 **	−0.589 **	−0.406 *	−0.483 **	−0.494 **	−0.375 *	−0.442 **	−0.462 **	−0.483 **	−0.445 **
RH	−0.621 **	−0.663 **	−0.671 **	−0.681 **	−0.739 **	−0.759 **	−0.717 **	−0.724 **	−0.741 **	−0.715 **	−0.636 **
VPD	0.562 **	0.655 **	0.711 **	0.704 **	0.699 **	0.727 **	0.664 **	0.707 **	0.717 **	0.707 **	0.623 **

* and ** represent the correlation coefficient exceeding 95% and 99% of the significance tests, respectively.

The combined correlation results for *P. schrenkiana* were analyzed with the meteorological elements obtained from each meteorological station. This analysis shows that the tree-ring $\delta^{13}C_{corr}$ chronologies of *P. schrenkiana* correlated well with mean temperature and relative humidity from July of the previous year to June of the current year. They also correlated well with precipitation from September of the previous year to August of the current year. The analysis of meteorological data from the Balikun meteorological station revealed a poor correlation between *P. schrenkiana* and the mean maximum temperature.

As a result, further analysis with the mean minimum temperature was conducted to identify the climate variable with the greatest influence on stable carbon isotope fractionation. The results showed that the *P. schrenkiana* $\delta^{13}C_{corr}$ tree-ring chronologies were significantly correlated with the mean minimum temperature from June to October, with a correlation coefficient of −0.692 (n = 58, $p < 0.01$) (Table 4). Additionally, a first-order difference correlation analysis of the mean minimum temperature data for all months, combined with the $\delta^{13}C_{corr}$ chronologies, showed that all data failed the 99% significance test, except for those from July to August of the previous year. Therefore, the correlation between the $\delta^{13}C_{corr}$ chronologies of *P. schrenkiana* and the mean minimum temperature showed only a trend correlation.

Table 4. Correlations between various meteorological variables observed at Balikun meteorological station and the stable carbon isotope chronology of TSS during the growth season. p7c6 represents the combination from July to June of the previous year; c6c10 represents the combination from June to October of the current year.

	p7c6	P8c7	p9c8	p10c9	c5c8	c5c9	c6c7	c6c8	c6c9	c6c10	c7c8
T	−0.544 **	−0.526 **	−0.514 **	−0.491 **	−0.436 **	−0.474 **	−0.477 **	−0.492 **	−0.513 **	−0.52 **	−0.459 **
Tmin	−0.638 **	−0.634 **	−0.627 **	−0.635 **	−0.656 **	−0.667 **	−0.678 **	−0.671 **	−0.677 **	−0.692 **	−0.653 **
P	−0.422 **	−0.531 **	−0.55 **	−0.5 **	−0.432 **	−0.48 **	−0.378 **	−0.377 **	−0.424 **	−0.452 **	−0.267 *
RH	0.474 **	0.437 **	0.41 **	0.338 **	0.324 *	0.357 **	0.26 *	0.351 **	0.376 **	0.391 **	0.383 **
VPD	−0.597 **	−0.536 **	−0.488 **	−0.453 **	−0.403 **	−0.432 **	−0.373 **	−0.434 **	−0.456 **	−0.466 **	−0.431 **

* and ** represent the correlation coefficient exceeding 95% and 99% of the significance tests, respectively.

The first-order difference analysis revealed that the combined correlation data passed the 99% significance test. This analysis pertained to the combined correlation data of other meteorological elements with the *P. schrenkiana* $\delta^{13}C_{corr}$ chronologies. Additionally, it showed a significantly positive correlation of these chronologies with the mean temperature and the mean maximum temperature from the end of the previous growing season to the current growing season. The first-order difference results showed that *P. schrenkiana* $\delta^{13}C_{corr}$ chronologies exhibited a significantly negative correlation with precipitation from September of the previous year to August of the current year, with a correlation coefficient of −0.502 (n = 57, $p < 0.01$). They also demonstrated a significant negative correlation with relative humidity from May to July of the current year, with a correlation coefficient of −0.549 (n = 57, $p < 0.01$).

After comparing the data from the meteorological stations in Qincheng and Balikun, we found that the *P. schrenkiana* tree-ring $^{13}C_{corr}$ chronologies were positively correlated with the mean temperature and mean minimum temperature of the meteorological station in Balikun and negatively correlated with the mean temperature and mean minimum temperature in Qincheng. The data for relative humidity also had opposite correlations. To study this difference, we determined the 21-year sliding correlation between the better-combined correlation data from the Balikun meteorological station and chronology; this could reflect the change in correlation between the two sets of data over time.

The sliding correlation analysis results showed that the combined correlation data from the Balikun meteorological station and the 21-year sliding correlation results of the TSS chronology underwent a significant change from negative to positive and positive to negative over time (Figure 8). In contrast, data from the Qincheng meteorological station were of a shorter age, which stopped in 1996. The time scale of these data and those of the Balikun meteorological station only overlapped for a short period of time, so the overall correlations were positive and negative in opposite directions. We then performed a correlation analysis of the first-order differences using the single-month meteorological data from the two stations and the first-order differences using the chronology data. We found that the two were consistent regarding their changes at high frequencies. The above correlations indicated that the main climatic factors affecting $\delta^{13}C$ fractionation in *P. schrenkiana* were precipitation from the end of the previous growing season and throughout the current growing season and the mean growing season temperature.

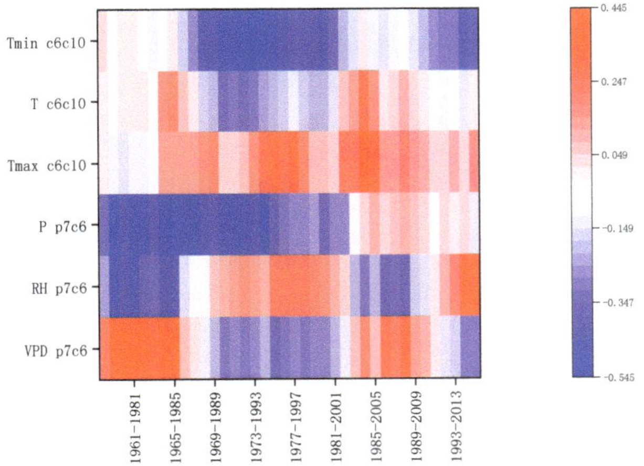

Figure 8. Sliding correlation analysis between the *P. schrenkiana* $\delta^{13}C_{corr}$ chronology and meteorological variables. P7c6 represents the combination from July to June of the previous year;.c6c10 represents the combination from June to October of the current year.

The results of the single-factor correlation analysis indicated that the growing season precipitation and relative humidity were significantly negatively correlated with *L. sibirica* tree-ring $\delta^{13}C_{corr}$ chronologies. On the other hand, *P. schrenkiana*'s $^{13}C_{corr}$ chronologies were somewhat influenced by the growing season temperature, precipitation, and relative humidity. The stable carbon isotope fractionation in tree rings was probably influenced by a combination of climatic factors. To further investigate these relationships, we performed a collinearity analysis between the main climatic factors and the tree-ring $\delta^{13}C_{corr}$ chronologies. The results showed that the mean growing season temperature, precipitation, and relative humidity explained 58.6% of the variance in the BFL tree-ring stable carbon isotope chronologies. For TSS chronologies, the growing season mean temperature, precipitation from September of the previous year to August of the current year, and relative humidity from July of the previous year to June of the current year explained 56.6% of the variance. The climatic factors influencing the stable carbon isotope fractionation were higher in *P. schrenkiana* tree rings than those of *L. sibirica*.

The main factors limiting stable carbon isotope fractionation in *L. sibirica* included relative humidity during the growing season and the mean temperature. The *P. schrenkiana* tree-ring stable carbon isotope fractionation process was synergistically influenced by precipitation from the previous growing season to the current growing season and by the mean growing season temperature. These variables accounted for more than 30% of the variance (Figure 9).

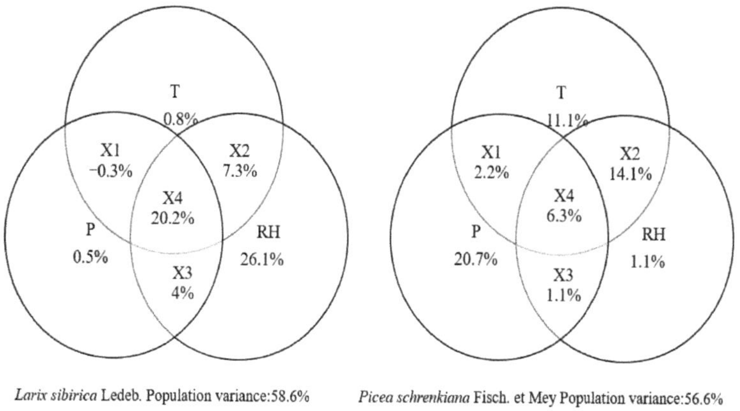

Figure 9. Collinearity analysis between the $\delta^{13}C_{corr}$ and meteorological factors. T is the mean temperature combination from June to August at Qingcheng meteorological station; P is the precipitation–growing season combination; RH is the relative humidity growing season combination. Only the combinations with the best correlations with the $\delta^{13}C_{corr}$ of *L. sibirica* are shown. T is the mean temperature combination from June to October (c6c10) at Balikun meteorological station; P is the precipitation combination from September of the previous year to August of the current year(p9c8); RH is the relative humidity combination from July of the previous year to June of the current year(p7c6). Only the combinations with the best correlations with the $\delta^{13}C_{corr}$ of *P. schrenkiana* are shown. X1 is the combined effect of the mean temperature and precipitation. X2 is the combined effect of the mean temperature and relative humidity. X3 is the combined effect of precipitation and relative humidity. X4 is the combined effect of all three climate factors. The corresponding percentage values represent the explained variance of the parameters.

4. Discussion

A series of enzymatic reactions took place during the CO_2 assimilation process and during photosynthesis. Changes in the climate can impact both processes. The climatic factors that had a significant influence on organic matter synthesis could be divided into two categories: moisture and light [45]. Relative humidity and precipitation fell into the first category, whereas temperature and light intensity fell into the second. Changes in precipitation and relative humidity affected the humidity gradients and pressure, which altered the stomatal opening sizes and CO_2 concentrations inside the cells. These, in turn, altered the isotopic composition of the synthesized organic matter. The production and activity of photosynthetic enzymes were primarily influenced by temperature and light intensity, which affected the speed and effectiveness of photosynthesis, and, thus, the isotopic composition of the synthesized organic matter. Additionally, changes in precipitation and relative humidity affected the saturated water vapor pressure, which drove stomatal conductance [46]. Both the rates of CO_2 uptake by plants and changes in leaf stomatal conductance fluxes affected the intra-leaf intercellular CO_2 concentration (C_i) or the ratio of intra-leaf intercellular CO_2 concentration to the atmospheric CO_2 concentration (C_i/C_a). Climatic factors such as temperature, humidity, and light all affected $\delta^{13}C$ by influencing the distribution effects of the stomata and photosynthetic carboxylases on carbon isotopes and intercellular CO_2 concentration [43].

Stable carbon isotopes record the balance between the photosynthetic rate and stomatal conductance. In arid regions, precipitation, relative humidity, and soil moisture status are the dominant factors that cause changes in stable carbon isotope signals. In humid regions, summer irradiance and temperature are the primary drivers of these changes [45]. *L. sibirica* is a deciduous species that grows on sunny slopes and has a deep root system. As a result, it is drought-tolerant and is less affected by precipitation compared to *P. schrenkiana*. *L. sibirica* tree-ring stable carbon isotope chronologies are significantly negatively correlated with precipitation and relative humidity, indicating how stomatal conductance has an important effect on the *L. sibirica* tree-ring stable carbon isotope fractionation [47]. Because the meteorological record at the Qincheng station, which was used for *L. sibirica* analysis, was shorter, the meteorological variables showed higher overall correlations with *P. schrenkiana* chronologies. *L. sibirica's* stable carbon isotopes were most responsive to relative humidity during the growing season, and relative humidity was the main factor controlling the fractionation of stable carbon isotopes in *L. sibirica*. In the Tianshan Mountains, the growing season of *L. sibirica* is from May to September. Growth is fastest in June and July, which is when the $\delta^{13}C_{corr}$ chronology can be most significantly correlated with temperature and relative humidity. This is also the period during which plants are most affected by water stress (low relative humidity or low precipitation) and/or high temperature. Plant stomata may close to avoid excessive water loss, reducing the CO_2 concentration in the plant [4]. The above analysis indicates that the main factor controlling $\delta^{13}C$ fractionation in *L. sibirica* tree rings during the growing period is relative humidity which is influenced by both temperature and precipitation. This is especially true during the growing season when the gas exchange with the atmospheric environment occurs, and higher temperatures lead to enhanced transpiration, which indirectly affects photosynthesis. Drought stress is accompanied by more intense light, higher temperatures, and high evaporation from July to August. Together, these factors result in stomatal closure, reduced conductance, and an increase in photosynthetically fixed ^{13}C. Given that isotope fractionation in *L. sibirica* is less influenced by precipitation, the temperature increase in summer decreases the relative humidity in *L. sibirica*'s growing environment. A lower relative humidity leads to an increase in diffusive evaporation, which trees regulate by decreasing their stomatal conductance in order to reduce water loss, thus affecting the stable carbon isotope fractionation of *L. sibirica*.

Saurer et al. [7] studied the relationship between $\delta^{13}C$ in the tree rings of French beech trees, as well as climatic variables, at sites with different moisture conditions. They found that the $\delta^{13}C$ values of beech tree rings were significantly correlated with relative air humidity from April to September. This finding is consistent with that of this study, i.e., relative humidity during the growing season had a significantly negative correlation with the *L. sibirica* tree-ring $\delta^{13}C$. Similarly, Alexander et al. [11] found that the stable carbon isotopes of *L. sibirica* tree rings in eastern Siberia were significantly and positively correlated with the mean maximum temperature from June to July ($r = 0.46$, $p < 0.01$) and were negatively correlated with precipitation in July ($r = -0.44$, $p < 0.05$). Jin Xiang et al. [48], in a study on Minjiang fir, found that the tree-ring $\delta^{13}C$ chronologies from a high-elevation zone were significantly and negatively correlated with the monthly mean relative humidity in July and August of the previous year and April of the current year. This result is consistent with the results of the combined correlation analysis presented in this paper.

However, a study on Linzhi fir trees in Tibet by Liu et al. [19] showed the opposite results to those discussed above. In their study, Linzhi fir tree-ring $\delta^{13}C$ values were affected by a strong lag effect and showed a significantly positive correlation with relative humidity from the previous September to November. This is due to the fact that Linzhi fir occurs in the humid zone along the middle reaches of the Yarlung Tsangpo River, where the relative humidity is not a growth-limiting factor. After September, with air humidity significantly decreased, relative humidity becomes the growth-limiting factor. This is also related to the influence of unique climatic conditions on the physiological and metabolic activities of trees at high altitudes on the Tibetan Plateau.

P. schrenkiana is an evergreen conifer species that grows on shady slopes and has a shallow root system. As a result, it has limited access to deep soil water resources and is more susceptible to drought stress during the growing season. The data and charts above show that *P. schrenkiana* $\delta^{13}C_{corr}$ chronologies are strongly correlated with precipitation at high and low frequencies, indicating that precipitation plays a key role in the stable carbon isotope fractionation of tree rings. In general, the optimum temperature for net photosynthesis in evergreen conifers is 10–25 °C [49]. At the sampling site, the mean monthly temperature in summer (June to August) was 17 °C; the summer season was when *P. schrenkiana* growth was at its peak [50]. Because evergreen trees continue to photosynthesize at the end of the growing season to maintain their above-ground biomass, they are also affected by climatic factors at the end of the previous year's growing season.

It is, therefore, unsurprising that the combined meteorological data (Balikun station) from the previous year's growing season to the current year's growing season correlated well with *P. schrenkiana* tree-ring stable carbon isotopes. Following the first-order difference analysis, we found that the first-order difference between the $\delta^{13}C$ values and the combined mean maximum temperature was significantly and positively correlated from the end of the previous growing season to the current growing season. *P. schrenkiana* grows on shady slopes with limited sunlight. As temperatures decrease at the end of the growing season, photosynthetic enzyme activity decreases, inhibiting photosynthesis. Thus, the mean maximum temperature affects high-frequency changes in tree-ring stable carbon isotope fractionation. Upon looking at the single-month correlation data, we found that the $\delta^{13}C$ values of *P. schrenkiana* were significantly and negatively correlated with the mean minimum temperature from June to July. From the combined correlation data, we found that the *P. schrenkiana* $\delta^{13}C$ values were most strongly correlated with the mean minimum temperature from June to October. Our plant physiology analysis indicated that June and July represented a period of rapid growth, but higher mean minimum temperatures outside these months could prolong the growing season and enhance photosynthetic efficiency.

Schleser et al. [51] suggested that the relationship between tree-ring $\delta^{13}C$ values and temperature is not simply linear but rather "bell-shaped". That is, the carbon isotope fractionation of plants is greatest when temperatures are optimal, which means that $\delta^{13}C$ values are at their lowest rate. If temperatures are lower than is optimal for plant growth, then tree-ring $\delta^{13}C$ values can be negatively correlated with temperature. If temperatures are higher than is optimal, then tree-ring $\delta^{13}C$ values can be positively correlated with temperature. However, if temperatures oscillate near the optimal temperature point, their response relationship is more complicated. The mean minimum temperature occurs at night and is lower than the optimal temperature for plant growth. As a result, tree-ring $\delta^{13}C$ values can be negatively correlated with temperature. Photosynthesis is extremely weak at night when there is little atmospheric turbulence and a significant accumulation of CO_2 under the canopy. At the beginning of the night, stomatal conductance is greater than photosynthesis, and the $\delta^{13}C$ fractionation of the carboxylation process is stronger. As the fractionation process is weakened due to the depletion of chemical energy accumulated during the daytime and respiration continues, the external CO_2 concentration increases.

The results of our collinearity analysis indicate that the tree-ring stable carbon isotopes of *P. schrenkiana* are synergistically influenced by the mean summer temperature, precipitation, and relative humidity. During the growing season (especially in summer), carbon isotope fractionation is jointly regulated by the photosynthetic rate (temperature) and stomatal conductance (precipitation and relative humidity). Previous studies on the responses of tree-ring stable carbon isotopes to the climate in Xinjiang suggest that tree-ring stable carbon isotope fractionation in the Altai Mountains may be controlled by a combination of temperature and moisture [30]. In the central Tianshan Mountains, fractionation may be primarily moisture-driven [52]. The findings for both regions are consistent with those of the present study. Hemming et al. [6] found that high-frequency changes in the tree-ring $\delta^{13}C$ values of beech, oak, and pine trees were negatively correlated with relative humidity from June to September. These findings are consistent with those of this study, which indicate a negative correlation between relative humidity and *P. schrenkiana* tree-ring $\delta^{13}C$ during the growing season.

5. Conclusions

The stable carbon isotope values, coefficients of variation, variance, and standard deviation of *P. schrenkiana* tree rings were marginally higher than those of *L.sibirica*, suggesting that interannual variation in the tree-ring stable carbon isotopes of *L. sibirica* were relatively more stable. *L. sibirica* tree-ring $\delta^{13}C_{corr}$ chronologies showed a significantly negative correlation with precipitation and relative humidity during the growing season and a significantly positive correlation with mean temperature, mean maximum temperature, and saturated water vapor pressure deficit during the growing season. The *P. schrenkiana* tree-ring $\delta^{13}C_{corr}$ was negatively correlated with mean temperature, mean minimum temperature, precipitation, and saturated water vapor pressure deficit from the previous growing season to the current year and positively correlated with relative humidity. The main climatic factor controlling stable carbon isotope fractionation in *L. sibirica* tree rings during the growing season was relative humidity. In contrast, stable carbon isotope fractionation in *P. schrenkiana* tree rings appeared to be jointly controlled by hydrothermal conditions.

Overall, climatic factors had a greater impact on the stable carbon isotope fractionation in *P. schrenkiana* compared to *L.sibirica*. Relative humidity during the growing season and the synergistic effect of the mean temperature and relative humidity were the main factors limiting stable carbon isotope fractionation in *L. sibirica* tree rings. In *P. schrenkiana* tree rings, the mean temperature and precipitation during the growing season had the strongest influence on the fractionation process.

This research improved our understanding of how stable carbon isotope fractionation in tree rings varies between species and geographical regions. It can also be used as a reference to forecast how *L. sibirica* and *P. schrenkiana* change as the climate changes. Finally, it provides evidence to support more intelligent forest management in arid lands and advances our understanding of how climate change impacts tree growth in extremely dry environments.

Author Contributions: Conceptualization, R.Z.; Data curation, R.Z., L.Q. and H.G.; Formal analysis, Y.J.; Funding acquisition, R.Z. and L.Q.; Investigation, R.Z., L.Q. and Z.P.; Methodology, R.Z., L.Q., Z.P. and R.A.; Resources, R.Z. and G.L.; Validation, Y.J.; Writing—original draft, Y.J. and R.Z.; Writing—review and editing, Y.J., L.Q. and R.Z. All authors have read and agreed to the published version of the manuscript.

Funding: This work was supported by the Natural Science Foundation of Xinjiang (2022D01A160), the regional collaborative innovation project of Xinjiang (2021E01022, 2022E01045), the National Natural Science Foundation of China (41975110), the Tianshan Talent Project of Xinjiang (2022TSY-CCX0003), and the Tianshan Youth Project of Xinjiang (2020Q026).

Institutional Review Board Statement: Not applicable.

Informed Consent Statement: Not applicable.

Data Availability Statement: The data presented in this study are available on request from the corresponding author.

Acknowledgments: We thank Tongwen Zhang, Huaming Shang, Yuting Fan, Heli Zhang, Youping Chen and Bin Liu for their help and support during the sample collection and experimental analysis process.

Conflicts of Interest: The authors declare no conflict of interest.

References

1. Robertson, I.; Switsur, V.R.; Carter, A.H.C.; Barker, A.C.; Waterhouse, J.S.; Briffa, K.R.; Jones, P.D. Signal strength and climate relationships in $^{13}C/^{12}C$ ratios of tree ring cellulose from oak in east England. *J. Geophys. Res. Atmos.* **1997**, *102*, 19507–19516. [CrossRef]
2. Leavitt, S.W.; Lone, A. Seasonal stable-carbon isotope variability in tree-ring: Possible paleoenvironmental signals. *Chem. Geol. Isot. Geosci. Sect.* **1991**, *87*, 59–70. [CrossRef]
3. Wilson, A.T.; Grinsted, M.J. $^{13}C/^{12}C$ in cellulose and lignin as palaeothermometers. *Nature* **1977**, *265*, 133–135. [CrossRef]
4. Francey, R.J.; Farquhar, G.D. An explanation of $^{13}C/^{12}C$ variations in tree-rings. *Nature* **1982**, *297*, 28–31. [CrossRef]
5. Saurer, M.; Siegenthaler, U.; Schweingruber, F. The climate-carbon isotope relationship in tree-ring and the significance of site conditions. *Tellus B* **1995**, *47*, 320–330. [CrossRef]
6. Hemming, D.I.; Switsur, V.R.; Waterhouse, J.S.; Heaton, T.H.E.; Carter, A.H.C. Climate variation and the stable carbon isotope composition of tree ring Cellulose: An intercomparison of Quercus robur, Fagus sylvatica and Pinus silvestris. *Tellus B Chem. Phys. Meteorol.* **1998**, *50*, 25–33. [CrossRef]
7. Saurer, M.; Siegenthaler, U. $^{13}C/^{12}C$ isotope ratios in trees are sensitive to relative humidity. *Dendrochronologia* **1989**, *7*, 9–13.
8. Barber, V.A.; Juday, G.P.; Finney, B.P.; Wilmking, M. Reconstruction of Summer Temperatures in Interior Alaska from Tree-Ring Proxies: Evidence for Changing Synoptic Climate Regimes. *Clim. Chang.* **2004**, *63*, 91–120. [CrossRef]
9. Hafner, P.; Levanič, T. Stable carbon isotopes in Norway Spruce (*Picea abies* (L.) Karst.) tree-ring at two sites in Slovenia. *Zb. Gozdarstva Lesar.* **2009**, *88*, 43–52.
10. Pawelczyk, S.; Pazdur, A.; Halas, S. Stable carbon isotopic composition of tree-ring from a Pine tree from Augustów Wilderness, Poland, as a temperature and local environment conditions indicator. *Isot. Environ. Health Stud.* **2004**, *40*, 145–154. [CrossRef]
11. Kirdyanov, A.V.; Treydte, K.S.; Nikolaev, A.; Helle, G.; Schleser, G.H. Climate signals in tree-ring width, density and $\delta^{13}C$ from larches in eastern Siberia (Russia). *Chem. Geol.* **2008**, *252*, 31–41. [CrossRef]
12. Sonninen, E.; Jungner, H. Stable carbon isotope in tree-rings of a Scots alpine pine (*Pinus sylvestris* L.) from northern Finland. *Paläoklimaforschung* **1995**, *15*, 121–128.
13. Sheu, D.D.; Kou, P.; Chiu, C.H.; Chen, M.J. Variability of tree-ring $\delta^{13}C$ in Taiwan fir: Growth effect and response to May–October temperatures. *Geochim. Cosmochim. Acta* **1996**, *60*, 171–177. [CrossRef]
14. Saurer, M.; Borella, S.; Schweingruber, F.; Siegwolf, R. Stable carbon isotopes in tree-ring of beech: Climatic versus site-related influences. *Trees* **1997**, *11*, 291–297. [CrossRef]
15. Liu, Y.; Wu, X.D.; Leavitt, S.W.; Hughes, M.K. Stable carbon isotope in tree-ring from Huangling, China and climatic Variation. *Sci. China Ser. D Earth Sci.* **1996**, *39*, 152–160.
16. Liu, X.H.; Shao, X.M.; Wang, L.L.; Zhao, L.J.; Wu, P.; Chen, T.; Qin, D.H.; Ren, J.W. Climatic significance of the stable carbon isotope composition of tree-ring cellulose: Comparison of Chinese hemlock (Tsuga chinensis Pritz) and alpine pine (Pinus densata Mast) in a temperate-moist region of China. *Sci. China Ser. D Earth Sci.* **2007**, *50*, 1076–1085. [CrossRef]
17. Wang, W.Z.; Liu, X.H.; Shao, X.M.; Leavitt, S.; Xu, G.B.; An, W.L.; Qin, D.H. A 200 years temperature record from tree ring $\delta^{13}C$ at the Qaidam Basin of the Tibetan Plateau after identifying the optimum method to correct for changing atmospheric CO_2 and $\delta^{13}C$. *J. Geophys. Res.* **2011**, *116*, G04022. [CrossRef]
18. Liu, X.H.; Qin, D.H.; Shao, X.M.; Chen, T.; Ren, J.W. Tibetan Himalayan fir annual rings $\delta^{13}C$ and climatic significance. *Sci. Bull.* **2003**, *48*, 1209–1213, (In Chinese, with English Abstract).
19. Liu, X.H.; Qin, D.H.; Shao, X.M.; Ren, J.W.; Wang, Y. Stable Carbon Isotope of Abies spectabibis from Nyingchi County of Tibet Autonomous Region and Its Response to Climate Change. *J. Glaciol. Geocryol.* **2002**, *24*, 574–578, (In Chinese, with English Abstract).
20. Liu, Y.; Wu, X.D.; Leavitt, S.W.; Hughes, M.K. Huangling tree rings stable C isotope and climate change. *Chin. Sci.* **1996**, *26*, 125–130, (In Chinese, with English Abstract).
21. Zhang, R.; Ermenbaev, B.; Zhang, H.; Shang, H.; Zhang, T.; Yu, S.; Chontoev, D.T.; Satylkanov, R.; Qin, L. Natural discharge changes of the Naryn River over the past 265 years and their climatic drivers. *Clim. Dyn.* **2020**, *55*, 1269–1281. [CrossRef]
22. Zhang, R.; Ermenbaev, B.; Zhang, T.; Ali, M.; Qin, L.; Satylkanov, R. The Radial Growth of Schrenk Spruce (Picea schrenkiana Fisch. et Mey.) Records the Hydroclimatic Changes in the Chu River Basin over the Past 175 Years. *Forests* **2019**, *10*, 223. [CrossRef]
23. Zhang, R.; Shang, H.; Yu, S.; He, Q.; Yuan, Y.; Bolatov, K.; Mambetov, B.T. Tree-ring-based precipitation reconstruction in southern Kazakhstan, reveals drought variability since A.D. 1770. *Int. J. Climatol.* **2017**, *37*, 741–750. [CrossRef]
24. Zhang, R.; Yuan, Y.; Gou, X.; He, Q.; Shang, H.; Zhang, T.; Chen, F.; Ermenbaev, B.; Yu, S.; Qin, L.; et al. Tree-ring-based moisture variability in western Tianshan Mountains since A.D. 1882 and its possible driving mechanism. *Agric. For. Meteorol.* **2016**, *218–219*, 267–276. [CrossRef]
25. Zhang, R.; Qin, L.; Shang, H.; Yu, S.; Gou, X.; Mambetov, B.T.; Bolatov, K.; Zheng, W.; Ainur, U.; Bolatova, A. Climatic change in southern Kazakhstan since 1850 C.E. inferred from tree rings. *Int. J. Biometeorol.* **2020**, *64*, 841–851. [CrossRef] [PubMed]
26. Zhang, R.; Wei, W.; Shang, H.; Yu, S.; Gou, X.; Qin, L.; Bolatov, K.; Mambetov, B.T. A tree ring-based record of annual mass balance changes for the TS.Tuyuksuyskiy Glacier and its linkages to climate change in the Tianshan Mountains. *Quat. Sci. Rev.* **2019**, *205*, 10–21. [CrossRef]
27. Qin, L.; Shang, H.M.; Zhang, T.W.; Liu, W.P.; Zhang, R.B. Response comparison of the tree-ring $\delta^{13}C$ to climate on the southern and northern slopes of Tianshan Mountains. *Acta Ecol. Sin.* **2021**, *41*, 5713–5724. (In Chinese, with English Abstract)

28. Qin, L.; Yuan, Y.J.; Yu, S.L.; Shang, H.M.; Chen, F.; Fan, Z.A.; Zhang, T.W.; Tong, Z. Tree-ring based monthly mean relative humidity reconstruction for the Eastern Tianshan Mountains during the last 443 years. *J. Earth Environ.* **2012**, *3*, 908–914. (In Chinese, with English Abstract)
29. Shang, H.M.; Wei, W.S.; Yuan, Y.J.; Yu, S.L.; Zhang, R.B.; Chen, X.J. Environmental Response of Tree-ring Width and Stable Carbon Isotope at North Slope of Eastern Tianshan Mountains. *Desert Oasis Meteorol.* **2010**, *4*, 6–10. (In Chinese, with English Abstract)
30. Zhang, R.B.; Yuan, Y.J.; Wei, W.S.; Shang, H.M.; Yu, S.L.; Zhang, T.W.; Chen, F.; Fan, Z.A.; Qin, L. Response of Stable Carbon Isotope of *Larix sibirica* Ledeb. Tree-rings to Climate Change. *Arid Zone Res.* **2012**, *29*, 328–334. (In Chinese, with English Abstract)
31. Peng, Z.B.; Li, X.J.; Zhang, R.B.; Qin, L.; Zhang, H.L.; Chen, Y.P.; Liu, R. The response of tree-ring chronologies of Schrenk spruce (*Picea schrenkiana* Fisch. et Mey.) to climate change at high- and low- elevations of the Eastern Tianshan Mountains, Xinjiang, using different detrending methods. *Acta Ecol. Sin.* **2019**, *39*, 1595–1604. (In Chinese, with English Abstract)
32. Holmes, R.L. Computer-assisted quality control in tree-ring dating and measurement. *Tree-Ring Bull.* **1983**, *43*, 69–78.
33. Cook, E.R. *A Time Series Analysis Approach to Tree Ring Standardization*; The University of Arizona: Tucson, AZ, USA, 1985.
34. Porté, A.; Loustau, D. Seasonal and interannual variations in carbon isotope discrimination in a maritime pine (Pinus pinaster) stand assessed from the isotopic composition of cellulose in annual rings. *Tree Physiol.* **2001**, *21*, 861–868. [CrossRef]
35. Gagen, M.; McCarroll, D.; Loader, N.; Robertson, I.; Jalkanen, R.; Anchukaitis, K. Exorcising the 'segment length curse': Summer temperature reconstruction since AD 1640 using non-detrended stable carbon isotope ratios from pine trees in northern Finland. *Holocene* **2007**, *17*, 435–446. [CrossRef]
36. Brendel, O.; Iannetta, P.; Stewart, D. A rapid and simple method to isolate pure alpha-cellulose. *Phytochem. Anal.* **2000**, *11*, 7–10. [CrossRef]
37. Crampton, E.W.; Maynard, L.A. The Relation of Cellulose and Lignin Content to the Nutritive Value of Animal Feeds. *J. Nutr.* **1938**, *15*, 383–395. [CrossRef]
38. Evans, M.N.; Reichert, B.K.; Kaplan, A.; Anchukaitis, K.J.; Vaganov, E.A.; Hughes, M.K.; Cane, M.A. A forward modeling approach to paleoclimatic interpretation of tree-ring data. *J. Geophys. Res.* **2006**, *111*, G03008. [CrossRef]
39. Coplen, T.B. Discontinuance of SMOW and PDB. *Nature* **1995**, *375*, 285. [CrossRef]
40. Leavitt, S.W.; Lara, A. South American tree-ring show declining $\mu^{13}C$ trend. *Tellus B Chem. Phys. Meteorol.* **1994**, *46*, 152–157. [CrossRef]
41. Rubino, M.; Etheridge, D.M.; Thornton, D.P.; Howden, R.; Allison, C.E.; Francey, R.J.; Langenfelds, R.L.; Steele, L.P.; Trudinger, C.M.; Spencer, D.A.; et al. Revised records of atmospheric trace gases CO_2, CH_4, N_2O, and $\delta^{13}C$-CO_2 over the last 2000 years from Law Dome, Antarctica. *Earth Syst. Sci. Data* **2019**, *11*, 473–492. [CrossRef]
42. Cernusak, L.A.; Ubierna, N. Carbon Isotope Effects in Relation to CO_2 Assimilation by Tree Canopies. In *Stable Isotopes in Tree Rings*; Springer International Publishing: Cham, Switzerland, 2022; pp. 291–310. [CrossRef]
43. Campbell, G.S.; Norman, J.M. *An Introduction to Environmental Biophysics*, 2nd ed.; Springer: New York, NY, USA, 1998. [CrossRef]
44. Speer, J.H. *Fundamentals of Tree-Ring Research*; The University of Arizona Press: Tucson, AZ, USA, 2010; p. 333.
45. McCarroll, D.; Loader, N.J. Stable isotopes in tree rings. *Quat. Sci. Rev.* **2004**, *23*, 771–801. [CrossRef]
46. McCarroll, D.; Pawellek, F. Stable carbon isotope ratios of Pinus sylvestris from northern Finland and the potential for extracting a climate signal from long Fennoscandian chronologies. *Holocene* **2001**, *11*, 517–526. [CrossRef]
47. Treydte, K.S.; Frank, D.C.; Saurer, M.; Helle, G.; Schleser, G.H.; Esper, J. Impact of climate and CO2 on a millennium-long tree-ring carbon isotope record. *Geochim. Cosmochim. Acta* **2009**, *73*, 4635–4647. [CrossRef]
48. Jin, X.; Xu, Q.; Liu, S.; Jiang, C.Q. Relationships of stable carbon isotope of Abies faxoniana tree-rings to climate in sub-alpine forest in Western Sichuan. *Acta Ecol. Sin.* **2014**, *34*, 1831–1840. (In Chinese, with English Abstract)
49. Yu, G.R.; Wang, Q.F. Physiological ecology of photosynthesis, transpiration, and water use in plants. *J. Nucl. Agric.* **2010**, *24*, 579. (In Chinese, with English Abstract)
50. Zhang, R.B.; Yuan, Y.J.; Gou, X.H.; Zhang, T.W.; Zou, C.; Ji, C.R.; Fan, Z.A.; Qin, L.; Shang, H.M.; Li, X.J. Intra-annual radial growth of Schrenk spruce (Picea schrenkiana Fisch. et Mey) and its response to climate on the northern slopes of the Tianshan Mountains. *Dendrochronologia* **2016**, *40*, 36–42. [CrossRef]
51. Schleser, G.H.; Frielingsdorf, J.; Blair, A. Carbon isotope behaviour in wood and cellulose during artifcial aging. *Chem. Geol.* **1999**, *158*, 121–130. [CrossRef]
52. Xu, G.B.; Liu, X.H.; Qin, D.H.; Chen, T.; Sun, W.Z.; An, W.L.; Wang, W.Z.; Wu, G.J.; Zeng, X.M.; Ren, J.W. Drought history inferred from tree ring $\delta^{13}C$ and $\delta^{18}O$ in the central Tianshan Mountains of China and linkage with the North Atlantic Oscillation. *Theor. Appl. Climatol.* **2014**, *116*, 385–401. [CrossRef]

forests

MDPI

Article

Regional Characteristics of the Climatic Response of Tree-Ring Maximum Density in the Northern Hemisphere

Shulong Yu , Yuting Fan *, Tongwen Zhang *, Li Qin , Huaming Shang, Heli Zhang, Kexiang Liu and Xiaoxia Gou

Key Laboratory of Tree-Ring Physical and Chemical Research of China Meteorological Administration, Key Laboratory of Tree-Ring Ecology of Xinjiang Uigur Autonomous Region, Institute of Desert Meteorology, China Meteorological Administration, Urumqi 830002, China; yushl@idm.cn (S.Y.); jiangsx1989@163.com (S.J.); river0511@163.com (R.Z.); qinli@idm.cn (L.Q.); shanghm@idm.cn (H.S.); zhangheli@idm.cn (H.Z.); liukx@idm.cn (K.L.); gouxiaox@idm.cn (X.G.)

* Correspondence: fanyt@idm.cn (Y.F.); zhangtw@idm.cn (T.Z.)

Abstract: The maximum latewood density (MXD) of tree rings can reflect the temperature of the growing season, but the timing of the response differs among regions. We selected 152 maximum latewood density chronologies from the Northern Hemisphere that showed a significant response to temperature. Based on a cluster analysis and the sensitivity of MXD to the monthly mean temperature, the chronologies were classified into six clusters. The clusters showed distinct regional characteristics, and the period and peak month of significant response of the chronologies in each cluster to temperature were different. Spatial synchronization of the MXDs revealed that the two clusters distributed in Europe showed the most consistency and the strongest response to the April–September monthly mean temperature compared with the other clusters. Temperature accounted for more than 40% of the total MXD variance in all clusters, whereas the effect of precipitation was much smaller. In addition to climatic factors, the random effect of the latitude and longitude of sampling sites, elevation, and tree species was a major factor contributing to the variance in MXD in each cluster. Latitude and longitude had the strongest influence among the three random effects, and tree species had the weakest influence, except at high latitudes. The MXD of each cluster showed sensitivity to temperature within a certain interval, with a positive linear response, and the sensitivity interval was greatest at high latitudes. Certain clusters showed a negative linear sensitivity to precipitation. The results provide a reference for studying the climatic threshold of large-scale tree-ring density formation.

Keywords: northern hemisphere; maximum density chronology; cluster analysis; random effect; sensitivity analysis

1. Introduction

The maximum latewood density (MXD) of tree rings holds considerable potential for global climate reconstruction owing to its greater sensitivity to temperature in the growing season compared with that of tree-ring width (RW) [1]. The MXD has a stronger and more robust climatic signal compared with the signal from tree-ring widths and is reasonably independent of species and site ecology [2]. Based on the existing results of tree-ring density analysis, Schweingruber [3] suggested that the maximum latewood density is suitable for large-scale climate reconstruction. Briffa et al. [4,5] explored in detail the nature of the temperature sensitivity of MXD across the Northern Hemisphere, dividing the network of sampling sites into nine regions according to chronological similarity and demonstrating the dominant common influence of April–September monthly mean temperatures on MXD variability. Frank and Esper [6] observed that the density chronologies explained more of the variance than ring-width chronologies and were significantly correlated with temperature in the April–September period based on a network of 31 MXD and 53 ring-width data sets distributed across the European Alps. However, Büntgen et al. [7] found that the growth response was strongest in the June–September mean temperatures based on 180 larches (*Larix decidua* Mill.) MXD series from the European Alps. Björklund et al. [8] compared the common variability and climate sensitivity of MXD chronologies for Scots pine (*Pinus sylvestris* L.) at five sites in the Scandinavian Mountains and determined that the MXD values showed the strongest and most stable response to the temperature from April to August. In addition to temperature, precipitation [9], the self-calibrated Palmer Drought Severity Index (scPDSI) [10], proxies of water availability [11], and other factors influence the MXD chronology in some areas. However, substantially fewer studies have evaluated the effects of such factors than the sensitivity of MXD to summer temperature. Although summer temperature is an important factor influencing the growth sensitivity of the MXD chronology, the months of the strongest responses differ among regions, comprising April–September, April–August, and June–September. Esper et al. [12] observed a good response to spring and autumn temperatures in the south of Spain, which revealed a notable difference from other regions.

Friedrichs et al. [13] used Varimax rotation principal component analysis, Ward's method, and the average-linkage method to conduct a cluster analysis to effectively classify 46 homogeneous oak (*Quercus* spp.) tree-ring width chronologies from Central and Western Germany. The different methods resulted in differences in classification for about 20% of the sample sites. In evaluating the trends and uncertainty of warming indicators in Siberia, Esper et al. [14] classified the tree-ring width and density in Siberia. Based on the similarities among the tree-ring chronologies, 78 tree-ring density chronologies in this region were divided into seven clusters. However, the influence of geographical location, elevation, and other factors, as well as the difference in sensitivity to climatic factors, have not been analyzed previously in the classification of MXD data sets from the Northern Hemisphere. Previous chronological classifications were based on the similarity between chronologies. Given that the tree-ring density shows a certain similarity in response to climate, the aims of the present study are (i) to classify tree-ring densities in the Northern Hemisphere by cluster analysis based on the response of MXD chronologies to monthly temperature in different regions; (ii) to identify the differences in response to climatic factors, especially mean temperature and precipitation, among the clustered MXD chronologies; and (iii) to analyze possible reasons for the differences among the clusters. Through classification and feature analysis, we have identified regional differences in tree-ring density and provided a reference for ecological studies and climate reconstruction research across large regions.

2. Materials and Methods

2.1. MXD Chronologies Network

All tree-ring MXD data sets, except the six MXD series for the Tienshan Mountains in China, were sourced from the International Tree-Ring Data Bank (ITRDB; https://www.ncei.noaa.gov/products/paleoclimatology/tree-ring (accessed on 23 October 2023)). Two types of data were obtained from the ITRDB. The first type was the standardized NOAA template chronology, which was used directly in the subsequent analysis. When no standardized NOAA template chronology was available, the MXD raw measurement data, provided by the ITRDB for cross-dating, were downloaded. ARSTAN 4.0 software [15] was used to detrend the series and establish the MXD chronologies. Sampling sites in Europe and North America greatly outnumber data sets from other locations. Given that imbalanced data sets may significantly increase the risk of overfitting [16] and given the uneven geographical distribution of the MXD chronologies, we reduced the number of sampling sites by randomly selecting the 164 MXD chronologies based on their spatial location. In selecting the sites, we considered the spatial distribution and response to climatic factors of the chronologies and excluded certain chronologies that did not show statistically significant correlations with the monthly mean temperature from January to December ($p < 0.05$). In addition, because there are few MXD chronologies for sites in Central Asia, we chose six new MXD chronologies from high elevations for this region. The sample size of the six chronologies in the Tienshan Mountains was more than 20 trees, and these were cross-dated using COFECHA [17]. In the present analysis, we used (a) cross-dating results of the raw tree-ring width and (b) series individually detrended using ARSTAN with a negative exponential curve or Spline function [15]. Ultimately, we selected 152 chronologies (Table S1) from 170 to generate a MXD network (Figure 1). Based on the Köppen–Geiger climate classification [18], the sampling sites were mostly located in the Cold Without Dry Season (Df), Arid Steppe (BS), Arid Desert (BW), Temperate Without Dry Season (Cf), and Temperate Dry Summer (Cs) climate types. The sites at low and mid latitudes were located in high-altitude mountains. Most of the chronologies used in the dendroclimatic correlation analysis began in 1901, from which the expressed population signal (EPS) [19] value of the chronologies is higher than 0.85. For some chronologies in which the EPS value did not attain 0.85 in 1901, the starting year was based on the year that the EPS was greater than 0.85.

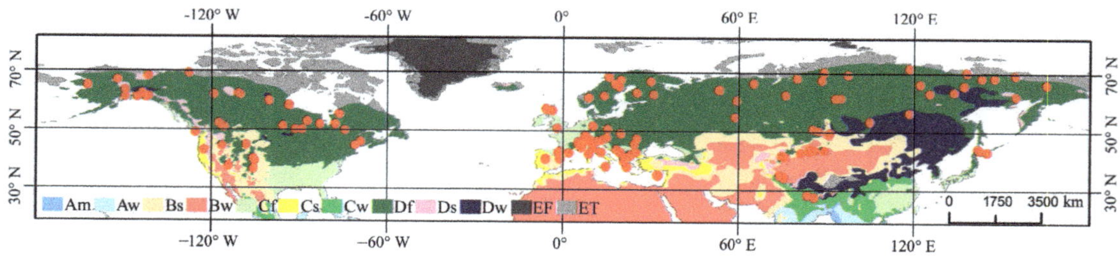

Figure 1. Location of the maximum latewood density chronologies in the Northern Hemisphere. The Köppen–Geiger climate classification map was derived from Peel et al. [18]. Climate types: Am = Tropical Monsoon; Aw = Tropical Savannah; Bs = Arid Steppe; Bw = Arid Desert; Cf = Temperate Without Dry Season; Cs = Temperate Dry Summer; Cw = Temperate Dry Winter; Df = Cold Without Dry Season; Ds = Cold Dry Summer; Dw = Cold Dry Winter; EF = Polar Frost; ET = Polar Tundra.

2.2. Climate Data

Dendroclimatic analysis of the tree-ring densities was based on monthly mean temperature and monthly precipitation data from the nearest Climatic Research Unit (CRU) $0.5^\circ \times 0.5^\circ$ grid cell (TS 4.04) [20,21] for each MXD chronology. We obtained the CRU data from the KNMI Climate Explorer (https://climexp.knmi.nl/ (accessed on 23 October 2023)). The CRU data set spans the period 1901–2019, but there are few meteorological stations providing long-term instrumental records in some regions. We screened all meteorological grid data using a 3 year sliding variance window from 1901 and observed that the variance of some gridded data equaled zero in the early portion of the data. Therefore, we discarded the early portion of the records and used data only for the period when the variance was greater than zero.

2.3. Methods

The partitioning around medoids (PAM) clustering method [22] was adopted for cluster analysis of the MXD data. In the PAM method, the algorithm searches the data set and selects k representative objects as the centrotypes (or medoids). After finding a group of centrotypes, the clusters are constructed by assigning each data point to the nearest representative object. Clustering performance was evaluated on the basis of compactness (CP), and the significance of the clustering results was assessed by calculating the p-value with SigClust [23]. The CP is a metric that uses the information inherent in a classified data set [23]. The CP measures the average distance between every pair of data points within the same cluster. The lower the CP value, the better the configuration of the cluster. The SigClust test uses a K-means-based algorithm to calculate a p-value to assess significance in a cluster analysis [24]. This approach sequentially identifies significant splits between groups of clusters and ultimately determines the significance of individual clusters or groups of clusters. Small p-values indicate a superior clustering structure.

We fitted a variance–covariance mixed model (CVM) [25,26] to analyze the spatial synchrony of climate sensitivities among clusters. Based on the diagonal positive-definite matrix structure, CVM calculates the common variance and estimated signal strength parameters for spatial synchrony from the outcome of the models for each cluster. A linear mixed-effects model (LMM) was developed to fit the distribution characteristics of the MXD data in terms of temperature, precipitation, latitude and longitude, elevation, and tree species for each cluster. The LMM is a statistical tool that examines the relationship between climatic factors and explanatory covariates based on the response between the two, taking into account these two types of variations. The coefficient of determination (R^2) [27,28], a summary statistic that describes the amount of variance explained, was calculated to quantify the goodness-of-fit of the LMM. Sensitivity analysis of the response of the MXD to meteorological variables was conducted using a treed constant model as a basic Bayesian treed Gaussian process model (TPM) [29,30]. The sensitivity indices of the models were estimated using a Latin hypercube sampling-based method [31], through prediction via one of the TPM models conditional on an observed set of locations, and the sensitivity indices were calculated at each Markov chain Monte Carlo iteration of the TPM fitting. Sensitivity analysis attempts to resolve the natural variability in responses. The indices comprised the main effects and first-order and total sensitivity indices. The main effects were the major contributions of variables to the TPM; the main-effects curves determine the way the variables contribute to the response. The first-order sensitivity is the proportion of the variability in the response attributable to the variable main effects. The total sensitivity measures the residual variance in the conditional expectation and represents all influences connected to a given variable. All statistical analyses were performed using R software version 4.2.0 [32].

3. Results

3.1. Classification of the MXD Chronologies

Briffa et al. determined that most MXD chronologies included a significant positive response to temperature in April–September [4,5]. Therefore, the monthly correlation coefficient from April to September was selected as the observed value, and these sampling points were selected as the sample points. The PAM clustering method was used for classification of the MXD chronologies. The 152 chronologies were classified into 4–9 clusters. The cumulative distribution function (Figure 2a), consensus matrix (Figure 2b), CP, and p-values (Table 1) were calculated to assess the credibility of the six classifications. The cumulative distribution function graph showed that the probability for more than six clusters was more than 0.75 for all consensus index values, and little difference was observed between six and nine clusters, indicating that there were differences in classification among about 25% of the sample points. In contrast, the difference between four and five clusters was more than 40%. Regarding CP, little difference was observed between the six classifications, and values ranged from 0.34 to 0.41. The lowest CP value was observed with eight clusters, but the difference compared with the value with six clusters was 0.018 at most. The lowest p-value (0.0006) was observed for six clusters and indicated extreme significance. Based on the clustering results, six clusters were selected for the final classification. A consensus matrix heatmap of the six clusters was generated (Figure 2b). The heatmap showed there were strong similarities in each classification of the MXD chronologies and the similarity between different clusters was close to zero.

Table 1. Test values for the clustering results.

Test	4	5	6	7	8	9
CP	0.4039	0.3754	0.3609	0.3528	0.3423	0.3519
p	0.0964	0.0055	0.0006	0.0051	0.0055	0.8088

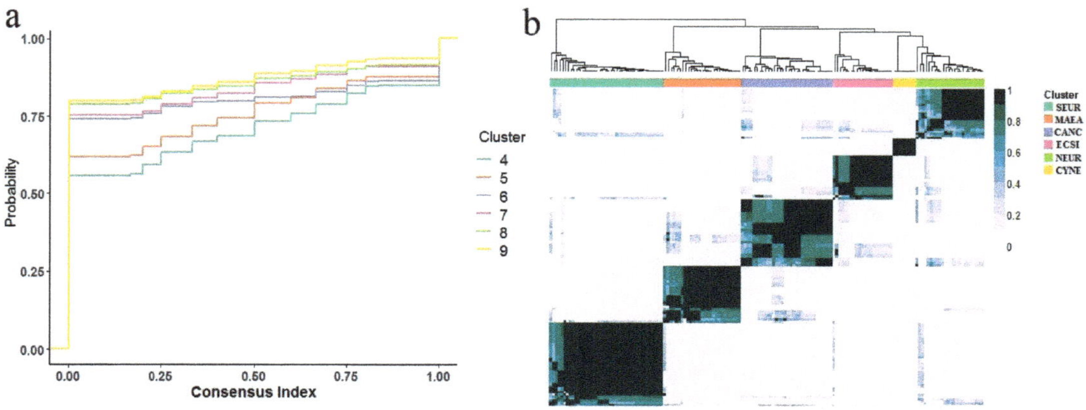

Figure 2. (**a**) Cumulative distribution function graph. (**b**) Consensus matrix heatmap of the six clusters. ECSI = East and Central Siberia; NEUR = Northern Europe; CANC = Central Asia and Northwestern Canada; MAEA = mid-latitude Asia, Europe, and America; SEUR = Southern Europe; CYNE = Cyprus and Nepal.

3.2. Characteristics of the Six Clusters

In the final classification of the MXD chronologies, the six clusters showed distinct spatial distribution characteristics. Chronologies in the same cluster did not occur entirely in the same climate zone but were located in relatively concentrated areas. According to the main distribution area of each cluster, the clusters were named from high to low latitude and comprised East and Central Siberia (ECSI), Northern Europe (NEUR), Central Asia and Northwestern Canada (CANC), mid-latitude Asia, Europe, and America (MAEA), Southern Europe (SEUR), and Cyprus and Nepal (CYNE) (Figure 3).

In the ECSI cluster, the sampling sites were mainly concentrated in the high-latitude region of Eastern and Central Siberia in Central Asia, with individual sites distributed in Southern Europe and Central North America. The sites were mostly located in the Cold Without Dry Season (Df) climate type and mountainous areas of the Temperate Dry Summer (Cs) type. A strong correlation between monthly temperature and MXD chronologies in this cluster was observed from June to September. The correlation coefficient for most chronologies was highest in June (Figure 4), with a second peak in September at some sites. No consistency in precipitation among all chronologies in this cluster was observed (Figure 5). Esper et al. [14] reported that the correlation coefficients with temperature in this region from June to August were 0.77–0.89, and the correlation was relatively stable in Central Siberia.

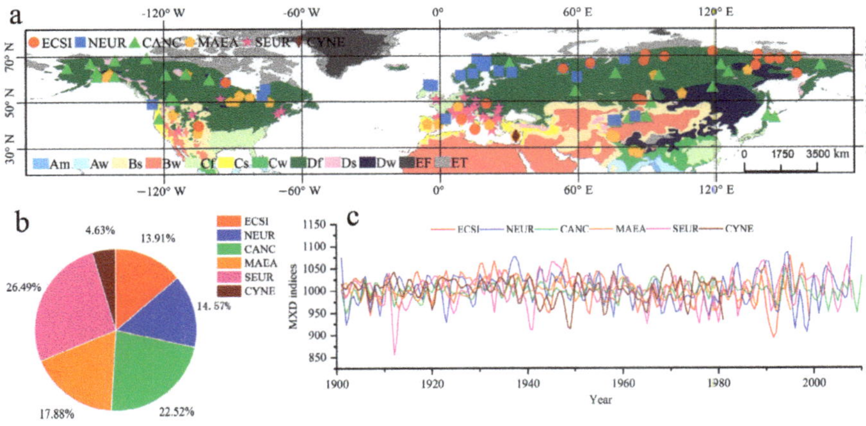

Figure 3. (**a**) Distribution of the six clusters and all MXD sampling sites in the Northern Hemisphere. Clusters: ECSI = East and Central Siberia; NEUR = Northern Europe; CANC = Central Asia and Northwestern Canada; MAEA = mid-latitude Asia, Europe, and America; SEUR = Southern Europe; CYNE = Cyprus and Nepal. (**b**) Pie chart of the cluster proportions. (**c**) Weighted average line for each cluster based on correlation coefficients.

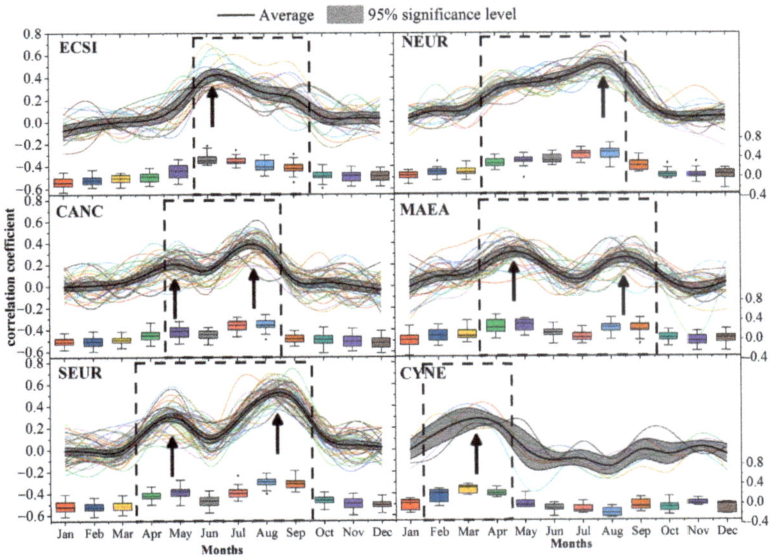

Figure 4. Line and box plots of the correlation coefficient between MXD chronologies and monthly mean temperature in each cluster. The dotted line indicates the period of strong mean correlations in each cluster.

The second cluster at high latitudes was NEUR. Most sites were located in Northern Europe, but some were also distributed in Central Asia and Central Siberia. Similar to the ECSI cluster, most sites were located in the Df climate type. A significant response to the monthly mean temperature was observed from April to August, with the highest correlation coefficient in August. Most sites showed a negative response, with precipitation from July to August. These correlations differed notably from those for the ECSI cluster. The MXD chronology assemblage in Northern Europe showed a correlation coefficient of 0.76 with June and August temperatures [33]. Duthorn [34] noted that the MXD in Northern

Finland was most strongly correlated with the temperature in August. The strongest correlation with the monthly mean temperature in the two clusters of high-latitude sites differed: June for ECSI and August for NEUR. The secondary peak value was variable in the NEUR cluster, so the mean monthly coefficient showed only one peak.

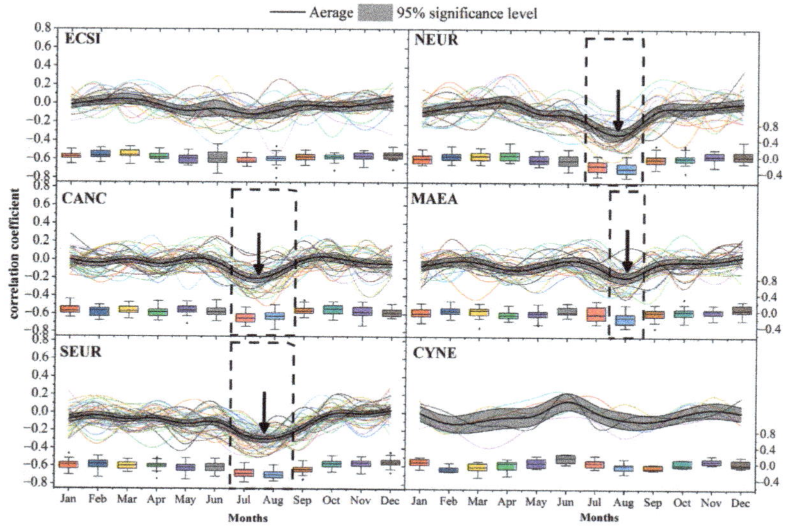

Figure 5. Lines and boxplots of the correlation coefficient between MXD chronologies and monthly precipitation for each cluster. The dotted line indicates the period of strong mean correlations in each cluster.

The sites in the CANC cluster were located at high- and mid-latitudes, mainly in Central Asia and Northwestern Canada, but some sites on the west coast of the Pacific Ocean were also classified in this cluster. The response characteristics of MXD chronologies to temperature revealed a strong correlation from May to July. Two peaks were observed in May and July, and the July peak was higher in this cluster. A significant effect of precipitation was observed in July and August, which was later than that for temperature.

The MAEA cluster was distributed slightly further to the south than the CANC cluster. The sampling sites in this cluster were mostly located in mid-latitudes and were distributed in Central Asia, South Asia, Central Europe, and Central North America. The cluster was characterized by the influence of temperature in April to May and August to September, and the influence in spring was greater than that in autumn. The influence of precipitation was mainly observed in August, which was slightly earlier or synchronous with the influence of temperature from August to September.

The SEUR cluster included sites mainly located in Southern Europe. The cluster was influenced by temperature from April to May and August to September, but, in contrast to the MAEA cluster, the effect was stronger in August to September than in April to May. Most of the sampling sites were affected by precipitation in July and August, and some in September. The NEUR and SEUR sites in Europe were located in similar taxonomic regions to Briffa's classification [4,5], but there were significant differences in other clusters.

The CYNE cluster included the fewest sampling sites among all clusters, comprising one site in Nepal in South Asia and several sites in the Mediterranean region. The characteristics of the CYNE cluster were quite different from those from other regions. The correlation with temperature was highest from February to April, and individual sites were also strongly correlated in September. Correlations with precipitation showed no consistency. Similar results have been reported in Southern Spain in the Northern Mediterranean region [9].

Discrepancies in the response of MXD chronologies to temperature among clusters were observed in the period and peak of high correlation. The clusters at high latitudes were unimodal, mid-, and low-latitude clusters were bimodal, and the month of the peak value and the maximum value were not identical. A portion of the MXD chronologies showed a negative response to precipitation late in the growing season.

3.3. Spatial Synchrony between the Six Clusters

We calculated the correlation coefficients and spatial synchronization of the CVM model between the MXD chronologies for all clusters and April–September monthly mean temperatures to compare differences in climate sensitivity among the clusters (Figure 6). The NEUR and SEUR clusters in the European region showed the strongest correlation with April–September temperatures. The correlation coefficients for the NEUR cluster were slightly higher than those for the SEUR cluster. The maximum coefficient value in the NEUR cluster was more than 0.75. Except for the CYNE cluster, no significant difference in sensitivity to temperature was observed among the other three clusters, and the correlation coefficients ranged from 0.2 to 0.6.

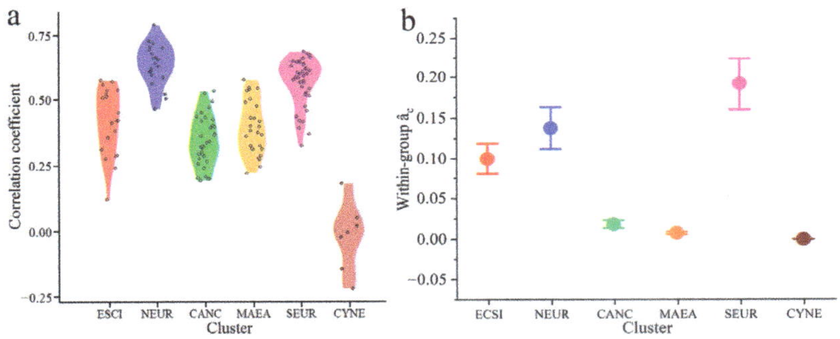

Figure 6. (**a**) Violin plot of correlation coefficients and (**b**) spatial synchrony between MXD chronologies and April–September mean monthly temperature.

The spatial synchronization results showed some differences from the correlation analysis. The SEUR cluster showed the most consistent response to temporal variation in temperature, followed by NEUR. The two clusters were more concentrated than the other clusters, which may be the reason for the high consistency. The ECSI cluster also showed a good response to temperature; the sites in this cluster were mainly located in the high-latitude region. Whereas climate had a greater influence on the spatial variability of the CANC and MAEA clusters, in Europe and Northern Asia, climate warming is leading to temporal coherence in MXD records, whereas in other parts of Asia and the Americas in the Northern Hemisphere, the MXD can represent consistent changes in temperature over a large range. The poor synchronization of MXD responses may be linked to global shifts in the timing of plant phenological events [35] and drought stress at mid-latitudes [36,37], among other factors.

3.4. Influence of Climatic and External Factors

Taking April–September temperature and July–August precipitation as independent variables, latitude and longitude, elevation, and tree species as random variables, and MXD of each cluster as a dependent variable, we established LMMs and calculated R^2 for each variable to evaluate their influence on each cluster (Figure 7a–d). Because insufficient sites were classified in the CYNE cluster to enable a LMM model to be established, the influence of climatic and external factors on this cluster is not discussed. When the three random variables were included in the LMMs (Figure 7a, ALL), temperature accounted for more than 40% of the MXD variance in all clusters. Among the five clusters, temperature had the greatest influence on ESCI. Precipitation only had a small influence, explaining approximately 1% of the total chronology variation. The random variables contributed to 29%–49% of the MXD variance and had the least influence on the CANC cluster. Temperature, precipitation, and the random variables explained similar proportions of the total variance for the NEUR and SEUR clusters. Precipitation had the strongest influence on these two clusters compared with the other clusters, accounting for approximately 2%–3% of the MXD variation.

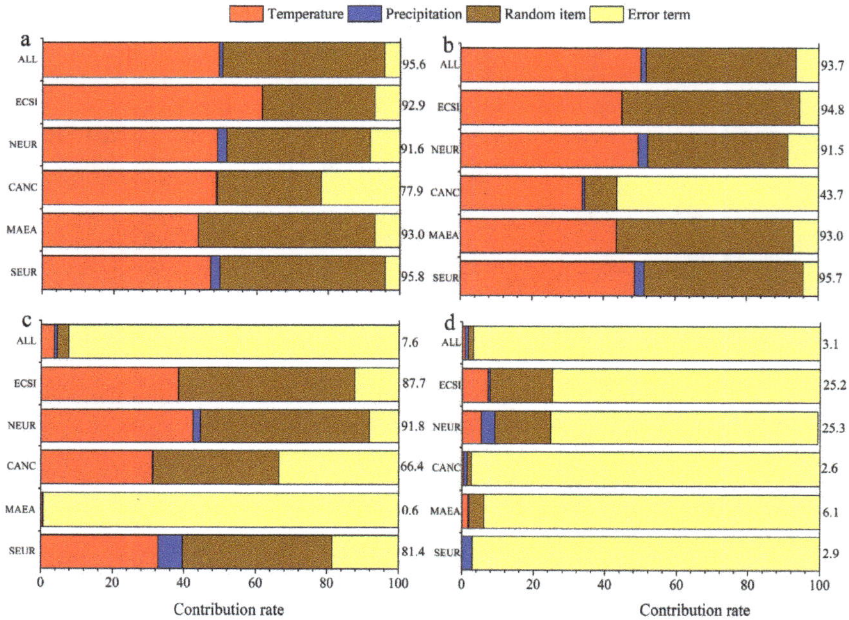

Figure 7. Contribution rate of factors in the mixed-effect models. The fixed-effect factors were April–September mean monthly temperature and July–August monthly precipitation. The random-effect factors were: (**a**) latitude and longitude, elevation, and tree species; (**b**) latitude and longitude; (**c**) elevation; and (**d**) tree species. The values on the right of each plot are R^2 values for each mixed model.

In addition, mixed models incorporating the three random variables individually were established (Figure 7b–d). Latitude and longitude were the main contributors to the random effects for all MXD chronologies, explaining about 41.9% of the total variance. Excluding CANC, the latitude and longitude of each cluster explained 39%–49% of the variance in the model. Elevation did not significantly affect the model for all chronologies. The ECSI, NEUR, and SEUR clusters each had a relatively restricted spatial distribution and were significantly affected by elevation, whereas the MAEA cluster was not obviously influenced by elevation. This may be due to the broad geographic distribution of the MAEA cluster, and thus the influence of elevation is on a comparatively small scale. In contrast to the other clusters, elevation was the main influencing factor contributing to the random effects for CANC. This may be because this cluster is mainly distributed in the Cold Without Dry Season (Df) climatic area, and some mid-latitude sites are distributed in the Arid Steppe (Bs) area. The trees in this cluster all grow in cold mountainous regions at elevations above 2200 m at mid-latitudes, and the elevation decreases with an increase in latitude. Therefore, the change in elevation is a major influencing factor for this cluster. Tree species is not an important factor influencing the climatic response of the MXD chronologies in each cluster. However, tree species may contribute more than 15% of the random effect at high latitudes, which indicates that tree species at high latitudes may affect the climate response of the MXD.

3.5. Sensitivity to Climatic Factors

The treed constant model was selected as a basic Bayesian treed Gaussian process model, and the sensitivity of all MXD chronologies and each cluster to the climatic factors was analyzed. The model factors for all chronologies and those of five clusters (ECSI, NEUR, CANC, MAEA, and SEUR) were anomalies of April–September mean monthly temperature and July–August monthly precipitation, and for CYNE anomalies of February–April temperature and precipitation. The main-effect sensitivity to temperature of the MXD chronologies in each cluster resembled a near-linear positive response within a certain temperature range (Figure 8), which was consistent with the results of the correlation analysis, but the sensitivity intensity and sensitivity range differed among each cluster. The range in temperature response sensitivity of all sites was similar to the maximum sensitivity, approximately −1.6 to −1.9 °C, but the slope was low and the intensity was weak. Among all clusters, ECSI and NEUR (comprising sites distributed at high latitudes) showed a greater range in sensitivity to temperature changes than most sites at mid and low latitudes, and the sensitivity intensity was greater at −1.6 to −1.7 °C and −1.9 to −1.6 °C, respectively. The slope of the main-effect sensitivity index for SEUR (comprising sites at mid- and low-latitudes) was the steepest, temperature had the greatest influence on MXD, and the sensitivity range was −1.5 to −0.8 °C. The slope of the main-effect sensitivity index for CANC was much gentler than that of the other clusters, and the sensitivity range was −1.5 to −1.1 °C. The sensitivity range for MAEA was the narrowest among the clusters except CYNE, and the sensitivity range was −1.3 to −0.5 °C. The sensitivity of the MXD chronologies to low temperatures was stronger than the sensitivity to high temperatures at mid and low latitudes.

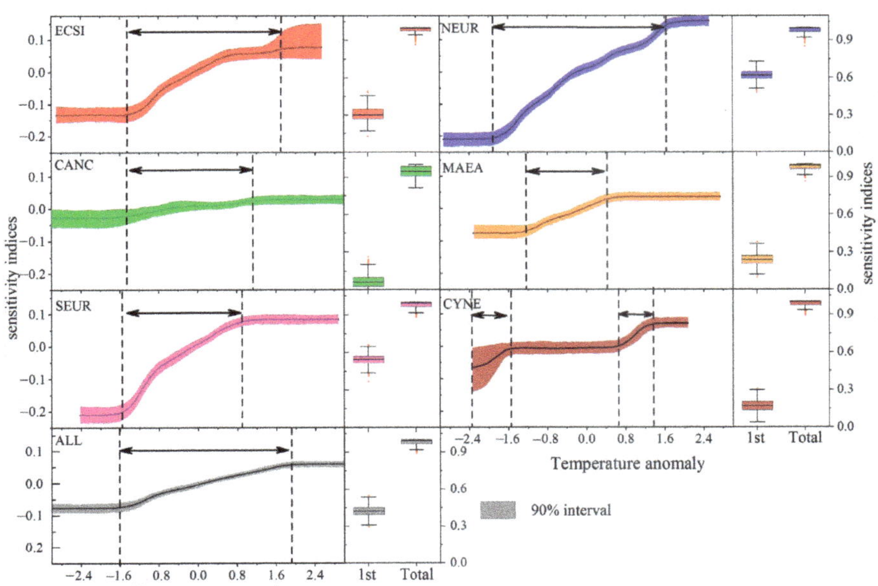

Figure 8. Main effects, first-order, and total sensitivity indices for temperature of Bayesian treed Gaussian process models. The independent variable for the ECSI, NEUR, CANC, MAEA, SEUR, and ALL clusters is July–August monthly precipitation, and that for the CYNE cluster is February–April mean monthly temperature.

Regarding the first-order sensitivity index, temperature had the strongest influence on the main effect of NEUR and SEUR, followed by ECSI. The total sensitivity index revealed that the overall sensitivity to temperature and precipitation of MXD was high.

The main-effect sensitivity of MXD chronologies of all clusters to precipitation mainly showed a negative linear response within a certain precipitation interval (Figure 9). The ECSI and CYNE clusters showed no sensitivity to precipitation, which was consistent with the correlation analysis between each cluster and July–August monthly precipitation. The main-effect sensitivity of CANC was predominantly between −140 and −40 mm. The first-order and total sensitivities of this cluster were the highest among all clusters, indicating that the precipitation of the region had a significant influence on the main effect. The MXD responded negatively to precipitation change in the period of low precipitation, whereas change in precipitation had no influence on the MXD in the period of severely low precipitation or when the precipitation was sufficiently high to meet the needs for tree growth. The main-effect sensitivities of the NEUR, SEUR, and MAEA clusters responded to precipitation in the region of zero precipitation anomaly. Their first-order sensitivity index values were low, indicating that precipitation had little influence on the main effect. However, a certain degree of total sensitivity indicated that precipitation may influence MXD sensitivity to temperature, but temperature limitation creates additional sensitivity to precipitation [33].

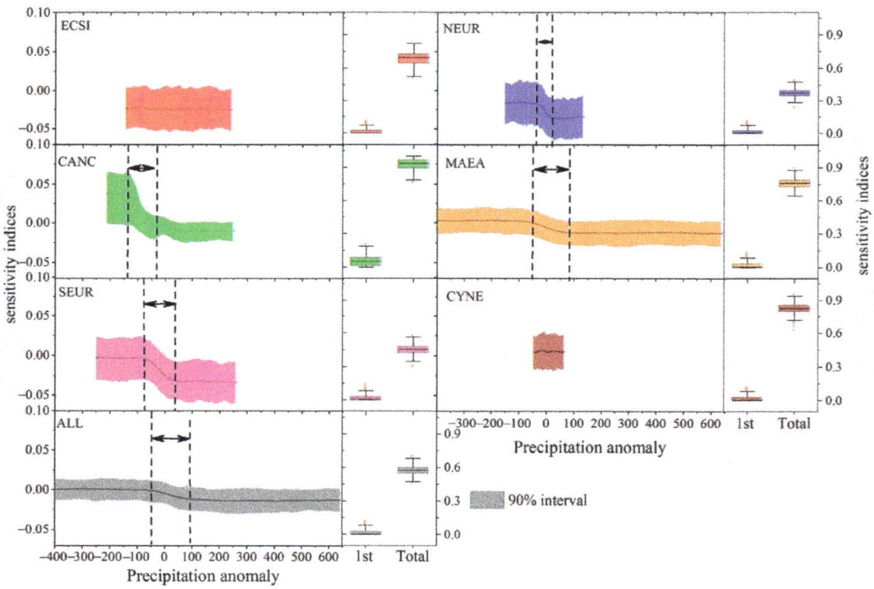

Figure 9. Main effects, first-order, and total sensitivity indices for precipitation of Bayesian treed Gaussian process models. The independent variable for the ECSI, NEUR, CANC, MAEA, SEUR, and ALL clusters is July–August monthly precipitation, and that for the CYNE cluster is February–April monthly precipitation.

4. Discussion

4.1. Comparison with Previous Classifications

Although the six clusters in the present classification were fewer than the number resolved by Briff et al. [4,5], there are similarities in the clusters between the two classifications. The NEUR and SEUR clusters were essentially consistent, with the samples located in the same region, and the highest correlation coefficient was between MXD and monthly temperature in August. The correlation coefficient of NEUR from April to August shows a trend of increasing with the month, while the relationship between SEUR and temperature shows a lower peak in April and May than in August [5].

In the Briff et al. classification [4,5], Central and Western Siberia are mainly divided into two clusters, whereas Esper et al. [12] classified the region into eight clusters based on the correlation between chronologies. The present classification divides the sampling points equally into the ESCI cluster. The response characteristics of the ESCI cluster to temperature were more similar to those of Northern Siberia (NSIB) [5], with high correlation coefficients occurring in May.

The CANC cluster includes the clusters of Northwestern North America (NWNA) and Central Asia (CAS) in the Briff et al. classification. The relevant peak months for NWNA are May and August, consistent with CANC, whereas in CAS the peaks are in August.

The Tibetan Plateau (TIBP) and Eastern and Central Canada (ECCA) clusters in the Briff et al. classification were mostly included in the MAEA cluster, and their correlation characteristics with temperature were also consistent. The highest correlation coefficients were in August–September, with April–May representing the secondary peak interval.

The CYNE cluster was not recognized in the classification of Briff et al., possibly because there were insufficient samples with the same temperature response characteristics. We distinguished the CYNE cluster because, in addition to the Cyprus and Nepal regions, the same response characteristics to January–April temperatures were observed in Southern Spain [9], which may represent the characteristics of low-latitude regions where local growth environments (such as differences in elevation) vary.

The present classification exhibited a degree of consistency with that of Briff et al., but some samples with relatively long geographical distances were grouped into a larger cluster based on their response to temperature. The Briff et al. classification places greater emphasis on geographical distribution and consistency between chronologies, whereas the present classification is focused more on the climate response characteristics of MXD.

4.2. Influence of Elevation on Tree-Ring Densities

In addition to geographical location, the elevation of each sampling point is an important factor that affects the temperature response of the MXD chronologies. Buckley et al. [38] observed that the tree-ring width chronologies at high elevation exhibit a strong, direct response to temperature for most growing-season months, whereas the chronologies at low elevation exhibit a weaker, direct response in Western Tasmania, Australia. In British Columbia, Canada, the tree-ring width and MXD chronologies response to climatic factors changed with elevation [39,40]; summer temperature limited tree-ring width at high elevations and MXD at low and high elevations, although the relationship was much stronger at high elevations [39]. In the Tatra Mountains of Slovenia [41], the peak response of MXD to temperature at high elevation is in May and September, which is consistent with the peak for the ESCI cluster. Whereas the low-elevation sampling site only showed a peak response to temperature in May and had a lower correlation coefficient than the high-elevation sampling site. In addition, precipitation in July is an important influencing factor. The mixed-effects model (Figure 7b) also showed that elevation had a significant impact on different clusters in addition to the CANC cluster.

4.3. Threshold and Sensitivity to Climatic Factors

There is a certain physiological range for tree growth to adapt to ecological factors, such as temperature; that is, there is a threshold value [42]. For example, in Northwestern Canada, the critical threshold for tree growth of white spruce at the timberline is 11.3 °C (the temperature in July) [43]. The threshold range for the negative response of Engelmann spruce to July temperature is 13.7–16.8 °C, and when the temperature exceeds 16.8 °C, the tree enters the positive response threshold range [44]. There are reports of similar temperature thresholds in other regions [45,46]. The climatic threshold of tree-ring width is based on the divergence pitfalls [43–47], but no similar study has been undertaken on the climatic threshold of tree-ring density growth. The sensitivity to temperature shown in Figure 8 is a positive linear correlation, whereas precipitation shows a negative correlation in some regions, which is consistent with previous research in the Northern Hemisphere [1–5,48–50]. Therefore, the results of the present sensitivity analysis may represent the linear response of tree-ring density to temperature in the study area from April to September. However, because of the presence of common intervals in the samples and differences in tree species (Table S1), further research is needed to determine whether the results of the sensitivity analysis represent climatic thresholds.

5. Conclusions

Taking the monthly correlation coefficient between MXD and April to September as the observation value, 152 MXD chronologies in the Northern Hemisphere were grouped into six clusters. Each cluster had distinct regional characteristics, and the differences among the clusters were mainly the time period of the significant response to temperature and the month of the peak response. Temperature contributed to more than 40% of the MXD variance. The location of the site was the main contributor to the random effect of the mixed-effect model, followed by elevation. Tree species had a small impact at high latitudes. The main-effect sensitivity to temperature of each cluster of MXD chronologies showed a near-linear response within a certain temperature range. High-latitude MXD chronologies had a broader range in temperature sensitivity, and the intensity was greater. The MXD was more sensitive to low temperatures. The MXD of some clusters was not sensitive to changes in precipitation. The main-effect sensitivity of other clusters to precipitation showed a negative linear response within a certain precipitation interval.

Supplementary Materials: The following supporting information can be downloaded at: https://www.mdpi.com/article/10.3390/f14112122/s1, Table S1: Information on the 152 maximum latewood density (MXD) chronologies selected for the study.

Author Contributions: Conceptualization and research design, S.Y., T.Z., Y.F. and R.Z.; data analysis, S.J. and L.Q.; contributed materials, H.S. and H.Z.; manuscript—writing, K.L. and X.G. All authors have read and agreed to the published version of the manuscript.

Funding: This research was supported by the Key Laboratory Opening Subject of Xinjiang Uigur Autonomous Region (2020D004040) and the National Science Foundation of China (Nos. 42375196, 41975095, and U1803245).

Data Availability Statement: Data will be made available upon reasonable request to the corresponding author.

Conflicts of Interest: The authors declare that the research was conducted in the absence of any commercial or financial relationships that could be construed as a potential conflict of interest.

References

1. D'Arrigo, R.; Davi, N.; Jacoby, G.; Wilson, R.; Wiles, G. *Dendroclimatic Studies: Tree Growth and Climate Change in Northern Forests*; John Wiley & Sons Inc: Hoboken, NJ, USA, 2014; p. 88.
2. Frank, D.; Esper, J. Temperture reconstruction and comparisons with instrumental data from a tree-ring network for the European Alps. *Int. J. Climatol.* **2005**, *25*, 1437–1454. [CrossRef]
3. Schweingruber, F.H. Measurement of densitometric properties of wood. In *Climate from Tree Rings*; Hughes, M.K., Kelly, P.M., Pilcher, J.R., LaMarche, V.C., Eds.; Cambridge University Press: New York, NY, USA, 1982; pp. 8–12.
4. Briffa, K.R.; Osborn, T.J.; Schweingruber, F.H.; Jones, P.D.; Shiyatov, S.G.; Vaganov, E.A. Tree-ring width and density data around the Northern Hemisphere (Part 2): Spatio-temporal variability and associated climate patterns. *Holocene* **2002**, *12*, 759–789. [CrossRef]
5. Briffa, K.R.; Osborn, T.J.; Schweingruber, F.H. Large-scale temperature inferences from tree rings: A review. *Glob. Planet Change* **2004**, *40*, 11–26. [CrossRef]
6. Frank, D.; Esper, J. Characterization and climate response patterns of a high elevation, multi species tree-ring network in the European Alps. *Dendrochronologia* **2005**, *22*, 107–121. [CrossRef]
7. Büntgen, U.; Frank, D.C.; Nievergelt, D.; Esper, J. Summer temperature variations in the European Alps, AD 755-2004. *J. Clim.* **2006**, *19*, 5606–5623. [CrossRef]

8. Björklund, J.A.; Gunnarson, B.E.; Krusic, P.J.; Grudd, H.; Josefsson, T.; Östlund, L.; Linderholm, H.W. Advances towards improved low-frequency tree-ring reconstructions, using an updated *Pinus sylvestris* L. MXD network from the Scandinavian Mountains. *Theor. Appl. Climatol.* **2013**, *113*, 697–710. [CrossRef]
9. Rocha, E.; Gunnarson, B.E.; Holzkämper, S. Reconstructing Summer Precipitation with MXD Data from *Pinus sylvestris* Growing in the Stockholm Archipelago. *Atmosphere* **2020**, *11*, 790. [CrossRef]
10. Yang, B.; He, M.; Yang, L.; Wang, F.; Ljungqvist, F.C. Pine maximum latewood density in semi-arid northern China records hydroclimate rather than temperature. *Geophys. Res. Lett.* **2023**, *50*, e2023GL104362. [CrossRef]
11. Pompa-García, M.; Hevia, A.; Camarero, J.J. Minimum and maximum wood density as proxies of water availability in two Mexican pine species coexisting in a seasonally dry area. *Trees* **2021**, *35*, 597–607. [CrossRef]
12. Esper, J.; Hartl, C.; Tejedor, E.; de Luis, M.; Günther, B.; Büntgen, U. High-Resolution Temperature Variability Reconstructed from Black Pine Tree Ring Densities in Southern Spain. *Atmosphere* **2020**, *11*, 748. [CrossRef]
13. Friedrichs, D.A.; Neuwirth, B.; Winiger, M.; Löffler, J. Methodologically induced differences in oak site classifications in a homogeneous tree-ring network. *Dendrochronologia* **2009**, *27*, 21–30. [CrossRef]
14. Esper, J.; Frank, D.; Büntgen, U.; Verstege, A.; Hantemirov, R.M.; Kirdyanov, A.V. Trends and uncertainties in Siberian indicators of 20th century warming. *Glob. Chang. Biol.* **2010**, *16*, 386–398. [CrossRef]
15. Cook, E.R.; Kairiukstis, L.A. *Methods of Dendrochronology Applications in the Environmental Science*; Springer: Dordrecht, The Netherlands, 1990; Chapter 1.
16. Onel, M.; Kieslich, C.A.; Guzman, Y.A.; Pistikopoulos, E.N. Simultaneous Fault Detection and Identification in Continuous Processes via nonlinear Support Vector Machine based Feature Selection. *Comput. Aided Chem. Eng.* **2018**, *44*, 2077–2082.
17. Holmes, R.L. Computer-assisted quality control in tree-ring dating and measurement. *Tree-Ring Bulletin.* **1983**, *43*, 69–75.
18. Peel, M.C.; Finlayson, B.L.; McMahon, T.A. Updated world map of the Köppen-Geiger climate classification. Hydrol. *Earth Syst. Sci.* **2007**, *12*, 1633–1644. [CrossRef]
19. Wigley, T.M.L.; Briffa, K.R.; Jones, P.D. On the average value of correlated time series, with applications in dendroclimatology and hydrometeorology. *J. Appl. Meteorol. Climatol.* **1984**, *23*, 201–213. [CrossRef]
20. Harris, I.; Osborn, T.J.; Jones, P.; Lister, D. Version 4 of the CRU TS monthly high-resolution gridded multivariate climate dataset. *Sci. Data* **2020**, *109*, 7. [CrossRef]
21. Mitchell, T.D.; Jones, P.D. An improved method of constructing a database of onthly climate observations and associated high-resolutiongrids. *Int. J. Climatol.* **2005**, *25*, 639–712. [CrossRef]
22. Kaufman, L.; Rousseeuw, P.J. *Finding Groups in Data: An Introduction to Cluster Analysis*; John Wiley & Sons Inc.: New York, NY, USA, 1990; Chapter 1.
23. Liu, Y.; Hayes, D.N.; Nobel, A.; Marron, J.S. Statistical Significance of Clustering for High-Dimension, Low-Sample Size Data. *J. Am. Stat. Assoc.* **2008**, *103*, 1281–1293. [CrossRef]
24. Nguyen, N.; Caruana, R. Consensus Clusterings. In Proceedings of the Third IEEE International Conference on Data Mining, Omaha, NE, USA, 28–31 October 2007; pp. 607–612.
25. Shestakova, T.A.; Aguilera, M.; Ferrio, J.P.; Gutierrez, E.; Voltas, J. Unravelling spatiotemporal tree-ring signals in Mediterranean oaks: A variance-covariance modelling approach of carbon and oxygen isotope ratios. *Tree Physiol.* **2014**, *34*, 819–838. [CrossRef]
26. Shestakova, T.A.; Gutierrez, E.; Kirdyanov, A.V.; Camarero, J.J.; Genova, M.; Knorre, A.A.; Linares, J.C.; Resco de Dios, V.; Sanchez-Salguero, R.; Voltas, J. Forests synchronize their growth in contrasting Eurasian regions in response to climate warming. *Proc. Natl. Acad. Sci. USA* **2016**, *113*, 662–667. [CrossRef]
27. Nakagawa, S.; Schielzeth, H. A general and simple method for obtaining R^2 from generalized linear mixed-effects models. *Methods Ecol. Evol.* **2013**, *4*, 133–142. [CrossRef]
28. Nakagawa, S.; Johnson, P.C.; Schielzeth, H. The coefficient of determination R^2 and intra-class correlation coefficient from generalized linear mixed-effects models revisited and expanded. *J. R. Soc. Interface* **2017**, *14*, 20170213. [CrossRef]
29. Morris, R.D.; Kottas, A.; Taddy, M.; Furfaro, R.; Ganapol, B.D. A Statistical Framework for the Sensitivity Analysis of Radiative Transfer Models. *IEEE Trans. Geosci. Remote Sens.* **2008**, *46*, 4062–4074. [CrossRef]
30. Gramacy, R.B. *Surrogates: Gaussian Process Modeling, Design and Optimization for the Applied Sciences*; Chapman Hall/CRC: Boca Raton, FL, USA, 2020; Chapter 8.
31. Saltelli, A. Making best use of model evaluations to compute sensitivity indices. *Comput. Phys. Commun.* **2002**, *145*, 280–297. [CrossRef]
32. R Core Team. *R: A Language and Environment for Statistical Computing*; R Foundation for Statistical Computing: Vienna, Austria, 2023. Available online: https://www.R-project.org/ (accessed on 1 May 2023).
33. Esper, J.; Duthorn, E.; Krusic, P. Northern European summer temperature variations over the Common Era from integrated tree-ring density records. *J. Quat. Sci.* **2014**, *29*, 487–494. [CrossRef]
34. Düthorn, E.; Schneider1, L.; Günther, B.; Gläser, S.; Esper, J. Ecological and climatological signals in tree-ring width and density chronologies along a latitudinal boreal transect. *Scand. J. Forest Res.* **2016**, *31*, 750–757. [CrossRef]
35. Menzel, A.; Sparks, T.H.; Esterella, N.; Koch, E.; Aasa, A.; Ahas, R.; Alm-kübler, K.; Bissolli, P.; Braslauska, C.; Briede, A.; et al. European phenological response to climate change matches the warming pattern. *Glob. Chang. Biol.* **2006**, *12*, 1969–1976. [CrossRef]

36. Williams, A.P.; Allen, C.D.; Macalady, A.K.; Griffin, D.; Woodhouse, C.A.; Meko, D.M.; Swetnam, T.W.; Rauscher, S.A.; Seager, R.; Grission-Mayer, H.D.; et al. Temperature as a potent driver of regional forest drought stress and tree mortality. *Nat. Clim. Chang.* **2013**, *3*, 292–297. [CrossRef]
37. Macias, M.; Andreu, L.; Bosch, O.; Camarero, J.; Gutiérrez, E. Increasing aridity is enhancing silver fir (*Abies alba* Mill.) water stress in its southwestern distribution limit. *Clim. Chang.* **2006**, *79*, 289–313. [CrossRef]
38. Buckley, B.; Cook, E.; Peterson, M.; Barbetti, M. Changing temperature response with elevation for *Lagarostrobos franklinii* in Tasmania, Australia. *Clim. Chang.* **1997**, *36*, 477–498. [CrossRef]
39. Bernhard, S.; Dobry, J.; Klinka, K. Tree-ring characteristics of subalpine fir (*Abies lasiocarpa* (Hook.) Nuttv.) in relation to elevation and climatic fluctuations. *Ann. Forest Sci.* **2000**, *57*, 89–100.
40. Zhang, Q.; Hebda, R. Variation in radial growth patterns of *Pseudotsuga menziesii* on the central coast of British Columbia, Canada. *Can. J. Forest Res.* **2004**, *34*, 1946–1954. [CrossRef]
41. Klippel, L.; Büntgen, U.; Konter, O.; Kyncl, T.; Esper, J. Climate sensitivity of high- and low-elevation Larix decidua MXD chronologies from the Tatra Mountains. *Dendrochronologia* **2020**, *60*, 125674. [CrossRef]
42. Kramer, P.; Kozlowski, T. *Physiology of Woody Plants*; Academic Press: New York, NY, USA, 1979; pp. 546–627.
43. D'Arrigo, R.; Kaufmann, R.; Davi, N. Thresholds for warming-induced growth decline at elevational tree line in the Yukon Territory, Canada. *Glob. Biogeochem. Cycles* **2004**, *18*, GB3021. [CrossRef]
44. Hart, S.; Laroque, C. Searching for thresholds in climateradial growth relationships of Engelmann spruce and subalpine fir, Jasper National Park, Alberta, Canada. *Dendrochronologia* **2013**, *31*, 9–15. [CrossRef]
45. Wilmking, M.; Juday, G.; Barber, V. Recent climate warming forces contrasting growth responses of whitespruce at treeline in Alaska through temperature thresholds. *Glob. Chang. Biol.* **2004**, *10*, 1724–1736. [CrossRef]
46. Peltier, D.; Ogle, K. Tree growth sensitivity to climate is temporally variable. *Ecol. Lett.* **2020**, *23*, 13575. [CrossRef]
47. Esper, J.; Frank, D. Divergence pitfalls in tree-ring research. *Clim. Chang.* **2009**, *94*, 261–266. [CrossRef]
48. Grudd, H. Torneträsk tree-ring width and density AD 500–2004: A test of climatic sensitivity and a new 1500-year reconstruction of north Fennoscandian summers. *Clim. Dynam.* **2008**, *31*, 843–857. [CrossRef]
49. Cerrato, R.; Salvatore, M.C.; Gunnarson, B.E.; Linderholm, H.W.; Carturan, L.; Brunetti, M.; De Blasi, F.; Baroni, C. A *Pinus cembra* L. tree-ring record for late spring to late summer temperature in the Rhaetian Alps, Italy. *Dendrochronologia* **2019**, *53*, 22–31. [CrossRef]
50. Nagavciuc, V.; Roibu, C.-C.; Ionita, M.; Mursa, A.; Cotos, M.-G.; Ionel, P. Different climate response of three tree ring proxies of *Pinus sylvestris* from the Eastern Carpathians, Romania. *Dendrochronologia* **2019**, *54*, 56–63. [CrossRef]

forests

MDPI

Article

Tree Ring Blue Intensity-Based August Temperature Reconstruction for Subtropical Central China

Yonghong Zheng [1,*] and Rob Wilson [2]

[1] School of Resource and Environmental Sciences, Wuhan University, Wuhan 430079, China
[2] School of Earth and Environmental Sciences, University of St. Andrews, St. Andrews KY16 9TS, UK; rjsw@st-andrews.ac.uk
* Correspondence: zhengyh@whu.edu.cn

Abstract: Tree-ring blue intensity (BI) has the potential to provide information on past summer temperatures of a similar quality to that of tree-ring maximum latewood density and at a substantially reduced cost. To explore the applicability of BI in subtropical regions, the inverted BI for the earlywood, latewood, and the delta BI (DBI) parameters, together with tree-ring width of subalpine fir (*Abies fargesii* Franch.) in the Shennongjia area of China, were measured, and the corresponding chronologies were developed. The relationships between these chronologies and monthly precipitation, monthly mean minimum temperature, monthly mean temperature, and monthly mean maximum temperature were explored through correlation analysis. Results show that the DBI chronology is closely related to the temperature in August of the current year, indicating that BI, specifically delta BI, data are suitable for use in dendroclimatology studies in subtropical areas. The resultant mean temperature reconstruction for August, based on DBI, explains 40.8% of the temperature variance and is robustly validated using independent periods from the calibration. This pilot study not only highlights the potential of DBI for temperature reconstruction in China but also offers valuable insights into historical climate variations in the Shennongjia region. Moreover, it shows the potential for utilizing such tree-ring data from low-latitude regions to derive past climate data in subtropical warm-humid zones.

Keywords: tree-ring blue intensity; Shennongjia; *Abies fargesii* Franch.; temperature reconstruction

1. Introduction

According to recent research [1], 2023 was the warmest northern hemisphere extratropical summer over the past 2000 years exceeding the 95% confidence range of natural climate variability by more than 0.5 °C. Climate change is now not only a scientific challenge but also a political issue [2]. Long climate records are essential for evaluating the extent of natural fluctuations in the climate system and placing recent warming in a historical context. However, instrumental meteorological records are often insufficiently long for this purpose, particularly in regions where meteorological measurements commenced later (e.g., East Asia). Tree-ring (TR) data are a valuable high-resolution and exactly-dated proxy for reconstructing the climate over the past hundreds [3–5] and even thousands [6–8] of years, as they provide important information about the amplitudes of natural climate fluctuations that extend beyond the observational period. However, in general, ring-width data have been shown to be a poor proxy of past summer temperatures even when trees are sampled from temperature-limited tree-line locations [9,10]. Over the last 30 years, however, ring density parameters have provided a robust data source for reconstructing past summer temperatures across many regions of the northern hemisphere [11,12]. Unfortunately, due to the high costs involved in generating density data, including equipment, software, technician training, analytical time, and maintenance costs, only a few tree-ring laboratories around the world have the capability to measure tree-ring density [13]. Tree-ring blue intensity (BI), a method that measures the amount of blue light reflected from tree-ring cores, has been shown to be a robust alternative to tree-ring density and is gaining increasing attention in dendroclimatic studies [13–20]. Unlike tree-ring density, BI is easy and inexpensive to measure. Previous studies demonstrate that when tree-ring width (RW) serves as a weak temperature predictor due to complex growth–climate relationships at lower latitudes (<40° N) [5,21,22], BI can still be strongly representative of local to regional summer temperatures. Although there has been an increasing number of BI-based summer temperature reconstruction studies [5,20,21,23–26], there is still a paucity of TR blue intensity records for SE Asia. Herein, we highlight the potential of BI to substantially improve on RW-based reconstructions of past temperatures.

In this study, we examine the use of BI data from cross-dated subalpine fir (*Abies fargesii* Franch.) tree-ring samples in the Shennongjia region of subtropical China (Figure 1). This area is significant as it serves as an ecological green belt in the world-renowned Three Gorges Reservoir region and is a crucial component of the South-to-North Water Transfer Project in China. *Abies fargesii* Franch. is one of the more widely distributed Chinese firs. It extends from the eastern margin of the Tibet–Qinghai Plateau to southeastern Gansu, southern Shaanxi, western Henan, western Hubei, and northern Hunan. Shennongjia is located at the eastern edge of its distribution range, which spans a vertical distribution from 2000 to 3100 m and dominates the forests above 2500 m. Building on previous research [27] based on tree-ring width data, our objectives are as follows: (1) to investigate the feasibility of using BI data from fir trees, specifically *Abies fargesii* Franch., in subtropical dendroclimatic research and (2) to extend climate information before the instrument record. This work is an important exploration of using BI in dendroclimatology for subtropical regions and could serve as a reference for reconstructing past climates in similar climatic regions around the globe.

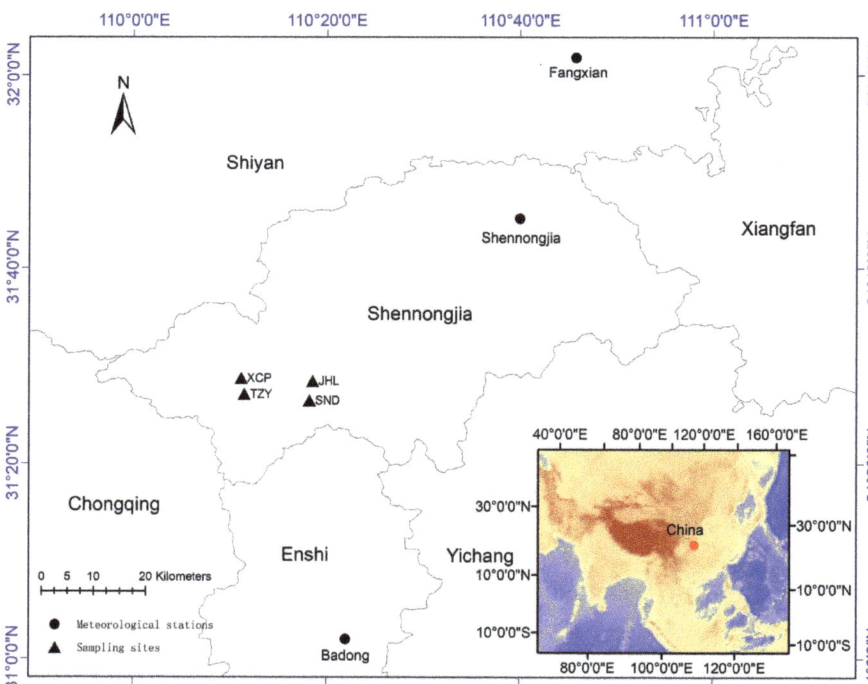

Figure 1. Locations of tree-ring sampling sites (black triangle) and meteorological stations (black dots).

2. Materials and Methods

2.1. Study Region and Sampling Information

The Shennongjia region, situated along the upper Yangtze River, has an average elevation of 1700 m. Its climate is significantly influenced by the southeast subtropical monsoon, resulting in ample precipitation and moderate mean air temperatures. According to 30-year climate records from the Shennongjia meteorological station (1991–2020), the area experiences an average annual precipitation of 897 mm and a mean monthly temperature of 12.6 °C. Using the CRU TS 4.07 data [28] for the 0.5° grid relevant to the study sites, the actual precipitation at the sampling sites is approximately 2000 mm. The highest temperatures typically occur in July and August, peaking around 23 °C. Tree-ring cores utilized in this study were obtained every few years between 2012 and 2021 from healthy dominant or co-dominant mature trees at breast height (1.3 m above ground surface) in various locations around the Shennongjia region—specifically, Taiziya (TZY), Xiaocongping (XCP), Jinhouling (JHL), and Shennongding (SND), as illustrated in Figure 1 and detailed in Table 1 (excluding the final year, which corresponds to the sampling year). An earlier version of JHL data was used in a pilot experiment study [22]. Sample collection was primarily performed using a 5.15 mm Haglöf Sweden increment borer with a small portion sampled using a 10 mm Haglöf Sweden increment borer. In general, we sampled two cores from each tree parallel to the contour of the slope to reduce abnormalities in tree-ring growth. In a few cases, one core may be sampled when the second core is dangerous or difficult to sample. Sometimes, three cores may be sampled when the other two cores seem to not be in good condition for further analysis.

Table 1. Sample information.

Sampling Site	XCP	TZY	JHL	SND
Latitude	31.478° N	31.451° N	31.473° N	31.440° N
Longitude	110.187° E	110.192° E	110.310° E	110.304° E
Elevation (m a.s.l.)	2722	2494	2564	2876
number of trees/cores	25/49	37/64	54/98	10/20
Total period (CE)	1774–2011	1811–2018	1830–2020	1846–2011

2.2. Climate Data

There are three meteorological stations around the sampling sites (Figure 1): Shennongjia Meteorological Station, Fangxian Meteorological Station, and Badong Meteorological Station—with meteorological records starting in 1975, 1958, and 1953, respectively. However, none of these stations are directly adjacent to the sampling sites; the closest distance from the sampling sites to the nearest meteorological station is ~45 km. Data from the Badong meteorological station, including monthly precipitation (PRE), monthly mean minimum temperature (TMN), monthly mean temperature (TMP), and monthly mean maximum temperature (TMX), were utilized due to the longer time coverage data compared with the other two stations.

2.3. Chronology Development

All samples were processed following recommended procedures [13], such as air-drying, mounting on wooden holders, and sanding with progressively finer sandpaper (up to 800 grit). Resin extraction was not performed, as the samples from fir do not exhibit a visible heartwood–sapwood color change. Before scanning, the scanners were calibrated using the SilverFast 9 scanner software and an IT8 color card target (IT8.7/2) to ensure consistency between laboratories [25]. The mounted cores were then scanned at 3200 DPI using a flatbed Epson V850 Pro scanner, which was covered by a box with matt black side walls to minimize bias from external ambient light and internal box reflections of light [13]. Tree-ring width and tree-ring blue intensity data were generated using CooRecorder software (9.8.1) [29] in combination with CDendro software (9.8.1). Inverted latewood blue intensity (LWBI), inverted earlywood blue intensity (EWBI), and delta blue intensity (DBI) [18] data were generated using frame specification parameters with a "window" (width-offset-limiting-depth-margin) of 300-3-0-500-0.2 and 200-3-5-200-0.2, respectively. Percentile extraction values 85% and 15%, which have been shown to produce optimal results in previous studies [13,22], were employed to extract light and dark wood reflection intensity information for EWBI and LWBI. All tree-ring data were detrended using ARSTAN with an age-dependent spline with positive trend retention. An initial spline flexibility of 50 years was used, which more naturally tracks the juvenile, while the increasing incremental changes in the spline window tracks the long-term radial growth trajectory better than more rigidly defined approaches, such as negative exponential functions.

2.4. Correlation Analysis and Reconstruction Model Development

Initially, the climate signal in the tree-ring parameters was evaluated by computing Pearson's correlation coefficients between the tree-ring chronologies and monthly/seasonal PRE, TMN, TPM, TMX from Badong meteorological station. Considering that the climate conditions in the previous year may affect the growth in the current year, the monthly climate from previous September to current October were employed in the analysis. Then, based on the result of the correlation analysis, a simple linear regression analysis was utilized to generate a transfer function to transform the tree-ring data to estimates of past climate. The period from 1953 to 2020 was used for an assessment of the reconstruction. The dataset was first split into a calibration period from 1953 to 1986 and a verification period from 1987 to 2020. Then, the analysis was repeated with a reversed split: the calibration period from 1987 to 2020 and the verification period from 1953 to 1986. To assess the accuracy of the model, the R-squared (R^2), the F value (F), and the root mean square error (RMSE) were calculated for the calibration period. Then, R^2, RMSE, reduction in error (RE), and coefficient of efficiency (CE) [30] were calculated for the verification period. The Durbin–Watson value (DW) was calculated for the full period.

Secondly, the reconstructed series, using the original and the first differenced transform, were compared with the corresponding CRU TS4.07 temperature data using spatial correlation analysis to assess its robustness and spatial representativeness.

3. Results and Discussion

3.1. Chronology Characteristics

The periods covered by the tree-ring chronologies (XCP, TZY, JHL, and SND) are as follows: XCP spans from 1774 to 2011, TZY from 1811 to 2018, JHL from 1830 to 2020, and SND from 1846 to 2011 (Table 1, Figure 2). Notably, both XCP and TZY extend over more than 200 years, presenting a significant opportunity to potentially extend our climate knowledge prior to the instrumental record through dendrochronological analysis. The chronologies exhibit a consistent growth pattern for the same tree-ring parameter (Figure 2). This consistency suggests a coherence in environmental response across these distinct temporal intervals. Combining data from these four sites enables us to develop more replicated robust and longer reconstruction of past climate variability, contributing to our understanding of the complex interactions driving climatic changes on both local and regional scales.

Figure 3 provides a comprehensive overview of the statistical characteristics of the parameter tree-ring chronologies. Across all sites, RW consistently shows higher coefficient of variation (CV) values, high first-order autocorrelation (AC1), and strong common signal strength (Rbar). For AC1, the order is RW, EWBI, LWBI, and DBI, showing a substantial difference in persistence properties between RW and DBI. Rbar values are low for EWBI, indicating a rather weak signal strength and need more cores to attain a robust chronology. However, the Rbar value of 0.172 to 0.199 for LWBI and DBI shows that for these parameters, one needs to sample about 25 cores to attain an expressed population signal (EPS) value $\geq$ 0.85 [31].

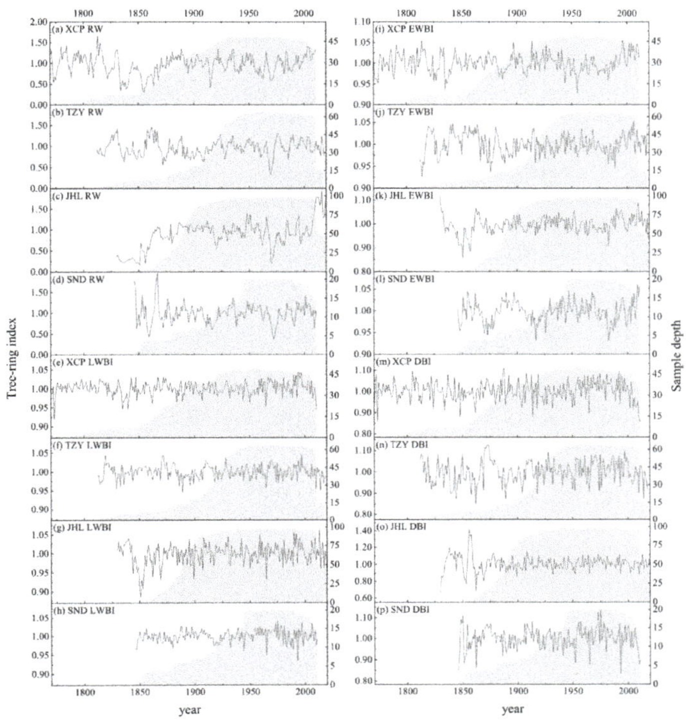

Figure 2. Chronologies for XCP, TZY, JHL, and SND in Shennongjia.

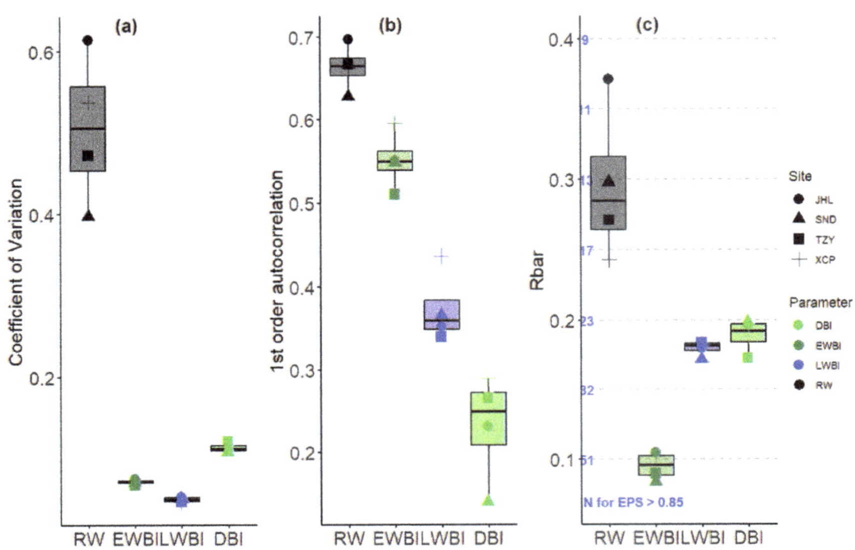

Figure 3. Statistics of the tree-ring detrended parameter chronologies. Coefficient of variation (**a**), first-order autocorrelation (**b**), and Rbar (**c**).

The higher CV and Rbar values for RW compared to EWBI and LWBI have been observed in previous studies [24,32], and it has been found that more samples are needed to maintain a 0.85 threshold [31] for BI chronologies than for RW chronologies in general. The higher and similar AC1 values for RW and EWBI compared to LWBI and DBI suggest why RW and EWB can express pre-growing season climate while LWBI and DBI typically reconstruct current growing season climate [5]. Previous studies [33,34], including this one, have demonstrated that the different signals expressed by the different tree-ring parameters present a chance to uncover additional climatic information.

For the period 1901–2010 represented by high replication for each site, the pairwise correlation coefficients among various tree-ring parameters for all sites predominantly indicate low or negative values between RW and other tree-ring parameters such as EWBI, LWBI, and DBI. Previous studies [14] have also identified distinct signals of the different TR parameters, which generally represent different aspects of climate influence on growth. Furthermore, a consistent negative correlation is observed between EWBI and DBI across all sites. Conversely, weak but positive correlation coefficients are consistently evident between EWBI and LWBI, while strong correlations between LWBI and DBI are noted across all sites (Figure 4). It should be noted that the patterns between TR parameters in different studies may differ, with one reason being the inconsistent interpretations and understanding of the terminology for BI [18].

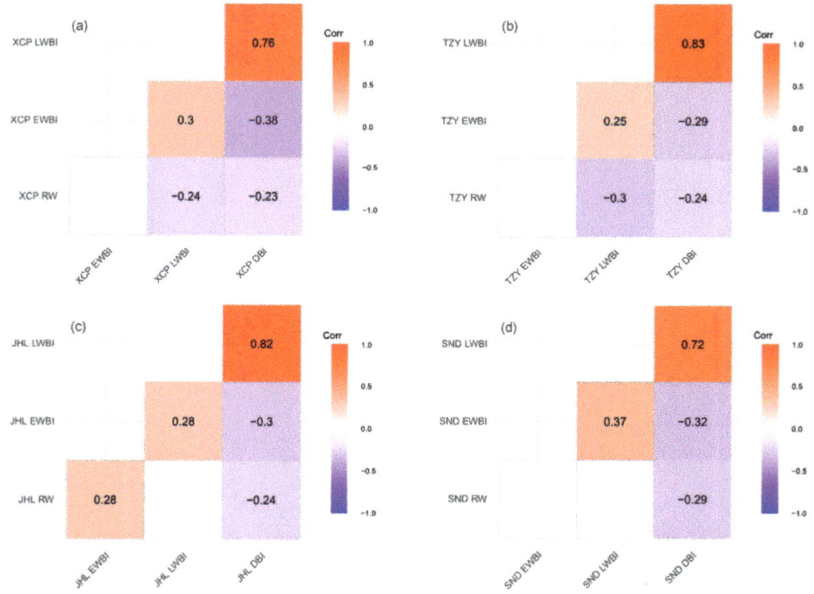

Figure 4. Pairwise correlation between different tree-ring parameters for XCP (**a**), TZY (**b**), JHL (**c**), and SND (**d**) for the period 1901–2010. The blank square indicates that the correlation did not reach the significance level of 0.05.

Overall, comparing the same tree-ring parameters between sites, there is a strong common signal between the sites (Figure 5), although this is weaker for RW (range: 0.23–0.57) followed by EWBI (0.49–0.66), LWBI (0.6–0.73) and DBI (0.6–0.78). Principal component analyses were performed for each tree-ring parameter to examine common patterns across the four locations (Table 2). This parameter-specific common variance is further validated over the period 1901–2010, where the variance explained by PC1 from a PCA for RW, EWBI, LWBI and DBI is 58.1%, 68.1%, 73.9%, and 76.4%, respectively.

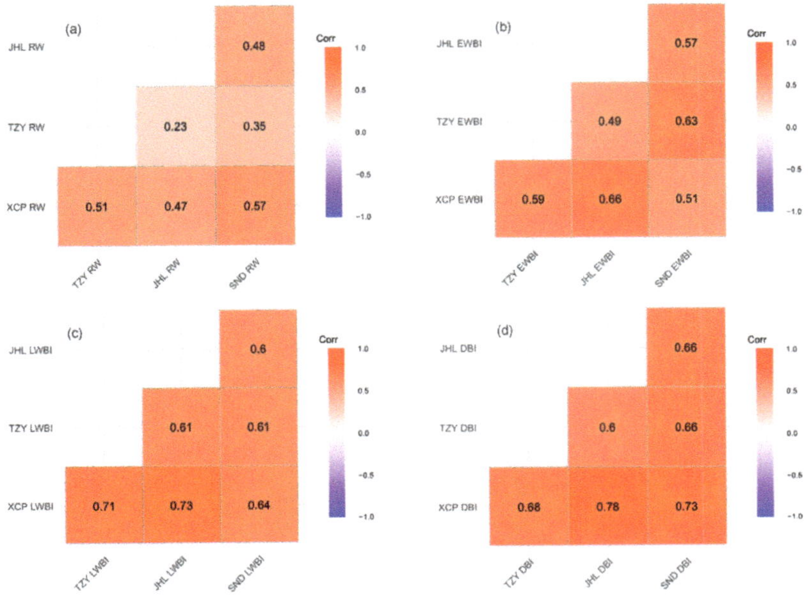

Figure 5. Pairwise correlation between different sites for RW (**a**), EWBI (**b**), LWBI (**c**), and DBI (**d**) for the period 1901–2010. The number and color of the box indicate the correlation coefficient.

Table 2. Results of principal component analysis.

Parameters		PC1	PC2	PC3	PC4
RW	Eigenvalue	2.325	0.798	0.499	0.378
	Proportion	0.581	0.199	0.125	0.095
	Cumulative	0.581	0.781	0.905	1.000
EWBI	Eigenvalue	2.724	0.569	0.438	0.269
	Proportion	0.681	0.142	0.109	0.067
	Cumulative	0.681	0.823	0.933	1.000
LWBI	Eigenvalue	2.955	0.421	0.386	0.239
	Proportion	0.739	0.105	0.096	0.060
	Cumulative	0.739	0.844	0.940	1.000
DBI	Eigenvalue	3.056	0.419	0.319	0.207
	Proportion	0.764	0.105	0.080	0.052
	Cumulative	0.764	0.869	0.948	1.000

Despite the weaker within-site common signal of the BI parameters compared to the RW parameters (Figure 3), which has been previously noted [32–35], the BI parameters exhibit a higher inter-site correlation than the RW parameters for *Abies fargesii* Franch. in the Shennongjia region (Figure 5). Hence, due to the strong common signal between the sites, it is more logical to develop a composite regional chronology for each tree-ring parameter. Therefore, we derived four regional composite chronologies for each parameter (Figure 6) but excluded cores that did not exhibit a significant correlation with the master series at the 0.05 level. Ultimately, we retained 226, 209, 226, and 220 cores out of 231 for RW, EWBI, LWBI, and DBI, respectively. EPS calculations were performed using a window length of 31 years with a 30-year overlap. The results show that in the early stages of all chronologies, when the sample size is small, the EPS value is low and unstable, indicating that the chronologies from this period are not reliable.

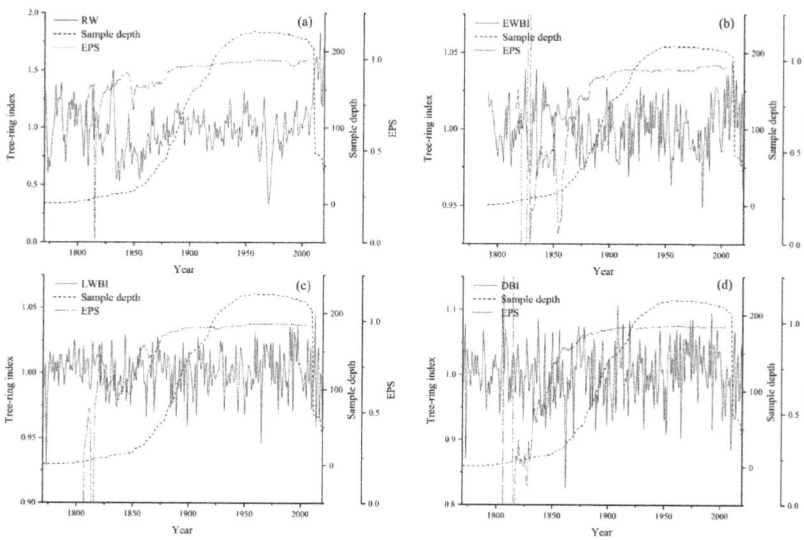

Figure 6. Composite chronologies for RW (**a**), EWBI (**b**), LWBI (**c**), and DBI (**d**) in Shennongjia.

3.2. Relationship between Chronologies and Climate Variables

The correlations between the chronologies and climate data (1953–2020) revealed that all chronologies express a temperature response and show weaker relationships with precipitation (Figure 7). As all tree-ring samples were collected at relatively high elevations (Table 1), where there is abundant rainfall, it is not surprising that precipitation does not limit tree growth. Previous studies have demonstrated that tree rings collected at high elevations are significantly influenced by temperature variations due to temperature limitations in the growing season [36].

On the whole, RW exhibits a weak correlation with climate data with relatively higher correlations observed for the mean TMN from the previous November to the current March ($r = 0.39$) and March–April TMX ($r = 0.49$) (Figure 7). However, after applying the first difference to both RW and climate data, the previously mentioned correlation results between them did not reach significance at the 0.05 level (Figure 8). This reaffirms that although tree-ring width data may express strong common signals (within and between sites), they may not express a strong climate signal and in fact, the correlations presented in Figure 7 represent the influence of trend on the correlation values rather than a true climate signal. Therefore, the timely and selective incorporation of other tree-ring parameters, such as tree-ring isotopes, tree-ring density, and tree-ring blue intensity, may significantly enhance the depth and breadth of our understanding of past climate data.

Typically, in the original series, EWBI demonstrates a weak correlation with the aforementioned four climate variables (Figure 7). This result explains why EWBI is a less-explored parameter in dendroclimatic research [14]. However, it is noteworthy that June, June–July, and January–July TMX show significant correlations with EWBI, with values of 0.59, 0.54 and 0.54, respectively, after applying the first difference to both the chronology and climate data (Figure 8). These correlations are weaker for the unfiltered correlations (Figure 7). However, these results may present an opportunity to understand the historical high-frequency TMX fluctuations in aforesaid month and seasons. The mean TMX for the previous December–October and January–October also showed a significant influence on EWBI. However, this may be meaningless because the tree-ring earlywood is not formed so later.

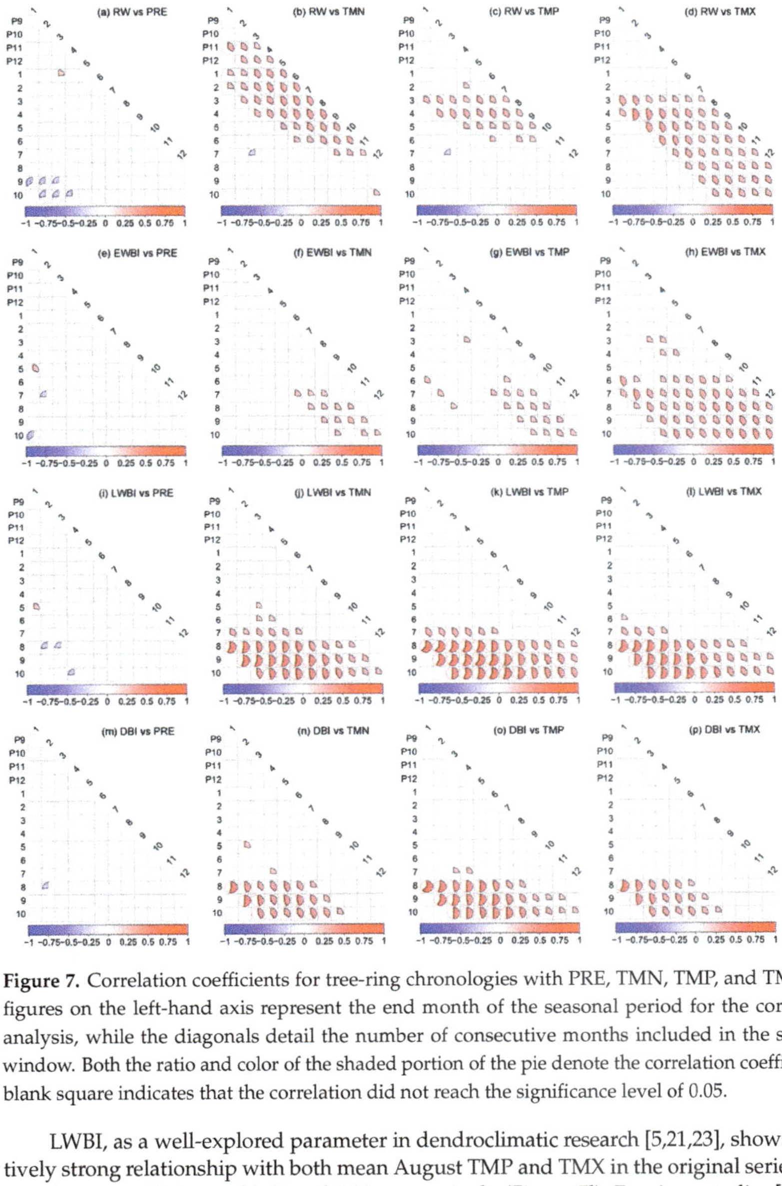

Figure 7. Correlation coefficients for tree-ring chronologies with PRE, TMN, TMP, and TMX. The figures on the left-hand axis represent the end month of the seasonal period for the correlation analysis, while the diagonals detail the number of consecutive months included in the seasonal window. Both the ratio and color of the shaded portion of the pie denote the correlation coefficient. A blank square indicates that the correlation did not reach the significance level of 0.05.

LWBI, as a well-explored parameter in dendroclimatic research [5,21,23], shows a relatively strong relationship with both mean August TMP and TMX in the original series, with correlation coefficients of 0.63 and 0.59, respectively (Figure 7l). Previous studies [5,21,23] have also confirmed that LWBI effectively reflects summer temperature variations. In fact, after applying the first difference to both the LWBI chronology and climate data, the correlation coefficients for TMN, TMP, and TMX significantly improve (Figure 8j–l). For example, August TMP and August TMX rank as the two highest correlations with coefficients of 0.76 and 0.75, respectively.

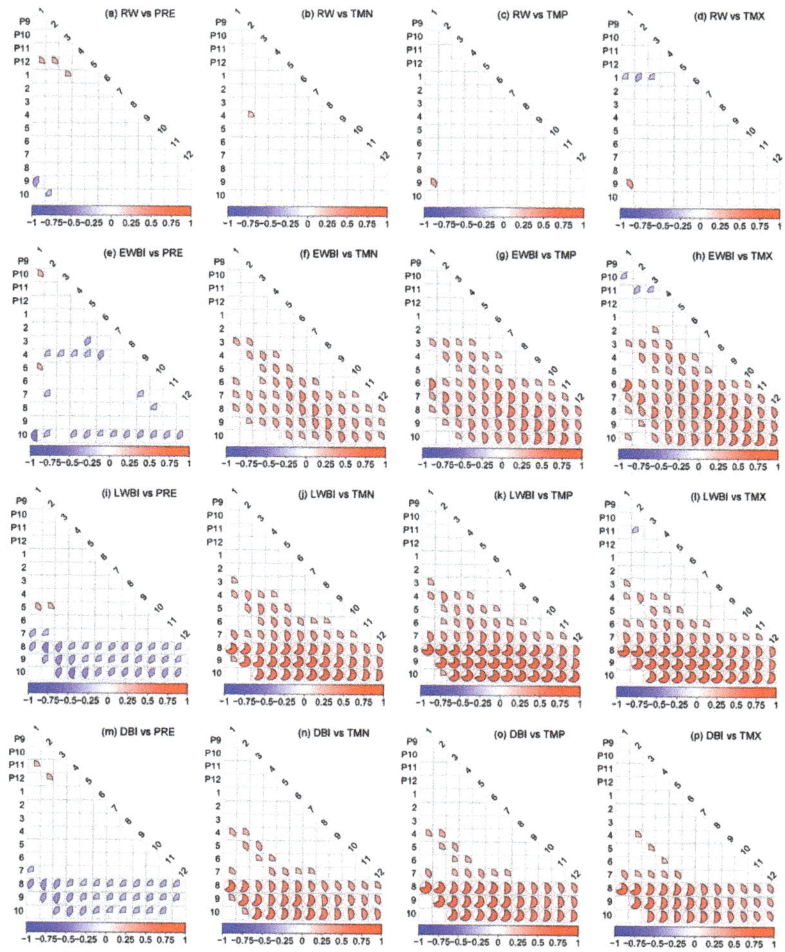

Figure 8. As Figure 7, but after the tree-ring and climate data have been transformed to 1st differences.

For DBI, the unfiltered correlation coefficients with August TMN, TMP and TMX rank highest at 0.54, 0.64, and 0.57, respectively. Similar to LWBI, these coefficients also show significant improvement after applying the first difference to both the DBI chronology and climate data, increasing to 0.65, 0.73, and 0.71. As we can see, DBI exhibits slightly better performance than LWBI in capturing low-frequency signals, while it is slightly less effective in high-frequency signals. This reaffirms the potential of both DBI and LWBI in tree-ring research [20,32]. Clearly, LWBI and DBI in this study demonstrate a different and more robust relationship with temperature compared to RW. This finding provides an opportunity to deepen our understanding of temperature variability in subtropical areas and represents a significant advancement beyond traditional research, which has mainly focused on higher-latitude regions [20,32]. As has been shown before, the strong correlations noted above show that despite the weaker signal strength in the BI parameters compared to RW, this does not reflect that these data express a weak climate signal [32–35,37]. The discrepancy in correlations between RW and climate versus BI and climate (Figure 8), whether species-specific or regional-specific, should be further explored.

3.3. Temperature Reconstruction

Based on the correlation analysis results, a regression analysis was performed to establish the transfer function between the DBI chronology and the August TMP (1953–2020). The derived regression equation is as follows:

$$TMP_{Aug} = 8.009 \times DBI + 19.56 \quad (1)$$

TMP_{Aug} represents the estimated August mean temperature (in degrees Celsius), while DBI denotes the corresponding tree-ring DBI value for the same year. As shown in Table 3, the reconstruction accounts for 40.81% of the temperature variance over the full calibration period from 1953 to 2020. The F-value is 45.506 with a corresponding p-value of 0.000, indicating high statistical significance. Additionally, the RMSE value stands at 1.05, suggesting the extent of deviation between actual and predicted values. The DW value of 1.719 suggests no significant (99%) autocorrelation in the model residuals. RE and CE, for both validation periods, are consistently positive, indicating an acceptable amount of model fidelity.

Table 3. Statistics of the August temperature model.

	R^2	F	DW	RMSE	RE	CE
Calibration 1953–1986	0.362	18.131	-	1.057	-	-
Verification 1987–2020	0.402	-	-	1.046	0.435	0.402
Calibration 1987–2020	0.407	22.011	-	1.041	-	-
Verification 1953–1986	0.356	-	-	1.062	0.448	0.356
Full period 1953–2020	0.408	45.506	1.719	1.050	-	-

The DBI based August temperature reconstruction covers the period from 1770 to 2020 with enhanced reliability observed since 1865 where EPS > 0.85 (Figure 9c). The average estimated August TMP for the reliable period spanning from 1865 through 2020 was 27.667 °C. An 11-year moving average of the reconstructed temperature series highlights, since 1865, warm periods were predominantly observed from 1869 to 1883 and from 1956 to 1999, while cold periods occurred mainly from 1884 to 1904, 1925 to 1955, and from 2000 to 2020. A year was classified as very warm if it exceeded the mean + 1σ and as cool if it fell below the mean—1σ; the reconstructed series indicated 29 very warm years and 26 very cold years. The cold summers, such as those in 1884, and 1965, may be related to volcanic eruptions, namely Krakatoa in 1883, and Mount Agung in 1963–1964. By contrast, some cold summers, such as those in 1869 and 1889, occurred without any associated volcanic eruption events. Therefore, in addition to the volcanic cooling effect [38], other factors should be given attention in future studies and may reflect larger-scale dynamical processes impacting season climate conditions.

As shown in Figure 9a,b, the reconstructed TMP closely aligns with the observed TMP over the common period of 1953–2020, and there is a stronger significant correlation ($r = 0.732$, $p < 0.001$) between the two sequences after the 1st difference transform. This indicates that our reconstructed sequence is very robust at high frequencies, but the fact that the reconstruction model passes RE and CE validation and there is no autocorrelation in the residuals suggests that the reconstruction models the lower frequency trends quite well also.

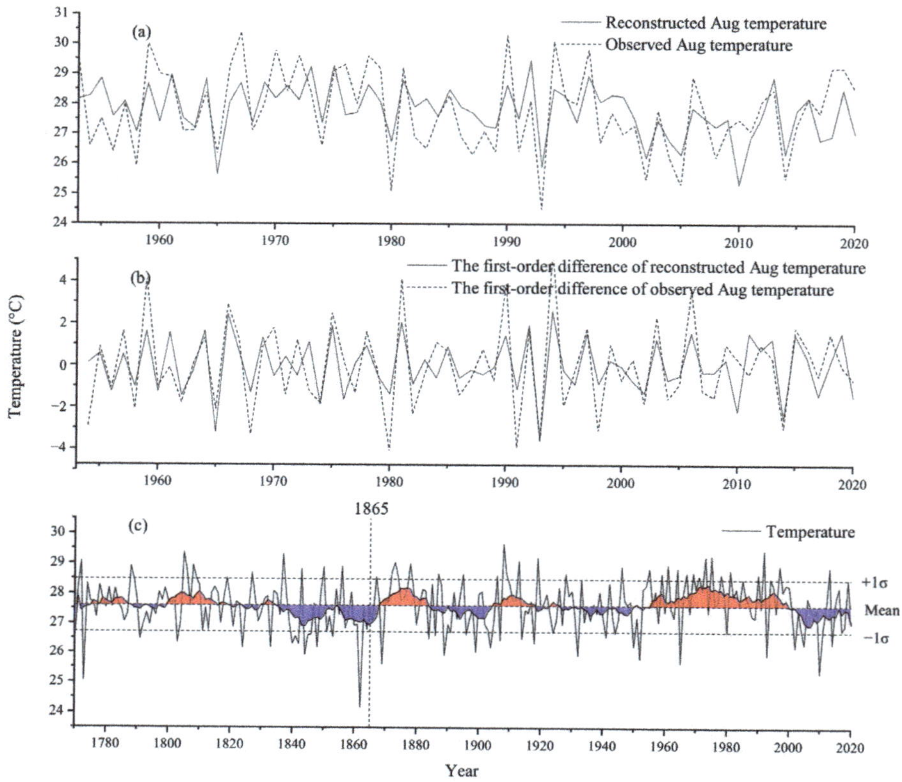

Figure 9. Comparison of reconstructed and observed August temperatures; unfiltered (**a**), 1st differenced (**b**), unfiltered reconstructed August temperatures with a 11-year moving average (**c**). The red and blue shadows represent warm and cool periods, respectively, based on the 11-year moving average temperature.

Previous research conducted in the region [27] showed an early growing season temperature signal in RW. However, this temperature signal primarily manifests in low-frequency variations rather than high-frequency variations. In contrast, as has been observed in other BI-based studies in Europe [20], DBI has a substantially stronger temperature signal at interannual timescales than the corresponding RW chronology. This study, unlike previous tree-ring width-based temperature reconstructions [39,40] that show a warming trend in recent decades, shows a gentle temperature change. This difference may result from the reconstructed seasons being different; additionally, the study areas are also not the same.

As well as being well correlated with local temperatures (Figure 10a), the reconstruction coheres strongly with a much larger region especially after 1st differencing (Figure 10b). This latter observation suggests that the interannual signal is robust over an extremely large area across most of China and even as far east as the Korean Peninsula. This indicates that the variability of August temperatures in the Shennongjia area is representative of not just local temperatures but also of larger-scale regional climate patterns. This reaffirms that BI can effectively capture high-frequency temperature changes, although further work is needed to assess whether the weaker correlation for the non-transformed data is related to trend distortion in the DBI records or possible inhomogeneity issues in the instrumental data themselves.

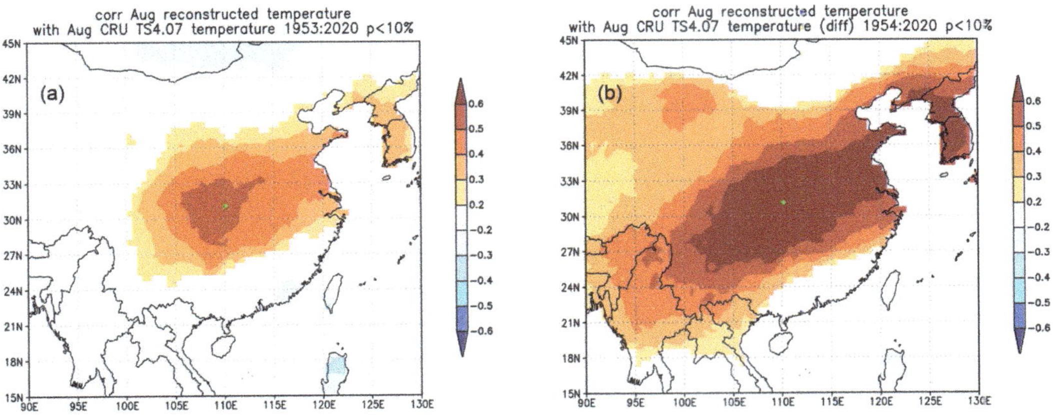

Figure 10. The field-correlation between the reconstructed August temperature and the CRU TS 4.07 dataset for the period from 1953 to 2020, using both unfiltered data (**a**) and the first difference (**b**). Correlation fields were produced using KNMI Climate explorer "http://climexp.knmi.nl (accessed on 23 July 2024)". The green circle shows the position of the Badong meterological station.

4. Conclusions

In this study, we examined the relationship between RW, EWBI, LWBI, and DBI chronologies with climate to determine if BI can express more climate information than RW data from *Abies fargesii* Franch. trees in subtropical China. The results are encouraging and suggest that BI is a superior parameter for extracting temperature information from tree rings in subtropical areas. Our findings indicate that DBI, which explains 40.8% of the variance in August temperature, is a promising parameter for reconstructing past climate variability in subtropical regions. Reconstructed results are particularly important for this region with limited historical climate data, as they can provide foundational data for studies related to climate change. Given the longevity of *Abies fargesii* Franch., exceeding 200 years, and the absence of heartwood/sapwood color differentiation, it is a favorable species for future dendroclimatic research in this region of Asia. By further exploration and network expansion, the temperature reconstruction can be significantly improved in terms of length and spatial coverage with the inclusion of new material [41]. This study serves as a reference for understanding past climates in subtropical, warm–humid regions. It is worth noting that except for fir and spruce, most tree species exhibit visible heartwood–sapwood color changes, which can affect the relationship between tree-ring chronology and climate. Effectively eliminating this color change is crucial and requires further study. Additionally, tree species with very tiny rings may not produce a sharply focused image with a scanner; this issue may be addressed using camera mode.

Author Contributions: Conceptualization, data curation, formal analysis, resource acquisition, visualization, and original draft preparation: Y.Z.; Review and editing: R.W. All authors have read and agreed to the published version of the manuscript.

Funding: This research was funded by the Key Regional Joint Research Program of the National Natural Science Foundation of China, grant no. U23A2021, and the National Natural Science Foundation of China, grant no. 41771227. Rob Wilson was further funded by the NSF/NERC project, grant no. NE/W007223/1—Understanding Trans-Hemispheric Modes of Climate Variability: A Novel Tree-Ring Data Transect spanning the Himalaya to the Southern Ocean.

Data Availability Statement: The data presented in this study are available on request from the corresponding author.

Acknowledgments: We extend our gratitude to Shuai Yuan, Yongdong Qi, Dan Wang, and Zhengsheng Hu for their assistance in sampling tree rings. Yue Wang, Shuai Yuan, and Fanxi Kong conducted initial exploratory investigations.

Conflicts of Interest: The authors declare no conflicts of interest.

References

1. Esper, J.; Torbenson, M.; Büntgen, U. 2023 summer warmth unparalleled over the past 2000 years. *Nature* **2024**, *631*, 94–97. [CrossRef]
2. IPCC. Summary for Policymakers. In *Climate Change 2023: Synthesis Report. Contribution of Working Groups I, II and III to the Sixth Assessment Report of the Intergovernmental Panel on Climate Change*; Core Writing Team, Lee, H., Romero, J., Eds.; IPCC: Geneva, Switzerland, 2023; pp. 1–34.
3. Cook, E.; Krusic, P.; Anchukaitis, K.; Buckley, B.; Nakatsuka, T.; Sano, M. Tree-ring reconstructed summer temperature anomalies for temperate East Asia since 800 C.E. *Clim. Dyn.* **2012**, *41*, 2957–2972. [CrossRef]
4. Shah, S.K.; Berkelhammer, M.; Li, Q.; Mehrotra, N.; Thomte, L.; Shell, R.; Pandey, U.; Gaire, N.P.; Kathayat, G.; Sinha, A. Regional tree-ring oxygen isotope deduced summer monsoon drought variability for Kumaun-Gharwal Himalaya. *Quat. Sci. Rev.* **2023**, *301*, 107927. [CrossRef]
5. Heeter, K.J.; Harley, G.L.; Maxwell, J.T.; Wilson, R.J.; Abatzoglou, J.T.; Rayback, S.A.; Rochner, M.L.; Kitchens, K.A. Summer temperature variability since 1730 CE across the low-to-mid latitudes of western North America from a tree ring blue intensity network. *Quat. Sci. Rev.* **2021**, *267*, 107064. [CrossRef]
6. Wang, J.; Yang, B.; Ljungqvist, F.C. A millennial summer temperature reconstruction for the eastern tibetan plateau from tree-ring width. *J. Clim.* **2015**, *28*, 5289–5304. [CrossRef]
7. Wilson, R.; Anchukaitis, K.; Briffa, K.R.; Büntgen, U.; Cook, E.; D'Arrigo, R.; Davi, N.; Esper, J.; Frank, D.; Gunnarson, B.; et al. Last millennium northern hemisphere summer temperatures from tree rings: Part I: The long term context. *Quat. Sci. Rev.* **2016**, *134*, 1–18. [CrossRef]
8. Salzer, M.W.; Bunn, A.G.; Graham, N.E.; Hughes, M.K. Five millennia of paleotemperature from tree-rings in the Great Basin, USA. *Clim. Dyn.* **2014**, *42*, 1517–1526. [CrossRef]
9. Huang, R.; Zhu, H.; Liang, E.; Liu, B.; Shi, J.; Zhang, R.; Yuan, Y.; Grießinger, J. A tree ring-based winter temperature reconstruction for the southeastern Tibetan Plateau since 1340 CE. *Clim. Dyn.* **2019**, *53*, 3221–3233. [CrossRef]
10. Keyimu, M.; Li, Z.; Liu, G.; Fu, B.; Fan, Z.; Wang, X.; Wu, X.; Zhang, Y.; Halik, U. Tree-ring based minimum temperature reconstruction on the southeastern Tibetan Plateau. *Quat. Sci. Rev.* **2021**, *251*, 106712. [CrossRef]
11. Fan, Z.-X.; Bräuning, A.; Yang, B.; Cao, K.-F. Tree ring density-based summer temperature reconstruction for the central Hengduan Mountains in southern China. *Glob. Planet. Chang.* **2009**, *65*, 1–11. [CrossRef]
12. Briffa, K.; Osborn, T.; Schweingruber, F.H.; Harris, I.C.; Jones, P.D.; Shiyatov, S.G.; Vaganov, E. Low-frequency temperature variations from a northern tree ring density network. *J. Geophys. Res. Space Phys.* **2001**, *106*, 2929–2941. [CrossRef]
13. Rydval, M.; Larsson, L.; McGlynn, L.; Gunnarson, B.E.; Loader, N.J.; Young, G.H.; Wilson, R. Blue intensity for dendroclimatology: Should we have the blues? Experiments from Scotland. *Dendrochronologia* **2014**, *32*, 191–204. [CrossRef]
14. Cao, X.; Hu, H.; Kao, P.; Buckley, B.M.; Dong, Z.; Chen, X.; Zhou, F.; Fang, K. Improved spring temperature reconstruction using earlywood blue intensity in southeastern China. *Int. J. Clim.* **2022**, *42*, 6204–6220. [CrossRef]
15. McCarroll, D.; Pettigrew, E.; Luckman, A.; Guibal, F.; Edouard, J.L. Blue reflectance provides a surrogate for latewood density of high-latitude pine tree rings. *Arct. Antarct. Alp. Res.* **2002**, *34*, 450–453. [CrossRef]
16. Campbell, R.; McCarroll, D.; Loader, N.J.; Grudd, H.; Robertson, I.; Jalkanen, R. Blue intensity in *Pinus sylvestris* tree-rings: Developing a new palaeoclimate proxy. *Holocene* **2007**, *17*, 821–828. [CrossRef]
17. Kaczka, R.J.; Spyt, B.; Janecka, K.; Beil, I.; Büntgen, U.; Scharnweber, T.; Nievergelt, D.; Wilmking, M. Different maximum latewood density and blue intensity measurements techniques reveal similar results. *Dendrochronologia* **2018**, *49*, 94–101. [CrossRef]
18. Björklund, J.; Seftigen, K.; Kaczka, R.J.; Rydval, M.; Wilson, R. A definition and standardised terminology for Blue Intensity from Conifers. *Dendrochronologia* **2024**, *85*, 126200. [CrossRef]
19. Kaczka, R.J.; Wilson, R. I-BIND: International Blue intensity network development working group. *Dendrochronologia* **2021**, *68*, 125859. [CrossRef]
20. Fuentes, M.; Salo, R.; Björklund, J.; Seftigen, K.; Zhang, P.; Gunnarson, B.; Aravena, J.-C.; Linderholm, H.W. A 970-year-long summer temperature reconstruction from Rogen, west-central Sweden, based on blue intensity from tree rings. *Holocene* **2018**, *28*, 254–266. [CrossRef]
21. Heeter, K.J.; Harley, G.L.; Maxwell, J.T.; McGee, J.H.; Matheus, T.J. Late summer temperature variability for the Southern Rocky Mountains (USA) since 1735 CE: Applying blue light intensity to low-latitude Picea engelmannii Parry ex Engelm. *Clim. Chang.* **2020**, *162*, 965–988. [CrossRef]
22. Zheng, Y.; Shen, H.; Abernethy, R.; Wilson, R. Experiments of the efficacy of tree ring blue intensity as a climate proxy in central and western China. *Biogeosciences* **2023**, *20*, 3481–3490. [CrossRef]
23. Björklund, J.; Gunnarson, B.E.; Seftigen, K.; Zhang, P.; Linderholm, H.W. Using adjusted Blue Intensity data to attain high-quality summer temperature information: A case study from Central Scandinavia. *Holocene* **2015**, *25*, 547–556. [CrossRef]

24. Wilson, R.; Rao, R.; Rydval, M.; Wood, C.; Larsson, L.; Luckman, B.H. Blue Intensity for dendroclimatology: The BC blues: A case study from British Columbia, Canada. *Holocene* **2014**, *24*, 1428–1438. [CrossRef]
25. Wilson, R.; Anchukaitis, K.; Andreu-Hayles, L.; Cook, E.; D'arrigo, R.; Davi, N.; Haberbauer, L.; Krusic, P.; Luckman, B.; Morimoto, D.; et al. Improved dendroclimatic calibration using blue intensity in the southern Yukon. *Holocene* **2019**, *29*, 1817–1830. [CrossRef]
26. Buckley, B.M.; Hansen, K.G.; Griffin, K.L.; Schmiege, S.; Oelkers, R.; D'arrigo, R.D.; Stahle, D.K.; Davi, N.; Nguyen, T.Q.T.; Le, C.N.; et al. Blue intensity from a tropical conifer's annual rings for climate reconstruction: An ecophysiological perspective. *Dendrochronologia* **2018**, *50*, 10–22. [CrossRef]
27. Zheng, Y.; Shao, X.; Lu, F.; Li, Y. February–May temperature reconstruction based on tree-ring widths of Abies fargesii from the Shennongjia area in central China. *Int. J. Biometeorol.* **2016**, *60*, 1175–1181. [CrossRef]
28. Harris, I.; Osborn, T.J.; Jones, P.; Lister, D. Version 4 of the CRU TS monthly high-resolution gridded multivariate climate dataset. *Sci. Data* **2020**, *7*, 19. [CrossRef]
29. Maxwell, R.S.; Larsson, L.-A. Measuring tree-ring widths using the CooRecorder software application. *Dendrochronologia* **2021**, *67*, 125841. [CrossRef]
30. Cook, E.R.; Briffa, K.R.; Jones, P.D. Spatial regression methods in dendroclimatology: A review and comparison of two techniques. *Int. J. Climatol.* **1994**, *14*, 379–402. [CrossRef]
31. Wigley, T.M.; Briffa, K.R.; Jones, P.D. On the average value of correlated time series, with applications in dendroclimatology and hydrometeorology. *J. Appl. Meteorol. Climatol.* **1984**, *23*, 201–213. [CrossRef]
32. Wilson, R.; Allen, K.; Baker, P.; Boswijk, G.; Buckley, B.; Cook, E.; D'Arrigo, R.; Druckenbrod, D.; Fowler, A.; Grandjean, M.; et al. Evaluating the dendroclimatological potential of blue intensity on multiple conifer species from Tasmania and New Zealand. *Biogeosciences* **2021**, *18*, 6393–6421. [CrossRef]
33. Blake, S.A.; Palmer, J.G.; Björklund, J.; Harper, J.B.; Turney, C.S. Palaeoclimate potential of New Zealand Manoao colensoi (silver pine) tree rings using Blue-Intensity (BI). *Dendrochronologia* **2020**, *60*, 125664. [CrossRef]
34. Cao, X.; Fang, K.; Chen, P.; Zhang, P.; Björklund, J.; Pumijumnong, N.; Guo, Z. Microdensitometric records from humid subtropical China show distinct climate signals in earlywood and latewood. *Dendrochronologia* **2020**, *64*, 125764. [CrossRef]
35. Wiles, G.C.; Charlton, J.; Wilson, R.J.; D'arrigo, R.D.; Buma, B.; Krapek, J.; Gaglioti, B.V.; Wiesenberg, N.; Oelkers, R. Yellow-cedar blue intensity tree-ring chronologies as records of climate in Juneau, Alaska, USA. *Can. J. For. Res.* **2019**, *49*, 1483–1492. [CrossRef]
36. Frank, D.; Esper, J. Characterization and climate response patterns of a high-elevation, multi-species tree-ring network in the European Alps. *Dendrochronologia* **2005**, *22*, 107–121. [CrossRef]
37. Buras, A. A comment on the expressed population signal. *Dendrochronologia* **2017**, *44*, 130–132. [CrossRef]
38. Yin, H.; Li, M.-Y.; Huang, L. Summer mean temperature reconstruction based on tree-ring density over the past 440 years on the eastern Tibetan Plateau. *Quat. Int.* **2021**, *571*, 81–88. [CrossRef]
39. Cai, Q.; Liu, Y.; Duan, B.; Sun, C. Regional difference of the start time of the recent warming in Eastern China: Prompted by a 165-year temperature record deduced from tree rings in the Dabie Mountains. *Clim. Dyn.* **2018**, *50*, 2157–2168. [CrossRef]
40. Yuan, Y.; Deng, Y.; Gao, L.; Fu, X.; Xu, Y.; Gou, X. Unprecedented recent warming as recorded by tree-ring in the western Qinling Mountains, China. *Palaeogeogr. Palaeoclim. Palaeoecol.* **2024**, *644*, 112202. [CrossRef]
41. Dang, H.; Zhang, K.; Zhang, Y.; Tong, X.; Zhang, Q. Regeneration dynamics of subalpine fir (Abies fargesii) forest across the altitudinal range in the Shennongjia Mountains, central China. *J. Plant Ecol.* **2013**, *6*, 36–47. [CrossRef]

forests

MDPI

Article

Anomalous Warm Temperatures Recorded Using Tree Rings in the Headwater of the Jinsha River during the Little Ice Age

Chaoling Jiang, Haoyuan Xu, Yuanhe Tong and Jinjian Li *

School of Atmospheric Sciences, Plateau Atmosphere and Environment Key Laboratory of Sichuan Province, Chengdu Plain Urban Meteorology and Environment Observation and Research Station of Sichuan Province, Sichuan Meteorological Disaster Prediction and Early Warning Engineering Laboratory, Chengdu University of Information Technology, Chengdu 610225, China; 2021013020@stu.cuit.edu.cn (C.J.); 2022013194@stu.cuit.edu.cn (H.X.); 2021013057@stu.cuit.edu.cn (Y.T.)

* Correspondence: ljj@cuit.edu.cn

Abstract: As a feature of global warming, climate change has been a severe issue in the 21st century. A more comprehensive reconstruction is necessary in the climate assessment process, considering the heterogeneity of climate change scenarios across various meteorological elements and seasons. To better comprehend the change in minimum temperature in winter in the Jinsha River Basin (China), we built a standard tree-ring chronology from *Picea likiangensis* var. *balfouri* and reconstructed the regional mean minimum temperature of the winter half-years from 1606 to 2016. This reconstruction provides a comprehensive overview of the changes in winter temperature over multiple centuries. During the last 411 years, the regional climate has undergone seven warm periods and six cold periods. The reconstructed temperature sensitively captures the climate warming that emerged at the end of the 20th century. Surprisingly, during 1650–1750, the lowest winter temperature within the research area was about 0.44 °C higher than that in the 20th century, which differs significantly from the concept of the "cooler" Little Ice Age during this period. This result is validated by the temperature results reconstructed from other tree-ring data from nearby areas, confirming the credibility of the reconstruction. The Ensemble Empirical Mode Decomposition method (EEMD) was adopted to decompose the reconstructed sequence into oscillations of different frequency domains. The decomposition results indicate that the temperature variations in this region exhibit significant periodic changes with quasi-3a, quasi-7a, 15.5-16.8a, 29.4-32.9a, and quasi-82a cycles. Factors like El Niño–Southern Oscillation (ENSO), Pacific Decadal Oscillation (PDO), and solar activity, along with Atlantic Multidecadal Oscillation (AMO), may be important driving forces. To reconstruct this climate, this study integrates the results of three machine learning algorithms and traditional linear regression methods. This novel reconstruction method can provide valuable insights for related research endeavors. Furthermore, other global climate change scenarios can be explored through additional proxy reconstructions.

Keywords: dendrochronology; tree-ring width; temperature reconstruction; winter half-year; climate change; Tibetan Plateau

1. Introduction

Climate change poses a long-term challenge to human society and the Earth's environment. A substantial temperature rise will result in a multitude of extensive and profound consequences for both human society and ecosystems [1]. Several studies have demonstrated evident asymmetry in the rise in the minimum and maximum temperatures during global climate change [2]. Additionally, the seasonal distribution of warming exhibits asymmetry, with warming often being more pronounced in winter than in summer [3,4]. This climatic driver, in addition to directly impacting surface snow cover, exerts influence over species adaptation and the structure and function of terrestrial ecosystems. Furthermore, it affects virtually all biochemical processes on Earth through carbon cycling mechanisms and inter-layer interactions [5–7]. In light of this, a more comprehensive reconstruction is necessary for the climate assessment process, considering the heterogeneity of climate change scenarios across various meteorological elements and seasons [8].

The Jinsha River (China) originates from the Tibetan Plateau, which is often called the "Roof of the World" and the "Third Pole," and feeds the upper regions of the Yangtze River. Its watershed encompasses the eastern part of the Tibetan Plateau and the Hengduan Mountain Range area, extending southward to the northern Yunnan Plateau and eastward to the southwestern edge of the Sichuan Basin. It serves as an important ecological reserve and hydropower energy base in China [9]. In summer, due to the influence of the oceanic southwest monsoon and southeast monsoon, the basin experiences substantial rainfall, with the precipitation levels diminishing from southeast to northwest. In contrast, during winter, the basin is primarily impacted by a westerly airflow, leading to continental sunny and dry weather. Therefore, gaining a comprehensive understanding and elucidating the local long-term climate change pattern can facilitate the rational planning of resource development and utilization in the basin, holding significance for predicting future climate change and assessing the climate risks not only in the Qinghai–Tibet Plateau, but across the entire Yangtze River region on a broader scale. Several scholars have conducted extensive research focused on climate change in the Jinsha River Basin. The spatio-temporal characteristics of the monthly average atmospheric temperature at the surface and the levels of precipitation were investigated at 20 meteorological stations from 1961 to 2010 [10]. Furthermore, the daily rainfall patterns across four distinct sub-areas within the Jinsha River Basin were scrutinized [11], and four extreme precipitation indices were employed, as formulated by the Expert Team on Climate Change Detection and Indices (ETCCDIs). This will help us to analyze the temporal and spatial fluctuations in extreme precipitation occurrences. In addition, methodologies such as the dual-mass curve method, the change in slope, and the water balance method have been employed to examine how human activities and the contribution of anthropogenic climate alteration affected the discharge dynamics within the Jinsha River Basin from 1961 to 2010 [12].

The Little Ice Age is characterized by the conditions of general coldness, increased precipitation, and glacier expansion in many parts of the world, significantly influenced both the world and human society, and thus has become a focus in modern research [13–17]. The Northern Hemisphere experienced a temperature drop of about 0.5 °C during this period, and cold weather led to massive famines, wars, and epidemics [18,19]. Since the Little Ice Age, large-scale climate anomalies have occurred simultaneously in the southeastern zones of the Qinghai–Tibet Plateau and the Himalayas [20]. However, given the paucity of long-term instrumental climatic archives for the area, the limited observational data currently available are insufficient for a thorough comprehension of the enduring climate change patterns in the Jinsha River Basin since the Little Ice Age. Therefore, it is urgent for high-resolution climate proxies to be used to reconstruct the historical climate changes in this area, thus contributing to a deeper comprehension of the interaction between modern natural and anthropogenic climate variability [21]. Among the various climate proxies, tree rings have been extensively used for reconstructing ancient climatic alterations due to their reliable dating, precise measurements, continuity, and high resolution [22–25]. The Tibetan Plateau is a prominent area for dendroclimatological studies in China [26–31]. For instance, *Abies squamata* Mast. trees were leveraged to reconstruct early summer temperatures in the Shalv Mountains within the southeastern reaches of the Tibetan Plateau in the last 563 years [32]. The winter temperatures were reconstructed across the preceding 351-year timeframe in the Gongga Mountains using tree rings [33]. In addition, *Juniperus tibetica* tree rings were employed to reveal the annual runoff changes in the Lhasa River over 472 years in the southern Tibetan Plateau [34]. The smallest temperature changes in the last 564 years in the east-central Tibetan Plateau were investigated [35]. Furthermore, the smallest winter temperature changes in the southeastern Tibetan Plateau were determined [27].

To date, several studies have been conducted to estimate climate shifts in the Jinsha River and its neighboring basins during the historical period. These studies have primarily focused on reconstructing the runoff, drought, and summer temperatures [36–40], which have significantly contributed to understanding the configurations of climate alteration within the region across multiple centuries. Nevertheless, these studies have yet to adequately characterize the winter half-year temperature changes in the Jinsha River Basin. Consequently, this study will address the winter temperature changes using tree-ring data from the Jinsha River Basin. This research will construct the average minimum temperature of the winter half-year (November–April) in the last 400 years using a tree-ring width chronology. In addition, the characteristics of the reconstructed temperature changes and the influencing factors will be analyzed, thus enriching the historical climate change data in the Jinsha River Basin, providing scientific data support for understanding the response to global temperature changes and estimating the future temperature change trends.

The primary and specific aims of this research are as follows: (a) to construct the tree-ring width chronology; (b) to reconstruct the minimum winter temperatures of the last four centuries; and (c) to discuss the possible elements driving climate changes.

2. Materials and Methods

2.1. Study Area Overview and Chronological Development

The data set of tree-ring width employed in this research was obtained from the eastern slopes of the upper Jinsha River in Baiyu County, Sichuan Province, China (98.83° E, 31.25° N; Figure 1a). The sampling location is situated at an altitude ranging from 4000 to 4100 m. The species selected for sampling is *Picea likiangensis* var. *balfouriana*, commonly known as western Sichuan spruce, the dominant species in this dense forest. The understory of this forest is characterized by the presence of dwarf shrubs such as alpine rhododendrons, which coexist beneath the canopy. The forest exhibits a high level of canopy closure, indicating a complex and well-established ecological community.

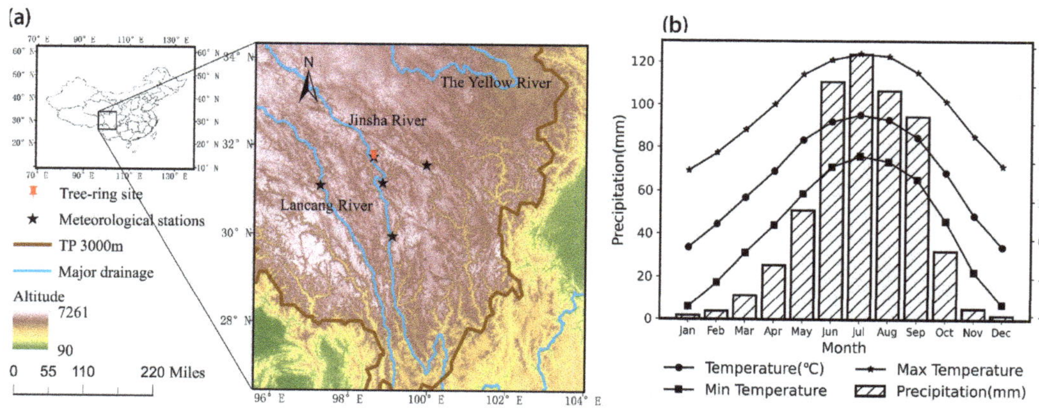

Figure 1. (**a**) The specific location of the tree-ring sampling site and meteorological stations utilized in this research, along with (**b**) the average climatic factors (including the minimum temperature, the average temperature, the maximum temperature, and precipitation) for each month acquired at five meteorological stations from 1961 to 2016.

The sampling site is situated at the junction of Sichuan Province and the Tibet Autonomous Region within the southeastern Qinghai–Tibetan Plateau, along the Jinsha River (i.e., the section of the Yangtze River between Zhimenda and Pingshan; see Figure 1a). This region has a unique highland climate formed by the Tibetan Plateau's topography and the synergistic effects of the South and East Asian monsoons and westward winds of the mid-latitudes. The extreme changes in climate are particularly evident here, with both the maximum and average temperatures increasing at a slower rate than the minimum temperature, and the warming rate increases with elevation [41]. It seems there is an increasingly positive correlation between extreme precipitation and overall rainfall, and precipitation intensifies in volume and frequency over time [42]. The increasing frequency of drought and frost phenomena and changes in the meteorological driving factors such as extreme weather have significantly affected vegetation growth in the study area [43]. Climate data from the study area (1961–2017) indicate an annual average temperature of 7.7 °C, with the highest average temperature of 16.2 °C in July and the lowest temperature of −1.1 °C in January. The average yearly precipitation was 568.1 mm, distributed in a single peak primarily between May and September, constituting 85.8% of the entire precipitation. The peak wet condition is in July, reaching up to 123.9 mm (Figure 1b). During sample collection, the collection object was older live trees. Increments borers were used to obtain two tree cores near the chest-high position of the trees, and one or three tree cores were collected for some trees. Ultimately, 61 cores were gathered from 37 trees.

All the sampled cores underwent preliminary processing following the globally recognized standard procedure for analyzing annual tree rings. With a measurement error rate of 0.01 mm, the LinTab™ system can be utilized to gauge the tree ring widths. The LinTab system is widely used in tree-ring width measurement due to its low data storage requirements and high resolution [44,45]. The cross-dating of rings was performed employing the COFECHA software [46]. In order to eliminate non-climatic data embedded within the original tree-ring widths, while preserving the maximum amount of climatic data, detrending was carried out using the ARSTAN program [47,48]. After multiple growth trend fittings using different methods, it was determined that the first-lowest filter method provided a better fit with the growth trend. This process yielded three standardized chronologies: Residual (RES), ARSTAN (ARS), and Standard (STD) chronologies. During the synthesis of the chronology, the individual sample cores that significantly deviated from the main series were excluded. The STD chronology was primarily employed for analyses in this research (all the following references to chronology are to the STD chronology, unless otherwise stated). The chronology spanned 663 years (1355–2017), with an average sensitivity of 0.188. The intra-tree and inter-tree correlation coefficients in the common interval between 1900 and 2000 were 0.562 and 0.327, respectively. Moreover, the signal-to-noise ratio (SNR), overall representativeness (expressed population signal, EPS), and proportion of variation described by the initial primary element were 0.964, 26.575, and 41.6%, respectively. These statistical characteristics imply that the chronology has a high average sensitivity and may represent climate changes in the past. To validate the reconstruction's integrity, an EPS criterion of 0.85 was established to identify the period of reliable chronologies [49]. When the number of sample cores reached 12, the EPS exceeded 0.85, indicating a reliable time period from 1605 to 2017 (Figure 2).

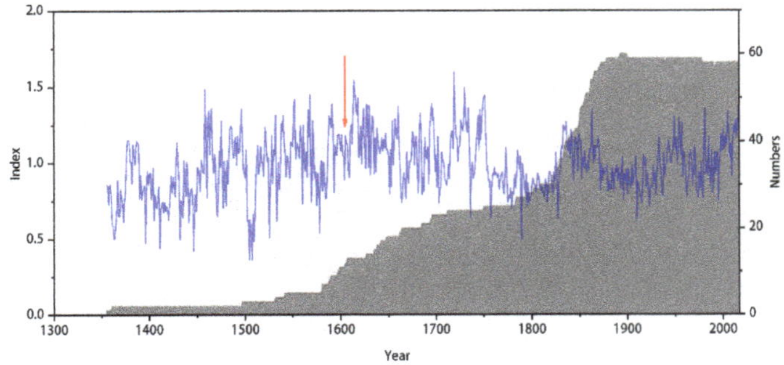

Figure 2. The tree ring index and their count shows a red arrow highlighting the first year where the EPS exceeds a value of 0.85.

2.2. Climate Data

The climate data were sourced from five distinct stations: Dege County (98°35′ E, 31°48′ N, elevation 3184 m), Bainyu County (98°50′ E, 31°13′ N, elevation 3260 m), Batang County (99°6′ E, 30° N, elevation 2589.2 m), Changdu County (91°10′ E, 31°9′ N, elevation 3315 m), and Ganzi County (100° E, 31°37′, elevation 3393.5 m). These stations are geographically close to the sampling site (Figure 1a) and provided the minimum (T_{min}) temperature, monthly mean (T_{mean}), maximum (T_{max}), and precipitation (P) data from 1961 to 2016. In addition, the average values of these climatic factors from these sites were utilized to characterize the area's weather. The global sea surface temperature (SST) data set (Extended Reconstructed Sea Surface Temperature V5, ERSST V5), with a spatial resolution of 2° × 2°, offered by NOAA, supplied the SST information. The sea surface temperature was used to examine the coupled air–sea interaction between the climate and the global SST in the study area.

2.3. Reconstruction and Analysis Methods

Pearson analysis could be employed to evaluate the linear correlations between the STD chronology and different climatic factors within the study region and to choose the primary climatic variables for reconstruction. While the traditional reconstruction methods primarily rely on constructing linear regression equations [50–52], the interplay between tree growth and climatic factors is highly complex and exhibits non-linear characteristics [53]. The previous studies have shown the efficacy of machine learning techniques in capturing non-linear regression relationships between tree growth and climatic factors [54–56]. Therefore, this study used three machine learning algorithms—K-nearest neighbors (KNNs), support vector machines (SVMs), and random forest (RF)—for modeling and reconstruction purposes. During the reconstruction process, chronological data from both the current and preceding years were employed for model training. This approach acknowledges the temperature's lag effect on tree rings [57], indicating that the temperature in year t impacts the tree-ring widths both in that year and the next (year $t + 1$). The construction of machine learning algorithms was completed using the Python library package known as sklearn.

The accuracy and dependability of the reconstruction were evaluated through various statistical metrics, such as the correlation coefficient (R), the coefficient of determination (R^2), the adjusted coefficient of determination (R^2_{adj}), error reduction (RE), the variance test (F), the product mean test, the sign test (ST), and the root-mean-square error (RMSE). To better understand extreme temperature variations, the temperature years were classified as high or low based on being one standard deviation (SD) above or below the mean. Years with reconstructed temperatures above the mean exceeding the SD (−4.08 °C) were identified as extremely high-temperature years, while years with reconstructed temperatures below the mean exceeding the SD (−5.2 °C) were classified as extremely low-temperature years. To further explain the decadal characteristics of the temperature shifts during the previous 411 years, analysis was conducted using the 11-year mean moving of the reconstruction. The phases in which the reconstruction results were lower or higher than the average for 10 consecutive years or more were defined as cold/warm phases.

The reconstructed and instrumental climate series were subjected to a spatial correlation analysis with the Climatic Research Unit (CRU TS 4.07) data set to gauge the spatial representativeness of the reconstructed minimum winter temperatures for the research region during the observational period from AD 1962 to 2005. To extract multi-timescale fluctuations from the reconstruction, ensemble empirical modal decomposition (EEMD) [58] was applied. EEMD decomposed the time series into various intrinsic modal functions (IMFs) with actual meanings, which extracted both time and frequency domain information from the amalgamated changes and trends [59]. Utilizing global sea surface temperature data, spatial correlation analysis was performed on the reconstruction and instrumental information to verify the influence of driving factors. The computation of the parameters for testing and the calculation of correlation coefficients were performed using both Python and SPSS software. The implementation of the Ensemble Empirical Mode Decomposition (EEMD) approach was also achieved using Python, and plotting was carried out using the matplotlib library in Python.

3. Results

3.1. Relationship between Tree Growth and Climate Factors

Figure 3 depicts the relationship between the tree-ring chronology and the regional climatic factors. The results imply that the radial growth of trees exhibits the most significant correlation with March and April precipitation, with correlation coefficients of 0.312 and 0.300, respectively ($p < 0.05$). Meanwhile, the correlation coefficient linking the trees' radial growth and the T_{mean} in January, August, and November of the previous year, as well as the T_{max} in February and August, were 0.287, 0.425, 0.314, 0.271, and 0.269, respectively ($p < 0.05$). A better correlation ($p < 0.01$) with the T_{mean} between this August and last December and between the T_{max} in November and December of the previous year was found, with respective correlation coefficients of 0.425, 0.474, 0.391, and 0.433.

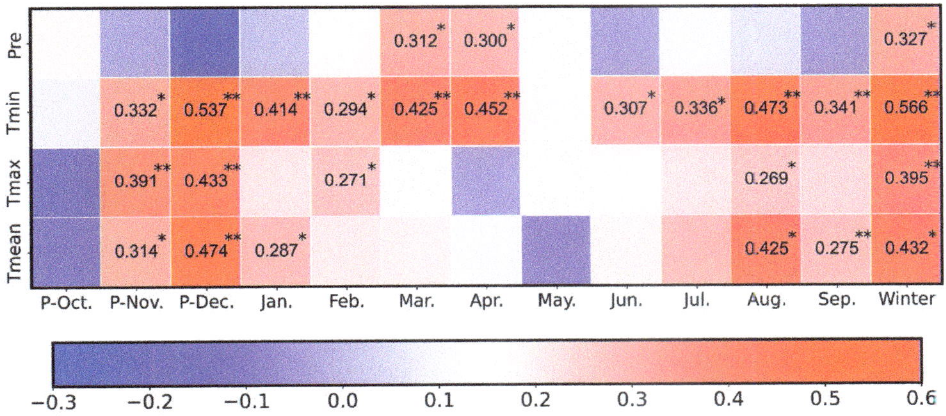

Figure 3. The coefficients of the correlation between the tree-ring index and the climatic factors (pre, T_{min}, T_{max}, and T_{mean} indicate the monthly precipitation, minimum temperature, maximum temperature, and mean temperature, respectively). The statistical relationship of the tree-ring index with the climatic indicators, namely monthly precipitation (pre), and the trio of temperatures (minimum (T_{min}), maximum (T_{max}), and mean (T_{mean})) is delineated by their correlation coefficients. Here, "Winter" represents the winter half-year (from last November to this April), and the correlation coefficients used are Pearson correlation coefficients. The symbols ** and * display the significance levels of 0.01 and 0.05, respectively.

The analysis of correlations for each individual month indicated that tree radial expansion was significantly associated ($p < 0.05$) with the minimum temperatures, with the exception of May in the current year. The most substantial correlation was observed with the average minimum temperature of the preceding December, boasting a correlation coefficient of 0.537 ($p < 0.01$). Through further correlation analysis of the climate elements of different month combinations, the correlation between the mean minimum temperature and the tree-ring chronology was found to pass the significance test at a level of significance of 0.01 in spring (March–April), high summer (August–September), and winter (last November–December and January in this year), and especially during the winter half-year (from last November to this April). The correlation coefficient of the winter half-year was the most significant, with its correlation coefficient reaching 0.566.

3.2. Minimum Temperature Reconstruction

In light of the aforementioned analysis, the average minimum temperature of the Jinsha River Basin during the winter half-year was selected as the object of reconstruction. Given the advantages of machine learning methods in modeling non-linear relationships, the present research incorporated three machine learning models, namely K-nearest neighbor (KNN), SVMs, and RF, in addition to the traditional regression equations for model construction.

The regression equation Lr ($T_{\text{min}} = 2.282 \times \text{std}_t + 1.08\ \text{std}_{(t+1)} - 7.994$) and the leave-one-out cross-validation (LOOCV) approach were employed to adjust the parameter of the machine learning model, and then the KNN, RF, and SVM models for temperature reconstruction were established. Ultimately, the reconstructed sequences obtained using the four methods were weighted and averaged with equal weights. Thus, an integrated reconstruction was obtained [60]. The correlation coefficient between the rebuilt and actual temperatures was 0.73 (with a sample size of 56, $p < 0.001$), and after accounting for degrees of freedom, the variance explained was 52.4% (as depicted in Figure 4a).

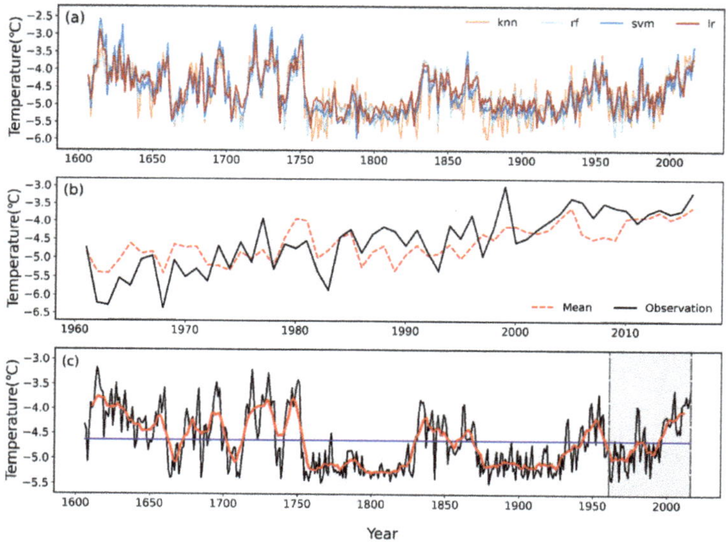

Figure 4. (**a**) Comparison of four reconstruction results; (**b**) the constructed and observed temperatures during the calibration period of 1960–2016; and (**c**) the average minimum temperatures for the winter months reconstructed using the chronological data ranging from 1606 to 2016. The red line shows the moving average for 11 years, while the horizontal line signifies the average temperature across the entire reconstructed timeline.

To assess the robustness and dependability of the reconstruction model, a split-sample calibration/verification methodology was deployed for the ensemble reconstruction model. The segmented verification outcomes were consolidated and are presented in Table 1. Sign tests were conducted on both the initial data (ST) and the first-differenced data (ST1), and they successfully met the significance threshold at a level of 0.01, suggesting that the model has excellent reliability for the construction of both high- and low-frequency changes. The most stringent statistical reduction in error (RE) and coefficient of efficiency (CE) were positive; the RE values of the split calibration/verification test were much greater than 0, at 0.79 and 0.80, respectively.

Table 1. Statistics of split calibration/verification test results for the tree-ring construction.

Calibration					Verification							
Period	R	R^2	R^2adj	F	Period	R	R^2	R^2adj	RE	ST	ST1	CE
1961~1988	0.568 **	32.3%	29.7%	12.409 **	1989~2016	0.611 **	37.4%	35.0%	0.79	23 **	21 **	0.18
1989~2016	0.611 **	37.4%	35.0%	15.521	1961~1988	0.568 **	32.3%	29.7%	0.80	21 **	20 **	0.02
1961~2016	0.730 **	53.2%	52.4%	61.481 **								

Note: ** $p < 0.01$. CE, coefficient of efficiency; RE, reduction in error. ST, sign tests of the original data; ST1, first-differenced data (ST1).

This suggests that the reconstruction outcomes are both stable and trustworthy, thereby validating their utility in reconstructing the average minimum winter half-year temperatures for the headwater of Jinsha River from 1606 to 2016. Furthermore, the reconstructed values of the average minimum winter half-year temperature exhibit a strong correlation with the observed values (Figure 4b) and the correlation coefficient of 0.730 ($n = 56$, $p < 0.01$).

3.3. Characteristic Changes in Minimum Temperature

As shown in Figure 4c, the fluctuations in the average minimum temperature during the wintertime in the year exhibited marked variations during the last 411 years (between 1606 and 2016), and the mean of the reconstructed minimum temperature (−4.64 °C) was lower than the observed temperature in the instrumental period (−4.57 °C, 1961–2016). The reconstructed temperature varied from −5.47 °C (in 1880) to −3.17 °C (in 1719), with a standard deviation (SD) of 0.56 °C. In the last 411 years, there have been 80 years of exceptionally high temperatures and 92 years of exceptionally low temperatures, constituting 19.4% and 22.4%, respectively. Notably, the periods of 1755–1759, 1775–1779, 1795–1805, and 1813–1817 were characterized by consecutive and persistent extremely low-temperature phases lasting five or more years, while the periods of 1610–1618, 1626–1630, 1693–1697, 1725–1734, 1742–1751, and 2010–2016 experienced sustained extreme high temperatures for five or more consecutive years.

In the last 411 years, the study area has experienced six cold and seven warm phases. The cold periods occurred during 1663–1672, 1702–1712, 1754–1829, 1869–1942, 1959–1979, and 1985–1994. The warm periods, on the other hand, were observed in 1611–1662, 1673–1701, 1713–1753, 1830–1852, 1858–1868, 1943–1958, and 1995–2011.

4. Discussion

4.1. Response of the Radial Growth of Tree Rings to Meteorological Elements

The examination of the relationship between the tree-ring width chronologies and meteorological variables showed the radial tree growth within the study region was more responsive to temperature fluctuations compared to fluctuations in the rainfall levels. The T_{min} value of the winter half-year might be the primary influence on tree growth. This finding aligns with the outcomes of previous research conducted in neighboring regions [33,61], the lower Yangtze River [62], and southern Poland [63]. An excessive T_{min} in a given winter half-year would result in the condensation of ice crystals, disrupting the vesicles and leading to cellular dehydration, which would cause permanent damage to the tree root tissues, subsequently affecting tree growth in the following growing season [64]. Additionally, colder winters may lead to permafrost layer thickening, which would impede water uptake and cause winter drying.

In conclusion, the prospects for tree growth in the ensuing year are anticipated to decrease [65]. Furthermore, a thicker permafrost delays the snowmelt date as well as the onset of the season of tree growing season, limiting their growth and development [28]. Conversely, warmer winters could reduce the consumption of stored carbohydrates by trees to maintain physiological activities. This favorable condition promotes tree growth and development in the upcoming year. Specifically, spruce might exhibit significant photosynthesis during warm winters, which ensures sufficient nutrient availability for the subsequent growing season, thus facilitating the growth of spring timber [23]. These explanations shed light on why the radial growth of spruce in the Jinsha River Basin is affected by low winter temperatures.

4.2. Temporal Representativeness and Reconstruction Reliability

To further evaluate the accuracy of this reconstruction on longer timescales before the instrumental period and the representativeness of large spatial scales utilizing the reconstructed and observed minimum winter temperatures, we conducted spatial correlation analysis with the Climatic Research Unit time series (CRU) gridded data set (TS 4.07). The spatial correlation outcomes revealed analogous patterns between the observed (Figure 5a) and reconstructed (Figure 5b) temperature data. Although the observed data exhibited stronger correlations, both data sets demonstrated significant correlations with the CRU gridded surface temperature data within the area. This indicates that the reconstructed results are representative of the climate fluctuations within the study region and its adjacent regions.

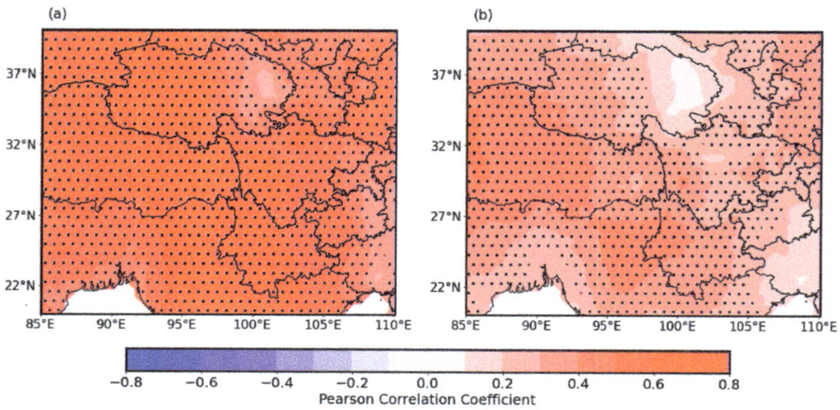

Figure 5. The spatial correlations between the observed (**a**) and reconstructed (**b**) minimum winter temperatures of the southeastern Tibetan Plateau were assessed against the winter gridded temperatures provided by the Climatic Research Unit (CRU TS 4.07) for the period from 1962 to 2016. The stippled areas indicate statistical significance at the $p < 0.01$ level.

The reconstructed series were juxtaposed with the temperature reconstructions from nearby places, encompassing winter temperatures of the southeastern Tibetan Plateau during the last 1340 years [27], the reconstruction of warm season temperatures in the southeastern Tibetan Plateau for the previous 449 years [30], and the reconstructed minimum winter temperatures in the western Sichuan Plateau from 1650 to 1994 [66]. The reconstructed temperatures exhibit a significant level of agreement with each other, as indicated by the correlation coefficients of 0.298 ($n = 359$, $p < 0.01$), 0.279 ($n = 407$, $p < 0.01$), and 0.304 ($n = 345$, $p < 0.01$) in the common periods. The warm and cold changes were comparatively analyzed using an 11-year sliding average and *Z*-score normalization for all the sequences (Figure 6). Despite variations across the reconstructed temporal segments and variables studied, the synchronization of warm and cold changes among the four temperature reconstructions remained consistent, further evidencing the reliability of the results. The reconstructed temperature results share warm and cold periods, including a span of warmer conditions in the 1610s, an interval characterized by cooler temperatures in the 1740s, a stretch of colder climatic conditions in the early 1900s, a phase of elevated thermal conditions in the 1950s, and an interval marked by increased warmth in the late 1900s. In addition, the cold phase of the early 18th century, the warm phase of the 1710s, the cold phase of the early 19th century, the cold phase of the 1850s, and the cold phase of the 1960s are represented among several reconstructions (including the reconstruction of the present study). Some of the disparities observed in the warm and cold phases across the four reconstructions can likely be ascribed to a variety of causes. These include the specific objects of reconstruction (varying meteorological elements), the distinct time periods chosen for the reconstructions, the methodologies applied for detrending the data, as well as the geographical demarcation and categorization of the study area. The interplay of these elements may jointly contribute to the differences noted in the reconstruction outcomes.

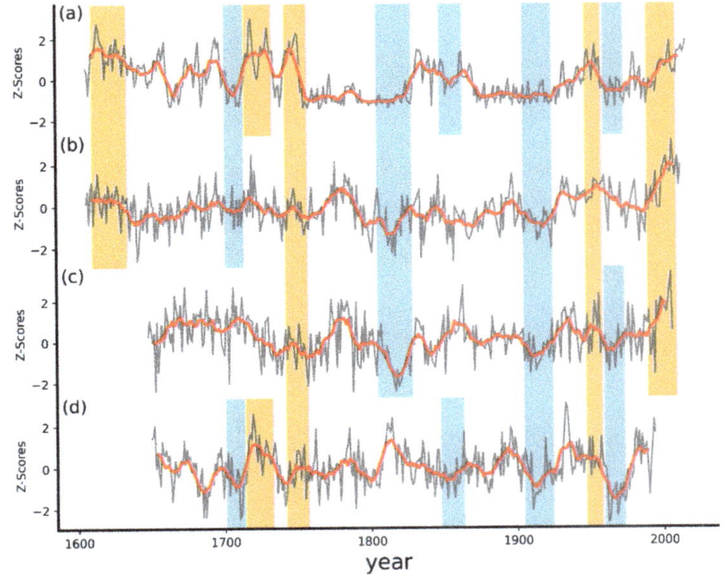

Figure 6. Comparison of the winter minimum temperature reconstruction in this research (**a**), the winter temperature reconstruction of the southeastern Tibetan Plateau [27] (**b**), the reconstruction of the warm period temperatures in the southeastern Tibetan Plateau [30] (**c**), and the reconstructed minimum winter temperatures in the western Sichuan Plateau [66] (**d**). The regions with shading signify the durations when the reconstructions align in their trend movements.

In addition, the reconstructed temperature that occurred during the cold phase in the 1900s was confirmed in the work on temperature reconstruction using ice cores taken from the Tibetan Plateau [67]. The reconstructed temperature aligns well with the cold winters of 1875–1880, 1890, 1920, and 1980, as indicated by the European instrumental record [63]. The oscillations in the reconstruction correspond to the advancement and retreat of glaciers in the Tibetan Plateau region. Notably, the reconstructed temperatures experienced a period of significant cooling and persistent cold from 1748 to 1767, coinciding with the largest glacier advance event since the Little Ice Age in the Middushu Ice, located in the western Tibetan Plateau. This result is corroborated by the validation of the most extensive and peripheral moraines, which occurred around 1767 [68]. In addition, the reconstructed temperatures demonstrated a notable cooling trend before 1880, which corresponds to a clear progression of glacial activity in the Trans-Himalayan region [69]. Furthermore, the reconstruction between 1790 and 1820 indicates a significant period of low temperatures, which is called the Dalton Minimum, at the transition between the 18th and 19th centuries. This cold period was affected by changes in solar irradiance, volcanic activity (e.g., the 1815 eruption of the Tambora volcano in Indonesia), and rising CO_2 concentrations. These factors resulted in a sudden drop in temperatures not only in the Northern Hemisphere, but even globally, with reports of a summer-free year [70,71]. Moreover, a sustained cooling trend from 1860 to 1880 was documented in the Western Himalayan region [24]. The reconstruction captured signs of rapid warming on the Tibetan Plateau during the last fifty years in the 1900s, indicating a direct temperature increase at a rate of 0.2 °C during the last five decades. These findings match the conclusions drawn from climate reconstruction efforts using tree rings [72–74]. It merits particular attention that some paleoclimate findings in the Northern Hemisphere show temperature discrepancies, which are attributed to issues related to tree-ring dispersion [75]. However, the reconstructions in this study retain temperature sensitivity, and the increase in the reconstructed temperatures align with the warming trends observed in the instrumental data.

It is intriguing to observe that the relatively "warm period" of the study area on the centennial timescale is clearly different from the notion of a "cool period" (Little Ice Age) during this time, with the T_{min} of the half-year wintertime in the research region during 1650–1750 being superior to those of the 20th century by approximately 0.44 °C. Despite being in the Maunder minimum, apparent warming in the 17th century was not an isolated phenomenon, and this has been reported in climate reconstructions using tree rings and ice cores in the Tibetan Plateau region [27,76,77]. The reconstruction findings of the cold season temperature changes in the northeastern Qinghai–Tibet Plateau using alkene ketone also indicate that the chilly seasons of the Little Ice Age were not the coldest periods, aligning with the findings of this research [78]. This phenomenon is not unique to some regions of the Tibetan Plateau, but also occurred in North America across the ocean [25,79,80].

A plausible rationale for this phenomenon is that an augmented pressure gradient between south-central Asia and northern Eurasia could potentially intensify the westerlies in the mid-latitude regions, preventing cold air from moving southward toward the Tibetan Plateau [81].

4.3. Reconstructing Multiscale Fluctuations in Temperature

Considering that long-term climate evolution is usually non-stationary and non-linear and occurs at multiple timescales, to extract the available climate change signals and explore the cyclic changes in the extrinsic forcing of the minimum temperature during the winter semester, the authors of this paper used the Ensemble Empirical Mode Decomposition (EEMD) approach to break down the reconstruction at multiple scales (Figure 7). A total of six IMFs (IMFs 1–6) were obtained using EEMD that could characterize the fluctuations in the reconstruction from high-frequency to low-frequency scales, and each mode contains information with an actual physical meaning (Table 2).

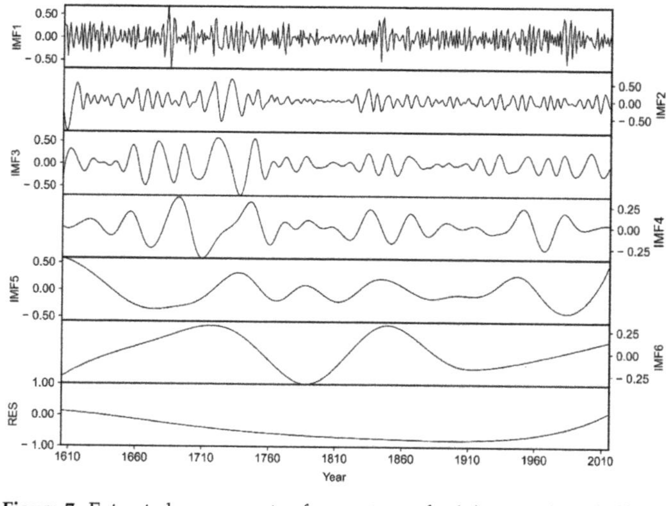

Figure 7. Extracted components of reconstructed minimum winter half-year temperatures using Ensemble Empirical Mode Decomposition.

Table 2. Contributions to the variance and coefficient of reconstruction after the EEMD of IMFs and trends (RES).

Variable	IMF1	IMF2	IMF3	IMF4	IMF5	IMF6	RES
Major cycle (year)	3.3–3.4	7.1–7.4	15.5–16.8	29.4–32.9	82.4	164.8	
Contribution (%)	13.17	12.72	16.95	6.67	20.18	13.60	16.71

The reconstructed temperatures had relatively invariant intervals characterized by high-frequency temporal dynamics, with major periods of 3.3–3.4 and 7.1–7.4 years for IMF1 and IMF2, respectively. The results align with the periods (2–8a) of the El Niño–Southern Oscillation (ENSO), suggesting that large-scale coupled sea–air oscillations have an important influence on the variation in the winter minimum temperatures within the purview of this research locale, which has been documented in the existing body of research [82]. During the El Niño years of the ENSO cycle, an anomalous anticyclone located in the western troposphere of the Pacific Ocean weakened the strength of the East Asian Winter Monsoon (EAWM). This could cause significantly higher-than-average winter surface temperatures in most parts of China, while in the La Niña years, the situation was reversed, and such temperature anomalies would persist into the following spring or even early summer [83]. Nevertheless, the spatial correlation assessment of the reconstructed and empirically obtained temperatures during the global winter half-year (P11C4) SST (Figure 8a,b) indicates that the correlation between the SST and the reconstructed and observed temperatures in the El Niño Sea area is insignificant ($p > 0.05$). However, there is better agreement between the Indian Ocean and the western and central parts of the mid-latitude Pacific Ocean. The bridging anomalous anticyclone of ENSO influencing the winter temperatures in East Asia is profoundly influenced by the evolution of SST anomalies in the equatorial western Pacific Ocean, and the anomalous cooling of SSTs in this region directly affect the Rossby waves within the lower stratum of the troposphere, which, in turn, affect the strength of the anomalous anticyclone [84]. Meanwhile, the influence of ENSO on the climate of East Asia is also regulated by the SST of the Indian Ocean, and the anomalous enhancement of the anticyclone anomaly in the western North Pacific Ocean occurs when the El Niño phenomenon exists concurrently with the affirmative phase of the Indian Ocean dipole sub-mode [85]. The aforementioned mechanism helps elucidate the distribution of the regions where the winter half-year temperature and the global SST were significantly correlated.

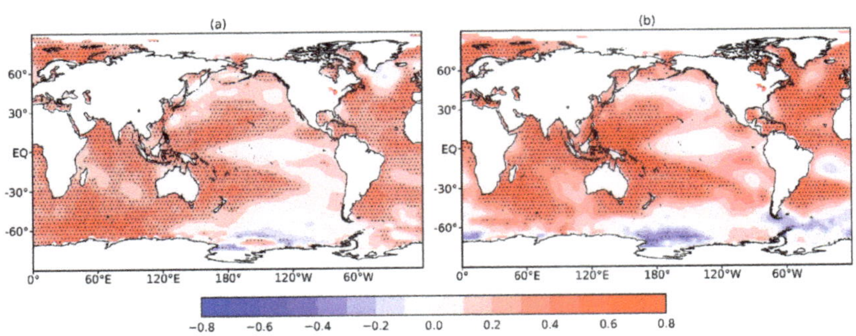

Figure 8. Spatial correlations of reconstructed (**a**) and observed (**b**) minimum winter half-year temperatures with SST data during the period of 1845–2016 and during the period of 1961–2016, respectively. The 95% confidence interval is shown by the dotted region.

The variation in the 15.5–16.8-year cycle of IMF3 might be indicative of Pacific Decadal Oscillation (PDO), which is an important component of the North Pacific SST anomalies, and its influence covers the North Pacific SST and even the global climate [86]. The correlation coefficient between the reconstruction and the PDO was 0.295 (1606–1996, $n = 391$, $p < 0.001$). The interplay between Pacific Decadal Oscillation (PDO) and the atmospheric circulation patterns suggests a robust inverse pressure correlation between these two elements over the North Pacific Ocean and eastern China. When the PDO entered its positive (negative) phase, there was a corresponding decrease (increase) in sea level pressure within the mid- to high-latitude regions of East Asia and the North Pacific. The potency of the Siberian high-pressure system decreased (elevated), and the upper-level latitudinal winds weakened (strengthened), which, in turn, led to warmer (colder) winters in China [87]. The sunspot cycle (Schwabe) was around 11 years, which corresponds to the cycle of IMF4 in this paper (multiples of 11a). The effect of solar activity on anthropogenic climate alteration has been verified in numerous climate reconstructions employing tree-ring width chronologies [30,31,88].

The quasi-82a period of IMF5 could potentially correlate with Atlantic Multidecadal Oscillation (AMO), an inter-decadal variation in the SST over a period of 60–80 years, as observed in various studies conducted in the Tibetan Plateau region [29,61]. The IMF6 component, which represents temperature variation in the reconstructed sequence on a multi-hundred-year scale, had a weaker oscillation and corresponds to an extrinsic driver, which might be the same as the AMO (quasi-multiples of 80 years). The correlation coefficient between the IMF5 and the AMO index during 1871–2016 is 0.518 ($n = 146$, $p < 0.001$). As an oceanic model, the AMO regulates the surface air pressure and temperature through atmospheric conduits extending from the Atlantic Ocean, encompassing westerly winds and the associated wave train. The westerlies are influenced by the large topography of the Tibetan Plateau and separated into several branches, whose southern branch reaches southwestern China and moderates the winter air temperatures in the region [89]. Meanwhile, the AMO is strongly related to the interdecadal variations of the East Asian Monsoon. Research has found that the positive (warm) phase of the AMO aligns with a period of weak EAWM in most of China, having relatively warm winters, and vice versa, which corresponds to winters with cooler temperatures [90].

Spatial correlation analysis was conducted on both the reconstructed temperature sequence from 1845 to 2016 and the instrumental record from 1961 to 2016, and each correlate with the global sea surface temperatures (SSTs) from the previous November to this April, as illustrated in Figure 8a,b. This analysis effectively validated the role of sea–air coupling patterns as driving factors within the climate system. The differences in the measured lowest temperature at regional meteorological stations are both synchronized with the fluctuations in the SST, in which there are more regions of significant positive correlation ($p < 0.05$) between the reconstructed sequences and the SST than instrumental sequences. The overall correlation between the observed temperature and the SST is slightly higher than that of the reconstructed temperature. The synergistic relationship between the minimum temperature during the winter semester and the SSTs in the western Pacific, Indian, North Atlantic, South Atlantic, and Arctic Oceans in the study area further demonstrate the association between thermal fluctuations within the region and the synergistic factors of the air and sea. The specific mechanisms of the influence warrant further research.

5. Conclusions

A standardized chronology of tree-ring width was formulated utilizing spruce from the Jinsha River Basin, and the regional mean minimum temperatures recorded during the winter semester were reconstructed for the period from 1606 to 2016. The reconstruction recorded stronger decadal variations, reflecting six cold periods and seven warm periods. The reconstructed temperature, when compared with those from the neighboring areas, was discovered to consistently reveal warm and cold epochs. Additionally, the reconstruction results from other proxies, meteorological data recorded by early instruments, and historical data on the timing of glacial activity and natural disasters also support the dependability of the reconstruction results. The reconstruction sensitively captures the climate warming observed during the last 50 years in the 1900s. Unexpectedly, the recorded tree-ring data showed that the research region was relatively warm in the winter half-years from the 17th to 18th century, but should have been cold during the Little Ice Age. This phenomenon of the "warm winter of the Little Ice Age" is also supported by other winter temperature reconstructions. The winter half-year minimum temperatures could be affected by the El Niño–Southern Oscillation (ENSO), Pacific Decadal Oscillation (PDO), solar activity, and Atlantic Multidecadal Oscillation (AMO). In this study, the integration of several reconstruction methods was performed, and temperature's hysteresis influence on tree rings was considered, providing a new idea for reconstruction work. In future research, the reconstructed temperature should be supplemented with high-resolution meteorological data and sufficient sample sizes in more areas in order to more effectively address the challenges posed by global climate change.

Author Contributions: Conceptualization, J.L.; methodology, J.L. and C.J.; software, J.L., C.J. and Y.T.; validation, J.L. and H.X.; formal analysis, J.L.; investigation, J.L.; resources, J.L.; data curation, J.L. and C.J.; writing—original draft preparation, C.J. and H.X.; writing—review and editing, J.L.; visualization, C.J. and Y.T.; supervision, J.L.; project administration, J.L.; funding acquisition, J.L. All authors have read and agreed to the published version of the manuscript.

Funding: This research was jointly funded by the Second Tibetan Plateau Scientific Expedition and Research (STEP) program (No. 2019QZKK0103 and 2019QZKK0608), the Natural Science Foundation of Sichuan Province (No. 2022NSFSC0215), and the Undergraduate Innovation and Entrepreneurship Training Program (No. 202310621007).

Data Availability Statement: The data of this study are available upon request to the first or corresponding author.

Conflicts of Interest: The authors declare no conflicts of interest.

References

1. Fang, J.; Zhu, J.; Wang, S.; Yue, C.; Shen, H. Global Warming, Human-Induced Carbon Emissions, and Their Uncertainties. *Sci. China Earth Sci.* **2011**, *54*, 1458–1468. [CrossRef]
2. Boyles, R.P.; Raman, S. Analysis of Climate Trends in North Carolina (1949–1998). *Environ. Int.* **2003**, *29*, 263–275. [CrossRef]
3. Liu, X.; Chen, B. Climatic Warming in the Tibetan Plateau during Recent Decades. *Int. J. Climatol.* **2000**, *20*, 1729–1742. [CrossRef]
4. Huang, J.; Guan, X.; Ji, F. Enhanced Cold-Season Warming in Semi-Arid Regions. *Atmos. Chem. Phys.* **2012**, *12*, 5391–5398. [CrossRef]
5. Schwartz, M.D.; Ahas, R.; Aasa, A. Onset of Spring Starting Earlier across the Northern Hemisphere. *Glob. Change Biol.* **2006**, *12*, 343–351. [CrossRef]
6. Piao, S.; Friedlingstein, P.; Ciais, P.; Viovy, N.; Demarty, J. Growing Season Extension and Its Impact on Terrestrial Carbon Cycle in the Northern Hemisphere over the Past 2 Decades. *Glob. Biogeochem. Cycles* **2007**, *21*, 2006GB002888. [CrossRef]
7. Brooks, P.D.; Grogan, P.; Templer, P.H.; Groffman, P.; Öquist, M.G.; Schimel, J. Carbon and Nitrogen Cycling in Snow-Covered Environments: Carbon and Nitrogen Cycling in Snow-Covered Environments. *Geogr. Compass* **2011**, *5*, 682–699. [CrossRef]
8. Cai, Q.; Liu, Y.; Wang, Y.; Ma, Y.; Liu, H. Recent Warming Evidence Inferred from a Tree-Ring-Based Winter-Half Year Minimum Temperature Reconstruction in Northwestern Yichang, South Central China, and Its Relation to the Large-Scale Circulation Anomalies. *Int. J. Biometeorol.* **2016**, *60*, 1885–1896. [CrossRef]
9. Xiong, Y.; Zhou, J.; Sun, N.; Jia, B.; Hu, G. Variation Trends of Precipitation and Runoff in the Jinsha River Basin, China: 1961–2015. *IOP Conf. Ser. Earth Environ. Sci.* **2020**, *440*, 052044. [CrossRef]
10. He, Y.; Zhang, Y.; Wu, Z. Analysis of Climate Variability in the Jinsha River Valley. *J. Trop. Meteorol.* **2016**, *22*, 243–251. [CrossRef]
11. Zhang, D.; Wang, W.; Liang, S.; Wang, S. Spatiotemporal Variations of Extreme Precipitation Events in the Jinsha River Basin, Southwestern China. *Adv. Meteorol.* **2020**, *2020*, 3268923. [CrossRef]
12. Liu, X.; Peng, D.; Xu, Z. Identification of the Impacts of Climate Changes and Human Activities on Runoff in the Jinsha River Basin, China. *Adv. Meteorol.* **2017**, *2017*, 4631831. [CrossRef]
13. Schwartz, S.A. An Interdisplinary Approach to the Little Ice Age and Its Implications for Global Change Research. Ph.D. Thesis, University of Michigan, Ann Arbor, MI, USA, 1994.
14. Wang, S.; Liu, J.; Zhou, J. The Climate of Little Ice Age Maximum in China. *Lake Sci.* **2015**, *15*, 369–376. [CrossRef]
15. Felis, T.; Ionita, M.; Rimbu, N.; Lohmann, G.; Kölling, M. Mild and Arid Climate in the Eastern Sahara-Arabian Desert During the Late Little Ice Age. *Geophys. Res. Lett.* **2018**, *45*, 7112–7119. [CrossRef]
16. Pitman, K.J.; Smith, D.J. Tree-Ring Derived Little Ice Age Temperature Trends from the Central British Columbia Coast Mountains, Canada. *Quat. Res.* **2012**, *78*, 417–426. [CrossRef]

17. Shen, D.; Li, S. Environment Change of Sihu Lake in the Past 450 Years, Southern China. *Environ. Earth Sci.* **2013**, *70*, 2953–2962. [CrossRef]
18. Jomelli, V.; Lane, T.; Favier, V.; Masson-Delmotte, V.; Swingedouw, D.; Rinterknecht, V.; Schimmelpfennig, I.; Brunstein, D.; Verfaillie, D.; Adamson, K.; et al. Paradoxical Cold Conditions during the Medieval Climate Anomaly in the Western Arctic. *Sci. Rep.* **2016**, *6*, 32984. [CrossRef]
19. Degroot, D. Climate Change and Society in the 15th to 18th Centuries. *WIREs Clim. Chang.* **2018**, *9*, e518. [CrossRef]
20. Yang, B.; Braeuning, A.; Liu, J.; Davis, M.E.; Yajun, S. Temperature Changes on the Tibetan Plateau during the Past 600 Years Inferred from Ice Cores and Tree Rings. *Glob. Planet. Change* **2009**, *69*, 71–78. [CrossRef]
21. Nesje, A.; Dahl, S.O.; Thun, T.; Nordli, O. The "Little Ice Age" Glacial Expansion in Western Scandinavia: Summer Temperature or Winter Precipitation? *Clim. Dyn.* **2008**, *30*, 789–801. [CrossRef]
22. Chen, F.; Yuan, Y.; Wei, W.; Yu, S.; Zhang, T. Tree Ring-Based Winter Temperature Reconstruction for Changting, Fujian, Subtropical Region of Southeast China, since 1850: Linkages to the Pacific Ocean. *Theor. Appl. Climatol.* **2012**, *109*, 141–151. [CrossRef]
23. Shah, S.K.; Pandey, U.; Mehrotra, N.; Wiles, G.C.; Chandra, R. A Winter Temperature Reconstruction for the Lidder Valley, Kashmir, Northwest Himalaya Based on Tree-Rings of Pinus Wallichiana. *Clim. Dyn.* **2019**, *53*, 4059–4075. [CrossRef]
24. Borgaonkar, H.P.; Ram, S.; Sikder, A.B. Assessment of Tree-Ring Analysis of High-Elevation Cedrus Deodara D. Don from Western Himalaya (India) in Relation to Climate and Glacier Fluctuations. *Dendrochronologia* **2009**, *27*, 59–69. [CrossRef]
25. Porter, T.J.; Pisaric, M.F.J.; Field, R.D.; Kokelj, S.V.; Edwards, T.W.D.; deMontigny, P.; Healy, R.; LeGrande, A.N. Spring-Summer Temperatures since AD 1780 Reconstructed from Stable Oxygen Isotope Ratios in White Spruce Tree-Rings from the Mackenzie Delta, Northwestern Canada. *Clim. Dyn.* **2014**, *42*, 771–785. [CrossRef]
26. Wang, L.; Duan, J.; Chen, J.; Huang, L.; Shao, X. Temperature Reconstruction from Tree-ring Maximum Density of Balfour Spruce in Eastern Tibet, China. *Int. J. Climatol.* **2010**, *30*, 972–979. [CrossRef]
27. Huang, R.; Zhu, H.; Liang, E.; Liu, B.; Shi, J.; Zhang, R.; Yuan, Y.; Grießinger, J. A Tree Ring-Based Winter Temperature Reconstruction for the Southeastern Tibetan Plateau since 1340 CE. *Clim. Dyn.* **2019**, *53*, 3221–3233. [CrossRef]
28. Gou, X.; Chen, F.; Jacoby, G.; Cook, E.; Yang, M.; Peng, J.; Zhang, Y. Rapid Tree Growth with Respect to the Last 400 Years in Response to Climate Warming, Northeastern Tibetan Plateau. *Int. J. Climatol.* **2007**, *27*, 1497–1503. [CrossRef]
29. Liang, E.; Shao, X.; Qin, N. Tree-Ring Based Summer Temperature Reconstruction for the Source Region of the Yangtze River on the Tibetan Plateau. *Glob. Planet. Change* **2008**, *61*, 313–320. [CrossRef]
30. Duan, J.; Zhang, Q.-B. A 449 Year Warm Season Temperature Reconstruction in the Southeastern Tibetan Plateau and Its Relation to Solar Activity: Temperature Reconstruction in the Tibet. *J. Geophys. Res. Atmos.* **2014**, *119*, 11578–11592. [CrossRef]
31. Gou, X.; Gao, L.; Deng, Y.; Chen, F.; Yang, M.; Still, C. An 850-year Tree-ring-based Reconstruction of Drought History in the Western Qilian Mountains of Northwestern China. *Int. J. Climatol.* **2015**, *35*, 3308–3319. [CrossRef]
32. Deng, Y.; Gou, X.; Gao, L.; Yang, T.; Yang, M. Early-Summer Temperature Variations over the Past 563 Yr Inferred from Tree Rings in the Shaluli Mountains, Southeastern Tibet Plateau. *Quat. Res.* **2014**, *81*, 513–519. [CrossRef]
33. Li, J.; Li, J.; Li, T.; Au, T.F. 351-Year Tree Ring Reconstruction of the Gongga Mountains Winter Minimum Temperature and Its Relationship with the Atlantic Multidecadal Oscillation. *Clim. Chang.* **2021**, *165*, 49. [CrossRef]
34. Chen, F.; Shang, H.; Panyushkina, I.P.; Meko, D.M.; Yu, S.; Yuan, Y.; Chen, F. Tree-Ring Reconstruction of Lhasa River Streamflow Reveals 472 Years of Hydrologic Change on Southern Tibetan Plateau. *J. Hydrol.* **2019**, *572*, 169–178. [CrossRef]
35. Li, T.; Li, J. A 564-Year Annual Minimum Temperature Reconstruction for the East Central Tibetan Plateau from Tree Rings. *Glob. Planet. Change* **2017**, *157*, 165–173. [CrossRef]
36. Li, Z.-S.; Zhang, Q.-B.; Ma, K. Tree-Ring Reconstruction of Summer Temperature for A.D. 1475–2003 in the Central Hengduan Mountains, Northwestern Yunnan, China. *Clim. Change* **2012**, *110*, 455–467. [CrossRef]
37. Gou, X.; Yang, T.; Gao, L.; Deng, Y.; Yang, M.; Chen, F. A 457-Year Reconstruction of Precipitation in the Southeastern Qinghai-Tibet Plateau, China Using Tree-Ring Records. *Chin. Sci. Bull.* **2013**, *58*, 1107–1114. [CrossRef]
38. Fan, Z.; Bräuning, A.; Cao, K. Tree-ring Based Drought Reconstruction in the Central Hengduan Mountains Region (China) since A.D. 1655. *Int. J. Climatol.* **2008**, *28*, 1879–1887. [CrossRef]
39. Yang, B.; Chen, X.; He, Y.; Wang, J.; Lai, C. Reconstruction of Annual Runoff since CE 1557 Using Tree-Ring Chronologies in the Upper Lancang-Mekong River Basin. *J. Hydrol.* **2019**, *569*, 771–781. [CrossRef]
40. Xiao, D.; Shao, X.; Qin, N.; Huang, X. Tree-Ring-Based Reconstruction of Streamflow for the Zaqu River in the Lancang River Source Region, China, over the Past 419 Years. *Int. J. Biometeorol.* **2017**, *61*, 1173–1189. [CrossRef]
41. Yao, L.; Lu, J.; Zhang, W.; Qin, J.; Zhou, C.; Ngoc, N.T.; Pinage, E.R. Spatiotemporal Analysis of Extreme Temperature Change on the Tibetan Plateau Based on Quantile Regression. *Earth Space Sci.* **2022**, *9*, e2022EA002571. [CrossRef]
42. Ge, G.; Shi, Z.; Yang, X.; Hao, Y.; Guo, H.; Kossi, F.; Xin, Z.; Wei, W.; Zhang, Z.; Zhang, X.; et al. Analysis of Precipitation Extremes in the Qinghai-Tibetan Plateau, China: Spatio-Temporal Characteristics and Topography Effects. *Atmosphere* **2017**, *8*, 127. [CrossRef]
43. Valjarevic, A.; Popovici, C.; Stilic, A.; Radojkovic, M. Cloudiness and Water from Cloud Seeding in Connection with Plants Distribution in the Republic of Moldova. *Appl. Water Sci.* **2022**, *12*, 262. [CrossRef]

44. Maes, S.L.; Vannoppen, A.; Altman, J.; Van Den Bulcke, J.; Decocq, G.; De Mil, T.; Depauw, L.; Landuyt, D.; Perring, M.P.; Van Acker, J.; et al. Evaluating the Robustness of Three Ring-Width Measurement Methods for Growth Release Reconstruction. *Dendrochronologia* **2017**, *46*, 67–76. [CrossRef]
45. Ziemiańska, M.; Kalbarczyk, R. Biometrics of Tree-Ring Widths of (Populus X Canadensis Moench) and Their Dependence on Precipitation and Air Temperature in South-Western Poland. *Wood Res.* **2018**, *63*, 2018.
46. Holmes, R.L. Computer-Assisted Quality Control in Tree-Ring Dating and Measurement. *Tree-Ring Bull* **1983**, *43*, 51–67. [CrossRef]
47. Cook, E.R. *A Time Series Analysis Approach to Tree Ring Standardization*; Dendrochronology, Forestry, Dendroclimatology, Autoregressive Process: Tucson, AZ, USA, 1985.
48. Cook, E.; Briffa, K. A Comparison of Some Tree-Ring Standardization Methods. In *Methods of Dendrochronology*; Cook, E.R., Kairiukstis, L.A., Eds.; Applications in the Environmental Sciences: Boston, MA, USA, 1990.
49. Choi, E.-B.; Sano, M.; Park, J.-H.; Kim, Y.-J.; Li, Z.; Nakatsuka, T.; Hakozaki, M.; Kimura, K.; Jeong, H.-M.; Seo, J.-W. Synchronizations of Tree-Ring δ18O Time Series within and between Tree Species and Provinces in Korea: A Case Study Using Dominant Tree Species in High Elevations. *J. Wood Sci.* **2020**, *66*, 53. [CrossRef]
50. Wang, Y.; Feng, Q.; Kang, X. Tree-Ring-Based Reconstruction of Temperature Variability (1445–2011) for the Upper Reaches of the Heihe River Basin, Northwest China. *J. Arid Land* **2016**, *8*, 60–76. [CrossRef]
51. Yang, B.; Kang, X.; Braeuning, A.; Liu, J.; Qin, C.; Liu, J. A 622-Year Regional Temperature History of Southeast Tibet Derived from Tree Rings. *Holocene* **2010**, *20*, 181–190. [CrossRef]
52. Tian, Q.; Gou, X.; Zhang, Y.; Peng, J.; Wang, J.; Chen, T. Tree-Ring Based Drought Reconstruction (A.D. 1855–2001) for the Qilian Mountains, Northwestern China. *Tree-Ring Res.* **2007**, *63*, 27–36. [CrossRef]
53. Khan, N.; Sachindra, D.A.; Shahid, S.; Ahmed, K.; Shiru, M.S.; Nawaz, N. Prediction of Droughts over Pakistan Using Machine Learning Algorithms. *Adv. Water Resour.* **2020**, *139*, 103562. [CrossRef]
54. Lin, Y.; Salekin, S.; Meason, D.F. Modelling Tree Diameter of Less Commonly Planted Tree Species in New Zealand Using a Machine Learning Approach. *Forestry* **2023**, *96*, 87–103. [CrossRef]
55. Jevšenak, J.; Džeroski, S.; Levanič, T. Predicting the Vessel Lumen Area Tree-Ring Parameter of Quercus Robur with Linear and Nonlinear Machine Learning Algorithms. *Geochronometria* **2018**, *45*, 211–222. [CrossRef]
56. Zhang, H.; Feng, Z.; Wang, S.; Ji, W. Disentangling the Factors That Contribute to the Growth of *Betula* spp. and *Cunninghami Lanceolata* in China Based on Machine Learning Algorithms. *Sustainability* **2022**, *14*, 8346. [CrossRef]
57. Briffa, K.R.; Jones, P.D.; Bartholin, T.S.; Eckstein, D.; Schweingruber, F.H.; Karlén, W.; Zetterberg, P.; Eronen, M. Fennoscandian Summers from Ad 500: Temperature Changes on Short and Long Timescales. *Clim. Dyn.* **1992**, *7*, 111–119. [CrossRef]
58. Shi, F.; Yang, B.; Von Gunten, L.; Qin, C.; Wang, Z. Ensemble Empirical Mode Decomposition for Tree-Ring Climate Reconstructions. *Theor. Appl. Climatol.* **2012**, *109*, 233–243. [CrossRef]
59. Yin, H.; Liu, H.; Linderholm, H.W.; Sun, Y. Tree Ring Density-Based Warm-Season Temperature Reconstruction since A.D. 1610 in the Eastern Tibetan Plateau. *Palaeogeogr. Palaeoclimatol. Palaeoecol.* **2015**, *426*, 112–120. [CrossRef]
60. Zhao, X.; Fang, K.; Chen, F.; Martín, H.; Roig, F.A. Reconstructed Jing River Streamflow from Western China A 399-Year Perspective for Hydrological Changes in the Loess Plateau. *J. Hydrol.* **2023**, *621*, 129573. [CrossRef]
61. Shi, S.; Li, J.; Shi, J.; Zhao, Y.; Huang, G. Three Centuries of Winter Temperature Change on the Southeastern Tibetan Plateau and Its Relationship with the Atlantic Multidecadal Oscillation. *Clim. Dyn.* **2017**, *49*, 1305–1319. [CrossRef]
62. Shi, J.; Cook, E.; Lu, H.; Li, J.; Wright, W.; Li, S. Tree-Ring Based Winter Temperature Reconstruction for the Lower Reaches of the Yangtze River in Southeast China. *Clim. Res.* **2010**, *41*, 169–176. [CrossRef]
63. Opała, M.; Mendecki, M.J. An Attempt to Dendroclimatic Reconstruction of Winter Temperature Based on Multispecies Tree-Ring Widths and Extreme Years Chronologies (Example of Upper Silesia, Southern Poland). *Theor. Appl. Climatol.* **2014**, *115*, 73–89. [CrossRef]
64. Körner, C. A Re-Assessment of High Elevation Treeline Positions and Their Explanation. *Oecologia* **1998**, *115*, 445–459. [CrossRef]
65. Körner, C. *Alpine Treelines*; Springer: Basel, Switzerland, 2012; ISBN 978-3-0348-0395-3.
66. Shao, X.; Fan, J. Past Climate on West Sichuan Plateau as Reconstructed from Ring-Widths of Dragon Spruce. *Quat. Sci.* **1999**, *19*, 81–89.
67. Yao, T.; Guo, X.; Thompson, L.; Duan, K.; Wang, N.; Pu, J.; Xu, B.; Yang, X.; Sun, W. δ 18O Record and Temperature Change over the Past 100 Years in Ice Cores on the Tibetan Plateau. *Sci. China Ser. D* **2006**, *49*, 1–9. [CrossRef]
68. Xu, P.; Zhu, H.; Shao, X.; Yin, Z. Tree Ring-Dated Fluctuation History of Midui Glacier since the Little Ice Age in the Southeastern Tibetan Plateau. *Sci. China Earth Sci.* **2012**, *55*, 521–529. [CrossRef]
69. Mayewski, P.A.; Pregent, G.P.; Jeschke, P.A.; Ahmad, N. Himalayan and Trans-Himalayan Glacier Fluctuations and the South Asian Monsoon Record. *Arct. Alp. Res.* **1980**, *12*, 171. [CrossRef]
70. Wagner, S.; Zorita, E. The Influence of Volcanic, Solar and CO_2 Forcing on the Temperatures in the Dalton Minimum (1790–1830): A Model Study. *Clim. Dyn.* **2005**, *25*, 205–218. [CrossRef]
71. Gao, C.; Gao, Y.; Zhang, Q.; Shi, C. Climatic Aftermath of the 1815 Tambora Eruption in China. *J. Meteorol. Res.* **2017**, *31*, 28–38. [CrossRef]
72. Jiang, Y.; Cao, Y.; Zhang, J.; Li, Z.; Shi, G.; Han, S.; Coombs, C.E.O.; Liu, C.; Wang, X.; Wang, J.; et al. A 168-Year Temperature Recording Based on Tree Rings and Latitude Differences in Temperature Changes in Northeast China. *Int. J. Biometeorol.* **2021**, *65*, 1859–1870. [CrossRef]

73. Lyu, S.; Li, Z.; Zhang, Y.; Wang, X. A 414-Year Tree-Ring-Based April–July Minimum Temperature Reconstruction and Its Implications for the Extreme Climate Events, Northeast China. *Clim. Past* **2016**, *12*, 1879–1888. [CrossRef]
74. Zhu, L.; Li, Z.; Zhang, Y.; Wang, X. A 211-year Growing Season Temperature Reconstruction Using Tree-ring Width in Zhangguangcai Mountains, Northeast China: Linkages to the Pacific and Atlantic Oceans. *Int. J. Climatol.* **2017**, *37*, 3145–3153. [CrossRef]
75. Zeng, A.; Zhou, F.; Li, W.; Bai, Y.; Zeng, C. Tree-Ring Indicators of Winter-Spring Temperature in Central China over the Past 200 Years. *Dendrochronologia* **2019**, *58*, 125634. [CrossRef]
76. Thompson, L.G.; Yao, T.; Davis, M.E.; Mosley-Thompson, E.; Wu, G.; Porter, S.E.; Xu, B.; Lin, P.-N.; Wang, N.; Beaudon, E.; et al. Ice Core Records of Climate Variability on the Third Pole with Emphasis on the Guliya Ice Cap, Western Kunlun Mountains. *Quat. Sci. Rev.* **2018**, *188*, 1–14. [CrossRef]
77. Wang, N.; Yao, T.; Pu, J.; Zhang, Y.; Sun, W. Climatic and Environmental Changes over the Last Millennium Recorded in the Malan Ice Core from the Northern Tibetan Plateau. *Sci. China Ser. Earth Sci.* **2006**, *49*, 1079–1089. [CrossRef]
78. Yao, Y.; Wang, L.; Li, X.; Cheng, H.; Cai, Y.; Vachula, R.S.; Liang, J.; Li, H.; Liu, G.; Zhao, J.; et al. Unexpected Cold Season Warming during the Little Ice Age on the Northeastern Tibetan Plateau. *Commun. Earth Environ.* **2023**, *4*, 182. [CrossRef]
79. Viau, A.E.; Gajewski, K. Reconstructing Millennial-Scale, Regional Paleoclimates of Boreal Canada during the Holocene. *J. Clim.* **2009**, *22*, 316–330. [CrossRef]
80. Houle, D.; Moore, J.-D.; Provencher, J. Ice Bridges on the St. Lawrence River as an Index of Winter Severity from 1620 to 1910. *J. Clim.* **2007**, *20*, 757–764. [CrossRef]
81. Zhu, H.; Huang, R.; Asad, F.; Liang, E.; Bräuning, A.; Zhang, X.; Dawadi, B.; Man, W.; Grießinger, J. Unexpected Climate Variability Inferred from a 380-Year Tree-Ring Earlywood Oxygen Isotope Record in the Karakoram, Northern Pakistan. *Clim. Dyn.* **2021**, *57*, 701–715. [CrossRef]
82. Zhang, Y.; Shao, X.M.; Yin, Z.-Y.; Wang, Y. Millennial Minimum Temperature Variations in the Qilian Mountains, China: Evidence from Tree Rings. *Clim. Past* **2014**, *10*, 1763–1778. [CrossRef]
83. Wang, B.; Wu, R.; Fu, X. Pacific–East Asian Teleconnection: How Does ENSO Affect East Asian Climate? *J. Clim.* **2000**, *13*, 1517–1536. [CrossRef]
84. Zhang, R.; Sumi, A.; Kimoto, M. Impact of El Niño on the East Asian Monsoon: A Diagnostic Study of the '86/87 and '91/92 Events. *J. Meteorol. Soc. Jpn. Ser. II* **1996**, *74*, 49–62. [CrossRef]
85. Kim, J.; An, S. Western North Pacific Anticyclone Change Associated with the El Niño–Indian Ocean Dipole Coupling. *Int. J. Climatol.* **2019**, *39*, 2505–2521. [CrossRef]
86. Ma, L.; Yin, Z. Possible Solar Modulation of Pacific Decadal Oscillation. *Sol. Syst. Res.* **2017**, *51*, 417–421. [CrossRef]
87. Xu, Y.; Li, T.; Shen, S.; Hu, Z. Assessment of CMIP5 Models Based on the Interdecadal Relationship between the PDO and Winter Temperature in China. *Atmosphere* **2019**, *10*, 597. [CrossRef]
88. He, M.; Yang, B.; Bräuning, A.; Wang, J.; Wang, Z. Tree-Ring Derived Millennial Precipitation Record for the South-Central Tibetan Plateau and Its Possible Driving Mechanism. *Holocene* **2013**, *23*, 36–45. [CrossRef]
89. Fang, K.; Guo, Z.; Chen, D.; Wang, L.; Dong, Z.; Zhou, F.; Zhao, Y.; Li, J.; Li, Y.; Cao, X. Interdecadal Modulation of the Atlantic Multi-Decadal Oscillation (AMO) on Southwest China's Temperature over the Past 250 Years. *Clim. Dyn.* **2019**, *52*, 2055–2065. [CrossRef]
90. Ding, Y.; Liu, Y.; Liang, S.; Ma, X.; Zhang, Y.; Si, D.; Liang, P.; Song, Y.; Zhang, J. Interdecadal Variability of the East Asian Winter Monsoon and Its Possible Links to Global Climate Change. *J. Meteorol. Res.* **2014**, *28*, 693–713.

forests

MDPI

Article

A 278-Year Summer Minimum Temperature Reconstruction Based on Tree-Ring Data in the Upper Reaches of Dadu River

Jinjian Li [1,2], Liya Jin [1,2,*] and Zeyu Zheng [1]

1 MOE Key Laboratory of Western China's Environmental System, College of Earth and Environmental Sciences, Lanzhou University, Lanzhou 730000, China; ljj@cuit.edu.cn (J.L.); zhengzy17@lzu.edu.cn (Z.Z.)
2 Plateau Atmosphere and Environment Key Laboratory of Sichuan Province, School of Atmospheric Sciences, Chengdu University of Information Technology, Chengdu 610225, China
* Correspondence: jinly@lzu.edu.cn

Abstract: In the context of global warming, climate change in river headwater regions and its drivers have attracted increasing attention. In this study, tree-ring width (TRW) chronology was constructed using tree-ring samples of fir (*Abies faxoniana*) in Dadu River Basin in the central part of the western Sichuan Plateau, China. Correlation analysis with climatic factors implies that the radial growth of trees in the region is mainly limited by temperature and has the highest correlation with the mean minimum temperature in summer (June and July) (R = 0.602, $p < 0.001$). On this basis, the TRW chronology was adopted to reconstruct variations in the mean minimum temperatures in summer from 1733 to 2010 in the upper reaches of Dadu River. The reconstruction equation was stable and reliable and offered a variance explanation rate of 36.2% in the observed period (1962~2010). In the past 278 years, the region experienced nine warm periods and ten cold periods. The warmest and coldest years occurred in 2010 and 1798, respectively, with values of 13.6 °C and 11.0 °C. The reconstruction was highly spatiotemporally representative and verified by temperatures reconstructed using other tree-ring data in surrounding areas. A significant warming trend was found in the last few decades. Moreover, the multi-taper method (MTM) analysis indicated significant periodic changes in quasi-2-year and 21–35-year periods, for which the El Niño Southern Oscillation (ENSO) and the Pacific decadal oscillation (PDO) could be the key controlling factors.

Keywords: tree ring; temperature reconstruction; up catch of river; climate change; China

1. Introduction

The short-term instrument measurements and sparse distribution of meteorological stations are the main factors that limit our understanding of regional climate change, particularly in alpine and plateau regions [1,2]. Additionally, to reveal the characteristics, causes, and trends of climate change at various temporal and spatial scales and forecast future climate change and its impacts, long-term climate records are needed [3,4]. Existing research and analysis of historical climate change generally rely on climate proxy data, particularly high-resolution climate proxy data. These data not only provide important basic information for verifying climate modes but also play a crucial role in understanding the characteristics and causes of climate and environmental changes, performing climatic predictions, and analyzing the relationship between human activities and climate change [5–7]. Numerous scholars have used climate proxy data such as ice cores, lacustrine sedimentary records, and historical literature to reconstruct climate change over the past several centuries and millennia [8–10]. Among diverse climate proxy data, tree-ring data are highly regarded due to their accurate dating and high time resolution. By using tree-ring data, multiple climate series were reconstructed, which revealed the trends and characteristics of historical climate change [11–14].

Recent research has shown that the Qinghai–Tibet Plateau (QTP) is the driver of climate change in China and the whole of East Asia [15,16]; as such, this area's climate-change characteristics have attracted increasing attention. Located between the main part of QTP and the Sichuan Basin, the Western Sichuan Plateau (WSP), which is scarcely populated and covered by rich vegetation, is a transition zone from the alpine climate zone of the QTP to the subtropical climate prevailing in the Sichuan Basin [17,18]. Climate characteristics are of important scientific significance for determining climate change in the QTP, the nearby Sichuan Basin, and even the entirety of southwestern China. Since the 1990s, many tree-ring researchers have sampled tree rings in the WSP to reconstruct the climate changes at numerous locations in the plateau using the tree-ring width (TRW), maximum density, isotopes, etc. These researchers have made great efforts to reveal the historical climate changes over the past 100~500 years on the WSP [18–23]. Rivers are densely distributed over the WSP and breed important tributaries of the Yangtze River, including the Minjiang River, Dadu River, and Yalong River. However, existing tree-ring-based climatological research in the WSP mainly focuses on alpine timberline with old trees and mainly represents plateau climates, while attention to climate change in river regions remains to be improved [3]. Against this background, *Abies faxoniana* samples were collected by setting sampling points on slopes along banks of the main tributaries upstream of Dadu River in the central part of the WSP to construct the TRW's chronology. Based on an analysis of the relationship between the TRW and climatic factors, temperature variations in the last several hundred in the upper reaches of Dadu River were reconstructed, thus revealing temperature variations and enabling us to explore the possible drivers. This research expands our understanding of long-term climate change in river headwater regions and provides a scientific basis for ecological conservation in these regions.

2. Materials and Methods

2.1. Study Area and Tree-Ring Data

The tree-ring samples used in this research were collected from the Lover Sea scenic area in Maori Township, Aba Tibetan, and Qiang Autonomous Prefecture, Sichuan Province, China (101°42′ E, 31°31′ N, Figure 1). *Abies faxoniana* tree rings were collected at sampling points established on the slope of main tributaries upstream of Dadu River at an altitude of 3000 m. This region is within the range of the QTP climate zone. Under joint action of the jet stream in the Southern Branch of the westerlies, southeast monsoon, and southwest monsoon, a mountain climate affording a cold winter, cool summer, and moderate rainfall is formed in the region [17]. The annual mean temperature (1961~2010) is 10.6 °C, and the monthly mean temperature changes during the year are single-peaked, with the lowest temperature (0.9 °C) in January and the highest temperature (18.5 °C) in July. The annual mean precipitation (1961~2010) is 634.5 mm, and the monthly precipitation distribution shows a bimodal pattern, with peaks, respectively, in June and September. Tree cores were collected within several square kilometers (not exceeding 0.2 square kilometers) in the sampling points where the difference in elevation did not exceed 100 m. The forest depression of the sampling site was up to about 0.4. The trees were mostly healthy living trees, and the understory vegetation was rich, with common plants, such as rhododendron and lonicera, a denser grass cover layer, and scattered mosses of 10~30 mm. An increment borer was used to collect two tree cores (three from some trees) at breast height. A total of 59 cores were collected from 24 old trees.

All core samples were pre-processed following the basic international procedures for tree-ring analysis [24,25]. A Lintab system with a precision of 0.01 mm was used to measure the TRW, and the COFECHA software [26] was adopted for cross-dating. Afterward, the ARSTAN program [27] was used for detrending to eliminate non-climatic information contained in the original TRW and retain as much climatic information as possible. By repeatedly fitting the growth trend using different methods, the spline function with a step of 50 years was found to perform best in fitting the growth trend. Finally, we obtained three tree-ring chronologies: standard (STD), residual (RES), and autoregressive standard (ARS) chronologies [27]. Several core samples with large differences from the main series were eliminated in the process of constructing the chronologies. STD chronology was mainly adopted for the analysis. The total length of the chronology was 336 years (1675~2010), and the average sensitivity was 0.158. In the common period of 1900~2000, the intra-tree and inter-tree correlation coefficients were, respectively, 0.604 and 0.208, and the expressed population signal (EPS), signal-to-noise ratio (SNR), and variance explained by the first principal component were, respectively, 0.943, 16.518, and 29.7%. These statistical characteristics of the chronologies indicate that the chronology has the potential to reflect past climate change. EPS > 0.85 was used to determine the reliable interval of the chronology [21,28]. After reaching 21 tree rings, the EPS exceeded 0.85, so the reliable period was deemed to be 1733~2010 (Table 1 and Figure 2).

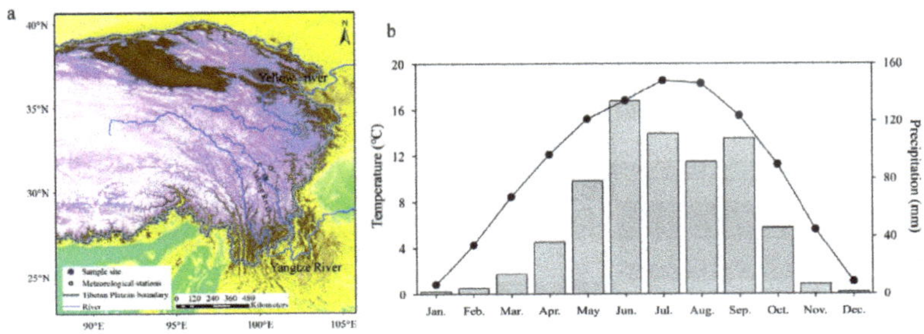

Figure 1. Locations of the sampling site, meteorological stations, (**a**) and monthly mean temperatures and total precipitation for the study area, which were calculated from 5 meteorological stations (Danba, Luhuo, Jinchuan, Xiaojin, and Daofu) during the period of 1962~2010 (**b**).

Table 1. Statistical characteristics of the chronology.

Statistical Characteristic	Values
Period	1675–2010
Common period	1900–2000
First-order autocorrelation (AC1)	0.773
Signal-to-noise ratio (SNR)	16.518
Mean sensitivity (MS)	0.158
Mean correlations within trees	0.604
Mean correlations between trees	0.208
Expressed population signal (EPS)	0.943
Period with EPS > 0.85 (minimum number of trees)	1733 (21)
Variance of first eigenvector (PC1)	29.7%

2.2. Meteorological Data

Meteorological data used in this study included monthly mean temperature, maximum temperature, minimum temperature, and precipitation from 1961 to 2010 recorded at five stations: Danba County (101°88′ E, 30°88′ N, at an altitude of 1950 m), Luhuo County (100°67′ E, 31°40′ N, at an altitude of 3250 m), Daofu County (101°12′ E, 30°98′ N, at an altitude of 2957 m), Jinchuan County (102°07′ E, 31°53′ N, at an altitude of 2169 m), and Xiaojin County (102°35′ E, 31° N, at an altitude of 2369 m). The mean values of climatic factors at the five stations were used to characterize the climatic characteristics of the study area. Sea surface temperatures were taken from the Extended Reconstructed Sea Surface Temperature V5 (ERSST V5) dataset with a spatial resolution of 2° × 2° provided by the National Oceanic and Atmospheric Administration (NOAA).

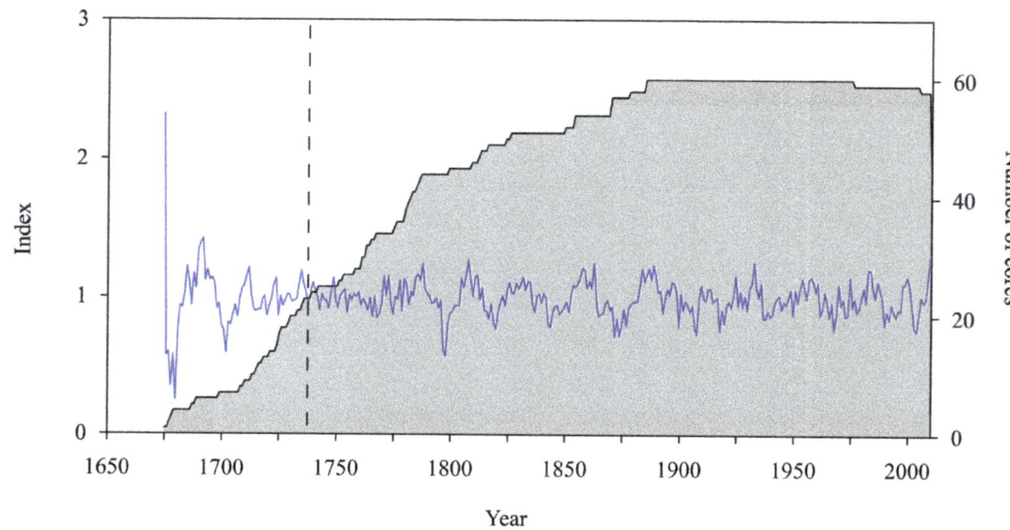

Figure 2. The tree ring width and number of tree rings. The red arrow indicates that the first year of EPS > 0.85.

2.3. Methods

Pearson's correlation analysis was used to identify the key climatic factors in this study. The method of leave-one-out cross-validation (LOOCV) was used to assess the stability and reliability of the reconstruction equation. The main statistic parameters were the correlation coefficient (R), variance explanation (R2), sign tests, sign tests of first-order difference, F value, mean product (t), and reduced error (RE) [24]. The standard deviation (SD) could reflect the degree of the dispersion of a data set; it was used to identify extreme climate events [18,21,29]. Here, a mean threshold of $\pm$ 1.5 times the standard deviation (SD) was used to judge years with extremely high and low temperatures. A year with a reconstructed temperature higher than the mean + 1.5SD was regarded as an extremely warm year; if the reconstructed temperature was lower than the mean −1.5SD, the year was considered to be an extremely cold year. To determine the decadal variations in cold and warm periods, time periods lower and higher than the mean temperature for more than nine successive years were defined as cold and warm periods, respectively, based on low-pass filtering of the reconstructed series for 11 years. The multi-taper method (MTM), which is a low-variance high-resolution spectral analysis method especially suitable for diagnostic analysis of quasi-periodic signals with short sequences and high noise background [30,31], was used to identify the cycle of reconstruction.

3. Results

3.1. Climate-Growth Response Analysis

Figure 3 illustrates correlations of the tree-ring index with regional climatic factors (including the mean temperature, maximum temperature, minimum temperature, and precipitation). Results showed that for correlations with precipitation, the radial growth of trees was significantly negatively correlated in the previous December (R = −0.346, $p < 0.05$) and significantly positively correlated with the current July (R = 0.321, $p < 0.05$). For correlations with the temperature, the radial growth of trees was significantly positively correlated with the mean and minimum temperatures in June and July of the current year (R = 0.528, 0.284, 0.503, and 0.489, respectively; $p < 0.05$) and the maximum temperatures in February and June of the current year (R = 0.283 and 0.5, respectively; $p < 0.05$). Therein, radial growth was the most significantly correlated with mean temperature in June, with a correlation coefficient of 0.528 ($p < 0.01$). Correlation analysis was also performed for climatic factors under different month combinations. The results indicated that the mean minimum temperatures in June and July were most significantly correlated with the tree-ring index, with the correlation coefficient reaching 0.601 ($p < 0.01$). Meanwhile, the 31-year moving correlations also showed that the tree-ring index had a stable relationship with the climatic factors (Figure 4).

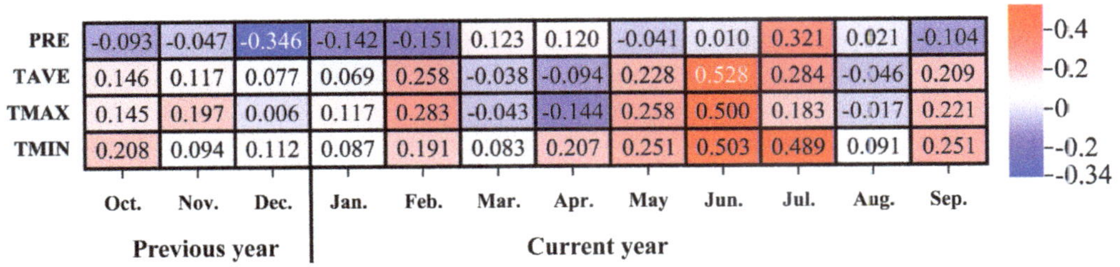

Figure 3. The correlation coefficients between the tree-ring index and climatic factors (PRE, TAVE, TMAX, and TMIN indicate monthly precipitation, mean temperature, maximum temperature, and minimum temperature, respectively).

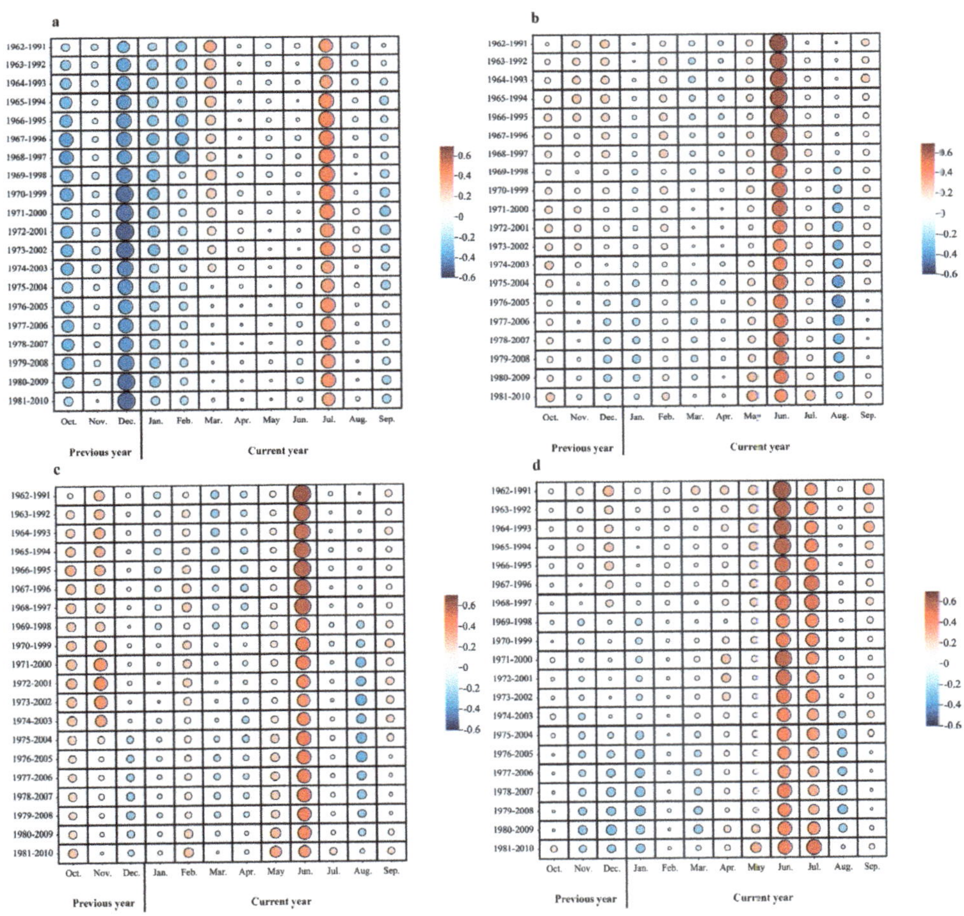

Figure 4. The 31-year moving correlations between the tree-ring index and climatic factors (**a**–**d**) indicate the relationship between TRW and monthly precipitation, mean temperature, maximum temperature, and minimum temperature, respectively).

3.2. Temperature Reconstruction

According to the above analysis, linear regression equation was established by taking the tree-ring index as the independent variable and the July–June average minimum temperature ($Tmin_{JJ}$) as the dependent variables:

$$T_{JJ} = 9.029 + 3.499 \times I$$

where T_{JJ} is the reconstructed June–July mean minimum temperature, and I is the growth index for the total period of calibration (1733–2010). The reconstruction equation explained 36.2% of the variance of the $Tmin_{JJ}$ series in the region, which was 34.9% after adjusting the degree of freedom, and the correlation coefficient was 0.602 (n = 49, $p < 0.001$).

The equation was tested through leave-one-out cross-validation (LOOCV), and the statistics in each test are summarized in Table 2. Here, the reduced error (RE = 0.36), mean product (t = 3.07), correlation coefficient (r = 0.602), and F-test statistic (F = 26.733) all reached a 0.01 significance level, indicating the stability of the reconstruction equation. Meanwhile, the sign tests and sign tests of the first-order difference between the reconstructed and measured values both reached a 0.01 significance level, suggesting a good reconstruction effect when using the reconstruction equation on both low-frequency and high-frequency variations. As shown in Figure 5a, the observed and reconstructed $Tmin_{JJ}$ values were very consistent in the instrumental period (1962–2010).

Table 2. Test statistics in leave-one-out cross-validation of the regression equation.

Time Period	Correlation Coefficient	Variance Explanation	F-Value	Sign Tests	Sign Tests of First-Order Difference	Mean Product (t)	Reduced Error (RE)
1962~2010	0.602	36.2%	26.733	36 (32 *, 34 **)	34 (32 *, 34 **)	3.07	0.36

Note: * and ** are the numbers of same signs needed for 0.05 and 0.01 significance levels, respectively.

3.3. Temperature Variation Characteristics

Figure 5b presents variations in the reconstructed $Tmin_{JJ}$ over 278 years (1733~2010). Here, the mean value of the reconstructed temperature series is 12.5 °C; the highest and lowest values of $Tmin_{JJ}$ are 11.0 °C (in 1798) and 13.6 °C (in 2010), respectively, with a standard deviation (SD) of 0.4 °C. Ultimately, 13 and 20 years, respectively, were found to be extremely warm and cold years over the past 278 years (Table 3), accounting for 4.3% and 7.2% of the period. Therein, 1858~1859 and 1983~1984 featured persistent extremely warm years for two or more years, while 1797~1799, 1819~1820, 1843~1844, 1874~1875, 1907~1908, and 2003~2004 featured persistent extremely cold years for two or more years.

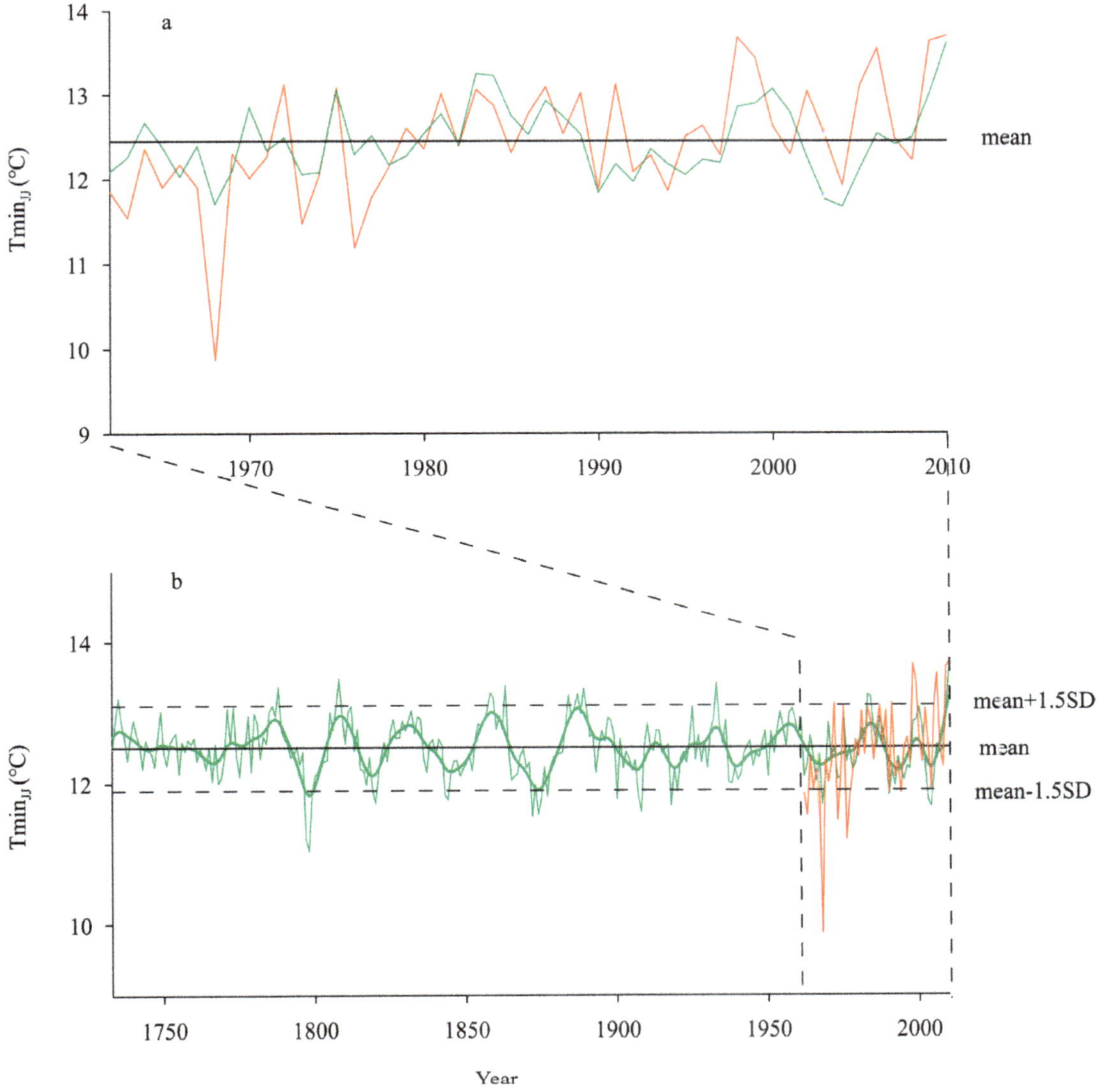

Figure 5. Comparisons between the reconstructed and observed July–June average minimum temperature ($Tmin_{JJ}$) during the period of 1962~2010 (**a**) and the reconstructed $Tmin_{JJ}$ in the reliable period of 1733~2010 (**b**). The thin orange and green lines represent the observed and reconstructed $Tmin_{JJ}$, respectively. The thick green line indicates 11-year low-pass filtering of the reconstruction.

Table 3. Extreme events during the past 278 years.

Year	The Value of Extremely High Temperature (°C)	Year	The Value of Extremely High Temperature (°C)
1735	13.2	1797	11.2
1788	13.4	1798	11.0
1808	13.5	1799	11.8
1858	13.3	1819	11.9
1859	13.2	1820	11.7
1863	13.4	1843	11.8
1884	13.2	1844	11.8
1887	13.3	1872	11.5
1889	13.3	1874	11.6
1933	13.4	1875	11.8
1983	13.2	1878	11.8
1984	13.2	1900	11.8
2010	13.6	1907	11.8
		1908	11.6
		1918	11.7
		1920	11.9
		1968	11.7
		1990	11.8
		2003	11.8
		2004	11.7

During 1733~2010, there were nine warm periods and ten cold periods. The warm periods occurred in 1733~1743, 1771~1791, 1804~1814, 1825~1837, 1853~1864, 1881~1899, 1924~1936, 1950~1962, and 1980~1988; the cold periods happened in 1759~1770, 1792~1803, 1815~1824, 1838~1852, 1865~1880, 1900~1911, 1915~1923, 1937~1949, 1963~1979, and 1989~1997.

As shown in Figure 6, the results of MTM at a confidence level of 95% indicated that the reconstructed series mainly features an interannual variation period of 2 years and a decadal oscillation period of 21 to 35 years.

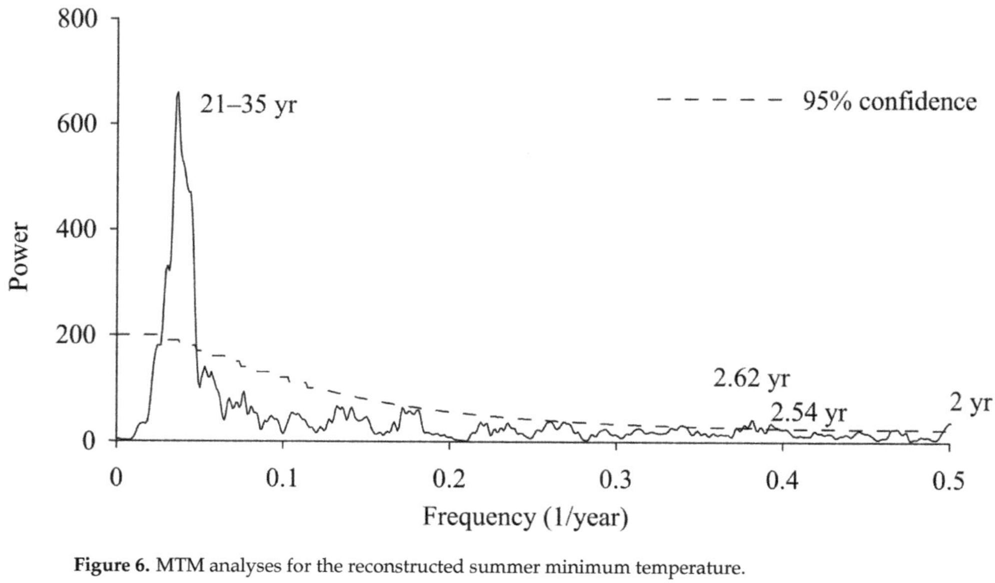

Figure 6. MTM analyses for the reconstructed summer minimum temperature.

4. Discussion

4.1. Climate Responses

As shown in Figures 3 and 4, the summer temperatures (June and July), particularly the minimum temperatures, are the leading limiting factor influencing the radial growth of trees in the region. The samples were established near the alpine timberline in the central part of WSP, where the climate is mainly driven by the East Asian monsoon (EAM) and South Asian monsoon (SAM). In summer, it becomes rainy and hot, which coincides with the most vigorous and rapid growth of trees in the region. The high temperatures in summer are favorable for the photosynthesis of trees. Particularly, a higher minimum temperature can promote the accumulation of nutrients produced by photosynthesis and accelerate the splitting of cambial cells, which is conducive to the production of wide tree rings [32,33]. Conversely, a lower minimum temperature in summer not only limits the accumulation of photosynthesis products but can also affect the root system development and water uptake of trees [34,35], thus forming narrow tree rings and even frost rings or missing rings. Therefore, the mean minimum temperature in summer is the main limiting factor that affects the growth of *Abies faxoniana* and has specific physiological significance. Similar climatic limiting factors were also found in others research focusing on high-altitude or high-latitude areas [32,36–38].

In addition, correlation analysis showed that precipitation also affected the radial growth of *Abies faxoniana*, with precipitation in December of the previous year and the current July showing significant negative and positive correlations with the tree-ring index, respectively. In winter, precipitation mainly occurs in the form of snow, and more precipitation in winter may lead to a longer duration of snow cover, thus delaying cambial activity and shortening the reproductive period [39–41], resulting in a narrower tree-ring width. In contrast, greater precipitation in summer compensates for the evaporation of soil water, which facilitates root activity and photosynthesis, thus promoting the radial growth of trees.

4.2. Verification of the Reconstructed Temperature

It is necessary to explore whether the regional representativeness of the reconstructed series is consistent with the observed series across a large range. To this end, spatial correlation analysis was conducted for the reconstructed and the observed $Tmin_{JJ}$ using the grid data of the Climate Research Unit (CRU TS4.06) in the same period. The results are shown in Figure 7. Here, the spatial correlations of observed (Figure 7a) and reconstructed $Tmin_{JJ}$ (Figure 7b) values with CRU data exhibit similar spatial patterns, suggesting high representativeness of the reconstructed $Tmin_{JJ}$ for the WSP and QTP. However, due to the possibility of the loss of some variation characteristics in the reconstruction, the correlations between the reconstructed $Tmin_{JJ}$ and CRU data were weaker than those of the observed and CRU data.

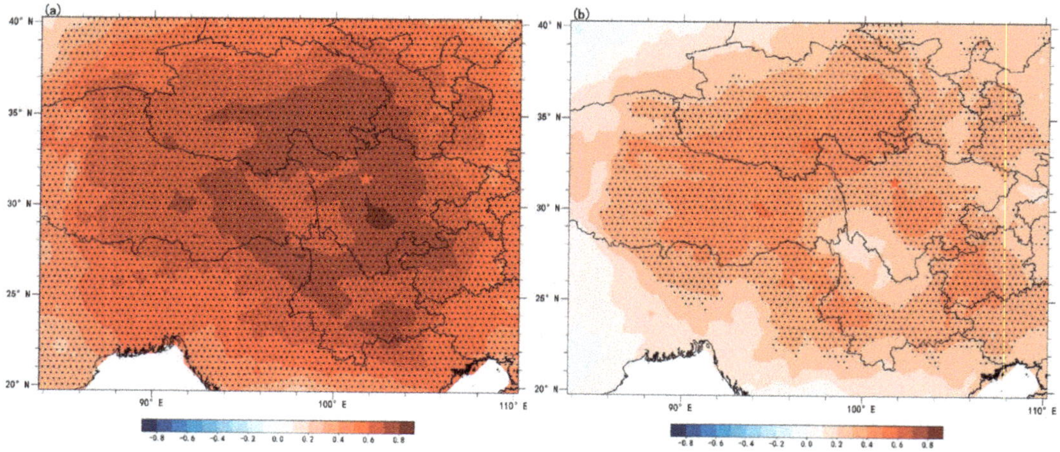

Figure 7. Spatial correlations of observed (**a**) and reconstructed (**b**) $Tmin_{JJ}$ with CRU data during the period of 1962~2010. The dotted area indicates 95% confidence.

To verify the reliability of the reconstructed series, the reconstructed series was compared with some reconstructed temperature series in surrounding areas (Figure 8). The reconstructed series for comparison and validation included the mean maximum temperature in May and June in the southeastern Qinghai–Tibet Plateau reconstructed by Zhang et al. [17], the anomaly series of mean minimum temperatures in summer in the Yangtze River headwater region reconstructed by Liang et al. [33], and the PC1 series reflecting mean minimum temperatures in the Western Sichuan Plateau reconstructed by Shao and Fan [42]. During the comparison, 11-year Gaussian low-pass filtering was performed for all series. As shown in Figure 8, despite certain differences in the reconstruction seasons and objects, the reconstructed $Tmin_{JJ}$ in the current research presented high consistency with cold and warm periods of the other three reconstructions. This result further validated the reliability of the reconstructed series. The cold and warm periods that were well reflected by all of the four reconstructions included warm periods in the 1780s, 1810s, 1830s, 1880s~1890s, 1950s, and 1980s and common cold periods in the 1790s, 1870s, and 1960s. Except for the aforementioned cold and warm periods, the warm periods in the early 1750s and 1850s and the cold periods in the 1760s and 1840s were also commonly reflected by three reconstructed series (including those in this study). In addition, the four reconstructions also differed in their views of cold and warm periods. These differences may arise from the discrepancy in local climate due to different reconstruction time periods, reconstruction objects (maximum or minimum temperatures), and habitats at different samples. These differences may occur because of the deviations caused by the use of different detrending methods when presenting the chronologies. Some other reconstructions also recorded the cold and warm periods reflected by the reconstructed $Tmin_{JJ}$ in this study. The continuous cold period from 1900~1923 was also reflected by the reconstructed temperature variations of southeastern QTP in winter [43], the mean temperatures in June and July in the WSP [18], the mean temperatures from June to August in the Eastern QTP [44], and the mean temperature field during May and June in the southeastern QTP [2]. The warm period from 1950 to 1962 was also well reflected by the reconstructed annual mean minimum temperatures in Hengduan Mountains [37] and the annual mean temperatures in Songpan [45].

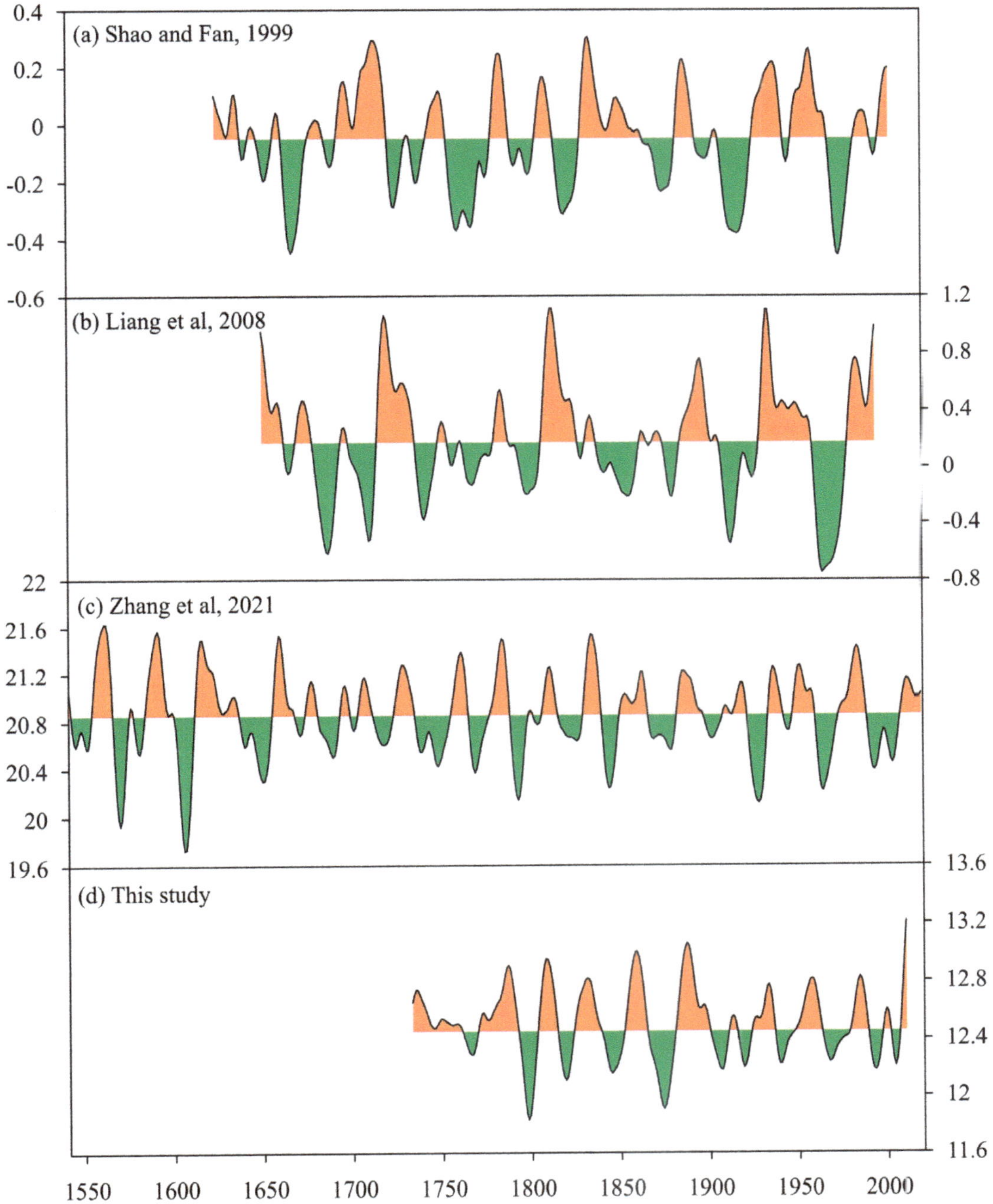

Figure 8. Comparisons between our reconstruction and other proxies: (**a**) PC1 series reflecting mean minimum temperatures in the WSP [42]; (**b**) the anomaly series of mean minimum temperatures in summer in the Yangtze River headwater region [33]; (**c**) the mean maximum temperature in May and June in the southeastern Qinghai–Tibet Plateau [17], and (**d**) the reconstructed mean summer minimum temperature of the present study. Here, 11-year Gaussian low-pass filtering was performed for all series.

Although few historical climatic events before 1950 can be determined, some clues can still be garnered from historical records to provide evidence for a reconstruction. According to some low-temperature events (snow disasters and freeze-related injuries) in the Sichuan Volume of the China Meteorological Calamity Grand Ceremony [46], there were few low-temperature events, indicating a high-temperature period. Correspondingly, low-temperature events were frequently recorded in the low-temperature period. For example, 1865~1880 was found to be a cold period in our reconstruction. Correspondingly, large-scale snow disasters happened frequently from the 1860s to early 1870s, particularly in 1871, in which snow disasters were most severe: "countless passers-by died, and most cattle and sheep froze to death" [46]. Thereafter, snow disasters occurred frequently from 1890 to 1930, and snow disasters were recorded to happen in counties and cities around the study area in 1913, 1917, and 1922 [46]. This period corresponds to the two continuous cold periods from 1900 to 1911 and 1915 to 1923 in the reconstructed series.

The glacial advance and retreat around the study area also conform to the reconstruction [47–49]. The continuous low-temperature periods reflected by the reconstructed series correspond to the stable or advanced stages of glaciers. Glacial recession or significant glacial retreat were observed when the reconstructed series showed a persistent trend toward increasing temperatures or when the temperature remained high.

4.3. Climate Responses

MTM spectral analysis of the reconstruction shows that the reconstructed series included significant periodic changes (Figure 6). Furthermore, wavelet analysis indicated the significant presence of decadal oscillations with a period of 21~35 y, nearly throughout the period analyzed. The interannual periodic changes of 2 y mainly correspond to high-frequency oscillations of the reconstructed series. On the one hand, this result is likely related to the quasi-two-year period commonly present in the climatic system; on the other hand, these changes are consistent with the main period of El Niño–Southern Oscillation (ENSO), indicating that ENSO may have a certain modulating effect on temperature variations in the region. Similar results were also obtained for the reconstructed temperature in the surrounding areas [17,21,50,51]. The decadal periodic oscillation is likely related to Pacific decadal oscillation (PDO), which was verified in numerous studies as one of the main factors controlling decadal climate variability in East Asia [17,52,53]. The reconstructed series has a correlation coefficient of 0.14 with the reconstructed PDO [54] (1733~2004, $p < 0.05$), indicating that PDO exerts an important influence on the temperature in the study area. Here, when PDO is in a warm phase, the sea temperature in the middle-east equator of the Pacific Ocean is higher, and the location of the ridge point of the western Pacific subtropical high (WPSH) is much more westward. The study area is likely to be controlled by WPSH, with strong downdrafts leading to higher temperatures. Conversely, if PDO is in a cold phase, it becomes conducive to the northward movement of warm and humid air flows, leading to frequent rain and relatively lower summer temperatures in the study area. Studies in the nearby area revealed that PDO is one of the main controlling factors [55,56]. Correlations between the reconstructed series and sea surface temperature (SST) of the Pacific also support this result (Figure 9).

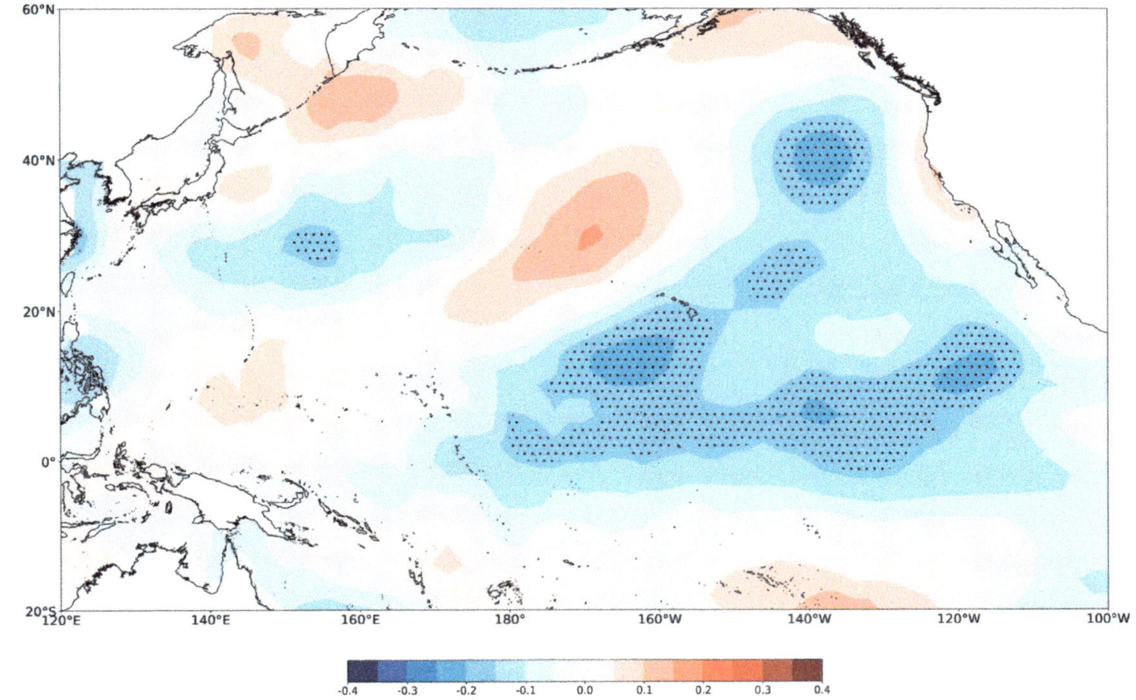

Figure 9. Spatial correlations between the reconstruction and sea surface temperature (SST) during the period of 1854–2010. The dotted area indicates 95% confidence.

5. Conclusions

TRW chronologies were constructed using tree-ring samples of *Abies faxoniana* collected on the Snowy Mountains in Dadu River Basin in the central part of WSP. The summer (June and July) mean minimum temperature was reconstructed for the period of 1733~2010. The reconstruction equation was found to be stable and reliable, with a variance explanation rate of 36.2% in the observed period (1962~2010). A total of 13 years of extremely high temperatures and 20 years of extremely low temperatures occurred in the past 278 years. In addition, there were 9 warm periods and 10 cold periods on a decadal scale. The correlation analysis shows that the reconstruction is highly spatiotemporally representative. Moreover, the reconstruction could be verified by other temperatures reconstructed using tree-ring data, historical meteorological disaster records, and glacial advance and retreat events in surrounding areas. The MTM results suggest that there were significant quasi-2-year and 21–35-year cycles in temperature variation in the study area. ENSO and PDO were possibly important controlling factors of temperature variations in the region. The reconstruction results indicate greater historical climate change at the western Sichuan Plateau and provide a reference for tackling climate change.

Author Contributions: Conceptualization, J.L. and L.J.; methodology, J.L.; software, J.L.; validation, J.L. and Z.Z.; formal analysis, J.L.; investigation, J.L. and Z.Z.; resources, L.J.; data curation, Z.Z.; writing—original draft preparation, J.L.; writing—review and editing, L.J. and J.L.; visualization, J.L.; supervision, L.J.; project administration, J.L. and L.J.; funding acquisition, J.L. and L.J. All authors have read and agreed to the published version of the manuscript.

Funding: This research was jointly funded by the Natural Science Foundation of Sichuan Province (No. 2022NSFSC0215), and the Sichuan Central Government guides local science and technology development project (No. 2021ZYD0019).

Data Availability Statement: The data of this study are available upon request to the first author or the corresponding author.

Acknowledgments: The authors would like to thank the editor and the two anonymous reviewers for their valuable and helpful comments and suggestions, which helped us to improve the quality of the manuscript.

Conflicts of Interest: The authors declare no conflict of interest.

References

1. Deng, Y.; Gou, X.; Gao, L.; Yang, M.; Zhang, F. Spatiotemporal Drought Variability of the Eastern Tibetan Plateau during the Last Millennium. *Clim. Dyn.* **2017**, *49*, 2077–2091. [CrossRef]
2. Li, J.; Jin, L.; Zheng, Z.; Qin, N. Variability of the Early Summer Temperature in the Southeastern Tibetan Plateau in Recent Centuries and the Linkage to the Indian Ocean Basin Mode. *Lithosphere* **2022**, *2022*, 9648384. [CrossRef]
3. Gou, X.; Deng, Y.; Chen, F.; Yang, M.; Fang, K.; Gao, L.; Yang, T.; Zhang, F. Tree Ring Based Streamflow Reconstruction for the Upper Yellow River over the Past 1234 Years. *Chin. Sci. Bull.* **2010**, *55*, 4179–4186. [CrossRef]
4. Li, J.; Xie, S.-P.; Cook, E.R.; Chen, F.; Shi, J.; Zhang, D.D.; Fang, K.; Gou, X.; Li, T.; Peng, J.; et al. Deciphering Human Contributions to Yellow River Flow Reductions and Downstream Drying Using Centuries-Long Tree Ring Records. *Geophys. Res. Lett.* **2019**, *46*, 898–905. [CrossRef]
5. Shi, F.; Sun, C.; Guion, A.; Yin, Q.; Zhao, S.; Liu, T.; Guo, Z. Roman Warm Period and Late Antique Little Ice Age in an Earth System Model Large Ensemble. *J. Geophys. Res. Atmos.* **2022**, *127*, e2021JD035832. [CrossRef]
6. Jiang, D.; Yu, G.; Zhao, P.; Chen, X.; Liu, J.; Liu, X.; Wang, S.; Zhang, Z.; Yu, Y.; Li, Y.; et al. Paleoclimate Modeling in China: A Review. *Adv. Atmos. Sci.* **2015**, *32*, 250–275. [CrossRef]
7. Parsons, L.A.; Amrhein, D.E.; Sanchez, S.C.; Tardif, R.; Brennan, M.K.; Hakim, G.J. Do Multi-Model Ensembles Improve Reconstruction Skill in Paleoclimate Data Assimilation? *Earth Space Sci.* **2021**, *8*, e2020EA001467. [CrossRef]
8. Pang, H.; Hou, S.; Zhang, W.; Wu, S.; Jenk, T.M.; Schwikowski, M.; Jouzel, J. Temperature Trends in the Northwestern Tibetan Plateau Constrained by Ice Core Water Isotopes Over the Past 7000 Years. *J. Geophys. Res. Atmos.* **2020**, *125*, e2020JD032560. [CrossRef]
9. Gyawali, A.R.; Wang, J.; Ma, Q.; Wang, Y.; Xu, T.; Guo, Y.; Zhu, L. Paleo-Environmental Change since the Late Glacial Inferred from Lacustrine Sediment in Selin Co, Central Tibet. *Palaeogeogr. Palaeoclimatol. Palaeoecol.* **2019**, *516*, 101–112. [CrossRef]
10. Chinese Academy of Meteorological Sciences; China Meteorological Administration. *Yearly Charts of Dryness/Wetness in China for the Last 500-Year Period*; SinoMaps Press: Beijing, China, 1981.
11. Yang, B.; Qin, C.; Bräuning, A.; Osborn, T.J.; Trouet, V.; Ljungqvist, F.C.; Esper, J.; Schneider, L.; Grießinger, J.; Büntgen, U.; et al. Long-Term Decrease in Asian Monsoon Rainfall and Abrupt Climate Change Events over the Past 6700 Years. *Proc. Natl. Acad. Sci. USA* **2021**, *118*, e2102007118. [CrossRef]
12. Liu, Y.; Song, H.; An, Z.; Sun, C.; Trouet, V.; Cai, Q.; Liu, R.; Leavitt, S.W.; Song, Y.; Li, Q.; et al. Recent Anthropogenic Curtailing of Yellow River Runoff and Sediment Load Is Unprecedented over the Past 500 y. *Proc. Natl. Acad. Sci. USA* **2020**, *117*, 18251–18257. [CrossRef]
13. PAGES 2k. Consortium Continental-Scale Temperature Variability during the Past Two Millennia. *Nat. Geosci.* **2013**, *6*, 339–346. [CrossRef]
14. Neukom, R.; Steiger, N.; Gómez-Navarro, J.J.; Wang, J.; Werner, J.P. No Evidence for Globally Coherent Warm and Cold Periods over the Preindustrial Common Era. *Nature* **2019**, *571*, 550–554. [CrossRef]
15. Mann, M.E.; Bradley, R.S.; Hughes, M.K. Global-Scale Temperature Patterns and Climate Forcing over the Past Six Centuries. *Nature* **1998**, *392*, 779–787. [CrossRef]
16. Li, H.; Liu, L.; Shan, B.; Xu, Z.; Niu, Q.; Cheng, L.; Liu, X.; Xu, Z. Spatiotemporal Variation of Drought and Associated Multi-Scale Response to Climate Change over the Yarlung Zangbo River Basin of Qinghai–Tibet Plateau, China. *Remote Sens.* **2019**, *11*, 1596. [CrossRef]
17. Zhang, Y.; Li, J.; Zheng, Z.; Zeng, S. A 479-Year Early Summer Temperature Reconstruction Based on Tree-Ring in the Southeastern Tibetan Plateau, China. *Atmosphere* **2021**, *12*, 1251. [CrossRef]
18. Zhang, Y.; Li, J.; Wang, S.; Shao, X.; Qin, N.; An, W. A Reconstruction of June–July Temperature since AD 1383 for Western Sichuan Plateau, China Using Tree-Ring Width. *Int. J. Climatol.* **2022**, *42*, 1803–1817. [CrossRef]
19. Li, J.; Li, J.; Li, T.; Au, T.F. 351-Year Tree Ring Reconstruction of the Gongga Mountains Winter Minimum Temperature and Its Relationship with the Atlantic Multidecadal Oscillation. *Clim. Change* **2021**, *165*, 49. [CrossRef]
20. Fang, K.; Guo, Z.; Chen, D.; Wang, L.; Dong, Z.; Zhou, F.; Zhao, Y.; Li, J.; Li, Y.; Cao, X. Interdecadal Modulation of the Atlantic Multi-Decadal Oscillation (AMO) on Southwest China's Temperature over the Past 250 Years. *Clim. Dyn.* **2019**, *52*, 2055–2065. [CrossRef]
21. Nie, W.; Li, M. July Mean Temperature Reconstruction for the Southern Tibetan Plateau Based on Tree-Ring Width Data during 1763–2020. *Forests* **2022**, *13*, 1911. [CrossRef]
22. Zhu, L.; Zhang, Y.; Li, Z.; Guo, B.; Wang, X. A 368-Year Maximum Temperature Reconstruction Based on Tree-Ring Data in the Northwestern Sichuan Plateau (NWSP), China. *Clim. Past* **2016**, *12*, 1485–1498. [CrossRef]

23. Xu, G.; Liu, X.; Zhang, Q.; Zhang, Q.; Hudson, A.; Trouet, V. Century-Scale Temperature Variability and Onset of Industrial-Era Warming in the Eastern Tibetan Plateau. *Clim Dyn* **2019**, *53*, 4569–4590. [CrossRef]
24. Fritts, H. *Tree Rings and Climate*; Elsevier: Amsterdam, The Netherlands, 2012.
25. Fritts, H.C. Dendroclimatology and Dendroecology. *Quat. Res.* **1971**, *1*, 419–449. [CrossRef]
26. Holmes, R.L. Computer-Assisted Quality Control in Tree-Ring Dating and Measurement. *Tree-Ring Bull.* **1983**, *43*, 51–67.
27. Cook, E.R. A Time Series Analysis Approach to Tree Ring Standardization. Ph.D. Thesis, University of Arizona, Tuscon, AZ, USA, 1985.
28. Wigley, T.M.L.; Briffa, K.R.; Jones, P.D. On the Average Value of Correlated Time Series, with Applications in Dendroclimatology and Hydrometeorology. *J. Appl. Meteorol. Climatol.* **1984**, *23*, 201–213. [CrossRef]
29. An, W.; Li, J.; Wang, S.; Xu, C.; Shao, X.; Qin, N.; Guo, Z. Hydrological Extremes in the Upper Yangtze River over the Past 700 Yr Inferred from a Tree Ring Δ18O Record. *J. Geophys. Res. Atmos.* **2022**, *127*, e2021JD036109. [CrossRef]
30. Babadi, B.; Brown, E.N. A Review of Multitaper Spectral Analysis. *IEEE Trans. Biomed. Eng.* **2014**, *61*, 1555–1564. [CrossRef]
31. Prieto, G.A.; Parker, R.L.; Vernon, F.L., III. A Fortran 90 Library for Multitaper Spectrum Analysis. *Comput. Geosci.* **2009**, *35*, 1701–1710. [CrossRef]
32. Huang, R.; Zhu, H.; Liang, E.; Asad, F.; Grießinger, J. A Tree-Ring–Based Summer (June–July) Minimum Temperature Reconstruction for the Western Kunlun Mountains since AD 1681. *Theor. Appl. Climatol.* **2019**, *138*, 673–682. [CrossRef]
33. Liang, E.; Shao, X.; Qin, N. Tree-Ring Based Summer Temperature Reconstruction for the Source Region of the Yangtze River on the Tibetan Plateau. *Glob. Planet. Change* **2008**, *61*, 313–320. [CrossRef]
34. Mayr, S. Limits in Water Relations. In *Trees at Their Upper Limit: Treelife Limitation at the Alpine Timberline*; Springer: Berlin/Heidelberg, Germany, 2007; pp. 145–162.
35. Wieser, G.; Tausz, M. *Trees at Their Upper Limit: Treelife Limitation at the Alpine Timberline*; Springer Science & Business Media: Berlin/Heidelberg, Germany, 2007; Volume 5.
36. Shi, C.; Wang, K.; Sun, C.; Zhang, Y.; He, Y.; Wu, X.; Gao, C.; Wu, G.; Shu, L. Significantly Lower Summer Minimum Temperature Warming Trend on the Southern Tibetan Plateau than over the Eurasian Continent since the Industrial Revolution. *Environ. Res. Lett.* **2019**, *14*, 124033. [CrossRef]
37. Keyimu, M.; Li, Z.; Zhang, G.; Fan, Z.; Wang, X.; Fu, B. Tree Ring–Based Minimum Temperature Reconstruction in the Central Hengduan Mountains, China. *Theor. Appl. Climatol.* **2020**, *141*, 359–370. [CrossRef]
38. Liang, E.; Wang, Y.; Piao, S.; Lu, X.; Camarero, J.J.; Zhu, H.; Zhu, L.; Ellison, A.M.; Ciais, P.; Peñuelas, J. Species Interactions Slow Warming-Induced Upward Shifts of Treelines on the Tibetan Plateau. *Proc. Natl. Acad. Sci. USA* **2016**, *113*, 4380–4385. [CrossRef]
39. Vaganov, E.A.; Hughes, M.K.; Kirdyanov, A.V.; Schweingruber, F.H.; Silkin, P.P. Influence of Snowfall and Melt Timing on Tree Growth in Subarctic Eurasia. *Nature* **1999**, *400*, 149–151. [CrossRef]
40. D'Arrigo, R.; Wilson, R.; Liepert, B.; Cherubini, P. On the 'Divergence Problem' in Northern Forests: A Review of the Tree-Ring Evidence and Possible Causes. *Glob. Planet. Change* **2008**, *60*, 289–305. [CrossRef]
41. Linderholm, H.W.; Chen, D. Central Scandinavian Winter Precipitation Variability during the Past Five Centuries Reconstructed from Pinus Sylvestris Tree Rings. *Boreas* **2008**, *34*, 43–52. [CrossRef]
42. Shao, X.; Fan, J. Past Climate on West Sichuan Plateau as Reconstructed from Ring-Widths of Dragon Spruce. *Quat. Sci.* **1999**, *1*, 81–89.
43. Huang, R.; Zhu, H.; Liang, E.; Liu, B.; Shi, J.; Zhang, R.; Yuan, Y.; Grießinger, J. A Tree Ring-Based Winter Temperature Reconstruction for the Southeastern Tibetan Plateau since 1340 CE. *Clim. Dyn.* **2019**, *53*, 3221–3233. [CrossRef]
44. Wang, J.; Yang, B.; Ljungqvist, F.C. A Millennial Summer Temperature Reconstruction for the Eastern Tibetan Plateau from Tree-Ring Width. *J. Clim.* **2015**, *28*, 5289–5304. [CrossRef]
45. Li, J.; Shao, X.; Li, Y.; Qin, N. Annual Temperature Recorded in Tree-Ring from Songpan Region. *Chin. Sci. Bull.* **2014**, *59*, 1446–1458.
46. Wen, K.; Zhan, Z. *China Meteorological Calamity Grand Ceremony: Sichuan*; Meteorological Press: Beijing, China, 2005.
47. Xu, X.; Yi, C. Little Ice Age on the Tibetan Plateau and Its Bordering Mountains: Evidence from Moraine Chronologies. *Glob. Planet. Change* **2014**, *116*, 41–53. [CrossRef]
48. Xu, P.; Zhu, H.; Shao, X.; Yin, Z. Tree Ring-Dated Fluctuation History of Midui Glacier since the Little Ice Age in the Southeastern Tibetan Plateau. *Sci. China Earth Sci.* **2012**, *55*, 521–529. [CrossRef]
49. Li, Z.; He, Y.; Jia, W. Response of"Glaciers-Runoff"System in a Typical Temperate-Glacier, Hailuogou Glacier in Gongga Mountain of China to Global Change. *Sci. Geogr. Sin.* **2008**, *28*, 229.
50. Liu, N.; Bao, G.; Liu, Y.; Linderholm, H.W. Two Centuries-Long Streamflow Reconstruction Inferred from Tree Rings for the Middle Reaches of the Weihe River in Central China. *Forests* **2019**, *10*, 208. [CrossRef]
51. Su, M.; Wang, H. Relationship and Its Instability of ENSO—Chinese Variations in Droughts and Wet Spells. *Sci. China Ser. D* **2007**, *50*, 145–152. [CrossRef]
52. Yin, H.; Liu, H.; Linderholm, H.W.; Sun, Y. Tree Ring Density-Based Warm-Season Temperature Reconstruction since a.D. 1610 in the Eastern Tibetan Plateau. *Palaeogeogr. Palaeoclimatol. Palaeoecol.* **2015**, *426*, 112–120. [CrossRef]
53. Yin, H.; Li, M.-Y.; Huang, L. Summer Mean Temperature Reconstruction Based on Tree-Ring Density over the Past 440 Years on the Eastern Tibetan Plateau. *Quat. Int.* **2021**, *571*, 81–88. [CrossRef]

54. D'Arrigo, R.; Wilson, R. Spring Pacific Decadal Oscillation Index Reconstruction. In *IGBP PAGES/World Data Center for Paleoclimatology Data Contribution Series*; World Data Center for Paleoclimatology: Boulder, CO, USA, 2006; Volume 95.
55. Qin, J.; Tian, Y.; Ren, J.; Wan, Y. Interdecadal Oscillation of Pacific and Interdecadal Change of Summer Temperature in Yunnan. *Plateau Meteor* **2004**, *23*, 69–76.
56. Yang, Y.; Zhao, D.; Chen, H. Variability of Bio-Climatology Indicators in the Southwest China under Climate Warming during 1961–2015. *Int. J. Biometeorol.* **2019**, *63*, 107–119. [CrossRef]

forests

MDPI

Article

Hydrological Variability in the El Cielo Biosphere Reserve, Mexico: A Watershed-Scale Analysis Using Tree-Ring Records

José Villanueva-Díaz [1], Arian Correa-Díaz [2], Jesús Valentín Gutiérrez-García [2], Claudia C. Astudillo-Sánchez [3,*] and Aldo R. Martínez-Sifuentes [1]

1 Centro Nacional de Investigación Disciplinaria en Relación Agua, Suelo, Planta, Atmósfera CENID-RASPA, INIFAP, Gómez Palacio 35140, Durango, Mexico; villanueva.jose@inifap.gob.mx (J.V.-D.); martinez.rafael@inifap.gob.mx (A.R.M.-S.)
2 Centro Nacional de Investigación Disciplinaria en Conservación y Mejoramiento de Ecosistemas Forestales CENID-COMEF, INIFAP, Ciudad de México 04010, Ciudad de México, Mexico; correa.arian@inifap.gob.mx (A.C.-D.); gutierrez.jesus@inifap.gob.mx (J.V.G.-G.)
3 Facultad de Ingeniería y Ciencias, Universidad Autónoma de Tamaulipas, Ciudad Victoria 87149, Tamaulipas, Mexico
* Correspondence: ccastudillo@docentes.uat.edu.mx

Abstract: The El Cielo Biosphere Reserve (CBR) stands as a vital forested region in eastern Mexico due to its high biodiversity in flora and fauna and provision of environmental services. This study established a network of 10 ring-width chronologies of different species within the CBR and adjacent watersheds. The objective was to analyze their climatic response and reconstruct the seasonal streamflow contribution of each sub-basin to the main stream, utilizing data from a gauge network of eight hydrological stations located at strategic locations of the CBR. With chronologies ranging from 116 to 564 years, most exhibited association with the accumulated streamflow between January and June. Based on the adjusted R^2, Akaike Information Criteria, and Variance Inflation Factor, the stepwise regression procedure was selected among different statistical methods for developing the reconstruction model. In spite of differences in the seasonal reconstructed periods, all the species showed potential to develop hydrological reconstructions as indicated by their common response to streamflow variability, as occurred in the wet years of 1976, 1993, 2000, and 2008, and dry years of 1980, 1982, 1996, and 2011. It was found that the response of the chronologies to gauge records increased as a function of the chronologies' interseries correlation, average mean sensitivity, and distance of the tree-ring series to the gauge station. Streamflow reconstructions at the sub-basin level allowed a better understanding of the hydroclimatic variability characterizing the CBR, but also suggested the need to increase the network of chronologies for some particular sub-basins lacking tree-ring series to improve the reconstructed models.

Keywords: biodiversity; droughts; stepwise model; streamflow reconstruction; tree rings; watershed

1. Introduction

Hydrological resources are some of the most essential environmental services provided by forest ecosystems [1]; however, in the last few decades, an increase in deforestation, overgrazing, logging, and other land-use changes, along with climate warming, has disrupted the hydrological cycle in most of the ecosystems worldwide [2,3]. These changes affect the amount and distribution of rainfall, increasing evapotranspiration rates and decreasing water quality for different human uses [4,5].

The El Cielo Biosphere Reserve (CBR) is northeastern Mexico's most important protected area. It was created in 1985 by the State government of Tamaulipas, Mexico, and recognized as a Biosphere Reserve by the United Nations in 1987 [6]. The CBR covers an area of 144,530 ha, and it is recognized for its high biodiversity in flora and fauna richness favored by the interaction of Neotropical and Nearctic ecosystems [7].

The importance of this protected area involves different ecological and socioeconomic issues, given that rural communities settled on the reserve made their way of living based on ecotourism services. However, ecotourism could have negative impacts on the reserve's biodiversity if the design of interfaces that promote knowledge exchange among different sectors, systems, and stakeholders is not considered [8].

The CBR has been the focus of a fair amount of ecological research studies, mainly biodiversity [9,10]; however, no studies are available to analyze the impact of climate variability on the water resources yielded in this watershed that could explain much of the biodiversity of the area and the use of water for productive activities.

At the same time, a long-term understanding of climate variability in the CBR is constrained by the availability of climatic and gauge records that are scarce for northeastern Mexico, short in length (usually <50 years), poor quality, and representativeness. For example, only one gauge station, located within the CBR limits, has records from the middle of the 20th century. An alternative approach to obtain climate and hydrological information back in time is the use of proxy records, where tree rings constitute one of the most suitable sources of information given their annual resolution and the potential to develop centuries-long time series that are helpful in reconstructing climate and streamflow volumes and determining their interannual and long-term variability, as well as potential trends [11].

Streamflow reconstructions from tree rings are essential to analyze periods larger than the available gauge records, to evaluate the amount of flow produced at a watershed level, and to determine the frequency and trends of extreme hydroclimatic events, the influence of major circulatory modes on climate variability, and the impact they have on the proper planning and management of available water resources, mainly in dry regions [12,13].

Long-term information on water availability helps in designing strategies for proper ecosystem management, preserving biodiversity, and developing conservation strategies for the forested areas with the highest water yield, mainly where downstream settlements depend on the water produced at the watershed headwaters.

The objective of this study was to reconstruct the streamflow variability of the El Cielo Biosphere Reserve at the basin level using a set of tree-ring-width chronologies. For this, we did not restrict our focus to the boundaries of the CBR, but instead included information from all sub-basins that, at some point, influenced the hydrological behavior of the CBR. Finally, the variables we examined were distance to the gauge station, tree species, and statistic quality of the tree-ring series, to determine their relationship with the strength of the correlation between streamflow and ring-width indices. The relationship between tree rings and streamflow is essential to analyze their interannual variability and the potential trends as affected by major atmospheric circulatory modes and climate change.

2. Materials and Methods

2.1. Site Description

The El Cielo Biosphere Reserve (CBR) is part of the Sierra Madre Oriental (Eastern Sierra Madre), located in the southeastern portion of the State of Tamaulipas, in the municipalities of Gomez Farias, Llera, Jaumave, and Ocampo in northeastern Mexico. Geographically, it encompasses the extreme coordinates from 22.925° to 23.431° N and −99.097° to −99.441° W, at an elevation that ranges from 200 to 2320 m a.s.l. (Figure 1).

Geologically, limestone rocks of sedimentary origin dominate the region and have formed a karstic topography with rocky outcrops, sinkholes, and caves. Dominant soils are lithosols, morphologically characterized by their poor development, low fertility, and limited water-holding capacity [14]. Most rainfall infiltrated appears at lower elevations as springs, representing the main water source for the Rio Sabinas and Rio Frio.

The CBR is part of a larger basin known as "The Bajo Rio Panuco" (Lower Panuco River), Hydrological Region 26 (RH-26), that covers an area of 15,256.5 km^2, where the main stream Rio Panuco (Panuco River) drains a mean annual volume of circa 200,000 × 106 m^3 into the Gulf of Mexico. The CBR encompasses four sub-basins: Guayalejo, Sabinas,

Comandante, and Drenaje Subterraneo, covering an area ranging from 160 to 5519 km^2. Notably, the CBR is characterized by its significant hydrological potential, contributing 72% of the water flow in the basin. This substantial water input originates from the Sabinas, Frio, and Comandante rivers and a portion of the Guayalejo River, the primary collector for the entire basin.

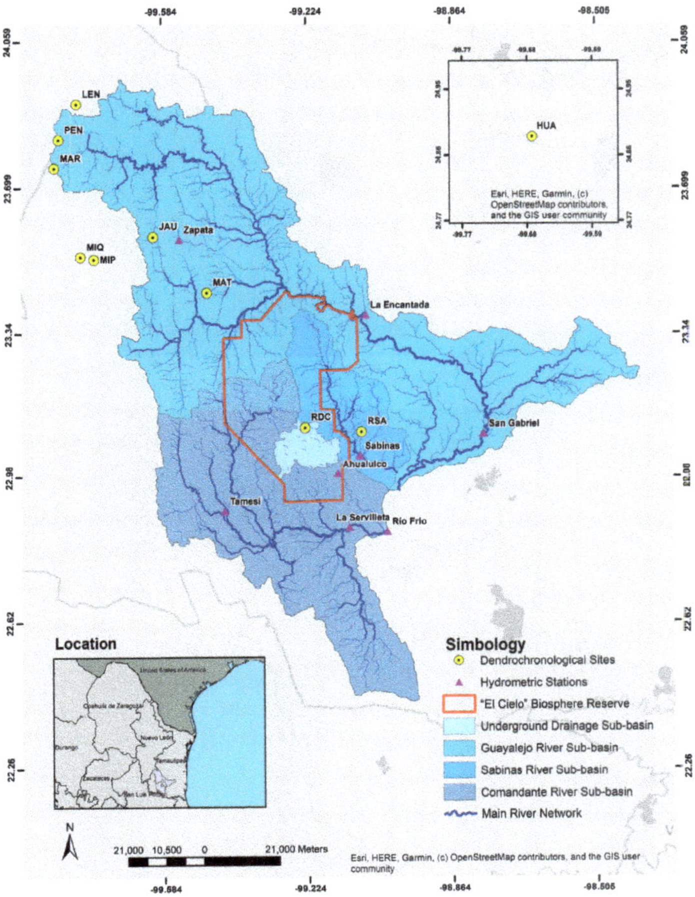

Figure 1. Localization of the El Cielo Biosphere Reserve (CBR) and the sub-basins that converge in the zone.

The climate of the reserve has different variations according to the elevation, with a dominance of warm–subhumid conditions at elevations from 300 to 800 m, where the lowest temperature for the coldest month is >18 °C with an annual precipitation range from 1500 to 2000 mm. Elevations between 700 and 2000 m dominate a semi-warm humid climate with a temperature of the coldest month between 15 and 18 °C and annual precipitation between 1000 and 1500 mm [15]. Most of the rainfall (55 to 65% of the annual total) occurs in the summer months (June to September), with the influence of the Bermuda–Azores (a large subtropical semi-permanent center of high atmospheric pressure favoring tropical cyclogenesis) [16]. In contrast, the rest of the precipitation occurs in the winter months, influenced by the entrance of cold fronts [17].

The variability in environmental conditions, topography, and elevation characterizing the reserve favors the presence of over ten vegetation types [18]; among them are tropical deciduous and semi-deciduous forest, cloud forest, mixed-conifer forest, oak–pine forest, desert shrublands, grasslands, and aquatic and subaquatic vegetation, where riparian vegetation is included [7].

2.2. Tree-Ring Chronologies

We employed ten dendrochronological sites distributed within and outside the CBR limits, considering the sub-basin distribution (Figure 1). Accordingly, we used tree-ring measurements of Montezuma bald cypress (*Taxodium mucronatum* Ten., HUA and RSA sites), vejar fir (*Abies vejarii* Martínez, LEN site), nut pine (*Pinus cembroides* Zucc., MIP and JAU sites), Montezuma pine (*Pinus montezumae* Lamb., RDC site), nelson pine (*Pinus nelsonii* Shaw, MIQ site), weeping pinyon pine (*Pinus pinceana* Gordon, MAT site), Mexican mountain pine (*Pinus hartwegii* Lindl., MAR site), and Douglas fir (*Pseudotsuga menziesii* Mirb. Franco, PEN site) (Figure 1). Most of our sites were developed by the authors, except for one being an updated version of the ring-width database developed by Stahle and collaborators for the Rio Sabinas [19] https://www.ncei.noaa.gov/access/paleo-search/study/4935 (accessed on 16 March 2024). To update this site, we sampled 66 bald cypress trees located along the Rio Sabinas stream banks (23.148° N, −99.15° W; 317 m a.s.l.). At all sites, we collected two to three increment cores per tree with a Pressler increment borer, in the opposite direction to the stem and avoiding sampling rotten wood.

We processed the cores from the rest of the sites following standard dendrochronological techniques [20]. The dating of the raw measurement file was conducted with the COFECHA program [21], and for detrending purposes, we used dplR software version 3.4.3 [22,23]. A ring-width index (RWI) was calculated by dividing the raw measurements with a fitted value according to a smoothing spline function with a frequency response of 0.50 at a wavelength of 0.67 series length. To build a site chronology, we averaged the RWI using Tukey's biweight robust mean [11]. Finally, to analyze the quality of the chronologies, we used the standard version of the chronology and considered different parameters such as series intercorrelation, average mean sensitivity, mean interseries correlation (Rbar) [24], and Expressed Population Signal (EPS) [25].

2.3. Streamflow Data and Imputation Methods

Gauge data were acquired through the National Water Commission (CONAGUA) hydrometric database, which provided access to historical records from gauge stations located in the sub-basins of interest [26]. We compiled data from eight gauge stations in the studied area (Table 1).

Table 1. Gauge stations located within or near the El Cielo Biosphere Reserve (CBR). The stations are arranged from low to high average streamflow.

No.	Gauge Identification	Sub-Basin	Information PERIOD	Latitude (N)	Longitude (W)
1	Zapata	Rio Guayalejo	1987–2014	23.2714°	99.5417°
2	Servilleta	Rio Comandante	1960–2014	22.8536°	99.1211°
3	Ahualulco	Rio Comandante	1945–2014	22.9869°	99.1489°
4	La Encantada	Rio Guayalejo	1960–2014	23.3855°	99.0791°
5	Sabinas	Rio Sabinas	1960–2014	23.0331°	99.0942°
6	Frio	Rio Tamesi	1960–2013	22.8431°	99.0261°
7	San Gabriel	Rio Guayalejo	1984–2014	23.0853°	98.7878°
8	Tamesi	Rio Comandante	1973–2014	22.8942°	99.4319°

The La Encantada, Sabinas, and La Servilleta gauge stations showed no missed data during the recorded periods. The Rio Frio and San Gabriel stations had less than 1% and 4% data gaps, respectively. The Tamesi station exhibited an 11% data gap for 2006–2010, while the Zapata station had a 15% absence in daily records for 1995–1998 and 2010–2013. To address this issue, we employed the imputTS package from R to impute missing data. This package offers several imputation methods suitable for univariate time series. The "na.interpolation" method with the "stine" option was selected due to its effectiveness in adjusting the curvature of the interpolation based on the slope of adjacent points, thereby providing more accurate estimations than simpler methods. Theoretical details and the implementation of this method are explained in the imputeTS manual [27].

2.4. Association between Ring-Width Chronologies and Streamflow Data and Reconstruction Models

The ring-width chronologies of the CBR were compared against accumulated monthly streamflow records from September of the previous year to October of the current year of growth. This comparison was conducted using a bootstrap Pearson correlation with the *treeclim* package in R [28], where we correlated all tree-ring sites with gauge records to determine the optimal period of association between variables.

We accumulated streamflow records of consecutive months with a significant association, and then compared them to the ring-width chronology to develop a seasonal reconstruction model for each gauge station. After identifying the optimal period for each station, we tested single dendrochronological sites, the first component of all dendrochronological sites (PC1), and stepwise regressions using site-level information for runoff reconstruction. We selected the best model according to the adjusted R^2, Akaike Information Criteria (AIC), and a Variance Inflation Factor (VIF) lower than ten, when appropriate. Residuals for all models were inspected for normality, trends, and autocorrelation. Afterwards, the gauged streamflow records were divided into two subperiods, one for calibration and one for verification, and if reconstruction statistics were adequate (reduction of error—RE; sign test—ST; correlation coefficient—r), the whole period of recorded data was used to build the final model.

For the final model, we followed the approach described by [13], using a set of 1000 bootstrap regressions. The bootstrap approach allows more accurate information to be obtained among predictors and is particularly useful when small databases are used and do not come from a normal distribution. In this case, we used bootstrapping with resampling, which results in multiple replicates of the original data [29]. This process was performed for each gauge station.

2.5. Influence of Explanatory Variables in the Strength Signal of Tree Rings and Streamflow Data

To assess the impact of various explanatory variables on the strength of the correlation between tree rings and streamflow data, we performed an analysis of covariance (ANCOVA) in R, version 4.1.0 [22]. This analysis considered categorical variables (tree species and land use) as well as continuous variables (distance between tree-ring sites and gauge stations, elevation, stream order, intercorrelation, and sensitivity).

To explore the common variability and detect periods of low and high streamflow at each gauge station, yearly records for both observed (annual cumulative runoff) and reconstructed periods were transformed into z-scores. This was conducted by subtracting the mean value from each record and dividing it by the standard deviation, allowing for comparison among gauge stations even if the streamflow differed. Furthermore, we used a principal component analysis to determine a common behavior among gauge reconstruction runoffs, using the common period from 1902 to 1995.

3. Results

3.1. Length and Quality of the Ring-Width Chronologies

The dendrochronological series spanned from 1450 to 2021, covering a 572-year period. The PEN site had the longest record (564 years, *Pseudotsuga menziesii*), while JAU had the shortest (116 years, *Pinus cembroides*) (Figure 2). All sites exhibited high interseries correlation values, ranging from 0.42 to 0.75, exceeding the minimum 0.328 threshold to be considered properly dated [21]. Additionally, all sites had high rbar and EPS values, indicating suitable characteristics of the tree-ring data. Spatially, only two sites were located at lower elevations, near the bottom of the sub-basins, while the rest were at higher elevations (Table 2).

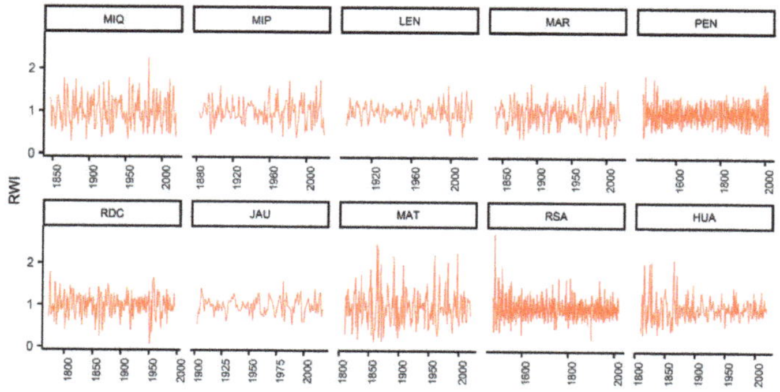

Figure 2. Ring-width index (RWI) chronologies for sampling sites within the influence area of the El Cielo Biosphere Reserve (CBR). For a full site description, see Table 2.

Table 2. Dendrochronological sites within the influence area of the El Cielo Biosphere Reserve (CBR). Sites are listed from the highest to the lowest intercorrelation.

Site	Species	Elevation m a.s.l.	Period	n (cores)	Series Intercorrelation	Average Mean Sensitivity	rbar	EPS
MIQ	*Pinus nelsonii*	1950	1846–2019	87	0.75	0.50	0.59	0.99
MIP	*Pinus cembroides*	2080	1882–2019	114	0.72	0.42	0.53	0.99
LEN	*Abies vejarii*	2700	1894–2021	295	0.69	0.34	0.51	0.99
MAR	*Pinus hartwegii*	3300	1838–2019	90	0.68	0.41	0.50	0.99
PEN	*Pseudotsuga menziesii*	2700	1450–2013	108	0.66	0.31	0.48	0.99
RDC	*Pinus montezumae*	1800	1772–1995	27	0.61	0.43	0.42	0.95
JAU	*Pinus cembroides*	1760	1902–2017	69	0.59	0.33	0.37	0.98
MAT	*Pinus pinceana*	1070	1810–2019	52	0.57	0.56	0.35	0.96
RSA	*Taxodium mucronatum*	317	1474–2017	118	0.44	0.44	0.19	0.97
HUA	*Taxodium mucronatum*	400	1810–2019	68	0.42	0.39	0.19	0.94

3.2. *Daily and Monthly Runoff for the Gauge Stations*

The gauge stations Gabriel and Tamesi recorded the highest streamflow (m^3 day^{-1}), while Zapata and Servilleta reported the lowest. Despite this, the region experienced common extreme events, such as a notorious peak observed in several stations in 1976 (Figure 3). Most sites begin their highest runoff season in June, at the onset of the rainy season, concluding by October, except for Zapata, which extends until November. We detected critical years with low annual runoff in 1973, 1993, 2000, and 2014. In contrast, years with high annual runoff were identified in 1974, 1991, and 1992.

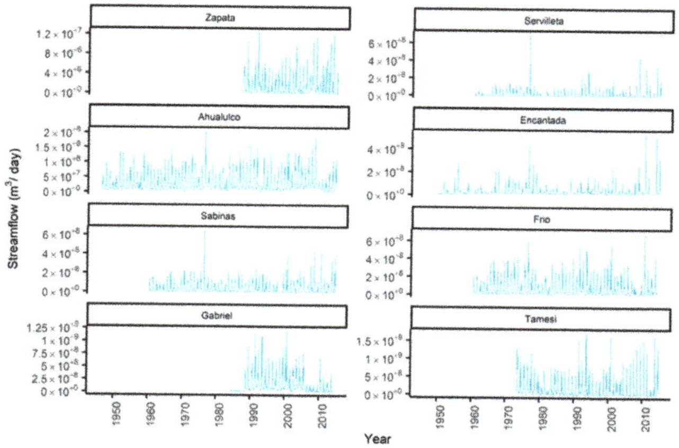

Figure 3. Accumulated daily streamflow from gauge stations located within or near the El Cielo Biosphere Reserve (CBR). The stations are arranged from low to high runoff volume.

3.3. Response of the Chronology to Monthly Streamflow Records

A comparison between the ring-width indices and monthly streamflow from the previous September to the current October indicates significant correlations ($p < 0.05$) across all gauge stations (Figure 4). However, some dendrochronological sites were less correlated with the gauge data. For instance, the Zapata station exhibited the highest correlation values (up to 0.62 in April, MIQ site), followed by Gabriel (r = 0.55 in May, HUA site) and Sabinas (r = 0.54 in April, PEN site). Generally, the April to June period had better correlations for most of the tree-ring sites. However, specific sites were also correlated with previous conditions (previous November to current May, LEN, MAR, and PEN with the Ahualulco station). Interestingly, the sign of the association changes from a positive correlation to a negative one starting in August, although it is not statistically significant.

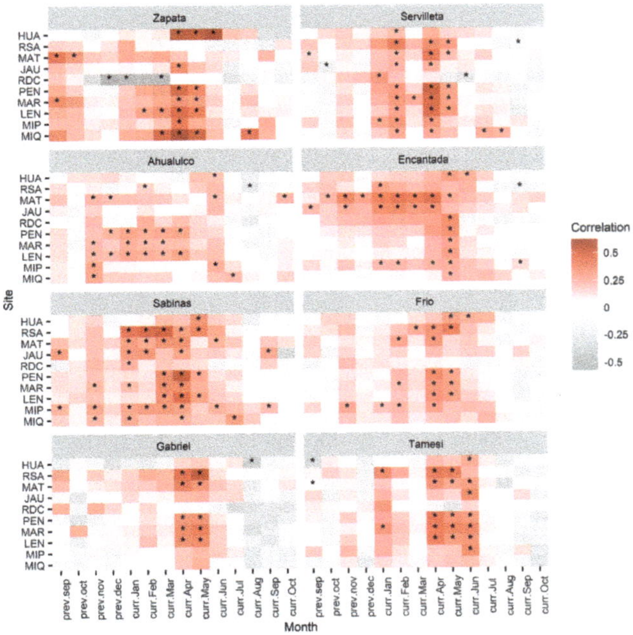

Figure 4. Heatmap of temporal correlations between tree-ring data and gauge stations. The analysis included data from the previous September to the current October. Correlation values with * are significant ($p < 0.05$). Dendrochronological sites are listed from the highest to the lowest intercorrelation. Site names can be found in Table 2.

3.4. Development of the Streamflow Models for Reconstruction Purposes

Based on the information from Figure 4, we selected the best correlated period at each gauge station for reconstruction purposes. Consequently, we chose to accumulate runoff from January to June for the Zapata station, February to April for Servilleta, January to April for Ahualulco, January to May for Encantada, January to June for Sabinas, February to June for Frio, April to May for Gabriel, and April to June for Tamesi.

After testing all models, they demonstrated statistical superiority over each station's pooled version, principal component model, or single version (Table 3). Additionally, these models exhibited homoscedasticity, normally distributed residuals, and VIF < 10 (Table S1 in Supplementary Materials). As mentioned earlier, some gauge stations were more suitable for reconstruction purposes, as indicated by high R^2 and $R^2_{adjusted}$ values (up to 0.93 and 0.85, respectively, for the Zapata station) and the lowest AIC. In contrast, Ahualulco exhibited the lowest performance, with R^2 and $R^2_{adjusted}$ values of 0.28 and 0.22, respectively (Table 3).

Table 3. Statistical comparison among different models explaining the runoff within or near the El Cielo Biosphere Reserve (CBR). Note that some statistics are given in intervals for single tree species because they represent values across sites. The symbol (+) indicates the best model for each gauge station based on R^2 and AIC.

Gauge Station	Period	Model	Variables	*p*-Value	R^2	$R^2_{adjusted}$	AIC
Zapata	January to June	Single tree species	HUA to MIQ	[0.02, 0.96]	[0.02, 0.18]	[0.01, 0.14]	[940.2, 945.6]
		Principal component model	PC1	0.54	0.02	0.01	945.6
		Pooled model	HUA to MIQ	0.45	0.94	0.56	327.7
		+Stepwise model	HUA, LEN, JAU, RSA	0.02	0.93	0.85	109.2
Servilleta	February to April	Single tree species	HUA to MIQ	[0.001, 0.44]	[0.02, 0.23]	[0.01, 0.22]	[1352.5, 1362.2]
		Principal component model	PC1	0.007	0.14	0.13	1355.6
		Pooled model	HUA to MIQ	0.08	0.48	0.25	852.5
		+Stepwise model	JAU, MAR, RSA	0.002	0.43	0.36	831.4
Ahualulco	January to April	Single tree species	HUA to MIQ	[0.01, 0.96]	[0.02, 0.13]	[0.01, 0.12]	[2468.3, 2478.3]
		Principal component model	PC1	0.05	0.05	0.04	2474.4
		Pooled model	HUA to MIQ	0.08	0.30	0.15	1792.5
		+Stepwise model	HUA, MAR, PEN, RSA	<0.001	0.28	0.22	1778.9
Encantada	January to May	Single tree species	HUA to MIQ	[0.001, 0.20]	[0.02, 0.23]	[0.02, 0.22]	[2435, 2450.4]
		Principal component model	PC1	0.002	0.14	0.12	2442.3
		Pooled model	HUA to MIQ	<0.001	0.51	0.39	1731.7
		+Stepwise model	JAU, MAR, MAT, PEN, MIQ, RSA	<0.001	0.51	0.43	1722.4
Sabinas	January to June	Single tree species	HUA to MIQ	[0.001, 0.96]	[0.01, 0.21]	[0.01, 0.20]	[2160.5, 2164.1]
		Principal component model	PC1	0.008	0.12	0.11	2160
		Pooled model	HUA to MIQ	0.003	0.60	0.44	1407.3
		+Stepwise model	HUA, MAR, MAT, MIP, MIQ, RSA, RDC	<0.001	0.56	0.45	1398.2
Frio	February to June	Single tree species	HUA to MIQ	[0.001,0.66]	[0.03, 0.13]	[0.02, 0.11]	[2207.6, 2212.4]
		Principal component model	PC1	0.036	0.08	0.06	2209.7
		Pooled model	HUA to MIQ	<0.01	0.55	0.39	1450.2
		+Stepwise model	HUA, MAT, MIP, MIQ, RSA, RDC	<0.001	0.53	0.44	1439.9
Gabriel	April to May	Single tree species	HUA to MIQ	[0.001,0.96]	[0.02,0.28]	[0.01,0.27]	[1223.4, 1229.4]
		Principal component model	PC1	0.03	0.15	0.12	1225.4
		Pooled model	HUA to MIQ	0.08	0.47	0.24	1239.9
		+Stepwise model	JAU, MAR, RSA	0.003	0.40	0.33	1220.1
Tamesi	April to June	Single tree species	HUA to MIQ	[0.02, 0.58]	[0.01, 0.17]	[0.01, 0.14]	[1353.7, 1358.6]
		Principal component model	PC1	0.03	0.14	0.11	1354.8
		Pooled model	HUA to MIQ	0.33	0.47	0.11	1003.6
		+Stepwise model	JAU, LEN, MAR, MIP	0.05	0.39	0.25	978.2

All stepwise models exhibited satisfactory reconstruction skills, as indicated by positive values for RE, ST, and R^2 during both the calibration and verification phases. For instance, the Pearson correlation coefficient ranged from 0.46 (Ahualulco) to 0.86 (Sabinas). Additionally, the RE ranged from 0.211 to 0.742, indicating that the performance exceeds that of simply using the mean of the records (Table S2 in Supplementary Materials). Thus, we used the whole period at each gauge station to conduct a set of 1000 replications. The dependent variable was the runoff in the selected period, ranging from January to June for the more extended season (Zapata and Sabinas) to the shorter period from April to May (Gabriel). The independent variables included the sites and, consequently, the species-level chronologies. The statistics reveal that tree-ring information explained the variance in the runoff in CBR, ranging from 21% to 64% (see Table 4).

Our results demonstrate that, irrespective of high or low runoff volumes, all gauge stations were correlated to tree-ring information, enabling the reconstruction of runoff events beyond the available records. In some cases, specific sites exhibited the capability to detect both extremely high runoff and years with low runoff, as exemplified by the Servilleta and Gabriel gauge stations (Figure 5b,g). Although some tree-ring chronologies spanned centuries (Table 2), merging them with shorter yet highly sensitive sites resulted in the oldest reconstructed runoff dating back to the mid-19th century, while the shortest spanned from the beginning of the 20th century onwards (see Figure 5). Furthermore, in specific instances, a lack of tree-ring information was observed despite the availability of gauge data, highlighting the importance of updated information, particularly in Sabinas and Frio (see Figure 5e,f).

Table 4. Bootstrap performance of selected periods at each gauge station. The statistics represent the median for 1000 replicates. See Table 2 for a complete reference of site names.

St	Pe	Final Model	R^2	$R^2_{adjusted}$	AIC
Za [1]	J-J	$Y_i = -2.97 \times 10^8 + 1.46 \times 10^7(HUA) + 3.97 \times 10^6(JAU) - 6.48 \times 10^6(LEN) - 4.01 \times 10^6(RSA)$	0.39	0.28	934.4
Se [2]	F-A	$Y_i = -3.84 \times 10^5 - 2.78 \times 10^5(JAU) + 3.32 \times 10^5(MAR) + 5.85 \times 10^5(RSA)$	0.36	0.32	1341.6
Ah [3]	J-A	$Y_i = 2.88 \times 10^7 - 1.64 \times 10^7(HUA) - 8.70 \times 10^6(MAR) + 3.51 \times 10^7(PEN) + 4.50 \times 10^6(RSA)$	0.21	0.16	2465.1
En [4]	J-M	$Y_i = 7.10 \times 10^6 + 1.85 \times 10^7(JAU) - 2.93 \times 10^7(MAR) + 3.44 \times 10^7(MAT) - 3.13 \times 10^7(MIQ) + 4.94 \times 10^7(PEN) + 3.30 \times 10^7(RSA)$	0.43	0.37	2421.2
Sa [5]	J-J	$Y_i = -1.44 \times 10^8 + 1.74 \times 10^8(HUA) - 7.18 \times 10^7(MAR) + 5.52 \times 10^7(MAT) + 9.08 \times 10^7(MIP) - 5.85 \times 10^7(MIQ)$ $-1.05 \times 10^8(PEN) + 1.72 \times 10^8(RSA)$	0.64	0.55	1386.9
Fr [6]	F-J	$Y_i = -1.68 \times 10^8 + 3.69 \times 10^8(HUA) + 1.05 \times 10^8(MAT) + 1.51 \times 10^8(MIP) - 1.48 \times 10^8(MIQ) - 2.60 \times 10^8(RDC) + 1.54 \times 10^8(RSA)$	0.60	0.52	1430.1
Ga [7]	A-M	$Y_i = -1.25 \times 10^8 - 1.29 \times 10^8(JAU) + 1.14 \times 10^8(MAR) + 2.50 \times 10^8(RSA)$	0.41	0.35	1213.9
Ta [8]	A-J	$Y_i = 2.45 \times 10^8 - 1.06 \times 10^9(JAU) - 6.83 \times 10^8(LEN) + 1.06 \times 10^9(MAR) + 7.84 \times 10^8(MIP)$	0.34	0.25	1350.3

St: station; Pe: period; [1]: Zapata (January–June); [2]: Servilleta (February–April); [3]: Ahualulco (January–April); [4]: Encantada (January–May); [5]: Sabinas (January–June); [6]: Frio (February–June); [7]: Gabriel (April–May); [8]: Tamesi (April–June).

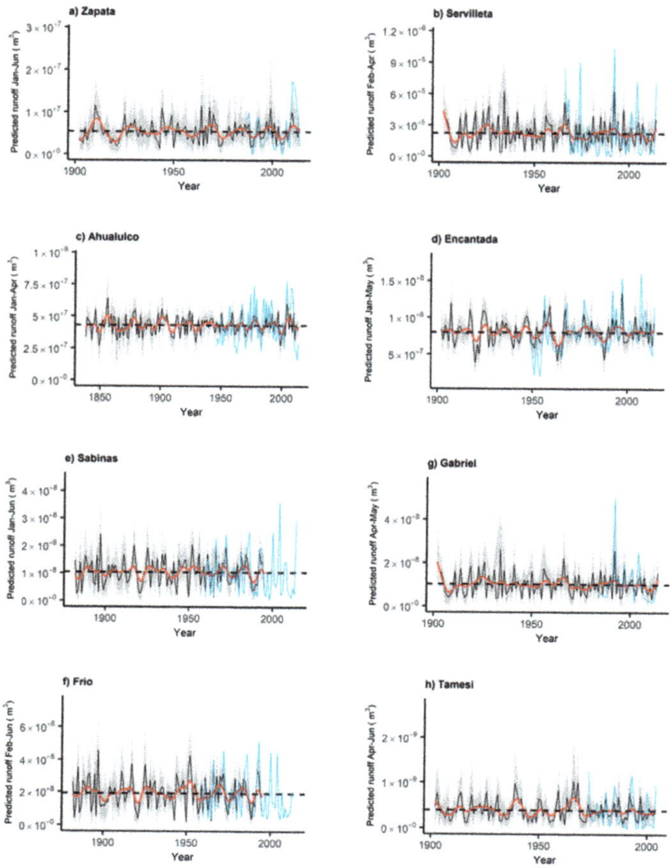

Figure 5. Seasonal reconstruction of runoff within the influence zone of the El Cielo Biosphere Reserve (CBR). The black line represents the mean, while the thin gray lines represent each of the 1000 bootstrapped reconstructions. The black dashed horizontal line corresponds to the average historical runoff, while the blue line represents gauge records. Additionally, the red line depicts a decadal spline, emphasizing low-frequency events such as wet and dry spells. Subfigures (**a**–**h**) represent the reconstructed seasonal streamflow contribution at each gauge station.

3.5. The Common Variability within the El Cielo Biosphere Reserve

Overall, all gauge stations exhibited a common variability in their behavior when considering the annual cumulative runoff. Consequently, years with high runoff volumes, such as 1976, 1993, 2000, and 2008, were observed. Similarly, years with low runoff were concurrent in 1980, 1982, 1996, and 2011 (see Figure 6a). However, it is important to note that in recent years, lower synchrony has been observed, particularly with some stations like Zapata.

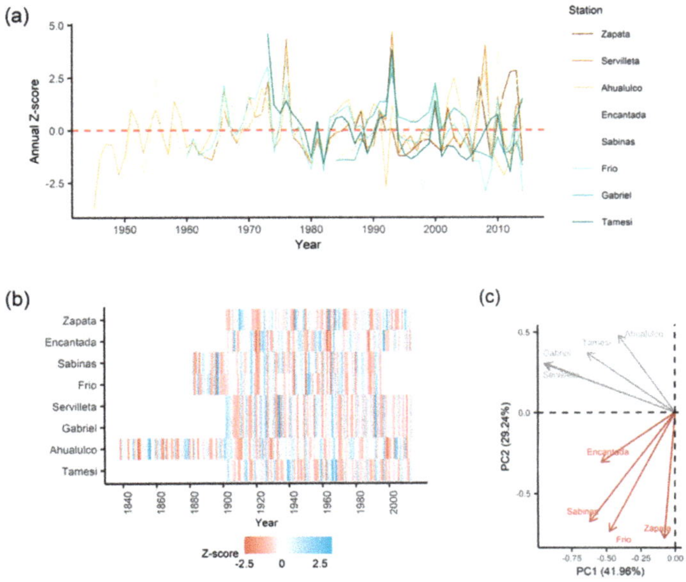

Figure 6. Reconstructions common variability within the El Cielo Biosphere Reserve. (**a**) Annual cumulative runoff registered at each gauge station represented as z-scores. (**b**) Heatmap of z-scores of seasonal reconstructions of runoff. Note that for (**a**), the runoff corresponds to the accumulated values from January to December, while for (**b**), it represents different months (See Table 4). (**c**) Principal component analysis using the seasonal reconstructed runoff data.

The common variability in their reconstructed versions showed mixed results, with few common periods concurrent among all gauge stations. For instance, the period 1907–1912 was observed in almost all stations except for Zapata, Frio, and Sabinas. Similar results were observed around 1920, 1940, and 1975. This apparent mismatch may originate from the fact that the reconstructions represent different periods of the year (e.g., months). For example, while Sabinas represents the accumulated runoff for six months, the Gabriel station represents only two months (Figure 6b). This difference was mirrored in two different groups according to a principal component analysis, with the first group involving Ahualulco, Tamesi, Gabriel, and Servilleta, and the second group comprising Encantada, Sabinas, Frio, and Zapata (See Figure 6c). However, the higher synchrony at the yearly level implies a common influence of climatic drivers at the watershed scale, such as those related to ocean–atmospheric modes.

3.6. The Effect of Species Composition on the Strength of Correlations

The variables influencing the strength of the correlation between tree-ring series and gauge records involved the tree species, series intercorrelation, average mean sensitivity, and the distance to the gauge stations (Table 5). Notably, *Taxodium mucronatum*, *Pseudotsuga menziesii*, and *Pinus hartwegii* exhibited higher correlations, whereas *Pinus montezumae* and *Pinus nelsonii* showed the lowest correlations. Regarding dendrochronological statistics, higher intercorrelation values—indicating similar climatic responses among trees within a site—and average mean sensitivity were associated with statistically significant correlations with runoff data. Similarly, a shorter distance between tree-ring data and gauge stations resulted in a higher correlation, highlighting the importance of having tree-ring information close to each one of the gauge stations. Notably, the amount of runoff, which can be indirectly related to the stream order, does not influence the strength of the correlation. Therefore, any station, regardless of its position in the watershed (highlands or outlets), has the potential to be related to the tree-ring records.

Table 5. Analysis of covariance (ANCOVA) to detect variables influencing the correlation between tree-ring series and gauge data.

Variable	F-Value	*p*-Value
Tree species	8.33	<0.001
Elevation	0.75	0.387
Series intercorrelation	8.95	0.002
Average mean sensitivity	25.78	<0.001
Stream order	0.01	0.905
Distance to gauge station	52.56	<0.001

4. Discussion

4.1. The Importance of Dendrochronological Networks for Hydrometric Reconstructions

This study has explored runoff variability in the CBR and adjacent sub-basins using tree-ring chronologies (Table 2). Our findings reveal significant patterns of hydrological variability, reflecting the complex interaction between climate and water resources in this region. Despite the spatial location of the gauge stations within the sub-basins, all were capable of being related to tree-ring records, albeit reflecting different seasons throughout the year. While some stations showed correlations for six months (e.g., the Zapata and Sabina stations), others only reflected a limited period over the year, such as Gabriel station, but with a strong correlation with the tree-ring data (Figure 4).

Research in dendrohydrology has been distinguished by implementing increasingly extensive networks of moisture-sensitive trees [30,31]. These networks have been important in developing proxy records that reflect historical regional hydroclimate and interannual variability and trends. A notable example is a series of studies that have reconstructed the paleohydrology of the Colorado River [32–35]. These studies have enriched our understanding due to the addition of new tree-ring chronologies from various sites in the river basin. Moreover, applying innovative statistical methods to tree-ring data and hydroclimatic information has enabled more profound and detailed insights. Here, we demonstrate the importance of certain tree-ring series that are highly sensitive to runoff records, such as RSA (i.e., *Taxodium mucronatum*, Table 4), which must be regularly updated to enhance our capability to understand the hydrological variability of critical regions such as the CBR. Furthermore, our results also highlighted the significance of the distance of tree-ring data from the gauge location. Thus, selecting sensitive tree species closer to the gauge locations may ensure a higher chance of reconstructing the historical runoff of a sub-basin; nevertheless, this approach will be based on a statistical relationship between secondary growth and runoff excluding physiological processes, whose understanding may contribute to explaining the influence of internal factors that trigger the size of the annual ring [36].

In the context of Mexico, the studies conducted by Osorio et al. [37] have been fundamental in understanding climate dynamics over several centuries. These researchers identified the recurrent presence of droughts at the beginning and at the end of each century, from the 16th to the 21st, based on a historical reconstruction of precipitation in the CBR from tree rings of *Taxodium mucronatum*. Furthermore, they highlighted the occurrence of cyclical droughts approximately every 50 to 60 years. Other findings with moisture-sensitive species distributed in this region like *Pseudotsuga menziesii* and *Pinus cembroides* indicate a winter–spring (January–April or January–May) precipitation response [38,39], similar to the one detected with *Taxodium mucronatum* in the CBR, which implies that they have a greater response to the cool season precipitation extending from the previous October to the current May or June. When comparing these findings with our results, we observed a synchrony between the periods of low precipitation (Table S3 in Supplementary Materials) and the reduction in runoff at the same time intervals, i.e., 1870s, 1900s, 1930s, 1950s, 1980s, 2010s (Figure 5). This pattern reinforces the idea that variations in precipitation, particularly during winter and spring, directly and significantly impact runoff levels in the CBR, underscoring the importance of these studies for water resource management and environmental planning in the region.

With climate change, precipitation patterns are expected to become more erratic, potentially leading to greater variability in runoff [40]. This could mean more prolonged periods of drought interspersed with extreme precipitation events. The alteration in runoff patterns would have significant implications for water resource management [41]. A more erratic water supply could affect water availability for agriculture, human consumption, and biodiversity conservation, especially in regions that depend on natural sources like the CBR.

Changes in runoff will also influence aquatic and terrestrial ecosystems. Species that depend on a specific water regime could face elevated risks, potentially leading to changes in species abundance, distribution, and possibly local extinctions [40].

4.2. The Runoff Variability beyond the Hydrometric Records in the El Cielo Biosphere Reserve

The correlation between tree rings and gauge records reveals that other environmental factors may become limiting in years with higher moisture, as noted by [11]. This dynamic could explain why certain high-flow events are not fully recorded in the tree-ring series, particularly occurring during the rainy season, which coincides with the highest runoff season (Figure S1 in Supplementary Materials). It is probable that these may have occurred during periods when trees are physiologically inactive or in soils with a low water-holding capacity or low infiltration rate. When developing an annual streamflow reconstruction, tree growth is limited during the winter season, making it difficult to capture the entire streamflow behavior in the reconstruction [42,43]. However, our results showed that specific sites had the ability to detect extreme positive flow peaks, as well as reflect years of low runoff. We observed this behavior at stations like Servilleta and Gabriel, where the recording capability of tree-ring chronologies is notable and coincides with the highest association between runoff and precipitation (Figure 5b,g, Table S3 in Supplementary Materials). Nevertheless, it is clear that even with 1000 replications through the bootstrap procedure, there were years with overestimated/underestimated runoff levels, which can limit management strategies in the region. This could be addressed by establishing new sampling sites close to the gauge stations and using moisture-sensitive species such as *Taxodium mucronatum* or *Pseudotsuga menziesii*. Additionally, exploring different detrending methods, such as signal-free or Bayesian detrending, could maximize the climatic signal contained in tree rings [44].

4.3. Outlook in the El Cielo Biosphere Reserve

Our findings significantly impact the management of water resources in the El Cielo Biosphere Reserve. Understanding long-term hydrological variability can assist in water conservation planning and adaptation to future climate changes, especially considering the climate change scenarios in precipitation. Pichardo et al. [45] forecasted a decrease in annual precipitation in the Guayalejo River sub-basin, which could result in modifications to the water balance.

A potential limitation of our study could be the geographic representativeness of the tree-ring samples. Future research could expand the sampling network to include more sub-basins, tree species, and sampling sites near gauge stations, thus improving the accuracy of our reconstructions. Additionally, integrating more climatic and environmental variables into the flow reconstruction models, as Goeking and Tarboton [5] suggested, would allow for a better understanding of the factors influencing hydrological variability, including common climatic conditions (mainly precipitation and evapotranspiration), which control tree growth and streamflow reconstruction [46,47].

Consequently, dendrochronology is an instrumental methodology for reconstructing historical climatic and hydrological patterns, providing indispensable insights for current decision-making processes and future strategic planning. Its implementation within the CBR, as well as in analogous regions, could play a fundamental role in formulating water management strategies that are both more effective and sustainable, adapted to the unique challenges and environmental conditions inherent to each locality.

5. Conclusions

Most of the published papers on dendrohydrological reconstruction are based on a representative network of tree-ring chronologies for a whole basin, but in this study, we emphasize the importance of considering partial reconstructions at the sub-basin level, as independent contributors to the total streamflow of the basin. This finding is important when the tree-ring series developed from different species correspond to a common period of streamflow record, which in this case corresponded to April, May, and June.

The use of different approaches (stepwise, bootstrap) allowed us to find a proper reconstruction model explaining over 50% of the seasonal streamflow variability for some single sub-basins.

In order to improve the chronologies' response to streamflow records, most of them should be developed closer to the gauge stations, and these data should be complemented with hydrological and ecological information of the sub-basin for a better interpretation of the results.

Supplementary Materials: The following supporting information can be downloaded at: https://www.mdpi.com/article/10.3390/f15050826/s1, Table S1: Testing of statistical model assumptions of linear regression models; Table S2: Verification statistics in the calibration and verification periods within the El Cielo Biosphere Reserve (CBR). Table S3: Correlation coefficient between reconstructed runoff and seasonal precipitation. Figure S1: Monthly average runoff for the gauge stations.

Author Contributions: Conceptualization, J.V.-D., A.C.-D. and J.V.G.-G.; methodology, J.V.-D., A.C.-D., J.V.G.-G., C.C.A.-S. and A.R.M.-S.; software, A.C.-D., J.V.G.-G. and A.R.M.-S.; validation, J.V.-D., A.C.-D. and C.C.A.-S.; formal analysis, A.C.-D., J.V.G.-G., J.V.-D., A.R.M.-S. and C.C.A.-S.; investigation, J.V.-D., C.C.A.-S. and A.R.M.-S.; writing—original draft preparation, J.V.-D. and A.C.-D.; writing—review and editing, J.V.-D., A.C.-D., C.C.A.-S. and J.V.G.-G.; visualization, C.C.A.-S. and A.R.M.-S.; funding acquisition, J.V.-D. and C.C.A.-S. All authors have read and agreed to the published version of the manuscript.

Funding: This research was funded by the Sectoral Research Fund for Education No. 283134/CB 2016-1 of the Mexican Fund Agency of Science and Technology, project "Red dendrocronológica mexicana: aplicaciones hidroclimáticas y ecológicas (Mexican dendrocronological network for dendroclimatic and ecological applications).

Data Availability Statement: The data presented in this study are available upon request from the corresponding author.

Acknowledgments: We acknowledge Fátima del Rocío Reyes-Camarillo for dating support. Leroy Soria-Díaz and Esteban Berrones helped with fieldwork.

Conflicts of Interest: The authors declare no conflicts of interest.

References

1. Vose, J.M.; Martin, K.L.; Luce, C.H. Forests, Water, and Climate Change. U.S. Department of Agriculture, Forest Service, Climate Change Resource Center. 2017. Available online: https://www.climatehubs.usda.gov/sites/default/files/Forests-Water-Climate%20Change_CCRC.pdf (accessed on 16 March 2024).
2. Foley, J.A.; DeFries, R.; Asner, G.P.; Barford, C.; Bonan, G.; Carpenter, S.R.; Chapin, F.S.; Coe, M.T.; Daily, G.C.; Gibbs, H.K.; et al. Global Consequences of Land Use. *Science* **2005**, *309*, 570–574. [CrossRef] [PubMed]
3. Hlásny, T.; Kočický, D.; Maretta, M.; Sitková, Z.; Barka, I.; Konôpka, M.; Hlavatá, H. Effect of deforestation on watershed water balance: Hydrological modelling-based approach. *Cent. Eur. For. J.* **2015**, *61*, 89–100. [CrossRef]
4. Marshall, E.; Randhir, T. Effect of climate change on watershed system: A regional analysis. *Clim. Change* **2008**, *89*, 263–280. [CrossRef]
5. Goeking, S.A.; Tarboton, D.G. Forests and water yield: A synthesis of disturbance effects on streamflow and snowpack in western coniferous forests. *J. For.* **2020**, *118*, 172–192. [CrossRef]
6. UNESCO. El Cielo Biosphere Reserve, Mexico. In Biosphere Reserves in Latin America and the Caribbean, October 2018. Available online: https://en.unesco.org/biosphere/lac/el-cielo (accessed on 16 March 2024).
7. González-Medrano, F. La vegetación. In *Historia Natural de la Reserva de la Biosfera El Cielo, Tamaulipas, México*; Sánchez-Ramos, G., Reyes-Castillo, P., Dirzo, R., Eds.; Universidad Autónoma de Tamaulipas: Ciudad Victoria, Mexico, 2005; pp. 88–105. Available online: https://libros.uat.edu.mx/index.php/librosuat/catalog/view/25/15/236-1 (accessed on 16 March 2024).

8. Caballero-Rico, F.C.; Roque-Hernández, R.V.; de la Garza Cano, R.; Arvizu-Sánchez, E. Challenges for the Integrated Management of Priority Areas for Conservation in Tamaulipas, México. *Sustainability* **2022**, *14*, 494. [CrossRef]
9. Vargas-Contreras, J.A.; Hernández-Huerta, A. Distribución Altitudinal de la Mastofauna en la Reserva de la Biosfera "El Cielo", Tamaulipas, México. *Acta Zool. Mex.* **2001**, *82*, 83–109. Available online: https://www.redalyc.org/pdf/575/57508205.pdf (accessed on 16 March 2024). [CrossRef]
10. Sánchez-Ramos, G.; Reyes-Castillo, P.; Dirzo, R. *Historia Natural de la Reserva de la Biósfera El Cielo, Tamaulipas, México*; Universidad Autónoma de Tamaulipas: Tamaulipas, Mexico, 2005; ISBN 968-7662-67-0. Available online: https://libros.uat.edu.mx/index.php/librosuat/catalog/view/25/15/236-1 (accessed on 16 March 2024).
11. Fritts, H.C. *Tree Rings and Climate*; Academic Press: London, UK, 1976.
12. Villanueva-Díaz, J.; Estrada-Ávalos, J.; Martínez-Sifuentes, A.R.; Correa-Díaz, A.; Meko, D.M.; Castruita-Esparza, L.U.; Cerano-Paredes, J. Historic Variability of the Water Inflow to the Lazaro Cardenas Dam and Water Allocation in the Irrigation District 017, Comarca Lagunera, Mexico. *Forests* **2022**, *13*, 2057. [CrossRef]
13. Villanueva-Díaz, J.; Correa-Díaz, A.; Castruita-Esparza, L.U.; Gutiérrez-García, J.V.; Martínez-Sifuentes, A.R.; Reyes-Camarillo, F.d.R. Tree Rings as Proxies of Historical Runoff in a National Park in Northern Mexico: A Major Ecosystem Service Provider. *Atmosphere* **2023**, *14*, 1199. [CrossRef]
14. INEGI (Instituto Nacional de Estadística, Geografía e Informática). Conjunto Nacional de Datos Vectorial Edafológico. Escala 1:250 000 Serie II (Digital). 2000. Available online: https://www.inegi.org.mx/temas/edafologia/ (accessed on 15 March 2024).
15. García, E.; Sánchez-Santillán, N. Análisis climático de la Reserva de la Biosfera El Cielo. *Rev. Geofís.* **1996**, *45*, 181–199. Available online: https://biblat.unam.mx/es/revista/revista-geofisica/articulo/analisis-climatico-de-la-reserva-de-la-biosfera-el-cielo (accessed on 16 March 2024).
16. Tang, B.H.; Fang, J.; Bentley, A.; Kilroy, G.; Nakano, M.; Park, M.-S.; Rajasree, V.P.M.; Wang, Z.; Wing, A.A.; Wu, L. Recent advances in research on tropical cyclogenesis. *Trop. Cyclone Res. Rev.* **2020**, *9*, 87–105. [CrossRef]
17. Sánchez-Santillán, N.; Binnqüist, G.S.; Garduño, R. Sequía intraestival en La Reserva de la Biosfera El Cielo y su entorno, Tamaulipas, México. *Cuad. Geogr. Rev. Colomb. Geogr.* **2018**, *27*, 146–163. [CrossRef]
18. Rzedowski, J. *La Vegetación de México*; Limusa: Ciudad de México, Mexico, 1978.
19. Stahle, D.W.; Cleaveland, M.K.; Therrell, M.D.; Paull, G. Stahle—Rio Sabinas—TAMU—ITRDB MEXI035. NOAA Study Page. 2002. Available online: https://www.ncei.noaa.gov/access/paleo-search/study/4935 (accessed on 16 March 2024).
20. Stokes, M.A.; Smiley, T.L. *An Introduction to Tree-Ring Dating*; University of Chicago Press: Chicago, IL, USA, 1968.
21. Holmes, R.L. Computer-Assisted Quality Control in Tree-Ring Dating and Measurement. *Tree-Ring Bull.* **1983**, *43*, 69–78. Available online: https://repository.arizona.edu/handle/10150/261223 (accessed on 17 March 2024).
22. R Development Core Team. *R: A Language and Environment for Statistical Computing*; R Foundation for Statistical Computing: Vienna, Austria; Available online: http://www.R-project.org/ (accessed on 16 March 2024).
23. Bunn, A.; Korpela, M.; Biondi, F.; Campelo, F.; Mérian, P.; Qeadan, F.; Zang, C.; Buras, A.; Cecile, A.; Mudelsee, M.; et al. Dendrochronology Program Library in R. 2023. Available online: https://cran.r-project.org/web/packages/dplR/dplR.pdf (accessed on 16 March 2024).
24. Speer, J.H. *Fundamentals of Tree-Ring Research*; University of Arizona Press: Tucson, AZ, USA, 2010.
25. Wigley, T.M.; Briffa, K.R.; Jones, P.D. On the average value of correlated time series, with applications in dendroclimatology and hydrometeorology. *J. Appl. Meteorol. Climatol.* **1984**, *23*, 201–213. [CrossRef]
26. CONAGUA (Comisión Nacional del Agua, México). Portal de Sistemas de Información del Agua. *Sistema de Información Hidrológica. Estaciones Hidrométricas.* 2024. Available online: https://sih.conagua.gob.mx/hidros.html (accessed on 16 March 2024).
27. Moritz, S.; Gatscha, S.; Wang, E.; Hause, R. Package Imputets Version 3.3. 2022. Available online: https://cran.r-project.org/web/packages/imputeTS/imputeTS.pdf (accessed on 16 March 2024).
28. Zang, C.; Biondi, F. Treeclim: An R package for the numerical calibration of proxy-climate relationships. *Ecography* **2015**, *38*, 431–436. [CrossRef]
29. Faraway, J.J. *2016. Extending the Linear Model with R: Generalized Linear, Mixed Effects and Nonparametric Regression Models*, 2nd ed.; Chapman and Hall/CRC: New York, NY, USA, 2016. [CrossRef]
30. Martin, J.T.; Pederson, G.T. Streamflow reconstructions from tree rings and variability in drought and surface water supply for the Milk and St. Mary River basins. *Quat. Sci. Rev.* **2022**, *288*, 107574. [CrossRef]
31. Littell, J.S.; Pederson, G.T.; Martin, J.T.; Gray, S.T. Networks of tree-ring based streamflow reconstructions for the Pacific Northwest, U.S.A. *Water Resour. Res.* **2023**, *59*, e2023WR035255. [CrossRef]
32. Hidalgo, H.G.; Dracup, J.A.; MacDonald, G.M.; King, J.A. Comparison of tree species sensitivity to high and low extreme hydroclimatic events. *Phys. Geogr.* **2001**, *22*, 115–134. [CrossRef]
33. Woodhouse, C.A.; Lukas, J.J. Multi-century tree-ring reconstructions of Colorado streamflow for water resources planning. *Clim. Change* **2006**, *78*, 293–315. [CrossRef]
34. Woodhouse, C.A.; Gray, S.T.; Meko, D.M. Updated streamflow reconstructions for the Upper Colorado River Basin. *Water Resour. Res.* **2006**, *42*, W05415. [CrossRef]
35. Timilsena, J.; Piechota, T.C.; Hidalgo, H.; Tootle, G. Five hundred years of hydrological drought in the upper Colorado River basin. *J. Am. Water Resour. Assoc.* **2007**, *43*, 798–812. [CrossRef]

36. Garcia-Forner, N.; Vieira, J.; Nabais, C.; Carvalho, A.; Martínez-Vilalta, J.; Campelo, F. Climatic and physiological regulation of the biomodal xylem formation pattern in *Pinus pinaster* saplings. *Tree Physiol.* **2019**, *39*, 2008–2018. [CrossRef] [PubMed]
37. Osorio-Osorio, J.A.; Astudillo-Sánchez, C.C.; Villanueva-Díaz, J.; Soria-Díaz, L.; Vargas-Tristán, V. Reconstrucción histórica de la precipitación en la Reserva de la Biosfera El Cielo, México, mediante anillos de crecimiento en *Taxodium mucronatum* (Cupressaceae). *Rev. Biol. Trop.* **2020**, *68*, 818–832. [CrossRef]
38. Villanueva-Diaz, J.; Stahle, D.W.; Luckman, B.H.; Cerano-Paredes, J.; Therrell, M.D.; Cleaveland, M.K.; Cornejo-Oviedo, E. Winter-spring precipitation reconstructions from tree rings for northeast Mexico. *Clim. Change* **2007**, *83*, 117–131. [CrossRef]
39. Constante-García, V.; Villanueva, J.; Cerano, J.; Cornejo, E.H.; Valencia, S. Dendrocronología de *Pinus cembroides* Zucc. y reconstrucción de precipitación estacional para el sureste de Coahuila. *Rev. Mex. Cien. For.* **2009**, *34*, 17–39. Available online: https://cienciasforestales.inifap.gob.mx/index.php/forestales/article/view/685 (accessed on 17 March 2024).
40. IPCC. *2022: Climate Change 2022: Impacts, Adaptation and Vulnerability. Contribution of Working Group II to the Sixth Assessment Report of the Intergovernmental Panel on Climate Change*; Pörtner, H.-O., Roberts, D.C., Tignor, M., Poloczanska, E., Mintenbeck, K., Alegría, A., Craig, M., Langsdorf, S., Löschke, S., Möller, V., et al., Eds.; Cambridge University Press: Cambridge, UK; New York, NY, USA, 2022. [CrossRef]
41. Miller, O.L.; Putman, A.L.; Alder, J.; Miller, M.; Jones, D.K.; Wise, D.R. Changing climate drives future streamflow declines and challenges in meeting water demand across the southwestern United States. *J. Hydrol. X* **2021**, *11*, 100074. [CrossRef]
42. Tootle, G.; Oubeidillah, A.; Elliott, E.; Formetta, G.; Bezak, N. Streamflow Reconstructions Using Tree-Ring-Based Paleo Proxies for the Sava River Basin (Slovenia). *Hydrology* **2023**, *10*, 138. [CrossRef]
43. Li, K.; Qin, X.; Plunkett, G.; Brown, D.; Xu, B.; Zhang, L.; Gu, Z.; Mu, G.; Jia, H.; Yin, Z.; et al. Hydrological fluctuations in the Tarim Basin, northwest China, over the past millennium. *Geology*, 2024; *advance online publication*. [CrossRef]
44. Schofield, M.R.; Barker, R.J.; Gelman, A.; Cook, E.R.; Briffa, K.R. A Model-Based Approach to Climate Reconstruction Using Tree-Ring Data. *J. Am. Stat. Assoc.* **2016**, *111*, 93–106. [CrossRef]
45. Pichardo, R.; Treviño, J.; Rolón, J.C. *Tópicos Selectos de Ingeniería y Ciencias Ambientales*; Universidad Autónoma de Tamaulipas: Ciudad Victoria, Mexico, 2022. [CrossRef]
46. Loaiciga, H.A.; Haston, L.; Michaelsen, J. Dendrohydrology and long-term hydrologic phenomena. *Rev. Geophys.* **1993**, *31*, 151–171. [CrossRef]
47. Shekhar, M.; Ranhotra, P.S.; Bhattacharyya, A.; Singh, A.; Dhyani, R.; Singh, S. Tree-ring-based hydrological records reconstructions of the Himalayan Rivers: Challenges and opportunities. In *Springer Climate*; Rani, S., Kumar, R., Eds.; Springer Science and Business Media B.V.: Berlin/Heidelberg, Germany, 2022; pp. 47–72. [CrossRef]

forests

MDPI

Article

Total Streamflow Variation for the Upper Catchment of Bosten Lake Basin in China Inferred from Tree-Ring Width Records

Kexiang Liu [1,2,3], Tongwen Zhang [1,2,3,*], Huaming Shang [1,2,3], Yuting Fan [1,2,3], Shulong Yu [1,2,3], Shengxia Jiang [1,2,3], Weiyi Mao [1,2,3] and Xinchun Liu [1,2,3]

1 Institute of Desert Meteorology, China Meteorological Administration, Urumqi 830002, China
2 Key Laboratory of Tree-Ring Physical and Chemical Research of China Meteorological Administration, Urumqi 830002, China
3 Xinjiang Key Laboratory of Tree-Ring Ecology, Urumqi 830002, China
* Correspondence: zhangtw@idm.cn

Abstract: Bosten Lake Basin not only is a major source of drinking water for the residents of the surrounding area, but also maintains the ecological balance of the region. However, with the influence of climate change and human activities, the water level of Bosten Lake fluctuates sharply and has a great impact on the surrounding ecological environment. Therefore, the study of its historical water flow changes as a reference has become a focus of research. In this study, the radial growth of Schrenk spruces (*Picea schrenkiana* Fisch. et Mey.) significantly correlated with the tributary streamflow coming from the mountainous region near Bosten Lake Basin. On the basis of this good coherence, the tree-ring chronologies were used to reconstruct the streamflow for Huangshuigou River from the previous August to the present July ($r = 0.766$, $p < 0.0001$, $n = 50$). The reconstructed streamflow series matched observations well, explaining 63.3% of the variation in the observed streamflow of 1956–2005. Then, the sum of the streamflow reconstruction of Huangshuigou River and another two tree-ring-based streamflow reconstructions (Kaidu River and Qingshui River) was used to represent the hydrological variation of the upper catchment of Bosten Lake Basin, and the reconstruction sequence was 306 years. The 10.7, 5.5, and 2.1 year cycles of the power spectrum and wavelet analysis revealed that the runoff series reconstructed from tree-ring hydrometeorology was related to solar activity. Some dry and wet years in the reconstructed streamflow series of the upper catchment of Bosten Lake Basin corresponded to the historical record. During the wet years, the Indian Ocean was probably the main source of precipitation.

Keywords: Schrenk spruces (*Picea schrenkiana*); tree ring; streamflow reconstruction; Huangshuigou River; Bosten Lake Basin

1. Introduction

Inland lakes, evaporating dishes in arid and semi-arid areas, are not only the sensitive factors for climate change but also indicators of ecosystems [1,2]. The temporal and spatial variation of water storage of lakes objectively reflects the process of regional water cycle and water balance [3]. These lakes provide valuable water resources for industrial production, agricultural irrigation, and animal husbandry development, and they play a crucial part in regional residents' drinking and wildlife breeding water supply [4,5]. Thus, the timely and accurate analyses of the distribution, area, storage, and water level of inland lakes are beneficial to comprehend the characteristics and mechanism of lake variation and to develop policies of regional water resource management.

China's largest inland freshwater lake, Bosten Lake, is located on the southern slope of the middle Tianshan Mountains and provides the basis of agricultural irrigation, fishery production, and municipal water for the around area. Zhu et al. [6] analyzed remote sensing data to indicate that there are two water level decreasing periods (1972–1990 and 2002–2010) and one increasing period (1990–2002) for Bosten Lake. While exceptionally high lake levels will result in the sinking of farmland, soil salinization, and an increase in the risk of flooding, continuous lake level decreases can cause the loss of lake wetlands, deterioration of flora, and harm to biodiversity and fishing resources [7]. These possible secondary disasters caused by the fluctuation of water storage in lakes have been research hotspots attracting much attention in recent years. Most of the previous studies on Bosten Lake Basin investigated the lake level and water storage, and analyzed the relationship between the climatic factors and the surface runoff of the watershed [6–9]. It is a pity that short-term observed data and fragmentary historical records make it difficult to satisfactorily describe and understand the variation characteristics of the environmental factors and their driving mechanism for Bosten Lake Basin and its surrounding area.

Tree rings, as important proxy data, are frequently used to reconstruct the hydrological variation in history. Gou et al. [10,11] reconstructed the streamflow of Yellow River in the past millennium and revealed the long-timescale variation pattern of Yellow River streamflow. Woodhouse et al. [12,13] conducted streamflow reconstruction of the upper Colorado River on a different timescale, which provided reliable data to support water resource management in the basin. Panyushkina et al. [14] reconstructed the past 235 years of streamflow variation in Lake Balkhash Basin using Schrenk spruce's tree-ring width. Liu et al. [15] used a moisture-sensitive tree-ring width chronology to reconstruct streamflow in the middle reaches of Yellow River and found that reduced streamflow resulted in a 58% reduction in sediment load in the upper Yellow River and a 29% reduction in the middle reaches. Ferrero et al. [16] reconstructed the streamflow variability of Río Bermejo in subtropical South America. Pederson et al. [17] reconstructed the streamflow of Kherlen River in northeastern Mongolia for the last 345 years using tree-ring width to explore possible links between regional streamflow variability and solar activity, among others. The coniferous forest widely distributed in the surrounding mountains of Bosten Lake Basin provides an excellent opportunity for dendrohydrological studies [18]. However, these studies mainly focused on the streamflow reconstruction for a given river, whereas less attention has been paid to the elaborate description of the total runoff from tributaries for the watershed. These dendrohydrological reconstructions for the main tributaries may provide a good background for researching and analyzing the hydrological variations of the whole watershed over the long-term period.

Therefore, this study established the following goals in order to fill these knowledge gaps and expand on the current knowledge base: (1) reconstruct Huangshuigou River's streamflow on the basis of connections among hydroclimatic factors, as well as tree-ring width, (2) develop the total streamflow variation for the upper catchment of Bosten Lake Basin from the reconstructed streamflow series of these tributaries, and (3) analyze the variation characteristics of the total streamflow of Bosten Lake Basin and investigate how it relates to other environmental factors.

2. Materials and Methods

2.1. Study Area

Bosten Lake is situated in the southeastern Yanqi Basin, bounded by 41°56′–42°14′ N and 86°40′–87°26′ E. The mean depth of this lake is 8 m, and the maximum depth is 16.2 m. The area of the lake is approximately 1000 km^2, and its storage is 8.8 km^3. This lake is between the Tianshan Mountains and the Kuluketage Mountains. The lake's surface is 1048 m a.s.l. and the watershed of the lake is close to 55,600 km^2 [19]. Bosten Lake, as a typical alternately filled lake, is the home of more than 10 rivers originating from the mountainous area of Yanqi Basin and the source of Kongque River [20]. The runoff of Kaidu River, Huangshuigou River, and Qingshui River accounts for above 90% of the Bosten Lake inflow [21].

2.2. Sampling and Tree-Ring Width Measurement

In the mountainous region around Bosten Lake Basin, there are a total of nine tree-ring sample locations. These tree-ring cores were obtained in August 2005 (two sites in the basin of Huangshuigou River), September 2010 (five sites in the basin of Kaidu River), and June 2015 (two sites in the basin of Qingshui River) (Figure 1). In these locations, forest stands are comparatively open. From west to east, the soil beneath the forest gradually becomes thinner and contains more sand and gravel. Schrenk spruces (*Picea schrenkiana* Fisch. et Mey.), a type of dominating and enduring tree species found in the Tianshan Mountains, which range in elevation from 1200 to 2800 m a.s.l., were collected from every tree-ring sampling site. In order to prevent non-climatic factors from influencing the radial growth of the tree trunk, we only chose healthy spruces that showed minimal signs of being harmed by human activity, pests and insects, landslides, earthquakes, or animal invasion. For the majority of the studied trees, we simultaneously obtained two cores from one tree, at breast height. Each tree's two cores were taken from opposing sides. In total, we collected 516 cores from 262 trees at nine sites using increment borers with a 10 mm diameter. The information of these sampling sites is listed in Table 1.

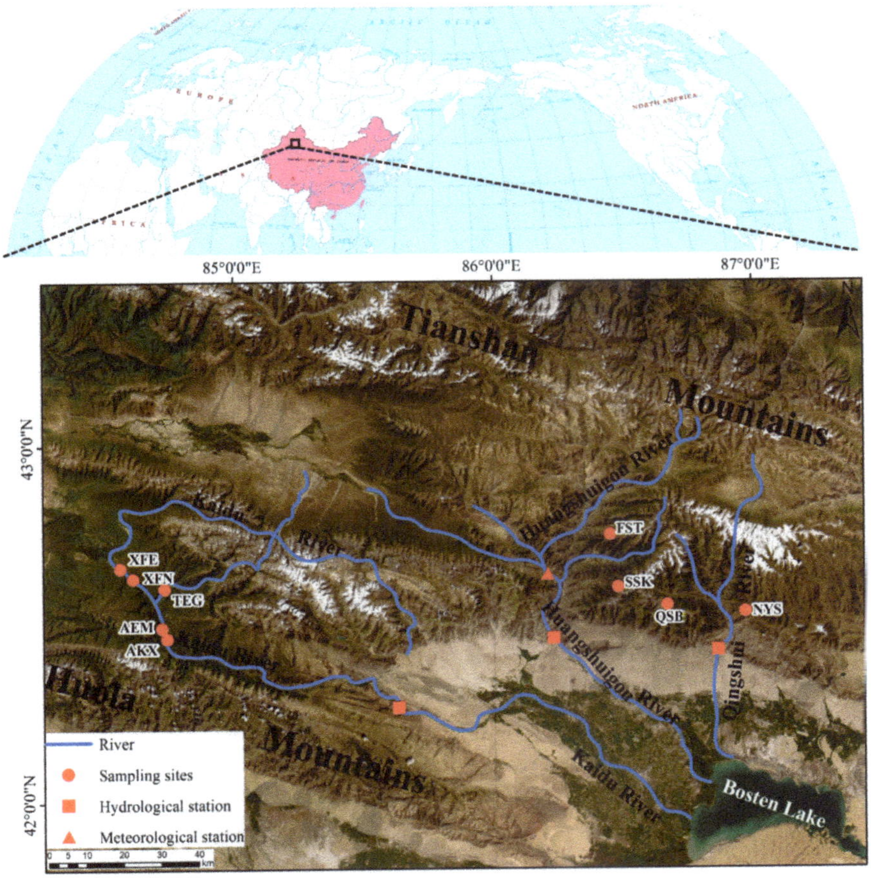

Figure 1. Location map of the study area, mainly showing the distribution of rivers, hydrological stations, meteorological stations, and sampling sites.

Table 1. The information of sampling sites in the upper catchment of Bosten Lake Basin.

Watershed	Site Name	Site Code	Latitude (N)	Longitude (E)	Trees/Cores	Elevation (m)	Aspect	Slope	Maximum Tree Age	The Rate of Absent Rings (%)
Kaidu River	Aermuxi	AEM	42.49°	84.72°	25/50	~2540	W	20°	395 (1616–2010)	0.552
	Akenxiaer	AKX	42.46°	84.74°	27/53	~2740	W	15°	378 (1633–2010)	0.289
	Taoergaotingxili	TEG	42.60°	84.73°	28/56	~2560	NW	20°	414 (1597–2010)	0.409
	Xingfumuchangdong	XFE	42.66°	84.56°	28/56	~2490	N	25°	316 (1694–2009)	0.329
	Xingfumuchangnan	XFN	42.63°	84.61°	29/58	~2650	N	35°	355 (1656–2010)	0.250
Qingshui River	Nayitigou	NYS	42.53°	86.97°	31/60	~2870	NNW	20°	405 (1611–2015)	0.994
	Qingshuihe	QSB	42.55°	86.67°	27/53	~2710	N	40°	580 (1436–2015)	0.787
Huangshuigou River	Fusitan	FST	42.75°	86.45°	34/66	~2600	N	40°	744 (1262–2005)	3.019
	Saseke	SSK	42.60°	86.48°	33/64	~2800	N	40°	439 (1567–2005)	0.666

Utilizing standard dendrochronological technologies, the sampled trees' tree-ring cores were dried naturally and placed on a grooved wooden plank [22]. Next, each core was sanded with abrasive paper of varying grades, and then marked with needles while viewed through a microscope. The LINTAB measuring system with 0.001 mm resolution was used to measure the width of each ring.

2.3. Development of Tree-Ring Chronologies

The curves of tree-ring width series for every core were displayed and contrasted using the LINTAB measuring system. The output of the COFECHA program enabled checking for missing tree-ring width series and false rings, as well as cross-dating quality control [23]. To obliterate the indications of tree ages influenced by variables other than climate, the ARSTAN program was chosen with the modified negative exponential curve option [24]. Because the common variations across various ring-width series are included in the standard tree-ring chronologies created by the program and since they maintain a low-throughput high-frequency common variance, they were utilized in the subsequent analysis and reconstruction [25]. The reliable length of tree-ring chronologies was assessed by the expressed population signal (EPS). These recently created tree-ring chronologies were given a reliable length by using an EPS $\geq$ 0.85 [26]. If a single sampling site's chronologies were found to be coherent, all of the tree-ring width series in the same watershed of a given tributary were incorporated to establish a regional chronology.

2.4. Hydrometeorological Data and Developed Streamflow Reconstructions

We choose the Baluntai meteorological station's monthly mean temperature, mean maximum temperature, mean minimum temperature, and precipitation data (42.67° N, 86.33° E, and 1752.5 m a.s.l.) and the monthly streamflow data of the Huangshuigou hydrological station (42.45° N, 86.23° E, and 1320.0 m a.s.l.) for correlation analyses and hydrological reconstruction. Climatic data from 1958 to 2013 and hydrologic data from 1955 to 2013 were from the China Meteorological Data Service Center (CMDC, https://data.cma.cn/, accessed on 6 September 2019) and Xinjiang Uygur Autonomous Region Water Authority, respectively. The meteorological station is located in the mountainous area, which can adequately describe the climatic conditions for the Huangshuigou River watershed. The hydrological station is located in the watershed of the Huangshuigou River's mountain pass. In the higher and intermediate basins above this hydrological gauging station, there is little herding activity and no dam; however, in the lower watershed, there is significant agricultural irrigation and human activity. Taking into account the impact of hydroclimatic conditions prior to the growing season, the association between tree growth and hydroclimatic conditions was revealed using various combinations of monthly hydrometeorological data from the previous year to the current year [27–29]. Furthermore, two tree-ring-based streamflow reconstructions for Kaidu River and Qingshui River were used in the process of total streamflow variation for the upper catchment of Bosten Lake Basin [18,30].

2.5. Statistical Analysis

The monthly and annual hydrological and meteorological data were analyzed as the hydroclimatic background for the study area. For the single sampling site, the consistency of these tree-ring chronologies was tested using the Gleichläufigkeit index (GLK), correlation coefficients in the different domains, and extreme value years [31–33]. The regional chronology was developed in accordance with a good coherence of growth of tree rings in the research area. The strength of the hydroclimatic signals present in these chronologies from spruces was measured using correlation analysis. Hydrometeorological data and tree-ring chronology were shown to have the strongest seasonal association; then, the linear and exponential regression models were employed for reconstruction. Leave-one-out cross-validation and split-sample calibration–verification tests were used to evaluate the statistical reliability of the reconstruction model [34,35]. The hydrological data period was divided into two continuous sections for calibration and verification, with the same length over the split-sample calibration–verification experiments. Several statistical parameters, including reduction in error (*RE*), coefficient of efficiency (*CE*), correlation coefficient (r_1), sign test (*S*), and product mean test (*t*), were computed to assess the data, both observed and estimated [36]. The total streamflow for the upper catchment of a given lake was calculated as the sum of the streamflow reconstructions of these tributaries originating from the mountainous area. The aforementioned analyses were carried out utilizing the Statistical Product and Service Solutions and Data Processing System. Reasonable periodicities in the rebuilt series were examined using power spectrum and wavelet analysis (http://climexp.knmi.nl, accessed on 27 September 2019) [31].

2.6. Characteristics of Hydrometeorological Data Variation

Figure 2a shows that the monthly mean temperature was higher than 0 °C from March to October in the watershed of Huangshuigou River, and that summer (June–August) had the greatest mean temperature, with July (19.1 °C) being the hottest. Most of the precipitation (147.2 mm), 70.6% of the yearly total, fell during the summer (Figure 2c). Figure 2e demonstrates that the yearly total runoff, which peaked in July, primarily constituted the monthly mean streamflow during the summer (25.438 m^3/s). The highest value of annual mean streamflow (from January to December) occurred in 2000 (20.126 m^3/s), while the lowest value occurred in 1985 (5.279 m^3/s) (Figure 2f). The partial correlation coefficient for annual mean streamflow and annual total precipitation (Figure 2d) with the controlled annual mean temperature series was 0.779 ($p < 0.001$, $n = 53$) in 1958–2013, while the partial correlation between annual mean streamflow and annual mean temperature (Figure 2b) with the controlled annual total precipitation series was clearly lower ($r = 0.406$, $p < 0.01$, $n = 53$). The aforementioned findings show that precipitation had a significant influence on the Huangshuigou River's streamflow.

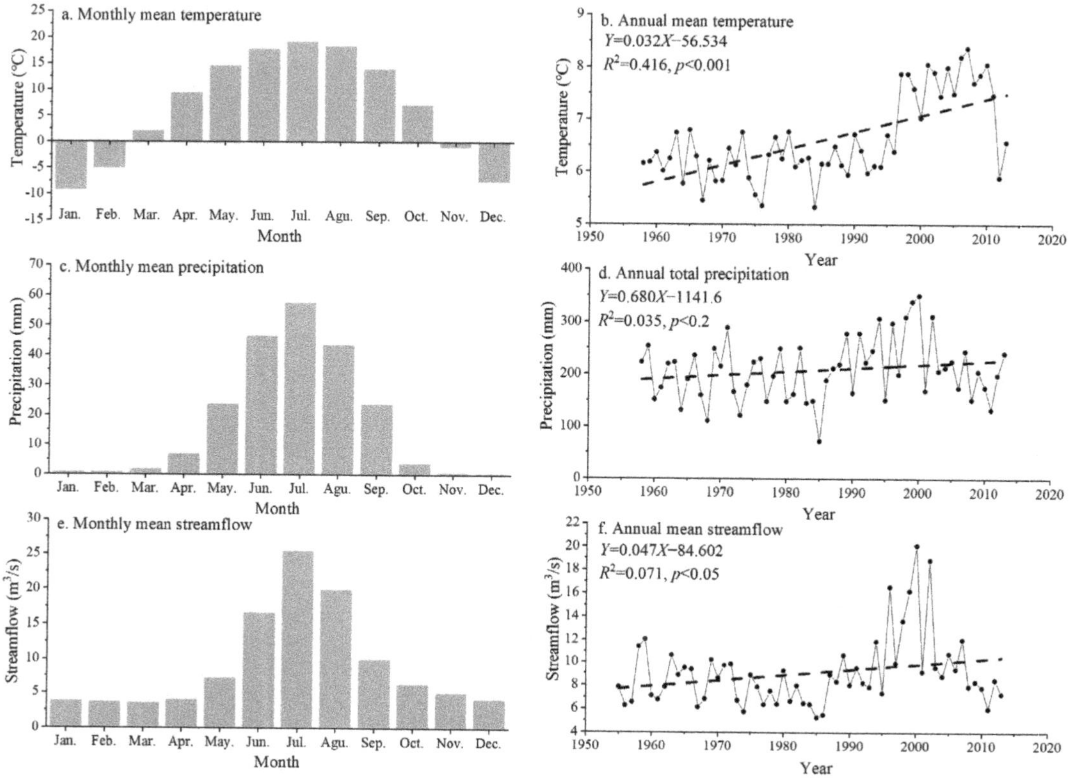

Figure 2. (**a**) Monthly mean temperature, (**b**) annual mean temperature, (**c**) monthly mean precipitation, (**d**) annual total precipitation at the Baluntai meteorological station. (**e**) Monthly mean streamflow and (**f**) annual mean streamflow at Huangshuigou hydrological station.

3. Results

3.1. Regional Chronologies Development and Analyses

The regional tree-ring chronologies for the watershed of Kaidu River and Qingshui River were developed on the basis of the good coherence between each tree-ring chronology for a single sampling site and used for streamflow reconstruction [18,30]. Thus, the regional tree-ring chronology for the watershed of Huangshuigou River was determined following the above principle. In the common period 1567–2005, the GLK value between the FST and SSK chronologies (79.7%) reached the 0.001 significance threshold. We examined two chronologies in the various domains using Pearson correlation coefficients in addition to the reciprocal filters. During the common period, the original domain correlation coefficient for the two chronologies was 0.700 ($p < 0.0001$, $n = 439$). In the high-frequency domain ($r = 0.713$, $n = 427$), the correlation coefficient between the two chronologies was greater than in the low-frequency domain ($r = 0.667$, $n = 427$). Furthermore, a comparison of the years with the 10 highest values and 10 lowest values of the FST and SSK chronologies in the common period revealed that there were four high-value years (1595, 1704, 1924, and 2002) and six low-value years (1776, 1878, 1918, 1951, 1957, and 1974) in total.

The above results and similar natural environments for the two sampling sites (Table 1) indicated that the spruce radial growth in the Huangshuigou River watershed was well synchronized. Consequently, the series of tree-ring widths in this watershed were combined to create the regional chronology HSR (Figure 3c). The information on tree-ring chronologies is listed in Table 2. The values of first-order autocorrelation (AC1) ranged from 0.485 to 0.555, indicating that the tree-ring width had low-frequency variation, which might have been influenced by the lag effects of climate and tree physiology. The high correlation coefficients among these ring-width series revealed a similar variation of tree growth in the study area. The values of signal-to-noise ratio (SNR) and EPS for the HSR chronology were obviously higher than the two chronologies for the single site due to the greater sample depth, indicating that the regional chronology might have contained more climatic signals. The reliable length of the HSR chronology was 536 years (1470–2005).

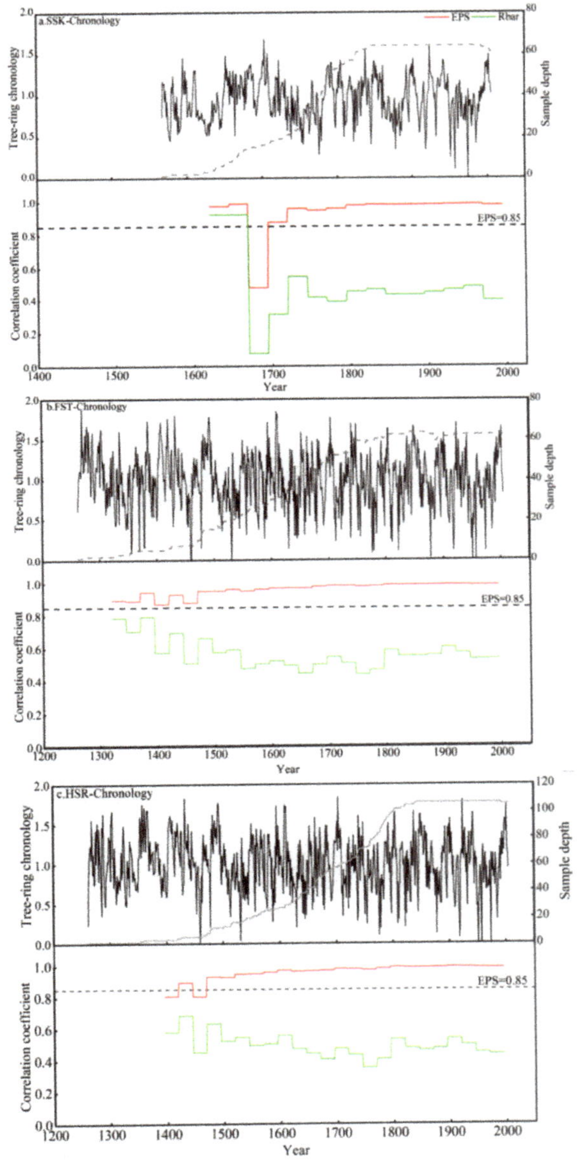

Figure 3. The upper part of each figure represents the chronology and sample depth, and the lower part is expressed population signal (EPS) and mean inter-series correlation (Rbar) of tree-ring width chronology.

Table 2. The statistical characteristics of SSK, FST, and synthetic regional chronology HSR.

Statistic	FST	SSK	HSR
Mean index (MI)	0.983	0.952	0.998
Standard deviation (SD)	0.357	0.264	0.340
Mean sensitivity (MS)	0.349	0.222	0.314
First-order autocorrelation (AC1)	0.485	0.555	0.508
Inter-series correlation between trees	0.426	0.356	0.384
Inter-series correlation among all series	0.432	0.364	0.388
Mean within-tree correlation	0.778	0.753	0.769
Signal-to-noise ratio (SNR)	43.327	30.294	62.029
Expressed population signal (EPS)	0.977	0.968	0.984
First year EPS $\geq$ 0.85	1320	1695	1470

3.2. Climate Response Analysis

A strong biological lag effect was indicated by the high AC1 of two tree-ring chronologies for the single sampling site and the regional chronology (Table 2). Therefore, in order to evaluate the impact of hydrometeorological factors on the radial growth of spruces, monthly climate and hydrological data were applied from July in the previous year to October in the current year (covering a 16 month period) (Figure 4). The HSR chronology showed a substantial positive association with the 5 month average temperature with a 0.01 significance (Figure 4b), while significantly positive correlations between this chronology and precipitation were found in 2 months (Figure 4a). Additionally, the findings of the correlation analysis showed that streamflow and the HSR chronology had a better link than precipitation and temperature. The monthly streamflow data and the chronology had 14 clearly positive relationships. After examining several combinations of months, the HSR chronology and the streamflow from the previous August to the current July showed the greatest association coefficient ($r = 0.766$, $p < 0.0001$, $n = 50$).

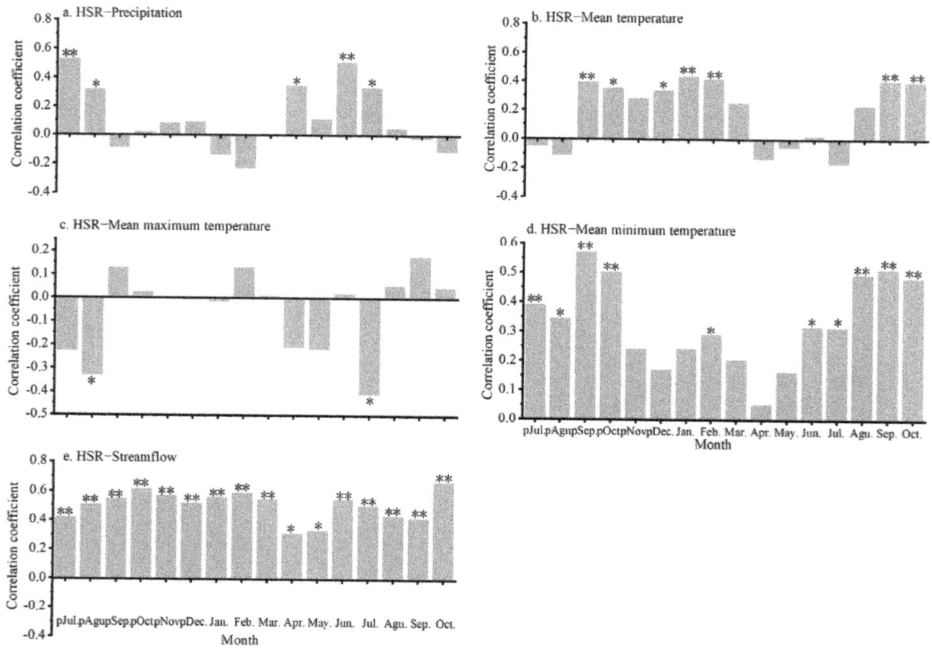

Figure 4. Pearson correlations between hydroclimatic data and the HSR chronology; * $p < 0.05$, ** $p < 0.01$.

The results for the correlation of streamflow to various meteorological factors in the period of 1958–2013 show that the yearly streamflow for Huangshuigou River positively correlated with mean temperature in January of the current year, as well as precipitation in April, June, and July of the current year, at the 0.01 significance level (Figure 5). After experimenting with various month combinations, the yearly streamflow and pSeptember–August precipitation (where p represents a month from the previous year) had the strongest connection ($r = 0.805$, $n = 55$).

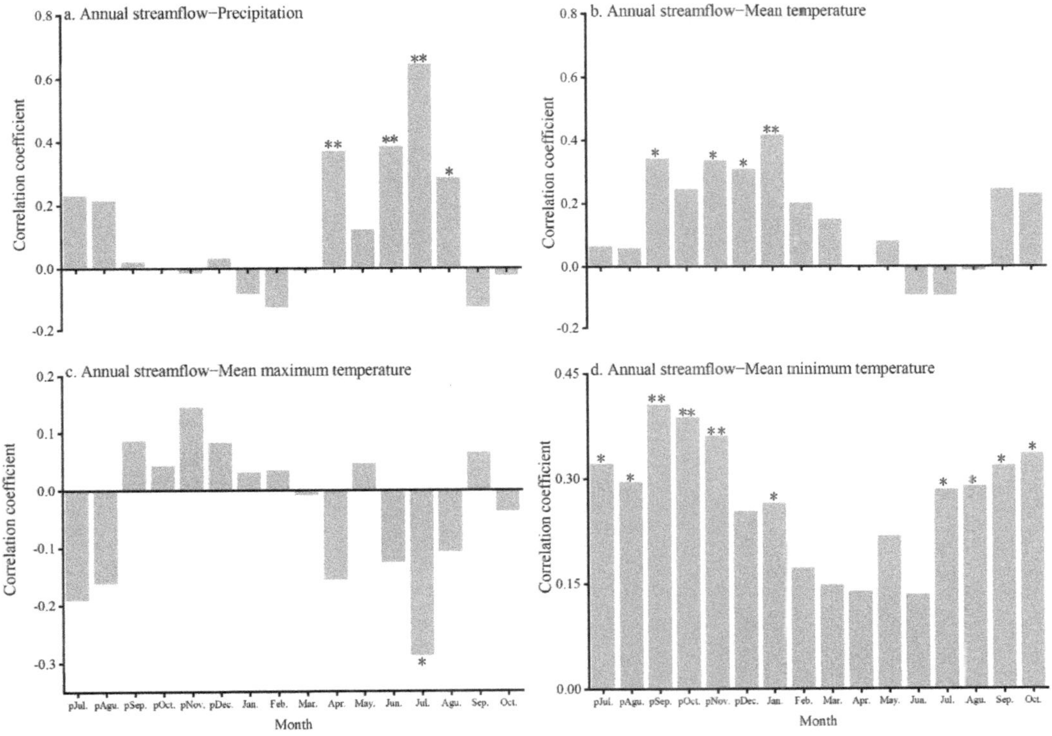

Figure 5. Pearson correlations between Huangshuigou River's yearly streamflow and climate information; * $p < 0.05$, ** $p < 0.01$.

3.3. *Streamflow Reconstruction and Its Reliability Tests*

Due to the strong correlation between streamflow and spruce radial growth, we firstly reconstructed the pAugust–July streamflow of the targeted river using the regional chronology. These relationships were represented using the linear and exponential regression models:

$$S_{pAug.\text{-}Jul.} = 2.803 + 6.583 \times HSR, \quad (1)$$

($n = 50$, $r_1 = 0.766$, $r_2 = 0.639$, $R^2 = 58.6\%$, $R^2_{adj} = 57.8\%$, SE = 1.930, and $F = 67.986$),

$$S_{pAug.\text{-}Jul.} = 4.565 \times e^{0.675 \times HSR}, \quad (2)$$

($n = 50$, $r_1 = 0.795$, $r_2 = 0.695$, $R^2 = 63.3\%$, $R^2_{adj} = 62.5\%$, SE = 0.179, and $F = 82.652$),

where $S_{pAug.\text{-}Jul.}$ is the streamflow of Huangshuigou River from August of last year through July of the current year, and the regional tree-ring chronology is referred to as HSR. In the HSR chronology during the 1956–2005 calibration era, two models explained 58.6% and 63.3% of the streamflow variance. The results of the leave-one-out test revealed that, although these test values were almost identical to the initial values in the two models

(Table 3), the correlation coefficients of Equation (2) were obviously higher than those of Equation (1). The results of split-sample calibration–verification tests (Table 4) revealed that the values of *RE* and *t* in the four verification periods all passed the test, while the values of Equation (2) were obviously higher. The value of *CE* of Equation (1) in 1956–1980 was negative, but the values of Equation (2) all exceeded 0. The streamflow series reconstructed by Equation (2) revealed a higher R^2 and improved test results after comparison. Therefore, we used Equation (2) to reconstruct Huangshuigou River's streamflow from August of last year through July of the current year in the years 1470–2005 (Figure 6c). Figure 6a,b reveal that the original and high-frequency domains of the reconstructed streamflow data closely match the measurement ($r_2 = 0.695$, $p < 0.001$, $n = 49$) domains. Furthermore, the streamflow reconstruction for Kaidu River significantly correlated with the streamflow observation from August of last year through July of the current year ($r = 0.797$, $p < 0.001$, $n = 38$). Thus, the streamflow reconstructions of Kaidu River, Qingshui River, and Huangshuigou River in the common period were added together to reconstruct the total streamflow series for the upper catchment of Bosten Lake Basin, yielding the following model:

$$TSB = \mathrm{SK} + \mathrm{SQ} + \mathrm{SH}, \tag{3}$$

where *TSB* is the total streamflow for the upper catchment of Bosten Lake Basin; SK, SQ, and SH respectively refer to the streamflow reconstructions of Kaidu River, Qingshui River, and Huangshuigou River. According to Equation (3), the streamflow series for the upper catchment of Bosten Lake Basin in the period of 1700–2005 is shown in Figure 6d.

Table 3. Results of the leave-one-out tests for the reconstruction of streamflow from August of last year through July of the current year using exponential and linear regression models.

Statistic	Linear Regression Model Mean (Range)	Exponential Regression Model Mean (Range)
Correlation coefficient (r_1)	0.766 (0.733–0.783)	0.795 (0.770–0.811)
Squared multiple correlation (R^2)	0.586 (0.538–0.613)	0.633 (0.593–0.657)
Adjusted squared multiple correlation (R^2_{adj})	0.577 (0.528–0.604)	0.625 (0.584–0.650)
Standard error (SE)	1.929 (1.771–1.950)	0.179 (0.174–0.181)
F-value (F)	66.671 (54.7230–74.299)	81.032 (68.388–90.222)

Table 4. Statistics of streamflow reconstruction from August of last year through July of the current year using linear and exponential regression models and split-sample calibration–verification tests; correlation coefficient (r_1, r_2), coefficient of determination (R^2), reduction in error (*RE*), coefficient of efficiency (*CE*), product mean test (*t*), and sign test (*S*).

Statistic	Linear Regression Model					Exponential Regression Model				
	Calibration (1956–1980)	Verification (1981–2005)	Calibration (1981–2005)	Verification (1956–1980)	Full Calibration (1956–2005)	Calibration (1956–1980)	Verification (1981–2005)	Calibration (1981–2005)	Verification (1956–1980)	Full Calibration (1956–2005)
r_1	0.621 ($p < 0.001$)	0.807 ($p < 0.001$)	0.841 ($p < 0.001$)	0.608 ($p < 0.001$)	0.766 ($p < 0.001$)	0.620 ($p < 0.001$)	0.847 ($p < 0.001$)	0.871 ($p < 0.001$)	0.604 ($p < 0.001$)	0.795 ($p < 0.001$)
r_2	0.792 ($p < 0.001$)	0.579 ($p < 0.01$)	0.653 ($p < 0.001$)	0.786 ($p < 0.001$)	0.639 ($p < 0.001$)	0.781 ($p < 0.001$)	0.636 ($p < 0.001$)	0.696 ($p < 0.001$)	0.774 ($p < 0.001$)	0.695 ($p < 0.001$)
R^2	0.386	0.652	0.706	0.370	0.586	0.385	0.717	0.759	0.365	0.633
RE	/	0.674	/	0.485	/	/	0.703	/	0.661	/
CE	/	0.599	/	–0.149	/	/	0.635	/	0.248	/
t	/	4.307	/	3.170	/	/	4.642	/	3.653	/
S	/	20 + /5–	/	15 + /10–	/	/	24 + /1–	/	16 + /9–	/

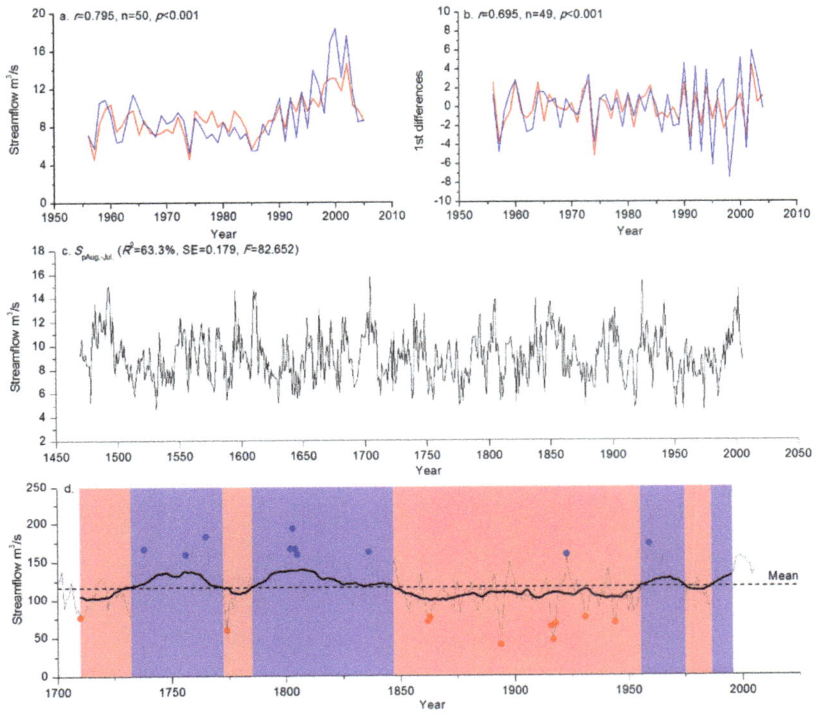

Figure 6. Information on streamflow reconstruction: (**a**) comparison between Huangshuigou River's reconstructed (red line) and observed (blue line) streamflow for pAugust–July; (**b**) comparison of the first differences (year-to-year changes) between the recorded (blue line) and reconstructed (red line) streamflow series; (**c**) reconstructed streamflow for Huangshuigou River from August of last year through July of the current year during the period of 1470–2005; (**d**) streamflow series for the upper catchment of Bosten Lake Basin in the period of 1700–2005, including Qingshui River, Kaidu River, and Huangshuigou River.

3.4. Characteristics of the Total Streamflow Series

The mean value of the total streamflow series for the upper catchment of Bosten Lake Basin was 117.083 m^3/s with a standard deviation (σ) of 22.791 m^3/s. For this series, according to the division approach proposed by Liu et al., we designated a wet year, defined as having values higher than the mean (139.873 m^3/s), and a dry year, defined as having values lower than the mean (94.292 m^3/s) [37]. The data showed that 45 years were dry, and 50 years could be classified as wet. The 211 years that remained were classified as normal. Table 5 contains the mean centenary values, as well as the extreme values of years/decades. As can be observed, there was a difference of 154.651 m^3/s between the wettest (1803) and driest (1894) years and a difference of 47.141 m^3/s between the wettest and driest decades (the 1800s and 1910s, respectively). For this total streamflow series, Table 5 also includes the long-term means and coefficients of variation.

In the total streamflow series of the upper catchment of Bosten Lake Basin, an analysis of decadal variability using a 21 year moving average was conducted (Figure 6d), which could be divided into four wet and four dry phases. The wet periods were 1730–1772, 1785–1846, 1955–1974, and 1987–1995 (above the series mean value). The dry periods were 1710–1729, 1773–1784, 1847–1954, and 1975–1986 (below the mean value). Power spectral examination of the whole duration of this series revealed significant periodicity at frequencies of 10.7 ($p < 0.05$), 5.5 ($p < 0.05$), and 2.1 ($p < 0.1$) years (Figure 7a). Figure 7b shows the results of wavelet analysis of the distinct cycle's temporal properties. The wavelet analysis showed a strong, approximately 11 year cycle from 1790 to 1850 and from 1880 to 1930.

Table 5. Summary of streamflow reconstruction characteristics from August of last year through July of the current year.

10 Wettest Years		10 Driest Years		10 Wettest Decades		10 Driest Decades		LONG-TERM		
Year	Value (m^3/s)	Year	Value (m^3/s)	Decade	Mean Value (m^3/s)	Decade	Mean Value (m^3/s)	Years	Mean Value (m^3/s)	Variation Coefficient
1803	193.803	1894	39.152	1800s	142.794	1910s	95.653	1700–1799	119.182	0.183
1765	182.419	1917	45.688	1750s	140.166	1860s	97.068	1800–1899	116.856	0.204
1959	173.156	1774	59.154	1990s	137.890	1710s	99.099	1900–1999	113.567	0.192
1804	166.177	1916	64.630	1810s	135.953	1930s	100.268	1700–2005	117.083	0.194
1802	166.169	1918	68.222	1760s	134.339	1870s	102.969			
1738	166.159	1944	69.927	1960s	133.622	1770s	103.196			
1836	161.934	1862	71.085	1790s	133.339	1940s	105.448			
1756	159.365	1931	76.214	1740s	126.968	1850s	106.451			
1805	159.044	1710	76.468	1920s	123.448	1900s	106.805			
1923	158.874	1863	76.551	1730s	123.260	1980s	107.755			

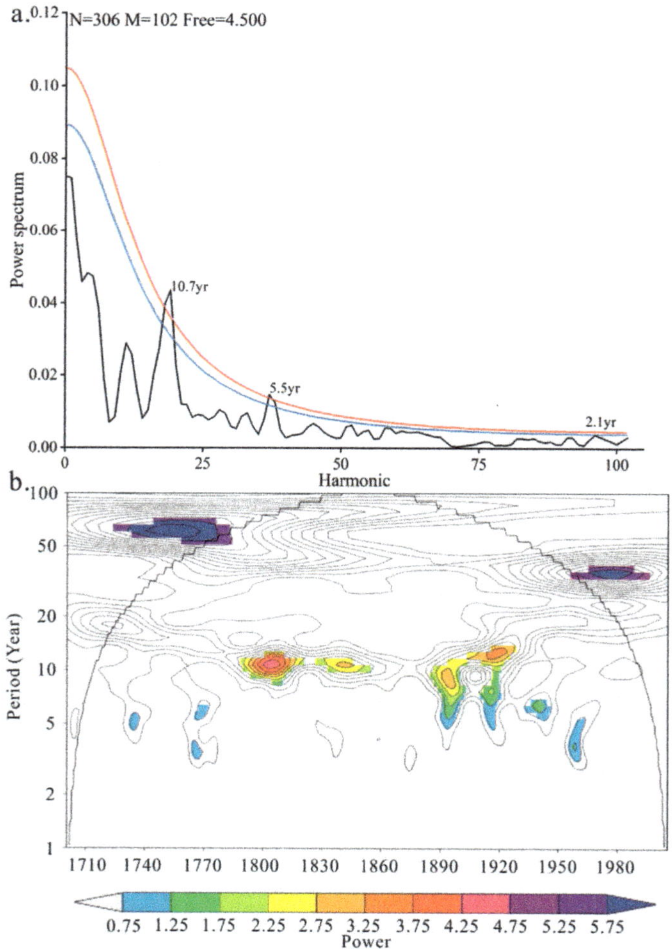

Figure 7. (**a**) Power spectrum and (**b**) wavelet analysis of the reconstruction of streamflow from August of last year through July of the current year. The red line indicates a 95% confidence level, whereas the blue line indicates a 90% confidence level.

4. Discussion

Positive responses for precipitation were found in the previous and current fast growth season, while the positive correlations between tree-ring chronology and mean temperature occurred in the end growth season and non-growth season in the previous and current year. Substantial rain during the fast growth season may, in the context of arid and semi-arid regions, encourage photosynthesis, lead to larger needles, buds, and roots, and store more nutrients for the following growth season [38]. Thus, more precipitation in the above two periods would benefit the formation of a wider tree ring in the current and next year. In addition, the mean minimum temperature has a facilitating effect on tree growth (Figure 4d), and the mean minimum temperature reflects the nighttime temperature to some extent. When the nighttime temperature increases, it can promote the division of forming layer cells to a certain extent. In contrast, the mean maximum temperature usually occurs during the day, and the high temperature limits the accumulation of plant organic matter and affects tree growth. Schrenk spruce is a kind of evergreen conifer species, and its roots are distributed 40–60 cm underground [9]. Furthermore, the sampling trees in this study are located in the high-elevation area because the sampling locations are above 2600 m in height (Table 1). Thus, the forceful low temperature in winter might easily damage the needles exposed to the air and the roots distributed in the relatively shallow soil layer, especially in the high-elevation area. Contrarily, higher temperature at the end of the growing season and during the dormant season may shield needles, roots, and cambial cells from harm caused by the cold [39]. Additionally, a warmer winter might lessen the depth of frozen soil and advance the start of the growth season the following year [40]. These results due to abundant rainfall and high temperature would benefit the formation of a wider tree ring. Numerous dendroclimatological studies were conducted from west to east along the Tianshan Mountains, as well as in the surrounding mountain system, revealing similar climatic responses for Schrenk spruce to those mentioned above [41–45].

Although several significant correlations occurred in different months, the results of correlation analysis of the yearly streamflow of Huangshuigou River and monthly climatic data as a whole were quite similar to the climatic reactions of spruce radial growth (Figure 5). More precipitation in the wet season may bring more water for the watershed and directly increase the runoff of Huangshuigou River [46]. The minimum streamflow period of Huangshuigou River occurs in winter, when this river is primarily supplied by subsurface water [47]. Higher temperature in this period can retreat the depth of frozen soil and increase underground water. Thus, the streamflow may increase under this climate condition. An indirect connection has been established between the radial growth of spruces and the fluctuations in streamflow of Huangshuigou River caused by climatic factors based on the correlation results mentioned above. Coherent variations between tree-ring chronology and runoff have been widely found in the Central Asia. Yuan et al. (2007) reconstructed the Manasi River runoff variation in the Tianshan Mountains using the chronology of Schrenk spruce tree-ring width [48]. Zhang et al. (2019) developed a regional tree-ring width chronology of Schrenk spruce and reconstructed annual runoff for Issly Lake over 345 years [49]. Jiang et al. (2016) recorded the precipitation variation in the upper Irtysh River region for the past 291 years according to Siberian spruce tree-ring width [50]. Zhao et al. (2022) reconstructed the Ishim–Tobol River Basin's hydrological changes over the past 299 years [51].

Solar radiation denotes the electromagnetic waves and the flow of particles emitted by the sun into the cosmic space. It is the energy source of Earth and the fundamental driving force of climatic formation and environmental evolution. It has been suggested that solar activity may have an impact on Earth's climate by way of direct radiation, ultraviolet rays, cosmic ray, and geomagnetic field [52]. In accordance with the Schwabe cycle, solar activity alternates between active and quite phases, with each phase lasting approximately 11 years [53]. The 10.7 year cycles appearing in the total streamflow series for the upper catchment of Bosten Lake Basin suggest a possible connection with the above cycle of solar activity. According to Fu et al. [54], El Niño is the first to be impacted by solar activity, before transferring this influence to streamflow. The approximate 5 year cycle in the streamflow series falls within the range of the El Niño Southern Oscillation (ENSO) [55]. The reconstruction streamflow series, which was detected by cycle analyses and the interannual period (2.1 years), includes the Quasi-Biennial Oscillation [56]. For the various regions of the Tianshan Mountains, the aforementioned interdecadal and interannual cycles have been widely extrapolated from other tree-ring-based hydrometeorological reconstructions [14,41,49,57].

Unlike the exhaustive historical records from southern China, the records of hydrometeorological disasters in northwestern China are sporadic and isolated due to the historically low levels of human habitation and the nomadic lifestyle. The limited historical records do not sufficiently describe the past hydrometeorological fluctuations, but they can verify the accuracy of this recently developed streamflow series. The streamflow reconstruction's wettest and driest year for the total streamflow series for the upper catchment of Bosten Lake Basin (Table 5) were compared to histories of hydrometeorological catastrophes in China's Xinjiang Province. Table 6 reveals that the six wettest (1803–1805, 1836, 1923, and 1959) and six driest (1894, 1916–1918, 1931, and 1944) years have been documented in historical records [58]. The aforementioned coherence demonstrates that the streamflow series reconstruction can capture some signs of drought and flood occurrences for the surrounding area. Whenever a drought lasts for a significant amount of time, famines are always worsened, and social unrest is always exacerbated. The historical records from the 1710s and the 1860s to 1870s contain several accounts of ferocious clashes between local secession regimes and the central ruler (Xinjiang Academy of Social Sciences 1980). The driest of the reconstructed series coincide with the above two periods (Table 5). The Junggar Tribe and the Qing Dynasty engaged in battle in the 1690s, which lasted about 70 years. During the years 1864–1877, the Qing Dynasty fought with Aguber, and Zuo Zongtang put an end to Aguber's rebellion, establishing Xinjiang Province.

Table 6. Comparison of streamflow reconstruction's wettest and driest years from August of last year through July of the current year with the records of hydrometeorological disasters.

	Short Description of Flood, Snowfall, Drought, or Locust Disasters for the Study Area and Surrounding Areas
Wettest Year	
1803–1805	1. Frequently intense rainfall in Ili Prefecture, which is in the northwest of the study area, led to grain reduction in 1803. 2. Flood occurred in Yarkant River, a river source of Tarim River, in July 1804. 3. Thousands of people suffered from snowfall disaster in Huocheng County of Ili Prefecture, in spring 1804.
1836	1. Flood occurred in Dihua City (Urumqi now), which is in the north of the study area, in May 1836.
1923	1. Snowfall and freezing disasters occurred in Aletai Prefecture, which is in the north of the study area, leading to a large number of people and livestock death in winter 1922. 2. Intense rainfall led to a flood in Baiyang River, which is in the north of the study area, in July 1923.
1959	1. Frequently intense rainfall occurred in Akesu Prefecture, Boertala Prefecture, Changji Prefecture, Hami Prefecture, Kashi Prefecture, and Shihezi Prefecture, which are around the study area, in spring and summer 1959. 2. Great snowfalls occurred in Ili Prefecture, Tacheng Prefecture, and Boertala Prefecture, which are in the northwest of the study area, leading to a large amount of livestock death in spring 1959.
Driest Year	
1894	1. Drought occurred in Dihua City and Changji Prefecture in 1893. 2. Plagues of locusts occurred in Wusu County of Tacheng Prefecture in June 1916.
1916–1918	1. Frequently great droughts occurred in Ili Prefecture and Kashi Prefecture in 1916–1918. People fled their homes, and 90% of rooms were empty. 2. Plagues of locusts occurred in Jinghe County of Boertala Prefecture, which is in the northwest of the study area, in June 1916.

Table 6. *Cont.*

	Short Description of Flood, Snowfall, Drought, or Locust Disasters for the Study Area and Surrounding Areas
Driest Year	
1931	1. Great drought occurred in Shule County of Kashi Prefecture, which is in the southwest of the study area, in 1930.
1944	1. Less snowfall in the previous year and lack of rainfall in the current year, resulting in great drought, occurred in northern Xinjiang Province in 1943. 2. Drought occurred in Hutubi County, which is in the northwest of the study area, in 1944. 3. Frequent plagues of locusts occurred in Wusu County in 1943–1945.

5. Conclusions

Water in the upper catchment of Bosten Lake Basin is mainly supplied by Kaidu River, Qingshui River, and Huangshuigou River. To construct streamflow sequences for the historical period in the region, we first established a region tree-ring chronological series for the Huangshuigou River watershed using a total of 130 tree-ring width sequences of 67 living healthy Schrenk spruces from two sampling sites. On the basis of the high correlation between the tree-ring chronological series and streamflow data, streamflow sequences were constructed for the Huangshuigou River watershed from August of the previous year to July of the current year, from 1470 to 2005. By combining the already constructed historical streamflow series of the Qingshui River and Kaidu River watersheds, the historical streamflow series of the upper catchment of Bosten Lake Basin from August of the previous year to July of the current year were also reconstructed for a total of 306 years. The streamflow variation for the study area may have been influenced by solar activity and the oscillations of land–atmospheric–ocean circulation systems according to the result of spectral analysis. The streamflow reconstruction precisely captured the wettest (1803–1805, 1836, 1923, and 1959) and driest (1894, 1916–1918, 1931, and 1944) years noted in historical documents for the surrounding areas. Furthermore, two turbulent periods in Xinjiang corresponded to the reconstruction in the 1710s and 1860s–1870s. The frequent drought after 1700 might have been one of the main reasons for the conflicts increasing.

Author Contributions: Conceptualization, K.L., T.Z. and H.S.; methodology, K.L. and S.Y.; software, K.L., Y.F. and S.J.; writing—original draft preparation, K.L.; supervision, W.M. and X.L. All authors have read and agreed to the published version of the manuscript.

Funding: This research was supported by the National Natural Science Foundation of China (41975095 and U1803245) and the China Desert Meteorological Science Research Foundation (Sqj2020003).

Data Availability Statement: Data can be provided upon reasonable request.

Conflicts of Interest: The authors declare no conflict of interest.

References

1. Stanev, E.V. Temporal and Spatial Patterns of Sea Level in Inland Basins: Recent Events in the Aral Sea. *Geophys. Res. Lett.* **2004**, *31*, L15505. [CrossRef]
2. Bai, J.; Chen, X.; Li, J.; Yang, L.; Fang, H. Changes in the Area of Inland Lakes in Arid Regions of Central Asia during the Past 30 Years. *Environ. Monit. Assess* **2011**, *178*, 247–256. [CrossRef] [PubMed]
3. Harris, A. Time Series Remote Sensing of a Climatically Sensitive Lake. *Remote Sens. Environ.* **1994**, *50*, 83–94. [CrossRef]
4. Zuo, Q.; Dou, M.; Chen, X.; Zhou, K. Physically-Based Model for Studying the Salinization of Bosten Lake in China. *Water Energy Abstr.* **2007**, *16*, 5. [CrossRef]
5. Huang, X.Z.; Chen, F.H.; Fan, Y.X.; Yang, M.L. Dry Late-Glacial and Early Holocene Climate in Arid Central Asia Indicated by Lithological and Palynological Evidence from Bosten Lake, China. *Quat. Int.* **2009**, *194*, 19–27. [CrossRef]
6. Zhu, C.M.; Li, J.L.; Zhang, X.; Luo, J.C. Bosten water resource dynamic detection and feature analysis in recent 40 years by remote sensing. *J. Nat. Resour.* **2015**, *30*, 106–114. (In Chinese)
7. Guo, M.; Wu, W.; Zhou, X.; Chen, Y.; Li, J. Investigation of the Dramatic Changes in Lake Level of the Bosten Lake in Northwestern China. *Theor. Appl. Climatol.* **2015**, *119*, 341–351. [CrossRef]

8. Sun, Z.D.; Opp, C.; Wang, R.; Gao, Q.Z. Response of land surface flow to climate change in the mountain regions of Bosten Lake valley, Xinjiang, China. *J. Mt. Sci.* **2010**, *28*, 206–211. (In Chinese) [CrossRef]
9. Luo, C.; Wu, S.L. Analysis on the changes of temperature and precipitation in the Bosten Lake basin, Xinjiang. *J. Beihua Univ. (Nat. Sci.)* **2013**, *14*, 472–475. (In Chinese)
10. Gou, X.; Chen, F.; Cook, E.; Jacoby, G.; Yang, M.; Li, J. Streamflow Variations of the Yellow River over the Past 593 Years in Western China Reconstructed from Tree Rings: Streamflow variations of the Yellow River. *Water Resour. Res.* **2007**, *43*. [CrossRef]
11. Gou, X.; Deng, Y.; Chen, F.; Yang, M.; Fang, K.; Gao, L.; Yang, T.; Zhang, F. Analysis of tree-ring reconstruction and change characteristics of the flow of the upper Yellow River in the past 1234 years. *Chin. Sci. Bull.* **2010**, *55*, 3236–3243. (In Chinese) [CrossRef]
12. Woodhouse, C.A.; Lukas, J.J. Multi-Century Tree-Ring Reconstructions of Colorado Streamflow for Water Resource Planning. *Clim. Change* **2006**, *78*, 293–315. [CrossRef]
13. Woodhouse, C.A.; Gray, S.T.; Meko, D.M. Updated Streamflow Reconstructions for the Upper Colorado River Basin. *Water Resour. Res.* **2006**, *42*. [CrossRef]
14. Panyushkina, I.P.; Meko, D.M.; Macklin, M.G.; Toonen, W.H.J.; Mukhamadiev, N.S.; Konovalov, V.G.; Ashikbaev, N.Z.; Sagitov, A.O. Runoff Variations in Lake Balkhash Basin, Central Asia, 1779–2015, Inferred from Tree Rings. *Clim. Dyn.* **2018**, *51*, 3161–3177. [CrossRef]
15. Liu, Y.; Song, H.; An, Z.; Sun, C.; Trouet, V.; Cai, Q.; Liu, R.; Leavitt, S.W.; Song, Y.; Li, Q.; et al. Recent Anthropogenic Curtailing of Yellow River Runoff and Sediment Load Is Unprecedented over the Past 500 y. *Proc. Natl. Acad. Sci. USA* **2020**, *117*, 18251–18257. [CrossRef]
16. Ferrero, M.E.; Villalba, R.; De Membiela, M.; Hidalgo, L.F.; Luckman, B.H. Tree-Ring Based Reconstruction of Río Bermejo Streamflow in Subtropical South America. *J. Hydrol.* **2015**, *525*, 572–584. [CrossRef]
17. Pederson, N.; Jacoby, G.C.; D'Arrigo, R.D.; Cook, E.R.; Buckley, B.M.; Dugarjav, C.; Mijiddorj, R. Hydrometeorological Reconstructions for Northeastern Mongolia Derived from Tree Rings: 1651–1995. *J. Climate* **2001**, *14*, 872–881. [CrossRef]
18. Zhang, T.; Liu, Y.; Zhang, R.; Yu, S.; Fan, Y.; Shang, H.; Jiang, S.; Qin, L.; Zhang, H. Tree-Ring Width Based Streamflow Reconstruction for the Kaidu River Originating from the Central Tianshan Mountains since AD 1700. *Dendrochronologia* **2020**, *61*, 125700. [CrossRef]
19. Chen, F.; Huang, X.; Yang, M.; Yang, X.; Fan, Y.; Zhao, H. Westerly dominated Holocene climate model in arid central Asia-Case study on Bosten Lake Basin, Xinjiang, China. *Quat. Sci.* **2006**, *26*, 881–887. (In Chinese)
20. Wang, H.; Pan, Y.; Chen, Y. Impacts of Regional Climate and Teleconnection on Hydrological Change in the Bosten Lake Basin, Arid Region of Northwestern China. *J. Water Clim. Change* **2018**, *9*, 74–88. [CrossRef]
21. Li, Y.; Chen, Y.; Zhang, Q.; Fang, G. Analysis of the change in water level and its influencing factors on Bosten Lake Basin from 1960 to 2018. *Arid. Zone Res.* **2021**, *38*, 48–58. (In Chinese) [CrossRef]
22. Speer, J.H. *Fundamentals of Tree-Ring Research*; University of Arizona Press: Tucson, AZ, USA, 2010.
23. Holmes, R.L. *Computer-Assisted Quality Control in Tree-Ring Dating and Measurement*; Tree-Ring Society: Tucson, AZ, USA, 1983.
24. Cook, E.R.; Krusic, P.J. *Program ARSTAN: A Tree-Ring Standardization Program Based on Detrending and Autoregressive Time Series Modeling, with Interactive Graphics*; Lamont-Doherty Earth Observatory, Columbia University: Palisades, NY, USA, 2005.
25. Cook, E.R. A Time Series Analysis Approach to Tree Ring Standardization. Ph.D. Thesis, University of Arizona Tucson, Tucson, AZ, USA, 1985.
26. Wigley, T.M.; Briffa, K.R.; Jones, P.D. On the Average Value of Correlated Time Series, with Applications in Dendroclimatology and Hydrometeorology. *J. Appl. Meteorol. Climatol.* **1984**, *23*, 201–213. [CrossRef]
27. D'Arrigo, R.; Mashig, E.; Frank, D.; Wilson, R.; Jacoby, G. Temperature Variability over the Past Millennium Inferred from Northwestern Alaska Tree Rings. *Clim. Dyn.* **2005**, *24*, 227–236. [CrossRef]
28. Liu, Y.; Wang, C.; Hao, W.; Song, H.; Cai, Q.; Tian, H.; Sun, B.; Linderholm, H.W. Tree-Ring-Based Annual Precipitation Reconstruction in Kalaqin, Inner Mongolia for the Last 238 Years. *Chin. Sci. Bull.* **2011**, *56*, 2995–3002. [CrossRef]
29. Opala, M.; Mendecki, M.J. An Attempt to Dendroclimatic Reconstruction of Winter Temperature Based on Multispecies Tree-Ring Widths and Extreme Years Chronologies (Example of Upper Silesia, Southern Poland). *Theor. Appl. Climatol.* **2014**, *115*, 73–89. [CrossRef]
30. Zhang, T.; Shang, H.; Fan, Y.; Yu, S.; Zhang, R.; Qin, L.; Jiang, S. A 475-Year Tree-Ring-Width Record of Streamflow for the Qingshui River Originating in the Southern Slope of the Central Tianshan Mountains, China. *Geogr. Ann. Ser. A Phys. Geogr.* **2020**, *102*, 247–266. [CrossRef]
31. Fritts, H. *Tree Rings and Climate*; Academic Press, Inc. Ltd.: London, UK; New York, NY, USA, 1976.
32. Schweingruber, F.H.; Briffa, K.R.; Nogler, P. A Tree-Ring Densitometric Transect from Alaska to Labrador. *Int. J. Biometeorol.* **1993**, *37*, 151–169. [CrossRef]
33. Yuan, Y.; Zhang, T.; Wei, W.; Nievergelt, D.; Verstege, A.; Yu, S.; Zhang, R.; Esper, J. Development of Tree-Ring Maximum Latewood Density Chronologies for the Western Tien Shan Mountains, China: Influence of Detrending Method and Climate Response. *Dendrochronologia* **2013**, *31*, 192–197. [CrossRef]
34. Michaelsen, J. Cross-Validation in Statistical Climate Forecast Models. *J. Appl. Meteorol. Climatol.* **1987**, *26*, 1589–1600. [CrossRef]
35. Meko, D.; Graybill, D.A. Tree-Ring Reconstruction of Upper Gila Rwer Discharge 1. *JAWRA J. Am. Water Resour. Assoc.* **1995**, *31*, 605–616. [CrossRef]

36. Cook, E.R.; Meko, D.M.; Stahle, D.W.; Cleaveland, M.K. Drought Reconstructions for the Continental United States. *J. Clim.* **1999**, *12*, 1145–1162. [CrossRef]
37. Liu, Y.; Zhang, X.; Song, H.; Cai, Q.; Li, Q.; Zhao, B.; Liu, H.; Mei, R. Tree-Ring-Width-Based PDSI Reconstruction for Central Inner Mongolia, China over the Past 333 Years. *Clim. Dyn.* **2017**, *48*, 867–879. [CrossRef]
38. Liang, E.; Shao, X.; Hu, Y.; Lin, J. Dendroclimatic Evaluation of Climate-Growth Relationships of Meyer Spruce (*Picea Meyeri*) on a Sandy Substrate in Semi-Arid Grassland, North China. *Trees* **2001**, *15*, 230–235. [CrossRef]
39. Pederson, N.; Cook, E.R.; Jacoby, G.C.; Peteet, D.M.; Griffin, K.L. The Influence of Winter Temperatures on the Annual Radial Growth of Six Northern Range Margin Tree Species. *Dendrochronologia* **2004**, *22*, 7–29. [CrossRef]
40. Zhang, T.; Yuan, Y.; Liu, Y.; Wei, W.; Zhang, R.; Chen, F.; Yu, S.; Shang, H.; Qin, L. A Tree-Ring Based Temperature Reconstruction for the Kaiduhe River Watershed, Northwestern China, since AD 1680: Linkages to the North Atlantic Oscillation. *Quat. Int.* **2013**, *311*, 71–80. [CrossRef]
41. Li, J.; Gou, X.; Cook, E.R.; Chen, F. Tree-Ring Based Drought Reconstruction for the Central Tien Shan Area in Northwest China. *Geophys. Res. Lett.* **2006**, *33*. [CrossRef]
42. Zhang, T.; Zhang, R.; Yuan, Y.; Gao, Y.; Wei, W.; Diushen, M.; He, Q.; Shang, H.; Wang, J. Reconstructed Precipitation on a Centennial Timescale from Tree Rings in the Western Tien Shan Mountains, Central Asia. *Quat. Int.* **2015**, *358*, 58–67. [CrossRef]
43. Jiao, L.; Jiang, Y.; Wang, M.; Kang, X.; Zhang, W.; Zhang, L.; Zhao, S. Responses to Climate Change in Radial Growth of *Picea schrenkiana* along Elevations of the Eastern Tianshan Mountains, Northwest China. *Dendrochronologia* **2016**, *40*, 117–127. [CrossRef]
44. Chen, F.; He, Q.; Bakytbek, E.; Yu, S.; Zhang, R. Reconstruction of a Long Streamflow Record Using Tree Rings in the Upper Kurshab River (Pamir-Alai Mountains) and Its Application to Water Resources Management. *Int. J. Water Resour. Dev.* **2017**, *33*, 976–986. [CrossRef]
45. Shang, H.M.; Mamattuersun, K.; Li, H.; Yu, S.; Zhang, R.; Zhang, T.; Chen, F.; Lu, H.; Wei, W.; Yuan, Y. Evaluation on the climatic information of tree-ring width of *Picea schrenkiana* from Yarkand River Basin. *Desert Oasis Meteorol.* **2017**, *11*, 17–24. (In Chinese)
46. Zhou, J.; Abulimiti, A.; Mao, W.; Jiao, M.; An, M.; Shen, Y. The responses of the runoff processes to climate in the Qingshuihe River watershed on the southern slope of Tianshan Mountains. *J. Glaciol. Geocryol.* **2014**, *36*, 685–690. (In Chinese)
47. Zhou, J.; Gao, Y.; Shen, Y.; Han, T.; Mao, W. Response of extreme hydrological events to climate change in cold watersheds of Huangshuigou River and Qingshuihe River on the southern slope of Tianshan. *J. Glaciol. Geocryol.* **2014**, *36*, 1042–1048. (In Chinese)
48. Yuan, Y.; Shao, X.; Wei, W.; Yu, S.; Gong, Y.; Trouet, V. The Potential to Reconstruct Manasi River Streamflow in the Northern Tien Shan Mountains (NW China). *Tree Ring Res.* **2007**, *63*, 81–93. [CrossRef]
49. Zhang, T.; Diushen, M.; Bakytbek, E.; Shang, H.; Gao, Y.; Huang, L.; Zhang, R.; Chen, F.; Yu, S.; Yin, Z. Tree Ring Record of Annual Runoff for Issyk Lake, Central Asia. *J. Water Clim. Change* **2019**, *10*, 610–623. [CrossRef]
50. Jiang, S.; Yuan, Y.; Chen, F.; Shang, H.; Zhang, T.; Yu, S.; Qin, L.; Zhang, R. A 291 year precipitation reconstruction in the upper Irtyshn river basin based on tree-ring width. *Acta Ecol. Sin.* **2016**, *36*, 2866–2875. (In Chinese)
51. Zhao, X.; Zhang, R.; Chen, F.; Bagila, M.; Vitaliy, K.; Bulkair, M.; Yu, S.; He, Q.; Daniyar, D.; Nurzhan, K. Reconstructed summertime (June–July) streamflow dating back to 1788 CE in the Kazakh Uplands as inferred from tree rings. *J. Hydrol. Reg. Stud.* **2022**, *40*, 101007. [CrossRef]
52. Xiao, Z. Impact of solar activity on Earth's climate. *Chin. J. Nat.* **2021**, *43*, 408–419. (In Chinese)
53. Nagovitsyn, Y.A. A Nonlinear Mathematical Model for the Solar Cyclicity and Prospects for Reconstructing the Solar Activity in the Past. *Astron. Lett.* **1997**, *23*, 742–748.
54. Fu, C.; James, A.L.; Wachowiak, M.P. Analyzing the Combined Influence of Solar Activity and El Niño on Streamflow across Southern Canada. *Water Resour. Res.* **2012**, *48*. [CrossRef]
55. Allan, R.; Lindesay, J.; Parker, D. *El Niño Southern Oscillation and Climatic Variability*; CSIRO Publishing: Collingwood, Australia, 1996.
56. Baldwin, M.P.; Gray, L.J.; Dunkerton, T.J.; Hamilton, K.; Haynes, P.H.; Randel, W.J.; Holton, J.R.; Alexander, M.J.; Hirota, I.; Horinouchi, T. The Quasi-Biennial Oscillation. *Rev. Geophys.* **2001**, *39*, 179–229. [CrossRef]
57. Chen, F.; Yuan, Y. Streamflow Reconstruction for the Guxiang River, Eastern Tien Shan (China): Linkages to the Surrounding Rivers of Central Asia. *Environ. Earth Sci.* **2016**, *75*, 1–9.
58. Wen, K.G.; Shi, Y.G.; Ren, Y.Y. *Documents of Chinese Meteorological Disaster: Volume of Xinjiang*; Meteorological Publishers: Bejing, China, 2006. (In Chinese)

MDPI AG
Grosspeteranlage 5
4052 Basel
Switzerland
Tel.: +41 61 683 77 34

Forests Editorial Office
E-mail: forests@mdpi.com
www.mdpi.com/journal/forests

www.ingramcontent.com/pod-product-compliance
Lightning Source LLC
Chambersburg PA
CBHW061820210926
51850CB00006B/759